Volume Properties
Liquids, Solutions and Vapours

Volume Properties
Liquids, Solutions and Vapours

Edited by

Emmerich Wilhelm
Institute of Physical Chemistry, University of Vienna, Vienna, Austria
Email: Emmerich.Wilhelm@univie.ac.at

Trevor M. Letcher
University of KwaZulu-Natal, Durban, South Africa
Email: trevor@letcher.eclipse.co.uk

ROYAL SOCIETY OF CHEMISTRY

IUPAC

International Association of Chemical Thermodynamics

Print ISBN: 978-1-84973-899-6
PDF eISBN: 978-1-78262-704-3

A catalogue record for this book is available from the British Library

Published by The Royal Society of Chemistry,
Thomas Graham House, Science Park, Milton Road,
Cambridge CB4 0WF, UK

Registered Charity Number 207890

Visit our website at www.rsc.org/books

Printed and bound by CPI Group (UK) Ltd, Croydon, CR0 4YY

THE QUEEN'S AWARDS FOR ENTERPRISE: INTERNATIONAL TRADE 2013

Preface

The majority of chemical processes of interest occur in the fluid state. Thus, gases or vapours, pure liquids and liquid solutions are of prime scientific and engineering importance, and many of the most significant developments in physical chemistry and chemical physics (if they are, indeed, fields apart), biophysical chemistry and chemical engineering are based on contributions originating from chemical thermodynamics as applied to fluid systems.[1–3] The most profitable approach for practical applications as well as theoretical advances is based on a combination of chemical thermodynamics with molecular theory and statistical mechanics, thereby creating the field of *molecular thermodynamics* (this term was coined by Prausnitz more than four decades ago). The continuously increasing number of articles reporting experimental data on thermodynamic properties of fluids (pure and mixed) and on fluid phase equilibria, as well as on novel experimental techniques, improved data reduction methods, advances in molecular theory and computer simulation, demonstrate the unabated growth of this field. Of particular note are activities in biophysical chemistry aimed to broaden our understanding of the thermodynamic basis of physico-chemical phenomena associated with biological processes.

Volume properties belong to the most important thermodynamic/thermophysical properties, and play a central role in the pure sciences as well as in chemical engineering and industrial applications. For instance, virial coefficients and their temperature derivatives allow the description of many thermophysical properties of real gases, an approach thoroughly grounded in statistical mechanics. For gases/vapours as well as dense liquids, *PVTx* data or $P\rho_nTx$ data provide the basic input for the development of *equations of state* (EOS) valid for both pure fluids and fluid mixtures which are based on the pioneering ideas of van der Waals.[4] Here, P is the pressure, V is the molar volume, T is the temperature, $\rho_n = 1/V$ is the

Volume Properties: Liquids, Solutions and Vapours
Edited by Emmerich Wilhelm and Trevor M. Letcher

Published by the Royal Society of Chemistry, www.rsc.org

amount-of-substance density and x is the mole fraction. There can be little doubt that the most popular class of EOS is the one comprising the so-called cubic EOS, which evolved from the famous equation proposed by van der Waals in his PhD dissertation at the University of Leiden in 1873:

$$P = \frac{RT}{V-b} - \frac{a}{V^2}.$$

Here, R denotes the (molar) gas constant, and a and b are positive parameters characterising attraction and repulsion, respectively, in a particular fluid. Over the past five decades, numerous improved modifications were developed (including extensions to multicomponent mixtures *via* semiempirical mixing rules and combination rules), and these *generalised* van der Waals equations remain indispensable for applied research and development in chemical engineering. In contradistinction, modern thermodynamic property formulations for fluids are based on *fundamental equations* that are usually explicit in the Helmholtz energy as a function of *temperature* and *density*.[5] These semiempirical fundamental "equations of state" allow the calculation of any thermodynamic fluid property *via* differentiation only. For developing such an accurate multi-parameter Helmholtz energy-based EOS, besides extensive density data, perfect-gas state heat capacities as a function of temperature are essential, and vapour pressures. In addition, high-precision data of the *thermodynamic* speed of ultrasound (that is sound speeds at sufficiently low frequencies, well below the dispersive region) are becoming increasingly important. However, the required amount of experimental input data and state-of-the-art data treatment is considerable, and the equations available for selected *pure liquids* still have benchmark character.[6,7]

Liquid mixtures are usually more easily dealt with *via* properties that measure their deviation from *ideal-solution* behaviour, that is to say, *via excess properties* as a function of temperature, *pressure* and composition.[8] *Excess molar volumes* $V^{\mathrm{E}}(T,P,\{x_i\}) = (\partial G^{\mathrm{E}}(T,P,\{x_i\})/\partial P)_{T,\{x_i\}}$ are of particular importance, and Chapter 7 is devoted to this topic; here, $\{x_i\}$ denotes the set of compositional variables (mole fractions) and G^{E} is the excess molar Gibbs energy. Finally, we would like to point out that, during the last two decades, research activities have been rapidly extended to include volumetric properties of multicomponent systems of increased complexity, including systems and processes of biochemical interest,[9] such as proteins in aqueous solutions,[10,11] and the protein folding problem *in vitro* and in the cellular environment.[12,13]

In this monograph the reader will find 22 contributions dealing with volume properties and related thermodynamic properties of liquid systems and gases/vapours, both pure and mixed. The topics are approached from different angles representing the varying research background of the respective authors. This book has its origin in committee meetings of the *International Association of Chemical Thermodynamics* (IACT) which is an associate organisation of the *International Union of Pure and Applied Chemistry* (IUPAC). The IACT developed from the IUPAC *Commission on Thermodynamics* and has continued to play an active role in the definition and maintenance of standards in all aspects of thermodynamics and in the development of the subject and the encouragement of young scientists to take up research careers in chemical thermodynamics (further information can be found at www.IACTweb.org).

This book, entitled *Volume Properties: Liquids, Solutions and Vapours*, is published under the auspices of both IACT and IUPAC. On the one hand, it continues the topic-oriented approach represented by the recent monograph entitled *Heat Capacities: Liquids, Solutions and Vapours*, edited by E. Wilhelm and T. M. Letcher, and published in 2010 (Royal Society of Chemistry, Cambridge), and on the other hand it may be considered as another volume in the tradition of the long line of books that started in 1956 with *Experimental Thermochemistry*, Volume I (Interscience Publishers, Inc., New York) edited by F. D. Rossini, and *Experimental Thermochemistry*, Volume II, edited by H. A. Skinner, and published in 1962 (Interscience-Wiley, New York). Following these two monographs, over the years a series of books on various aspects of *experimental thermodynamics* (in a broad sense) have been added. So far, eight volumes have been published: (a) Volume I, *Calorimetry of Non-Reacting Systems*, edited by J. P. McCullough and D. W. Scott (Butterworths, London, 1968); (b) Volume II, *Experimental Thermodynamics of Non-Reacting Systems,* edited by B. LeNeindre and B. Vodar (Butterworths, London, 1975); (c) Volume III, *Measurement of the Transport Properties of Fluids*, edited by W. A. Wakeham, N. Nagashima and J. V. Sengers (Blackwell Science Publications, Oxford, 1991); (d) Volume IV, *Solution Calorimetry*, edited by K. N. Marsh and P. A. G. O'Hare (Blackwell Science Publications, Oxford, 1994); (e) Volume V, *Equations of State for Fluids and Fluid Mixtures*, edited by J. V. Sengers, R. F. Kayser, C. J. Peters and H. J. White, Jr. (Elsevier, Amsterdam, 2000); (f) Volume VI, *Measurement of the Thermodynamic Properties of Single Phases*, edited by A. R. H. Goodwin, K. N. Marsh and W. A. Wakeham (Elsevier, Amsterdam, 2003); (g) Volume VII, *Measurement of the Thermodynamic Properties of Multiple Phases*, edited by R. D. Weir and T. W. de Loos (Elsevier, Amsterdam, 2005); and (h) Volume VIII, *Applied Thermodynamics of Fluids*, edited by A. R. H. Goodwin, J. V. Sengers and C. J. Peters (Royal Society of Chemistry, Cambridge, 2010).

In true IUPAC spirit, the authors represent some of the most important names in their respective fields and come from many countries around the world, including: Australia, Austria, Belarus, Canada, Czech Republic, France, Germany, Israel, Japan, Poland, Russia, South Africa, Spain, United Kingdom and the United States of America.

One of the objectives of the book is to bring together research from disparate disciplines which have a bearing on *volume properties* of fluids. Cross-links between these chapters, we believe, will lead to new ways of looking at volume property-related issues, and thus to new ways of solving associated problems in physics, chemistry and engineering. Underlying this philosophy is our inherent belief that a book is still an important vehicle for the dissemination of knowledge.

Two features are of paramount importance in monographs like this one: the timeliness of the topic and the coverage and critical evaluation of the pertinent publications. In fact, important features of this book include the underlying theory, some of the most important experimental techniques, modelling and computer simulation, as well as significant and new results related to volume properties. The authors have endeavoured to cover the relevant literature up to 2013. This book is meant for researchers in chemical thermodynamics, either from academia or from chemical industry, and provides an overview of the progress recently achieved. Its success ultimately rests with the 32 authors and we, the editors, would like to thank all of them for their cooperation and enthusiastic contributions which are highly valued. We would also like to thank Professor Ron Weir who, on behalf of the IUPAC subcommittee, the *Interdivisional Committee on Terminology, Nomenclature and Symbols* (ICTNS), checked each chapter for the correct usage of thermodynamic quantities, units and symbols, always exercising the liberal spirit invoked in the *Green Book* of IUPAC.[14] Finally we wish to thank the Royal Society of Chemistry, whose representatives were helpful and patient in producing this monograph on *Volume Properties*.

Emmerich Wilhelm
Institute of Physical Chemistry,
University of Wien (Vienna),
Wien (Vienna), Austria

Trevor M. Letcher
University of KwaZulu-Natal,
Durban,
South Africa

References

1. J. H. Hildebrand, J. M. Prausnitz and R. L. Scott, *Regular and Related Solutions. The Solubility of Gases, Liquids, and Solids*, Van Nostrand Reinhold Company, New York, USA, 1970.
2. J. S. Rowlinson and F. L. Swinton, *Liquids and Liquid Mixtures*, Butterworth Scientific, London, UK, 3rd edn, 1982.
3. J. M. Prausnitz, R. N. Lichtenthaler and E. G. de Azevedo, *Molecular Thermodynamics of Fluid-Phase Equilibria*, Prentice Hall PTR, Upper Saddle River, NJ, USA, 3rd edn, 1999.
4. J. D. van der Waals, *On the Continuity of the Gaseous and Liquid States*, ed. (and with an Introduction by) J. S. Rowlinson, Dover Publications, Inc., Mineola, NY, USA, 2004.
5. (a) R. T. Jacobsen, S. G. Penoncello, E. W. Lemmon and R. Span, Multiparameter Equations of State, in *Equations of State for Fluids and*

Fluid Mixtures: Experimental Thermodynamics, Vol. V, ed. J. V. Sengers, R. F. Kayser, C. J. Peters and H. J. White Jr., Elsevier/IUPAC, Amsterdam, The Netherlands, 2000, ch. 18, pp. 849–881; (b) E. W. Lemmon and R. Span, Multi-parameter Equations of State for Pure Fluids and Mixtures, in *Applied Thermodynamics of Fluids*, ed. A. R. H. Goodwin, J. V. Sengers and C. J. Peters, Royal Society of Chemistry/IUPAC, Cambridge, UK, 2010, ch. 12, pp. 394–432.
6. *International Thermodynamic Tables of the Fluid State, IUPAC: Vol. 1, Argon*, ed. S. Angus, B. Armstrong and K. M. de Reuck, Butterworths, London, UK, 1972 through Vol. 13, Methane, ed. W. Wagner and K. M. de Reuck, Blackwell Scientific, Oxford, UK, 1996.
7. W. Wagner and A. Pruß, The IAPWS Formulation 1995 for the Thermodynamic Properties of Ordinary Water Substance for General and Scientific Use, *J. Phys. Chem. Ref. Data*, 2002, **31**, 387–535.
8. J. P. O'Connell and J. M. Haile, *Thermodynamics: Fundamentals for Applications*, Cambridge University Press, New York, NY, USA, 2005.
9. M. Gruebele and D. Thirumalai, Perspective: Reaches of chemical physics in biology, *J. Chem. Phys.*, 2013, **139**, 121701-1–121701-9.
10. *Frontiers in High Pressure Biochemistry and Biophysics*, ed. C. Balny, P. Masson and K. Heremans, Elsevier Science B. V., Amsterdam, The Netherlands, 2002.
11. T. V. Chalikian, Volumetric Properties of Proteins, *Annu. Rev. Biophys. Biomol. Struct.*, 2003, **32**, 207–235.
12. K. A. Dill, S. B. Ozkan, M. S. Shell and T. R. Weikl, The Protein Folding Problem, *Annu. Rev. Biophys*, 2008, **37**, 289–316.
13. D. Thirumalai, E. P. O'Brien, G. Morrison and C. Hyeon, Theoretical Perspectives on Protein Folding, *Annu. Rev. Biophys*, 2010, **39**, 159–183.
14. *International Union of Pure and Applied Chemistry: Quantities, Units and Symbols in Physical Chemistry*, RSC Publishing/IUPAC, Cambridge, UK, 3rd edn, 2007.

Contents

Volume Properties: Liquids, Solutions and Vapours
Edited by Emmerich Wilhelm and Trevor M. Letcher

Published by the Royal Society of Chemistry, www.rsc.org

CHAPTER 1

Volumetric Properties: Introduction, Concepts and Selected Applications

EMMERICH WILHELM

Institute of Physical Chemistry, University of Wien, Währinger Strasse 42, A-1090, Wien (Vienna), Austria
Email: emmerich.wilhelm@univie.ac.at

Science is not a collection of truths. It is a continuing exploration of mysteries......an unending argument between a great multitude of voices.

Freeman Dyson, How we know, *The New York Review of Books*, **58**(4), 10 March 2011.

1.1 Introduction

This monograph is concerned with *volumetric properties* of fluids and their role in the physicochemical description of liquid and gaseous systems, pure and mixed, that is to say, of systems ranging from pure rare gases to proteins in solution. Only *non-reacting equilibrium* systems of uniform temperature T and pressure P (*i.e.* systems in thermal, mechanical and diffusional equilibrium) characterised by the essential absence of surface effects and of extraneous influences, such as electric or magnetic fields, will be considered. Note, however, that the influence of the earth's gravitational field, while usually ignored, will become important near a critical point. Volumetric properties of fluids are of pivotal importance in physics, physical chemistry and chemical engineering, and have thus received due attention in all

Volume Properties: Liquids, Solutions and Vapours
Edited by Emmerich Wilhelm and Trevor M. Letcher

Published by the Royal Society of Chemistry, www.rsc.org

modern monographs/textbooks dealing with fluids,[1–24] the most profitable approach being that based on a "marriage" of chemical thermodynamics with molecular theory and statistical mechanics, effectively anchoring it in the field of *molecular thermodynamics*. This term was coined by Prausnitz more than four decades ago.[9,25] It is an interdisciplinary field of great academic fascination and an indispensable part of chemical engineering. The impressive growth of molecular thermodynamics has been stimulated by the continuously increasing need for thermodynamic property data and phase equilibrium data[26–51] in the applied sciences, and has greatly profited by unprecedented advances in experimental techniques,[10,18,52–61] by advances in the theory of liquids in general, and by advances in computer simulations of reasonably realistic model systems.[62–72]

In this introductory Subsection 1.1, I will present something like a rough "road-map" of the major scientific areas covered by this monograph, though its aim and scope will only be crudely outlined by commenting on just a few selected topics close to my own research interests in physical chemistry of *non-electrolyte* fluids. The following Subsection 1.2 is essentially a concise review of pertinent parts of chemical thermodynamics, and provides the basis for most of the thermodynamic relations used in the other chapters. In addition, however, ramifications into neighbouring disciplines will be indicated, and occasionally historically significant contributions will be included. For details and additional aspects the interested reader should consult a textbook close to his/her taste, perhaps one of those listed in references 1 through 25. Concluding remarks and a brief outlook will be given in Subsection 1.3, while Subsection 1.4 will provide a glossary of symbols used, together with a few critical remarks concerning nomenclature in chemical thermodynamics.

Of course, true to the title of this book, pride of place will be given to the molar volume V and the mass density ρ characterising a single-phase fluid (either pure or a multicomponent mixture/solution), and their derivatives with respect to temperature and pressure. These two material properties are related by

$$\rho(T,P,\{x_i\}) = m/(nV(T,P,\{x_i\})), \tag{1.1}$$

where, $m=\sum_i m_i = \sum_i n_i m_{\mathrm{m},i}$ is the total mass of the phase, m_i is the mass of component i ($i=1, 2,\ldots$) with molar mass $m_{m,i}$, that is $m_i = n_i m_{m,i}$, n_i denotes the amount of substance of component i, $n=\sum_i n_i$ is the total amount of substance, nV is the total volume of the phase, $x_i = n_i/\sum_i n_i$ is the mole fraction of i, and $\{x_i\}$ is the set of compositional variables with $\sum_i x_i = 1$; for a pure fluid $x_i = 1$. The precise experimental determination of V or ρ over wide ranges of temperature and pressure, and in the case of mixtures/solutions also as a function of composition, yields information on the fluid state which is eminently useful at the experimental, practical level as well as at the theoretical level. The same is true for material properties characterising the temperature and pressure dependence of V or ρ. For instance, for a *constant-composition* fluid, the *isobaric expansivity* is defined by

$$\alpha_P \equiv \frac{1}{V}\left(\frac{\partial V}{\partial T}\right)_P = -\frac{1}{\rho}\left(\frac{\partial \rho}{\partial T}\right)_P, \text{ const }\{x_i\}, \tag{1.2}$$

the *isothermal compressibility*† is defined by

$$\beta_T \equiv -\frac{1}{V}\left(\frac{\partial V}{\partial P}\right)_T = \frac{1}{\rho}\left(\frac{\partial \rho}{\partial P}\right)_T, \text{ const }\{x_i\}, \tag{1.3}$$

and the *isentropic compressibility* (often loosely called the adiabatic compressibility) is defined by

$$\beta_S \equiv -\frac{1}{V}\left(\frac{\partial V}{\partial P}\right)_S = \frac{1}{\rho}\left(\frac{\partial \rho}{\partial P}\right)_S, \text{ const }\{x_i\}, \tag{1.4}$$

where S denotes the molar entropy. In passing we note that while β_S of liquids has been determined by purely mechanical methods (*PVT* methods),[73–77] *i.e.* by directly measuring the volume increase on sudden decompression, this method is considerably less common and less accurate than that based on ultrasonics.[55,78–81] For small sound wave amplitudes and sufficiently *low* frequencies well below any region where sound speed dispersion due to relaxation effects is observed,[82,83] to an excellent approximation

$$v_0^2 = \left(\frac{\partial P}{\partial \rho}\right)_S = \frac{1}{\rho\beta_S} = \frac{\kappa}{\rho\beta_T}, \text{ const }\{x_i\}, \tag{1.5}$$

where $v_0(T,P,\{x_i\})$, the low-frequency speed of ultrasound, may be treated as a thermodynamic equilibrium property, and as shown in detail in subsection 1.2

$$\kappa \equiv \frac{\beta_T}{\beta_S} = \frac{C_P}{C_V}, \text{ const }\{x_i\}. \tag{1.6}$$

Here, C_P denotes the molar heat capacity at constant pressure (molar isobaric heat capacity), and C_V is the molar heat capacity at constant volume (molar isochoric heat capacity).

The response of the system pressure to a temperature change at constant volume is represented by the *isochoric thermal pressure coefficient*

$$\gamma_V \equiv \left(\frac{\partial P}{\partial T}\right)_V, \text{ const }\{x_i\}. \tag{1.7}$$

---

†In this chapter the isothermal compressibility is represented by the symbol β_T and not by κ_T, as was recently recommended by IUPAC. Similarly, the isentropic compressibility is represented by the symbol β_S and not by κ_S. For their ratio, the symbol $\kappa \equiv \beta_T/\beta_S$ is used.

Since the three mutual derivatives of P, V and T satisfy the triple product rule

$$\left(\frac{\partial V}{\partial P}\right)_T \left(\frac{\partial P}{\partial T}\right)_V \left(\frac{\partial T}{\partial V}\right)_P = -1, \text{ const } \{x_i\}, \tag{1.8}$$

the three *mechanical coefficients* are related as follows:

$$\gamma_V = \frac{\alpha_P}{\beta_T}, \text{ const } \{x_i\}, \tag{1.9}$$

and

$$\left(\frac{\partial \alpha_P}{\partial P}\right)_T = -\left(\frac{\partial \beta_T}{\partial T}\right)_P, \text{ const } \{x_i\}. \tag{1.10}$$

As already pointed out, the experimental determination of volumetric properties of fluids occupies a central position in physics, physical chemistry and chemical engineering, and many distinguished scientists have contributed to this subject, that is to say, they contributed to the development of pressure–volume–temperature–composition relations which will eventually lead to reliable *PVTx equations of state* (EOS), applicable to both gaseous and liquid phases. These equations of state relate the variables in either a *pressure-explicit* form $P = f(T, nV, n_1, n_2, \ldots)$, or in a *volume-explicit* form $nV = f(T, P, n_1, n_2, \ldots)$, though most realistic equations of state are pressure-explicit, *i.e.* T, V, and the composition are the independent variables. In particular, experimental vapour phase or gas phase *PVTx*-data at low densities/low pressures have provided a large body of second virial coefficients B and third virial coefficients C,[26a,b] and have thus contributed enormously to our knowledge of intermolecular interaction.[84–88] For a *pure* fluid with a spherically symmetric potential-energy function $u(r)$ for a *pair* of molecules, where r is the distance between the molecules, the second virial coefficient is given by

$$B(T) = -2\pi L \int_0^\infty [\exp(-u(r)/k_B T) - 1] r^2 \mathrm{d}r. \tag{1.11}$$

Here, k_B is the Boltzmann constant, the quantity $f(r) \equiv e^{-u(r)/k_B T} - 1$ is commonly known as the Mayer f-function, and L is the Avogadro constant: the letter L is used in honor of Josef Loschmidt, pioneering Austrian physicist 1821–1895, who, in 1865, provided the first reasonable estimate for the number of particles N in a given volume of gas at ambient conditions, *i.e.* for the number density

$$\rho_N \equiv \frac{N}{nV} = \frac{L}{V}. \tag{1.12}$$

Equation (1.11) provides access to the *fundamentally* important potential-energy function $u(r)$,[86] which is frequently approximated by a Mie (n,m)-type function, introduced in 1903,[89,90]

$$u(r) = \left(\frac{n}{n-m}\right)\left(\frac{n}{m}\right)^{m/(n-m)} \varepsilon\left[\left(\frac{\sigma}{r}\right)^n - \left(\frac{\sigma}{r}\right)^m\right]. \tag{1.13}$$

The positive constants n and m $(n>m)$ are associated with molecular repulsion and attraction, respectively, ε is an intermolecular energy parameter characterising the well-depth of the interaction energy function, *i.e.* $u(r_{\min}) = -\varepsilon$, and σ is an intermolecular distance parameter characterised by $u(\sigma)=0$. Special cases of the Mie (n,m) function were introduced by Lennard-Jones in 1924 and connected with gas viscosities,[91] the equations of state of real gases,[92] X-ray measurements on crystals,[93] and quantum mechanics.[94] The most common form of the Lennard-Jones (12,6) function is[94]

$$u(r) = 4\varepsilon\left[\left(\frac{\sigma}{r}\right)^{12} - \left(\frac{\sigma}{r}\right)^{6}\right], \tag{1.14}$$

where $\sigma = 2^{-1/6} r_{\min}$.

On the practical side, when available over extended density/pressure regions, reliable equations of state may be used to calculate *residual functions*[10,18] which, together with *perfect-gas heat capacities*,[61] allow the convenient calculation of property changes of single-phase, constant-composition fluids for any change of state. Residual properties represent the difference between a real fluid property and the corresponding perfect-gas state value.

However, because of the more complex behaviour of *liquid* mixtures/solutions, their thermodynamic description *via* equations of state is frequently *not* satisfactory. In fact, many experimental data are more informative and easier to handle when expressed relative to some conveniently selected model behaviour closer to reality.[1–3,5,8] For instance, instead of determining total properties of mixtures, it is frequently useful to compare mixture properties to the properties of the *pure* constituents, which are indicated by a superscript asterisk (*). Denoting any *intensive mixture* property, such as molar Gibbs energy, molar enthalpy, molar volume, or molar isobaric heat capacity by $M(T,P,\{x_i\})$ (*excluding*, of course, T, P and $\{x_i\}$ which are used advantageously to describe the state of homogeneous equilibrium fluids), and the corresponding *intensive pure-fluid* property of component i by $M_i^*(T,P)$, a new class of thermodynamic functions, known as *molar property changes on mixing* may be defined, *i.e.*

$$\Delta M \equiv M - \sum_i x_i M_i^*. \tag{1.15}$$

They also depend on T, P and $\{x_i\}$. The corresponding *partial molar property change on mixing* is defined by

$$\Delta M_i \equiv \left(\frac{\partial(n\Delta M)}{\partial n_i}\right)_{T,P,n_{j\neq i}} = M_i - M_i^*, \tag{1.16}$$

where $M_i(T,P,\{x_i\})$ denotes a partial molar property, *i.e.*

$$M_i \equiv \left(\frac{\partial(nM)}{\partial n_i}\right)_{T,P,n_{j\neq i}}, \tag{1.17}$$

and nM is a total property of the solution phase.

Euler's theorem asserts that if a function f is homogeneous of degree k in the variables $z_1,z_2,\ldots,z_p$, that is, if it satisfies for any value of the parameter λ

$$f(\lambda z_1, \lambda z_2, \ldots, \lambda z_p) = \lambda^k f(z_1, z_2, \ldots, z_p), \tag{1.18}$$

it must also satisfy

$$kf(z_1, z_2, \ldots, z_p) = \sum_{i=1}^{p} z_i(\partial f/\partial z_i)_{z_{j\neq i}}. \tag{1.19}$$

In thermodynamics only homogeneous functions of degree $k=0$ and $k=1$ are important. The former are known as *intensive* variables, such as temperature and pressure, and the latter are known as *extensive* variables, such as mass, amount of substance and volume. The *ratio* of two extensive variables is again intensive, hence the density and the molar volume are intensive quantities. Since the total property nM of a phase is a homogeneous function of the first degree in the amounts of substance $\{n_i\}$, Euler's theorem yields

$$nM = \sum_i n_i M_i, \tag{1.20a}$$

or after division by the total amount of substance $n = \sum_i n_i$

$$M = \sum_i x_i M_i, \tag{1.20b}$$

and correspondingly

$$\Delta M = \sum_i x_i \Delta M_i. \tag{1.21}$$

A partial molar property is an intensive property of the mixture/solution and not of the particular component i. It has to be evaluated for each mixture/solution and, generally, $M_i \neq M_i^*$. However, a partial molar property *defined* by Equation (1.17) can always be used to provide a systematic *formal* subdivision of the extensive property nM into a sum of contributions of the individual species i in solution constrained by Equation (1.20a), or a systematic *formal* subdivision of the intensive property M into a sum of contributions of the individual species i in solution constrained by Equation (1.20b). Thus one may use partial molar properties as though they possess property values referring to the individual constituent species in solution. Such a formal subdivision may also be based on *mass* instead of amount of substance, in which case partial *specific* properties are obtained with similar physical significance.

The exact differential of the extensive property $nM = f(T,P,n_1,n_2,\ldots,n_p)$ is

$$\begin{aligned} d(nM) &= \left(\frac{\partial(nM)}{\partial T}\right)_{P,n} dT + \left(\frac{\partial(nM)}{\partial P}\right)_{T,n} dP + \sum_i \left(\frac{\partial(nM)}{\partial n_i}\right)_{T,P,n_{j \neq i}} dn_i \\ &= n\left(\frac{\partial M}{\partial T}\right)_{P,\{x_i\}} dT + n\left(\frac{\partial M}{\partial P}\right)_{T,\{x_i\}} dP + \sum_i M_i dn_i, \end{aligned} \tag{1.22}$$

where the subscript $n \equiv \{n_i\}$ indicates that all amounts of substance, and thus the composition, are held constant. From Equation (1.20a), for a general differential change of nM, we obtain

$$d(nM) = \sum_i M_i dn_i + \sum_i n_i dM_i, \tag{1.23}$$

hence comparison with Equation (1.22) immediately yields

$$\begin{aligned} 0 &= n\left(\frac{\partial M}{\partial T}\right)_{P,\{x_i\}} dT + n\left(\frac{\partial M}{\partial P}\right)_{T,\{x_i\}} dP - \sum_i n_i dM_i \\ &= \left(\frac{\partial M}{\partial T}\right)_{P,\{x_i\}} dT + \left(\frac{\partial M}{\partial P}\right)_{T,\{x_i\}} dP - \sum_i x_i dM_i. \end{aligned} \tag{1.24}$$

Equation (1.24) is the *most general* form of the Gibbs–Duhem equation for a homogeneous phase (see also below in Section 1.2): *all* changes in T, P and M_i must satisfy this equation. The Gibbs–Duhem equation is of fundamental importance in solution thermodynamics.

To reiterate, the following general system of notation will be used throughout:

- molar properties of multicomponent solutions, such as the molar volume, will be represented by the symbol M;
- molar pure-substance properties will be characterised by a superscript asterisk and identified by a subscript, *i.e.* M_i^*, $i = 1,2,\ldots$;
- partial molar properties referring to a component i in solution will be identified by a subscript alone, *i.e.* M_i, $i = 1,2,\ldots$

Additional superscripts/subscripts will be attached by definition.

Perhaps the most popular property-estimation techniques for *real* liquid (fluid) solutions/mixtures are based on the *ideal-solution model* (ideal-mixture model), indicated by a superscript id, for which the property changes on mixing are given by particularly simple mathematical expressions. The quantities that measure deviations from ideal-solution behaviour constitute still another class of thermodynamic functions, and are called *excess molar properties.* They are defined by

$$M^{\mathrm{E}}(T,P,\{x_i\}) \equiv M(T,P,\{x_i\}) - M^{\mathrm{id}}(T,P,\{x_i\}), \tag{1.25}$$

i.e. M^{E} is the difference between the property value of the real solution and the value calculated for an ideal solution at the same temperature, pressure and composition. The corresponding *excess partial molar property* for component i in solution is defined by

$$M_i^{\mathrm{E}} \equiv \left(\frac{\partial(nM^{\mathrm{E}})}{\partial n_i}\right)_{T,P,n_{j\neq i}} = M_i - M_i^{\mathrm{id}}, \tag{1.26}$$

where M_i^{id} is a partial molar property of component i in an ideal solution, and

$$M^{\mathrm{E}} = \sum_i x_i M_i^{\mathrm{E}}. \tag{1.27}$$

Thus the excess molar properties are also the differences between the real changes of properties on mixing and the ideal-solution changes of properties on mixing (that is the excess property changes on mixing),

$$\begin{aligned}\Delta M^{\mathrm{E}}(T,P,\{x_i\}) \equiv\ & \Delta M(T,P,\{x_i\}) \\ & - \Delta M^{\mathrm{id}}(T,P,\{x_i\}) = M^{\mathrm{E}}(T,P,\{x_i\}),\end{aligned} \tag{1.28a}$$

with a similar relation holding for the partial molar quantities:

$$\begin{aligned}\Delta M_i^{\mathrm{E}}(T,P,\{x_i\}) \equiv\ & \Delta M_i(T,P,\{x_i\}) \\ & - \Delta M_i^{\mathrm{id}}(T,P,\{x_i\}) = M_i^{\mathrm{E}}(T,P,\{x_i\}).\end{aligned} \tag{1.28b}$$

The terms excess molar property and excess molar property change on mixing may both be used interchangeably and are indeed found in the literature. Excess properties and residual properties (not discussed here at all) are, of course, related.

Focusing now on the excess molar volume, *i.e.* $M = V$, for an ideal solution, by definition

$$V_i^{\mathrm{id}}(T,P,\{x_i\}) = V_i^*(T,P), \tag{1.29}$$

and

$$V^{\mathrm{id}} = \sum_i x_i V_i^* \tag{1.30}$$

for all temperatures, pressures and compositions, hence

$$V^{\mathrm{E}} = \Delta V. \tag{1.31}$$

I emphasise that the definition Equation (1.25) is not restricted to any phase, though excess molar properties in general, and excess molar volumes in particular, are predominantly used for *liquid* mixtures/solutions. In fact, the single-phase thermodynamic property most frequently measured is the excess volume or the volume of mixing, see above. Note that

$$V^{\mathrm{E}}(T,P,\{x_i\}) = \left(\frac{\partial G^{\mathrm{E}}(T,P,\{x_i\})}{\partial P}\right)_{T,\{x_i\}}, \tag{1.32}$$

where G^{E} is the excess molar Gibbs energy. Because of its role in the thermodynamic treatment of phase equilibria, $G^{\mathrm{E}}(T,P,\{x_i\})$ has received particular attention, experimental and theoretical. This topic will be touched upon in the next section.

PVTx-measurements in all variants have a long history, and the nature and size of the area make it virtually impossible to cover the entire subject in one book. Fortunately, in recent years considerable effort has been invested by the *International Union of Pure and Applied Chemistry* (IUPAC) and the *International Association of Chemical Thermodynamics* (IACT) to review experimental techniques as well as the corresponding thermodynamic formalism, with emphasis on progress in equations of state research.[10,18,52,53,56,58–61] In the present monograph the focus is on topics not treated so far, and on topics where recent developments make it desirable to revisit them. Any omission is not to be taken as a measure of its importance, but is essentially a consequence of space limitations. Thus, although the book is not comprehensive, it is intended to present state-of-the-art overviews and to discuss advances in many of the currently active fields of *PVTx*-research.

The individual specialised chapters have again[61] been written by internationally renowned thermodynamicists/thermophysicists. Because of their topical diversity, in this introductory chapter I shall try to summarise concisely most of the important basic thermodynamic relations relevant for the discussion of volumetric properties of fluid systems that will be used in other chapters, to clarify, perhaps, some points occasionally obscured or overlooked, to indicate cross-fertilisation with neighbouring disciplines, and to point out a few less familiar yet potentially interesting problems. Because of the fundamental character of thermodynamics, a certain parallelism with the introductory chapter[95] of our recent monograph on heat capacities is, however, unavoidable.

Thermodynamics rests on an experiment-based *axiomatic* fundament. Experiments, together with theory and computer simulation, are the *pillars of science*, and Figure 1.1 (the "knowledge triangle"[96,97]) indicates what may be learned from a comparison of respective results under idealised conditions. It may be used to illustrate the process of *inductive* reasoning in science, also known informally as *bottom-up* reasoning, which amplifies and generalises our experimental observations, eventually leading to theories and new knowledge. In contradistinction, *deduction*, informally known as *top-down* reasoning, orders and explicates already existing knowledge, thereby leading to predictions which may be corroborated by experiment, or, in principle, falsified (see Popper[98]). Classical thermodynamics is a highly

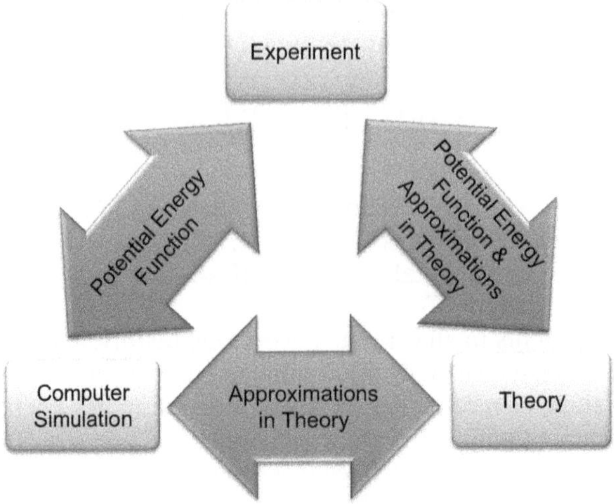

Figure 1.1 The *three pillars of science*: experiment, theory and computer simulation. The double-headed arrows indicate possible fundamentally important comparisons which will contribute to a deeper understanding of the role of approximations concerning interaction energies and theoretical models (after E. Wilhelm, Determination of caloric quantities of dilute liquid solutions, *Thermochim. Acta*, 1987, **119**, 17–33).[256]

formalised scientific discipline of enormous generality, providing a mathematical framework of equations (and a few inequalities) from a small number of fundamental postulates, which yields exact *relations* between macroscopically observable thermodynamic equilibrium properties of matter and restricts the course of any natural process. The central feature of thermodynamics is its independence from considerations of microscopic, molecular phenomena. None of the derived relations has in fact ever been shown experimentally to be false. In the sense that mathematics is an exact science, thermodynamics is an exact science, and the validity of the derived relations depends only on the validity of these fundamental postulates. Indeed, the role of mathematics in physical theories in general is an important topic in contemporary philosophy and physics.[99–103]

1.2 Thermodynamics: Fundamentals and Applications

When focusing on *PVTx* systems, any differential change from one *equilibrium state* to another in a *closed multiphase, multicomponent* system is described by the *basic* differential equation combining the *first* and the *second laws* of thermodynamics, which relates the three primary thermodynamic properties internal energy, entropy and volume either in the energy representation

$$\mathrm{d}(nU) = T\mathrm{d}(nS) - P\mathrm{d}(nV), \tag{1.33}$$

or, alternatively, in the entropy representation

$$d(nS) = \frac{1}{T}d(nU) + \frac{P}{T}d(nV). \tag{1.34}$$

Here, U, S, and V represent *overall* molar properties of a fluid system and n is the total amount of substance. For the special case of a *homogeneous solution*, *i.e.* a *single-phase* multicomponent system, either open or closed,[1,104,105] we assume that $nU = f(nS, nV, \{n_i\})$ or $nS = f(nU, nV, \{n_i\})$, respectively, where U is the molar internal energy, S is the molar entropy, V is the molar volume of the phase, and n_i designates the amount of substance of the chemical species i. The corresponding *fundamental equations for a change of the state of a phase*, also known as the *fundamental property relations*, or the *differential forms of the fundamental equations*, or the *Gibbs equations* are

$$d(nU) = Td(nS) - Pd(nV) + \sum_i \mu_i dn_i, \tag{1.35}$$

and equivalently,

$$d(nS) = \frac{1}{T}d(nU) + \frac{P}{T}d(nV) - \sum_i \frac{\mu_i}{T}dn_i. \tag{1.36}$$

The intensive parameter furnished by the first-order partial derivatives of the internal energy with respect to the amount of substance of component i,

$$\mu_i \equiv \left[\frac{\partial(nU)}{\partial n_i}\right]_{nS,nV,n_{j\neq i}}, \tag{1.37}$$

is called the *chemical potential* of component i. Its introduction extends the scope to the general case of a single-phase system in which the n_i may *vary*, either by exchanging matter with its surroundings (open system) or by changes in composition occurring as a result of chemical reactions within (reactive system) or both. Chemical-reaction equilibria, however, will *not* be considered here.

Analogously, from Equation (1.36) we obtain

$$\frac{\mu_i}{T} = \left[\frac{\partial(nS)}{\partial n_i}\right]_{nU,nV,n_{j\neq i}}. \tag{1.38}$$

Corresponding to Equations (1.35) and (1.36), the *primary thermodynamic functions* (or *integrated forms of the fundamental equations for a change of the state of a phase*, or *fundamental equations*, or *cardinal functions*, or *thermodynamic potentials*, or *Euler equations*) are

$$nU = T(nS) - P(nV) + \sum_i \mu_i n_i \tag{1.39}$$

in the energy representation, and

$$nS = \frac{1}{T}(nU) + \frac{P}{T}(nV) - \sum_i \frac{\mu_i}{T}n_i \tag{1.40}$$

in the entropy representation. They may be obtained by integrating Equation (1.35) and (1.36), respectively, at constant values of the intensive properties T, P and the μ_i, or $\frac{1}{T}, \frac{P}{T}$ and the $\frac{\mu_i}{T}$, respectively. Alternatively, Equations (1.39) and (1.40) can be regarded as a consequence of Euler's theorem: the internal energy is a homogeneous function of degree one in terms of the extensive properties nS,nV and $\{n_i\}$, *i.e.* $nU = f(nS,nV,\{n_i\})$, while the entropy is a homogeneous function of degree one in terms of the extensive properties nU,nV and $\{n_i\}$, *i.e.* $nS = f(nU,nV,\{n_i\})$. These sets of variables are called the corresponding *canonical* (or natural) variables. *All* thermodynamic equilibrium properties of any system can be derived from these functions, and it is for this reason that they are called primary functions or cardinal functions or fundamental functions.

In both the energy and entropy representations the extensive quantities are the mathematically independent variables, while the intensive parameters are derived, a situation which does *not* conform to experimental practice. The choice of nS and nV as independent extensive variables in the fundamental property relation in the energy representation is *not* convenient, and Equation (1.39) suggests the definition of useful *alternative* energy-based primary functions related to nU and with total differentials consistent with Equation (1.35), but with a set of canonical variables different from $\{nS, nV, \{n_i\}\}$. The appropriate method for generating them *without loss of information* is the *Legendre transformation*.[82,105,106] The most popular additional equivalent primary functions are the *enthalpy*

$$nH \equiv nU + P(nV), \tag{1.41}$$

the Helmholtz energy

$$nF \equiv nU - T(nS), \tag{1.42}$$

and the *Gibbs energy* (a double transform)

$$nG \equiv nU + P(nV) - T(nS) = nH - T(nS), \tag{1.43}$$

where H, F and G are the respective *molar* quantities. The alternative energy-based fundamental property relations for the enthalpy, the Helmholtz energy and the Gibbs energy are thus

$$\mathrm{d}(nH) = T\mathrm{d}(nS) + (nV)\mathrm{d}P + \sum_i \mu_i \mathrm{d}n_i, \tag{1.44}$$

$$\mathrm{d}(nF) = -(nS)\mathrm{d}T - P\mathrm{d}(nV) + \sum_i \mu_i \mathrm{d}n_i, \tag{1.45}$$

and

$$\mathrm{d}(nG) = -(nS)\mathrm{d}T + (nV)\mathrm{d}P + \sum_i \mu_i \mathrm{d}n_i. \tag{1.46}$$

Equation (1.46) is of central importance in solution thermodynamics.

The integrated forms of the fundamental property relations Equations (1.44), (1.45) and (1.46), *i.e.* the corresponding primary functions, or alternative fundamental equations, or alternative thermodynamic potentials, are

$$nH = T(nS) + \sum_i \mu_i n_i, \tag{1.47}$$

$$nF = -P(nV) + \sum_i \mu_i n_i, \tag{1.48}$$

and

$$nG = \sum_i \mu_i n_i. \tag{1.49}$$

These alternative groupings may also be obtained from Equations (1.41), (1.42) and (1.43), respectively, by substituting for nU according to Equation (1.39).

Since Equations (1.44), (1.45) and (1.46) are equivalent to Equation (1.35), we have

$$\mu_i \equiv \left[\frac{\partial(nU)}{\partial n_i}\right]_{nS,nV,n_j \neq i} = \left[\frac{\partial(nH)}{\partial n_i}\right]_{nS,P,n_j \neq i} = \left[\frac{\partial(nF)}{\partial n_i}\right]_{T,nV,n_j \neq i} = \left[\frac{\partial(nG)}{\partial n_i}\right]_{T,P,n_j \neq i}. \tag{1.50}$$

Division of Equations (1.39), (1.47), (1.48) and (1.49) by n yields the corresponding molar functions:

$$U = TS - PV + \sum_i x_i \mu_i, \tag{1.51}$$

$$H = TS + \sum_i x_i \mu_i, \tag{1.52}$$

$$F = -PV + \sum_i x_i \mu_i, \tag{1.53}$$

$$G = \sum_i x_i \mu_i. \tag{1.54}$$

The four fundamental property relations/primary functions presented so far are *equivalent*, however each one is associated with a different set of canonical (natural) *variables*: $\{nS, nV, \{n_i\}\}$, $\{nS, P, \{n_i\}\}$, $\{T, nV, \{n_i\}\}$ and $\{T, P, \{n_i\}\}$, respectively. The selection of any primary thermodynamic function/fundamental property relation depends on which set of independent variables best simplifies the problem to be solved. In physical chemistry by far the most useful set of independent variables is $\{T,P,\{n_i\}\}$, because, in principle, temperature, pressure and composition are easily measured and controlled. Thus, for chemists, the Gibbs energy $nG = f(T,P,\{n_i\})$ is of central importance. Since all the fundamental property relations are equivalent, alternative expressions for the chemical potential are possible [see Equation (1.50)], of which

$$\mu_i \equiv \left[\frac{\partial(nG)}{\partial n_i}\right]_{T,P,n_{j\neq i}} \tag{1.55}$$

is the preferred one. We recognise that the chemical potential of component i is just the partial molar Gibbs energy of component i, that is

$$\mu_i = G_i. \tag{1.56}$$

This property holds a key position in solution thermodynamics.

A primary function which arises naturally in statistical mechanics is the *grand canonical potential.* It is the double Legendre transform of the internal energy nU when simultaneously the extensive entropy is replaced by its conjugate intensive variable, the temperature, and the extensive amount of substance by its conjugate intensive variable, the chemical potential:

$$nJ \equiv nU - T(nS) - \sum_i \mu_i n_i, \tag{1.57}$$

and the corresponding fundamental property relation is

$$\mathrm{d}(nJ) = -(nS)\mathrm{d}T - P\mathrm{d}(nV) - \sum_i n_i \mathrm{d}\mu_i, \tag{1.58}$$

with the associated canonical variables $(T,nV,\{\mu_i\})$. The integrated, alternative form is

$$nJ = -P(nV), \tag{1.59}$$

and the corresponding molar function is given by

$$J = -PV. \tag{1.60}$$

The *complete* Legendre transform vanishes identically for any system. The complete transform of the internal energy replaces *all* extensive canonical variables by their conjugate intensive variables, thus yielding the *null-function*

$$0 = nU - T(nS) + P(nV) - \sum_i \mu_i n_i \tag{1.61}$$

as the final alternative primary function in the energy representation, and

$$0 = -(nS)\mathrm{d}T + (nV)\mathrm{d}P - \sum_i n_i \mathrm{d}\mu_i \tag{1.62}$$

as the corresponding alternative form of the fundamental property relation. Division by $n = \sum_i n_i$ yields

$$0 = -SdT + VdP - \sum_i x_i d\mu_i. \tag{1.63}$$

Equations (1.62) and (1.63) are frequently used forms of the *Gibbs–Duhem* equation. They represent important relations between the intensive parameters T, P and μ_i of the system and show that they are *not* independent of each other.

For an exact differential expression containing p independent variables there exist $2^p - 2$ *partial* Legendre transforms, while the *complete* Legendre transform vanishes identically for any system. Thus treating the sum $\sum_i \mu_i n_i$ in the energy representation Equation (1.39) as a single term, the *total* number of equivalent primary functions and therefore the *total* number of equivalent fundamental property relations for a thermodynamic *PVTx*-system is eight ($= 2^3$). That is nU, Equation (1.39), plus seven equivalent primary functions (including the *null-function*), or alternatively Equation (1.35) plus seven equivalent fundamental property relations (including the Gibbs–Duhem equation). Of the equivalent primary functions, five have already been treated above: nH, nF, nG, nJ and the null-function. The remaining two,

$$nX = nU - \sum_i \mu_i n_i, \tag{1.64}$$

and

$$nY = nU + P(nV) - \sum_i \mu_i n_i, \tag{1.65}$$

with the alternative forms

$$nX = T(nS) - P(nV), \tag{1.66}$$

and

$$nY = T(nS), \tag{1.67}$$

respectively, are rarely used and have not received separate symbols or names. The corresponding fundamental property relations are

$$d(nX) = Td(nS) - Pd(nV) - \sum_i n_i d\mu_i, \tag{1.68}$$

and

$$d(nY) = Td(nS) + (nV)dP - \sum_i n_i d\mu_i. \tag{1.69}$$

Equations (1.35) and (1.44) through (1.46) are exact differential expressions, hence application of the reciprocity relation yields Maxwell-type equations. For instance, focusing on the Helmholtz function, Equation (1.45), which is of primary importance in statistical mechanics, we find

$$nS = -\left(\frac{\partial(nF)}{\partial T}\right)_{nV,n}, \tag{1.70}$$

$$P = -\left(\frac{\partial(nF)}{\partial(nV)}\right)_{T,n}, \tag{1.71}$$

and the Maxwell-type equation

$$\left(\frac{\partial(nS)}{\partial(nV)}\right)_{T,n} = \left(\frac{\partial P}{\partial T}\right)_{nV,n}, \tag{1.72}$$

where the subscript n signifies that all amounts of substance $n = \{n_i\}$ are kept constant. As already pointed out, for physical chemists the Gibbs function is of central importance. Equation (1.46) yields

$$nS = -\left(\frac{\partial(nG)}{\partial T}\right)_{P,n}, \tag{1.73}$$

$$nV = \left(\frac{\partial(nG)}{\partial P}\right)_{T,n}, \tag{1.74}$$

and the Maxwell-type equations

$$-\left(\frac{\partial(nS)}{\partial P}\right)_{T,n} = \left(\frac{\partial(nV)}{\partial T}\right)_{P,n}, \tag{1.75}$$

$$\left(\frac{\partial \mu_i}{\partial T}\right)_{P,n} = -\left(\frac{\partial(nS)}{\partial n_i}\right)_{T,P,n_{j\neq i}} = -S_i, \tag{1.76}$$

$$\left(\frac{\partial \mu_i}{\partial P}\right)_{T,n} = \left(\frac{\partial(nV)}{\partial n_i}\right)_{T,P,n_{j\neq i}} = V_i, \tag{1.77}$$

and

$$\left(\frac{\partial \mu_i}{\partial n_j}\right)_{T,P,n_{k\neq j}} = \left(\frac{\partial \mu_j}{\partial n_i}\right)_{T,P,n_{k\neq i}}. \tag{1.78}$$

Here, S_i denotes the partial molar entropy.

We note that a differential in p variables,

$$\mathrm{d}f(z_1, z_2, \ldots, z_p) = \sum_{i=1}^{p} Q_i(z_1, z_2, \ldots, z_p)\mathrm{d}z_i, \tag{1.79}$$

is *exact* if, and only if, the reciprocity relations

$$\left(\frac{\partial Q_i}{\partial z_j}\right)_{z_{i\neq j}}=\left(\frac{\partial Q_j}{\partial z_i}\right)_{z_{j\neq i}},i,j=1,2,\ldots,p, \tag{1.80}$$

are fulfilled for each of the $p(p-1)/2$ pairs of conjugate variables (Q_i,z_i) and (Q_j,z_j). If df is an exact differential, then $\int_a^b \sum Q_i \mathrm{d}z_i$ is independent of the integration path, and in thermodynamics such a function f is called a state function.

The Maxwell equations, Equations (1.72) and (1.75), are particularly useful in EOS applications, because entropy derivatives are replaced by derivatives involving the directly measurable quantities P, V, and T. Since the subscript n signifies that all amounts of substance are held constant, for a constant-composition PVT system they simplify to

$$\left(\frac{\partial S}{\partial V}\right)_T=\left(\frac{\partial P}{\partial T}\right)_V=\gamma_V,\ \text{constant composition}, \tag{1.81}$$

and

$$\left(\frac{\partial S}{\partial P}\right)_T=-\left(\frac{\partial V}{\partial T}\right)_P=-\alpha_P V,\ \text{constant composition}. \tag{1.82}$$

Maxwell equations form part of the thermodynamic basis of the relatively new experimental technique known as *scanning transitiometry* (see Chapter 14).[107,108]

Legendre transformations of the primary function in the *entropy representation*, $nS=f(nU,nV,\{n_i\})$, Equation (1.40), resulting in the replacement of one or more extensive variables by the corresponding conjugate intensive variable(s) $1/T$, P/T and μ_i/T, respectively, yield new primary functions known as *Massieu–Planck functions*, whose total differentials are compatible with Equation (1.36). Again, treating the sum $\sum_i \mu_i n_i$ as a single term, the *total* number of equivalent primary functions and therefore the *total* number of equivalent fundamental property relations in the entropy representation for a thermodynamic $PVTx$ system is eight: nS, Equation (1.40), plus seven alternatives (including the *null-function*), or alternatively Equation (1.36), plus seven alternatives, respectively (including the Gibbs–Duhem equation). For instance, *the* Massieu function is defined by

$$n\Psi \equiv nS-\frac{1}{T}(nU), \tag{1.83}$$

with its alternative form

$$n\Psi=\frac{P}{T}(nV)-\sum_i \frac{\mu_i}{T}n_i, \tag{1.84}$$

and the corresponding form of the entropy-based fundamental property relation is

$$\mathrm{d}(n\Psi)=-(nU)\mathrm{d}\frac{1}{T}+\frac{P}{T}\mathrm{d}(nV)-\sum_i \frac{\mu_i}{T}\mathrm{d}n_i. \tag{1.85}$$

A second-order Legendre transformation yields *the* Planck function

$$n\Phi \equiv nS - \frac{1}{T}(nU) - \frac{P}{T}(nV), \tag{1.86}$$

with its alternative form

$$n\Phi = -\sum_i \frac{\mu_i}{T} n_i, \tag{1.87}$$

and the corresponding entropy-based fundamental property relation reads

$$\mathrm{d}(n\Phi) = -(nU)\mathrm{d}\frac{1}{T} - (nV)\mathrm{d}\frac{P}{T} - \sum_i \frac{\mu_i}{T}\mathrm{d}n_i. \tag{1.88}$$

Note that

$$\Psi = -\frac{F}{T}, \tag{1.89}$$

and

$$\Phi = -\frac{G}{T}. \tag{1.90}$$

Another second-order Legendre transform is the Kramer function

$$n\Omega \equiv nS - \frac{1}{T}(nU) + \sum_i \frac{\mu_i}{T} n_i. \tag{1.91}$$

Its alternative form is

$$n\Omega = \frac{P}{T}(nV), \tag{1.92}$$

hence for the molar Kramer function we obtain

$$\Omega = -\frac{J}{T}. \tag{1.93}$$

The corresponding form of the entropy-based fundamental property relation is

$$\mathrm{d}(n\Omega) = -(nU)\mathrm{d}\frac{1}{T} + \frac{P}{T}\mathrm{d}(nV) + \sum_i n_i \mathrm{d}\left(\frac{\mu_i}{T}\right). \tag{1.94}$$

Again, the complete Legendre transform is identical to zero, yielding the null-function

$$0 = nS - \frac{nU}{T} - \frac{P(nV)}{T} + \sum_i \frac{\mu_i}{T} n_i \tag{1.95}$$

as the alternative primary function in the entropy representation. The corresponding fundamental property relation that might be called a version of the *entropy-based* Gibbs–Duhem equation reads

$$0 = (nU)\mathrm{d}\left(\frac{1}{T}\right) + (nV)\mathrm{d}\left(\frac{P}{T}\right) - \sum_i n_i \mathrm{d}\left(\frac{\mu_i}{T}\right), \tag{1.96}$$

thus showing that the intensive parameters $1/T$, P/T and μ_i/T in the entropy representation are also *not* independent of each other.

Focusing now on *constant-composition* systems, and thus dropping the subscripts n, $\{x_i\}$, *etc.*, whenever unambiguously permissible, for *one mole* of a homogeneous fluid the following four energy-based fundamental property relations apply:

$$dU = TdS - PdV, \tag{1.97}$$

$$dH = TdS + VdP, \tag{1.98}$$

$$dF = -SdT - PdV, \tag{1.99}$$

$$dG = -SdT + VdP. \tag{1.100}$$

It follows that

$$T = (\partial U/\partial S)_V = (\partial H/\partial S)_P, \tag{1.101}$$

$$P = -(\partial U/\partial V)_S = -(\partial F/\partial V)_T, \tag{1.102}$$

$$V = (\partial H/\partial P)_S = (\partial G/\partial P)_T, \tag{1.103}$$

$$S = -(\partial F/\partial T)_V = -(\partial G/\partial T)_P. \tag{1.104}$$

These relations establish the link between the natural independent variables S, V, P, T and the energy-based functions U, H, F, G. In view of the definitions of F and G, and Equation (104), the *Gibbs–Helmholtz* equations

$$U = F - T(\partial F/\partial T)_V \tag{1.105a}$$

and

$$H = G - T(\partial G/\partial T)_P, \tag{1.105b}$$

are obtained. Simple mathematical transformations lead to the following alternative forms:

$$\left(\frac{\partial(F/T)}{\partial T}\right)_V = -\frac{U}{T^2}, \tag{1.106}$$

$$\left(\frac{\partial(F/T)}{\partial(1/T)}\right)_V = U, \tag{1.107}$$

$$\left(\frac{\partial(G/T)}{\partial T}\right)_P = -\frac{H}{T^2}, \tag{1.108}$$

$$\left(\frac{\partial(G/T)}{\partial(1/T)}\right)_P = H. \tag{1.109}$$

Equation (1.108) may be used to develop an alternative form of the fundamental property relation, Equation (1.46), with the canonical variables T, P and $\{n_i\}$:

$$d\left(\frac{nG}{RT}\right) = -\frac{nH}{RT^2}dT + \frac{nV}{RT}dP + \sum_i \frac{\mu_i}{RT}dn_i, \tag{1.110a}$$

and

$$\frac{G}{RT} = \sum_i x_i \frac{\mu_i}{RT}. \tag{1.110b}$$

Here, $R = Lk_B$ denotes the molar gas constant. This equation is of considerable utility, with all terms having the dimension of amount-of-substance and, in contrast to Equation (1.46), the enthalpy rather than the entropy appears in the first term of the right-hand side of Equation (1.110a).

We note that the *parallelism* existing between equations valid for constant-composition solutions, and for the corresponding partial molar quantities in such solutions, greatly facilitates the formulation of equations by analogy. This approach is valid whenever the properties appearing in any equation referring to a constant-composition solution are *linearly* related. For instance, the analogue to Equation (1.100) is

$$d\mu_i \equiv dG_i = -S_i dT + V_i dP, \tag{1.111}$$

and

$$G_i = H_i - TS_i, \tag{1.112}$$

where

$$H_i \equiv \left(\frac{\partial(nH)}{\partial n_i}\right)_{T,P,n_{j\neq i}} = U_i + PV_i, \tag{1.113}$$

and

$$U_i \equiv \left(\frac{\partial(nU)}{\partial n_i}\right)_{T,P,n_{j\neq i}}. \tag{1.114}$$

Similarly,

$$F_i \equiv \left(\frac{\partial(nF)}{\partial n_i}\right)_{T,P,n_{j\neq i}} = U_i - TS_i, \tag{1.115}$$

hence Helmholtz-type equations analogous to Equations (1.106) and (1.108) result,

$$\left(\frac{\partial(F_i/T)}{\partial T}\right)_{V,\{x_i\}} = -\frac{U_i}{T^2} \tag{1.116}$$

and

$$\left(\frac{\partial(G_i/T)}{\partial T}\right)_{P,\{x_i\}} = -\frac{H_i}{T^2}, \tag{1.117}$$

and so forth.

The volume dependence of the internal energy and the pressure dependence of the enthalpy are conveniently derived through differentiating the appropriate Gibbs–Helmholtz equations, Equations (1.105a) and (1.105b):

$$(\partial U/\partial V)_T = -P + T(\partial P/\partial T)_V = -P + T\gamma_V, \tag{1.118}$$

$$(\partial H/\partial P)_T = V - T(\partial V/\partial T)_P = V - TV\alpha_P. \tag{1.119}$$

Note that both equations can be contracted to yield

$$\left(\frac{\partial U}{\partial V}\right)_T = T^2\left(\frac{\partial(P/T)}{\partial T}\right)_V = -\left(\frac{\partial(P/T)}{\partial(1/T)}\right)_V, \tag{1.120}$$

$$\left(\frac{\partial H}{\partial P}\right)_T = -T^2\left(\frac{\partial(V/T)}{\partial T}\right)_P = \left(\frac{\partial(V/T)}{\partial(1/T)}\right)_P. \tag{1.121}$$

The molar heat capacity at constant volume (molar isochoric heat capacity) is defined by

$$C_V \equiv \left(\frac{\partial U}{\partial T}\right)_V = T\left(\frac{\partial S}{\partial T}\right)_V = -T\left(\frac{\partial^2 F}{\partial T^2}\right)_V, \tag{1.122}$$

And, from Equation (1.118), its volume dependence is given by

$$(\partial C_V/\partial V)_T = T(\partial^2 P/\partial T^2)_V = T(\partial\gamma_V/\partial T)_V. \tag{1.123}$$

The molar heat capacity at constant pressure (molar isobaric heat capacity) is defined by

$$C_P \equiv \left(\frac{\partial H}{\partial T}\right)_P = T\left(\frac{\partial S}{\partial T}\right)_P = -T\left(\frac{\partial^2 G}{\partial T^2}\right)_P, \tag{1.124}$$

and from Equation (1.119) its pressure dependence is given by

$$(\partial C_P/\partial P)_T = -T(\partial^2 V/\partial T^2)_P = -TV\left[\alpha_P^2 + (\partial\alpha_P/\partial T)_P\right]. \tag{1.125}$$

And the isothermal compressibility may also be expressed as

$$\beta_T = \left(V\left(\frac{\partial^2 F}{\partial V^2}\right)_T\right)^{-1} = -\frac{(\partial^2 G/\partial P^2)_T}{(\partial G/\partial P)_T}. \tag{1.126}$$

In high-pressure research,[82,95,107–115] Equations (1.123) and (1.125) are important. For instance, the *pressure* dependence of the isobaric heat capacity of a constant-composition fluid may be either determined from *PVT* data alone or by high-pressure calorimetry or by transitiometry or by measuring the speed of ultrasound v_0 as a function of P and T (see below), and the consistency of the respective experimental results has to be ascertained by Equation (1.125).

We are now in the position to present the functional dependence of the molar internal energy and the molar entropy of a homogeneous fluid at constant composition on T and V, and the functional dependence of the molar enthalpy and the molar entropy of such a fluid on T and P. Starting with

$$\mathrm{d}U = \left(\frac{\partial U}{\partial T}\right)_V \mathrm{d}T + \left(\frac{\partial U}{\partial V}\right)_T \mathrm{d}V \tag{1.127}$$

and

$$\mathrm{d}S = \left(\frac{\partial S}{\partial T}\right)_V \mathrm{d}T + \left(\frac{\partial S}{\partial V}\right)_T \mathrm{d}V, \tag{1.128}$$

respectively, and replacing the partial derivatives using Equations (1.81), (1.118) and (1.122), we obtain

$$\mathrm{d}U = C_V \mathrm{d}T + (T\gamma_V - P)\mathrm{d}V \tag{1.129}$$

and

$$\mathrm{d}S = \frac{C_V}{T}\mathrm{d}T + \gamma_V \mathrm{d}V. \tag{1.130}$$

If temperature and pressure are selected as the independent variables, an entirely analogous procedure using Equations (1.82), (1.119) and (1.124) gives

$$\mathrm{d}H = C_P \mathrm{d}T + V(1 - T\alpha_P)\mathrm{d}P \tag{1.131}$$

and

$$\mathrm{d}S = \frac{C_P}{T}\mathrm{d}T - V\alpha_p \mathrm{d}P. \tag{1.132}$$

Thus we obtain

$$\left(\frac{\partial S}{\partial V}\right)_P = \frac{C_V}{TV\alpha_P} + \gamma_V = \frac{C_P}{TV\alpha_P}, \tag{1.133}$$

$$\left(\frac{\partial S}{\partial P}\right)_V = \frac{C_P}{T\gamma_V} - V\alpha_P = \frac{C_V}{T\gamma_V}, \tag{1.134}$$

and

$$\mathrm{d}S = \frac{C_P}{TV\alpha_P}\mathrm{d}V + \frac{C_V}{T\gamma_V}\mathrm{d}P, \tag{1.135}$$

These equations complement Equations (1.81), (1.82), (1.130) and (1.132), respectively. Equations (1.130), (1.132) and (1.135) are equivalent.

The *difference* between the heat capacities C_P and C_V depends on *volumetric* properties *only*, and may be derived as follows. According to Equation (1.128),

$$\left(\frac{\partial S}{\partial T}\right)_P = \left(\frac{\partial S}{\partial T}\right)_V + \left(\frac{\partial S}{\partial V}\right)_T \left(\frac{\partial V}{\partial T}\right)_P, \tag{1.136}$$

which yields, with Equations (1.2), (1.7), (1.81), (1.122) and (1.124),

$$C_P - C_V = TV\alpha_P\gamma_V. \tag{1.137}$$

Alternatively, we obtain [*cf.* Equation (1.9)]

$$C_P - C_V = TV\alpha_P^2/\beta_T \tag{1.138}$$

and

$$C_P - C_V = TV\beta_T\gamma_V^2. \tag{1.139}$$

Since the compression factor is defined by

$$Z \equiv \frac{PV}{RT}, \tag{1.140}$$

alternatively,[82,95]

$$C_P - C_V = R\frac{\left[Z + T(\partial Z/\partial T)_P\right]^2}{Z - P(\partial Z/\partial P)_T}. \tag{1.141}$$

The *ratio* of the heat capacities, $\kappa \equiv C_P/C_V$, is accessible *via* Equations (1.122) and (1.124):

$$\kappa \equiv \frac{C_P}{C_V} = \frac{(\partial S/\partial T)_P}{(\partial S/\partial T)_V} = \frac{(\partial S/\partial V)_P(\partial V/\partial T)_P}{(\partial S/\partial P)_V(\partial P/\partial T)_V}. \tag{1.142}$$

According to the triple product rule

$$\frac{(\partial S/\partial V)_P}{(\partial S/\partial P)_V} = -\left(\frac{\partial P}{\partial V}\right)_S \tag{1.143}$$

and

$$\frac{(\partial V/\partial T)_P}{(\partial P/\partial T)_V} = -\left(\frac{\partial V}{\partial P}\right)_T = V\beta_T, \tag{1.144}$$

hence

$$\kappa \equiv \frac{C_P}{C_V} = -\left(\frac{\partial P}{\partial V}\right)_S V\beta_T. \tag{1.145}$$

With Equation (1.4) we now obtain the important Equation (1.6), *i.e.* $\kappa \equiv C_P/C_V = \beta_T/\beta_S$, thereby establishing a connection with ultrasonics.[55,78–81] Using Equation (1.138) in conjunction with Equation (1.5) leads to

$$\kappa = 1 + \frac{Tm_{\mathrm{m}}\alpha_P^2 v_0^2}{C_P}, \tag{1.146}$$

which is one of the *most important* equations in thermophysics. At low temperatures, where γ_V of liquids is *very* large, the *direct* calorimetric determination of C_V of *liquids* is not easy and requires sophisticated instrumentation, as evidenced by the careful work of Magee at NIST,[116a,b] though it becomes more practicable near the critical point where γ_V is much smaller.

From either one of the equations for the difference $C_P - C_V$ it follows that

$$C_V = TV\frac{\alpha_P^2}{\beta_T - \beta_S}\frac{\beta_S}{\beta_T} \tag{1.147a}$$

and

$$C_P = TV\frac{\alpha_P^2}{\beta_T - \beta_S}. \tag{1.147b}$$

Perhaps most noteworthy in the present context is that heat capacities may be determined *without* measuring a quantity of heat.

Combining the Equations (1.147a,b) with Equation (1.138) yields

$$\frac{C_V}{C_P - C_V} = \frac{\beta_S}{\beta_T - \beta_S} \tag{1.148a}$$

and

$$\frac{C_P}{C_P - C_V} = \frac{\beta_T}{\beta_T - \beta_S}. \tag{1.148b}$$

Equation (1.148a) establishes a connection with *Rayleigh–Brillouin light scattering.*[83] For simple liquids (liquid noble gases), the ratio of the integrated intensity of the central, unshifted Rayleigh peak, I_R, and the integrated intensity of the *two* Brillouin peaks, $2I_B$, is given by the *Landau–Placzek* ratio, *i.e.*

$$\frac{I_R}{2I_B} = \frac{C_P - C_V}{C_V} = \kappa - 1. \tag{1.149}$$

For *molecular* (normal) liquids, the ratio of the integrated intensity of the central, unshifted components of the scattered light (Rayleigh and Mountain) to the integrated intensity of the Brillouin peaks is a rather complicated expression,[117a,b] and the ratio is greater[83] than $(\kappa - 1)$. Evidently, when $C_P \approx C_V$ it is rather difficult to observe the central Rayleigh peak. Such a case exists for liquid water, at temperatures around that of the density maximum.

For glass-forming liquids, such as toluene, α-picoline, ethanol or glycerol, good agreement with theory is only observed at elevated temperatures, while at low temperatures the Landau–Placzek ratio becomes significantly larger.[118]

Most of the isochoric heat capacity data for liquids reported in the literature have been obtained *indirectly* through use of Equation (1.146) *via*

$$C_V = C_P/\kappa, \tag{1.150a}$$

that is to say from experimentally determined molar isobaric heat capacities, isobaric expansivities and ultrasonic speeds[55,81,119–123] at sufficiently low frequencies. With modern equipment, these three quantities can be measured accurately and speedily, thereby making the indirect method for determining C_V of liquids quite attractive. In addition, Equation (1.146) also provides a valuable *alternative* to the direct method for determining β_T (where hydrostatic pressure is applied to the fluid and the resulting volume change is measured), since

$$\beta_T = \kappa\beta_S. \tag{1.150b}$$

This approach is known as the *indirect* method for determining isothermal compressibilities and usually yields highly accurate results. From Equation (1.146) the difference between β_T and β_S may be expressed by

$$\beta_T - \beta_S = TV\alpha_P^2/C_P, \tag{1.151}$$

while

$$\beta_S^{-1} - \beta_T^{-1} = TV\gamma_V^2/C_V. \tag{1.152}$$

The most important use of β_S data obtained *via* speed-of-sound measurements is to calculate C_V and/or β_T using the appropriate equations presented above.

Isentropic changes on the *PVT* surface are described in terms of three quantities analogous to the isobaric expansivity, the isochoric thermal pressure coefficient and the isothermal compressibility:

$$\alpha_S \equiv \frac{1}{V}\left(\frac{\partial V}{\partial T}\right)_S, \tag{1.153}$$

$$\gamma_S \equiv \left(\frac{\partial P}{\partial T}\right)_S, \tag{1.154}$$

and β_S, already defined by Equation (1.4). Expressing them in terms of derivatives deducible from the thermal equation of state, the following relations are obtained:

$$\alpha_S = -\frac{1}{V}\frac{(\partial S/\partial T)_V}{(\partial S/\partial V)_T} = -\frac{C_V}{TV\gamma_V} = -\frac{C_V\beta_T}{TV\alpha_P} \tag{1.155a}$$

$$= \alpha_P - \frac{C_P}{TV\gamma_V} = \alpha_P - \frac{C_P\beta_T}{TV\alpha_P}, \tag{1.155b}$$

and

$$\gamma_S = -\frac{(\partial S/\partial T)_P}{(\partial S/\partial P)_T} = \frac{C_P}{TV\alpha_P} = \frac{\alpha_P}{\beta_T - \beta_S} \tag{1.156a}$$

$$= \gamma_V + \frac{C_V}{TV\alpha_P} = \gamma_V + \frac{\alpha_P}{\beta_T - \beta_S}\frac{1}{\kappa}, \tag{1.156b}$$

where the triple product rule was invoked and the appropriate Maxwell equation, together with Equations (1.138) and (1.147b). On the basis of Equation (1.156a), heat capacities C_P of liquid benzene and toluene were determined by Burlew[124–126] between 281 K and the normal boiling point:

$$C_P = T\frac{(\partial V/\partial T)_P}{(\partial T/\partial P)_S}. \tag{1.156c}$$

His method involves two independent sets of measurements for the determination of the two differential quotients appearing in Equation (1.156). Because the two principal measurements yielding the isentropic temperature–pressure coefficient $(\partial T/\partial P)_S \equiv 1/\gamma_S$ are those of T and P, he called it the piezo-thermometric method. For the determination of $(\partial V/\partial T)_P$ he devised a new type of weight dilatometer.

The three isentropic coefficients are related by

$$\gamma_S = -\frac{\alpha_S}{\beta_S}. \tag{1.157}$$

Since C_P is always positive, according to Equation (1.156a) the rate of change of temperature with pressure in an isentropic process,

$$\left(\frac{\partial T}{\partial P}\right)_S = \frac{TV}{C_P}\alpha_P, \tag{1.158}$$

has the same sign as the isobaric expansivity. Usually α_P is positive, and the temperature of the system increases upon isentropic compression and decreases upon isentropic expansion. However, liquid water and a few other substances, such as bismuth, show an anomalous temperature dependence of the density in certain temperature ranges, that is the density *increases* with temperature (at constant pressure). For water at 100 kPa, the expansivity is negative between 0 °C and 4 °C, hence in this range the temperature of water will decrease upon isentropic compression.

As pointed out by Rowlinson and Swinton,[4] the so-called mechanical coefficients α_P, β_T and γ_V (and the related quantities along the saturation curve) are determined, to a high degree of accuracy, solely by the intermolecular forces, whilst the isentropic coefficients, to which they are related through the thermal coefficients, *i.e.* the heat capacities, and the thermal coefficients themselves depend also on the *internal molecular* properties.

The most basic characteristic of liquids is that they possess short-range order, as opposed to the long-range periodicity of crystalline solids. The pair distribution function $g(r;T,\rho_N)$ is important for the description of the structure and the properties of equilibrium fluids,[19] and can be determined experimentally by X-ray scattering or neutron scattering. In the present context, perhaps the statistical–mechanical *compressibility equation*[19,127]

$$k_B T\left(\frac{\partial \rho_N}{\partial P}\right)_T = k_B T \rho_N \beta_T = \frac{RT}{V}\beta_T = 1 + 4\pi\rho_N \int_0^\infty [g(r;\, T, \rho_N) - 1] r^2 \mathrm{d}r \quad (1.159)$$

is particularly appropriate for linking macroscopic thermodynamics with microscopic fluid structure. The function $h(r;T,\rho_N) \equiv g(r;T,\rho_N) - 1$ is known as the *total correlation function*; it approaches zero as $r \to \infty$, reflecting the absence of any positional correlation between molecules. The derivation of Equation (1.159) does not require the assumption of a pairwise additive intermolecular potential-energy function $u(r)$, as is the case for the *pressure equation*

$$Z \equiv \frac{PV}{RT} = \frac{P}{\rho_N k_B T} = 1 - \frac{2\pi\rho_N}{3k_B T}\int_0^\infty \frac{\mathrm{d}u(r)}{\mathrm{d}r} g(r;\, T, \rho_N) r^3 \mathrm{d}r. \quad (1.160)$$

The compressibility equation is therefore more general than the pressure equation. For hard spheres with diameter σ_{hs}, the pressure equation can be integrated analytically, and becomes

$$Z_{\mathrm{hs}} = 1 + \frac{2\pi\sigma_{\mathrm{hs}}^3}{3}\rho_N g(\sigma_{\mathrm{hs}}+), \quad (1.161)$$

where $g(\sigma_{\mathrm{hs}}+)$ is the value of $g(r)$ at $\sigma_{\mathrm{hs}} + \varepsilon$, where $\varepsilon \to 0$ from the positive direction.

So far, the focus has been on *homogeneous* constant-composition fluids, of which pure fluids are special cases. I will now briefly consider the case where a *pure* liquid (L) is in equilibrium with its vapour (V) at the *vapour pressure* $P_\sigma(T)$. Such a situation is encountered, for instance, in *classical adiabatic calorimetry*, where the calorimeter vessel is *incompletely* filled with liquid in order to accommodate the thermal expansion of the sample (usually, the vapour space volume is comparatively small). One now has a closed two-phase single-component system. The heat capacity of such a system is closely related to C_σ^{L*}, *i.e.* the molar heat capacity of a pure liquid in *equilibrium*, at all temperatures, with an *infinitesimal* amount of vapour.[4,95,128] The mechanical coefficient that is most easily determined is the expansivity α_σ^{L*} which is obtained from measured densities of the saturated pure liquid (orthobaric liquid densities)

$$\rho^{L*}(T, P_\sigma(T)) = m_{\mathrm{m}} / V^{L*}(T, P_\sigma(T)) \quad (1.162)$$

over a reasonably large temperature range. Henceforth quantities pertaining to *saturation* (*orthobaric*) conditions will be indicated by the subscript $\boldsymbol{\sigma}$, and for the sake of a more compact notation, the superscript asterisk, indicating a pure-substance property, will be omitted.

The three mutual derivatives along the *saturation* (*orthobaric*) curve on a *PVT* surface of a pure liquid are related to each other by

$$\left(\frac{\partial V^{\mathrm{L}}}{\partial P}\right)_{\sigma}\left(\frac{\partial P}{\partial T}\right)_{\sigma}\left(\frac{\partial T}{\partial V^{\mathrm{L}}}\right)_{\sigma}=1, \tag{1.163}$$

hence the following relations for the expansivity $\alpha_{\sigma}^{\mathrm{L}}$ of a liquid in contact with its vapour equilibrium phase are obtained:

$$\alpha_{\sigma}^{\mathrm{L}} \equiv \frac{1}{V^{\mathrm{L}}}\left(\frac{\partial V^{\mathrm{L}}}{\partial T}\right)_{\sigma}=\alpha_{P}^{\mathrm{L}}-\beta_{T}^{\mathrm{L}}\gamma_{\sigma}, \tag{1.164}$$

$$=\alpha_{P}^{\mathrm{L}}\left(1-\frac{\gamma_{\sigma}}{\gamma_{V}^{\mathrm{L}}}\right), \tag{1.165}$$

Below the normal boiling point the difference $\alpha_{P}^{\mathrm{L}}-\alpha_{\sigma}^{\mathrm{L}}$ is usually *very* small. Here,

$$\gamma_{\sigma}\equiv(\partial P/\partial T)_{\sigma} \tag{1.166}$$

is the slope of the vapour-pressure curve $\mathrm{d}P_{\sigma}/\mathrm{d}T$, which in turn is related, *via* the *exact Clapeyron* equation, to the molar enthalpy of vaporisation $\Delta_{\mathrm{vap}}H$ and the molar volume change on vaporisation $\Delta_{\mathrm{vap}}V\equiv V^{\mathrm{V}}(T,P_{\sigma})-V^{\mathrm{L}}(T,P_{\sigma})$:

$$\gamma_{\sigma}=\frac{\Delta_{\mathrm{vap}}H}{T\Delta_{\mathrm{vap}}V}. \tag{1.167}$$

This area was carefully and comprehensively surveyed by Majer and collaborators.[28,129] The difference to the isochoric thermal pressure coefficient is given by

$$\gamma_{V}^{\mathrm{L}}-\gamma_{\sigma}=\alpha_{\sigma}^{\mathrm{L}}/\beta_{T}^{\mathrm{L}}. \tag{1.168}$$

At the critical point γ_{V} and γ_{σ} become equal, *i.e.*

$$\gamma_{V}^{\mathrm{L}}(T_{\mathrm{c}},P_{\mathrm{c}})=\gamma_{V}^{\mathrm{V}}(T_{\mathrm{c}},P_{\mathrm{c}})=\gamma_{\sigma}(T_{\mathrm{c}},P_{\mathrm{c}}), \tag{1.169}$$

and remain *finite*, whereas both α_{P}^{L} and β_{T}^{L} *diverge*. Here, T_{c} is the critical temperature and P_{c} is the critical pressure. Denoting the critical molar volume by V_{c}, the critical compression factor is given by

$$Z_{\mathrm{c}} \equiv \frac{P_{\mathrm{c}}V_{\mathrm{c}}}{RT_{\mathrm{c}}}. \tag{1.170}$$

This quantity is often used in correlations based on the extended corresponding states principle (see below).

The molar heat capacity C_{σ}^{L} of the liquid at saturation is defined by

$$C_{\sigma}^{\mathrm{L}} \equiv T\left(\partial S^{\mathrm{L}}/\partial T\right)_{\sigma}, \tag{1.171}$$

hence one obtains, for instance,

$$C_\sigma^{\rm L} = C_P^{\rm L} + T\left(\frac{\partial S^{\rm L}}{\partial P}\right)_T \left(\frac{\partial P}{\partial T}\right)_\sigma \tag{1.172}$$

$$= C_P^{\rm L} - TV^{\rm L}\alpha_P^{\rm L}\gamma_\sigma, \tag{1.173}$$

$$= C_P^{\rm L} - TV^{\rm L}(\alpha_P^{\rm L} - \alpha_\sigma^{\rm L})\gamma_V^{\rm L}, \tag{1.174}$$

$$= C_V^{\rm L} + T\left(\frac{\partial S^{\rm L}}{\partial V}\right)_T \left(\frac{\partial V^{\rm L}}{\partial T}\right)_\sigma, \tag{1.175}$$

$$= C_V^{\rm L} + TV^{\rm L}\gamma_V^{\rm L}\alpha_\sigma^{\rm L}, \tag{1.176}$$

$$= C_V^{\rm L} + TV^{\rm L}\alpha_P^{\rm L}(\gamma_V^{\rm L} - \gamma_\sigma). \tag{1.177}$$

Neither $C_P^{\rm L}$ nor $C_\sigma^{\rm L}$ is equal to the change of enthalpy with temperature along the saturation curve, which is given by

$$\left(\frac{\partial H^{\rm L}}{\partial T}\right)_\sigma = C_P^{\rm L} + V^{\rm L}(1 - T\alpha_P^{\rm L})\gamma_\sigma, \tag{1.178}$$

$$= C_\sigma^{\rm L} + V^{\rm L}\gamma_\sigma. \tag{1.179}$$

Since $U = H - PV$,

$$\left(\frac{\partial U^{\rm L}}{\partial T}\right)_\sigma = C_\sigma^{\rm L} - P_\sigma V^{\rm L}\alpha_\sigma^{\rm L}, \tag{1.180}$$

$$= C_V^{\rm L} + V^{\rm L}\alpha_P^{\rm L}(T\gamma_V^{\rm L} - T\gamma_\sigma - P_\sigma). \tag{1.181}$$

Thus, for the pure saturated *liquid* at temperatures well below the critical temperature, when $0 < T\alpha_P^{\rm L} < 1$, the following sequence is obtained:

$$\left(\frac{\partial H^{\rm L}}{\partial T}\right)_\sigma > C_P^{\rm L} > C_\sigma^{\rm L} > \left(\frac{\partial U^{\rm L}}{\partial T}\right)_\sigma > C_V^{\rm L}. \tag{1.182}$$

The differences between the first four quantities are generally *much* smaller than the difference between $(\partial U^{\rm L}/\partial T)_\sigma$ and $C_V^{\rm L}$. At low vapour pressures, the difference between $C_\sigma^{\rm L}$ and $C_P^{\rm L}$ is frequently negligible (see above), but at higher vapour pressures corrections in the spirit of Equation (1.172) have to be applied.

Entirely analogous equations may be written for the coexisting *pure saturated vapour* (superscript V). However, here the difference $C_\sigma^{\rm V} - C_P^{\rm V}$ is always significant since $\alpha_P^{\rm V}V^{\rm V}$ is always large. In fact, for vapours of substances consisting of *small* molecules, such as argon, carbon dioxide, ammonia and water (steam), $\alpha_P^V V^V$ may be large enough to even make C_σ^V *negative.*

Equation (1.146) is a suitable starting point for a discussion of the temperature dependence of $\kappa^L \equiv C_P^L / C_V^L = \beta_T^L / \beta_S^L$ along the orthobaric curve:[95,130,131]

$$\left(\frac{\partial \kappa^L}{\partial T}\right)_\sigma = \left(\kappa^L - 1\right)\left[\frac{1}{T} + \frac{2}{\alpha_P^L}\left(\frac{\partial \alpha_P^L}{\partial T}\right)_\sigma + \frac{2}{v_0^L}\left(\frac{\partial v_0^L}{\partial T}\right)_\sigma - \frac{1}{C_P^L}\left(\frac{\partial C_P^L}{\partial T}\right)_\sigma\right]. \tag{1.183}$$

Usually, the second term in parenthesis on the right-hand side of Equation (1.182) is positive and the third term is negative; the fourth term may contribute positively or negatively. Thus, for some liquids $\kappa^L(T,P_\sigma(T))$ may increase with temperature, while for others $\kappa^L(T,P_\sigma(T))$ may decrease with temperature.

The resolution of the variation of the molar heat capacity at constant volume, C_V^L, of a pure liquid along the *orthobaric curve*, *i.e.* along states with $V^L(T,P_\sigma(T))$, into the contribution due to the increase of *temperature* and the contribution due to the increase of *volume*, respectively, is a very interesting problem. It is important to realise that because of the close packing of molecules in a liquid, even a rather small change of the average volume available for their motion may have a considerable impact on the molecular dynamics, in particular on the (hindered) overall molecular rotation: volume changes may become *more* important in influencing molecular motion in the liquid state than temperature changes. In fact, careful analysis of the heat capacity at constant volume of liquids provides valuable information on the *hindered rotation* of the molecules as a whole.[74-76,82,95,97,130-133] Since

$$\left(\frac{\partial C_V^L}{\partial T}\right)_\sigma = \left(\frac{\partial C_V^L}{\partial T}\right)_V + \left(\frac{\partial C_V^L}{\partial V}\right)_T V^L \alpha_\sigma^L, \tag{1.184}$$

in the absence of calorimetrically determined values of $\left(\partial C_V^L / \partial T\right)_V$, evaluation of this quantity requires knowledge of the *second* term of the right-hand side of Equation (1.184). At temperatures below the normal boiling point, the saturation expansivity α_σ^L is practically equal to α_P^L of the liquid [see Equations (1.164) or (1.165)], and is frequently used instead. In principle, the quantity $\left(\partial C_V^L / \partial V\right)_T$ is accessible *via* precise *PVT* measurements, see Equation (1.123), but measurements of $(\partial^2 P/\partial T^2)_V$ are not plentiful. Available data indicate that it is small and *negative* for simple organic liquids at ordinary temperatures, that is to say, C_V^L *decreases* with *increasing* molar volume. Alternatively, one may use[95,97,130-132]

$$\left(\frac{\partial C_V^L}{\partial V}\right)_T = \frac{T}{\beta_T^L}\left[\left(\frac{\partial \alpha_P^L}{\partial T}\right)_P - 2\frac{\alpha_P^L}{\beta_T^L}\left(\frac{\partial \beta_T^L}{\partial T}\right)_P - \left(\frac{\alpha_P^L}{\beta_T^L}\right)^2\left(\frac{\partial \beta_T^L}{\partial P}\right)_T\right]. \tag{1.185}$$

Frequently, the last term in parenthesis on the right-hand side of Equation (1.185) is not known and has to be estimated. This may advantageously be done by means of a *modified Tait equation* (MTE),[130–140] that is

$$V^{\mathrm{L}}(T,P)/V^{\mathrm{L}}(T,P_{\mathrm{ref}}) = \left[1 + m_{\mathrm{MTE}}(P - P_{\mathrm{ref}})\beta_T^{\mathrm{L}}(T,P_{\mathrm{ref}})\right]^{-1/m_{\mathrm{MTE}}}, \qquad (1.186)$$

where $V^{\mathrm{L}}(T,P_{\mathrm{ref}})$ and $\beta_T^{\mathrm{L}}(T,P_{\mathrm{ref}})$ denote the molar volume and the isothermal compressibility of the pure liquid, respectively, at a reference pressure P_{ref}. For convenience, this reference pressure is often taken either as P_σ or 100 kPa, and m_{MTE} is a pressure-independent parameter. This equation is a versatile wide-pressure range equation of state for liquids.[134,135] From Equation (1.186) it follows that

$$\frac{1}{\beta_T^{\mathrm{L}}(T,P)} = \frac{1}{\beta_T^{\mathrm{L}}(T,P_{\mathrm{ref}})} + m_{\mathrm{MTE}}(P - P_{\mathrm{ref}}), \qquad (1.187)$$

which is identical to the tangent-modulus equation.[136] Indeed, at constant T and for pressures up to several hundred bars (bar = 100 kPa), $1/\beta_T^{\mathrm{L}}$ for any liquid is essentially a *linear* function of pressure. Thus, for the pressure dependence of the isothermal compressibility in this moderate pressure region

$$\left(\frac{\partial \beta_T^{\mathrm{L}}}{\partial P}\right)_T = -\, m_{\mathrm{MTE}}\left(\beta_T^{\mathrm{L}}\right)^2. \qquad (1.188)$$

For many liquid nonelectrolytes, experimental values of m_{MTE} cluster around

$$m_{\mathrm{MTE}} \approx 10, \qquad (1.189)$$

with only a very small temperature dependence. Note that

$$\frac{\beta_T^{\mathrm{L}}(T,P)}{\beta_T^{\mathrm{L}}(T,P_{\mathrm{ref}})} = \left(\frac{V^{\mathrm{L}}(T,P)}{V^{\mathrm{L}}(T,P_{\mathrm{ref}})}\right)^{m_{\mathrm{MTE}}}. \qquad (1.190)$$

This remarkably simple behaviour was successfully modelled by using a van der Waals-type equation of state.[137,138] As suitable starting point

$$\frac{PV^{\mathrm{L}}}{RT} = Z_{\mathrm{hs}} - \frac{a}{RTV^{\mathrm{L}}} = \frac{1 + y + y^2 - y^3}{(1-y)^3} - \frac{a}{RTV^{\mathrm{L}}}, \qquad (1.191)$$

was selected, with Z_{hs} being Carnahan and Starling's[141] expression for the compression factor of a hard sphere (hs) fluid. The packing density (compactness) is characterised by

$$y \equiv \frac{L\pi\sigma_{\mathrm{hs}}^3}{6V^{\mathrm{L}}}, \qquad (1.192)$$

where σ_{hs} denotes an appropriate *effective* hard sphere diameter,[142,143] and $-a/RTV^{\mathrm{L}}$ is a term reflecting a uniform background cohesive energy.[144–147]

For liquid tetrachloromethane[131] at $T = 298.15$ K and $V^{\mathrm{L}}(T,P_\sigma)$, the calculated value of $(\partial C_V^{\mathrm{L}}/\partial V)_T$ amounts to -0.48 J K^{-1} cm^{-3} (see also reference 132), for cyclohexane[131] -0.57 J K^{-1} cm^{-3} is obtained, and for 1,2-dichloroethane[130] $(\partial C_V^{\mathrm{L}}/\partial V)_T = -0.60$ J K^{-1} cm^{-3}. These results indicate a *substantial* contribution of $(\partial C_V^{\mathrm{L}}/\partial V)_T V^{\mathrm{L}}(T,P_\sigma)\alpha_\sigma^{\mathrm{L}}$ to the change of C_V^{L} with temperature along the orthobaric curve.

Definitely the most widely used equation for correlating compressed liquid volumes or densities is the Tait equation,[140,148] which can be written as

$$V^{\mathrm{L}}(T,P) - V_{\mathrm{ref}}^{\mathrm{L}}(T,P_{\mathrm{ref}}) = -C_{\mathrm{TV}} \ln \frac{B_{\mathrm{TV}} + P}{B_{\mathrm{TV}} + P_{\mathrm{ref}}}, \tag{1.193}$$

where $V_{\mathrm{ref}}^{\mathrm{L}}$ is a suitably selected reference volume, and P_{ref} is the corresponding reference pressure; alternatively, one may write

$$\rho^{\mathrm{L}}(T,P) = \frac{\rho_{\mathrm{ref}}^{\mathrm{L}}(T,P_{\mathrm{ref}})}{1 - C_{\mathrm{TD}} \ln \dfrac{B_{\mathrm{TD}} + P}{B_{\mathrm{TD}} + P_{\mathrm{ref}}}}, \tag{1.194}$$

where $\rho_{\mathrm{ref}}^{\mathrm{L}}$ is a suitably selected reference density. Frequently, the reference state is the *saturation* state at temperature T, hence $P_{\mathrm{ref}} = P_\sigma(T)$, and the substance specific parameters B_{TV} and C_{TV}, and B_{TD} and C_{TD} are usually temperature dependent. The Tait equation in the form of Equation (1.194) was used by Cibulka, Takagi and collaborators in their recent systematic, comprehensive and critical compilation of $P\rho T$ data of liquids.[149–156] Liquid *heptane* (n-C_7H_{16}) is a key hydrocarbon liquid in thermophysics. Using a new apparatus for simultaneous measurements of the density and viscosity of liquids at high temperatures (from room temperature to 500 K) and high pressures (up to 250 MPa) based on hydrostatic weighing and falling-body techniques, respectively, Sagdeev *et al.*[157] recently presented a comprehensive experimental study of this liquid.

An attractive alternative to the direct experimental route to *high-pressure* PVT data and heat capacities C_P^{L} and C_V^{L} of liquids is to measure the thermodynamic *speed of ultrasound* v_0 as a function of P and T (at constant composition),[123] and to combine these results, in the spirit of Equations (1.3)–(1.5) and (1.151), with data at ordinary pressure, say at $P_{\mathrm{ref}} = 10^5$ Pa, *i.e.* with $\rho^{\mathrm{L}}(T,P_{\mathrm{ref}})$ and $C_P^{\mathrm{L}}(T,P_{\mathrm{ref}})$. Specifically, according to Equation (1.151) the isothermal pressure dependence of the density may be expressed as

$$\left(\frac{\partial \rho^{\mathrm{L}}}{\partial P}\right)_T = \frac{1}{v_0^2} + \frac{T m_{\mathrm{m}} \left(\alpha_P^{\mathrm{L}}\right)^2}{C_P^{\mathrm{L}}}, \tag{1.195}$$

which gives, upon integration,[113,123,158–164]

$$\rho^{\mathrm{L}}(T,P) = \rho^{\mathrm{L}}(T,P_{\mathrm{ref}}) + \int_{P_{\mathrm{ref}}}^{P} v_0^{-2} \mathrm{d}P + T m_{\mathrm{m}} \int_{P_{\mathrm{ref}}}^{P} \left(\left(\alpha_P^{\mathrm{L}}\right)^2 / C_P^{\mathrm{L}}\right) \mathrm{d}P. \tag{1.196}$$

The first integral is evaluated directly by fitting the isothermal ultrasonic speed data with suitably selected polynomials or Padé approximants, and for the second integral several successive integration algorithms have been devised. The simplicity, rapidity and precision of this method makes it highly attractive for the determination of the density, isobaric expansivity, isothermal compressibility, isobaric heat capacity and isochoric heat capacity of liquids at high pressures. Its application to *Room Temperature Ionic Liquids* (RTILs)[165] has first been reported by Gomes de Azevedo and co-workers.[166] Details may be found in the original literature. Concerning the wide-temperature range/wide-pressure range results reported by Biswas *et al.*, for instance, on the isobaric heat capacity and the isobaric expansivity of heptane and toluene,[113] the unequivocal proof of the existence of *minima* of the *isotherms* $C_P^L = C_P^L(P)$ at elevated pressures, and of a substance-specific *crossing "point"* (small crossing region?) of the *isotherms* of the isobaric expansivity $\alpha_P^L(P)$ at elevated pressures are particularly interesting (and intriguing). For heptane, this crossing "point" is found at *ca.* 120 MPa, at which pressure $\partial\alpha_P^L/\partial T \approx 0$. Thus for any given pressure *lower* than 120 MPa, α_P^L of heptane *increases* with temperature, while at *higher* pressures α_P^L *decreases* with temperature, *i.e.* $\partial\alpha_P^L/\partial T$ becomes *negative*. As evidenced by Equation (1.125), the pressure dependence of C_P^L is directly influenced by the temperature dependence of α_P^L, that is, a minimum of the function C_P^L *vs.* P at constant T is observed for that pressure where $(\alpha_P^L)^2 + (\partial\alpha_P^L/\partial T)_P = 0$.

Equation (1.125) suggests still another method for obtaining heat capacities of liquids at high pressures using volumetric data. Upon integration

$$C_P^L(T,P) = C_P^L(T,P_{\mathrm{ref}}) - T\int_{P_{\mathrm{ref}}}^{P} V^L(T,P)\left[(\alpha_P^L(T,P))^2 + \left(\frac{\partial\alpha_P^L(T,P)}{\partial T}\right)_P\right]\mathrm{d}P, \text{ constant } T, \tag{1.197}$$

is obtained, with

$$V^L(T,P) = V^L(T_{\mathrm{ref}},P)\exp\left[\int_{T_{\mathrm{ref}}}^{T}\alpha_P^L(T,P)\mathrm{d}T\right], \text{ constant } P. \tag{1.198}$$

Here, the isobaric expansivity $\alpha_P^L(T,P)$ is the *main* experimental property to be measured as a function of T and P. The other experimental properties are the molar volume $V^L(T_{\mathrm{ref}},P)$ as a function of pressure at a convenient low reference temperature T_{ref} and the molar isobaric heat capacity $C_P^L(T,P_{\mathrm{ref}})$ as a function of temperature at a convenient low reference pressure P_{ref}. With a *scanning transitiometer*[107,108,114] it is possible to measure $\alpha_P^L(T,P)$ over wide ranges of temperature and pressure with an uncertainty of about 1% to 3%, the reference volume isotherm with an uncertainty of about 0.6%, and the reference heat capacity isobar with an uncertainty of about 0.3%. Thus, the overall uncertainty of the heat capacities of liquids at high pressures obtained by scanning transitiometry is estimated to be about 2%. Perhaps the most interesting result is the confirmation, for some simple organic liquids, of the existence of (shallow) *minima* of the *isotherms* of the isobaric heat capacity C_P^L at elevated pressures. Concerning the experimental results for the isobaric expansivity obtained with this technique, we note that again, say for liquid hexane, for the *isotherms* $\alpha_P^L = \alpha_P^L(P)$ a *crossing "point"* (a crossing region?) at about (65 ± 2) MPa is found, where $(\partial \alpha_P^L / \partial T)_P \approx 0$. However, the α_P^L isotherms of liquid hexan-1-ol do *not* exhibit a crossing region and, correspondingly, the C_P^L isotherms of this hydrogen-bonded liquid do *not* show any minima in the pressure range accessible so far. Similar work has also been reported by Romaní's group[167] in Spain.

Densities of liquids at ambient pressure are important input parameters in many process design calculations and are fairly easy to measure (for an extensive data compilation see Daubert *et al.*[168]). However, saturated liquid densities over large temperature ranges have been determined much less frequently, and a number of estimation techniques have thus been developed for practical use. The most popular prediction methods are based on the Rackett equation[169–172]

$$V(T, P_\sigma(T)) = V_c Z_c^{(1-T/T_c)^{2/7}}. \tag{1.199}$$

The *PVT* behaviour of *real gases* (pure or mixed) at low to moderate densities has been of interest in physics and physical chemistry for more than a century, and is of considerable practical and theoretical importance.[26a,b,173,174] At first, I will focus on *pure* gases and vapours. Historically, the virial expansion of the compression factor at constant temperature (and constant composition) was an empirical method of fitting experimental *PVT* data by either the *density* or the *pressure* forms, that is

$$Z = 1 + B\rho_n + C\rho_n^2 + D\rho_n^3 + \cdots \tag{1.200}$$

or

$$Z = 1 + B'P + C'P^2 + D'P^3 + \ldots, \tag{1.201}$$

but was later shown to evolve naturally from statistical mechanics.[84–88,173,175] Here,

$$\rho_n \equiv \frac{1}{V} = \frac{\rho_N}{L} = \frac{\rho}{m_m} \tag{1.202}$$

denotes the amount-of-substance density. For the sake of simplicity, the superscript V, indicating the vapour or gas state, has been omitted. According to their definition, the virial coefficients of *pure* fluids are *only* functions of *T* and, in the case of *mixtures*, they are *only* functions of *T* and $\{x_i\}$. The virial series is not convergent for all densities, and the radius of convergence is not established theoretically.

The virial coefficients B',C',... of the *infinite* pressure series are rigorously related to the virial coefficients B, C,... of the *infinite* series in amount-of-substance density. For instance,

$$B' = \frac{B}{RT}, \tag{1.203}$$

$$C' = \frac{C - B^2}{(RT)^2}, \ etc. \tag{1.204}$$

These two virial equations are known to converge at different rates, hence for a given degree of truncation they provide different extents of approximation to the true volumetric behaviour. The two most popular approximations corresponding to truncations[176] to two terms and to three terms, respectively, are

$$Z = 1 + \frac{BP}{RT}, \tag{1.205}$$

and

$$Z = 1 + B\rho_n + C\rho_n^2. \tag{1.206}$$

As already indicated in the Introduction [see Equation (1.11)], the theoretical importance of the virial expansion lies in the fact that its coefficients are explicitly related to the intermolecular potential-energy functions. Specifically, the second virial coefficient B depends on pairwise interaction between molecules, the third virial coefficient C depends on the interaction in clusters of three molecules, and so forth. Equation (1.11) can be transformed through integration by parts into[7,19,86,175]

$$B(T) = -\frac{2\pi L}{3k_{\mathrm{B}}T}\int_0^\infty \exp[-u(r)/k_{\mathrm{B}}T]\frac{\mathrm{d}u(r)}{\mathrm{d}r}r^3\mathrm{d}r. \tag{1.207}$$

Alternatively, one might expand the pair distribution function as a Taylor series in ρ_N about $\rho_N = 0$, and substitute it for $g(r;T,\rho_N)$ in Equation (1.160), *i.e.* the pressure equation, to get the virial coefficients in general and $B(T)$ in particular. For a very dilute gas/vapour, where the influence of other molecules on a certain pair can be neglected, we have

$$g(r;\,T,\rho_N)_{\rho_N=0} = \mathrm{e}^{-u(r)/k_{\mathrm{B}}T}, \tag{1.208}$$

which is the limiting relation as $\rho_N \to 0$ for the *potential of mean force* $w^{(2)}(r;T,\rho_N)$ defined by

$$g(r;\,T,\rho_N) \equiv \mathrm{e}^{-w^{(2)}(r;T,\rho_N)/k_{\mathrm{B}}T}. \tag{1.209}$$

$w^{(2)}(r;T,\rho_N)$ is the potential function that is responsible for the interaction between two molecules held at a fixed distance r while the other $N - 2$ molecules of the fluid are averaged over all configurations. It is a good measure of the *solvent induced* interactions between two particles. At constant T and V, $w^{(2)}(r)$ corresponds to the change of the Helmholtz energy accompanying the process of bringing two molecules from infinite separation ($r=\infty$) to a distance r. Note that $w^{(2)}(r;T,\rho_N)$ of a *dense* fluid has minima where the corresponding $g(r;T,\rho_N)$ has maxima and *vice versa*. When the density becomes very small, $w^{(2)}(r;T,\rho_N)\rightarrow u(r)$.

Another equivalent expression for the second virial coefficient is[19,86,177]

$$B(T) = \exp(\varepsilon/k_\mathrm{B}T)\frac{2\pi L}{3k_\mathrm{B}T}\int_0^\infty \Delta(\varphi)\exp[-\varphi/k_\mathrm{B}T]\mathrm{d}\varphi, \tag{1.210}$$

where ε is the depth of the potential energy well [see Equation (1.15)], $\varphi(r)=u(r)+\varepsilon$, and $\Delta(\varphi)$ is the *well-width function*. This quantity is defined by

$$\begin{aligned}\Delta(\varphi) &= r_\mathrm{l}^3(\varphi) - r_\mathrm{r}^3(\varphi) \quad \text{for } \varphi \leq \varepsilon \\ &= r_\mathrm{l}^3(\varphi) \quad \text{for } \varphi > \varepsilon\end{aligned} \tag{1.211}$$

with r_l and r_r denoting the inner (subscript l for "left") and outer (subscript r for "right") coordinate of the potential well at energy φ, respectively. That is, these values characterise the intermolecular separation for which the energy is φ. Note that

$$\mathscr{L}[\Delta(\varphi)] = \int_0^\infty \Delta(\varphi)\exp[-\varphi/k_\mathrm{B}T]\mathrm{d}\varphi, \tag{1.212}$$

where $\mathscr{L}[\ ]$ is the *Laplace transform operator*. Thus, in principle, if accurate $B(T)$ data have been determined over a range of temperatures, *inversion* of the Laplace transform in Equation (1.210) will yield $u(r)$ *only* for $\varphi(r)>\varepsilon$, *i.e.* for $u(r)>0$, whilst for $\varphi(r)\leq\varepsilon$, *i.e.* for $u(r)\leq 0$, only information on the well width as a function of the well depth is obtained. However, for monatomic gases other than helium, where the repulsive potential energy between two helium atoms was obtained by an inverse Laplace transform of $B(T)$ data in the range 300 K–1500 K,[178] rather sophisticated, less formal, inversion techniques have been devised that allow the extraction of the complete potential-energy function from experimental second virial coefficients.[86,179–181] For *molecular* gases it is *not* possible to write Equation (1.11) as a Laplace transform, that is a formal inversion procedure does not exist in this case.

The majority of virial coefficient data has been obtained by *PVT* (or $P\rho T$) measurements,[182–198] of which the Burnett technique[182–188] is perhaps the most commonly used. For subcritical temperatures, adsorption is a major problem, which has been, however, successfully resolved by Wagner and collaborators[191–195] by introducing a new compensation method based on *Archimedes' buoyancy principle*: two sinkers of identical mass and surface area but significantly different volumes were used.[198] Virial coefficients will be discussed in detail in Chapter 6.

The thermodynamic speed of ultrasound v_0, *i.e.* the sound speed at frequencies well below any dispersion region, is related to the equation of state, and thus, for real gases, to the *virial coefficients.*[55,81,199,200] That is to say, at constant temperature the density dependence of v_0^2 in a gas may be expressed as

$$v_0^2 = \frac{\kappa^{\rm pg} RT}{m_{\rm m}}\left(1 + B_{\rm ac}V^{-1} + C_{\rm ac}V^{-2} + \cdots\right), \tag{1.213}$$

where $\kappa^{\rm pg} \equiv C_P^{\rm pg}/C_V^{\rm pg}$, $B_{\rm ac}$ is the second *acoustic* virial coefficient, $C_{\rm ac}$ is the third *acoustic* virial coefficient, and so forth. Thus, from the zero-density limit of the speed of ultrasound the molar perfect-gas state heat capacities are accessible *via*

$$\frac{C_V^{\rm pg}}{R} = \frac{1}{\kappa^{\rm pg} - 1} \tag{1.214}$$

or

$$\frac{C_P^{\rm pg}}{R} = \frac{\kappa^{\rm pg}}{\kappa^{\rm pg} - 1}, \tag{1.215}$$

that is *without* measuring a quantity of heat (see also above). Acoustic measurements are quite rapid and represent one of the most precise methods for determining virial coefficients *and* $C_V^{\rm pg} = C_P^{\rm pg} - R$, and the results are essentially *free* of errors due to gas adsorption.

For pure fluids, the *acoustic* virial coefficients $B_{\rm ac}$, $C_{\rm ac}$, *etc.* are only functions of temperature. They are, of course, rigorously related to the ordinary *PVT* virial coefficients. For instance,[55]

$$B_{\rm ac} = 2B + 2(\kappa^{\rm pg} - 1)T\frac{{\rm d}B}{{\rm d}T} + \frac{(\kappa^{\rm pg} - 1)^2}{\kappa^{\rm pg}}T^2\frac{{\rm d}^2B}{{\rm d}T^2}, \tag{1.216}$$

and so forth. Thus, measurement of the speed of ultrasound as function of amount-of-substance density (or pressure) will yield information on B together with its *first* and *second* temperature derivative[201–206] *in addition* to $\kappa^{\rm pg}$ obtained from the zero-density limit of v_0^2. For gases and vapours at low to moderate pressures not too close to saturation, the spherical resonator provides the highest precision for the measurement of the speed of ultrasound, a technique which was pioneered by Moldover *et al.*[201,202]

Calorimetric determinations of the molar enthalpy of vaporisation of pure liquids in conjunction with vapour pressure $P_\sigma(T)$ and orthobaric liquid density data $\rho^{\rm L}(T,P_\sigma(T)) = m_{\rm m}/V^{\rm L}(T,P_\sigma(T))$ are a useful source of information on second virial coefficients of vapours *well below* the critical temperature, since for $T < T_{\rm C}$ experimental difficulties of conventional techniques greatly diminish accuracy. Starting with the exact Clapeyron equation, Equation (1.167), after rearrangement one obtains

$$V^{\rm V}(T,P_\sigma) = \frac{\Delta_{\rm vap}H(T)}{T({\rm d}P_\sigma/{\rm d}T)} + V^{\rm L}(T,P_\sigma), \tag{1.217a}$$

and assuming negligible influence of virial coefficients other than $B(T)$,

$$V^{\mathrm{V}}(T,P_\sigma)=\left(1+\frac{B(T)}{V^{\mathrm{V}}(T,P_\sigma)}\right)\frac{RT}{P_\sigma} \tag{1.217b}$$

is also valid. Equating Equations (1.217a) and (1.217b), solving for $B(T)/V^{\mathrm{V}}(T,P_\sigma)$ and then expressing $V^{\mathrm{V}}(T,P_\sigma)$ *via* Equation (1.217a) finally yields $B(T)$. Alternatively, one may use

$$V^{\mathrm{V}}(T,P_\sigma)=\left(1+\frac{B(T)P_\sigma}{RT}\right)\frac{RT}{P_\sigma}, \tag{1.217c}$$

resulting in

$$B(T)=\frac{\Delta_{\mathrm{vap}}H(T)}{T(\mathrm{d}P_\sigma/\mathrm{d}T)}+\frac{m_{\mathrm{m}}}{\rho^{\mathrm{L}}(T,P_\sigma)}-\frac{RT}{P_\sigma}. \tag{1.217d}$$

Virial coefficients evaluated *via* Equations (1.217a) and (1.217b)[207–210] were checked calorimetrically for consistency through use of

$$\lim_{P\to 0}\left(\frac{\partial C_P^{\mathrm{V}}}{\partial P}\right)_T=-T\frac{\mathrm{d}^2B}{\mathrm{d}T^2}. \tag{1.218}$$

For many substances such an approach has provided the only information on second virial coefficients at subcritical temperatures.

The refractive index $n(T,\rho_n,\lambda)$ of a gas depends on temperature, amount-of-substance density (molar density) $\rho_n=V^{-1}$ and wavelength λ, but is independent of the amount used, thus providing an alternative to *PVT* methods where adsorption may cause systematic errors. The molar refractivity R_{LL} (or Lorentz–Lorenz function) is defined by

$$R_{\mathrm{LL}}\equiv\frac{n^2-1}{n^2+2}\frac{1}{\rho_n}, \tag{1.219}$$

and can be expanded in powers of the density ρ_n, *i.e.*

$$R_{\mathrm{LL}}=A_{\mathrm{R}}+B_{\mathrm{R}}\rho_n+C_{\mathrm{R}}\rho_n^2+\cdots, \tag{1.220}$$

where A_{R}, B_{R}, C_{R},...denote the first, second, third, *etc.*, refractivity (or optical) virial coefficient.[211–215] The first refractivity virial coefficient is proportional to the mean polarisability volume $L\alpha/4\pi\varepsilon_0$, where α is the electric polarisability of a molecule, and ε_0 is the permittivity of vacuum. Replacing the density in Equation (1.220) with use of the virial equation in density, one obtains:

$$\frac{n^2-1}{n^2+2}\frac{RT}{P}=A_{\mathrm{R}}(T,\lambda)+[B_{\mathrm{R}}(T,\lambda)-B(T)A_{\mathrm{R}}(T,\lambda)]\frac{P}{RT}+\cdots. \tag{1.221}$$

Appropriate fitting procedures then yield the refractivity virial coefficients of the expansion of the Lorentz–Lorenz function, and the second (and possibly the third) *PVT* virial coefficient.

Measurements of the density dependence of the relative permittivity $\varepsilon_r(T,\rho_n)$ of a pure gas, or more precisely, expanding the Clausius–Mossotti function P_{CM} (total molar polarisation) in powers of the density ρ_n yields the dielectric analogue of Equation (1.220). Note, that for a gas at optical frequencies, the equation $n^2=\varepsilon_r/\mu_r$ applies, with the relative permeability μ_r being very close to unity. Thus,

$$P_{CM} \equiv \frac{\varepsilon_r - 1}{\varepsilon_r + 2}\frac{1}{\rho_n} = A_P + B_P\rho_n + C_P\rho_n^2 + \cdots. \tag{1.222}$$

In combination with the virial equation in density, this expansion yields an equation similar to Equation (1.221), providing information on the dielectric virial coefficients A_P, B_P, C_P,... which, for pure gases, are functions of temperature alone, and the second *PVT* virial coefficient.[216–220]

Although a large number of experimental second virial coefficients of pure substances are at our disposal,[26a] it is frequently necessary to *estimate* B because the compound has not been investigated or because B is needed in a temperature range not covered by experiment. *A fortiori* this is the case with third virial coefficients C, where available experimental results[26a] are much less reliable. The most successful correlations for estimating virial coefficients take advantage of the *extended corresponding states principle* (ECSP), which is grounded in experiment as well as in statistical mechanics. The *three-parameter* corresponding-states correlations, pioneered by Pitzer,[221–223] have been capable of predicting, satisfactorily, the *PVT* behaviour of normal, nonassociating fluids with not too large molecular dipole moments. The compression factor Z of such a fluid may be satisfactorily expressed as

$$Z = Z(T_r, P_r, \omega) = Z^{(0)}(T_r, P_r) + \omega Z^{(1)}(T_r, P_r), \tag{1.223}$$

where

$$\omega \equiv -1 - \log_{10}(P_{\sigma,r})_{T_r=0.7} \tag{1.224}$$

is Pitzer's *acentric factor*, a parameter that quantifies the extent of anisotropic molecular interaction. In fact, this method is a thermodynamic perturbation approach where the Taylor series expansion of Z is truncated after the term linear in ω. Here, $T_r \equiv T/T_c$ is the reduced temperature, $P_r = P/P_c$ is the reduced pressure, and $P_{\sigma,r} \equiv P_\sigma/P_c$ is the reduced vapour pressure of the liquid evaluated at $T_r = 0.7$. $Z^{(0)}$ and $Z^{(1)}$ are *universal* functions of T_r and P_r, hence, according to Lee and Kesler,[224] for the critical compression factor

$$Z_c = 0.2901 - 0.0879\omega \tag{1.225}$$

is obtained. Pitzer and Curl[223] developed a correlation for the reduced second virial coefficient of the form

$$\frac{BP_c}{RT_c} = B^{(0)}(T_r) + \omega B^{(1)}(T_r), \tag{1.226}$$

where $B^{(0)}$ and $B^{(1)}$ are *universal* dimensionless functions of T_r. Extensions and refinements of these expressions have been developed by Tsonopoulos,[225] amongst others, to permit estimation of B for more complex hydrocarbons and polar and hydrogen-bonded substances by adding additional terms. The Tsonopoulos correlation is one of the most reliable, and several revisions and extensions have appeared over the years.[226] An extensive recent one is due to Meng *et al.*[227,228] However, further experimental data on strongly polar or hydrogen-bonded fluids, in particular at *low* reduced temperatures $T_r<1$, are clearly needed.

Experimental data on third virial coefficients are not plentiful, and in addition, for $T_r<1$ the experimental uncertainty increases significantly. For *non-polar* gases, the temperature dependence of C has been correlated by the Orbey and Vera model,[229] and third virial coefficients of gases consisting of small polar molecules (halogenated methanes and ethanes) by a model of Weber,[230] who adopted ideas, with some simplification, of Kohler *et al.*[231] A modified form of the Weber correlation for the third virial coefficients was presented by Meng *et al.*[227] Using an extended version of the Hossenlopp–Scott[207–210] method, Equations (1.217a and b), Ramos-Estrada *et al.*[232] recently also used the Clapeyron equation to calculate third virial coefficients at low reduced temperatures, *i.e.* in the range $0.8<T_r<1$. The calculated values are in good agreement with directly measured third virial coefficients, and appear to be simple functions of the critical compression factor Z_c. Additional approaches may be found in the comprehensive monograph on the properties of gases and liquids by Poling *et al.*[37] In summary, advances in this area will depend crucially on the availability of more reliable experimental data on third virial coefficients, in particular at low reduced temperatures $T_r<1$.

As already pointed out, the virial equation of state is firmly grounded in statistical mechanics. When applying Equation (1.200) to a multicomponent mixture with mole fractions $x_1,x_2,\ldots,x_p$, the mixture virial coefficients are now also composition dependent. Theory provides exact *mixing* rules, *i.e.* prescriptions which express the composition dependence of the mixture virial coefficients:

$$B(T,\{x_i\})=\sum_i^p\sum_j^p x_ix_jB_{ij}(T), \tag{1.227}$$

$$C(T,\{x_i\})=\sum_i^p\sum_j^p\sum_k^p x_ix_jx_kC_{ijk}(T),\ etc. \tag{1.228}$$

Here, the unsubscripted virial coefficients on the left-hand side of Equations (1.227) and (1.228) refer to the mixture. On the right-hand side, the quantities B_{ii},C_{iii}, *etc.* with identical subscripts are the second, and the third virial coefficient, respectively, of pure component i, while the quantities B_{ij},C_{ijk}, *etc.* with mixed subscripts are known as the *interaction* virial coefficients or *cross* virial coefficients; they characterise unlike molecular interactions. Note that

$$B_{ij} = B_{ji}, \tag{1.229a}$$

$$C_{ijk} = C_{ikj} = C_{jik} = C_{jki} = C_{kij} = C_{kji},\ etc. \tag{1.229b}$$

Many of the methods used for measuring virial coefficients of pure fluids are also applicable to mixtures, and have indeed been used to obtain mixture virial coefficients and in turn interaction virial coefficients based on Equations (1.227) and (1.228), respectively. However, experimental techniques to obtain cross virial coefficients more directly have obvious advantages: gas–liquid chromatography is such a method.[233–235]

Estimation and correlation of interaction virial coefficients follows ECSP-based routes essentially similar to those used for pure fluids. The basis is, for instance, the assumption that Equation (1.226) may also be used for the second cross virial coefficients, that is

$$\frac{B_{ij}P_{cij}}{RT_{cij}} = B^{(0)}(T_{rij}) + \omega_{ij}B^{(1)}(T_{rij}). \tag{1.230}$$

$B^{(0)}$ and $B^{(1)}$ are the same universal functions as used with pure fluids, $T_{rij} \equiv T/T_{cij}$, and the *interaction parameters* T_{cij}, P_{cij} and ω_{ij} are obtained through use of appropriate *combination* rules,[9,37,226,236] thereby linking them to pure-substance properties. Again, the Tsonopoulos correlation is most popular and for nonpolar or weakly polar binaries it uses

$$T_{cij} = (1 - k_{ij})(T_{ci}T_{cj})^{1/2}, \tag{1.231}$$

$$\frac{P_{cij}}{RT_{cij}} = \frac{Z_{cij}}{V_{cij}}, \tag{1.232}$$

$$Z_{cij} = \frac{1}{2}(Z_{ci} + Z_{cj}), \tag{1.233}$$

$$V_{cij} = \frac{1}{8}\left(V_{ci}^{1/3} + V_{cj}^{1/3}\right)^3, \tag{1.234}$$

and

$$\omega_{ij} = \frac{1}{2}(\omega_i + \omega_j). \tag{1.235}$$

The most sensitive combining rule is that for T_{cij}, and customarily another interaction parameter k_{ij}, characteristic of the *ij* binary, is introduced here to improve estimation quality. The updated correlation of Meng *et al.*[227] has recently been extended to binary mixtures,[237] and using the extensive data base provided by Dymond *et al.*,[26b] optimal k_{ij} values for 268 nonpolar mixtures were determined and various methods for estimating k_{ij} were tested.

In summary, the volumetric properties represented by virial coefficients (and by their temperature derivatives) provide a wealth of information on the thermophysical behaviour of fluids (pure and mixed) and are indispensable for the extraction of many thermophysical properties from experimental results. Undoubtedly, the most important use of the virial equation of state is in the description of vapour–liquid equilibria (VLE) of mixtures/solutions in the *low* to *medium* pressure range (well below the critical pressure). The primary thermodynamic value of such VLE data is that they allow the determination of the *excess molar Gibbs energy* referring to the liquid mixture, a process known as data reduction.[5,9,12,13,21] G^{E} is defined by Equation (1.25) with $M = G$, and by conventional definition an *ideal solution* is one for which the partial molar Gibbs energy is given by

$$G_i^{\mathrm{id}} \equiv G_i^* + RT\ln x_i, \tag{1.236}$$

and

$$G^{\mathrm{id}} \equiv \sum_i x_i G_i^* + RT\sum_i x_i \ln x_i. \tag{1.237}$$

All other ideal-solution properties follow from these two equations, such as

$$S_i^{\mathrm{id}} = S_i^* - R\ln x_i, \tag{1.238}$$

and

$$S^{\mathrm{id}} = \sum_i x_i S_i^* - R\sum_i x_i \ln x_i, \tag{1.239}$$

and Equations (1.29) and (1.30). Real mixtures are nonideal, and the extent to which they deviate from ideal behaviour is most conveniently expressed through use of the activity coefficients $\gamma_i(T,P,\{x_i\})$ of each component i, using the *symmetric* convention. For convenience, this may be expressed as

$$\frac{G_i^{\mathrm{E}}}{RT} = \frac{\mu_i^{\mathrm{E}}}{RT} = \ln\gamma_i \tag{1.240}$$

and

$$\frac{G^{\mathrm{E}}}{RT} = \sum_i x_i \frac{\mu_i^{\mathrm{E}}}{RT} = \sum_i x_i \ln\gamma_i, \tag{1.241}$$

and in analogy to Equation (110a), the *fundamental excess-property relation* is given by

$$\mathrm{d}\left(\frac{nG^{\mathrm{E}}}{RT}\right) = -\frac{nH^{\mathrm{E}}}{RT^2}\mathrm{d}T + \frac{nV^{\mathrm{E}}}{RT}\mathrm{d}P + \sum_i \ln\gamma_i \mathrm{d}n_i. \tag{1.242}$$

The corresponding Gibbs–Duhem equation reads

$$-\frac{H^{\mathrm{E}}}{RT^2}\mathrm{d}T+\frac{V^{\mathrm{E}}}{RT}\mathrm{d}P-\sum_i x_i\mathrm{d}\ln\gamma_i=0, \tag{1.243}$$

which, at constant T and P, becomes

$$\sum_i x_i\mathrm{d}\ln\gamma_i=0. \tag{1.244}$$

Inspection of Equation (1.242) leads directly to Equation (1.32) since

$$\left(\frac{\partial(G^{\mathrm{E}}/RT)}{\partial P}\right)_{T,\{x_i\}}=\frac{V^{\mathrm{E}}}{RT}, \tag{1.245}$$

to

$$\left(\frac{\partial(G^{\mathrm{E}}/RT)}{\partial T}\right)_{P,\{x_i\}}=-\frac{H^{\mathrm{E}}}{RT^2}, \tag{1.246}$$

and to

$$\ln\gamma_i=\left(\frac{\partial(nG^{\mathrm{E}}/RT)}{\partial n_i}\right)_{T,P,n_{j\neq i}}. \tag{1.247}$$

The latter equation demonstrates that $\ln\gamma_i$ is a partial molar quantity with respect to G^{E}/RT. The partial molar analogues of Equations (1.245) and (1.246) are

$$\left(\frac{\partial\ln\gamma_i}{\partial P}\right)_{T,\{x_i\}}=\frac{V_i^{\mathrm{E}}}{RT}, \tag{1.248}$$

and

$$\left(\frac{\partial\ln\gamma_i}{\partial T}\right)_{P,\{x_i\}}=-\frac{H_i^{\mathrm{E}}}{RT^2}. \tag{1.249}$$

Furthermore, from the reciprocity relation the pressure dependence of the excess enthalpy is found to be [see also Equation (1.119)]

$$\left(\frac{\partial H^{\mathrm{E}}}{\partial P}\right)_{T,\{x_i\}}=V^{\mathrm{E}}-T\left(\frac{\partial V^{\mathrm{E}}}{\partial T}\right)_{P,\{x_i\}}. \tag{1.250}$$

Flow calorimeters[56,58] allow excess enthalpy measurements at elevated temperatures and pressures, and the results can be checked for consistency by measuring volumetric quantities as indicated in Equation (1.250).

To conclude this subsection I would like to point out that, in the conventional treatment of solution/mixture properties based on excess properties, all the relevant quantities are measurable: H^{E} by calorimetry, V^{E} by dilatometry or densimetry, and γ_i by VLE measurements. Provided an equation for G^{E}/RT in terms of temperature, pressure and composition is available, the fundamental excess-property relation Equation (1.242) will give *complete* information on excess properties. Equation (1.28a) presents the relation between property changes on mixing and excess properties, *i.e.* $M^{\mathrm{E}} = \Delta M - \Delta M^{\mathrm{id}}$, while Equation (1.28b) presents the relation between the respective partial properties. *i.e.* $M_i^{\mathrm{E}} = \Delta M_i - \Delta M_i^{\mathrm{id}}$. In conjunction with Equation (1.237) this leads to

$$G^{\mathrm{E}} = \Delta G - RT \sum_i x_i \ln x_i, \tag{1.251}$$

$$S^{\mathrm{E}} = \Delta S + R \sum_i x_i \ln x_i, \tag{1.252}$$

$$H^{\mathrm{E}} = \Delta H, \tag{1.253}$$

and

$$V^{\mathrm{E}} = \Delta V. \tag{1.31}$$

Unfortunately, no general theory exists that adequately describes the composition dependence of excess quantities, and the equations commonly used are empirical or semiempirical at best.[9] Focusing on binary mixtures, perhaps the most popular functional form is due to Redlich and Kister,[238] that is

$$\frac{M^{\mathrm{E}}}{x_1 x_2} = A_0 + \sum_{m=1}^{a} A_m (x_1 - x_2)^m, \tag{1.254}$$

with the excess partial molar property values at *infinite-dilution*

$$M_i^{\mathrm{E}\infty} = \lim_{x_i \to 0} M_i^{\mathrm{E}}, \; T \,\text{and}\, P \,\text{constant}, \tag{1.255}$$

being given by

$$M_1^{\mathrm{E}\infty} = A_0 + \sum_{m=1}^{a} A_m (-1)^m \tag{1.256a}$$

and

$$M_2^{\mathrm{E}\infty} = A_0 + \sum_{m=1}^{a} A_m. \tag{1.256b}$$

However, for highly skewed data the use of more than four terms may cause spurious oscillations in derived excess partial molar properties and, in particular, may lead to unreliable limiting values at infinite dilution. The flexibility required to fit strongly unsymmetrical curves $M^{\mathrm{E}} = M^{\mathrm{E}}(x_1)$ is provided by *Padé approximants*[239] of order $[a/b]$

$$\frac{M^{\mathrm{E}}}{x_1 x_2} = \frac{A_0 + \sum_{m=1}^{a} A_m (x_1 - x_2)^m}{1 + \sum_{n=1}^{b} B_n (x_1 - x_2)^n}, \tag{1.257}$$

where the denominator must never become zero.

When the number of components in a mixture reaches three and beyond, the experimental effort in determining excess properties increases sharply. This explains the relative scarcity of accurate excess property data for multicomponent mixtures in general, and excess volume data in particular, a situation which is paralleled by less reliable empirical or semiempirical correlating functions to represent their composition dependence. The frequently used prediction of multicomponent behaviour from data of the constituent binaries alone, without ternary (or higher) terms, is an approximate result arising from the model assumptions made.

At constant temperature and pressure, the intensive molar properties of mixtures are usually presented as functions of the mole fractions of the p components, *i.e.* $M = M(x_1, x_2, \ldots, x_p)$, and the direct use of the definition in Equation (1.17) for obtaining partial molar properties may not be convenient. Taking into account that not all mole fractions are independent variables but are constrained by $\sum_{k=1}^{p} x_k = 1$, Van Ness and Abbott[5] showed that the partial molar property of component i in a multicomponent solution may be obtained from

$$M_i = M - \sum_{k \neq i} x_k \left(\frac{\partial M}{\partial x_k} \right)_{T,P,x_{j \neq i,k}} \tag{1.258}$$

where the summation is over all species except i. In differentiating with respect to, say, mole fraction x_k, all the others except x_i and x_k are to be kept constant. However, Equation (1.258) is not the most useful for the numerical treatment of data. As suggested by Rowlinson and Swinton,[4] one may formally introduce differential operators indicating differentiation with respect to, say, x_k, in which *all* other mole fractions are treated as independent variables and are thus held constant. With this convention

$$M_i = M + \left(\frac{\partial M}{\partial x_i} \right)_{T,P,x_{j \neq i}} - \sum_{k=1}^{p} x_k \left(\frac{\partial M}{\partial x_k} \right)_{T,P,x_{j \neq k}}, \tag{1.259}$$

where the summation is over all constituents including i, and the summation term is the same for all i. The problem of obtaining partial molar properties from molar properties of multicomponent solutions has also been addressed in somewhat different ways by Haase,[2] by Brown[240] and by others. This topic has been recently reviewed by Näfe.[241]

I conclude this section by introducing *apparent molar* properties. They are primarily used for binary liquid solutions, since in many cases their values are readily obtained by experiment. The apparent molar quantity M_2^{app} of solute 2 dissolved in solvent 1 is defined by

$$M_2^{\mathrm{app}} \equiv \frac{nM - n_1 M_1^*}{n_2}, \tag{1.260}$$

where $n = n_1 + n_2$. Note that occasionally the symbol ${}^{\phi}M_2$, or similar notations, are found in the literature. From this definition, we have, for instance, for the volume of a binary system

$$nV = n_1 V_1^* + n_2 V_2^{\mathrm{app}}, \tag{1.261}$$

that is, the molar volume of the solvent is regarded as being independent of the composition of the solution, and *all* the volume change is attributed to the solute *via* the *apparent molar volume* V_2^{app}. By differentiating Equation (1.261) with respect to n_2, for the partial molar volume of the solute we obtain

$$V_2 = V_2^{\mathrm{app}} + n_2 \left(\frac{\partial V_2^{\mathrm{app}}}{\partial n_2} \right)_{T,P,n_1}, \tag{1.262}$$

and for the partial molar volume of the solvent

$$V_1 = V_1^* + n_2 \left(\frac{\partial V_2^{\mathrm{app}}}{\partial n_1} \right)_{T,P,n_2} = V_1^* - \frac{n_2^2}{n_1} \left(\frac{\partial V_2^{\mathrm{app}}}{\partial n_2} \right)_{T,P,n_1} \tag{1.263}$$

Thus, if V_2^{app} is known as a function of n_2, the partial molar volumes of both components can be determined. Clearly, in the limit of infinite dilution of the solute, Equation (1.262) yields the partial molar volume at infinite dilution, that is, at constant T and P

$$\lim_{n_2 \to 0} V_2^{\mathrm{app}} = V_2^{\infty}. \tag{1.264}$$

There are, of course, many additional interesting, important details and fascinating topics I have not touched upon at all, ranging from critical behaviour of fluids to EOS research to volumetric properties of proteins in solution, a fact which is amply evidenced by the contributions to be found in this monograph. I trust you'll enjoy reading them all!

1.3 Concluding Remarks, Outlook and Acknowledgements

PVT measurements, calorimetry and vapour–liquid equilibrium determinations (and liquid–liquid equilibrium determinations and solid–liquid equilibrium determinations) are the oldest and most fundamental experimental disciplines in physical chemistry. They provide quantitative information on properties to be used for theoretical advances on the one hand,

and to improve the practical application of science, that is chemical engineering, on the other. Although simple in principle, enormous effort and ingenuity has gone into designing the vast array of apparatus now at our disposal for the determination of *PVT*, ultrasonic and caloric properties of pure and mixed fluids over large ranges of temperature and pressure, and for vapour–liquid, liquid–liquid and solid–liquid equilibrium studies. In this introductory chapter, I did not cover experimental details at all—the reader is referred to the relevant chapters of this book, and to pertinent articles and monographs quoted in the references I provided, in particular to the three recent volumes of the *Experimental Thermodynamics* series, *i.e. Volume IV*,[56] *Volume VI*,[58] and *Volume VII*,[59] and to the monograph on heat capacities[61] published by RSC/IACT. Let it suffice to say that continuing advances in instrumentation (including automation and miniaturisation) leading to increased precision, accuracy and speed of measurement, as well as the ever widening ranges of application (higher temperatures, higher pressures, smaller concentrations and smaller sample sizes) and improved methods of data management, data storage and data transfer provide the impetus for experimental *PVT* work on fluids, pure and mixed, to remain an active, developing discipline. Parallel advances in the statistical–mechanical treatment of fluids, and increasingly sophisticated computer simulation techniques provide new insights and stimulating connections at the microscopic level. Without doubt, *PVTx* properties belong to the thermodynamic quantities occupying the *centre* of the stage called scientific endeavour. Cross-fertilisation with other disciplines, notably with biophysics, molecular biology and molecular medicine, will become increasingly important. Indeed, a rich research area lies ahead with many new and fascinating vistas of importance for mankind.[242–249]

The field has grown so big that covering it in a reasonably sized monograph is essentially impossible. However, it is hoped that the various topics treated here in the 22 chapters will provide a feeling for the scope of the field, its position in the development of science, and its potential as a whole. In this connection, it is a pleasure to acknowledge the many years of fruitful scientific collaboration with my former PhD advisor, Friedrich Kohler[‡], with Rubin Battino, now Emeritus Professor at Wright State University, Dayton, Ohio, USA, with Jean-Pierre E. Grolier, now Emeritus Professor at Université Blaise Pascal, Clermont-Ferrand, France, with Henry V. Kehiaian[‡], with Augustinus Asenbaum, now Retired Professor of Experimental Physics at the University of Salzburg, Salzburg, Austria, and about 80 colleagues, postdoctoral fellows and students from 17 countries. Without them, many projects would have been difficult to carry out.

Related to this more practical aspect is a *fundamental philosophical* question: *Is there an end to the age of scientific discovery?* Russell Stannard[250] answers in the affirmative. In his opinion, there will remain questions that

---

[‡]Deceased.

we shall never be able to answer, and he offers three reasons why this might be so. One is connected with the architecture of our brains: we may simply reach the limits of what we can imagine. The second reason is of a practical, economical nature: the experimental apparatus needed to test and improve theories, say, in particle physics, will eventually exceed our means; and a scientific theory must be testable. The third reason is the suspicion that there might be questions that are, for us, intrinsically unanswerable, though this aspect is perhaps part of reason number one. One of the recent *Austrian Science Discussions* (Österreichische Wissenschaftstage), organised by the Österreichische Forschungsgemeinschaft, was devoted to the quite general topic of *Image and Reality in Science*. Of particular relevance for the question raised above was the contribution by Jürgen Mittelstraß,[251] entitled *Thinking the Unthinkable*, where he discussed the so-called *sphere of knowledge* suspended in a universe of *non*-knowledge, denoted by ¬knowledge in Figure 1.2. This concept addresses the well-known situation that, with every scientific problem solved, new problems emerge. Research-generated knowledge, *optimistically* represented by the sphere's *volume* V_k, grows as $dV_k = 4\pi r^2 dr$, yet this growth of the sphere of knowledge causes a continuously increasing exposure to ¬knowledge because of the concomitantly growing *surface*, *i.e.* $dA_k = 8\pi r dr$. However, when focusing on the freshly perceived new problems, *i.e.* ¬knowledge contained in the *differential spherical shell* adjacent to $r + dr$, it is seen that the rate of increase of research-generated knowledge is *matched* by the increase of ¬knowledge to *first order*, which is distinctly *less* optimistic than the view advocated by Mittelstraß. According to him, independent of the interpretation selected, there are *no* theoretical but only practical limits to scientific discovery.

Whatever the arguments, I find the statement by Gilbert Newton Lewis (1875–1946) on the *practical* philosophy of science most encouraging. Although I quoted it in our recent monograph on heat capacities, the insight contained makes it appropriate to repeat it here:

> *The scientist is a practical man and his are practical aims. He does not seek the ultimate but the proximate. He does not speak of the last analysis but rather of the next approximation....On the whole, he is satisfied with his work, for while science may never be wholly right it certainly is never wholly wrong; and it seems to be improving from decade to decade.*

1.4 Glossary of Symbols

In almost all cases I have adhered to the nomenclature/symbols suggested by IUPAC (see Green Book, *Quantities, Units and Symbols in Physical Chemistry*[252]). Deviations are due to my desire to present a concise, unequivocal and logically consistent notation in compliance with usage preferred by the scientific community interested in this review's topics, *i.e.* by physical chemists, physicists and chemical engineers. Such an approach is in accord with the spirit of the Green Book, expressed so admirably by Martin Quack in

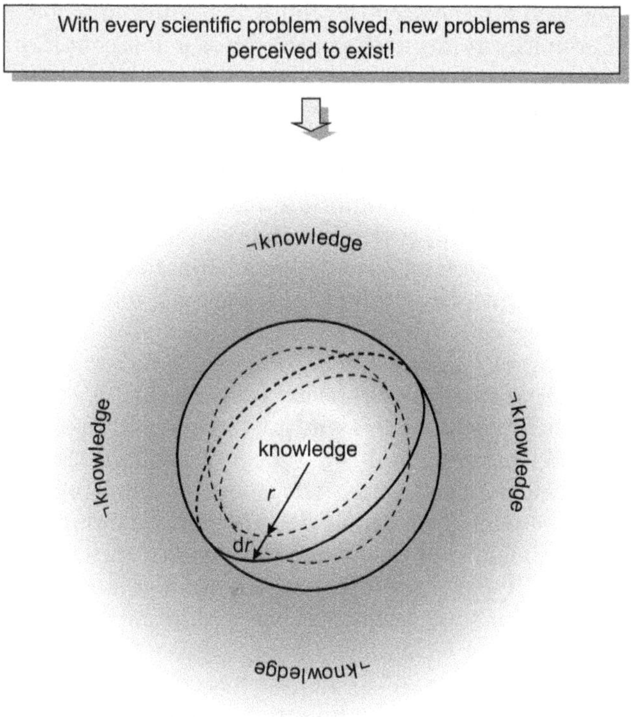

Figure 1.2 The *sphere of knowledge* with radius r suspended in the universe of *non*-knowledge (symbolised here by ¬knowledge) according to J. Mittelstraß,[250] who was stimulated by ideas expressed by Blaise Pascal (1623, Clermont–1662, Paris, France). Research-generated knowledge is *optimistically* represented by the sphere's *volume* $V_k(r)$; it *grows* as $dV_k(r) = 4\pi r^2 dr$, yet this growth of the sphere of knowledge causes a continuously increasing exposure to newly perceived problems, *i.e.* ¬knowledge, because of the concomitantly growing *surface*. When focusing on these freshly emerging problems, that is ¬knowledge contained in the differential spherical shell adjacent to the sphere with radius $r + dr$, it is seen that the rate of increase of research-generated knowledge is *matched* by the increase of ¬knowledge to *first order* (after E. Wilhelm, The art and science of solubility measurements: what do we learn?, *Netsu Sokutai*, 2012, **39** (2), 61–86).[257]

his *Historical Introduction* on p. XII of its 3rd edition, 2007: "*It is not the aim to present a list of recommendations in form of commandments. Rather we have always followed the principle that this manual should help the user in what may be called "good practice of scientific language*". The quantities in particular I would like to single out to comment on are pressure, compressibility and the thermal pressure coefficient. The symbol P for pressure is *now* accepted by IUPAC as an alternative to p, as indicated in Tables 2.2 and 2.10 of the Green Book. The reason why I (and many others) always preferred P is the following. Temperature and pressure are *both* intensive quantities, and together with the composition they form a set of basic thermodynamic variables advantageously used for *homogeneous* fluids in equilibrium states. They are not perceived primarily as properties of the fluids but as conditions imposed on/exhibited by them with the valuable bonus of being (in principle) easily measured and controlled. In other words, temperature and pressure are quantities of "*equal rank*", which fact should be reflected in the symbols we use, that is to say *capital T* and *capital P*. For heterogeneous *PVT* systems consisting of several phases in equilibrium with each other, temperature and pressure are identical in the coexisting phases. We note that Griffiths and Wheeler[253] call such variables *fields* (in contradistinction to variables that are in general *not* equal in coexisting phases, such as volume, enthalpy and entropy, which they call *densities*). For the isothermal compressibility, Rowlinson and Swinton,[4] amongst many others, use the symbol β_T. Together with the isobaric expansivity α_P and the isochoric thermal pressure coefficient γ_V, these mechanical coefficients form an mnemonic triple in $\alpha_P/\beta_T=\gamma_V$, Equation (1.9); and writing them this way, *i.e.* by indicating *via subscript* what quantity is to be held constant, is advantageous in general, and in particular when discussing the related isentropic or orthobaric quantities.

Some of the symbols listed below may be *modified* further, with obvious meaning, by adding appropriate subscripts, such as σ (saturation or orthobaric condition), and/or superscripts, such as * (pure substance), ∞ (infinite dilution), and L (liquid) or V (vapour). The *capital* superscript letters are used because (I) they are easy to read, (II) they are frequently used in the chemical engineering literature, including important monographs (for instance, Prausnitz *et al.*[9], Sandler,[13] and Poling *et al.*[37]) and volumes published under the auspices of the International Union of Pure and Applied Chemistry (IUPAC),[59] and (III) vapour–liquid equilibrium is usually abbreviated to VLE, and *not* to vle. The values for some fundamental physical constants listed in the glossary below are CODATA recommended values.[254,255]

$-a/RTV$	cohesive contribution to Z in a generalised van der Waals equation of state, Equation (1.191)
A_m	$m=0, 1, 2, \ldots, a$: coefficients used in the Redlich–Kister expansion for describing the composition dependence of an excess molar property M^{E}, see Equation (1.254), or in the Padé approximant, Equation (1.257)
A_{P}	first dielectric virial coefficient
$A_{\mathrm{R}}(T,\lambda)$	first refractivity (or optical) virial coefficient
B, B_{ii}	second virial coefficient of a pure gas or a vapour when using either the pressure-explicit virial equation in amount-of-substance density ρ_n, Equation (1.200), or the volume-explicit truncated two-term Equation (1.205)
B'	second virial coefficient of a pure gas or a vapour when using the volume-explicit virial equation in pressure, Equation (1.201)
$B(T,\{x_i\})$	second virial coefficient of a gaseous mixture
B_{ac}	second acoustic virial coefficient

B_{ij}	second interaction or cross virial coefficient
B_n	$n=1, 2,\ldots, b$: coefficients used in the Padé approximant, Equation (1.257)
B_{P}	second dielectric virial coefficient
$B_{\mathrm{R}}(T,\lambda)$	second refractivity (or optical) virial coefficient
B_{TD}	parameter of the Tait equation for correlating compressed liquid densities
B_{TV}	parameter of the Tait equation for correlating compressed liquid volumes
$B^{(0)}$	simple-fluid contribution to BP_c/RT_c according to Pitzer and Curl, Equation (1.230), refined by Tsonopoulos[225,226] and Meng *et al.*[227,228]
$B^{(1)}$	non-simple fluid contribution to $BP_{\mathrm{C}}/RT_{\mathrm{C}}$ according to Pitzer and Curl, Equation (1.230), refined by Tsonopoulos[225,226] and Meng *et al.*[227,228]
C,C_{iii}	third virial coefficient of a pure gas or vapour when using the pressure-explicit virial equation in amount-of-substance density ρ_n, Equation (1.200)
C'	third virial coefficient of a pure gas or a vapour when using the volume-explicit virial equation in pressure, Equation (1.201)
$C(T,\{x_i\})$	third virial coefficient of a gaseous mixture
C_{ac}	third acoustic virial coefficient
C_{ijk}	third interaction or cross virial coefficient
$C_P^{\mathrm{L}}, C_P^{\mathrm{V}}$	molar heat capacity at constant pressure (molar isobaric heat capacity) of a liquid (L) or of a vapour (V), respectively
$C_V^{\mathrm{L}}, C_V^{\mathrm{V}}$	molar heat capacity at constant volume (molar isochoric heat capacity) of a liquid (L) or of a vapour (V), respectively
C_σ^{L}	molar heat capacity of a liquid (L) maintained at all temperatures in equilibrium with an infinitesimal amount of vapour (or shorter: molar heat capacity of a liquid at saturation)
C_σ^{V}	molar heat capacity of a vapour at saturation
$C_P^{\mathrm{pg}}(T)$	molar heat capacity at constant pressure of a fluid in the perfect-gas (ideal-gas) state
$C_V^{\mathrm{pg}}(T)$	$=C_P^{\mathrm{pg}}(T)-R$, molar heat capacity at constant volume of a fluid in the perfect-gas (ideal-gas) state
C_P^{E}	excess molar heat capacity at constant pressure
C_{TD}	parameter of the Tait equation for correlating compressed liquid densities
C_{TV}	parameter of the Tait equation for correlating compressed liquid volumes
$f(r)$	Mayer *f*-function
F	molar Helmholtz energy
$g(r;T,\rho_N)$	pair distribution function
G	molar Gibbs energy

G_i	$=\mu_i$, partial molar Gibbs energy of component i of a mixture/solution, chemical potential of i
G^E	excess molar Gibbs energy
G_i^E	$=\mu_i^E=RT\ln\gamma_i$, excess partial molar Gibbs energy of component i of a mixture/solution, also known as excess chemical potential μ_i^E; γ_i denotes the activity coefficient based on the symmetric Lewis–Randall convention
H	molar enthalpy
H_i	partial molar enthalpy
H^E	excess molar enthalpy
H_i^E	excess partial molar enthalpy
$\Delta_{vap}{}^H$	molar enthalpy of vaporisation
I_B	integrated intensity of *one* Brillouin peak of a Rayleigh–Brillouin spectrum
I_R	integrated intensity of the *central*, unshifted Rayleigh peak of a Rayleigh–Brillouin spectrum
J	molar grand canonical potential
k_B	$= R/L=1.380\ 650\ 4(24)\times10^{-23}$ J K^{-1}, Boltzmann constant
k_{ij}	binary interaction parameter used in the combination rule Equation (1.231)
L	$= 6.022\ 141\ 79(30)\times10^{23}$ mol^{-1}, Avogadro constant
$\mathscr{L}[\ ]$	Laplace transform operator: $f(s)=\mathscr{L}[F(t)]=\int_0^\infty e^{-st}F(t)dt$
m	mass
m, n	Mie parameters, Equation (1.13); when $m=6$ and $n=12$ they are called Lennard-Jones parameters, Equation (1.14)
m_m	$=m/n$, molar mass
m_{MTE}	dimensionless parameter in the modified Tait equation (MTE), Equation (1.186); for many liquid nonelectrolytes at ordinary temperatures $m_{MTE}\approx10$
M	$= M(T,P,\{x_i\})$, intensive macroscopic property of a homogeneous *PVT* fluid mixture in equilibrium; for instance a *molar* quantity such as G, V or H of a mixture/solution
M_i^*	intensive macroscopic property of a pure homogeneous *PVT* fluid in equilibrium; for instance a *molar pure-substance* quantity such as G_i^*, V_i^* or H_i^*
M_i	partial molar property of component i of a mixture/solution, such as G_i, V_i or H_i
M_2^{app}	apparent molar quantity of solute 2 dissolved in solvent 1
M^E	excess molar property
M_i^E	excess partial molar property of component i of a mixture/solution
ΔM	molar property change on mixing
ΔM_i	partial molar property change on mixing
n	$=\sum_i n_i$, total amount of substance

n_i	amount of substance of component i of a mixture/solution
$n(T,\rho_n,\lambda)$	refractive index
N	number of molecules
p	number of components of a mixture/solution
P	pressure
P_c	critical pressure
P_r	reduced pressure
P_σ	vapour pressure
P_{ref}	suitably selected reference pressure, frequently 10^5 Pa or $P_\sigma(T)$
P_{CM}	total molar polarisability, Clausius–Mossotti function
r	distance between molecules
R	$=Lk_B=8.314\ 472(15)\ \mathrm{J\ K^{-1}\ mol^{-1}}$, (molar) gas constant
R_{LL}	molar refractivity, Lorentz–Lorenz function
S	molar entropy
S_i	partial molar entropy of component i of a mixture/solution
T	thermodynamic temperature
T_c	critical temperature
T_r	reduced temperature
T_{ref}	suitably selected reference temperature
$u(r)$	pair-potential-energy function
U	molar internal energy
v_0	thermodynamic speed of ultrasound (at low frequencies)
V	molar volume
V_c	critical molar volume
V_i	partial molar volume
V_2^{app}	apparent molar volume of solute 2 dissolved in solvent 1
V^E	excess molar volume
V_i^E	excess partial molar volume
$\Delta_{vap}V$	molar volume change on vaporisation
$w^{(2)}(r;T,\rho_N)$	potential of mean force
x_i	$=n_i/\sum_i n_i$, mole fraction of component i of a mixture/solution; $\sum_i x_i=1$
X	$\equiv U-\sum_i x_i\mu_i$, unnamed equivalent primary function in the energy representation, Equation (1.64)
Y	$\equiv U+PV-\sum_i x_i\mu_i$, unnamed equivalent primary function in the energy representation, Equation (1.65)
Z	$\equiv PV/RT$, compression factor
Z_c	critical compression factor
Z_{hs}	compression factor of a hard sphere fluid
$Z^{(0)}$	simple-fluid contribution to Z according to Pitzer, Equation (1.223), refined by Lee and Kesler[224]
$Z^{(1)}$	non-simple fluid contribution to Z, Equation (1.223), refined by Lee and Kesler[224]

Greek Letters

α electric polarisability of a molecule; $L\alpha/4\pi\varepsilon_0$ is the mean polarisability volume

α_P $\equiv V^{-1}(\partial V/\partial T)_P$, isobaric expansivity

$\alpha_\sigma^{\mathrm{L}}$ $\equiv (V^{\mathrm{L}})^{-1}(\partial V^{\mathrm{L}}/\partial T)_\sigma$, expansivity of a pure liquid (L) in contact with its vapour (saturation expansivity)

α_S $\equiv V^{-1}(\partial V/\partial T)_S$, isentropic expansivity

β_S $\equiv -V^{-1}(\partial V/\partial T)_S$, isentropic compressibility

β_T $\equiv -V^{-1}(\partial V/\partial T)_T$, isothermal compressibility

γ_V $\equiv (\partial V/\partial T)_V = \alpha_P/\beta_T$, isochoric thermal pressure coefficient

γ_S $\equiv (\partial P/\partial T)_S = -\alpha_S/\beta_S$, isentropic thermal pressure coefficient

γ_σ $\equiv (\partial P/\partial T)_\sigma$, slope of the vapour pressure curve, $\mathrm{d}P_\sigma/\mathrm{d}T$

γ_i $= \gamma_i(T,P,\{x_i\})$, activity coefficient of component i of a mixture/solution based on the symmetric Lewis–Randall convention

ε intermolecular energy parameter characterising the well-depth of the interaction energy function, say, of the Mie intermolecular pair-potential energy function, see Equation (1.13), or of the Lennard-Jones function, Equation (1.14)

ε_r $= \varepsilon/\varepsilon_0$, relative permittivity, formerly called dielectric constant; the permittivity ε is defined by $\boldsymbol{D} = \varepsilon\boldsymbol{E}$, where $\boldsymbol{D}$ is the electric displacement, and $\boldsymbol{E}$ is the electric field strength, and ε_0 is the permittivity of vacuum (electric constant)

ε_0 $= 8.854187\ldots \times 10^{-12}$ F m^{-1}, permittivity of vacuum (electric constant)

κ $\equiv C_P/C_V = \beta_T/\beta_S$, ratio of heat capacities or compressibilities

λ wavelength of light

μ_i chemical potential of component i of a mixture/solution

μ_i^{E} $= G_i^{\mathrm{E}} = RT\ln\gamma_i$, excess chemical potential of component i of a mixture/solution, Lewis–Randall convention

μ_r $= \mu/\mu_0$, relative permeability; the permeability μ is defined by $\boldsymbol{B} = \mu\boldsymbol{H}$, where $\boldsymbol{B}$ is the magnetic induction (magnetic flux density), and $\boldsymbol{H}$ is the magnetic field strength, and μ_0 is the permeability of vacuum (magnetic constant)

μ_0 $= 4\pi\times10^{-7}\mathrm{H\cdot m^{-1}}$ (defined); note that $c_0 = 1/\sqrt{\varepsilon_0\mu_0}$, where the vacuum speed of light is defined as $c_0 = 299\,792\,458\ \mathrm{m\cdot s^{-1}}$

ρ $\equiv m_\mathrm{m}/V = m/(nV)$, mass density

ρ_n $\equiv 1/V = \rho_N/L$, amount-of-substance density

ρ_N $\equiv N/nV = L/V$, number density

σ intermolecular distance parameter, say, of the Mie intermolecular pair-potential energy function, see Equation (1.13), or of the Lennard-Jones function, Equation (1.14); it is characterised by $u(\sigma) = 0$

σ_hs hard sphere diameter

Φ molar Planck function

Ψ molar Massieu function

Ω molar Kramer function

Superscripts

app	indicates an apparent molar property
E	excess quantity
id	ideal solution property
L	liquid phase
pg	perfect-gas state (ideal-gas state)
V	vapour phase
*	indicates a pure-substance property
∞	infinite dilution

Subscripts

c	critical property
i, j, k	general indices; usually i denotes a component of a mixture/solution
p	general index; usually p denotes the number of variables
r	reduced quantity
σ	saturation (orthobaric) condition

References

1. I. Prigogine and R. Defay, *Chemical Thermodynamics, translated and revised by D. H. Everett*, Longmans, Green and Co, London, UK, 1954.
2. R. Haase, *Thermodynamik der Mischphasen*, Springer-Verlag, Berlin, Germany, 1956.
3. K. Denbigh, *The Principles of Chemical Equilibrium*, Cambridge University Press, Cambridge, UK, 4th edn, 1981.
4. J. S. Rowlinson and F. L. Swinton, *Liquids and Liquid Mixtures*, Butterworth Scientific, London, UK, 3rd edn, 1982.
5. H. C. Van Ness and M. M. Abbott, *Classical Thermodynamics of Nonelectrolyte Solutions*, McGraw-Hill, New York, USA, 1982.
6. A. Kreglewski, *Equilibrium Properties of Fluids and Fluid Mixtures*, Texas A&M University Press, College Station, Texas, USA, 1984.
7. C. G. Gray and K. E. Gubbins, *Theory of Molecular Fluids. Vol. 1: Fundamentals*, Clarendon Press, Oxford, UK, 1984.
8. S. E. Wood and R. Battino, *Thermodynamics of Chemical Systems*; Cambridge University Press, Cambridge, UK, 1990.
9. J. M. Prausnitz, R. N. Lichtenthaler and E. G. de Azevedo, *Molecular Thermodynamics of Fluid-Phase Equilibria*, Prentice Hall PTR, Upper Saddle River, NJ, USA, 3rd edn, 1999.
10. *Equations of State for Fluids and Fluid Mixtures. Experimental Thermodynamics, Vol. V*, ed. J. V. Sengers, R. F. Kayser, C. J. Peters and H. J. White, Elsevier Science/IUPAC, Amsterdam, The Netherlands, 2000.
11. J.-L. Barrat and J.-P. Hansen, *Basic Concepts for Simple and Complex Liquids*, Cambridge University Press, Cambridge, UK, 2003.

12. J. P. O'Connell and J. M. Haile, *Thermodynamics: Fundamentals for Applications*, Cambridge University Press, New York, USA, 2005.
13. S. I. Sandler, *Chemical, Biochemical, and Engineering Thermodynamics*, Wiley, New York, USA, 4th edn, 2006.
14. J.-P. Hansen and I. R. McDonald, *Theory of Simple Liquids*, Academic Press/Elsevier, London, UK, 3rd edn, 2006.
15. A. Ben-Naim, *Molecular Theory of Solutions*, Oxford University Press, Oxford, UK, 2006.
16. K. Lucas, *Molecular Models for Fluids*, Cambridge University Press, New York, USA, 2007.
17. J. Z. Wu, Density Functional Theory for Liquid Structure and Thermodynamics, in *Molecular Thermodynamics of Complex Systems. Structure and Bonding, Vol. 131*, ed. X. Lu and Y. Hu, Springer-Verlag, Berlin, Heidelberg, Germany, 2009, pp. 1–73.
18. *Applied Thermodynamics of Fluids*, ed. A. R. H. Goodwin, J. V. Sengers and C. J. Peters, The Royal Society of Chemistry/IUPAC & IACT, Cambridge, UK, 2010.
19. C. G. Gray, K. E. Gubbins and C. G. Joslin, *Theory of Molecular Fluids. Vol. 2: Applications*, Oxford University Press, Oxford, UK, 2011.
20. U. K. Deiters and T. Kraska, *High-Pressure Fluid Phase Equilibria: Phenomenology and Computation*, Elsevier, Oxford, UK, 2012.
21. J. Gmehling, B. Kolbe, M. Kleiber and J. Rarey, *Chemical Thermodynamics for Process Simulation*, Wiley-VCH, Weinheim, Germany, 2012.
22. Y. Marcus, *Ion Solvation*, Wiley, Chichester, UK, 1985.
23. J. M. G. Barthel, H. Krienke and W. Kunz, *Physical Chemistry of Electrolyte Solutions: Modern Aspects*, Steinkopff, Darmstadt, Germany, and Springer, New York, USA, 1998.
24. L. L. Lee, *Molecular Thermodynamics of Electrolyte Solutions*, World Scientific, Singapore, 2008.
25. J. M. Prausnitz, *Molecular Thermodynamics of Fluid-Phase Equilibria*, Prentice-Hall, Englewood Cliffs, New Jersey, USA, 1969.
26. (a) J. H. Dymond, K. N. Marsh, R. C. Wilhoit and K. C. Wong, *Virial Coefficients of Pure Gases, in Landolt–Börnstein, Numerical Data and Functional Relationships in Science and Technology, New Series; Group IV: Physical Chemistry, Virial Coefficients of Pure Gases and Mixtures, Vol. 21A*, ed. M. Frenkel and K. N. Marsh, Springer-Verlag, Heidelberg, Germany, 2002; (b) J. H. Dymond, K. N. Marsh and R. C. Wilhoit, *Virial Coefficients of Mixtures, in Landolt–Börnstein, Numerical Data and Functional Relationships in Science and Technology, New Series; Group IV: Physical Chemistry, Virial Coefficients of Pure Gases and Mixtures, Vol. 21B*, ed. M. Frenkel and K. N. Marsh, Springer-Verlag, Heidelberg, Germany, 2003.
27. M. Frenkel, K. N. Marsh, K. J. Kabo, A. C. Wilhoit and G. N. Roganov, *Thermodynamics of Organic Compounds in the Gas State*, Thermodynamics Research Center, The Texas A&M System, College Station, Texas, USA, 1994, vol. I and II.

28. V. Majer and V. Svoboda, *Enthalpies of Vaporization of Organic Compounds. A Critical Review and Data Compilation*, Blackwell Scientific Publications/IUPAC, Oxford, UK, 1985.
29. L. Haar, J. S. Gallagher and G. S. Kell, *NBS/NRC Steam Tables (NSRDS). Thermodynamic and Transport Properties and Computer Programs for Vapor and Liquid States of Water in SI Units*, Hemisphere Publishing Corporation, New York, USA, 1984.
30. P. G. Hill and R. D. C. MacMillan, Virial Equation for Light and Heavy Water, *Ind. Eng. Chem. Res.*, 1988, **27**, 874–882.
31. W. Wagner and A. Pruß, The IAPWS formulation 1995 for the thermodynamic properties of ordinary water substance for general and scientific use, *J. Phys. Chem. Ref. Data*, 2002, **31**, 387–535.
32. M. Zábranský, V. Růžička Jr., V. Majer and E. S. Domalski, *Heat Capacity of Liquids, Volumes I and II: Critical Review and Recommended Values, J. Phys. Chem. Ref. Data, Monograph No. 6*, American Chemical Society and American Institute of Physics, 1996.
33. M. Zábranský, V. Růžička Jr. and E. S. Domalski, Heat Capacity of Liquids: Critical Review and Recommended Values, Supplement I, *J. Phys. Chem. Ref. Data*, 2001, **30**, 1199–1689.
34. M. W. Chase Jr., NIST-JANAF Thermochemical Tables, Parts I and II, *J. Phys. Chem. Ref. Data, Monograph No. 9*, American Chemical Society and American Institute of Physics, 4th edn, 1998.
35. *International DATA Series, SELECTED DATA ON MIXTURES, Series A*, published by the Thermodynamics Research Center, Texas A&M University, College Station, TX, from 1973 through 1994.
36. *Solubility Data Series (IUPAC)*: Vol. 1: Pergamon Press, Oxford, UK, 1979 and subsequent volumes; *Solubility Data Series (IUPAC-NIST)*: Vol. 66: *J. Phys. Chem. Ref. Data*, 1998, **27**, 1289–1470, and subsequent volumes.
37. B. E. Poling, J. M. Prausnitz and J. P. O'Connell, *The Properties of Gases and Liquids*, McGraw-Hill, New York, USA, 5th edn, 2001.
38. A. Kazakov, C. D. Muzny, R. D. Chirico, V. V. Diky and M. Frenkel, Web Thermo Tables – an On-Line Version of the TRC Thermodynamic Tables, *J. Res. Natl. Inst. Stand. Technol.*, 2008, **113**, 209–220.
39. *NIST SDR 4. NIST Thermophysical Properties of Hydrocarbon Mixtures Database: Version 3.2*, NIST, Boulder, Colorado, USA: http://www.nist.gov./srd/nist4.cfm (Last updated: June 19, 2012).
40. *NIST SDR 10. NIST/ASME Steam Properties Database: Version 2.22*, NIST, Boulder, Colorado, USA: http://www.nist.gov./srd/nist10.cfm (Last updated: February 28, 2012).
41. *NIST SDR 23: NIST Reference Fluid Thermodynamic and Transport Properties Database (REFPROP): Version 9.0*. NIST, Boulder, Colorado, USA: http://www.nist.gov./srd/nist23.cfm (Last updated: February 22, 2013).
42. *NIST SDR 103b: NIST ThermoData Engine Version 7.0 – Pure Compounds, Binary Mixtures, Ternary Mixtures, and Chemical Reactions*, NIST, Boulder, Colorado, USA: http://www.nist.gov./srd/nist103b.cfm (Last updated: February 14, 2013).

43. *NIST SDR 203. NIST Web Thermo Tables (WTT) – Professional Edition*, NIST, Boulder, Colorado, USA: http://www.nist.gov./srd/nistwebsub3.cfm (Last updated: January 11, 2013).
44. *Dortmund Data Bank Software and Separation Technology*: http://www.ddbst.de.
45. (a) R. C. Wilhoit, K. N. Marsh, X. Hong, N. Gadalla and M. Frenkel, Densities of Aliphatic Hydrocarbons: Alkanes, in *Landolt-Börnstein, Numerical Data and Functional Relationships in Science and Technology, New Series; Group IV: Physical Chemistry, Vol. 8, Subvolume B*, ed. K. N. Marsh, Springer-Verlag, Berlin, Heidelberg, 1996; (b) R. C. Wilhoit, K. N. Marsh, X. Hong, N. Gadalla and M. Frenkel, *Densities of Aliphatic Hydrocarbons: Alkenes, Alkadienes, Alkynes, and Miscellaneous Compounds, ibid., Subvolume C*, ed. K. N. Marsh, Springer-Verlag, Berlin, Heidelberg, 1996; (c) R. C. Wilhoit, X. Hong, M. Frenkel and K. R. Hall, *Densities of Monocyclic Hydrocarbons, ibid., Subvolume D*, ed. K. R. Hall and K. N. Marsh, Springer-Verlag, Berlin, Heidelberg, 1997; (d) R. C. Wilhoit, X. Hong, M. Frenkel and K. R. Hall, *Densities of Aromatic Hydrocarbons, ibid., Subvolume E*, ed. K. R. Hall and K. N. Marsh, Springer-Verlag, Berlin, Heidelberg, 1998; (e) R. C. Wilhoit, X. Hong, M. Frenkel and K. R. Hall, *Densities of Polycyclic Hydrocarbons, ibid., Subvolume F*, ed. K. R. Hall and K. N. Marsh, Springer-Verlag, Berlin, Heidelberg, 1998; (f) M. Frenkel, X. Hong, R. C. Wilhoit and K. R. Hall, *Densities of Alcohols, ibid., Subvolume G*, ed. K. R. Hall and K. N. Marsh, Springer-Verlag, Berlin, Heidelberg, 1998; (g) M. Frenkel, X. Hong, R. C. Wilhoit and K. R. Hall, *Densities of Esters and Ethers, ibid., Subvolume H*, ed. K. R. Hall and K. N. Marsh, Springer-Verlag, Berlin, Heidelberg, 2001; (h) M. Frenkel, X. Hong, Q. Dong, X. Yan and R. D. Chirico, *Densities of Phenols, Aldehydes, Ketones, Carboxylic Acids, Amines, Nitriles, and Nitrohydrocarbons, ibid., Subvolume I*, ed. K. R. Hall and K. N. Marsh, Springer-Verlag, Berlin, Heidelberg, 2002; (i) M. Frenkel, X. Hong, Q. Dong, X. Yan and R. D. Chirico, *Densities of Halohydrocarbons, ibid., Subvolume J*, ed. M. Frenkel and K. N. Marsh, Springer-Verlag, Berlin, Heidelberg, 2003.
46. J.-P. E. Grolier, C. J. Wormald, J.-C. Fontaine, K. Sosnkowska-Kehiaian and H. V. Kehiaian, Heats of Mixing and Solution, in *Landolt–Börnstein, Numerical Data and Functional Relationships in Science and Technology, New Series; Group IV: Physical Chemistry, Vol. 10A*, Springer-Verlag, Berlin, Heidelberg, 2004.
47. (a) I. Wichterle, J. Linek, Z. Wagner, J.-C. Fontaine, K. Sosnkowska-Kehiaian and H. V. Kehiaian, Vapor–Liquid Equilibrium in Mixtures and Solutions, in *Landolt–Börnstein, Numerical Data and Functional Relationships in Science and Technology, New Series; Group IV: Physical Chemistry, Vol. 13A1*, Springer-Verlag, Berlin, Heidelberg, 2007; (b) I. Wichterle, J. Linek, Z. Wagner, J.-C. Fontaine, K. Sosnkowska-Kehiaian and H. V. Kehiaian, Vapor–Liquid Equilibrium in Mixtures and Solutions, in *Landolt–Börnstein, Numerical Data and Functional Relationships*

in Science and Technology, New Series; Group IV: Physical Chemistry, Vol. 13A2, Springer-Verlag, Berlin, Heidelberg, 2008.

48. (a) R. Lacmann and C. Synowietz, *Nonaqueous Systems and Ternary Aqueous Systems, in Landolt–Börnstein, Numerical Data and Functional Relationships in Science and Technology, New Series; Group IV: Macroscopic and Technical Properties of Matter, Vol. 1, Densities of Liquid Systems, Part a*, ed. Kl. Schäfer, Springer-Verlag, Berlin, Heidelberg, 1974; (b) J. D'Ans, H. Surawski and C. Synowietz, *Densities of Binary Aqueous Systems and Hear Capacities of Liquid Systems, in Landolt–Börnstein, Numerical Data and Functional Relationships in Science and Technology, New Series; Group IV: Macroscopic and Technical Properties of Matter, Vol. 1, Densities of Liquid Systems and their Heat Capacities, Part b*, ed. Kl. Schäfer, Springer-Verlag, Berlin, Heidelberg, 1977; (c) I. Cibulka, L. Hnědkovský, J.-C. Fontaine, K. Sosnkowska-Kehiaian and H. V. Kehiaian, Binary Liquid Systems of Nonelectrolytes, in *Landolt–Börnstein, Numerical Data and Functional Relationships in Science and Technology, New Series; Group IV: Physical Chemistry, Vol. 23, Volumetric Properties of Mixtures and Solutions, Subvolume A*, ed. H. V. Kehiaian, Springer-Verlag, Berlin, Heidelberg, 2009.
49. I. Cibulka, J.-C. Fontaine, K. Sosnkowska-Kehiaian and H. V. Kehiaian, Heats of Mixing, Vapor–Liquid Equilibrium, and Volumetric Properties of Mixtures and Solutions, in *Landolt–Börnstein, Numerical Data and Functional Relationships in Science and Technology, New Series; Group IV: Physical Chemistry, Vol. 26A, Binary Liqid Systems of Nonelectrolytes I*, Springer-Verlag, Berlin, Heidelberg, 2011.
50. I. Cibulka, J.-C. Fontaine, K. Sosnkowska-Kehiaian and H. V. Kehiaian, Heats of Mixing, Vapor–Liquid Equilibrium, and Volumetric Properties of Mixtures and Solutions, in *Landolt–Börnstein, Numerical Data and Functional Relationships in Science and Technology, New Series; Group IV: Physical Chemistry, Vol. 26B, Binary Liqid Systems of Nonelectrolytes II*, Springer-Verlag, Berlin, Heidelberg, 2012.
51. I. Cibulka, J.-C. Fontaine, K. Sosnkowska-Kehiaian and H. V. Kehiaian, Heats of Mixing, Vapor–Liquid Equilibrium, and Volumetric Properties of Mixtures and Solutions, in *Landolt–Börnstein, Numerical Data and Functional Relationships in Science and Technology, New Series; Group IV: Physical Chemistry, Vol. 26C, Binary Liqid Systems of Nonelectrolytes III*, Springer-Verlag, Berlin, Heidelberg, 2012.
52. *Experimental Thermodynamics, Vol. I: Calorimetry of Non-reacting Systems*, ed. J. P. McCullough and D. W. Scott, Butterworths/IUPAC, London, UK, 1968.
53. *Experimental Thermodynamics, Vol. II: Experimental Thermodynamics of Non-reacting Fluids*, ed. B. Le Neindre and B. Vodar, Butterworths/IUPAC, London, UK, 1975.
54. W. Hemminger and G. Höhne, Calorimetry. *Fundamentals and Practice*, Verlag Chemie, Weinheim, Germany, 1984.
55. J. P. M. Trusler, *Physical Acoustics and Metrology of Fluids*, Hilger, Bristol, UK, 1991.

56. *Solution Calorimetry: Experimental Thermodynamics, Vol. IV*, ed. K. N. Marsh and P. A. G. O'Hare, Blackwell Scientific Publications/IUPAC, Oxford, UK, 1994.
57. G. Höhne, W. Hemminger and H.-J. Flammersheim, *Differential Scanning Calorimetry. An Introduction for Practitioners*, Springer, Berlin, Germany, 1996.
58. *Measurement of the Thermodynamic Properties of Single Phases: Experimental Thermodynamics, Vol. VI*, ed. A. R. H. Goodwin, K. N. Marsh and W. A. Wakeham, Elsevier/IUPAC, Amsterdam, The Netherlands, 2003.
59. *Measurement of the Thermodynamic Properties of Multiple Phases: Experimental Thermodynamics, Vol. VII*, ed. R. D. Weir and Th. W. de Loos, Elsevier/IUPAC, Amsterdam, The Netherlands, 2005.
60. *Development and Application in Solubility*, ed. T. M. Letcher, The Royal Society of Chemistry/IACT, Cambridge, UK, 2007.
61. *Heat Capacities: Liquids, Solutions and Vapours*, ed. E. Wilhelm and T. M. Letcher, The Royal Society of Chemistry/IUPAC & IACT, Cambridge, UK, 2010.
62. M. P. Allen and D. J. Tildesley, *Computer Simulation of Liquids*, Oxford University Press, New York, USA, 1989.
63. J. M. Haile, *Molecular Dynamics Simulation: Elementary Methods*, Wiley, New York, USA, 1997.
64. D. M. Heyes, *The Liquid State: Applications of Molecular Simulations*, Wiley & Sons, Chichester, UK, 1998.
65. R. J. Sadus, *Molecular Simulations of Fluids: Theory, Algorithms and Object-Orientation*, Elsevier Science, New York, USA, 1999.
66. A. Leach, *Molecular Modelling: Principles and Applications*, 2nd edn, Pearson Education, Essex, UK, 2001.
67. F. J. Vesely, *Computational Physics: An Introduction*, 2nd edn, Kluwer Academic/Plenum Publishers, New York, USA, 2001.
68. D. Frenkel and B. Smit, *Understanding Molecular Simulation*, 2nd edn, Academic Press, London, UK, 2002.
69. C. J. Cramer, *Essentials of Computational Chemistry: Theories and Models*, 2nd edn, John Wiley & Sons, Chichester, UK, 2004.
70. D. C. Rapaport, *The Art of Molecular Dynamics Simulation*, 2nd edn, Cambridge University Press, Cambridge, UK, 2004.
71. I. G. Kaplan, Intermolecular Interactions: Physical Picture, *Computational Methods and Model Potentials*, John Wiley & Sons, Chichester, UK, 2006.
72. M. E. Tuckerman, *Statistical Mechanics: Theory and Molecular Simulation*, Oxford University Press, New York, USA, 2010.
73. N. M. Philip, Adiabatic and isothermal compressibilities of liquids, *Proc. Indian Acad. Sci., Sect. A*, 1939, **9**, 109–129.
74. L. A. K. Staveley, K. R. Hart and W. I. Tupman, The heat capacities and other thermodynamic properties of some binary liquid mixtures, *Discuss. Faraday Soc.*, 1953, **15**, 130–150.

75. L. A. K. Staveley, W. I. Tupman and K. R. Hart, Some thermodynamic properties of the systems benzene + ethylene dichloride, benzene + carbon tetrachloride, and acetone + carbon disulphide, *Trans. Faraday Soc.*, 1955, **51**, 323–343.
76. D. Harrison and E. A. Moelwyn-Hughes, The heat capacities of certain liquids, *Proc. R. Soc. London, Ser. A*, 1957, **239**, 230–246.
77. J. Nývlt and E. Erdös, *P-V-T* relations in solutions of liquid non-electrolytes. I. Compressibility, *Collect. Czech. Chem. Commun.*, 1961, **26**, 485–499.
78. K. F. Herzfeld and T. A. Litovitz, *Absorption and Dispersion of Ultrasonic Waves*, Academic Press, New York, USA, and London, UK, 1959.
79. A. B. Bhatia, *Ultrasonic Absorption*, Oxford University Press, London, UK, 1967.
80. M. J. Blandamer, *Introduction to Chemical Ultrasonics*, Academic Press, London, UK, and New York, USA, 1973.
81. A. R. H. Goodwin and J. P. M. Trusler, Sound Speed, in *Measurement of the Thermodynamic Properties of Single Phases: Experimental Thermodynamics, Vol. VI*, ed. A. R. H. Goodwin, K. N. Marsh and W. A. Wakeham, Elsevier/IUPAC, Amsterdam, The Netherlands, 2003, ch. 6, pp. 237–323.
82. E. Wilhelm, What you always wanted to know about heat capacities, but were afraid to ask, *J. Solution Chem.*, 2010, **39**, 1777–1818.
83. E. Wilhelm and A. Asenbaum, Heat Capacities and Brillouin Scattering in Liquids, in *Heat Capacities: Liquids, Solutions and Vapours*, ed. E. Wilhelm and T. M. Letcher, The Royal Society of Chemistry/IUPAC & IACT, Cambridge, UK, 2010, ch. 11, pp. 238–263.
84. J. O. Hirschfelder, C. F. Curtiss and R. B. Bird, *Molecular Theory of Gases and Liquids*, John Wiley & Sons, New York, USA, 1954.
85. T. Kihara, *Intermolecular Forces*, John Wiley & Sons, Chichester, UK, 1978.
86. G. C. Maitland, M. Rigby, E. B. Smith and W. A. Wakeham, *Intermolecular Forces: Their Origin and Determination*, Clarendon Press, Oxford, UK, 1981.
87. J. Rowlinson, *Cohesion: A Scientific History of Intermolecular Forces*, Cambridge University Press, Cambridge, UK, 2004.
88. A. Stone, *The Theory of Intermolecular Forces*, Oxford University Press, Oxford, UK, 2nd edn, 2013.
89. G. Mie, Zur kinetischen Gastheorie der einatomigen Körper, *Ann. Phys.*, 1903, **11**, 657–697.
90. *The Mie Theory. Basics and Applications*, ed. W. Hergert and T. Wried, Springer Series in Optical Sciences, Vol. 169 , Springer-Verlag, Berlin, Germany, 2012.
91. J. E. Jones, On the determination of molecular fields. I. From the variation of the viscosity of a gas with temperature, *Proc. R. Soc. London, Ser. A*, 1924, **106**, 441–462.

92. J. E. Jones, On the determination of molecular fields. II. From the equation of state of a gas, *Proc. R. Soc. London, Ser. A*, 1924, **106**, 463–477.
93. J. E. Jones, On the determination of molecular fields. III. From crystal measurements and kinetic theory data, *Proc. R. Soc. London, Ser. A*, 1924, **106**, 709–718.
94. J. E. Lennard-Jones, Cohesion, *Proc. Phys. Soc.*, 1931, **43**, 461–482.
95. E. Wilhelm, Heat Capacities: Introduction, Concepts and Selected Applications, in *Heat Capacities: Liquids, Solutions and Vapours*, ed. E. Wilhelm and T. M. Letcher, The Royal Society of Chemistry/IUPAC & IACT, Cambridge, UK, 2010, ch. 1, pp. 1–27.
96. E. Wilhelm, Thermodynamics of Solutions, Especially Dilute Solutions of Nonelectrolytes, in *Molecular Liquids: New Perspectives in Physics and Chemistry*, ed. J. J. C. Teixeira-Dias, NATO ASI Series C: Mathematical and Physical Sciences, Vol. 379, Kluwer Academic Publishers, Dordrecht, The Netherlands, 1992, pp. 175–206.
97. E. Wilhelm, Chemical thermodynamics: A journey of many vistas, *J. Solution Chem.*, 2014, **43**, 525–576.
98. K. R. Popper, *The Logic of Scientific Discovery*, Routledge, London & New York, UK & USA, 2002.
99. E. P. Wigner, The unreasonable effectiveness of mathematics in the natural sciences, *Commun. Pure Appl. Math.*, 1960, **13**, 1–14.
100. R. W. Hamming, The unreasonable effectiveness of mathematics, *Am. Math. Mon.*, 1980, **87**, 81–90.
101. E. H. Lieb and J. Yngvason, The physics and mathematics of the second law of thermodynamics, *Phys. Rep.*, 1999, **310**, 1–96.
102. S. Sarukkai, Revisiting the "unreasonable effectiveness" of mathematics, *Curr. Sci.*, 2005, **88**, 415–423.
103. I. Grattan-Guinness, Solving Wigner's mystery: The reasonable (though perhaps limited) effectiveness of mathematics in the natural sciences, *Math. Intell.*, 2008, **30**(3), 7–17.
104. J. W. Gibbs, *The Scientific Papers of J. Willard Gibbs*, Vol. I, *Thermodynamics*, Longmans, Green and Co, London, UK, 1906.
105. H. B. Callen, *Thermodynamics and an Introduction to Thermostatistics*, John Wiley & Sons, New York, USA, 2nd edn, 1985.
106. IUPAC Committee on Legendre Transforms in Chemical Thermodynamics: Chairman: R. A. Alberty; Members: J. M. G. Barthel, E. R. Cohen, M. B. Ewing, R. N. Goldberg and E. Wilhelm: Use of Legendre Transforms in Chemical Thermodynamics (IUPAC Technical Report). *Pure Appl. Chem.*, 2001, **73**, 1349–1380.
107. S. L. Randzio, Scanning transitiometry, *Chem. Soc. Rev.*, 1996, **25**, 383–392.
108. S. L. Randzio, Scanning Transiometry and its Use to Determine Heat Capacities of Liquids at High Pressures, in *Heat Capacities: Liquids, Solutions and Vapours*, ed. E. Wilhelm and T. M. Letcher, The Royal

Society of Chemistry/IUPAC & IACT, Cambridge, UK, 2010, ch. 8, pp. 153–184.

109. H. Kashiwagi, T. Hashimoto, Y. Tanaka, H. Kubota and T. Makita, Thermal conductivity and density of toluene in the temperature range 273–373 K at pressures up to 250 MPa, *Int. J. Thermophys.*, 1982, **3**, 201–215.
110. M. J. P. Muringer, N. J. Trappeniers and S. N. Biswas, The effect of pressure on the sound velocity and density of toluene and n-heptane up to 2600 bar, *Phys. Chem. Liq.*, 1985, **14**, 273–296.
111. V. M. Shulga, F. G. Eldarov, Y. A. Atanov and A. A. Kuyumchev, Thermal conductivity and heat capacity of liquid toluene at temperatures between 255 and 400 K and at pressures up to 1000 MPa, *Int. J. Thermophys.*, 1986, **7**, 1147–1161.
112. A. Asenbaum, E Wilhelm and P. Soufi-Siavoch, Brillouin scattering in liquid toluene at high pressures, *Acustica*, 1989, **68**, 131–141.
113. T. F. Sun, S. A. R. C. Bominaar, C. A. ten Seldam and S. N. Biswas, Evaluation of the thermophysical properties of toluene and n-heptane from 180 to 320 K and up to 260 MPa from speed-of-sound data, *Ber. Bunsenges. Phys. Chem.*, 1991, **95**, 696–704.
114. S. L. Randzio, J.-P. E. Grolier, J. R. Quint, D. J. Eatough, E. A. Lewis and L. D. Hansen, n-Hexane as a model for compressed simple liquids, *Int. J. Thermophys.*, 1994, **15**, 415–441.
115. A. J. Easteal and L. A. Woolf, p, V, T and derived thermodynamic data for toluene, trichloromethane, dichloromethane, acetonitrile, aniline, and n-dodecane, *Int. J. Thermophys.*, 1985, **6**, 331–351.
116. (a) J. W. Magee, Molar heat capacity (C_V) for saturated and compressed liquid and vapor nitrogen from 65 to 300 K at pressures to 35 MPa, *J. Res. Natl. Inst. Stand. Technol.*, 1991, **96**, 725–740; (b) A. R. Perkins and J. W. Magee, Molar heat capacity at constant volume for isobutane at temperatures from (114 to 345) K and at pressures to 35 MPa, *J. Chem. Eng. Data*, 2009, **54**, 2646–2655.
117. (a) B. J. Berne and R. Pecora, *Dynamic Light Scattering*, Wiley, New York, USA, 1976; (b) J. B. Boon and S. Yip, *Molecular Hydrodynamics*, McGraw-Hill, New York, USA, 1980.
118. V. A. Popova and N. V. Surovtsev, Temperature dependence of the Landau-Placzek ratio in glass forming liquids, *J. Chem. Phys.*, 2011, **135**(134510), 1–7.
119. F. Eggers and U. Kaatze, Broad-band ultrasonic measurement techniques for liquids, *Meas. Sci. Tech.*, 1996, **7**, 1–20.
120. K. Lautscham, F. Wente, W. Schrader and U. Kaatze, High resolution and small volume automatic velocimeter for liquids, *Meas. Sci. Technol.*, 2000, **11**, 1432–1439.
121. U. Kaatze, T. O. Hushcha and F. Eggers, Ultrasonic broadband spectrometry of liquids. A research tool in pure and applied chemistry and chemical physics, *J. Solution Chem.*, 2000, **29**, 299–368.

122. U. Kaatze, F. Eggers and K. Lautscham, Ultrasonic velocity measurements in liquids with high resolution – techniques, selected applications and perspectives, *Meas. Sci. Technol.*, 2008, **19**(062001), p. 21.
123. T. Takagi and E. Wilhelm, Speed-of-Sound Measurements and Heat Capacities of Liquid Systems at High Pressure, in *Heat Capacities: Liquids, Solutions and Vapours*, ed. E. Wilhelm and T. M. Letcher, The Royal Society of Chemistry/IUPAC & IACT, Cambridge, UK, 2010, ch. 10, pp. 218–237.
124. J. S. Burlew, Measurement of the heat capacity of a small volume of liquid by the piezo-thermometric method. I. Apparatus for measuring $(\partial T/\partial P)_S$ and results for benzene and toluene, *J. Am. Chem. Soc.*, 1940, **62**, 681–689.
125. J. S. Burlew, Measurement of the heat capacity of a small volume of liquid by the piezo-thermometric method. II. Coefficient of thermal expansion of benzene and of toluene measured with a new type of weight dilatometer, *J. Am. Chem. Soc.*, 1940, **62**, 690–695.
126. J. S. Burlew, Measurement of the heat capacity of a small volume of liquid by the piezo-thermometric method. III. Heat capacity of benzene and toluene from 8 °C to the boiling point, *J. Am. Chem. Soc.*, 1940, **62**, 696–700.
127. D. A. McQuarrie, *Statistical Mechanics*, Harper & Row, New York, USA, 1976.
128. E. Wilhelm, Heat Capacities, Isothermal Compressibilities and Related Quantities of Fluids, in *Les Capacités Calorifiques des Systèmes Condensés*, ed. H. Tachoire, Société Française de Chimie, Marseille, France, 1987, pp. 138–163.
129. V. Majer, V. Svoboda and J. Pick, Heats of Vaporization of Fluids. *Studies in Modern Thermodynamics* 9, Elsevier, Amsterdam, The Netherlands, 1989.
130. E. Wilhelm, R. Schano, G. Becker, G. H. Findenegg and F. Kohler, Molar heat capacity at constant volume, *Trans. Faraday Soc.*, 1969, **65**, 1443–1455.
131. E. Wilhelm, M. Zettler and H. Sackmann, Molwärmen binärer Systeme aus Cyclohexan, Kohlenstofftetrachlorid, Siliziumtetrachlorid und Zinntetrachlorid, *Ber. Bunsenges. Phys. Chem.*, 1974, **78**, 795–804.
132. E. Wilhelm, The fascinating world of pure and mixed nonelectrolytes, *Pure Appl. Chem.*, 2005, **77**, 1317–1330.
133. E. A. Moelwyn-Hughes, *Physical Chemistry*, Pergamon Press, Oxford, UK, 2nd revised edn, 1961.
134. J. R. Macdonald, Some simple isothermal equations of state, *Rev. Mod. Phys.*, 1966, **38**, 669–679.
135. J. R. Macdonald, Review of some experimental and analytical equations of state, *Rev. Mod. Phys.*, 1969, **41**, 316–349.
136. A. T. J. Hayward, Compressibility equations for liquids: a comparative study, *Br. J. Appl. Phys.*, 1967, **18**, 965–977.

137. E. Wilhelm, Pressure dependence of the isothermal compressibility and a modified form of the Tait equation, *J. Chem. Phys.*, 1975, **63**, 3379–3381.
138. E. Wilhelm, Isothermal compressibility of liquids and a modified version of the Tait equation: a van der Waals approach, *Proc. 14th Intl. Conf. Chem. Thermodyn., Vol. II*, Montpellier, France, August 26–30, 1975, pp. 87–94.
139. E. Wilhelm, Calorimetry: its contribution to molecular thermodynamics of fluids, *Thermochim. Acta*, 1983, **69**, 1–44.
140. J. H. Dymond and R. Malhotra, The Tait equation: 100 years on, *Int. J. Thermophys.*, 1988, **9**, 941–951.
141. N. F. Carnahan and K. E. Starling, Equation of state for nonattracting rigid spheres, *J. Chem. Phys.*, 1969, **51**, 635–636.
142. E. Wilhelm and R. Battino, Estimation of Lennard-Jones (6,12) pair potential parameters from gas solubility data, *J. Chem. Phys.*, 1971, **55**, 4012–4017.
143. E. Wilhelm, On the temperature dependence of the effective hard sphere diameter, *J. Chem. Phys.*, 1973, **58**, 3558–3560.
144. H. C. Longuet-Higgins and B. Widom, A rigid sphere model for the melting of argon, *Mol. Phys.*, 1964, **8**, 549–556.
145. E. A. Guggenheim, The new equation of state of Longuet-Higgins and Widom, *Mol. Phys.*, 1965, **9**, 43–47.
146. E. A. Guggenheim, Variations on van der Waals' equation of state for high densities, *Mol. Phys.*, 1965, **9**, 199–200.
147. F. Kohler, E. Wilhelm and H. Posch, Recent advances in the physical chemistry of the liquid state, *Adv. Mol. Relax. Processes*, 1976, **8**, 195–239.
148. T. F. Sun, J. A. Schouten, P. J. Kortebeek and S. N. Biswas, Experimental equations of state for some organic liquids between 273 and 333 K and up to 280 MPa, *Phys. Chem. Liq.*, 1990, **21**, 231–237.
149. I. Cibulka and M. Zikova, Liquid densities at elevated pressures of 1-alkanols from C1 to C10: a critical evaluation of experimental data, *J. Chem. Eng. Data*, 1994, **39**, 876–886.
150. I. Cibulka and L. Hnědkovský, Liquid densities at elevated pressures of n-alkanes from C5 to C16: a critical evaluation of experimental data, *J. Chem. Eng. Data*, 1996, **41**, 657–668.
151. I. Cibulka, L. Hnědkovský and T. Takagi, *P-ρ-T* data of liquids: summarization and evaluation. 3. Ethers, ketones, aldehydes, carboxylic acids, and esters, *J. Chem. Eng. Data*, 1997, **42**, 2–26.
152. I. Cibulka, L. Hnědkovský and T. Takagi, *P-ρ-T* data of liquids: summarization and evaluation. 4. Higher 1-alkanols (C_{11}, C_{12}, C_{14}, C_{16}), Secondary, tertiary, and branched alkanols, cycloalkanols, alkanediols, alkanetriols, etheralkanols, and aromatic hydroxy derivatives, *J. Chem. Eng. Data*, 1997, **42**, 415–433.
153. I. Cibulka and T. Takagi, *P-ρ-T* data of liquids: summarization and evaluation. 5. Aromatic hydrocarbons, *J. Chem. Eng. Data*, 1999, **44**, 411–429.

154. I. Cibulka and T. Takagi, *P-ρ-T* data of liquids: summarization and evaluation. 6. Non-aromatic hydrocarbons (C_n, $n \geq 5$), *J. Chem. Eng. Data*, 1999, **44**, 1105–1128.
155. I. Cibulka, T. Takagi and K. Růžička, *P-ρ-T* data of liquids: summarization and evaluation. 7. Selected halogenated hydrocarbons, *J. Chem. Eng. Data*, 2001, **46**, 2–28; Correction: *J. Chem. Eng. Data*, 2001, **46**, 456.
156. I. Cibulka and T. Takagi, *P-ρ-T* data of liquids: summarization and evaluation. 8. Miscellaneous compounds, *J. Chem. Eng. Data*, 2002, **47**, 1037–1070.
157. D. I. Sagdeev, M. G. Fomina, G. Kh. Mukhamedzyanov and I. M. Abdulagatov, Experimental study of the density and viscosity of *n*-heptane at temperatures from 298 K to 470 K and pressure up to 245 MPa, *Int. J. Thermophys.*, 2013, **34**, 1–33.
158. L. A. Davies and R. B. Gordon, Compression of mercury at high pressure, *J. Chem. Phys.*, 1967, **46**, 2650–2660.
159. R. Vedam and G. Holton, Specific volumes of water at high pressures, obtained from ultrasonic-propagation measurements, *J. Acoust. Soc. Am.*, 1968, **43**, 108–116.
160. R. A. Fine and F. J. Millero, Compressibility of water as a function of temperature and pressure, *J. Chem. Phys.*, 1973, **59**, 5529–5536.
161. G. S. Kell and E. Whalley, Reanalysis of the density of liquid water in the range 0–150 °C and 0–1 kbar, *J. Chem. Phys.*, 1975, **62**, 3496–3503.
162. T. F. Sun, C. A. ten Seldam, B. J. Kortbeek, N. J. Trappeniers and S. N. Biswas, Acoustic and thermodynamic properties of ethanol from 273.15 to 333.15 K and up to 280 MPa, *Phys. Chem. Liq.*, 1988, **18**, 107–116.
163. E. Zorębski and M. Dzida, Study of the acoustic and thermodynamic properties of 1,2- and 1,3-butanediol by means of high-pressure speed of sound measurements at temperatures from (293 to 318) K and pressures up to 101 MPa, *J. Chem. Eng. Data*, 2007, **52**, 1010–1017.
164. M. Chorążewski and M. Skrzypek, Thermodynamic and acoustic properties of 1,3-dibromopropane and 1,5-dibromopentane within the temperature range from 293 K to 313 K at pressures up to 100 MPa, *Int. J. Thermophys.*, 2010, **31**, 26–41.
165. S. Aparicio, M. Atilhan and F. Karadas, Thermophysical properties of pure ionic liquids: review of present situation, *Ind. Eng. Chem. Res.*, 2010, **49**, 9580–9595.
166. R. Gomes de Azevedo, J. M. S. S. Esperança, V. Najdanovic-Visak, Z. P. Visak, H. J. R. Guedes, M. Nunes da Ponte and L. P. N. Rebelo, Thermophysical and thermodynamic properties of 1-butyl-3-methylimidazolium tetrafluoroborate and 1-butyl-3-methylimidazolium hexafluorophosphate over an extended pressure range, *J. Chem. Eng. Data*, 2005, **50**, 997–1008.
167. P. Navia, J. Troncoso and L. Romaní, Isobaric thermal expansivity for nonpolar compounds, *J. Chem. Eng. Data*, 2010, **55**, 2173–2179.

168. T. E. Daubert, R. P. Danner, H. M. Sibel and C. C. Stebbins, *Physical and Thermodynamic Properties of Pure Chemicals: Data Compilation*, Taylor & Francis, Washington, DC, USA, 1997.
169. H. G. Rackett, Equation of state for saturated liquids, *J. Chem. Eng. Data*, 1970, **15**, 514–517.
170. C. F. Spencer and R. P. Danner, Improved equation for prediction of saturated liquid density, *J. Chem. Eng. Data*, 1972, **17**, 236–241.
171. T. Yamada and R. D. Gunn, Saturated liquid molar volumes. The Rackett equation, *J. Chem. Eng. Data*, 1973, **18**, 234–236.
172. C. F. Spencer and S. B. Adler, A critical review of equations for predicting saturated liquid density, *J. Chem. Eng. Data*, 1978, **23**, 82–89.
173. E. A. Mason and T. H. Spurling, *The Virial Equation of State*, Pergamon Press, Oxford, UK, 1969.
174. J. P. M. Trusler, The Virial Equation of State, in *Applied Thermodynamics of Fluids*, ed. A. R. H. Goodwin, J. V. Sengers and C. J. Peters, The Royal Society of Chemistry/IUPAC & IACT, Cambridge, UK, 2010, ch. 3, pp. 33–52.
175. D. A. McQuarrie, *Statistical Mechanics*, Harper & Row, New York, USA, 1976.
176. R. L. Scott and R. D. Dunlop, On the determination of second virial coefficients, *J. Phys. Chem.*, 1962, **66**, 639–640.
177. H. L. Frisch and E. Helfand, Conditions imposed by gross properties on the intermolecular potential, *J. Chem. Phys.*, 1960, **32**, 269–270.
178. D. A. Jonah and J. S. Rowlinson, Direct determination of the repulsive potential between helium atoms, *Trans. Faraday Soc.*, 1966, **62**, 1067–1071.
179. G. C. Maitland and E. B. Smith, The direct determination of potential energy functions from second virial coefficients, *Mol. Phys.*, 1972, **24**, 1185–1201.
180. H. E. Cox, F. W. Crawford, E. B. Smith and A. R. Tindell, A complete iterative inversion procedure for second virial coefficient data. I. The method, *Mol. Phys.*, 1980, **40**, 705–712.
181. E. B. Smith, A. R. Tindell, B. H. Wells and F. W. Crawford, A complete iterative inversion procedure for second virial coefficient data. II. Applications, *Mol. Phys.*, 1981, **42**, 937–942.
182. E. S. Burnett, Compressibility determinations without volume measurements, *J. Appl. Mech.*, 1936, **3**, A136–A140.
183. M. Waxman and J. R. Hastings, A Burnett apparatus for the accurate determination of gas compressibility factors and second virial coefficients, and an evaluation of its capability based on some results for argon and carbon dioxide, *J. Res. Natl. Bur. Stand. Sect. C*, 1971, **75C**, 165–176.
184. M. R. Patel, L. L. Joffrion and P. T. Eubank, A simple procedure for estimating virial coefficients from Burnett PVT data, *AIChE J.*, 1988, **34**, 1229–1232.

185. W. Warowny and P. T. Eubank, Generalized equations of the Burnett *P-V-T* methods for adsorbing gases, *Fluid Phase Equilib.*, 1995, **103**, 77–95.
186. J. Yin and J. Wu, Gas phase PVT properties and second virial coefficients of dimethyl ether, *Fluid Phase Equilib.*, 2010, **298**, 298–302.
187. N. Sakoda, K. Shindo, K. Motomura, K. Shinzato, M. Kohno, Y. Takata and M. Fujii, Burnett method with absolute pressure transducer and measurements for PVT properties of nitrogen andhydrogen up to 473 K and 100 MPa, *Int. J. Thermophys.*, 2012, **33**, 6–21.
188. N. Sakoda, K. Shindo, K. Motomura, K. Shinzato, M. Kohno, Y. Takata and M. Fujii, Burnett PVT measurements of hydrogen and the development of a virial equation of state at pressures up to 100 MPa, *Int. J. Thermophys.*, 2012, **33**, 381–359.
189. B. Schramm and Ch. Weber, Measurement of the second virial coefficients of some new chlorofluorocarbons and of their mixtures at temperatures in the range from 230 K to 300 K, *J. Chem. Thermodyn.*, 1991, **23**, 281–292.
190. J. Millat, H Hendl and E. Bich, Quasi-isochoric $p\rho T$ measurements and second virial coefficients of n-heptane, *Int. J. Thermophys.*, 1994, **15**, 903–920.
191. W. Duschek, R. Kleinrahm and W. Wagner, Measurement and correlation of the (pressure, density, temperature) relation of nitrogen in the temperature range from 273.15 K to 323.15 K at pressures up to 8 MPa, *J. Chem. Thermodyn.*, 1988, **20**, 1069–1077.
192. G. Händel, R. Kleinrahm and W. Wagner, Measurement of the (pressure, density, temperature) relation of methane in the homogeneous gas and liquid regions in the temperature range from 100 K to 260 K and at pressures up to 8 MPa, *J. Chem. Thermodyn.*, 1992, **24**, 685–695.
193. M. Funke, R. Kleinrahm and W. Wagner, Measurement and correlation of the ($p\rho T$) relation of sulphur hexafluoride (SF_6). I. The homogeneous gas and liquid region in the temperature range from 225 K to 340 K at pressures up to 12 MPa, *J. Chem. Thermodyn.*, 2002, **34**, 717–734.
194. M. Funke, R. Kleinrahm and W. Wagner, Measurement and correlation of the ($p\rho T$) relation of ethane. I. The homogeneous gas and liquid regions in the temperature range from 95 K to 340 K at pressures up to 12 MPa, *J. Chem. Thermodyn.*, 2002, **34**, 2001–2015.
195. S. Glos, R. Kleinrahm and W. Wagner, Measurement of the ($p\rho T$) relation of propane, propylene, *n*-butane, and isobutene in the temperature range from (95 to 340) K at pressures up to 12 MPa using an accurate two-sinker densimeter, *J. Chem. Thermodyn.*, 2004, **36**, 1037–1059; Corrigendum: *J. Chem. Thermodyn.*, 2006, 38, 209.
196. M. O. McLinden and C. Lösch-Will, Apparatus for wide-ranging, high-accuracy fluid (p,ρ,T) measurements based on a compact two-sinker densimeter, *J. Chem. Thermodyn.*, 2007, **39**, 507–530.
197. M. R. Moldover and M. O McLinden, Using *ab initio* "data" to accurately determine the fourth density virial coefficient of helium, *J. Chem. Thermodyn.*, 2010, **42**, 1193–1203.

198. R. Kleinrahm and W. Wagner, Measurement and correlation of the equilibrium liquid and vapour densities and the vapour pressure along the coexistence curve of methane, *J. Chem. Thermodyn.*, 1986, **18**, 739–760.
199. J. P. M. Trusler, W. A. Wakeham and M. P. Zarari, Model intermolecular potentials and virial coefficients determined from the speed of sound, *Mol. Phys.*, 1997, **90**, 695–703.
200. A. R. H. Goodwin and J. P. M. Trusler, Speed of Sound Measurements and Heat Capacities of Gases, in *Heat Capacities: Liquids, Solutions and Vapours*, E. Wilhelm and T. M. Letcher, eds., The Royal Society of Chemistry/IUPAC & IACT, Cambridge, UK, 2010, ch. 9, pp. 185–217.
201. J. B. Mehl and M. R. Moldover, Precision acoustic measurements with a spherical resonator: Ar and C2H4, *J. Chem. Phys.*, 1981, **74**, 4062–4077.
202. M. R. Moldover, J. P. M. Trusler, T. J. Edwards, J. B. Mehl and R. S. Davis, Measurement of the universal gas constant R using a spherical acoustic resonator, *Phys. Rev. Lett.*, 1988, **60**, 249–252.
203. M. B. Ewing and J. P. M. Trusler, Speeds of sound in CF4 between 175 and 300 K measured with a spherical resonator, *J. Chem. Phys.*, 1989, **90**, 1106–1115.
204. J. P. M. Trusler and M. P. Zarari, Second and third acoustic virial coefficients of methane at temperatures between 125 K and 375 K, *J. Chem. Thermodyn.*, 1995, **27**, 771–778.
205. A. F. Estrada-Alexanders and J. P. M. Trusler, Thermodynamic properties of gaseous argon at temperatures between 110 and 450 K and densities up to 6.8 mol dm − 3 determined from the speed of sound, *Int. J. Thermophys.*, 1996, **17**, 1325–1347.
206. A. F. Estrada-Alexanders and J. P. M. Trusler, Speed of sound in carbon dioxide at temperatures between (220 and 450) K and pressures up to 14 MPa, *J. Chem. Thermodyn.*, 1998, **30**, 1589–1601.
207. S. S. Todd, I. A. Hossenlopp and D. W. Scott, Vapor-flow calorimetry of benzene, *J. Chem. Thermodyn.*, 1978, **10**, 641–648.
208. I. A. Hossenlopp and D. W. Scott, Vapor heat capacities and enthalpies of vaporization of six organic compounds, *J. Chem. Thermodyn.*, 1981, **13**, 405–414.
209. I. A. Hossenlopp and D. W. Scott, Vapor heat capacities and enthalpies of vaporization of five alkane hydrocarbons, *J. Chem. Thermodyn.*, 1981, **13**, 415–421.
210. I. A. Hossenlopp and D. W. Scott, Vapor heat capacities and enthalpies of vaporization of four aromatic and/or cycloalkane hydrocarbons, *J. Chem. Thermodyn.*, 1981, **13**, 423–428.
211. A. D. Buckingham, The molecular refraction of an imperfect gas, *Trans. Faraday Soc.*, 1956, **52**, 747–753.
212. H. Häusler and K. Kerl, Mean polarizabilities and second and third virial coefficients of the gases C_2H_4, C_2H_6, and SF_6, *Int. J. Thermophys.*, 1988, **9**, 117–130.

213. H. J. Achtermann, T. K. Bose and G. Magnus, Refractive virial coefficients and *P, ρ, T* data of ethylene, *Int. J. Thermophys.*, 1990, **11**, 133–144.
214. H. J. Achtermann, J. G. Hong, G. Magnus, R. A. Aziz and M. J. Slaman, Experimental determination of the refractivity virial coefficients of atomic gases, *J. Chem. Phys.*, 1993, **98**, 2308–2318.
215. U. Hohm and U. Trümper, Experimental determination of second (*p, V, T*) virial coefficients of xenon and chlorine at elevated temperatures, *J. Chem. Soc., Faraday Trans.*, 1995, **91**, 1277–1279.
216. A. D. Buckingham and J. A. Pople, The dielectric constant of an imperfect non-polar gas, *Trans. Faraday Soc.*, 1955, **51**, 1029–1035.
217. A. D. Buckingham, R. H. Cole and H. Sutter, Direct determination of the imperfect gas contribution to dielectric polarization, *J. Chem. Phys.*, 1970, **52**, 5960–5961.
218. A. Koschine and J. K. Lehmann, *Meas. Sci. Technol.*, 1992, **3**, 411–417.
219. T. Barão, C. Nieto de Castro, U. V. Mardolcar, R. Okambawa and J. M. St-Arnaud, Dielectric constants, dielectric virial coefficients, and dipole moments of 1,1,2,2-tetrafluoroethane, *J. Chem. Eng. Data*, 1995, **40**, 1242–1248.
220. M. B. Ewing and D. D. Royal, Relative permittivities and dielectric virial coefficients of nitrogen at $T=300$ K, *J. Chem. Thermodyn.*, 2002, **34**, 1089–1106.
221. K. S. Pitzer, The volumetric and thermodynamic properties of fluids. I. Theoretical basis and virial coefficients, *J. Am. Chem. Soc.*, 1955, **77**, 3427–3433.
222. K. S. Pitzer, D. Z. Lippmann, R. F. Curl, Jr., C. M. Huggins and D. E. Peterson, The volumetric and thermodynamic properties of fluids. II. Compressibility factor, vapor pressure, and entropy of vaporization, *J. Am. Chem. Soc.*, 1955, **77**, 3433–3440.
223. K. S. Pitzer and R. F. Curl, Jr., The volumetric and thermodynamic properties of fluids. III. Empirical equation for the second virial coefficient, *J. Am. Chem. Soc.*, 1957, **79**, 2369–2370.
224. B.-I. Lee and M. G. Kesler, A generalized thermodynamic correlation based on three-parameter corresponding states, *AIChE J.*, 1975, **21**, 510–527.
225. C. Tsonopoulos, An empirical correlation of second virial coefficients, *AIChE J.*, 1974, **20**, 263–272.
226. C. Tsonopoulos and J. H. Dymond, Second virial coefficients of normal alkanes, linear 1-alkanols (and water), alkyl ethers, and their mixtures, *Fluid Phase Equilib.*, 1997, **133**, 11–34.
227. L. Meng, Y.-Y. Duan and L. Li, Correlations for second and third virial coefficients of pure fluids, *Fluid Phase Equilib.*, 2004, **228**, 109–120.
228. L. Meng and Y.-Y. Duan, An extended correlation for second virial coefficients of associated and quantum fluids, *Fluid Phase Equilib.*, 2007, **258**, 29–33.

229. H. Orbey and J. H. Vera, Correlation for the third virial coefficient using Tc, Pc and ω as parameters, *AIChE J.*, 1983, **29**, 107–113.
230. L. A. Weber, Estimating the virial coefficients of small polar molecules, *Int. J. Thermophys.*, 1994, **15**, 461–482.
231. N. Van Nhu, G. A. Iglesias-Silva and F. Kohler, Correlation of third virial coefficients to second virial coefficients, *Ber. Bunsenges. Phys. Chem.*, 1989, **93**, 526–531.
232. M. Ramos-Estrada, G. A. Iglesias-Silva, K. R. Hall and F. Kohler, Estimation of third virial coefficients at low reduced temperatures, *Fluid Phase Equilib.*, 2006, **240**, 179–185.
233. A. J. B. Cruickshank, B. W. Gainey, C. P. Hicks, T. M. Letcher, R. W. Moody and C. L. Young, Gas-liquid chromatographic determination of cross-term virial coefficients using glycerol. Benzene + nitrogen and benzene + carbon dioxide at 50°C, *Trans. Faraday Soc.*, 1969, **65**, 1014–1031.
234. R. J. Laub and R. L. Pecsok, Determination of second-interaction virial coefficients by gas-liquid chromatography, *J. Chromatogr. A*, 1974, **98**, 511–526.
235. J. R. Conder and C. L. Young, *Physicochemical Measurements by Gas Chromatography*, Wiley, New York, USA, 1979.
236. C. Tsonopoulos, Second Virial Cross-Coefficients: Correlation and Prediction of kij, in *Equations of State in Engineering and Research, Advances in Chemistry Series, Vol. 182*, ed. K. C. Chao and R. L. Robinson, Jr., American Chemical Society, Washington, DC, USA, 1979, ch. 8, 143–162.
237. L. Meng and Y.-Y. Duan, Prediction of the second cross virial coefficients of nonpolar binary mixtures, *Fluid Phase Equilib.*, 2005, **238**, 229–238.
238. O. Redlich and A. T. Kister, Thermodynamics of nonelectrolyte solutions. Algebraic representation of thermodynamic properties and the classification of solutions, *Ind. Eng. Chem.*, 1948, **40**, 345–348.
239. K. N. Marsh, A general method for calculating the excess Gibbs free energy from isothermal vapour–liquid equilibria, *J. Chem. Thermodyn.*, 1977, **9**, 719–724.
240. W. B. Brown, The statistical thermodynamics of mixtures of Lennard-Jones molecules. I. Random mixtures, *Philos. Trans. R. Soc. A*, 1957, **250**, 175–220.
241. H. Näfe, Relationship between the partial molar and molar quantity of a thermodynamic state function in a multicomponent mixture – revisited, *J. Chem. Thermodyn.*, 2013, **61**, 138–145.
242. D. Chandler, Interfaces and the driving force of hydrophobic assembly, *Nature*, 2005, **437**, 640–647.
243. Y. Levy and J. N. Onuchic, Water mediation in protein folding and molecular recognition, *Annu. Rev. Biophys. Biomol. Struct.*, 2006, **35**, 389–415.

244. P. Ball, Water as an active constituent in cell biology, *Chem. Rev.*, 2008, **108**, 74–108.
245. I. O. Brovchenko and A. Oleinikova, Interfacial and Confined Water, Elsevier, Amsterdam, The Netherlands, 2008.
246. K. A. Dill, S. B. Ozkan, M. S. Shell and T. R. Weikl, The protein folding problem, *Annu. Rev. Biophys.*, 2008, **37**, 289–316.
247. I. Budin and J. W. Szostak, Expanding roles for diverse phenomena during the origin of life, *Annu. Rev. Biophys.*, 2010, **39**, 245–263.
248. J. L. England, Statistical physics of self-replication, *J. Chem. Phys.*, 2013, **139**(121923), 1–8.
249. D. R. Canchi and A. E. Garcia, Cosolvent effects on protein stability, *Annu. Rev. Phys. Chem.*, 2013, **64**, 273–293.
250. R. Stannard, *The End of Discovery*, Oxford University Press, Oxford, 2010.
251. J. Mittelstraß, Das Undenkbare denken. Über den Umgang mit dem Undenkbaren und Unvorstellbaren in der Wissenschaft, in *Virtualität und Realtät. Bild und Wirklichkeit in den Naturwissenschaften*, ed. K. Komarek and G. Magerl, Böhlau, Wien, Austria, 1998, pp. 1–25.
252. International Union of Pure and Applied Chemistry, *Quantities, Units and Symbols in Physical Chemistry*, RSC Publishing, Cambridge, UK, 2007.
253. R. B. Griffiths and J. C. Wheeler, Critical points in multicomponent systems, *Phys. Rev. A*, 1970, **2**, 1047–1064.
254. P. J. Mohr, B. N. Taylor and D. B. Newell, CODATA recommended values of the fundamental physical constants: 2006, *Rev. Mod. Phys.*, 2008, **80**, 633–730.
255. P. J. Mohr, B. N. Taylor and D. B. Newell, CODATA recommended values of the fundamental physical constants: 2012, *Rev. Mod. Phys.*, 2012, **84**, 1527–1605.
256. E. Wilhelm, Determination of caloric quantities of dilute liquid solutions, *Thermochim. Acta*, 1987, **119**, 17–33.
257. E. Wilhelm, The art and science of solubility measurements: what do we learn?, *Netsu Sokutai*, 2012, **39**(2), 61–86.

CHAPTER 2

Experimental Techniques 1: Direct Methods†

MARK O. McLINDEN

National Institute of Standards and Technology, Applied Chemicals and Materials Division, Boulder, Colorado 80305, USA
Email: markm@boulder.nist.gov

2.1 Introduction

Fluid density is, by definition, mass per unit volume, and the experimental techniques described in this chapter involve, in one way or another, a clearly defined mass and volume. For this reason, we describe them as "direct" methods. This is in contrast to the vibrating-tube method described in Chapter 3, in which neither mass nor volume is measured directly. (Instead the fluid density is calibrated in terms of the resonant frequency of the vibrating body.) The direct methods described here range from very simple and inexpensive devices to highly sophisticated instruments that have determined the most accurate fluid densities measured to date. This chapter focuses on general-purpose research instruments for wide-ranging measurements; specialized techniques, such as for partial molar volumes, are discussed in other chapters. Densimeters combining the Archimedes principle with a magnetic suspension coupling are increasingly used for wide-ranging, high-accuracy measurements, and such devices are extensively discussed. Older techniques are also described.

Volume Properties: Liquids, Solutions and Vapours
Edited by Emmerich Wilhelm and Trevor M. Letcher

Published by the Royal Society of Chemistry, www.rsc.org

2.2 Measurement by Solid Bodies

The determination of density using solid bodies relies on the Archimedes principle. The Archimedes principle dates to antiquity but remains among the simplest and most accurate techniques for the determination of density. It involves weighing an object or "sinker" of known volume V while it is immersed (fully or partially) in the fluid of interest. The balance reading W is the difference between the sinker mass m and the buoyancy force exerted by the fluid:

$$W = m_{\text{sinker}} - \rho_{\text{fluid}} V_{\text{sinker}}, \tag{2.1}$$

which is rearranged to yield the fluid density ρ:

$$\rho_{\text{fluid}} = \frac{m_{\text{sinker}} - W}{V_{\text{sinker}}}. \tag{2.2}$$

When the sinker is fully immersed in the fluid, the V in Equation (2.2) is the volume of the entire sinker, while for a partially-immersed sinker (such as a hydrometer, Section 2.2.1.1) the V is the immersed volume.

2.2.1 Near-Neutral Buoyancy

Densimeters employing a measuring element with a density near that of the fluid to be measured are simple, very common and widely used to monitor a process fluid in industry or to measure a quantity (such as alcohol concentration) that can be related to density. They are used less frequently in research applications where the objective is to determine the thermodynamic properties of a fluid.

2.2.1.1 Hydrometers

A hydrometer is typically made of glass and consists of a weighted bulb topped by a slender stem on which a scale is marked (see Figure 2.1). It is placed in the liquid sample and allowed to come to an equilibrium position, such that the downward force of gravity acting on the hydrometer's mass m, together with a surface-tension force at the liquid interface, is balanced by the buoyancy force of the liquid on the submerged portion and the (small) buoyancy force of air on the portion of the stem above the fluid:

$$mg + \pi D\gamma\cos\theta = g\rho_{\text{liq}}(V_{\text{bulb}} + AL_1) + g\rho_{air}AL_2, \tag{2.3}$$

where g is the acceleration of gravity, D is the diameter of the stem, γ is the surface tension, θ is the wetting angle of the liquid with the stem, A is the cross-sectional area of the stem, and L_1 and L_2 are the stem lengths below and above the liquid surface.

Hydrometers are used at a specified temperature, and they can be calibrated with an uncertainty of 0.01% or less, although surface tension and other effects typically increase the obtained uncertainty to 0.1% or more.

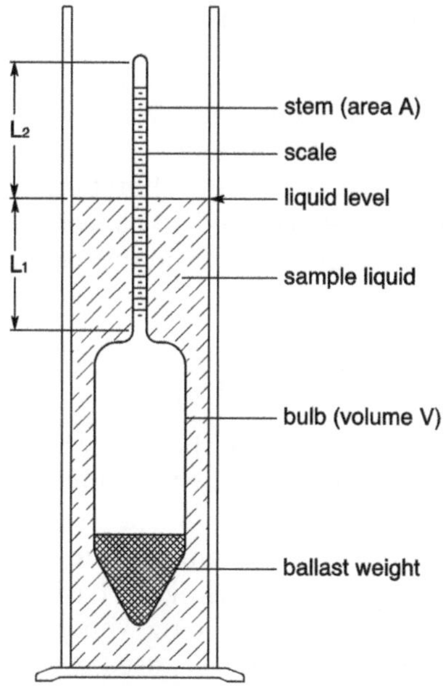

Figure 2.1 Elements of a hydrometer.

(Throughout this chapter expanded combined uncertainties ($k=2$) are given.) Wright *et al.*[1] thoroughly describe the calibration and also the theory of hydrometers.

Hydrometers are widely used in industry for testing fluids, and many hydrometers are calibrated with scales specific to a particular application, such as "degrees Baumé" for sugar solutions, "degrees API" for hydrocarbon fluids, and "proof" for alcohol solutions.

2.2.1.2 Density Floats

A particularly simple density transducer comprises a sinker (often very small and often termed a "float") with a density near that of the fluid to be measured. Typically a number of such sinkers, covering a range of density, are used. They are placed in the fluid, and the density is bounded by the densities of the most-dense sinker that floats and the least-dense sinker that remains at the bottom. This is the basis of simple, hardware-store testers to measure the "strength" of battery acid or automotive antifreeze. A more sophisticated embodiment employs a set of carefully calibrated glass density floats that are placed in a partially filled vial of fluid. The vial is sealed and the temperature is varied so that the density of the fluid matches the density of one of the floats, such that it neither sinks nor rises. The saturated liquid density at that temperature is the density of the corresponding float. The advantages of this method are its simplicity and the requirement for only a few milliliters of sample. For example, Basu and Wilson[2] have used this technique to measure the density of refrigerant 134a with an uncertainty of 0.05%.

2.2.2 Direct Archimedes Techniques

The straightforward application of the Archimedes principle has yielded some of the most accurate fluid densities ever measured. The method is an absolute technique, meaning that, as long as the sinker volume can be determined independently (see Chapter 4), calibration fluids are not required. In practice, two weighings are required to cancel the effect of the "pan" needed to lift the sinker and suspend it from the balance, and Equation (2.2) becomes

$$\rho_{\text{fluid}} = \frac{m_{\text{sinker}} - (W - W_{\text{tare}})}{V_{\text{sinker}}}, \tag{2.4}$$

where W_{tare} is the "tare" weighing with the sinker lifted off the pan; surface-tension forces on the suspension wire are (in principle) the same for the two weighings and cancel out.

For example, Masui *et al.*[3] have determined the density of water at 16 °C with an uncertainty of 0.6 parts in 10^6 using the basic Archimedes technique. They used a spherical sinker of fused quartz whose volume was determined by interferometry. Patterson and Morris[4] used a similar approach with a hollow glass sphere of ultra-low-expansion glass, whose volume was also determined by interferometry, to measure the specific volume of water over the temperature range 1 °C to 40 °C; their uncertainty ranged from 0.64 to 1.4 parts in 10^6.

Richter *et al.*[5] describe a "standard densimeter" for the special purpose of measuring the density of natural gases at standard conditions ($T = 273.15$ K, $p = 0.101325$ MPa). They use two sinkers: a stainless-steel ring ($m = 200$ g, $V = 25$ cm^3) and a hollow stainless-steel cylinder ($m = 200$ g, $V = 500$ cm^3); both sinkers are electropolished and gold plated. The sinkers have an equal surface area and the same surface finish and this largely cancels the effects of gas adsorption on the sinker surfaces. The basic approach is that of the two-sinker densimeters described in section 2.2.3.2, except that no magnetic suspension coupling is used. They achieve a relative combined uncertainty of 0.02%, which is remarkable considering that a typical density of natural gas at standard conditions is 0.8 kg m^{-3}.

2.2.3 Densimeters With Magnetic Suspension Coupling

The direct Archimedes experiment in which a sinker of known volume is suspended below a balance *via* a fine filament is limited to the narrow range of conditions tolerable by the balance. For wide-ranging measurements, another approach is needed, and the most successful has been the combination of the Archimedes principle with magnetic suspension. Beams *et al.*[6] described a magnetic suspension balance in 1955. Their group applied the technique to densimetry (Beams and Clarke[7]) and the simultaneous measurement of density and viscosity (Hodgins and Beams[8]). In these first generation instruments, the sinker (or "buoy") was either a permanent magnet or was made of a ferromagnetic material. The buoy was placed in a nonmagnetic cell (typically made of glass) filled with the fluid of interest. The cell was located in a magnetic field generated by one or more magnets. Three basic approaches were used. Beams and co-workers, along with Haynes *et al.*,[9] suspended the buoy using several electromagnets (see Figure 2.2). The electrical current needed to keep the buoy suspended at a fixed position yielded the buoyancy force and thus the density. This approach essentially integrated a force transducer of the type found in modern electronic balances with the magnetic suspension. It yielded good results but involved very tedious measurements and required very careful calibrations at each temperature measured. A variation on this approach was recently used by Wolf *et al.*[10] in a "magnetic flotation apparatus" to measure the density of water with an uncertainty of 0.8 parts in 10^6.

The second approach among the early instruments was to suspend the sample cell from a balance and weigh it with the magnet in suspension and at rest. This approach avoided many of the instrumental complications of the original technique, but it required a lightweight cell, and the filling line could influence the weighings. In the third approach, the electromagnetic coil was suspended from the balance, and it was weighed with and without the buoy in suspension. This approach was first described by Masui *et al.*[11] in 1984, who has recently refined the technique.[12] All three approaches suffer from various problems. The requirement for a magnetic buoy restricts the choice of materials. The position feedback needed to keep the buoy in stable suspension was typically optical, which often brought sealing difficulties. Masui[12] had difficulties with magnetic particles that would stick to the buoy (which was a permanent magnet) and change its mass.

The current generation of instruments incorporates a magnetic suspension coupling that is separate from the sinker, together with an analytical

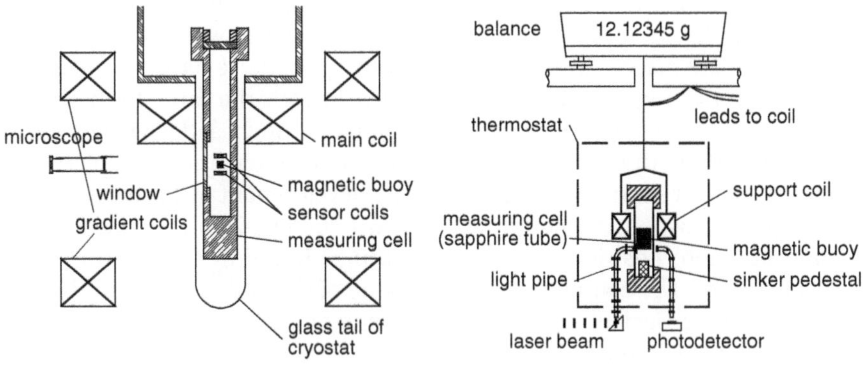

Figure 2.2 Early approaches for magnetic suspension densimeters: (left) levitated "buoy" approach of Beams and Clarke[7] and Haynes *et al.*;[9] the force needed to suspend the buoy is deduced from the current in the electromagnets; (right) levitated buoy approach of Masui *et al.*;[11,12] the electromagnet is weighed with and without the buoy in suspension.

balance; this removes the requirement that the sinker be magnetic, thus allowing a wider choice of sinker materials. The use of a separate analytical balance takes advantage of the extraordinary precision of modern balances, avoiding the need to develop complex electronics for the force measurement. Such instruments, which are sometimes called "hydrostatic balance densimeters," will be the focus of the remaining discussion of this section.

The central elements of the coupling are two magnets, one on each side of a nonmagnetic, pressure-separating wall. The top magnet, which is an electromagnet with a ferrite core, is outside the measuring cell and is suspended from the under-pan weighing hook of the balance. The bottom magnet, which is a permanent magnet, is inside the measuring cell; it is completely immersed in the fluid of interest and is held in a freely suspended state by the electromagnet. A device to lift the sinker is attached to the permanent magnet. A position sensor is also part of the coupling, and a stable suspension is maintained by means of a feedback-control circuit making fine adjustments in the electromagnet current, which is zero on average, to avoid heating and subsequent convection. (The bulk of the lifting force is provided by the attraction between the permanent magnet and the ferrite core of the electromagnet.) The weight of the sinkers is thus transmitted to the balance.

2.2.3.1 *Single-Sinker Densimeters*

The essential elements of a modern single-sinker, magnetic-suspension densimeter are a balance, the magnetic-suspension coupling and associated position sensor and control circuit, the sinker, and a mechanism to pick up the sinker for weighing ("measuring position") or place it on a rest to obtain a "tare" or "zero-position" weighing. (Systems for controlling and measuring the temperature and pressure are also required, but they are outside the scope of this discussion.) Many densimeters also have two "compensation masses" that are placed directly on the balance pan; these have masses such that the total loading on the balance is approximately the same for both the measuring-position and zero-position weighings. Such a design was first developed by Brachthäuser *et al.*[13,14] in 1993 and is schematically depicted in Figure 2.3. Key for such instruments was the development by Lösch *et al.*[15,16] of a magnetic suspension coupling with two stable states (*i.e.*, measuring position and zero position); previous couplings had only "measuring position" and "off." The coupling and decoupling of the sinker is accomplished simply by the vertical motion of the permanent magnet, and this greatly simplifies the mechanical design.

For such an instrument, the fluid density is given by

$$\rho_{\text{fluid}} = \frac{m_{\text{s}} - [(W_{\text{MP}} - m_{\text{c1}}) - (W_{\text{ZP}} - m_{\text{c2}})] - \rho_{\text{air}}(V_{\text{c1}} - V_{\text{c2}})}{V_{\text{s}}}, \qquad (2.5)$$

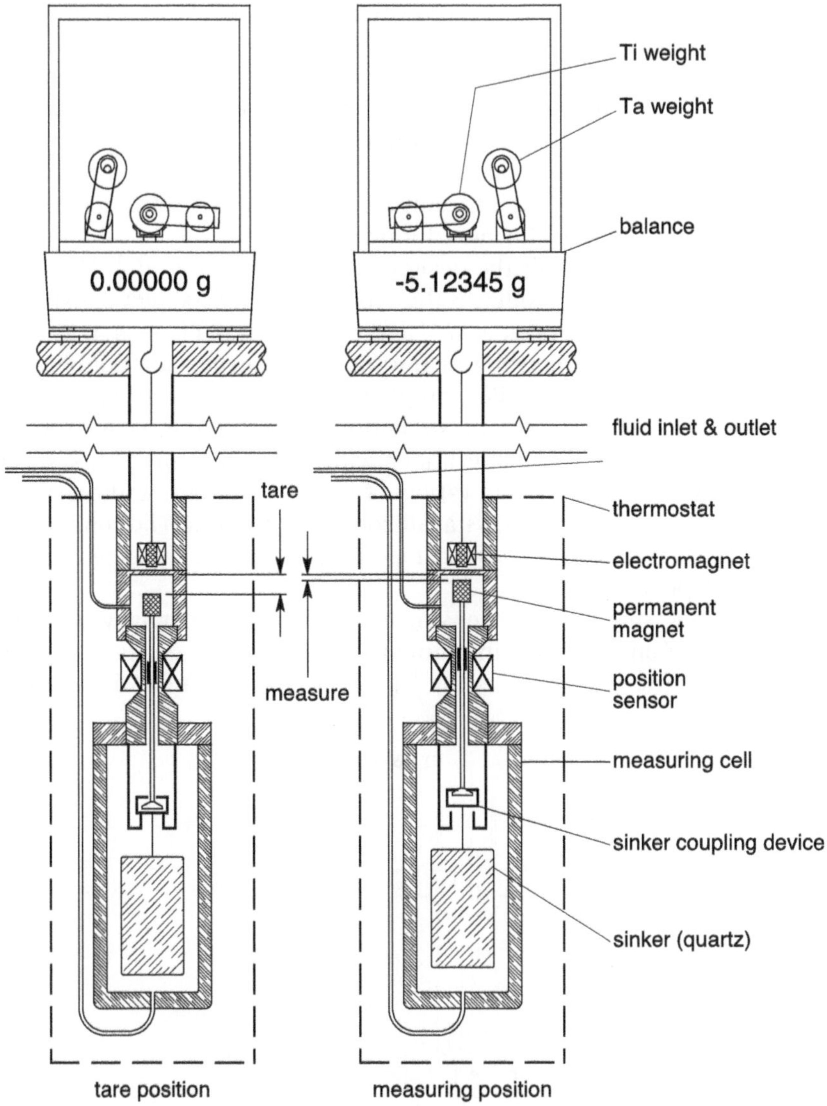

Figure 2.3 Principle of the single-sinker densimeter: (left) tare or "zero position" weighing; (right) sinker weighing ("measuring position"); figure adapted from Brachthäuser *et al.*[13,14]

where m is mass, W a balance reading, V are volumes, and the subscript s refers to the sinker, and c1 and c2 refer to the compensation masses. The compensation masses are typically made of different materials, so that $V_{c1} \approx V_{c2}$ and the air buoyancy term is zero. If the balance is tared when the sinker is on its rest, the balance reading W_{ZP} is zero. Note that corrections for the "force transmission error" should be applied to Equation (2.5), and these are discussed in Section 2.2.3.4.

The instrument described by Brachthäuser *et al.*[13,14] operates over a temperature range of 233 K to 523 K, with pressures up to 30 MPa. The measuring cell and coupling housing are made of beryllium copper, a nearly nonmagnetic material. The cylindrical quartz sinker ($m = 60$ g, $V = 26.5$ cm^3) is weighed with a balance having a resolution of 10 μg. The uncertainty is 0.02% for densities greater than 20 m^3 kg^{-1}. It has been used to measure a variety of fluids, including argon,[17] nitrogen,[17] methane,[17] carbon dioxide,[17] ethene,[18] ethane,[18] and sulfur hexafluoride.[18]

A noteworthy evolution of the design of Brachthäuser *et al.*,[13,14] and the most common instrument today, is the "compact" single-sinker densimeter, which was described by Docter *et al.*[19] and is depicted in Figure 2.4. Here the measuring cell and the housing of the magnetic suspension coupling are combined, resulting in a smaller sample volume (in the order of 100 cm^3). The position sensor is located in the bottom flange of the measuring cell. The sinker is a right circular cylinder with a central bore. This simple geometry facilitates the use of a variety of sinker materials, and these have included glass, quartz, single-crystal silicon, and titanium. The compact cell also facilitates the design of instruments operating at extremes of temperature and pressure.

The compact single-sinker densimeter is now commercially available, and a variety of single-sinker instruments based on this design have been developed. Wagner and Kleinrahm[20] present a summary of instruments

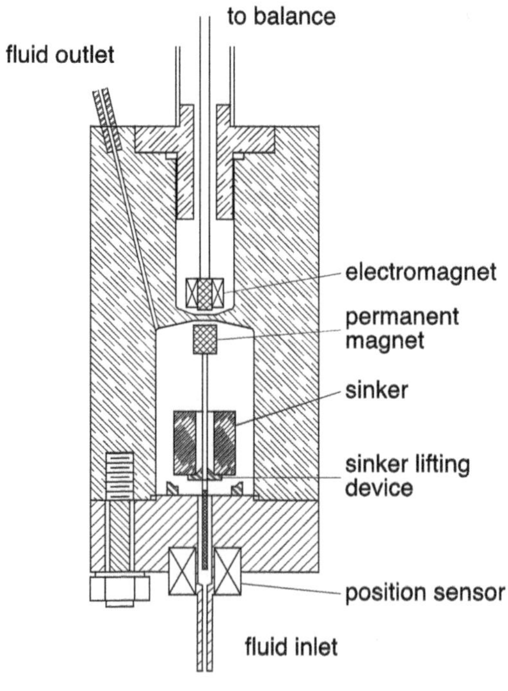

Figure 2.4 "Compact" single-sinker densimeter.

existing as of 2004. Notable examples include the high-pressure system at Texas A&M University,[21] with a maximum operating pressure of 200 MPa; its uncertainty is 0.05% to 0.10%. The system at the National Metrology Institute of Japan[22] has been used to measure several liquid density standards with uncertainties of about 0.002%, as discussed in Section 4.4.3. The cryogenic densimeter of Richter *et al.*[23] at the Ruhr-Universität Bochum operates over the temperature range of 90 K to 290 K, with pressures to 12 MPa; it is the first system to operate the magnetic suspension coupling at cryogenic temperatures. (Earlier cryogenic instruments had the coupling located outside the cryostat.) Gómez Mellado *et al.*[24] describe a portable system that is intended to calibrate *in situ* industrial densimeters used on natural-gas pipelines. Additional examples are discussed in Section 2.2.3.3.

2.2.3.2 Two-Sinker Densimeters

The accuracy of the Archimedes technique, especially at low densities, can be improved by the use of two sinkers. This technique was developed by Kleinrahm and Wagner.[25,26] For the two-sinker technique, the density is given by

$$\rho = \frac{(m_1 - W_1) - (m_2 - W_2)}{(V_1 - V_2)}, \tag{2.6}$$

where the subscripts refer to the two sinkers, which have the same mass, same surface area, and same surface material, but very different volumes (either by use of different materials or by employing solid and hollow sinkers of the same material). Systematic errors, including gas adsorption on the sinker surface, balance nonlinearity, and the influence of magnetic materials on the coupling are greatly reduced by the use of two sinkers and the resulting differential weighing.

Although the development of the two-sinker instrument predates the single-sinker instruments described in the previous section, only a handful of two-sinker instruments have been built to date. The greater accuracy of the two-sinker approach compared to a single-sinker instrument comes at the expense of greater complexity, and many applications can be satisfied by a simpler single-sinker instrument. The original two-sinker instrument as well as those known to be still in operation are briefly described.

The two-sinker densimeter developed at the Ruhr-Universität Bochum by Kleinrahm and Wagner[25,26] represented a major advance in density measurement over wide ranges of temperature and pressure. It operated at temperatures from 60 K to 340 K, with pressures to 12 MPa; the density range was 1 kg m^{-3} to 2000 kg m^{-3}. The sinkers were a gold-plated quartz sphere ($m = 54$ g, $V = 24.5$ cm^3) and a gold disk ($m = 54$ g, $V = 2.8$ cm^3). These were alternately placed on the "pan" for weighing by a mechanism involving a small winch outside the main thermostat (but immersed in the sample fluid), which actuated wires connected to devices which lowered the sinkers onto the pan. The magnetic suspension coupling was separate from the measuring cell, and it was thermostatted near ambient temperature. The system incorporated additional features for measurements near saturation as described in Section 2.2.3.3. The uncertainty of this instrument was 0.01% to 0.02%, depending on the measurement conditions. It was used to measure, among other fluids, the densities of methane,[27] carbon dioxide,[28,29] refrigerants 12 and 22,[30] argon,[31] ethene,[32,33] nitrogen,[34,35] and ethane.[36] The resulting data were key for the development of high-accuracy equations of state for these fluids, as discussed in Chapter 5. This instrument has been dismantled.

The two-sinker densimeter of Pieperbeck *et al.*[37] employs the same type of coupling and sinker changing mechanism as the Kleinrahm and Wagner[25,26] instrument. It was designed primarily for the measurement of natural-gas mixtures and operates over the temperature and pressure range of natural-gas pipelines, namely 273 K to 323 K, with pressures to 12 MPa; the density range is 1 kg m^{-3} to 1000 kg m^{-3}. The sinkers are a hollow sphere ($m = 123$ g, $V = 107$ cm^3) and a solid ring ($m = 123$ g, $V = 15.6$ cm^3); both are of stainless steel with electropolished and gold-plated surfaces. The large volume of the hollow sinker increases the sensitivity for low-density gas-phase measurements.

The two-sinker densimeter at the National Institute of Standards and Technology described by McLinden and Lösch-Will[38] combines the advantages of the two-sinker technique with those of the compact design of many single-sinker instruments. It operates over the temperature range of 210 K to 505 K with pressures to 40 MPa. The sinkers are titanium ($m = 60$ g, $V = 13.3$ cm^3) and tantalum ($m = 60$ g, $V = 3.6$ cm^3). The sinkers are alternately picked up for weighing by a mechanism that rotates the electromagnet, which in turn induces a matching rotation of the permanent magnet; two sets of "lifting forks" attached to the permanent magnet engage "pins" on the sinkers (see Figure 2.5). The motor operating this mechanism is located in ambient air, *i.e.*, not in the sample fluid like earlier designs; this avoids moving parts in the sample cell and possible condensation of sample in a "remote" mechanism. This densimeter also employs external masses placed on the balance pan, and while these are very similar to the compensation masses used with single-sinker densimeters as described in Section 2.2.3.1, they are used instead to calibrate the balance as a part of each density measurement and in the determination of the force-transmission error, as discussed in Section 2.2.3.4. The weighing design for this instrument involves two weighings of each of the four objects (two sinkers and two external masses) in a time-symmetric design that largely cancels any small drift in the temperature or pressure in the measuring cell.[39] This instrument has uncertainties ranging from 0.004% at near-ambient conditions to 0.02% at the extreme of temperature and pressure.[40] It has been used to measure both pure fluids and mixtures including density standards,[40] helium,[41] propane,[42] refrigerants,[43,44] and natural-gas mixtures.[39]

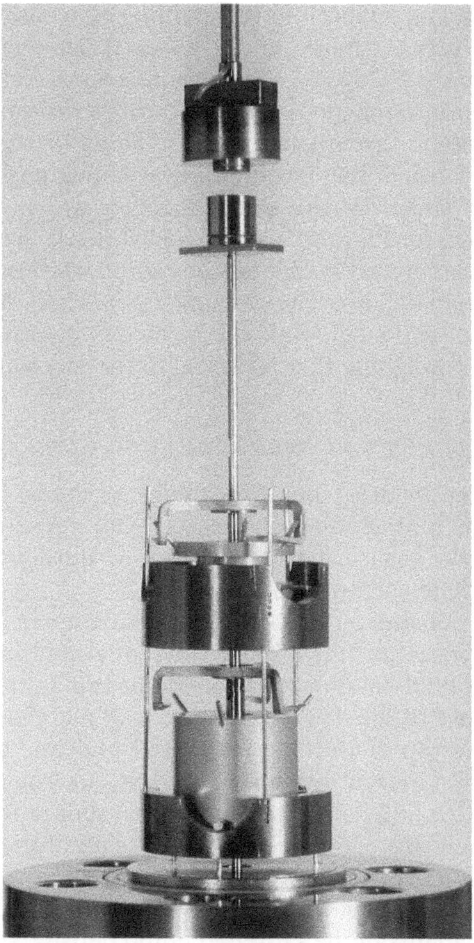

Figure 2.5 Sinkers and magnetic-suspension coupling in the densimeter of McLinden and Lösch-Will.[38] The electromagnet at the top hangs from the balance and levitates the permanent magnet below it. In this photo, the top (tantalum) sinker is being weighed, while the bottom (titanium) sinker sits on its rest.

Kayukawa *et al.*[45] at the National Metrology Institute of Japan describe a two-sinker system with sinkers of single-crystal silicon ($m=61$ g, $V=26.2$ cm^3) and single-crystal germanium ($m=60$ g, $V=11.3$ cm^3). The temperature range is 278 K to 323 K, with pressures to 20 MPa. An uncertainty in density of 1 part in 10^6 (not including uncertainties associated with the temperature and pressure measurement) is anticipated from this system due to the combination of single-crystal sinkers (which are very stable and well-characterized materials), a very detailed finite-element analysis of the force transmission error (see Section 2.2.3.4), and a coupling that maintains a constant position of the permanent magnet for the different weighings (as opposed to maintaining zero-current in the electromagnet, as is the case for most current systems). Thus far, preliminary results with tridecane at $T = 273.15$ K have yielded an uncertainty of 0.0005%.

A related type of instrument is designed for the measurement of density simultaneously with a thermogravimetric analysis or the investigation of adsorption. Here the coupling has multiple weighing positions such that it can lift a density sinker or the sinker together with a sample "basket" containing a sample for sorption or thermogravimetric investigation. Such a system was modified by May *et al.*[46] to study sorption effects in densimeters, and while this system had two "density sinkers" it is best thought of as being the combination of two single-sinker densimeters because it did not carry out the differential weighing that is central to the two-sinker principle.

2.2.3.3 Measurements at Near-Saturation Conditions

Measurements near the dew point or bubble point can be difficult because a small variation in temperature or pressure can cause a phase change. Modifications to the measuring cell or separate auxillary cells have been developed to facilitate such measurements.

The two-sinker densimeter of Kleinrahm and Wagner[25,26] was designed, in particular, to measure pure fluids close to saturation, and it implemented two additional small auxiliary cells for this purpose. Vapor-phase measurements close to the dew line made use of a "reference cell," which was located at the top of the measuring cell and was at the same temperature as the main measuring cell (see Figure 2.6). The reference cell was filled with the pure liquid being measured. The vapor pressure of the liquid in the reference cell was compared to the pressure of the vapor sample in the main cell with a sensitive differential pressure indicator. This allowed adjustment of the main cell pressure to closely approach the dew-point pressure. Measurements of the saturated liquid density were accomplished by partially fillling the measuring cell. Liquid-phase measurements at pressures slightly above the bubble-point pressure made use of a "pressure-adjusting cell." This small cell was again partially filled with the liquid being measured, and its vapor space was connected to the top of the main measuring cell. Its temperature could be controlled independently from the main cell, and by setting a temperature higher than the main cell, the vapor pressure thus generated served to pressurize the liquid sample in the measuring cell.

For the measurement of saturated liquid densities, Wendland and Saleh[47] used a measuring cell with the electromagnet in a re-entrant well (see Figure 2.6) such that a vapor space could be maintained at the top of the cell while keeping the permanent magnet immersed in the liquid phase. Blanke *et al.*[48,49] simultaneously measured both saturated liquid and (nearly) saturated vapor densities with two interconnected single-sinker densimeters. The pressure in both was that of the vapor pressure of the liquid; the vapor densimeter was maintained at a slightly higher temperature than the liquid

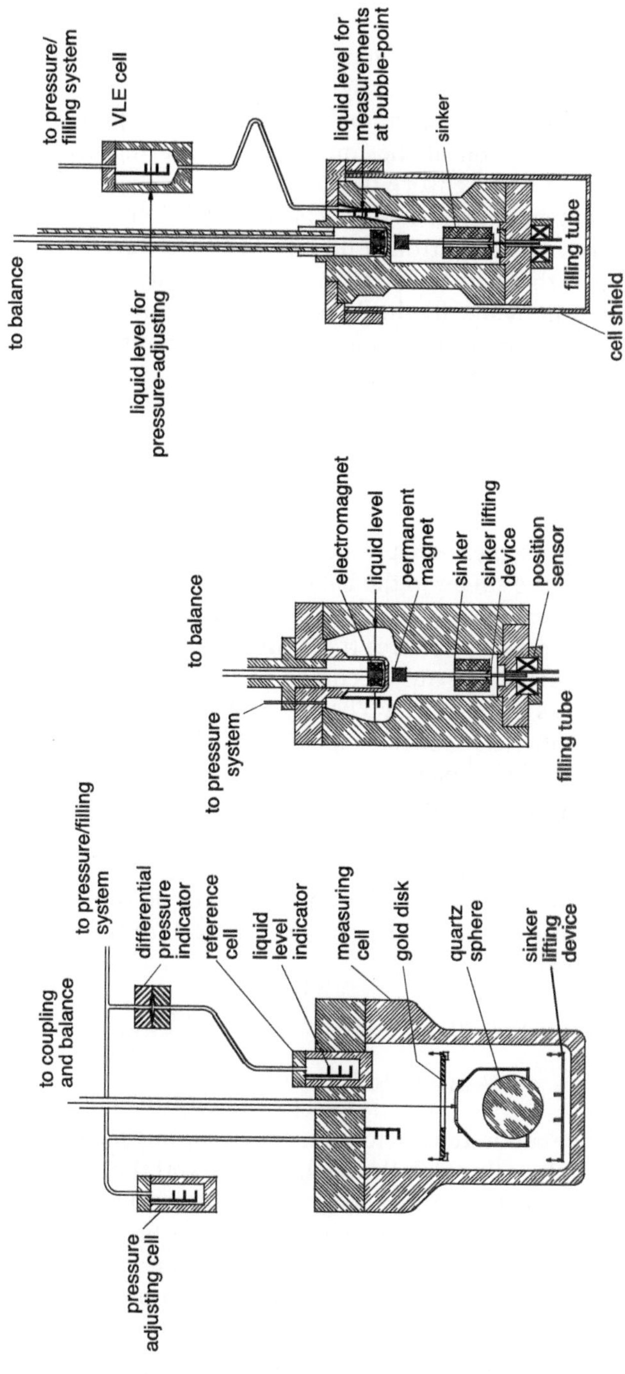

Figure 2.6 Modifications to facilitate measurements at near-saturation conditions: (left) reference cell and pressure-adjusting cell of Kleinrahm and Wagner;[25,26] (centre) re-entrant coupling of Wendland and Saleh,[47] (right) VLE cell of Richter *et al.*[23]

densimeter to avoid condensation. The sinkers were quartz, with a solid sphere for the liquid and a hollow sphere for the vapor, thus optimizing the sinker densities for each phase.

The cryogenic densimeter of Richter *et al.*[23] employs similar devices, but, unlike the above examples, it is designed primarily for measurements on mixtures. For measurements at the bubble point, the liquid level is maintained in a small extension at the top of the measuring cell (which also incorporates a liquid level sensor); the surface area of the liquid is small to minimize diffusion and demixing effects. A "VLE cell" above the main measuring cell provides a controlled liquid–vapor interface to avoid demixing when filling with a supercritical mixture; it is also used to generate pressure for measurements in the homogeneous liquid phase. The temperature of this auxilliary cell is controlled independently from the measuring cell to generate the desired pressure.

2.2.3.4 *Force-Transmission Errors in Magnetic-Suspension Densimeters*

A properly designed magnetic-suspension coupling very efficiently transmits the forces acting on the sinker to the balance, but it is slightly influenced by nearby magnetic materials, external magnetic fields, and the fluid being measured. These give rise to a "force transmission error" (FTE) that must be accounted for to realize the full accuracy of magnetic-suspension densimeters. The FTE was discussed by Lösch[15] and Klimeck *et al.*[17] Kuramoto *et al.*[22] presented a physical model for the FTE, but it is complex and requires detailed knowledge of the magnetic properties of both the apparatus and fluid, which may not be available, and would be applicable only to a specific instrument, in any case. We summarize here the empirical model of McLinden *et al.*[50]

The analysis of McLinden *et al.*[50] for a two-sinker densimeter starts with writing out the forces acting on the balance for each of the weighings. For the weighing of sinker 1:

$$\begin{aligned} W_1 = \alpha[\,\phi\{m_1 + m_{\text{p-mag}} - \rho_{\text{fluid}}(V_1 + V_{\text{p-mag}})\} \\ + (m_{\text{e-mag}} - \rho_{air} V_{\text{e-mag}}) + W_{\text{zero}}], \end{aligned} \tag{2.7}$$

where α is a balance calibration factor, ϕ is the "coupling factor", W_{zero} is the balance reading with nothing on the balance pan, and the subscripts p-mag and e-mag refer to the permanent magnet (in the fluid and including the lifting device) and to the electromagnet (in air), respectively. The key assumptions implicit in Equation (2.7) are that (a) the force transmitted to the balance by the magnetic suspension coupling is proportional to the suspended load; this is characterized by the coupling factor ϕ; (b) all quantities are constant over the time needed for a complete density determination; and (c) the balance reading is linear with the applied load. The electromagnet and (permanent magnet + lifting device) are always weighed, and the W_{zero} is the same for each weighing, so that these can be lumped together:

$$\beta = \phi\{m_{\text{p-mag}} - \rho_{\text{fluid}} V_{\text{p-mag}}\} + (m_{\text{e-mag}} - \rho_{\text{air}} V_{\text{e-mag}}) + W_{\text{zero}}. \tag{2.8}$$

Similar equations can be written for the weighing of the second sinker and also for the separate weighings of "tare" and "cal" masses directly on the balance pan (with the coupling in zero position). The result is a system of four equations with the four unknowns ρ_{fluid}, α, β, and the coupling factor ϕ, which yield:

$$\alpha = \frac{W_{\text{cal}} - W_{\text{tare}}}{(m_{\text{cal}} - m_{\text{tare}}) - \rho_{\text{air}}(V_{\text{cal}} - V_{\text{tare}})}, \tag{2.9}$$

$$\beta = \frac{W_{\text{cal}}}{\alpha} - (m_{\text{cal}} - \rho_{\text{air}} V_{\text{cal}}), \tag{2.10}$$

$$\rho_{\text{fluid}} = \left[(m_1 - m_2) - \frac{(W_1 - W_2)m_1}{W_1 - \alpha\beta}\right] \Big/ \left[(V_1 - V_2) - \frac{(W_1 - W_2)V_1}{W_1 - \alpha\beta}\right], \tag{2.11}$$

and

$$\phi = \frac{(W_1/\alpha) - \beta}{m_1 - \rho_{\text{fluid}} V_1}. \tag{2.12}$$

McLinden *et al.*[50] go on to demonstrate that the ϕ can be divided into apparatus and fluid-specific contributions. ϕ_0 is the value of ϕ in vacuum and is the apparatus contribution; it is obtained by weighing the sinkers in an evacuated measuring cell. The fluid-specific effect is proportional to the fluid density and the specific magnetic susceptibility of the fluid χ_s, with a proportionality constant ε_ρ:

$$\phi = \phi_0 + \varepsilon_\rho \chi_s \rho_{\text{fluid}}. \tag{2.13}$$

The coupling factor is shown in Figure 2.7 as a function of density for measurements on propane (a diamagnetic fluid with $\chi_s = -1.10\times10^{-8}\,\text{m}^3\,\text{kg}^{-1}$) and air (a paramagnetic mixture with χ_s varying with temperature from $35.74\times10^{-8}\,\text{m}^3\,\text{kg}^{-1}$ at $T = 250$ K to $19.12\times10^{-8}\,\text{m}^3\,\text{kg}^{-1}$ at $T = 460$ K). Note that these results are for a particular densimeter and illustrate only the general trends and the magnitude of the effect.

A similar analysis can be carried out for a single-sinker densimeter, except that there are only two or three equations, but still four unknowns. The balance calibration can be integrated with the sinker weighings (using the compensation masses) or carried out separately to obtain α. Weighings in vacuum yield the ϕ_0, but the fluid-specific effect (as quantified by the parameter ε_ρ) must be obtained by a separate determination. McLinden *et al.*[50] descibe two methods for obtaining ε_ρ. The first involves measurements with a fluid at the same (p, ρ, T) at different times using two different sinkers;

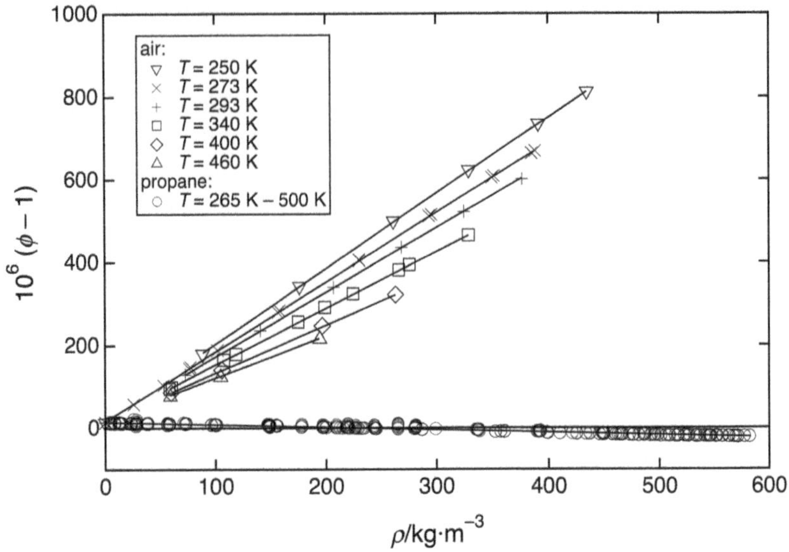

Figure 2.7 Coupling factor as a function of density for measurements on air (with $T = 250$ K to 460 K) and propane; figure adapted from McLinden *et al.*[50]

thus creating a sort of "sequential" two-sinker densimeter that allows an analysis by Equation (2.7)–(2.13). A simpler, but less accurate, approach requires measurements on a fluid with a well-characterized magnetic susecptibility, such as oxygen, and comparison with an accurate equation of state.

The magnitude of ϕ is typically 1 ± 0.000020. Ignoring the FTE corrections in a two-sinker densimeter will result in a relative (percentage) error that is small (typically less than 0.01%), except for strongly paramagnetic fluids. For a single-sinker densimeter, however, ignoring the FTE introduces an absolute error that is proportional to (ρ_s–ρ_{fluid}). This is most pronounced at low densities. Taking the example of a 60 g sinker of silicon with a volume $V_s = 25.76$ cm^3, a force transmission error of 1.2 mg (*i.e.*, 60 g $\times$ $(1 - 1.000020)$) would result in an error in density of 0.047 kg m^{-3}. For a liquid density of 1000 kg m^{-3}, this would be 0.0047%, but for a gas at 20 kg m^{-3} the error would be 0.16%.

2.3 Measurement by Calibrated Volumes

The techniques discussed in this section include the most direct measurement of fluid density, *i.e.*, a measurement of the mass of fluid required to fill a container of known volume. Although direct, the accuracies of these methods are often lower than those of the Archimedes methods discussed in Section 2.2. The volume must be calibrated with well-characterized fluids, and, thus, they cannot be an absolute determination of density. They are also often cumbersome to implement. For these reasons, such methods are falling out of favor in thermophysical properties research, but they remain important for measurements at extreme conditions, and the simpler devices are used in testing laboratories.

The amount of fluid can be determined by direct weighing (as in pycnometers and isochoric p–ρ–T instruments) or by one or more expansions into a larger volume (as in the Burnett technique). The volume itself can be fixed or variable. Systems with a variable volume are used to determine the compressibility of a fluid; such devices are also called "piezometers." Mercury was often used as a piston to form a variable volume, but because of current restrictions on the laboratory use of mercury due to its toxicity, such systems are largely obsolete and will not be discussed here. A metal bellows is now used for variable-volume systems.

2.3.1 Pycnometers

The simplest embodiment of a pycnometer consists of a glass bottle with a vented stopper (see Figure 2.8); the liquid volume is typically between 2 cm^3 and 100 cm^3. To determine density, the empty bottle is first weighed; it is then filled with the liquid of interest and the stopper is inserted, causing excess liquid to be expelled out of the top of the vent hole in the stopper. The bottle must then be carefully dried of excess liquid and weighed again. The mass of fluid is given by the difference in the weighings, and the volume is typically determined by weighing with water and taking the density of water

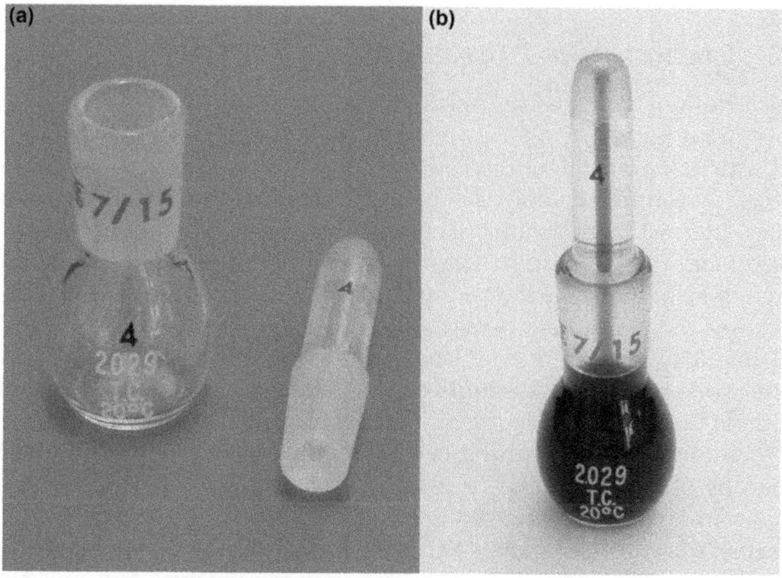

Figure 2.8 Simple glass pycnometer with a volume of 2 cm^2: (a) the pycnometer bottle and stopper; (b) filled with liquid that has been dyed to aid visualization; note that the liquid extends to the top of the stopper.

as known. This device is limited to near-ambient conditions, and the uncertainty is typically 0.1%, although this can be bettered by careful technique and calibration. These simple pycnometers are also called "specific-gravity bottles," and they are common in chemistry laboratories. A variety of designs for specific purposes, such as the measurement of very viscous liquids, are available; these are sometimes specified by standard test methods (see, for example, ASTM[51]). Pycnometers are calibrated at a specific temperature (usually 20 °C or 25 °C), and determining the temperature of the fluid is a major source of uncertainty; pycnometers with a built-in liquid-in-glass thermometer are available.

The concept of the pycnometer has been adapted to measurements over wide ranges of temperature and pressure by the "continously weighed pycnometer" in which the pycnometer is a small pressure vessel suspended from an analytical balance. The pycnometer is contained within a thermostat and is connected to a filling and pressure system by a fine, flexible capillary tube. Such systems have been developed at Cornell University by Machado and Streett[52] and at Texas A&M University, as described by Lau *et al.*[53] The volume of the pycnometer was determined by calibration with well-characterized fluids, and the mass of sample was determined directly by comparing the weight of the vessel filled with sample and when evacuated. The systems at Cornell and Texas A&M operated over temperature and pressure ranges of (298 K to 489 K, 104 MPa) and (100 K to 450 K, 200 MPa), respectively. Both claimed uncertainties in density of about 0.05%, although they were often higher in practice.

2.3.2 Isochoric p–ρ–T Instrument

The isochoric technique has much in common with the pycnometers described in Section 2.3.1. A pressure vessel (measuring cell) of fixed volume is filled with a known quantity of sample and sealed with a valve; the temperature is controlled, and the pressure is measured when the system reaches thermal equilibrium. The temperature is changed (usually increased), and the pressure is measured again. The process is repeated to a desired maximum temperature or until the maximum temperature or pressure of the instrument is reached. The major advantage of this method is its simplicity and ability to (at least partially) automate the measurements. The method can be used for liquids, gases, and supercritical fluids, and it is applicable to extremes of temperature and pressure.

The volume of the measuring cell can be determined at the time of fabrication by filling with water or mercury and weighing or *in situ* by a calibration with a well-characterized fluid (see Section 4.4). "Parasitic" volumes associated with the filling tubes and pressure-measuring device must also be accounted for. The expansion of the cell with temperature and pressure is often computed from the thermal and mechanical properties of the cell material, and this can be a source of uncertainty.

The amount of sample can be determined by expanding the entire sample into a much larger volume, such as the "gasometer" method of Goodwin[54] described in Section 2.3.3. The amount of sample can also be determined gravimetrically. A supply cylinder is weighed, connected to the isochoric system to fill the cell, and then weighed again. Alternately, the cell is filled to a desired starting pressure and the measurment commences; at the end of the isochore, the sample is condensed into a weighing cylinder cooled in liquid nitrogen. This approach has several advantages over the first: the weighing cylinder is small and easily weighed with good precision, and at the conclusion of the process, the cell and filling lines are filled with gas at the (usually very low) pressure corresponding to the vapor pressure of the sample fluid at liquid nitrogen temperature, minimizing corrections for fluid remaining in the system. In a variation on this method,[55] the system is filled to the highest density and the first isochore measured; a portion of the sample is condensed into the weighing cylinder and weighed; the next isochore is then measured. The process is repeated until, after the lowest density isochore is measured, all of the remaining sample is condensed into the weighing cylinder. The mass of sample for each of the isotherms is calculated from the sequential weighings. This method requires a minimum quantity of sample, but runs the risk that a leak or an error in any of the weighings can affect all of the isochores.

Numerous isochoric instruments have been built. As an example, the isochoric system of Goodwin[54] is shown in Figure 2.9. This system was developed in the late 1950s at the National Bureau of Standards (now NIST) for measurements on hydrogen at temperatures from 14 K to 100 K, with an upper pressure limit of 35 MPa. The uncertainty was 0.05% to 0.10%. It was subsequently modified to extend its temperature range to 400 K (see Magee and Ely[55]) and used to measure numerous fluids of industrial interest, including natural-gas systems[56] and refrigerants.[57,58] It remained in nearly continous operation for 40 years.

2.3.3 Expansion Techniques (Burnett Method)

The basic principle behind expansion techniques is to expand the fluid sample (which is gaseous or supercritical) into a larger (evacuated) volume one or more times (all at a fixed temperature) until the final state is at a sufficiently low pressure such that it is described by a simple virial equation. The best-known such technique was described by Burnett[59] in 1936. The advantage of this method is that neither volumes nor the amount of sample needs to be directly measured (in fact, the title of Burnett's original paper was "Compressibility determinations without volume measurements"). Only the temperature and pressures are measured, and the analysis yields densities and virial coefficients. These are molar quantities, and this is an advantage for mixture measurements—the Burnett method is less sensitive to

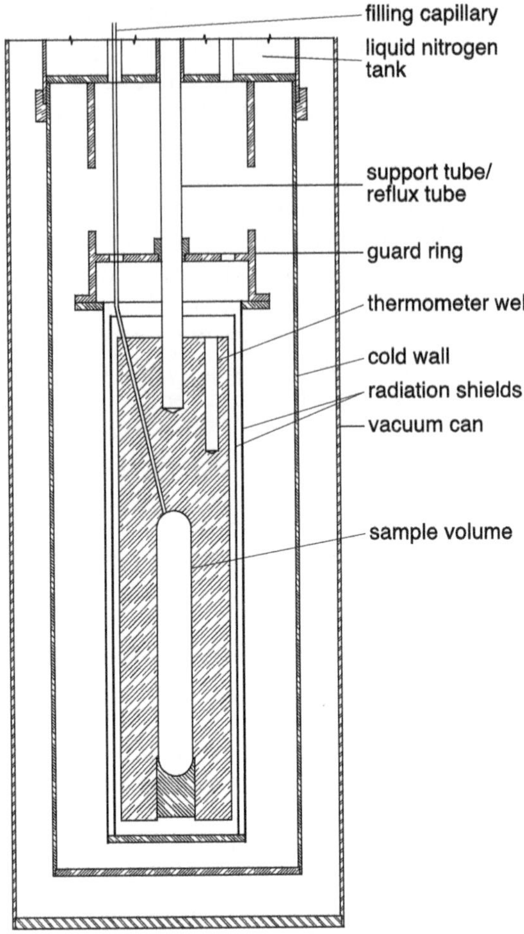

Figure 2.9 Isochoric p–ρ–T instrument of Goodwin.[54]

the mixture composition than when mass densities are measured. The ratio of the densities before and after an expansion is given by:

$$\frac{\rho_{i-1}}{\rho_i} = \frac{(V_a + V_b)_i}{(V_a)_i} \equiv r_i, \tag{2.14}$$

where V_a is the "sample volume," V_b is the volume of the "expansion cell," and the subscripts refer to the ith expansion. The ratio of the volumes r_i is termed the cell "constant" (although it varies slightly with pressure, thus the subscript i); it is typically about 1.1 to 2.0 in the Burnett method.

Several methods for analyzing Burnett expansion experiments are available (see Hall and Canfield[60]), and Kell[61] presents an analysis of the errors in the technique. Burnett[59] originally presented a graphical method.

Summarized here is the analytical method of Britt and Luecke.[62] The density after any expansion is related to the density after the final expansion ρ_m by:

$$\rho_i = \rho_m \prod_{j=i+1}^{m} r_j = \rho_m \prod_{j=i+1}^{m} r_0 \left[\frac{1 + \gamma_{ab} p_j}{1 + \gamma_a p_{j-1}} \right], \tag{2.15}$$

where the γ_a and γ_{ab} account for the pressure dilation effect on the volumes, and r_0 is the volume ratio in the limit of zero pressure; these parameters are typically determined by a calibration with helium. The virial coefficients are determined by a least-squares fit of the pressure data, where the basic virial equation (see Chapter 6)

$$p = RT\rho[1 + B_2\rho + B_3\rho^2 + B_4\rho^3 + \ldots], \tag{2.16}$$

is combined with Equation (2.15) to yield the objective function of the fit:

$$p_i = RT \sum_{k=1}^{n} B_k \left\{ \rho_m r_0^{m-1} \prod_{j=i+1}^{m} \left[\frac{1 + \gamma_{ab} p_j}{1 + \gamma_a p_{j-1}} \right] \right\}^k, \tag{2.17}$$

where the B_k are the virial coefficients, which, along with ρ_m, are the fitted parameters; the summation is taken to the fourth virial or higher.

The Burnett method has been used by numerous investigators, and some of these are referenced in the compilation of virial data in Section 6.6. An example are the virial coefficients of Blancett *et al.*[63] for helium and argon, which remain among the most accurate for these fluids even 40 years after their publication. The experimental system of Waxman and Hastings[64] is shown in Figure 2.10 as an example. This system integrated both volumes into a single block to facilitate thermostatting and minimize temperature gradients. Other systems used two spherical vessels; an approach that simplified corrections for the pressure expansion of the vessels.

The Burnett method has often been combined with the isochoric technique to increase the efficiency of data collection, see, for example, Linsky *et al.*[65] After each Burnett expansion the fluid sample in volume V_a is varied over a range of temperatures. The density is constant (except for small corrections for the temperature and pressure dilation of the cell) and is given by Equations (2.15) to (2.17); the temperature and pressure are measured at each point to yield p–ρ–T data. The temperature is then returned to the starting value, and the next Burnett expansion is carried out.

In the Burnett method, the volume ratio of the expansion is typically about 1.5. A much larger volume ratio (in the range of 100 to 1000) will yield, in a single expansion, a pressure sufficiently low to calculate the quantity of sample by a virial equation. Goodwin[54] used such a method, which he called a "gasometer" to determine the quantity of sample in his isochoric instrument. A similar approach has been applied in a commercial instrument termed a "Z-meter," [where "Z" refers to the compression factor $Z = p/(RT\rho)$]; see, for example, Jaeschke *et al.*[66]

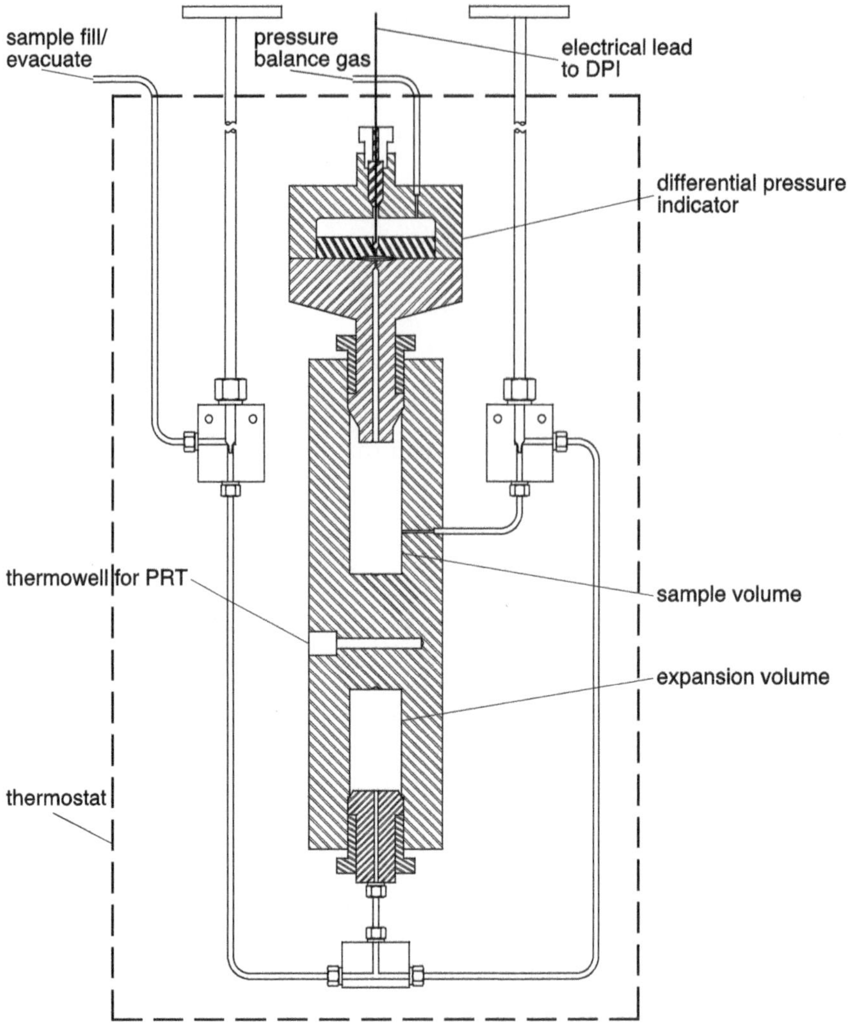

Figure 2.10 Burnett apparatus of Waxman and Hastings.[64]

2.3.4 Bellows Volumometer

In a bellows volumometer the fluid sample is contained in a metal bellows, which, in turn, is contained within a pressure vessel. A pressure-transmitting fluid on the outside of the bellows compresses the fluid sample, changing its volume. The classic implementation of this technique measures only the isothermal compressibility of a fluid, defined as:

$$\beta_T = \frac{1}{V}\frac{\partial V}{\partial p}, \tag{2.18}$$

where V is the specific (or molar) volume. The compressibility is then intregrated over pressure to yield density; this requires an independent measurement of the density at some reference pressure, often 0.1 MPa.

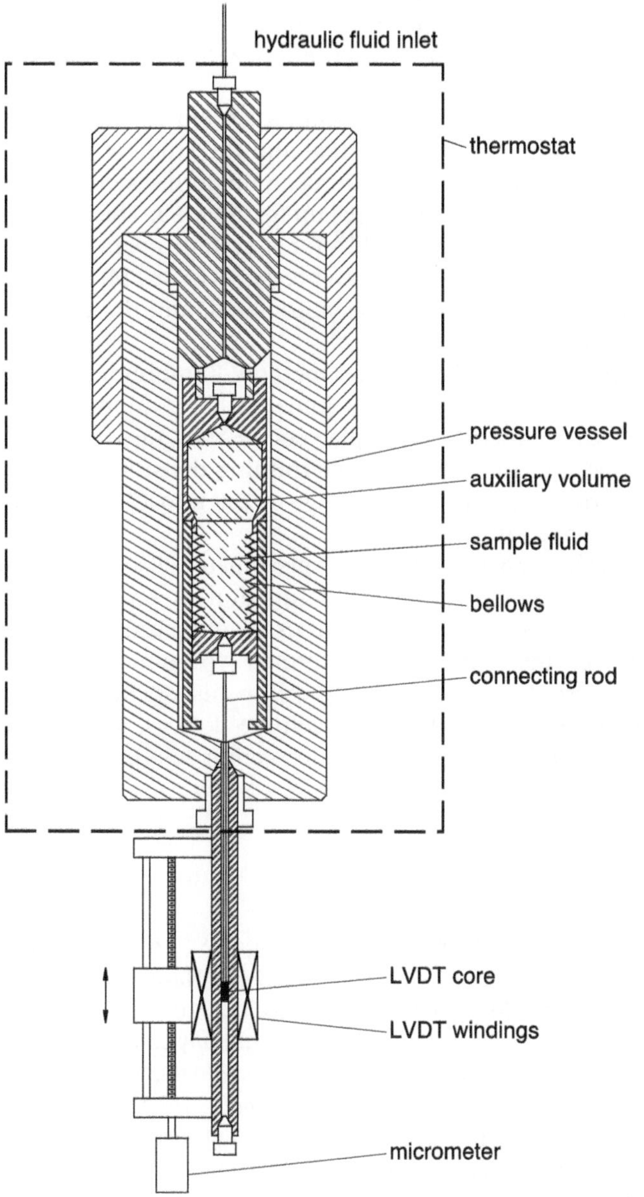

Figure 2.11 Bellows volumometer of Malhotra and Woolf.[69] The bellows assembly is removed from the pressure vessel for sample loading, and the auxiliary volume increases sensitivity to changes in pressure.

The volume change is determined by the cross-sectional area of the bellows and the change in bellows length. The linear displacement of the bellows is typically measured with a linear variable differential transformer (LVDT) with the sensing element of the LVDT connected by a rod to the end of the bellows and contained in a tubular extension of the pressure vessel. The cross-sectional area of the bellows must be determined as a function of temperaure and pressure by a calibration with a well-characterized fluid (typically water).

Bridgman[67] described the first bellows system in 1931; his instrument operated at pressures up to 1200 MPa, and the study of fluids at very high pressures is a strength of this technique. A modern system was described by Back *et al.*;[68] it operated at pressures up to 250 MPa. The same basic design was used by Malhotra and Woolf[69] for a system operating over the temperature range of 278 K to 338 K, with pressures to 380 MPa (see Figure 2.11). Pressure control in this system was automated, and the uncertainties were 0.05% to 0.10%.

In a variation on the volumometer method, Kamimura *et al.*[70] and Kabata *et al.*[71,72] have fully calibrated the volume (rather than incremental volume) of their cell as a function of displacement; they measure the amount of sample loaded into the cell, thus determining densities directly. Their system operates over the temperature range of 298 K to 473 K, with pressures to 200 MPa; the uncertainty is 0.1%.

A bellows volumometer is a type of piezometer. The measurement of the change in volume of a sample as a function of pressure (*i.e.*, the compressibility) corresponds to the current usage of "piezometry." In the past, a wider range of instruments were termed "piezometers." For example, Goodwin[54] referred to the measuring cell in his isochoric instrument as a piezometer. Holste[73] included fixed volume devices, such as the continuously weighed pycnometer (discussed in Section 2.3.1), in his review entitled "Piezometer."

References

1. J. D. Wright, V. E. Bean and J. Aguilera, *NIST calibration services for hydrometers, NIST Special Publication 250-78*, National Institute of Standards and Technology, Gaithersburg, MD, 2010.
2. R. S. Basu and D. P. Wilson, *Int. J. Thermophys.*, 1989, **10**, 591.
3. R. Masui, K. Fujii and M. Takenaka, *Metrologia*, 1995, **32**, 333.
4. J. B. Patterson and E. C. Morris, *Metrologia*, 1994, **31**, 277.
5. M. Richter, R. Kleinrahm, S. Glos, W. Wagner, R. Span, P. Schley and M. Uhrig, *Int. J. Thermophys.*, 2010, **31**, 680.
6. J. W. Beams, C. W. Hulbert, W. E. Lotz and R. M. Montague, *Rev. Sci. Instrum.*, 1955, **26**, 1181.
7. J. W. Beams and A. M. Clarke, *Rev. Sci. Instrum.*, 1962, **33**, 750.
8. M. G. Hodgins and J. W. Beams, *Rev. Sci. Instrum.*, 1971, **42**, 1455.

9. W. M. Haynes, M. J. Hiza and N. V. Frederick, *Rev. Sci. Instrum.*, 1976, **47**, 1237.
10. H. Wolf, H. Bettin and A. Gluschko, *Meas. Sci. Technol.*, 2006, **17**, 2581.
11. R. Masui, W. M. Haynes, R. F. Chang, H. A. Davis and J. M. H. Levelt Sengers, *Rev. Sci. Instrum.*, 1984, **55**, 1132.
12. R. Masui, *Int. J. Thermophys.*, 2002, **23**, 921.
13. K. Brachthäuser, R. Kleinrahm, H. W. Lösch and W. Wagner, *Entwicklung eines neuen Dichtemeßverfahrens und Aufbau einer Hochtemperatur-Hochdruck-Dichtemeßanlage (Development of a new density measurement method and construction of a high temperature-high pressure densimeter)*, Reihe 8, No. 371 edn, VDI-Verlag, Düsseldorf, 1993.
14. W. Wagner, K. Brachthäuser, R. Kleinrahm and H. W. Lösch, *Int. J. Thermophys.*, 1995, **16**, 399.
15. H. W. Lösch, *Entwicklung und Aufbau von neuen Magnetschwebewaagen zur berührungsfreien Messung vertikaler Kräfte*, VDI-Verlag, Düsseldorf, 1987.
16. H. W. Lösch, R. Kleinrahm and W. Wagner, *Chem. Ing. Tech.*, 1994, **66**, 1055.
17. J. Klimeck, R. Kleinrahm and W. Wagner, *J. Chem. Thermodyn.*, 2001, **33**, 251.
18. P. Claus, R. Kleinrahm and W. Wagner, *J. Chem. Thermodyn.*, 2003, **35**, 159.
19. A. Docter, H. W. Lösch and W. Wagner, *Int. J. Thermophys.*, 1999, **20**, 485.
20. W. Wagner and R. Kleinrahm, *Metrologia*, 2004, **41**, S24.
21. M. Atilhan, S. Aparicio, S. Ejaz, D. Cristancho and K. R. Hall, *J. Chem. Engr. Data*, 2011, **56**, 212.
22. N. Kuramoto, K. Fujii and A. Waseda, *Metrologia*, 2004, **41**, S84.
23. M. Richter, R. Kleinrahm, R. Span and P. Schley, in *International Gas Union Research Conference*, Seoul, Republic of Korea, 2011.
24. B. Gómez Mellado, R. Kleinrahm, H. W. Lösch, W. Wagner, O. Brandt and R. Boden, *GWF-Gas/Erdgas*, 2001, **142**, 282.
25. R. Kleinrahm and W. Wagner, *Entwicklund und Aufbau einer Dichtemeßalnage zur Messung der Siede- und Taudichten reiner fluider Stoffe auf der gesamten Phasengrenzkurve*, Fortschr.-Ber. VDI, Reihe 3, No. 92 edn, VDI-Verlag, Düsseldorf, 1984.
26. R. Kleinrahm and W. Wagner, *J. Chem. Thermodyn.*, 1986, **18**, 739.
27. G. Händel, R. Kleinrahm and W. Wagner, *J. Chem. Thermodyn.*, 1992, **24**, 685.
28. W. Duschek, R. Kleinrahm and W. Wagner, *J. Chem. Thermodyn.*, 1990, **22**, 827.
29. W. Duschek, R. Kleinrahm and W. Wagner, *J. Chem. Thermodyn.*, 1990, **22**, 841.
30. G. Händel, R. Kleinrahm and W. Wagner, *J. Chem. Thermodyn.*, 1992, **24**, 697.
31. R. Gilgen, R. Kleinrahm and W. Wagner, *J. Chem. Thermodyn.*, 1994, **26**, 383.

32. P. Nowak, R. Kleinrahm and W. Wagner, *J. Chem. Thermodyn.*, 1996, **28**, 1423.
33. P. Nowak, R. Kleinrahm and W. Wagner, *J. Chem. Thermodyn.*, 1996, **28**, 1441.
34. P. Nowak, R. Kleinrahm and W. Wagner, *J. Chem. Thermodyn.*, 1997, **29**, 1137.
35. P. Nowak, R. Kleinrahm and W. Wagner, *J. Chem. Thermodyn.*, 1997, **29**, 1157.
36. M. Funke, R. Kleinrahm and W. Wagner, *J. Chem. Thermodyn.*, 2002, **34**, 2001.
37. N. Pieperbeck, R. Kleinrahm and W. Wagner, *J. Chem. Thermodyn.*, 1991, **23**, 175.
38. M. O. McLinden and C. Lösch-Will, *J. Chem. Thermodyn.*, 2007, **39**, 507.
39. M. O. McLinden, *J. Chem. Eng. Data*, 2011, **56**, 606.
40. M. O. McLinden and J. D. Splett, *J. Res. Natl. Inst. Stand. Technol.*, 2008, **113**, 29.
41. M. R. Moldover and M. O. McLinden, *J. Chem. Thermodyn.*, 2010, **42**, 1193.
42. M. O. McLinden, *J. Chem. Eng. Data*, 2009, **54**, 3181.
43. M. O. McLinden, M. Thol and E. W. Lemmon, in *International Refrigeration and Air Conditioning Conference*, Purdue University, W. Lafayette, IN, July 12–15, 2010, paper 2189, http://docs.lib.purdue.edu/iracc/1041/.
44. M. Richter, M. O. McLinden and E. W. Lemmon, *J. Chem. Eng. Data*, 2011, **56**, 3254.
45. Y. Kayukawa, Y. Kano, K. Fujii and H. Sato, *Metrologia*, 2012, **49**, 513.
46. E. F. May, R. C. Miller and Z. Shan, *J. Chem. Eng. Data*, 2001, **46**, 1160.
47. M. Wendland and B. Saleh, *J. Chem. Eng. Data*, 2005, **50**, 429.
48. W. Blanke, G. Klingenberg and R. Weiss, *Int. J. Thermophys.*, 1995, **16**, 1143.
49. W. Blanke, G. Klingenberg and R. Weiss, *J. Chem. Eng. Data*, 1996, **41**, 576.
50. M. O. McLinden, R. Kleinrahm and W. Wagner, *Int. J. Thermophys.*, 2007, **28**, 429.
51. ASTM International, *ASTM E 570 – 09e1 Standard Test Method for Density of Semi-Solid Bituminous Materials (Pycnometer Method)*, ASTM International, West Conshohocken, PA, 2008.
52. J. R. S. Machado and W. B. Streett, *J. Chem. Eng. Data*, 1983, **28**, 218.
53. W. R. Lau, C.-A. Hwang, H. B. Brugge, G. A. Iglesias-Silva, H. A. Duarte-Garza, W. J. Rogers, K. R. Hall, J. C. Holste, B. E. Gammon and K. N. Marsh, *J. Chem. Eng. Data*, 1997, **42**, 738.
54. R. D. Goodwin, *J. Res. Natl. Bur. Stand., Sect. C*, 1961, **65C**, 231.
55. J. W. Magee and J. F. Ely, *Int. J. Thermophys.*, 1988, **9**, 547.
56. J. W. Magee and W. M. Haynes, *J. Chem. Thermodyn.*, 1997, **29**, 1439.
57. J. W. Magee and W. M. Haynes, *Int. J. Thermophys.*, 2000, **21**, 113.
58. J. W. Magee, *Int. J. Thermophys.*, 1996, **17**, 803.
59. E. S. Burnett, *J. Appl. Mechanics*, 1936, **3**, 136.

60. K. R. Hall and F. B. Canfield, *Physica*, 1970, **47**, 99.
61. G. S. Kell, *Physica A*, 1981, **105**, 536.
62. H. I. Britt and R. H. Luecke, *Technometrics*, 1973, **15**, 233.
63. A. L. Blancett, K. R. Hall and F. B. Canfield, *Physica*, 1970, **47**, 75.
64. M. Waxman and J. R. Hastings, *J. Res. Natl. Bur. Stand., Sect. C*, 1971, **75C**, 165.
65. D. Linsky, J. M. H. Levelt Sengers and H. A. Davis, *Rev. Sci. Instrum.*, 1987, **58**, 817.
66. M. Jaeschke, S. Audibert, P. van Caneghem, A. E. Humphreys, R. Janssen-van Rosmalen and Q. Pellei, *Int. J. Thermophys.*, 1990, **11**, 157.
67. P. W. Bridgman, *Proc. Am. Acad. Arts Sci.*, 1931, **66**, 185.
68. P. J. Back, A. J. Easteal, R. L. Hurle and L. A. Woolf, *J. Phys. E: Sci. Instrum.*, 1982, **15**, 360.
69. R. Malhotra and L. A. Woolf, *Int. J. Thermophys.*, 1993, **14**, 1021.
70. T. Kamimura, A. Iso, Y. Higashi, M. Uematsu and K. Watanabe, *Rev. Sci. Instrum.*, 1989, **60**, 3055.
71. Y. Kabata, S. Yamaguchi, M. Takada and M. Uematsu, *J. Chem. Thermodyn.*, 1992, **24**, 1019.
72. Y. Kabata, S. Yamaguchi, M. Takada and M. Uematsu, *J. Chem. Thermodyn.*, 1992, **24**, 785.
73. J. C. Holste, in *Measurement of the Thermodynamic Properties of Single Phases*, ed. A. R. H. Goodwin, K. N. Marsh and W. A. Wakeham, International Union of Pure and Applied Chemistry, Amsterdam, 2003, Experimental Thermodynamics Series, Volume VI.

CHAPTER 3

Experimental Techniques 2: Vibrating Tube Densimetry

DIEGO GONZÁLEZ-SALGADO,* JACOBO TRONCOSO AND LUIS ROMANI

Department of Applied Physics, University of Vigo, Ourense 32004, Spain
*Email: dgs@uvigo.es

3.1 Introduction

Density is, perhaps, the most important property in the context of the thermophysical characterization of fluids due to its key role in computing other important quantities. Measurements of density as a function of temperature and pressure allow for the determination of the isobaric thermal expansivity and the isothermal compressibility.[1] The combination of this information together with isobaric heat capacity data under the same thermodynamic conditions allows one to calculate the isoentropic compressibility and the isochoric heat capacity.[2] In addition, molar volumes and kinematic viscosities can be evaluated using density values with the aid of the molar mass and the dynamic viscosity, respectively. On the other hand, the values of density at several pressures and temperatures provide the crucial input data for the parameterization of equations of state or simple molecular models.[3,4] Equations of state are commonly used in industrial process design which includes optimizing exploitation conditions of petroleum products and the design of refrigerant equipment or heat pumps.

A number of different experimental techniques[5] have been developed for the measurement of densities including: bellows volumetry, piezometers, isochoric methods, densimeters based on vibrating elements (forks, wires,

Volume Properties: Liquids, Solutions and Vapours
Edited by Emmerich Wilhelm and Trevor M. Letcher

Published by the Royal Society of Chemistry, www.rsc.org

etc.), buoyancy densimeters (hydrostatic balance), and magnetic float and magnetic suspension densimeters. In 1969, Kratky *et al.* [6] presented a novel indirect method for the accurate measurement of the density of a sample that is based on the relation between the resonating vibration period of a tube filled with the sample and its density. The popularity of this method (named the vibrating tube densimeter VTD) has increased enormously since that date and it has become the most widely used method for density measurement. Its main advantages are its simple operation, high accuracy, small sample requirement, and the possibility to systematise rapid measurement procedures; an automated system can easily be implemented. The development of this type of densimeter has been carried out by several companies (Anton Paar,[7] Sodev Inc.,[8] Kyoto,[9] *etc.*) that provide products for both scientific and applied usage. In particular, VTDs are used industrially for the analysis of the quality of different substances such as beer, wine, flavours, asphalt, sulfuric acid in batteries, *etc.*

The main aim of this chapter is to give a useful and simple guide for scientists interested in the density measurement of fluids using VTDs. It includes the experimental set-up, the measurement principle, calibration methods, and the main sources of uncertainty and some ways of improving precision.

3.2 Experimental Set-up

A variety of designs and implementations[10–17] has been considered in the development of VTDs. In this section, we have not given an exhaustive revision of all possible types of equipment but have presented those designs that, in our opinion, are not only the most commonly used but also the most pedagogical. To this end, in Figure 3.1 the diagrams of selected designs are given.

The measuring cell in a VTD is a hollow tube bent at its central part to form a U-tube (a V-tube was also proposed). Its length is usually quite small (a few cm long). The values of the inner radius r_i and outer radius r_e depend on the stiffness requirements of the equipment; for instance, the (r_i, r_e) values for the Anton Paar DMA 602 atmospheric pressure cell are (2.0 mm, 2.2 mm) and for the Anton Paar DMA 512P high pressure cell (1.073 mm, 1.588 mm) respectively. They are usually constructed of borosilicated glass (only for atmospheric pressure measurements) or Hastelloy or Inconel stainless steel (for moderate high pressure). The cell is anchored at both ends to a relatively large counter mass to allow a harmonic vibration of the tube without interferences from other parts of the equipment. With the same aim, the counter mass/tube system is sometimes suspended from appropriate springs.

The harmonic vibration of the tube and the measurement of vibration period can be obtained through different methods, some of which are represented in Figure 3.1. In all cases, there is a drive element that makes the tube vibrate harmonically and a pick-up element for measuring the vibration

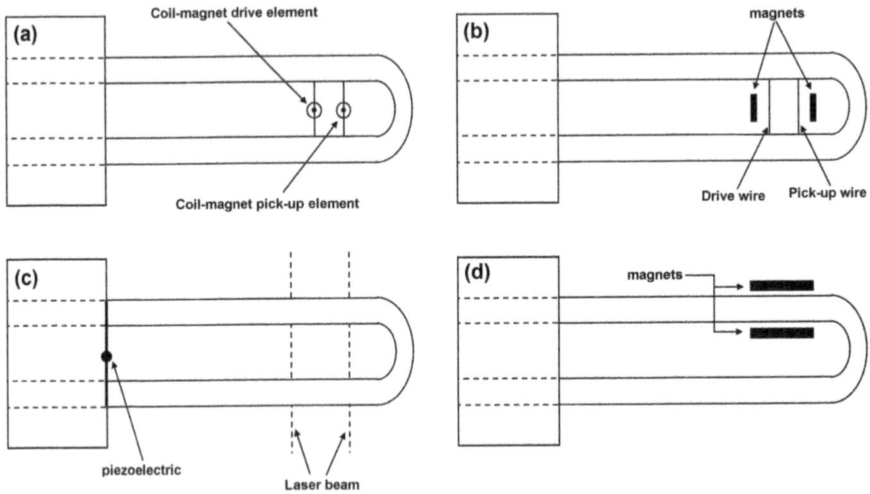

Figure 3.1 Schematic diagrams of several vibrating tube densimeters. See text for details.

period. Both elements take part in an electronic joint system that modulates the perturbation in the drive element in order to seek the vibration of the tube at the resonant frequency. In Figure 3.1(a), two magnets, each surrounded by one coil, are fixed to the end part of the tube. In the drive coil, an applied AC voltage provokes a periodic translational movement in the magnet that acts as a hammer on the tube. The periodic movement induced in the tube produces an alternating current in the pick-up coil (induced by the movement of the second magnet inside the coil) whose frequency is detected using a frequency counter. In Figure 3.1(b), a drive wire and a pick-up wire are attached across the tube with ceramic cement and electrically insulated from the vibration tube. Both wires are placed orthogonally to the magnetic field created by two magnets. A variable voltage is applied to the drive wire in order to induce a harmonic vibration in the tube; this vibration induces an AC current in the pick-up wire whose frequency is measured using an appropriate circuit. In Figure 3.1(c), the harmonic vibration is generated by a piezoelectric element located on one of the bearings at the end of the tube. The determination of the vibration period is made through a light beam located orthogonally in the plane of the tube. As the oscillators move, the light beam is modulated by the movement of the tube and this is detected by phototransistors which produce sinusoidal signals that are a function of the oscillation displacement. Finally, in apparatus shown in Figure 3.1(d), an alternating current is passed through the tube and a harmonic vibration is induced as a result of the force created by the magnetic field in the current. The vibration period is obtained by determining the effect of the induced voltage on the wire over the output of the exciting signal.

The oscillation of the tube in a VTD is carried out in a gas in order to reduce, to negligible values, the friction between the tube and its environment. The temperature control is made through a metal jacket around the tube (or the tube plus the counter mass system) using Peltier elements or a temperature controlled fluid bath. Thus, it is very important that the gas in contact with the tube and the metallic jacket has a high thermal conductivity. The temperature control is a crucial point in the design since all the physical properties of the materials forming the densimeter vary strongly with temperature. It is generally considered that the temperature changes during measurements must not be greater than 0.01 K. As for pressure, since it only affects the liquid and the tube, it is of less importance but it must be stable to within ± 0.01 MPa during measurements.

The harmonic vibration must be sustained for a certain length of time in order to measure the vibration period with a high degree of precision. For instance, when using a 1 MHz counter it is necessary to measure at least in the order of 10^8 periods to obtain the vibration period to nine digits. Thus the harmonic vibration must be maintained for at least 100 s. Taking into account that the frequency of a vibrating tube is typically around 300 Hz, the tube must oscillate about 30 000 times. These numbers indicate, again, that temperature control must be very precise.

3.3 Principle of Measurement

The mechanical model (Figure 3.2) commonly used to describe the behaviour of a VTD is a spring–mass system oscillating in the presence of damping and under an external periodic force. This oscillation can be described by the following differential equation:

$$\frac{d^2y}{dt^2} + \frac{b}{m}\frac{dy}{dt} + \frac{K}{m}y = \frac{F_0}{m}\cos(\omega t), \tag{3.1}$$

where y is the displacement with respect to the equilibrium position, b is the damping constant, K is the spring constant, F_0 is the amplitude of the external periodic force, ω the external frequency, m the mass, and t the time.

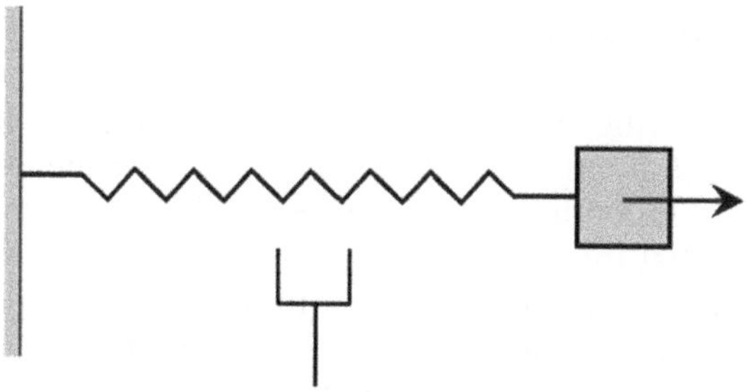

Figure 3.2 Scheme of the simple mechanical model used in the description of the behaviour of a vibrating tube densimeter.

It is easy to relate these terms to the physical parts of the designs given in Figure 3.1. Thus, using the spring–mass model for the U-tube clamped at both ends, the external force is that generated using the drive coil–magnet system, the piezoelectric element, the drive wire or the tube (acting as a wire); and the damping comes from the friction of the tube with the air, the sample, or an induction term. The steady-state solution for this equation is:

$$y(t) = A_0 \cos(\omega t - \delta), \tag{3.2}$$

where A_0 is the amplitude of the oscillation and δ is the phase shift between the external force and the oscillation $y(t)$. Most of the experimental equipment is based on *resonance of energy* for which the resonant frequency is the natural frequency of the oscillator, $\omega = \omega_0 = \sqrt{\frac{K}{m}}$ (this choice corresponds with $\delta = \pi/2$). In this situation, the damping force is completely balanced with the external force; both are of the same magnitude with a phase shift of π. Other equipment operates in a *resonance of amplitude* mode, for which $\omega^2 = \omega_0^2 - 2(b/2m)^2$ with $b/2m = \omega_0/2Q$, where Q is the quality factor of the resonator. However, in most designs and for low viscosity samples, the quality factor in the VTD is so large that the approximation $\omega^2 = \omega_0^2$ holds. From the equation $\omega_0 = \sqrt{\frac{K}{m}}$, the relation between ω_0 with natural vibration period τ, $\omega_0 = \frac{2\pi}{\tau}$, and the total mass m (where m is the sum of the evacuated tube mass m_0 and the mass of the sample ρV), the following equation can be deduced:

$$\rho = \frac{K}{4\pi^2 V}\tau^2 - \frac{m_0}{V} = A\tau^2 + B \tag{3.3}$$

where ρ is the density of the sample, V the inner tube volume, and A and B are the calibration constants. Equation (3.3) is the working equation commonly used for VTD measurements. Calibration of the equipment is needed for the computation of the constants A and B at the thermodynamic conditions of each experiment. Thus, once A and B are known, the density of a sample can easily be obtained by measuring the vibration period of the tube filled with the sample.

It is clear that the spring–mass system is an idealised model for the U-tube. One important limitation is that it cannot predict the existence of the higher modes of vibration (eigenfrequencies) of the U-tube. Some VTDs operate in such a way that the resonant frequency coincides with an eigenfrequency higher than the fundamental one.[10] In this case, the spring–mass model must be substituted for a more complex model with a bar or rod clamped at one or both ends,[10,14,18,19] as can be seen in Figure 3.3. For these more realistic models, Equation 3.3 also applies when the resonant frequency coincides with the fundamental one. On the other hand, these models can

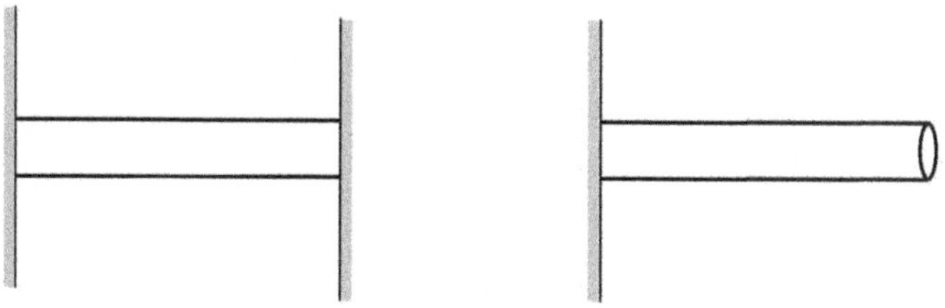

Figure 3.3 Two quasi-realistic models describing the behaviour of a vibrating tube densimeter: (left) rod clamped at both ends and (right) rod clamped at one end (cantilever).

be very useful in obtaining a relation between the *A* and *B* constants and the physical properties of the tube. For instance, in a cantilever model,[19] the spring constant *K* and the inner volume *V* are expressed as:

$$K = \frac{3E}{L^3}\frac{\pi}{4}(r_e^4 - r_i^4) \tag{3.4}$$

$$V = \pi r_i^2 L \tag{3.5}$$

where *E* is the Young's modulus, r_e and r_i are the outer and inner radius of the hollow tube, and *L* is the total length. Note that, since *E*, r_e, r_i and *L* depend on *T* and *p*, the *A* and *B* parameters of Equation (3.3) also depend on *T* and *p*. Therefore, one must not forget that the *A* and *B* values must be evaluated at the thermodynamic conditions of the measurement experiment.

3.4 Sources of Uncertainty in a VTD

Over the last twenty years, the Labor für Messtechnik (Dr. Hans Stabinger), the calibration laboratories H&D Fitzgerald Ltd. and PTB Germany worked jointly to identify the main uncertainty sources in density measurement using VTDs. They found that, when high precision is necessary, errors resulting from non-linearity, sample viscosity, and effects of temperature are the most relevant.[20] Some of these can only be overcome by improving the design of the equipment; however, others can be minimised by calibration. In this section, a brief explanation of these uncertainties is given with some possible solutions involving the VTD design. Alternative solutions using calibrations are given in Section 3.5.

3.4.1 Non-linearity

The linear relation between the density of the sample and the square of the vibration period [Equation (3.3)] does not hold over a wide range of density. This is produced by three phenomena related to (i) the mass distribution along the vibration tube, (ii) the dependence of the nodal points on the mass in the tube, and (iii) the parasitic resonances of ancillary components.

(i) It has been shown that the linear relation is violated when the mass is not uniformly distributed along the tube. In a well constructed evacuated tube, one can consider that the mass distribution is uniform, and this must also be true in a tube filled with a liquid or a gas. However, magnets or piezoelectric elements fastened along the tube [see Figure 3.1(a)] result in a non-uniform mass distribution of the system. In addition, this singularity around the magnet is different in magnitude depending on the sample being a gas (higher singularity) or a liquid (lower singularity), as is shown in Figure 3.4. These differences induce different deviations of behaviour with respect to Equation (3.3). This makes the VTD working equation for low density samples different from that of high density samples, *i.e.*, it is no longer valid for samples of very different densities. It is recommended nowadays, in order to eliminate this effect, that VTDs be designed in such a way that excitation of the vibrating tube is at the nodal points (tube points where no vibration occurs) [see Figure 3.1(c)].

(ii) The linear relation is also disrupted as a result of the dependence of the nodal position on the sample mass. As was explained in Section 3.2, the tube is clamped to a counter mass with a mass much greater than that of the tube. In the idealised model proposed in Section 3.3, the mass of the counter mass is considered as infinity, thus reducing the problem to only the movement of the U-tube (modelled as a spring–mass system). However, this is an idealised situation, since the tube vibration also induces a counter mass movement, albeit very small. A simple model describing this phenomenon consists of a spring linking the counter mass M and the tube + sample mass m, as is shown in Figure 3.5. The whole system oscillates keeping the centre of mass position invariable. However, if the small mass increases (introducing liquid inside the tube), the centre of mass position also changes. In the model it corresponds to the nodal point of the vibrating tube and then it depends on the liquid mass. Taking into account that the nodal point is the initial point needed to define the U-tube (and then its length and volume), it is clear that the calibration constants through Equations (3.3) to (3.5) depend on this effect. This problem can be solved by using an extremely heavy counter mass. If this is not feasible, one possibility is to use the

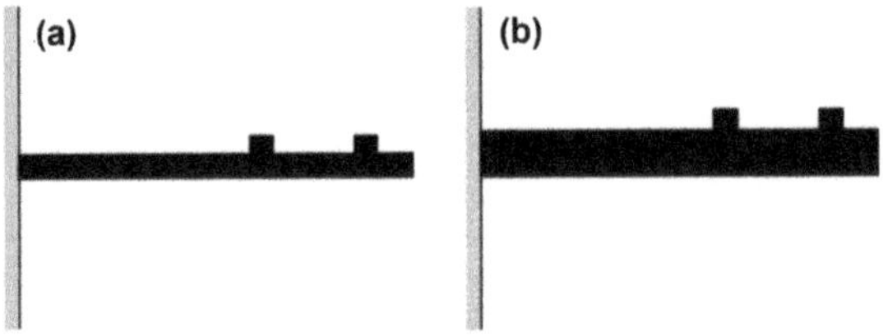

Figure 3.4 Mass distribution along a U-tube with two magnets fastened to a tube filled with a (a) gas and (b) liquid.

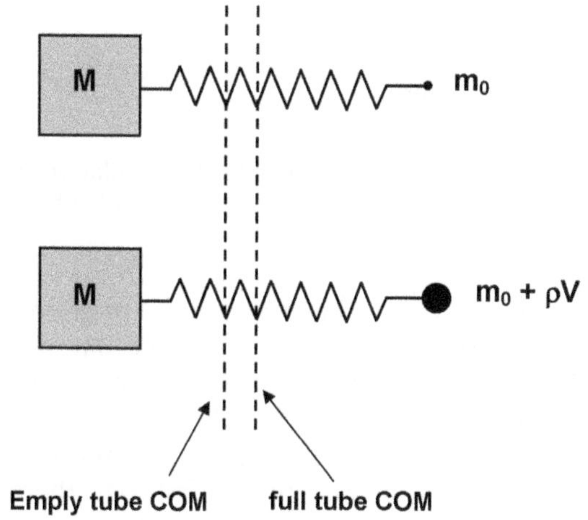

Figure 3.5 Diagram of the {counter mass–spring–mass} model and position of the center of mass COM in two situations: (up) evacuated tube and (down) tube filled with a sample.

counter mass–spring–mass model and apply an equation which relates the harmonic natural vibration frequency and the density of the sample,[12,15] instead of Equation (3.3).

(iii) The last phenomenon that requires attention involves parasitic resonant effects. The counter mass–vibrating tube system is not completely isolated. A number of other elements that form the equipment are connected to the counter mass. These components have their own eigenfrequencies and they can significantly disturb the density measurement if their values coincide with the resonant frequencies of the oscillator. These non-desired effects can be avoided by the proper implementation of ancillary elements.

3.4.2 Viscosity-induced Errors

The density of high viscosity samples obtained with VTDs is higher than its actual value; in other words, the method fails for high viscosity samples.[20–26] The question is: what is the difference regarding the vibrating tube behaviour between a low viscosity sample and a high viscosity sample, both with the same density? A simple explanation of this problem was given by Stabinger[20] and can be understood by looking at Figure 3.6, where the behaviour of a segment of liquid mass in the tube for a low, medium, and high viscosity sample is shown. In the first case, the movement of the liquid is only translational; in the second case, the liquid slides and also rotates; and in the last case, the liquid only rotates and behaves as a solid. In answering the question, the main difference is the presence of a rotational movement in high viscosity samples. This rotational movement comes from the shear forces between the liquid and the tube wall. This movement depends

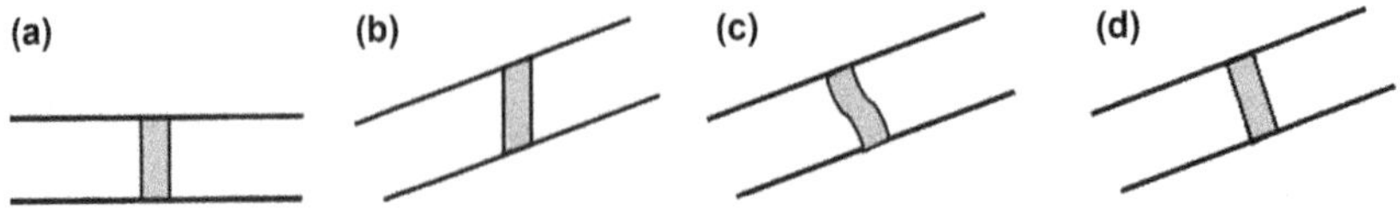

Figure 3.6 Behaviour of a slice of fluid during the oscillation in the equilibrium position (a) and out of equilibrium for a sample of (b) low, (c) medium, and (d) high viscosity.

strongly on the sample viscosity. The effect on the measured density comes from the inertia momentum of the rotated section that, when added to the inertia force of the translational movement, simulates a higher mass with respect to the volume and, thus, a higher density. This effect acts in both types of VTD apparatus, *i.e.*, those working with amplitude resonance and those working with energy resonance. In the first case, there is also another factor to be taken into account: the b term (or the quality factor Q) is no longer zero (or no longer infinity) for high viscosity samples.[11]

3.4.3 Thermal Effects

The last important source of error in the VTD is the ageing and thermal relaxation effects of the tube material.[20] As can be seen in Equation (3.3), the calibration constants depend on the inner volume of the tube and the spring constant which can be related to the Young's modulus of the material (see Equation (3.4) for the cantilever model). Both quantities can be affected by ageing and thermal relaxation effects of the tube. Therefore, these parameters, although obtained under the same thermodynamic conditions, can have different values depending on the time they are measured. To solve this problem, it is recommended that the measurement of the vibration period of a test fluid (or that of the evacuated tube) is checked to see if Equation (3.3) still holds. Stabinger and co-workers provide a design modification to solve this problem.[20] This involves working with two vibrating tubes, the usual U-tube and a reference tube, constructed of the same material. Therefore it is assumed that both tubes are affected in the same way by ageing and thermal relaxation effects. Thus, the ratio between the two vibration periods is a quantity quite insensitive to these factors. This ratio is used in Equation (3.3) instead of the usual vibration period.

3.5 Calibrations

3.5.1 Calibration for the Measurement of Low Viscosity Liquids

As was discussed in Section 3.3, the determination of the sample density using a VTD and Equation (3.3) requires the evaluation, *a priori*, of the constants A and B. The most popular procedure is the so-called "classical" calibration, which involves using two substances with known densities (*standards*). Both liquids (1 and 2) are inserted separately in the VTD in order to measure their vibration periods, τ_1 and τ_2, at specific thermodynamic conditions. Applying Equation (3.3) we have:

$$\rho_1 = A\tau_1^2 + B \tag{3.6}$$

$$\rho_2 = A\tau_2^2 + B \tag{3.7}$$

where ρ_1 and ρ_2 are the densities of the standards. From these equations, the calibration constants A and B are easily determined:

$$A = \frac{\rho_2 - \rho_1}{\tau_2^2 - \tau_1^2} \tag{3.8}$$

$$B = \frac{\rho_1\tau_2^2 - \rho_2\tau_1^2}{\tau_2^2 - \tau_1^2} \tag{3.9}$$

This procedure provides good results for density measurement if the following considerations are taken into account: (i) the density uncertainty of the standards is low; (ii) the densities of the standards are close to each other, and (iii) the sample density fits into the interval between the two standard densities ρ_1, ρ_2. The reason for the latter two considerations is that the linear relation between density and the square of the vibration period does not hold over a wide density interval and the involved interpolation or extrapolation could result in significant errors. Thus, since the "classical" calibration exploits this linearity, the densities of both standards must not be too different.

Therefore, the main problem of the "classical" calibration procedure is finding reliable density standards in the working density interval. For measurements at atmospheric pressure, it is relatively easy to find density standards. Good options for calibration standards are air,[27] nitrogen,[28] dodecane,[19,24,25] water,[29] and tetracholoroethylene.[19,24,25] Their densities cover the interval 0–1700 kg m^{-3} and their uncertainties are lower than 0.01 kg m^{-3} around room temperature. For measurements at high pressure, the situation is very different. Water must be chosen as one of the density standards since its densities are known with an uncertainty of lower than 0.03 kg m^{-3} up to 100 MPa. The selection of the second standard is, however, more difficult if a similar uncertainty is required. Common choices are nitrogen,[28] octane,[30] dichlorodifluoromethane,[31] and NaBr in water,[32] whose densities are known with an uncertainty of around 0.2 kg m^{-3} (one order of magnitude higher than in the case of water) for the same pressure interval. Furthermore there is little chance of finding a standard liquid which has a high density; tetrachloromethane is commonly used but its reliability is questioned nowadays.[19]

In order to solve the problem of the high uncertainty for the second density standard, alternative methods have been developed using one calibration fluid and a vacuum. The choice of a vacuum as standard is justified since vacuum bombs can be used coupled to a vibrating cell, with stated uncertainty in density lower than 10^{-3} kg m^{-3}. Equation (3.3) for a calibration experiment under vacuum then takes the form:

$$0 = \frac{K(T,0)}{4\pi^2 V(T,0)} \tau_v^2(T,0) - \frac{m_0}{V(T,0)}, \tag{3.10}$$

assuming that the density is zero inside the cell. In Equation (3.10), subscript v denotes vacuum. The thermodynamic conditions relating to temperature T and pressure (0 MPa) are included in parenthesis. With this information, Equation (3.3) can be transformed into the working equation:

$$\rho(T,p) = \frac{m_0}{V(T,p)} \left(\frac{K(T,p)}{K(T,0)} \frac{\tau^2(T,p)}{\tau_v^2(T,0)} - 1 \right) \tag{3.11}$$

Here, the quantities to be determined with the calibration fluid are $m_0/V(T, p)$ and the ratio $K(T,p)/K(T,0)$. Equation (3.11) for the second calibration experiment now becomes:

$$\rho_{ref}(T,p) = \frac{m_0}{V(T,p)} \left(\frac{K(T,p)}{K(T,0)} \frac{\tau_{ref}^2(T,p)}{\tau_v^2(T,0)} - 1 \right) \tag{3.12}$$

where the subindex *ref* denotes the reference calibration fluid. Unfortunately, the measurement of the vibration period of only one calibration fluid (with known density) does not allow one to obtain the quantities $m_0/V(T,p)$ and the ratio $K(T,p)/K(T,0)$. Therefore, additional considerations must be taken into account. A proposed solution is to make an additional hypothesis concerning the volume V and spring constant K pressure dependencies. Although a number of approximations have been reported,[18,33–38] the more popular proposals have been made by Sousa *et al.*,[33] Lagourette *et al.*,[34] and Bouchot and Richon.[18] Sousa *et al.*[33] considered that the spring constant does not depend on pressure $K(T,p)/K(T,0) = 1$, making the only unknown quantity to be $m_0/V(T,p)$. This quantity can easily be evaluated through Equation (3.12). Lagourette *et al.*[34] on the other hand, considered that both K and V depend on pressure but they depend on p in the same way; moreover, they assumed that $K(T,0) = K(T, 0.1\ \text{MPa})$. An analysis of the reliability of these approximations (and others) by Bouchot and Richon[18] resulted in the presentation of new approximations that significantly improved the previous ones. The idea was to consider the behaviour of the vibrating tube as a hollow thick tube whose response to the thermal and mechanical stress follows rigorous physical considerations. In this model, analytical expressions are provided for $V(T,p)$ and for the ratio $K(T,p)/K(T,0)$, which depend on only two parameters, m_0/L_{00} and γ. Here, m_0 is the tube mass, L_{00} its length under reference conditions and γ is related to the variation of the tube length with pressure. These three magnitudes must be computed at each temperature by fitting the density values of the calibration fluid at the same temperature and at several pressures.

The method proposed by Bouchot and Richon,[18] with water as the reference fluid, appears to be the best choice among the methods that use a vacuum as a standard in the interval 0–1000 kg m^{-3} for non-atmospheric pressure density measurements. This is because of the low uncertainty of the standard together with approximations related to the U-tube behaviour. However, this method has drawbacks for measurements of high density liquids because additional errors appear due to the need for extrapolation and the loss of the linear character of Equation (3.3). Some authors have replaced water by dichlorodifluoromethane.[31] Unfortunately this cannot be considered as the best solution since the uncertainty of the density of the standard increases the uncertainty by at least one order of magnitude. To solve this question, Sanmamed *et al.*[19] proposed two modifications to the method proposed by Bouchot and Richon.[18] The first point is to consider the U-tube as a thick hollow tube, as done by Bouchot and Richon, but with the additional constraint that it is clamped at one end, forming a cantilever. The working equation is obtained by substituting Equations (3.4) and (3.5) in Equation (3.3). These analytical expressions for the behaviour with T and p of E, r_e, r_i, and L have been proposed by Bouchot and Richon.[18] The parameters to be determined by calibration in these expressions are m_0, L_{00}, and γ, with these terms having the same meaning as given by Bouchot and Richon. The second point relates to the choice of calibration fluids. Since Sanmamed *et al.* were interested in the density interval 700–1600 kg m^{-3}, they selected dodecane ($\rho \sim 700$ kg m^{-3}) and tetrachloroethylene ($\rho \sim 1600$ kg m^{-3}) purchased from H&D Fitzgerald, whose densities at atmospheric pressure were certified with an uncertainty of 0.01 kg m^{-3}, and water ($\rho \sim 1000$ kg m^{-3}) with densities known with an uncertainty lower than 0.03 kg m^{-3} up to 100 MPa. The model parameters were determined as follows. The first quantities (m_0 and L_{00}) were computed at each temperature using two atmospheric pressure experiments for dodecane and tetrachloroethylene. After this, additional experiments at higher pressures were made with water. From literature density data and from experimentally determined vibration periods for this fluid, the γ parameter was fitted at each temperature. This method, by reducing the non-desirable effects of extrapolation, non-linearity, density standard uncertainties, together with approximations used for modelling the U-tube behaviour, resulted in the determination of densities which were reasonably accurate (uncertainty lower than 0.2 kg m^{-3}) in the working interval up to 70 MPa.

3.5.2 Calibration for the Measurement of High Viscosity Liquids

The methods proposed in Section 3.5.1 all resulted in significant errors in the density values for samples of high viscosity. This error was found to be dependent only on the sample viscosity and hence, using standard samples with certified densities and viscosities, corrections could be made.

The procedure was based on finding several standards whose viscosity values covered a wide range (between 0 and 500 mPa s or 1000 mPa s) and measuring their density through conventional calibration procedures such as those proposed in the previous sections. Thus, the viscosity-induced error $\Delta\rho$ was computed as the difference between the measured density and the certified density. This method was used by several authors[21,24–26,39,40] for different VTDs and some of these results are plotted in Figure 3.7. As can be seen, viscosity induced errors $\Delta\rho$ are different depending on the VTD, but the curve $\Delta\rho$ (η) shows the same shape for all of them: it increases strongly for low viscosity and reaches a plateau for high viscosity. It is worth noting that the errors are quite significant (they can be as much as 1.0 kg m^{-3}) and must always be taken into account for high viscosity samples. Moreover, this effect must also be considered if the density values are used for calculating other thermodynamic properties. For instance, it was shown than the curve of the isobaric thermal expansivity, α_p, against temperature, T, for room temperature ionic liquids can change from a parabolic relationship with a maximum (uncorrected) to a linear relationship, decreasing with T (corrected).[24]

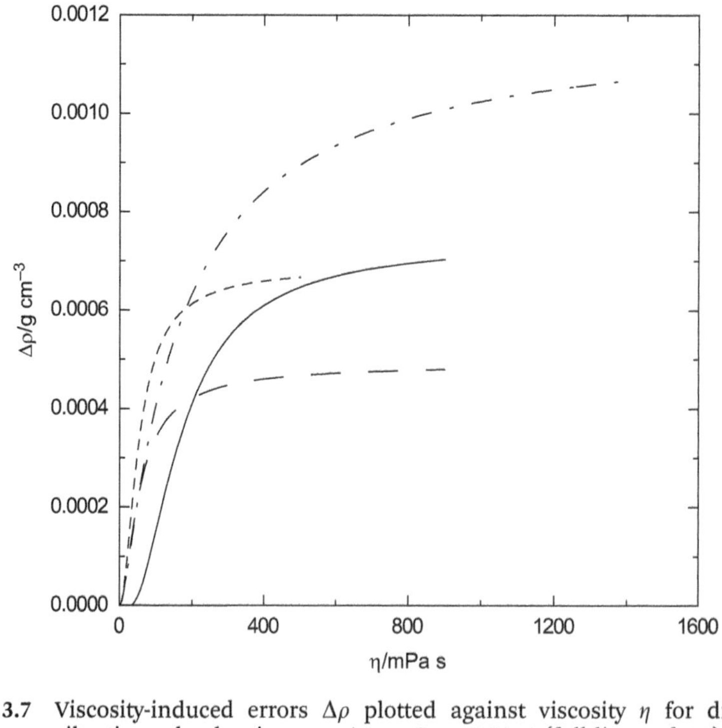

Figure 3.7 Viscosity-induced errors $\Delta\rho$ plotted against viscosity η for different vibrating tube densimeters. Anton Paar DSA48 (full line, ref. 24), Anton Paar DMA512P (long dashed line, ref. 25), Anton Paar DMA602 (short dashed line, ref. 40), and Anton Paar DMA HP (dashed dotted line, ref. 39).

It is worth noting that, although a correction curve $\Delta\rho$ (η) can be obtained, the viscosity of the sample must be known *a priori* for the calculation of $\Delta\rho$. Therefore, for determining the density of high viscosity samples, it is usually necessary to use a viscometer to determine viscosity. However the problem can be solved in another way by estimating the sample viscosity using a VTD, as proposed by Stabinger[20] and Fritz *et al.*[22] They found that the so-called "loss angle" in a VTD, that is, the ratio between the damping and spring force, can be related to the sample viscosity. This is done by means of two quantities. One of them is the vibration period of the tube filled with the sample in the first normal vibration mode ($n = 1$) under energy resonance ($\delta = \pi/2$); as was discussed in Section 3.3. In this case, the damping force is equal to the external force. The second one is the vibration period in the same vibration mode, but with $\delta = \pi/4$, *i.e.*, out of resonance; it is crucial, in this measurement, that the amplitude of the external force is also high enough to balance the damping force. Thus, such VTDs work in two steps: first, the vibration period in the fundamental ($n = 0$) overtone is measured; from it, the incorrect density (without any viscosity correction) is obtained. After this, the "loss angle" is determined and from this the viscosity is obtained. With this value, the viscosity-induced error can be calculated and the incorrect density corrected.

References

1. J. S. Rowlinson and F. L. Swinton, *Liquid and liquid mixtures*, Butterworth & Co., London, 3rd edn, 1982.
2. J. L. Valencia, D. González-Salgado, J. Troncoso, J. Peleteiro, E. Carballo and L. Romaní, *J. Chem. Eng. Data*, 2009, **54**, 904.
3. R. Span, *Multiparameter equations of state: an accurate source of thermodynamic property data*, Springer, Berlin, 1st edn, 2000.
4. J. V. Sengers, R. F. Kayser, C. J. Peters and H. J. White, *Equations of State for Fluids and Fluid Mixtures, vol. 5 (Experimental Thermodynamics)*, Elsevier, Amsterdam, 1st edn, 2000.
5. A. R. H. Goodwin, K. N. Marsh and W. A. Wakeham, *Measurement of the Thermodynamic Properties of Single Phases, vol. 6 (Experimental Thermodynamics)*, Elsevier, Amsterdam, 1st edn, 2003.
6. O. Kratky, H. Leopold and H. Stabinger, *Z. Angew. Phys.*, 1969, **27**(4), 273.
7. http://www.anton-paar.com/.
8. Sodev Inc., 1780 Rue Saint-Roch S, Rock Forest, WC, J1N 3B8, Canada.
9. http://www.kyoto-kem.com/.
10. R. F. Chang and M. R. Moldover, *Rev. Sci. Instrum.*, 1996, **67**(1), 251.
11. H. J. Albert and R. H. Wood, *Rev. Sci. Instrum.*, 1984, **55**(4), 589.
12. Y. Kayukawa, M. Hasumoto and K. Watanabe, *Rev. Sci. Instrum.*, 2003, **74**(9), 4134.
13. R. H. Wood, C. W. Buzzard and V. Majer, *Rev. Sci. Instrum.*, 1989, **60**(3), 493.

14. C. D. Holcomb and S. L. Outcalt, *Fluid Phase Equilibr.*, 1998, **150–151**, 815.
15. R. Laznickova and H. Huemer, *Meas. Sci. Technol.*, 1998, **9**, 719.
16. V. Hynek, L. Hnedkovský and I. Cibulka, *J. Chem. Thermodyn.*, 1997, **29**, 1237.
17. P. Picker, E. Tremblay and C. Jolicoeur, *J. Solution Chem.*, 1974, **3**, 377.
18. C. Bouchot and D. Richon, *Fluid Phase Equilibr.*, 2001, **191**, 189.
19. Y. A. Sanmamed, A. Dopazo-Paz, D. González-Salgado, J. Troncoso and L. Romani, *J. Chem. Thermodyn.*, 2009, **41**, 1060.
20. H. Stabinger, South Yorkshire Trading Standards Unit, Sheffield, 1994.
21. S. J. Ashcroft, D. R. Broker and J. C. R. Turner, *J. Chem. Soc. Faraday Trans.*, 1990, **86**(1), 145.
22. G. Fritz, G. Scherf and O. Glatter, *J. Phys. Chem. B*, 2000, **104**, 3463.
23. O. Glatter, *J. Phys. IV*, 1993, **3**, 27.
24. Y. A. Sanmamed, D. González-Salgado, J. Troncoso, C. A. Cerdeiriña and L. Romaní, *Fluid Phase Equilibr.*, 2007, **252**, 96.
25. Y. A. Sanmamed, D. González-Salgado, J. Troncoso, L. Romaní, A. Baylauq and C. Boned, *J. Chem. Thermodyn.*, 2010, **42**, 553.
26. H. Fitzgerald, D. Fitzgerald and G. Jones, *Pet. Rev.*, 1992, **46**, 544.
27. E. V. Ivanov and V. K. Abrosimov, *J. Chem. Eng. Data*, 2013, **58**, 1103.
28. L. A. Galicia-Luna, D. Richon and H. Renon, *J. Chem. Eng. Data*, 1994, **39**, 424.
29. T. Katrinak, L. Hnedkovsky and I. Cibulka, *J. Chem. Eng. Data*, 2012, **57**, 1152.
30. J. Troncoso, D. Bessieres, C. A. Cerdeiriña, E. Carballo and L. Romaní, *Fluid Phase Equilibr.*, 2003, **208**, 141.
31. C. Bouchot and D. Richon, *Ind. Chem. Eng. Res.*, 1998, **37**, 3295.
32. D. G. Archer, *J. Phys. Chem. Ref. Data*, 1992, **21**, 793.
33. A. T. Sousa, P. S. Fialho, C. A. Nieto de Castro, R. Tufeu and B. Le Neindre, *Fluid Phase Equilibr.*, 1992, **80**, 213.
34. B. Lagourette, C. Boned, H. Saint-Guirons, P. Xans and H. Zhou, *Meas. Sci. Technol.*, 1992, **3**, 699.
35. G. Morrison and D. K. Ward, *Fluid Phase Equilibr.*, 1991, **62**, 65.
36. V. G. Niesen, *J. Chem. Thermodyn.*, 1989, **21**, 915.
37. I. M. S. Lampreia and C. A. Nieto de Castro, *J. Chem. Thermodyn.*, 2011, **43**, 537.
38. M. J. P. Comuñas, J.-P Bazile, A. Baylaucq and C. Boned, *J. Chem. Eng. Data*, 2008, **53**, 986.
39. H. Guerrero, M. García-Mardones, P. Cea, C. Lafuente and I. Bandrés, *Thermochim. Acta*, 2012, **531**, 31.
40. P. Navia, J. Troncoso and L. Romaní, *J. Chem. Eng Data*, 2007, **52**, 1369.

CHAPTER 4

Density Standards and Traceability†

MARK O. McLINDEN

National Institute of Standards and Technology, Applied Chemicals and Materials Division, Boulder, Colorado 80305, USA
Email: markm@boulder.nist.gov

4.1 Introduction

Most instruments for measuring fluid density are not absolute instruments—they must be calibrated with one or more fluids of known density. The selection of appropriate calibration fluids is vital to achieving low uncertainty in a measurement. Also important is establishing the "traceability" of the calibration fluids (and thus the measurement) to established standards. This chapter discusses the concept of traceability and outlines the types and sources of available calibration fluids.

4.2 Traceability

Traceability, or more properly "metrological traceability", is the "property of a measurement result whereby the result can be related to a reference through a documented unbroken chain of calibrations, each contributing to the measurement uncertainty."[1] The Bureau International des Poids et Mesures (International Bureau of Weights and Measures, BIPM) considers the elements for confirming metrological traceability to be an "unbroken

---

†The US Government is authorized to reproduce and distribute reprints for Government purposes notwithstanding any copyright notation hereon.

Volume Properties: Liquids, Solutions and Vapours
Edited by Emmerich Wilhelm and Trevor M. Letcher

Published by the Royal Society of Chemistry, www.rsc.org

metrological traceability chain to an international measurement standard or a national measurement standard, a documented measurement uncertainty, a documented measurement procedure, accredited technical competence, metrological traceability to the SI, and calibration intervals."[1] While common usage is to refer to a measurement or instrument as being "traceable" to a national metrology institute (NMI), such as the National Institute of Standards and Technology (NIST) in the USA or the National Physical Laboratory (NPL) in the UK, the traceability is to a particular realization of an SI unit, which may have been carried out at an NMI. The implicit assumption is that a calibration performed at an NMI is indeed traceable to SI.

Establishing the traceability of a measurement does not necessarily imply that a measurement has a low uncertainty. Each step in the traceability chain has an uncertainty, and the chain can be long and/or the uncertainties in a step can be large.

Traceability is vital in research applications where the properties of a fluid are being measured for the purpose of developing an equation of state or other property model. Documentation of the traceability chain should be included in published results. This could be very detailed in the case of a standard promulgated by an NMI, but for a typical measurement, a brief summary is usually adequate. For some industrial applications, traceability may be a contractual requirement, while for others strict traceability may not be required; instead, conformance to some agreed-upon specification or density relative to a specified reference may be more important.

4.3 Solid Density Standards

Although this book is concerned with liquids and gases, solid density standards must be mentioned because virtually all liquid and gaseous density standards are traced back to solid standards. The current state of the art was reviewed by Fujii.[2] The diameters of nearly-perfect spheres of single-crystal silicon are measured by optical interferometry, thus defining the dimensions in terms of the SI definition of the metre. The mass of the sphere (often near 1 kg) is determined by a comparison weighing with national mass standards. Such methods have been employed by a collaboration among several NMIs[3,4] to achieve a density uncertainty of less than 1 part in 10^7. The reference spheres measured in the interferometer are then used to calibrate, by use of a hydrostatic comparator[5] or the "pressure of flotation" method,[6] the working sinkers employed in Archimedes-type densimeters. The Physikalisch-Technische Bundesanstalt (PTB, the NMI of Germany) offers a calibration service for the density of solid objects, including densimeter sinkers.[7] Earlier examples of this approach were applied to a glass cube, which was then used to determine the density of water;[8] a tungsten carbide cube used to determine the density of mercury;[9] steel spheres, which were used to calibrate silicon density standards;[10] and a quartz sphere used to measure the density of water.[11]

4.4 Calibration by "Known Fluids"

The typical calibration of a densimeter involves measuring fluid(s) of known density covering the temperature, pressure, and density range required by the tests to be carried out with the instrument. The instrument reading is compared to the known density, and the instrument or measurement is adjusted accordingly. The attributes of a good calibration fluid include:

stability over the temperature and pressure range of interest;
does not absorb air or atmospheric moisture (or at least the density is little affected);
density is not a strong function of temperature or pressure;
viscosity similar to test fluids of interest;
compatible with materials of construction of the instrument (including gaskets and seals); and
low hazard with regards to toxicity, flammability, and corrosivity.

Of course, the primary requirement is that the density of the fluid is known accurately. This can take the form of (1) a standard fluid that has been measured and certified by an accredited laboratory or (2) a high-purity material which has available a high-accuracy formulation of the density over the temperature and pressure range desired for the calibration. Such a formulation is possible only for single components of high purity, so that the replicate measurements by different laboratories needed to establish such a formulation are actually carried out on the same substance. The corallary to this requirement is that the fluid be readily available in high purity for calibration purposes.

The advantages of using a "known" high-purity fluid (compared to certified standards, which are discussed in Section 4.5) are a wider choice of fluids (especially when considering a wide range of temperature or pressure), ready availability, and lower cost (especially when large quantities are required). The disadvantages include uncertainties arising from purity effects and uncertainties in the density formulation that are often not well characterized (because the original data may not be well characterized, sometimes because of inappropriate calibrations in the original measurements). Thus, the actual uncertainty of the density of the material at a given temperature and pressure may be difficult to ascertain, and, as a result, strict traceability is difficult to establish.

4.4.1 Water

Water is the most commonly used calibration fluid, and an examination of its characteristics reveals why this is the case. Water is, first of all, very well characterized. Its density has been accurately determined over very wide ranges of temperature and pressure; these data are embodied in the equation of state of the International Association for the Properties of Water and Steam (IAPWS).[12] The density of liquid water is not a strong function of temperature or pressure, such that small errors in the measured conditions have a small effect on the density. Water presents low hazards in use. Water has a viscosity that is similar to many fluids that would be measured in industrial densimeters. However, water has characteristics that must be considered when using it as a calibration fluid, especially when the highest accuracy is sought. Its density is a function of dissolved atmospheric gases and its isotopic composition. The density of water saturated with air at atmospheric pressure is approximately 3 parts in 10^6 lower than air-free water.[13] The IAPWS equation of state is based on Vienna Standard Mean Ocean Water (VSMOW),[14] which refers to water having a particular isotopic composition. Water is H_2O, but there are two stable, naturally occuring isotopes of hydrogen (^{1}H and ^{2}H or deuterium), and three isotopes of oxygen (^{16}O, ^{17}O, and ^{18}O), with ^{16}O being the most abundant. Ocean water is enriched in the heavier isotopes compared to fresh water because evaporation (*i.e.*, the source of the water vapor that goes on to precipitate as rain water or snow) favors the lighter isotopes. Common fresh water will have a density approximately 3 parts in 10^6 lower than VSMOW. Combining isotopic effects with dissolved air can lead to a difference of 6 parts in 10^6 for a typical laboratory water compared to that tabulated by the IAPWS equation of state. Finally, it should be mentioned that water is corrosive to many materials at high temperatures.

Kell[13] has carefully considered the effects of isotopic variations and dissolved gases on the suitability of water as a density calibration standard; he summarized the situation in terms of the target uncertainty in the density as follows. For uncertainties of 0.1% "water is water, isotopic composition can be ignored, and temperature, pressure, and purity need not be controlled too closely." For uncertainties of 0.01% "the isotopic composition of ordinary water need not be considered, and all modern tables of the density of water at atmospheric pressure are equivalent." When working at the level of 0.001%, Kell[13] recommended that "the density of laboratory water should be checked at one temperature and pressure against a known standard…[and] the uncertainty of the isotopic composition of the material to be measured is important also." To realize uncertainties of 1 part in 10^6 would require strict control of the isotopic composition and accurate measurement of the temperature and pressure. Kell[13] characterized work at the level of 1 part in 10^7 "only within the competence of metrological laboratories that have specialized in this direction." Given that most laboratory calibrations use water from a local source (or "high-purity" water from a chemical vendor) of unknown isotopic composition and uncertain degree of gas saturation, it is important to distinguish between the very high precision that some instruments are capable of and the actual uncertainty of the density measurement.

4.4.2 Mercury

Next to water, the best-characterized liquid is elemental mercury. Some of the best density data for mercury, with uncertainties of less than 1 part in 10^6, were determined by Cook and Stone[9,15] in the late 1950s and early 1960s. Mercury is available in very high purity (99.999 99%), and while mercury has seven stable isotopes, the density of different samples varied by less than 1.7 parts in 10^6.[2] Mercury manometers are used by NMIs to realize the SI unit of pressure. The density of mercury (13 546 kg m^{-3} at $t = 20$ °C and $p = 0.1$ MPa) is much too high for the direct calibration of densimeters, although it has been used to determine the characteristic volume of certain types of densimeters, such as those described in Section 2.3. Mercury is also toxic, and its use is restricted in many laboratories.

4.4.3 Other Calibration Liquids

Beyond water and mercury, the applicability of pure liquids suitable for calibrations, based solely on their published density values, is limited. Kuramoto *et al.*[16] have measured the density of seven proposed reference liquids (*iso*-octane, *n*-nonane, *n*-tridecane, water, 2,4-dichlorotoluene, 3,4-dichlorotoluene, and bromobenzene) at 20.00 °C and atmospheric pressure using a single-sinker magnetic suspension densimeter (of the type described in Section 2.2.3.1). These fluids cover a density range of 692 to 1495 kg m^{-3}. The uncertainty of these data was very low, 13 to 23 parts in 10^6, but, apart from water, the purity of the fluids used was only 97% to 99%. Thus, these very accurate values may not be applicable to other samples of these fluids, or, at the very least, not with such low uncertainties.

Schilling *et al.*[17] measured four reference fluids (*n*-heptane, *n*-nonane, 2,4-dichlorotoluene, and bromobenzene); they also used a single-sinker densimeter, but considered a wide range of temperature (233.15 to 473.15 K) and pressure (up to 30 MPa). In a companion work, Sommer *et al.*[18] used the same instrument to measure cyclohexane, toluene, and ethanol over similar ranges of temperature and pressure. Together these fluids cover a density range of 718 to 1494 kg m^{-3}. The uncertainty of these data were 0.015% to 0.02%, increasing to 0.03% at the highest temperatures. The purities of these fluids were 98% for bromobenzene; 99% for *n*-nonane and bromobenzene; 99.3% for *n*-heptane; and 99.9% for toluene, cyclohexane, and ethanol. Both Schilling *et al.*[17] and Sommer *et al.*[18] compared their densities to literature values and found differences of up to 0.2% in density, indicating that the source and uncertainty of data must be carefully considered. Schilling *et al.*[17] compared their data to the high-accuracy data of Kuramoto *et al.*[16] for the three fluids they measured in common and noted differences of up to 0.068%; this is much more than the mutual uncertainties and was attributed to variations in the samples. Sommer *et al.*[18] also compared their own density measurements on two different batches of toluene from the same supplier (which were measured as received), together with the toluene data of McLinden and Splett[19] and found differences of up to 0.02%; thus, even for relatively high-purity material, batch-to-batch variations can be significant. This factor is circumvented when using certified density standards, as discussed in Section 4.5.

A related consideration is the stability of the sample, especially when working at high temperatures. Schilling *et al.*[17] observed decomposition of their samples of 2,4-dichlorotoluene and bromobenzene, and polymerization of the *n*-heptane; they note that these may have been catalyzed by the materials of their particular densimeter (*e.g.*, beryllium copper). Ethanol is also problematic because of its propensity to absorb water from the atmosphere.

4.4.4 Calibration Gases

There are several good choices for calibration with a gas. Research densimeters are more often calibrated with gases compared with industrial vibrating-tube densimeters. The best-characterized gases are helium, nitrogen, argon, propane, carbon dioxide, methane, ethane, ethylene, and sulfur hexafluoride. Wide-ranging equations of state are available for all of these gases; references and uncertainties are listed in Table 4.1. (All of the cited equations of state are implemented in the NIST REFPROP program[20] or on-line in the NIST Chemistry Web Book.[21]) Nitrogen and argon are perhaps the best choices because of the low hazard of these gases and the ready availability of high-purity material. (Research-grade samples with purities of 99.9999% are available.) Helium is also very well characterized (see Section 4.6), but its very low density results in a high relative uncertainty for most

Table 4.1 Calibration gases with relative molar mass, reference and publication year for equation of state, and uncertainty in density.

Gas	M_r	EOS reference	EOS year	Uncertainty[a]
Helium	4.0026	Ortiz-Vega[22]	2013	0.03% ($p<20$ MPa)
Methane	16.0428	Setzmann and Wagner[23]	1991	0.03% ($p<12$ MPa) 0.07% ($p<50$ MPa)
Nitrogen	28.0135	Span *et al.*[24]	2000	0.02%
Ethylene	28.0538	Smukala *et al.*[25]	2000	0.02%
Ethane	30.0690	Bücker and Wagner[26]	2006	0.02%–0.04%
Argon	39.948	Tegeler *et al.*[27]	1999	0.02% ($p<12$ MPa) 0.03% ($p<30$ MPa)
Propane	44.0096	Lemmon *et al.*[28]	2009	0.03%
Carbon dioxide	44.0098	Span and Wagner[29]	1996	0.03%–0.05%
Sulfur hexafluoride	146.0554	Guder and Wagner[30]	2009	0.02% ($T<340$ K) 0.03% ($T<500$ K)

[a]Uncertainties are expanded ($k=2$), or approximately 95% confidence interval, and apply for a temperature range of 240 to 500 K and pressures up to 30 MPa, unless noted. See the cited references for a more detailed discussion of uncertainty.

types of densimeters. Helium is also more sensitive to impurities than other gases. (The most common impurity when handling gases is air. The average molar mass of air is 7.3 times that of helium, but only 1.04 times that for nitrogen. Thus, an air impurity will have a larger relative effect on the density of helium.) Carbon dioxide and sulfur hexafluoride are relatively dense gases, but have critical temperatures near ambient (304.13 K and 318.72 K, respectively), and because of the high compressibility of a fluid at near-critical conditions, these gases may be less suitable for calibration purposes at some conditions.

When calibrating with a gas, the measurement of temperature and pressure is much more important than with a liquid calibration fluid. A target uncertainty of 0.01% in density would require a similar uncertainty in the pressure measurement of a gas and a temperature uncertainty of 0.03 K. In contrast, a calibration with water would require a temperature uncertainty of 0.5 K and a pressure uncertainty of 0.2 MPa to obtain a density uncertainty of 0.01%. While gases are usually considered to be low-density fluids, at high pressures their density can approach that of liquids; for example, argon at $p = 40$ MPa and $T = 293.15$ K has a density of 617 kg m^{-3}, nearly that of many light hydrocarbon liquids.

4.5 Certified Density Standards

Density standards that are distributed by NMIs and other accredited laboratories are not based on "known" properties of pure substances. Instead, the density of a particular lot of a fluid is determined in an absolute instrument, typically a densimeter operating on the Archimedes principle, as described in Section 2.2.2. The lot is divided into a large number of individual vials for sale. Some laboratories divide the lot first and carry out the density determination on a statistical sample of the individual vials and other laboratories measure the entire lot in a large apparatus and then divide it.

Because such standards consist of fluid that has actually been measured, the purity of the sample is not a factor—the calibration certificate applies only to that particular lot of material. (Although typically high-purity materials are used.) The traceability chain to the SI unit of density is, thus, clearly established. Such traceability is sometimes a contractual requirement when an instrument is used in the buying and selling of an industrial chemical. In contrast, the traceability of a "known" pure fluid is much more difficult to establish—it would require a detailed chemical analysis of impurities in the sample (with its own uncertainties and traceability chain) and an analysis of the effect of those impurities on the density.

The primary advantage of a certified standard is its low and clearly documented uncertainty. The disadvantages are high cost and a limited choice of fluids. Table 4.2 lists examples of available standards. These are termed "Standard Reference Materials (SRM)®" by NIST, "density reference liquids" by PTB, and "traceable liquid density standards" by H&D Fitzgerald,

Table 4.2 Examples of certified density standards.

Material	Density/kg m^{-3} (20 °C)	$U(\rho)$/kg m^{-3} ($k=2$)	t range/°C $p=0.1$ MPa	Cost
H&D Fitzgerald, UK, "Traceable Liquid Density Standards"				
Isooctane	692	0.01	15–25	£62.40
Dodecane	749	0.01	10–50	/10 mL
Water	998	0.015	20	
Dimethylphthalate	1191	0.02	15–25	
NaBr (aq)	1249	0.01	15–25	
Tetrachloroethylene	1623	0.015	10–50	
PTB, Germany, "Density Reference Liquids"				
n-Nonane	718	0.02–0.05	5–70	€290 (first)
Water	998	0.02–0.05	5–70	€190 (addt'l)
Polyalphaolefins (2)	795–816	0.02–0.05	5–70	/10 mL
NIST, USA, "Standard Reference Materials®"				
Isooctane	692	0.035	15–25	\$403–\$416
Toluene	867	0.060–0.098	-50–150 (p to 30 MPa)	/(4×5 mL)

a private laboratory in the UK accredited by NPL. Most of these standards are certified only at atmospheric pressure and near-ambient temperatures. Only the SRM of NIST based on toluene is certified over wide ranges of temperature and pressure. This SRM also considers the effect of dissolved air, which lowers the density by 0.0628 kg m^{-3} at $T=298.15$ K and $p=0.1$ MPa compared to the air-free material.[19]

4.6 *Ab Initio* Calculation of Fluid Properties

A relatively recent advance in the field is the ability to calculate fluid properties *ab initio*, or from first principles. These are based on solution of the equations of quantum mechanics, although the details are beyond the scope of this book. Among the properties that can be calculated are the virial coefficients, which provide the density of gases at relatively low pressures, as discussed in Chapter 6. To date, helium is the gas most thoroughly treated in this fashion, and argon is the subject of considerable current research.

The second and third virial coefficients of helium have been calculated essentially exactly[31,32] and Shaul *et al.*[33] argue that the fourth virial coefficient can be calculated better than any current experimental determination. The fifth virial coefficient has also been calculated, although not rigorously (such that it has a higher uncertainty and is valid only at temperatures above about 50 K).[33] The result is that the density of helium is known from first principles with an uncertainty less than 0.0025% at $T=223$ K and $p=38$ MPa, with smaller uncertainties at higher temperatures and lower pressures.

The second and third virial coefficients of argon have been calculated *ab initio* with low uncertainties.[34,35] Jäger *et al.*[34] have also computed the fourth (and higher) virial coefficients, although not including all effects. At a temperature of 234 K, the uncertainties in the theoretical and best experimental values of the third virial are comparable, with the experimental values having lower uncertainty at lower temperatures (where quantum effects complicate the theoretical calculations) and the *ab initio* virials having lower uncertainty at higher temperatures (where argon is more nearly an ideal gas). The result is that the density of argon is known from first principles with an uncertainty of approximately 0.01% at $T = 293$ K and $p = 5$ MPa and 0.02% at pressures up to 12 MPa. Which is to say that the theoretical values have uncertainties comparable to the equation of state at these conditions. Work continues to improve the *ab initio* values, and it is likely that theoretical densities for argon will be suitable for calibration purposes at higher pressures in the near future.

References

1. *BIPM, International vocabulary of metrology—Basic and general concepts and associated terms (VIM)*, Joint Committee for Guides in Metrology, Paris, 2012.
2. K. Fujii, *Metrologia*, 2004, **41**, S1.
3. G. Bartl, H. Bettin, M. Krystek, T. Mai, A. Nicolaus and A. Peter, *Metrologia*, 2011, **48**, S96.
4. R. A. Nicolaus and K. Fujii, *Meas. Sci. Technol.*, 2006, **17**, 2527.
5. H. A. Bowman, R. M. Schoonover and C. L. Carroll, *J. Res. Natl. Bur. Stand., Sect. A*, 1973, **78A**, 13.
6. H. Bettin and H. Toth, *Metrologia*, 2004, **41**, S52.
7. PTB Solide State Density Working Group 3.43 http://www.ptb.de/cms/nc/en/fachabteilungen/abt3/fb-34/ag-3430.html?print=1 (December 20, 2013).
8. C.-E. Guillaume, *Trav. Mem. Bur. Int. Poids Mes*, 1910, **14**, 1.
9. A. H. Cook and N. W. B. Stone, *Phil. Trans. R. Soc., A*, 1957, **250**, 279.
10. H. A. Bowman, R. M. Schoonover and C. L. Carroll, *Metrologia*, 1974, **10**, 117.
11. R. Masui, K. Fujii and M. Takenaka, *Metrologia*, 1995, **32**, 333.
12. W. Wagner and A. Pruß, *J. Phys. Chem. Ref. Data*, 2002, **31**, 387.
13. G. S. Kell, *J. Phys. Chem. Ref. Data*, 1977, **6**, 1109.
14. International Atomic Energy Agency Reference Sheet for International Measurement Standards. VSMOW2 & SLAP2. http://nucleus.iaea.org/rpst/ReferenceProducts/ReferenceMaterials/Stable_Isotopes/2H18O-water-samples/VSMOW2.htm (December 20, 2013).
15. A. H. Cook, *Phil. Trans. R. Soc., A*, 1961, **254**, 125.
16. N. Kuramoto, K. Fujii and A. Waseda, *Metrologia*, 2004, **41**, S84.
17. G. Schilling, R. Kleinrahm and W. Wagner, *J. Chem. Thermodyn.*, 2008, **40**, 1095.

18. D. Sommer, R. Kleinrahm, R. Span and W. Wagner, *J. Chem. Thermodyn.*, 2011, **44**, 117.
19. M. O. McLinden and J. D. Splett, *J. Res. Natl. Inst. Stand. Technol.*, 2008, **113**, 29.
20. E. W. Lemmon, M. L. Huber and M. O. McLinden, NIST Standard Reference Database 23, NIST Reference Fluid Thermodynamic and Transport Properties—REFPROP, version 9.1; Standard Reference Data Program, National Institute of Standards and Technology, Gaithersburg, MD, 2013.
21. NIST Chemistry Webbook. http://webbook.nist.gov/chemistry/fluid/.
22. D. O. Ortiz-Vega, PhD thesis, A new wide-range equation of state for helium-4, Texas A&M University, 2013.
23. U. Setzmann and W. Wagner, *J. Phys. Chem. Ref. Data*, 1991, **20**, 1061.
24. R. Span, E. W. Lemmon, R. T. Jacobsen, W. Wagner and A. Yokozeki, *J. Phys. Chem. Ref. Data*, 2000, **29**, 1361.
25. J. Smukala, R. Span and W. Wagner, *J. Phys. Chem. Ref. Data*, 2000, **29**, 1053.
26. D. Bücker and W. Wagner, *J. Phys. Chem. Ref. Data*, 2006, **35**, 205.
27. C. Tegeler, R. Span and W. Wagner, *J. Phys. Chem. Ref. Data*, 1999, **28**, 779.
28. E. W. Lemmon, W. Wagner and M. O. McLinden, *J. Chem. Eng. Data*, 2009, **54**, 3141.
29. R. Span and W. Wagner, *J. Phys. Chem. Ref. Data*, 1996, **26**, 1509.
30. C. Guder and W. Wagner, *J. Phys. Chem. Ref. Data*, 2009, **38**, 33.
31. W. Cencek, M. Przybytek, J. Komasa, J. B. Mehl, B. Jeziorski and K. Szalewicz, *J. Chem. Phys.*, 2012, **136**, 224–303.
32. G. Garberoglio, M. R. Moldover and A. H. Harvey, *J. Res. Natl. Inst. Stand. Technol.*, 2011, **116**, 729.
33. K. R. S. Shaul, A. J. Schultz, D. A. Kofke and M. R. Moldover, *Chem. Phys. Lett.*, 2012, **531**, 11.
34. B. Jäger, R. Hellman, E. Bich and E. Vogel, *J. Chem. Phys.*, 2011, **135**, 084308.
35. W. Cencek, G. Garberoglio, A. H. Harvey, M. O. McLinden and K. Szalewicz, *J. Phys. Chem.*, 2013, **117**, 7542.

CHAPTER 5

Volumetric Properties from Multiparameter Equations of State†

ROLAND SPAN*[a] AND ERIC W. LEMMON*[b]

[a] Thermodynamics, Ruhr-University Bochum, 44780 Bochum, Germany;
[b] Applied Chemicals and Materials Division, National Institute of Standards and Technology, 325 Broadway, Mailstop 647.08, Boulder, Colorado 80305-3337, USA
*Email: roland.span@thermo.rub.de; eric.lemmon@nist.gov

5.1 Introduction

Multiparameter equations of state are empirical property models. The experimental data they are based on are the primary reference for an accurate representation of thermodynamic properties of the corresponding fluid; in general a multiparameter equation of state cannot be more accurate than the most accurate experimental data available for the fluid. However, by principle, equations of state yield consistent results for all thermodynamic properties, which are considered just derivatives or combinations of derivatives of a common thermodynamic potential. Thus, the process of fitting a multiparameter equation of state to data for different thermodynamic properties measured by different groups using different experimental techniques implies a very rigid analysis of the available data sets. If the resulting equation of state is able to represent all reliable experimental data for

Volume Properties: Liquids, Solutions and Vapours
Edited by Emmerich Wilhelm and Trevor M. Letcher

Published by the Royal Society of Chemistry, www.rsc.org

different properties within their respective uncertainty, it is considered a "reference equation of state" for the corresponding fluid. Of course this status does not imply highest accuracy by itself—the accuracy of calculated properties still depends on the accuracy of the experimental data available for the fluid. However, in any case, a reference equation of state is a much better basis for all kinds of accurate thermodynamic property calculations than a simple fit to data for just one property measured by just one group of authors over a limited temperature and pressure range.

In technical applications high demands on the accuracy of calculated volumetric properties, or more specifically on calculated densities, are usually formulated for the calibration of secondary measuring devices, for the validation of new experimental techniques, and for custody transfer whenever a measured volume or volume flow has to be converted into a mass or a mass flow. In process simulations, as they are common, *e.g.*, in energy technologies and in chemical engineering, demands on the accuracy of volumetric properties are usually less advanced. However, in these calculations accurate values for caloric properties like heat capacities, enthalpies, and entropies are required. Due to the known relations between different thermodynamic properties, the real fluid or residual contributions to caloric properties can be calculated from the same equation of state that is used to accurately describe the volumetric behavior of a fluid (the ideal gas contribution has to be determined independently). Volumetric properties, like densities at homogeneous states and of the saturated liquid together with saturation pressures and, more recently, speeds of sound, usually form the most accurate part of the experimental basis for the development of multiparameter equations of state. Thus, accurate experimental data for volumetric properties are technically highly relevant, even if high demands on the accuracy of data for other properties are more obvious in many applications.

For a long time results from multiparameter equations of state were primarily used in the form of printed property tables and charts. Only a few specialized software tools contained algorithms that allowed for direct thermodynamic property calculations based on accurate multiparameter equations of state. In the late 1980s and early 1990s, reasonably user friendly software products with a graphical user interface designed for personal computers replaced property tables and charts in many applications. But, to integrate accurate property calculations from multiparameter equations of state into user defined algorithms, demanding and fault-prone links at source-code level were still required. This changed two decades ago when the required property algorithms were compiled into Dynamic Link Libraries (DLLs) for the first time. Today, such DLLs are easily available and can be called from most standard environments, for example from MS Excel® or Matlab®. Not only experts, but all natural scientists and engineers, and even undergraduate students, are able to integrate thermodynamic property calculations based on the most advanced multiparameter equations of state into their own algorithms—the calculation of accurate values for properties of many pure fluids became a matter of course.

Still, the term "accurate values for volumetric properties" is very vague—the accuracy that is actually achieved depends on the considered fluid and on the temperature and pressure range. For some "reference fluids" like, *e.g.*, nitrogen, argon, methane, ethane, propane, and carbon dioxide, densities can be calculated with an uncertainty of 0.02% to 0.03% over broad ranges of temperature and at pressures up to about 30 MPa. In the liquid phase much higher accuracies were achieved for water. For many other scientifically and technically important fluids, such as higher alkanes, refrigerants or working fluids for organic Rankine cycles, uncertainties in the range of 0.05% to 0.2% in density can be claimed over broad ranges of temperature and pressure. However, for some of the about 115 pure fluids described by multiparameter equations of state today, the uncertainty of calculated densities may be as high as 1%. Whenever high accuracy is relevant for an application the user cannot blindly use software tools; the user has to carefully assess the uncertainty of the equation of state, considering not only the fluid but also the relevant range of states. This information is usually not given by software tools—the original literature where the equation of state was published has to be consulted.

Multiparameter equations of state for pure fluids in combination with some kind of mixing rules have always been used to describe properties of mixtures as well. In the gas phase and at supercritical states some of these approaches have been quite successful, but in general properties of mixtures have always been described with significantly lower accuracy. In part, this effect can be explained by lower accuracy of the available experimental data—accurate experimental work on properties of mixtures has always been particularly demanding. Even if state of the art equipment is used and all mixture related effects are considered, the uncertainty of the composition of the investigated mixture increases the overall uncertainty beyond that which can be obtained for pure fluids. But the more limiting factor has been the lack of accurate mixture models based on multiparameter equations of state.

Over the last two decades a new approach has evolved, which allows for the development of accurate multiparameter equations of state for mixture properties. Even though the resulting property models may not be fully satisfactory from a theoretical point of view, accurate descriptions are now available over the whole range of fluid states of many technically relevant mixtures, in particular for refrigerant mixtures, for air, and for natural gas and natural gas like mixtures.

This chapter is intended to introduce multiparameter equations of state to readers interested in accurate calculation of volumetric properties. Section 5.2 will briefly introduce pressure explicit multiparameter equations of state. Even though this equation of state type is now considered obsolete, some information about them is relevant to understand the historical development and some details, which are still valid for modern fundamental equations of state. The way in which volumetric properties can be calculated from such fundamental equations of state will be discussed in Section 5.3. Section 5.4 will give examples of the uncertainty of volumetric properties calculated from multiparameter equations of state for pure fluids. And finally Section 5.5 will deal with recent multiparameter models for mixtures. The development of multiparameter equations of state and the setup of numerical algorithms for the calculation of different thermodynamic properties from multiparameter equations of state will not be discussed in detail—references to more detailed literature are given for interested readers.

5.2 Pressure Explicit Multiparameter Equations of State

To calculate data for volumetric properties of a pure fluid at homogeneous states, the specific volume v or the density ρ has to be expressed as a function of two independent variables. The most common independent variables are temperature T and pressure p, resulting in the relation

$$\rho = \rho(T,p) = \frac{1}{v(T,p)}. \tag{5.1}$$

However, when using temperature and pressure as independent variables, a discontinuity occurs at the saturation pressure $p_s = p_s(T)$ because the saturated liquid density ρ' and the saturated vapor density ρ'' are valid solutions at the same pressure,

$$\rho'(T) = \rho(T,p_s) \neq \rho''(T) = \rho(T,p_s). \tag{5.2}$$

Thus, an equation of state that is explicit in pressure cannot describe the whole range of fluid states with a closed mathematical expression. To overcome this problem, multiparameter equations of state that describe the whole fluid range have been formulated as

$$p = p(T,\rho) \quad \text{or} \quad p = p(T,v). \tag{5.3}$$

This way the equilibrium condition resulting from mechanical stability of the phase boundary,

$$p_S(T) = p(T,\rho') = p(T,\rho'') \tag{5.4}$$

with $\rho' \neq \rho''$ and $(\partial p/\partial \rho)_T > 0$ for $\rho = \rho''$ and for $\rho = \rho'$ can be satisfied if the function $p\ (T,\rho)$ is at least cubic in density.

Functions of the form $p\ (T,\rho)$ imply that the density has to be calculated iteratively if pressure and temperature are given. These iterative solutions in conjunction with the selection of suitable starting points are the main challenge when developing algorithms for property calculations based on multiparameter equations of state. The combination of temperature and density is not the only practically relevant set of independent variables. Pressure and enthalpy, or pressure and entropy, for example, are frequently used sets of independent variables in process simulations. Most software tools support at least these three sets of independent variables, see for example ref. 1–3.

The virial equation of state (see also chapter 14) is

$$\frac{p}{\rho RT}=Z(T,\rho)=1+B(T)\times\rho+C(T)\times\rho^2+D(T)\times\rho^3+\ldots \tag{5.5}$$

with the compression factor Z, the gas constant R, and the virial coefficients B, C, and D, which were introduced by Kamerlingh Onnes[4] in 1901, can be considered the basis of multiparameter equations of state explicit in pressure if the temperature dependence of the virial coefficients is empirically described by polynomials. However, simple virial expansions yield reasonable results only for gaseous or gas-like supercritical states. To describe liquid or liquid-like states, higher virial coefficients have to be introduced; the resulting equations have a very large number of adjustable coefficients (see, *e.g.*, ref. 5) and become numerically instable. In 1940 Benedict *et al.*[6] introduced an exponential function into a truncated virial expansion for the first time. The resulting "BWR-equation of state" can be written as

$$\frac{p}{\rho RT}=Z(T,\rho)=1+\sum_{i=1}^{6}n_iT^{t_i}\rho^{d_i}+\sum_{i=7}^{8}n_iT^{t_i}\rho^{d_i}\exp\left(-\left(\frac{\rho}{\rho_\mathrm{r}}\right)^2\right) \tag{5.6}$$

where the reducing parameter ρ_r is roughly equal to the critical density. The density exponents d_i and the temperature exponents t_i were selected with a trial and error procedure, and were considered universal for all fluids. With 8 fluid specific adjustable coefficients n_i, this formulation describes properties over the whole range of fluid states qualitatively correctly. The accuracy of calculated properties in Equation (5.6) is superior to cubic equations of state for pure fluids. However, Equation (5.6) does not satisfy high demands on accuracy and could never be considered a reference equation of state—in the 1940s the accuracy of the best experimental data was already much higher than the accuracy of the BWR-equation of state.

The BWR-equation, Equation (5.6), is the origin of the so-called "modified BWR-type" equations of state, which can be written either in a dimensional form as

$$\frac{p}{\rho RT}=Z(T,\rho)=1+\sum_{i=1}^{I_\mathrm{POL}}n_iT^{t_i}\rho^{d_i}+\sum_{i=I_\mathrm{POL}+1}^{I_\mathrm{POL}+I_\mathrm{EXP}}n_iT^{t_i}\rho^{d_i}\exp\left(-\left(\frac{\rho}{\rho_\mathrm{r}}\right)^2\right) \tag{5.7}$$

or in the more convenient reduced form as

$$\frac{p}{\rho RT}=Z(T,\rho)=1+\sum_{i=1}^{I_\mathrm{POL}}n_i^*\tau^{t_i^*}\delta^{d_i^*}+\sum_{i=I_\mathrm{POL}+1}^{I_\mathrm{POL}+I_\mathrm{EXP}}n_i^*\tau^{t_i^*}\delta^{d_i^*}\exp(-\gamma\delta^2) \tag{5.8}$$

with the reduced density $\delta=\rho/\rho_\mathrm{c}$ and the inverse reduced temperature $\tau=T_\mathrm{c}/T$. Equations (5.7) and (5.8) are equivalent with

$d_i^* = d_i$, $t_i^* = -t_i$, $n_i^* = n_i \times T_c^{t_i} \times \rho_c^{-d_i}$, and $\gamma = (\rho_c/\rho_r)^2$. The use of the critical parameters T_c and ρ_c as reducing parameters is common but not mandatory; other parameters can be used without disadvantages.

The simplest representation of the class of modified BWR-type equations of state are those by Strobridge,[7] Starling,[8] and Lee and Kessler.[9] More complex modifications found in the works of Bender[10] with 13 polynomial and 6 exponential terms and of Jacobsen and Stewart[11] with 19 polynomial and 13 exponential terms, which is referred to as *the MBWR equation* (Modified BWR equation) in literature, are still widely used.

Vapor liquid phase equilibria can be calculated consistently and without use of ancillary equations from pressure explicit multiparameter equations of state by evaluation of the phase equilibrium conditions $T' = T''$, $p' = p''$, and $g' = g''$. The equilibrium condition for the Gibbs enthalpy of both phases, $g' = g''$, can be replaced by the equality of the fugacities of both phases, $f'(T) = f(T, \rho') = f''(T) = f(T, \rho'')$, with

$$f(T,p) = p(T,\rho) \times \exp\left(\int_{p^0}^{p(T,\rho)} \left[\frac{v(T,p)}{RT} - \frac{1}{p} \right] \mathrm{d}p \Bigg|_{T=\mathrm{const.}} \right). \quad (5.9)$$

Even though the fugacity is commonly expressed in terms of pressure and $p' = p''$ holds, Equation (5.9) has to be evaluated independently for the saturated liquid and the saturated vapor because $p' = p(T,\rho')$ and $p'' = p(T,\rho'')$. For the saturated vapor the integration starts in the ideal gas and stops at the saturated vapor density. For the saturated liquid the integration has to be continued throughout the two-phase region until the liquid root for density is reached. In phase equilibrium, the contribution of the integral between saturated vapor and saturated liquid density becomes zero. Alternative approaches like the evaluation of the so-called "Maxwell criterion",

$$p_s(T_s) \times (v''(T_s) - v'(T_s)) = \int_{v'}^{v''} p(T_s, v)\mathrm{d}v \quad (5.10)$$

with $p(T_s, v') = p_s$ and $p(T_s, v'') = p_s$ for a given saturation temperature T_s can also be used. However, all approaches involve the evaluation of integral expressions and an iterative solution of the phase equilibrium conditions to determine p_s, v', and v''.

In contrast to cubic equations of state, multiparameter equations of state do not satisfy the phase equilibrium condition simply by their mathematical structure. Phase equilibrium conditions have to be considered when parameters of the equation are fitted. Independently of each other, Wagner,[12] Bender,[10] and McCarty[13] introduced corresponding algorithms in 1970. Wagner and Bender showed that phase equilibrium data can be considered in linear fits if data triplets with p_s, v', and v'' at the same temperature are given. In 1979, Ahrendts and Baehr[14–17] summarized the theory of simultaneous, nonlinear fits to data for different properties. This approach also enabled fits to vapor pressure and saturated liquid density data only—an important improvement because accurate experimental data for saturated vapor densities are available only for few fluids. The state of the art in the development of multiparameter equations of state is described, *e.g.*, in ref. 18 and 19.

The exponential term in the BWR-equation of state, Equation (5.6), and its modifications, Equation (5.8), makes the link between multiparameter and virial equations of state, Equation (5.5), less obvious. However, with the definitions

$$B(T) = \lim_{\rho\to 0}\left(\frac{\partial Z}{\partial \rho}\right)_T = \frac{1}{\rho_c}\times\lim_{\delta\to 0}\left(\frac{\partial Z}{\partial \delta}\right)_\tau, \tag{5.11}$$

$$C(T) = \frac{1}{2}\lim_{\rho\to 0}\left(\frac{\partial^2 Z}{\partial \rho^2}\right)_T = \frac{1}{2\rho_c^2}\times\lim_{\delta\to 0}\left(\frac{\partial^2 Z}{\partial \delta^2}\right)_\tau, \tag{5.12}$$

$$D(T) = \frac{1}{6}\lim_{\rho\to 0}\left(\frac{\partial^3 Z}{\partial \rho^3}\right)_T = \frac{1}{6\rho_c^3}\times\lim_{\delta\to 0}\left(\frac{\partial^3 Z}{\partial \delta^3}\right)_\tau, \tag{5.13}$$

and the Taylor expansion of the exponential terms in Equation (5.8),

$$\begin{aligned} n_i^*\delta^{d_i^*}\tau^{t_i^*}\exp(-\gamma\delta^2) &= n_i^*\delta^{d_i^*}\tau^{t_i^*}\times\sum_{n=0}^{\infty}\frac{\left(-\gamma\delta^2\right)^n}{n!} = n_i^*\delta^{d_i^*}\tau^{t_i^*} \\ &\times\left[1-\frac{\gamma\delta^2}{1}+\frac{\left(-\gamma\delta^2\right)^2}{2}+\ldots\right], \end{aligned} \tag{5.14}$$

it can be shown that polynomial and exponential terms with $d_i^* = 1$ contribute to the second virial coefficient $B(T)$, polynomial and exponential terms with $d_i^* = 2$ contribute to the third virial coefficient $C(T)$, polynomial terms with $d_i^* = 3$ and exponential terms with $d_i^* = 1$ and with $d_i^* = 3$ contribute to the fourth virial coefficient $D(T)$, and so on. As long as all exponents d_i^* are whole numbers with $d_i^* > 0$, virial coefficients can be derived from modified BWR-equations of state. The exponents t_i^* can be whole or real numbers without constraints on the algebraic sign. However, only the first few virial coefficients of multiparameter equations, typically B and C, come close to the values theoretically expected from a non-truncated virial expansion with precise representation of the temperature dependence of all virial coefficients. Higher virial coefficients are influenced by almost arbitrary contributions of the exponential terms. Virial coefficients calculated from multiparameter equations of state should be considered "practical virial coefficients", which yield an accurate description of $Z(T,\rho)$ up to very high (even liquid-like) densities based on a truncated virial expansion with a rather limited number of adjustable parameters.

5.3 Volumetric Properties Calculated from Fundamental Equations of State

The pressure explicit equations of state presented in Section 5.2 have two relevant disadvantages. In order to calculate caloric properties, an independent equation of state for the caloric behavior of the ideal gas is required. The residual part of the caloric properties (the difference between the properties of the hypothetical ideal gas at T, ρ and of the real fluid at T, ρ) includes integral expressions like the one in Equation (5.9). The need to integrate the equation of state for the calculation of caloric properties results in limitations on the mathematical structure.

The use of fundamental equations of state avoids these disadvantages. Once a correlation, in this case a multiparameter equation of state, has been established for one of the fundamental equations of state known in thermodynamics, all other thermodynamic properties can be calculated by combinations of derivatives of this function. For the reasons discussed in Section 5.2, temperature and density are the preferred independent variables of multiparameter equations of state. Thus, fundamental equations are formulated in the Helmholtz energy, $a(T,\rho)$, which is split into a contribution of the ideal gas, $a^{\mathrm{o}}(T,\rho)$, and a residual contribution, $a^{\mathrm{r}}(T,\rho)$, reflecting the difference between properties of the ideal gas and of the real fluid at the same temperature and density. To derive a dimensionless formulation, the Helmholtz energy is commonly reduced by RT and the reduced density $\delta=\rho/\rho_{\mathrm{c}}$ and the inverse reduced temperature $\tau=T_{\mathrm{c}}/T$ are used instead of ρ and T:

$$\alpha(\tau,\delta)=\alpha^{\mathrm{o}}(\tau,\delta)+\alpha^{\mathrm{r}}(\tau,\delta)=\frac{a^{\mathrm{o}}(T,\rho)}{RT}+\frac{a^{\mathrm{r}}(T,\rho)}{RT} \tag{5.15}$$

Tillner-Roth[20] pointed out that α corresponds to a reduced Massieu function because the Helmholtz energy is reduced by temperature and not by a constant reducing temperature, see also ref. 21. Although this argument is formally correct, α is still referred to as reduced Helmholtz energy in international literature. The use of T_{c} and ρ_{c} as reducing parameters of temperature and density is common but not mandatory.

From a fundamental equation of state in terms of the reduced Helmholtz energy the compression factor can be calculated according to

$$Z(T,\rho)=\frac{p}{\rho RT}=\frac{-(\partial a/\partial v)_T}{\rho RT}=\delta\times\left(\frac{\partial\alpha}{\partial\delta}\right)_{\tau}=1+\delta\times\left(\frac{\partial\alpha^{\mathrm{r}}}{\partial\delta}\right)_{\tau}. \tag{5.16}$$

Since the density dependence of $\alpha^{\mathrm{o}}(\tau,\delta)$ is simply given by $\ln(\delta)$, the ideal gas contribution in Equation (5.16) becomes unity ($Z=1$ is the thermal equation of state of the ideal gas!). The ideal gas part of the reduced Helmholtz energy, $\alpha^{\mathrm{o}}(\tau,\delta)$, is not required for the calculation of thermal properties. For details on common formulations for the ideal gas part of the reduced Helmholtz energy see, *e.g.*, ref. 18.

Simple fundamental equations of state use a functional form similar to the reduced modified BWR form, Equation 5.8, to describe the residual contribution to the reduced Helmholtz energy:

$$\frac{a^{\mathrm{r}}(T,\rho)}{RT} = \alpha^{\mathrm{r}}(\tau,\delta) = \sum_{i=1}^{I_{\mathrm{POL}}} n_i \tau^{t_i} \delta^{d_i} + \sum_{i=I_{\mathrm{POL}}+1}^{I_{\mathrm{POL}}+I_{\mathrm{EXP}}} n_i \tau^{t_i} \delta^{d_i} \exp(-\delta^{p_i}). \tag{5.17}$$

Thus, the compression factor becomes

$$Z(T,\rho) = 1 + \delta \times \left[\sum_{i=1}^{I_{\mathrm{POL}}} n_i d_i \tau^{t_i} \delta^{d_i-1} + \sum_{i=I_{\mathrm{POL}}+1}^{I_{\mathrm{POL}}+I_{\mathrm{EXP}}} n_i \tau^{t_i} \times \left(d_i \delta^{d_i-1} - p_i \delta^{p_i-1} \times \delta^{d_i} \right) \times \exp(-\delta^{p_i}) \right]. \tag{5.18}$$

An early example of this type of multiparameter fundamental equation of state is the equation of state by Schmidt and Wagner.[22] This equation was also the first example of a multiparameter equation of state with a systematically optimized functional form—the parameters I_{POL}, I_{EXP}, t_{i}, d_{i}, and p_{i} were not chosen by trial and error but by an evolutionary optimization algorithm.[23] More recently the simple functional form given in Equation (5.17) has been used to establish reference equations of state for organic working fluids, see *e.g.* refs. 24 and 25, and to describe a larger number of fluids with advanced technical accuracy,[26–29] but not necessarily with reference accuracy.

For the reference equation of state for methane, Setzmann and Wagner[30] adapted a term previously proposed by Haar *et al.*[31] in a similar form. These so-called "Gaussian bell shaped terms" read

$$\alpha^{\mathrm{r}}_{\mathrm{GBS},i}(\tau,\delta) = n_i \tau^{t_i} \delta^{d_i} \exp\left(-\eta_i(\delta-\varepsilon_i)^2 - \beta_i(\tau-\gamma_i)^2\right). \tag{5.19}$$

The first density derivative of the Gaussian bell shaped terms, which is required to evaluate Equation (5.16) for the compression factor, becomes

$$\left(\frac{\partial \alpha^{\mathrm{r}}_{\mathrm{GBS},i}}{\partial \delta}\right)_{\tau} = n_i \tau^{t_i} \delta^{d_i} \exp\left(-\eta_i(\delta-\varepsilon_i)^2 - \beta_i(\tau-\gamma_i)^2\right) \times \left[\frac{d_i}{\delta} - 2\eta_i(\delta-\varepsilon_i)\right]. \tag{5.20}$$

With ε_i and γ_i in the order of unity, the expression in the exponent roughly describes the distance from the critical point. The exponential expression dampens the influence of the terms further away from the critical point, allowing for an improved fit of the critical region that does not affect other regions. Using terms like this, Setzmann and Wagner[30] were able to describe highly accurate experimental $p\rho T$-data in the critical region with an uncertainty of 0.02% in pressure for the first time. Gaussian bell shaped terms of this kind have been used in all highly accurate reference equations of state since then, see *e.g.* ref. 19 and 32–34. More recently Gaussian bell shaped terms with different parameter sets were successfully introduced to improve the representation of properties in regions far from the critical point as well, see *e.g.* ref. 35.

Span and Wagner[36] supplemented the Gaussian bell shaped terms with a more advanced form for non-analytic critical-region terms. These very complex terms aim mostly to improve the description of caloric properties in the critical region and were used only for carbon dioxide[36] and water.[37] Non-analytic terms are not required for an accurate description of volumetric properties and will not be discussed here.

Densities as a function of temperature and pressure, and thermal properties at phase equilibrium, can be calculated with fundamental equations of state using the same iterative procedures as were described in Section 5.2, see also ref. 18. However, instead of the fugacity based criterion for phase equilibrium, the Gibbs energy criterion may be directly used as

$$\begin{aligned} g'(T) &= g(T,\rho') = RT\left(1+\alpha^{\mathrm{o}}+\alpha^{\mathrm{r}}+\delta\left(\frac{\partial\alpha^{\mathrm{r}}}{\partial\delta}\right)_{\tau}\right)\Bigg|_{\tau,\delta'} \\ &= g''(T) = g(T,\rho'') = RT\left(1+\alpha^{\mathrm{o}}+\alpha^{\mathrm{r}}+\delta\left(\frac{\partial\alpha^{\mathrm{r}}}{\partial\delta}\right)_{\tau}\right)\Bigg|_{\tau,\delta''}. \end{aligned} \tag{5.21}$$

As compared to Equation (5.9), this relation illustrates the advantage of fundamental equations of state: no integrations are required—all thermodynamic properties can be calculated from a combination of derivatives. Relations between the reduced Helmholtz energy and various thermodynamic properties including caloric properties are given, *e.g.*, in ref. 18, 19, 38 and 39.

Virial coefficients can be calculated from fundamental equations of state according to

$$B(T) = \lim_{\rho\to 0}\left(\frac{\partial Z}{\partial\rho}\right)_T = \frac{1}{\rho_{\mathrm{c}}}\times\lim_{\delta\to 0}\left(\frac{\partial\alpha^{\mathrm{r}}}{\partial\delta}\right)_{\tau}, \tag{5.22}$$

$$C(T) = \frac{1}{2}\lim_{\rho\to 0}\left(\frac{\partial^2 Z}{\partial\rho^2}\right)_T = \frac{1}{\rho_{\mathrm{c}}^2}\times\lim_{\delta\to 0}\left(\frac{\partial^2\alpha^{\mathrm{r}}}{\partial\delta^2}\right)_{\tau}, \tag{5.23}$$

$$D(T) = \frac{1}{6}\lim_{\rho\to 0}\left(\frac{\partial^3 Z}{\partial\rho^3}\right)_T = \frac{1}{2\rho_{\mathrm{c}}^3}\times\lim_{\delta\to 0}\left(\frac{\partial^3\alpha^{\mathrm{r}}}{\partial\delta^3}\right)_{\tau}. \tag{5.24}$$

As with pressure explicit multiparameter equations of state, the results have to be considered effective virial coefficients. Contributions from exponential and Gaussian bell shaped terms are implicitly considered by their contribution to the zero density limit of the corresponding derivative in Equations (5.22)–(5.24). The expansion of a fundamental equation of state explicit in the reduced Helmholtz energy into a virial series fails if all of the d_i in Equations (5.17) and (5.19) are not whole numbers with $d_i > 0$.

5.4 The Performance of Multiparameter Equations of State

An analysis of the performance of multiparameter equations of state has to consider two aspects. The obvious one is the representation of available experimental data; the less obvious one is the behavior of the equation of state in regions where little or no data are available.

5.4.1 The Representation of Experimental Data

Reference equations of state are supposed to represent all of the available experimental data for thermodynamic properties within their respective experimental uncertainty. Thus, in theory a simple comparison with experimental data is sufficient to decide whether the criterion for the status of a reference equation of state is met. In practice the decision is more complicated, because experimental data are sometimes less accurate than the authors claim. Detailed assessments of the uncertainty of relevant data sets have to be established by comparisons between different data sets, with the claimed uncertainty as the starting point of the assessment. Very often comparisons include an analysis of inconsistencies between data sets for different thermodynamic properties. By definition, equations of state yield consistent results for different properties—systematic deviations in an experimental data set for one property affect the ability of an equation of state to accurately represent this data set together with accurate data sets for other properties in the same temperature and density range. This way, systematic deviations in data sets can be discovered even if no other experimental data are available for the same property. In any case the accuracy of reference equations of state depends on the accuracy of the experimental data available for the corresponding fluid.

Multiparameter equations of state do not necessarily aim for reference status; to achieve a level of accuracy considered sufficient for a certain application is a legitimate scientific goal as well. The term "technical equations of state" is sometimes used to refer to this kind of multiparameter equation of state. Certain types of older pressure explicit formulations[7–10] may be considered technical equations of state, even though they aim for a level of accuracy that was not defined by the authors. At that point in time the development of reference equations of state was hardly possible due to numerical and methodical limitations. More recently a new generation of technical equations of state was published.[26–29] The accuracy goals specified for these equations of state aim at meeting demands of advanced technical applications. However, this new class of fundamental equations illustrates the dilemma of multiparameter equations of state—though they hardly come up to the specified level of accuracy for some fluids with very restricted experimental data sets, they can be considered reference equations of state for these fluids. The available experimental data do not allow for a more accurate description of thermodynamic properties. State of the art multiparameter equations of state are numerically more stable than older formulations and need less experimental data to be fitted to, but still systematic experimental campaigns are mandatory for an accurate description of the properties of a fluid. The fact that the number of laboratories working on accurate measurements of thermodynamic properties declines limits our ability to extend the group of fluids for which thermodynamic properties can be modeled with high accuracy.

To give an example of the accuracy that can be achieved for fluids with a very good database, Figure 5.1 shows the uncertainty of densities calculated from the current reference equation of state for nitrogen,[34] as claimed by the authors. In the temperature range from the melting line to 523 K and at pressures up to 30 MPa, the uncertainty of calculated densities is 0.05% or less. The lowest uncertainty is claimed in the temperature range 273 K to 353 K at pressures up to 12 MPa. In this typical "calibration range" the uncertainty of calculated nitrogen densities is 0.01%. At pressures above 30 MPa the uncertainty of calculated densities increases, but it remains within 0.6% up to 2200 MPa.

In the critical region it is hardly possible to define uncertainties in density. Due to the fact that $(\partial p/\partial \rho)_T$ becomes zero at the critical point, small

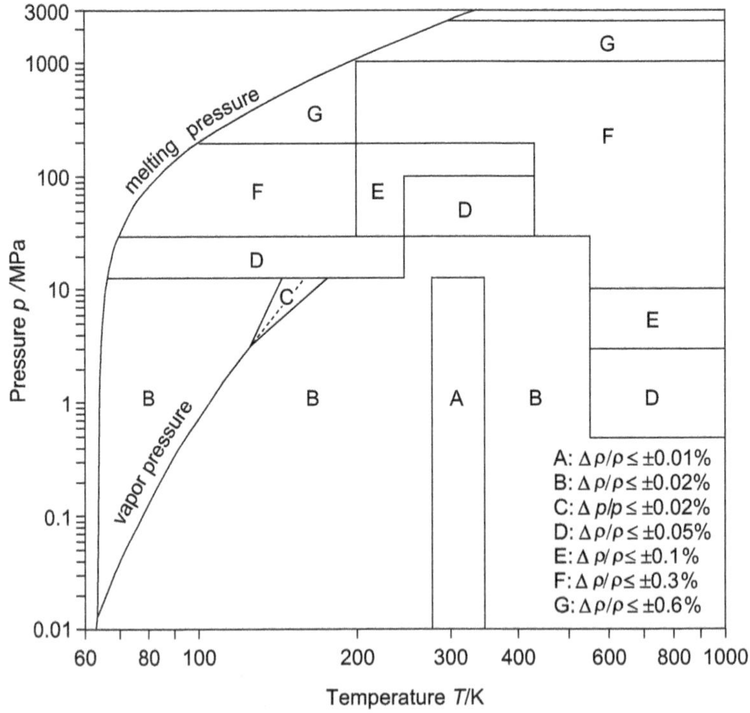

Figure 5.1 Uncertainty of densities calculated from the current reference equation of state for nitrogen.[34]

experimental uncertainties in pressure result in increasingly large uncertainties in density when approaching the critical point. The uncertainty in density is dominated by the distance to the critical point. However, an uncertainty in pressure can reasonably be defined in the critical region. Figure 1 gives an uncertainty of 0.02% in pressure for the extended critical region. The resulting uncertainty in density can be calculated by dividing the uncertainty in pressure through $(\partial p/\partial \rho)_T$ at the point of interest. Likewise, uncertainties in pressure become very large at a given temperature and density in the liquid region, because $(\partial p/\partial \rho)_T$ is very large in the liquid.

To illustrate typical deviations between measured and calculated densities, Figure 5.2 compares selected highly accurate density data with values calculated from different equations of state. The plotted data cover the range from liquid at a temperature of 66 K, which is just above the triple-point temperature, to the supercritical region at a temperature of 520 K, corresponding to more than four times the critical temperature. Pressures reach up to 30 MPa. The reference equation of Span *et al.*[34] represents the data well within the uncertainties discussed above. The older equations show significantly larger deviations, even though they were refitted to a data set containing the plotted data. The Bender[10]-type equation represents most of the data with an accuracy that can be considered sufficient for most technical applications. Starling[8]- and BWR[6]-type equations show unacceptably large deviations.

In this example the accuracy of the equations of state seems to depend directly on the number of fitted coefficients. It is true that rather large numbers of fitted coefficients are required to describe fluids with very high accuracy. However, the development of algorithms[23,50,51] for the optimization of functional forms (*i.e.*, for the selection of the parameters I_{POL}, I_{EXP}, t_{i}, d_{i}, p_{i},. . .) has decoupled the number of fitted coefficients and the accuracy of the resulting equation of state.

This fact is shown in Figure 5.3. The equations of state[44,45] developed by the group around Ely were considered reference equations of state for carbon dioxide, before a new generation of highly accurate experimental data became available. Span and Wagner[36] considered these new data when they developed the current reference equation of state for carbon dioxide. The observed increase in accuracy cannot be explained by a moderately larger number of fitted coefficients; the functional form of the equation by Span and Wagner was optimized using a modified version of the algorithm by Setzmann and Wagner.[50] The impact of this optimization process becomes obvious in the fourth diagram in Figure 5.3. With just 12 fitted coefficients the technical equation of state by Span and Wagner[28] describes the density of carbon dioxide as accurately as the older reference equations of state with 32 fitted coefficients. Using the latest fitting techniques,[19] which allow for a direct nonlinear fit of temperature exponents t_i, the number of terms in accurate equations of state can be reduced even further, see *e.g.* ref. 35.

Vapor pressures and densities of the saturated liquid and vapor can, in principle, be represented with the same accuracy as properties of the

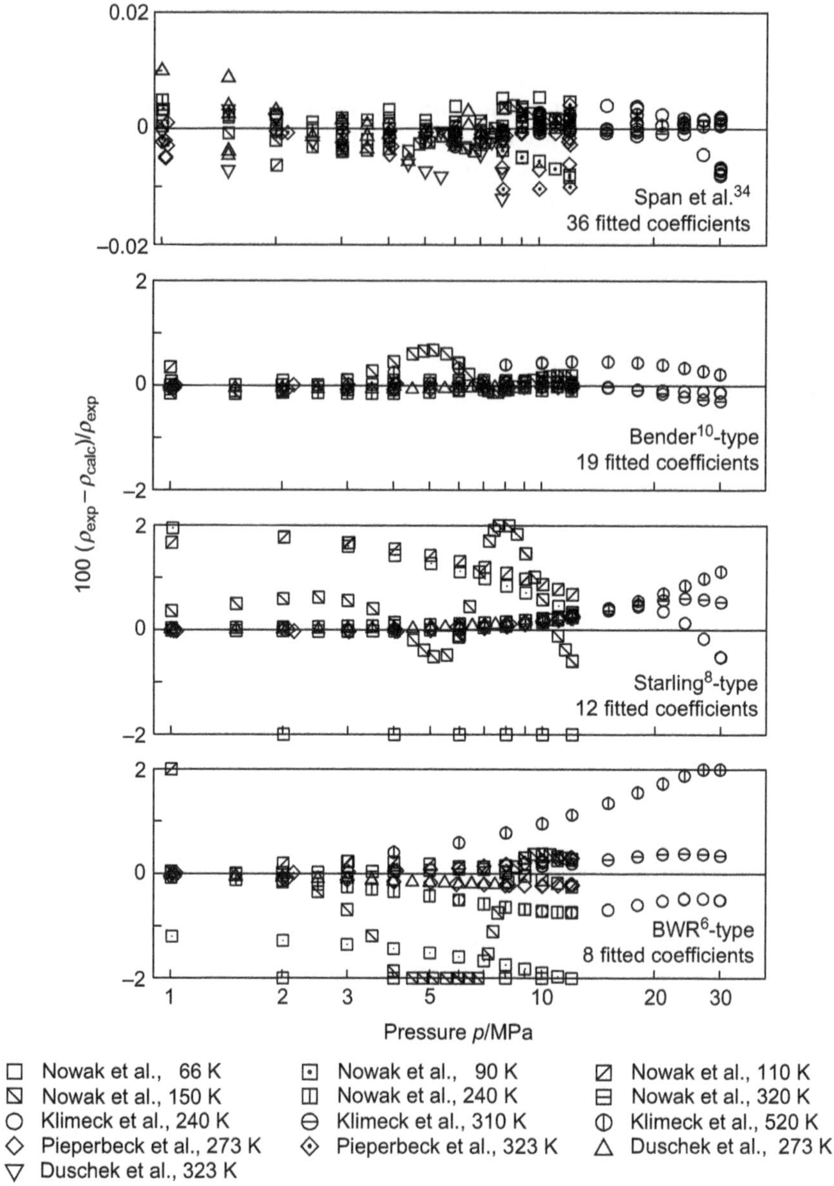

Figure 5.2 Deviations between highly accurate density data[40–43] for nitrogen and densities calculated from different equations of state, which were all fitted to the same data set.

homogeneous regions. Figure 5.4 shows the representation of highly accurate data for thermal properties on the phase boundary of carbon dioxide. The zero line corresponds to data calculated from the reference

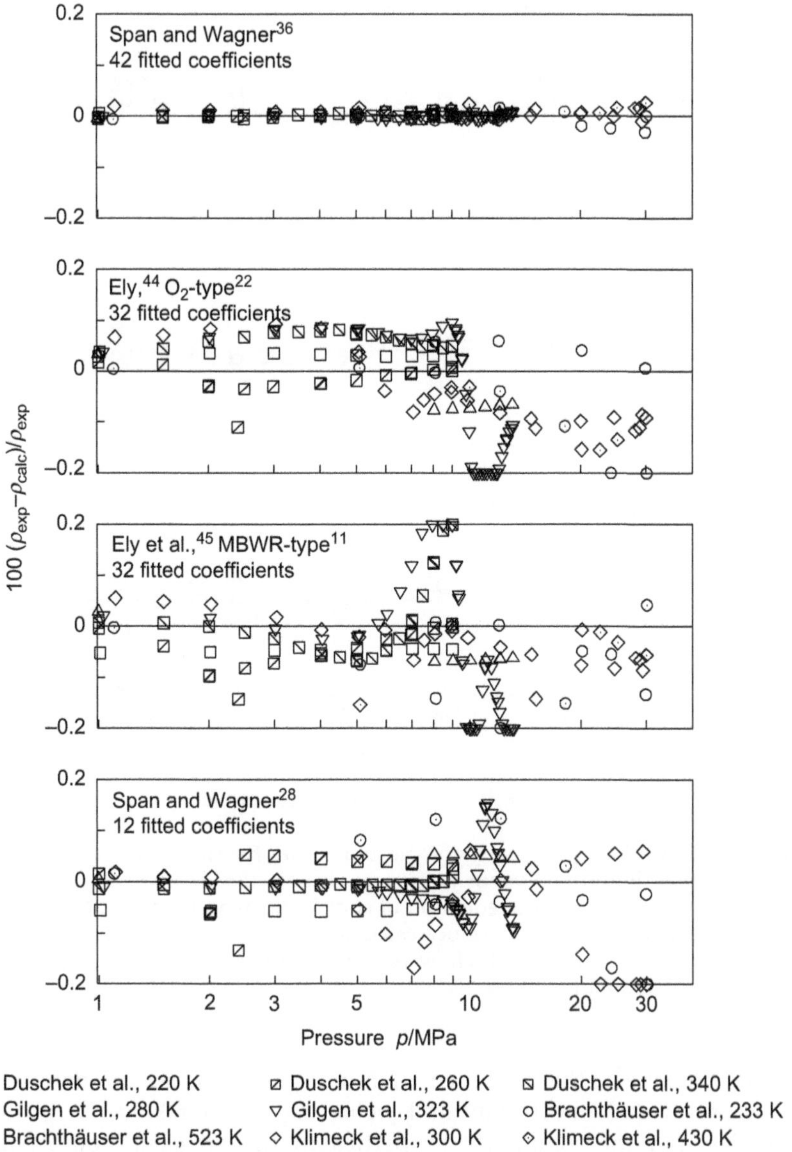

Figure 5.3 Deviations between highly accurate density data[46–49] for carbon dioxide and densities calculated from different equations of state.

equation of state for carbon dioxide[36] by solving the phase equilibrium condition. The dashed line represents values calculated from simple auxiliary equations,[36] which describe only the property they were fitted to as a simple empirical function of temperature. The average deviation between auxiliary equations and data is smaller than between the reference equation

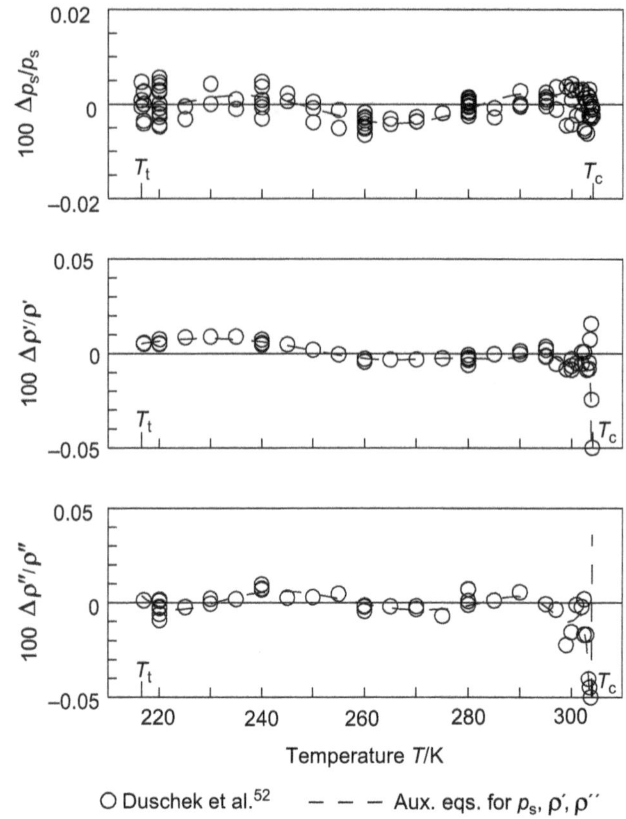

Figure 5.4 Deviations between highly accurate data for vapor pressure and saturated liquid and vapor density of carbon dioxide and values calculated from the reference equation of state by Span and Wagner.[36] The dashed lines correspond to values calculated from auxiliary equations for $p_s(T)$, $\rho'(T)$ and $\rho''(T)$, which were fitted directly to selected data for the corresponding property.[52]

and data. However, the reference equation still describes the data within their experimental uncertainty and yields consistent results for all three properties and for volumetric properties in the adjacent homogeneous regions. Whenever the consistency of results is relevant, phase equilibrium data should be calculated from equations of state and not from auxiliary equations.

5.4.2 The Extrapolation Behavior

The behavior of equations of state in regions where little or no experimental data are available is usually discussed as extrapolation behavior, even though problems may be related to interpolation or to the representation of derived properties as well. In the past, multiparameter equations of state were considered very unreliable in regions where they could not be fitted to a sufficient amount of accurate data. Unphysical behavior outside of the fitted range resulted in problems in calculations based on multiparameter equations of state. These problems were systematically addressed in the early 1990s.[53,54]

One important finding was that extrapolation behavior and results for derived properties like heat capacities or compressibility are closely related to the numerical stability of equations of state. Multiproperty fits and the use of optimized functional forms significantly improved extrapolation behavior and reliability of multiparameter equations of state. A typical example is shown in Figure 5.5. For *n*-octane reliable experimental data are available in the liquid and liquid-like supercritical region. The original Bender-type[10] equation of state published by Polt[55] shows obviously physically unreasonable behavior just outside of the range where experimental data are available. For derived properties unreasonable behavior occurs within the fitted range. A multiproperty refit of the 19 coefficients of the Bender-type equation

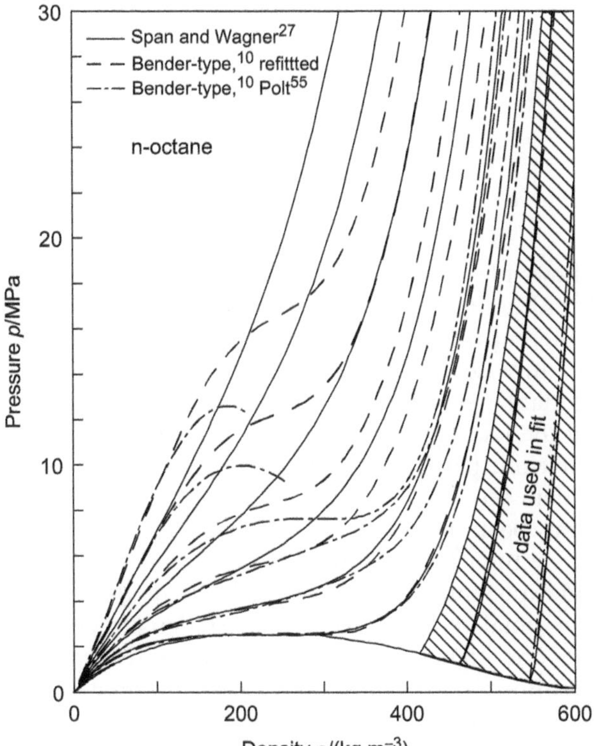

Figure 5.5 p,ρ-diagram of n-octane as calculated with three different equations of state. The dashed area corresponds to the p,ρ-range in which reliable experimental data are available.

avoids this problem, but still the equation shows unreasonable behavior for far supercritical isotherms. The equation of state by Span and Wagner describes non-polar fluids with somewhat higher accuracy than Bender-type equations of state with just 12 fitted coefficients. It uses a simultaneously optimized functional form,[26,51] which drastically reduces the intercorrelation between the different terms. Due to this increased numerical stability the equation yields physically reasonable results even far outside of the region where data are available.

Beside the numerical stability of functional forms the representation of so-called "ideal curves" of the compression factor was emphasized as an important tool for the validation of reasonable extrapolation behavior.[53,54] Along these curves the following conditions hold:

(Classical) Ideal curve $Z=1$

Boyle curve $$\left(\frac{\partial Z}{\partial \rho}\right)_T=0$$

Joule–Thomson inversion curve $$\left(\frac{\partial Z}{\partial \rho}\right)_p=0$$

Joule inversion curve $$\left(\frac{\partial Z}{\partial T}\right)_\rho=0$$

Figure 5.6 shows ideal curves of nitrogen as calculated with the equation by Span *et al.*[34] Although maxima and intersections with the axis $p=0$ are different for different fluids the general shape of the curves is similar and very sensitive to unreasonable curvatures—even equations of state that show physically reasonable behavior in a p,ρ-diagram may fail to properly represent the plot of the ideal curves. A reasonable shape of the ideal curves is a demanding criterion, particularly for equations of state describing fluids with a limited data set.

Finally, it has been pointed out[19,54] that the extrapolation behavior towards extremely high temperatures and densities can be traced back to coefficients and exponents of few or even of a single term in an equation of state. The polynomial term with the highest density power d_i is asymptotically leading for high densities; its temperature exponent t_i and coefficient n_i determine the behavior of an equation of state in the high density limit. If intercorrelations between different polynomial terms with high values of d_i (3 to 4, higher exponents should not be used at all) are avoided, equations of state can be constrained to reasonable extrapolation behavior up to arbitrary temperatures and pressures, which exceed the chemically stable and practically relevant range of states by far. As an example,

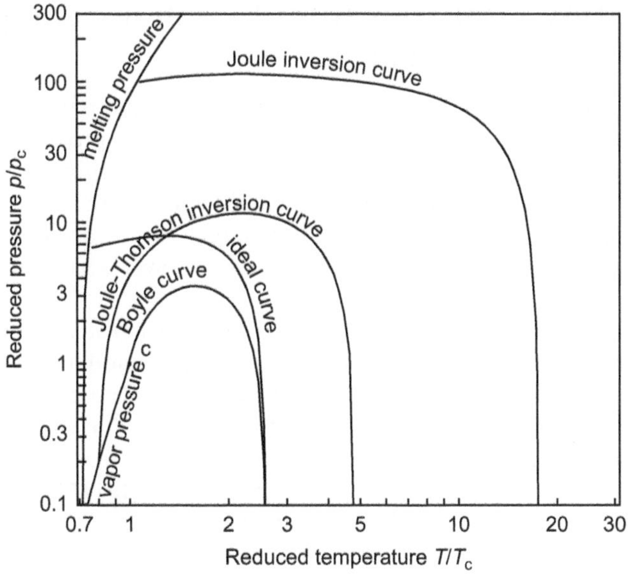

Figure 5.6 Ideal curves of nitrogen as calculated with the equation of state by Span *et al.*[34]

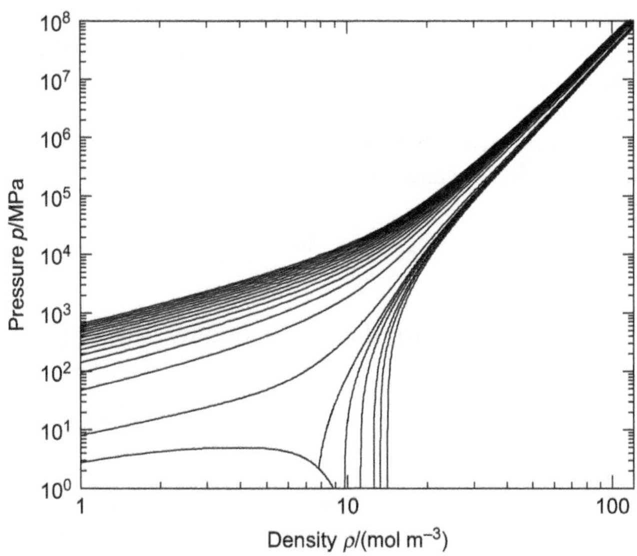

Figure 5.7 p,ρ-Diagram of benzene, as calculated with the equation of state by Thol *et al.*[56]

Figure 5.7 shows a p,ρ-diagram calculated with the benzene equation of state by Thol *et al.*[56] Experimental data are available up to 720 K for benzene; the highest isotherm shown in Figure 5.7 corresponds to 10^6 K.

5.5 Mixture Properties from Helmholtz Energy Equations of State

There are many practical models for calculating properties of mixtures of two or more fluids. A mixture equation of state should provide an accurate representation of the thermodynamic properties of the mixture over a wide range of compositions, including liquid and vapor properties. Virial equations of state derivable from statistical mechanics can be used to express the deviations from the perfect gas equation as a power series in density or volume.

Lemmon,[57] Lemmon and Jacobsen,[58] Tillner-Roth *et al.*,[59] and Lemmon and Tillner-Roth[60] developed generalized mixture models based on the equations of state for the pure fluids in the mixture and an excess function to account for the interaction between different species. The work of Lemmon and Jacobsen[61] documents the equations currently in use for mixtures of difluoromethane (R-32), pentafluoroethane (R-125), 1,1,1,2-tetrafluoroethane (R-134a), 1,1,1-trifluoroethane (R-143a), and 1,1-difluoroethane (R-152a). The work of Kunz *et al.*[62,63] expanded the earlier models of Lemmon[57] and coworkers[58,60] to include additional coefficients in the reducing parameters and equation of state, updated fits for mixtures of methane through butane or with nitrogen and carbon dioxide, and new fits for mixtures with alkanes of higher molar mass in order to precisely represent the thermodynamic properties of natural gas systems. Their work includes 21 components found in natural gas, including dilutants such as hydrogen, helium, and gaseous water. The equation can also be used to calculate the properties of moist air.

The different approaches have in common that the Helmholtz energy for mixtures of fluids can be calculated with the equation

$$a = a^{\text{cor.states}} + a^{\text{E}}, \tag{5.25}$$

where the Helmholtz energy for the corresponding states contribution is

$$a^{\text{cor.states}} = \sum_{i=1}^{n} x_i \left[a_i^0(\rho, T) + a_i^{\text{r}}(\delta, \tau) + RT \ln x_i \right]. \tag{5.26}$$

In this equation, n is the number of components in the mixture, a_i^0 is the ideal gas Helmholtz energy for component i, and a_i^{r} is the pure fluid residual Helmholtz energy of component i. Equation (5.26) represents a corresponding states approach rather than an ideal mixture because the reduced density δ and the inverse reduced temperature τ are formulated with a reducing temperature and a reducing density of the mixture (see Equations 5.30–5.33) and not with the T_{ci} and ρ_{ci} of the pure components.

The excess contribution to the Helmholtz energy from mixing is

$$\frac{a^{\mathrm{E}}}{RT}=\alpha^{\mathrm{E}}=\sum_{i=1}^{n-1}\sum_{j=i+1}^{n}x_i x_j F_{ij}\left[\sum_{k=1}^{K_{\mathrm{pol}}}N_k\delta^{d_k}\tau^{t_k}\right.$$
$$\left.+\sum_{k=K_{\mathrm{pol}}+1}^{K_{\mathrm{pol}}+K_{\mathrm{exp}}}N_k\delta^{d_k}\tau^{t_k}\exp\left(-\eta_k(\delta-\varepsilon_k)^2-\beta_k(\delta-\gamma_k)\right)\right], \tag{5.27}$$

where the coefficients and exponents were obtained from nonlinear regression to experimental mixture data. The parameter F_{ij} is used in a generalization to relate the excess properties of one binary mixture to those of another. With this parameter, the same set of mixture coefficients can be used for several binary mixtures in the model. Several binary pairs do not use the generalized parameter and instead have binary specific excess functions for the coefficients and exponents. These pairs include methane + nitrogen, methane + carbon dioxide, methane + ethane, methane + propane, methane + hydrogen, nitrogen + carbon dioxide, and nitrogen + ethane. The experimental data for these binary mixtures were sufficient such that individualized equations could be developed for each pair.

All single-phase thermodynamic properties can be calculated from the Helmholtz energy with the relations

$$\alpha^0=\sum_{i=1}^{n}x_i\left[\frac{a_i^0(\rho,T)}{RT}+\ln x_i\right] \tag{5.28}$$

and

$$\alpha^{\mathrm{r}}=\sum_{i=1}^{n}x_i\alpha_i^{\mathrm{r}}(\delta,\tau)+\alpha^{\mathrm{E}}(\delta,\tau,\mathbf{x}), \tag{5.29}$$

whereby derivatives are taken at constant composition. The reduced values of density and temperature for the mixture are

$$\delta=\rho/\rho_{\mathrm{r}} \tag{5.30}$$

and

$$\tau=T_{\mathrm{r}}/T, \tag{5.31}$$

where ρ and T are the mixture density and temperature, and ρ_{r} and T_{r} are the reducing values,

$$\frac{1}{\rho_{\mathrm{r}}(\mathbf{x})}=\sum_{i=1}^{n}\sum_{j=1}^{n}x_i x_j\beta_{v,ij}\gamma_{v,ij}\frac{x_i+x_j}{\beta_{v,ij}^2x_i+x_j}\frac{1}{8}\left(\frac{1}{\rho_{\mathrm{c},i}^{1/3}}+\frac{1}{\rho_{\mathrm{c},j}^{1/3}}\right)^3 \tag{5.32}$$

and

$$T_r(\mathbf{x}) = \sum_{i=1}^{n}\sum_{j=1}^{n} x_i x_j \beta_{T,ij}\gamma_{T,ij}\frac{x_i + x_j}{\beta_{T,ij}^2 x_i + x_j}\sqrt{T_{c,i}T_{c,j}}. \tag{5.33}$$

The parameters β and γ are used to define the shapes of the reducing temperature lines and reducing density lines. These reducing parameters are not the same as the critical parameters of the mixture and are determined simultaneously in the nonlinear fit of experimental data with the other parameters of the mixture model.

In principle, multiparameter mixture models of the described type allow for accuracies comparable to those of pure component reference equations of state. For natural gases, highly accurate experimental data are available in the temperature and pressure range relevant for pipelining applications. These data can be represented by state of the art mixture models[58,62,63] within $\Delta\rho/\rho \leq 0.1\%$; an example is shown in Figure 5.8. However, one has to be aware that highly accurate experimental data are available only for a few mixtures.

Beside the description of refrigerant mixtures and of mixtures like humid air and natural gases, the development of multiparameter mixture models

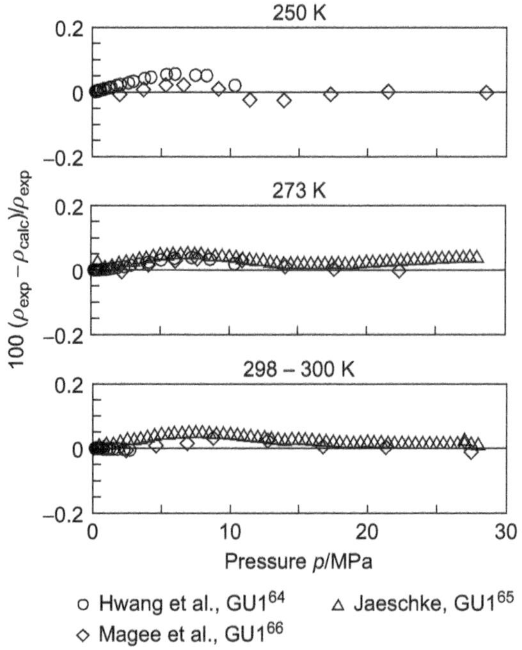

Figure 5.8 Deviations between highly accurate experimental data for a quintic natural gas-like mixture and values calculated from the model by Lemmon and Jacobsen.[58] For the composition of the mixture indicated as GU1 see the cited references 64–66.

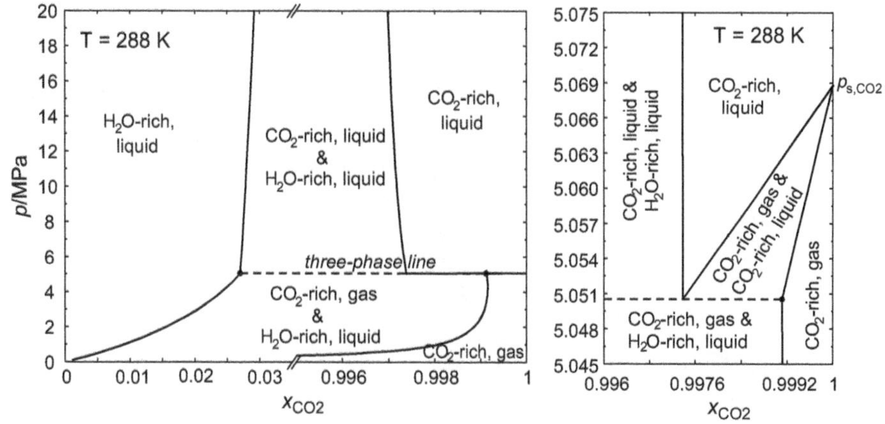

Figure 5.9 Phase equilibria in the system CO_2–H_2O as calculated by the mixture model of Gernert.[67]

has recently been driven by the need to accurately describe CO_2-rich mixtures for carbon capture and storage applications. This application implies the prediction of complex phase equilibria, including three-phase equilibria with two liquid phases (VLLE). Figure 5.9 gives an example for the description of three-phase equilibria in the system carbon dioxide–water. Multiparameter mixture models accurately predict such systems, even though they were fitted only to experimental data for homogeneous states and for vapor–liquid phase equilibria (VLE).

Phase equilibria in the system CO_2–H_2O become even more complex when solid phases (S) have to be considered at low temperatures. In principle, phase equilibria with solid phases (VSE, LSE) can be described with a closed formulation just as VLE and VLLE, provided an equation of state is also valid for the solid phase and satisfies the corresponding phase equilibrium conditions.[68] However, this approach results in high demands on the equations of state or the mixture model, respectively, and has not been realized in combination with accurate multiparameter models yet. Instead it is more convenient to describe the solid phase by an independent fundamental equation of state, which is typically formulated in terms of the Gibbs energy, $g(T,p)$, or the chemical potential of components, $\mu_i(T,p,\boldsymbol{x})$. If the models for the solid phases are constrained to the same reference states of enthalpy and entropy as the fluid model, phase equilibria can be calculated by intersection of the models. Figure 5.10 shows the resulting low temperature phase equilibria in the system CO_2–H_2O, as calculated using the model by Gernert[67] for the fluid phases, the model by Jäger *et al.*[69] for CO_2 hydrates, the model by Feistel and Wagner[70] for water ice, and the model by Jäger and Span[71] for dry ice (solid CO_2).

Although much has been achieved regarding the representation of mixture properties with multiparameter models, the models common today still have

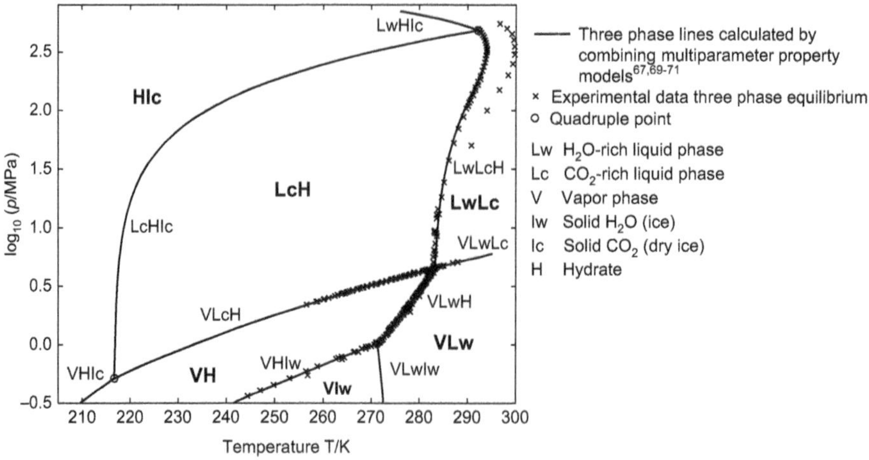

Figure 5.10 Low temperature phase equilibria in the system CO_2–H_2O. The plotted lines represent three-phase equilibrium lines. The indicated areas are two-phase regions. Composition dependences are omitted in this p,T-diagram.

shortcomings from a theoretical point of view. For most mixtures the available experimental data set is not sufficient to allow for the development of accurate empirical models. Models have to be developed that allow for a description of multicomponent mixtures based on a combination of accurate multiparameter approaches for well measured components and subsystems with numerically less demanding physically motivated approaches for less well measured components and sub-systems. It seems almost certain that the development of empirical multiparameter property models for mixtures has not been concluded yet.

References

1. E. W. Lemmon, M. L. Huber and M. O. McLinden, *REFPROP 9.1. NIST Standard Reference Database 23*, National Institute for Standards and Technology, 2013.
2. H. J. Kretzschmar, *Property Libraries for Calculating Heat Cycles, Boilers, Turbines and Refrigerators*. http://thermodynamik-zittau.de/, downloaded January 2014.
3. R. Span, T. Eckermann, J. Gernert, A. Jäger, M. Thol and S. Herrig, *TREND 1.1. Thermodynamic Reference and Engineering Data*, Ruhr-Universität Bochum, Bochum, Germany, 2013.
4. H. Kamerlingh Onnes, Expression of the equation of state of gases and liquids by means of series, in *Proceedings of the Royal Netherlands Academy of Arts and Sciences (KNAW)*, 4, 1901–1902, Amsterdam, 1902.
5. V. V. Altunin and O. G. Gadetskii, *Therm. Eng.*, 1971, **18**(3), 120(English translation).

6. M. Benedict, G. B. Webb and L. C. Rubin, *J. Chem. Phys.*, 1940, **8**, 334.
7. T. R. Strobridge, *The thermodynamic properties of nitrogen from 64 to 300° K between 0.1 and 200 atmospheres*. National Bureau of Standards, Technical Note 129, Washington, 1962.
8. K. E. Starling, *Fluid thermodynamic properties for light petroleum systems*. Gulf Publishing, Houston, 1973.
9. B. I. Lee and M. G. Kessler, *AIChE J.*, 1975, **21**, 510.
10. E. Bender, Equations of state exactly representing the phase behavior of pure substances, in *Proc. 5th Symp. Thermophys. Prop.*, ed. C. F. Bonila, ASME, New York, 1970, pp. 227–235.
11. R. T. Jacobsen and R. B. Stewart, *J. Phys. Chem. Ref. Data*, 1973, **2**, 757.
12. W. Wagner, *Eine thermische Zustandsgleichung zur Berechnung der Phasengleichgewichte flüssig-gasförmig für Stickstoff*. Dissertation, TU Braunschweig, Braunschweig, 1970.
13. R. D. McCarty, *Provisional thermodynamic functions for helium 4 for temperatures from 2 to 1500 K with pressures to 100 MN/m² (1000 atmospheres)*. Nat. Bur. Stand. Report 9762, Boulder CO, 1970.
14. J. Ahrendts and H. D. Baehr, *Forsch. Ingenieurwes.*, 1979, **45**, 1.
15. J. Ahrendts and H. D. Baehr, *Forsch. Ingenieurwes.*, 1979, **45**, 51.
16. J. Ahrendts and H. D. Baehr, *Int. Chem. Eng.*, 1981, **21**, 557.
17. J. Ahrendts and H. D. Baehr, *Int. Chem. Eng.*, 1981, **21**, 572.
18. R. Span, *Multiparameter equations of state – an accurate source of thermodynamic property data*, Springer, Berlin, 2000.
19. E. W. Lemmon and R. T Jacobsen, *J. Phys. Chem. Ref. Data*, 2005, **34**, 69.
20. R. Tillner-Roth, *Fundamental equations of state*, Shaker, Aachen, 1998.
21. ISO 31, *Quantities and units - Part 4: Heat*, International Organization for Standardization, Genève, 1992.
22. R. Schmidt and W. Wagner, *Fluid Phase Equilib.*, 1985, **19**, 175.
23. J. Ewers and W. Wagner, A method for optimizing the structure of equations of state and its application to an equation of state for oxygen, in *Proc. eighth symposium on thermophysical properties*, ed. J. V. Sengers, ASME, New York, 1982.
24. R. Tillner-Roth and H. D. Baehr, *J. Phys. Chem. Ref. Data*, 1994, **23**, 657.
25. R. Tillner-Roth and A. Yokozeki, *J. Phys. Chem. Ref. Data*, 1997, **26**, 1273.
26. R. Span and W. Wagner, *Int. J. Thermophys.*, 2003, **24**, 1.
27. R. Span and W. Wagner, *Int. J. Thermophys.*, 2003, **24**, 41.
28. R. Span and W. Wagner, *Int. J. Thermophys.*, 2003, **24**, 111.
29. E. W. Lemmon and R. Span, *J. Chem. Eng. Data*, 2006, **51**, 785.
30. U. Setzmann and W. Wagner, *J. Phys. Chem. Ref. Data*, 1991, **20**, 1061.
31. L. Haar, J. S. Gallagher and G. S. Kell, The anatomy of the thermodynamic surface of water: the formulation and comparison with data, in *Proc. eighth symposium on thermophysical properties*, ed. J. V. Sengers, ASME, New York, 1982.
32. C. Tegeler, R. Span and W. Wagner, *J. Phys. Chem. Ref. Data*, 1999, **28**, 779.

33. J. Smukala, R. Span and W. Wagner, *J. Phys. Chem. Ref. Data*, 2000, **29**, 1053.
34. R. Span, E. W. Lemmon, R. T Jacobsen, W. Wagner and A. Yokozeki, *J. Phys. Chem. Ref. Data*, 2000, **29**, 1361.
35. M. Richter, M. O. McLinden and E. W. Lemmon, *J. Chem. Eng. Data*, 2011, **56**, 3254.
36. R. Span and W. Wagner, *J. Phys. Chem. Ref. Data*, 1996, **25**, 1509.
37. W. Wagner and A. Pruß, *J. Phys. Chem. Ref. Data*, 2002, **31**, 387.
38. R. T. Jacobsen, S. G. Penoncello, E. W. Lemmon and R. Span, Multiparameter equations of state, in *Equations of state for fluids and fluid mixtures*, ed. J. V. Sengers, R. F. Kayser, C. J. Peters and H. J. White Jr., Elsevier, Amsterdam, 2000.
39. E. W. Lemmon and R. Span, Multiparameter Equations of State for Pure Fluids and Mixtures, in *Applied Thermodynamics of Fluids*, ed. A. R. H. Goodwin, J. V. Sengers and C. Peters, International Union of Pure and Applied Chemistry, Royal Society of Chemistry, Cambridge, 2010, ch. 12.
40. P. Nowak, R. Kleinrahm and W. Wagner, *J. Chem. Thermodyn.*, 1997, **29**, 1137.
41. J. Klimeck, R. Kleinrahm and W. Wagner, *J. Chem. Thermodyn.*, 1998, **30**, 1571.
42. N. Pieperbeck, R. Kleinrahm and W. Wagner, *J. Chem. Thermodyn.*, 1991, **23**, 175.
43. W. Duschek, R. Kleinrahm and W. Wagner, *J. Chem. Thermodyn.*, 1988, **20**, 1069.
44. J. F. Ely, An equation of state model for pure CO2 and CO2 rich mixtures, in *Proc. 65th annual convention of the Gas Processor Association*, San Antonio, TX, 1986.
45. J. F. Ely, J. W. Magee and B. C. Bain, *J. Chem. Thermodyn.*, 1989, **21**, 879.
46. W. Duschek, R. Kleinrahm and W. Wagner, *J. Chem. Thermodyn.*, 1990, **22**, 827.
47. R. Gilgen, R. Kleinrahm and W. Wagner, *J. Chem. Thermodyn.*, 1992, **24**, 1243.
48. K. Brachthäuser, R. Kleinrahm, H. W. Lösch and W. Wagner, *Fortschr.-Ber. VDI*, 8, vol. 371, VDI, Düsseldorf, 1993.
49. J. Klimeck, R. Kleinrahm and W. Wagner, *J. Chem. Thermodyn.*, 2001, **33**, 251.
50. U. Setzmann and W. Wagner, *Int. J. Thermophys.*, 1989, **10**, 1103.
51. R. Span, H. J. Collmann and W. Wagner, *Int. J. Thermophys.*, 1998, **19**, 491.
52. W. Duschek, R. Kleinrahm and W. Wagner, *J. Chem. Thermodyn.*, 1990, **22**, 841.
53. K. M. de Reuck, *Extrapolation of Accurate Equations of State Outside the Range of the Experimental Data, 1st draft*, personal communication to the participants of the 5th International Workshop on Equations of State, IUPAC Thermodynamic Tables Centre, London, 1991.
54. R. Span and W. Wagner, *Int. J. Thermophys.*, 1997, **18**, 1415.

55. A. Polt, *Zur Beschreibung thermodynamischer Eigenschaften reiner Fluide mit "Erweiterten BWR-Gleichungen"*, PhD Dissertation, Univ. Kaiserslautern, Kaiserslautern, 1987.
56. M. Thol, E. W. Lemmon and R. Span, *High Temp. – High Pressures*, 2012, **41**, 467.
57. E. W. Lemmon, *A Generalized Model for the Prediction of the Thermodynamic Properties of Mixtures Including Vapor–Liquid Equilibrium*, PhD dissertation, University of Idaho, Moscow, 1996.
58. E. W. Lemmon and R. T Jacobsen, *Int. J. Thermophys.*, 1999, **20**, 825.
59. R. Tillner-Roth, J. Li, A. Yokozeki, H. Sato and K. Watanabe, *Thermodynamic Properties of Pure and Blended Hydrofluorocarbon (HFC) Refrigerants*, Japan Society of Refrigerating and Air Conditioning Engineers, Tokyo, 1998.
60. E. W. Lemmon and R. Tillner-Roth, *Fluid Phase Equilib.*, 1999, **165**, 1.
61. E. W. Lemmon and R. T Jacobsen, *J. Phys. Chem. Ref. Data*, 2004, **33**, 593.
62. O. Kunz, R. Klimeck, W. Wagner and M. Jaeschke, The GERG-2004 Wide-Range Equation of State for Natural Gases and Other Mixtures, GERG TM15, *Fortschr.-Ber. VDI*, vol. 6, no. 557, 2007.
63. O. Kunz and W. Wagner, *J. Chem. Eng. Data*, 2012, **57**, 3032.
64. C.-A. Hwang, P. P. Simon, H. Hou, K. R. Hall, J. C. Holste and K. N. Marsh, *J. Chem. Thermodyn.*, 1997, **29**, 1455.
65. M. Jaeschke, *Results of refractive index measurements on natural gas like mixtures*, private communication, Ruhrgas AG, Dorsten, 1997.
66. J. W. Magee, W. M. Haynes and M. J. Hiza, *J. Chem. Thermodyn.*, 1997, **29**, 1439.
67. J. Gernert, *A New Helmholtz Energy Model for Humid Gases and CCS Mixtures*, PhD Dissertation, Ruhr-Universität Bochum, Bochum, 2013.
68. A. Yokozeki, *Fluid Phase Equilib.*, 2004, **222–223**, 55.
69. A. Jäger, V. Vinš, J. Gernert, R. Span and J. Hrubý, *Fluid Phase Equilib.*, 2013, **338**, 100.
70. R. Feistel and W. Wagner, *J. Mar. Res.*, 2005, **63**, 95.
71. A. Jäger and R. Span, *J. Chem. Eng. Data*, 2012, **57**, 590.

CHAPTER 6

Virial Coefficients

J. P. MARTIN TRUSLER

Department of Chemical Engineering, Imperial College London, South Kensington Campus, London SW7 2AZ, United Kingdom
Email: m.trusler@imperial.ac.uk

6.1 Introduction

The virial equation of state expresses the pressure p of a gas as a power-series expansion in the inverse of the molar volume V_m as follows:

$$p = (RT/V_m)(1 + B/V_m + C/V_m^2 + \ldots). \tag{6.1}$$

Here, R is the universal gas constant, T is the temperature, and 1, B, C, $\cdots$ are the first, second, third...virial coefficients. The leading term in this equation is, of course, the equation of state of the perfect gas; thus, the second and higher virial coefficients express the departure of the real gas properties from those of the perfect gas. Crucially, the virial coefficients depend only upon temperature and (in a mixture) the composition of the gas.

The convergence properties of the virial equation of state are such that it is useful only for gases. The radius of convergence is not generally known but, as a matter of practicality, the virial equation is usually truncated after the second or third virial coefficient, thereby limiting its range of applicability to densities well below the critical density.

Although the virial equation of state may be thought of simply as a power series expansion of the pressure about the perfect-gas limit, it has a more fundamental basis in statistical mechanics that yields expressions for the virial coefficients in terms of intermolecular potential-energy functions. These expressions show that B is related to interactions between isolated

Volume Properties: Liquids, Solutions and Vapours
Edited by Emmerich Wilhelm and Trevor M. Letcher

Published by the Royal Society of Chemistry, www.rsc.org

pairs of molecules, while C relates to interactions involving three isolated molecules, D to interactions between four molecules and so on. The statistical–mechanical analysis shows that the second and third virial coefficients of a mixture containing ν different components are given by

$$B = \sum_{i=1}^{\nu}\sum_{j=1}^{\nu} x_i x_j B_{ij} \tag{6.2}$$

$$C = \sum_{i=1}^{\nu}\sum_{j=1}^{\nu}\sum_{k=1}^{\nu} x_i x_j x_k C_{ijk}, \tag{6.3}$$

where x_i denotes the mole fraction of component i in the mixture. In Equation (6.2), B_{ii} is the second virial coefficient of pure i and B_{ij} is called the second interaction virial coefficient for the unlike pair of species i and j. Similarly, in Equation (6.3), C_{iii} is the third virial coefficient of pure component i, while C_{ijk} is an interaction third virial coefficient involving three species i, j and k. All of the virial coefficients appearing on the right-hand sides of Equations (6.2) and (6.3) are functions of temperature only. When the virial equation is truncated after the term in C, one has a closed-form analytical approximation to the equation of state that is cubic in both molar volume and mole fractions; and, when truncated after the term in B, the equation is quadratic in both molar volume and mole fractions.

As with other common equations of state, the independent variables in the virial equation are temperature, molar volume and (implicitly) composition. Thus, when temperature, pressure and composition are specified, it is necessary to solve Equation (6.1) for the corresponding molar volume. Alternatively, a power series for $V_m(T, p)$ may be written,

$$V_m = (RT/p)[1 + B'p + C'p^2 + \ldots], \tag{6.4}$$

and the coefficients B', C' $\cdots$ appearing in it may be expressed in terms of B, C, $\cdots$ by means of series inversion:

$$\left.\begin{aligned} B' &= B/(RT) \\ C' &= (C - B^2)/(RT)^2 \end{aligned}\right\}. \tag{6.5}$$

In the remainder of this chapter, some of the properties of the virial equation itself are expanded upon and the determination of the virial coefficients for both real and model systems is described.

6.2 Statistical Mechanical Analysis

The derivation of the virial equation of state is a standard problem in statistical mechanics and only the key results of such an analysis are discussed here; for further details, the reader is referred to the literature.[1,2] The problem may be approached at various levels of approximation, the most fundamental of which is an assumption of rigid molecules. Except for very light molecules such as H_2 and He, classical statistical mechanics is applicable, possibly with the addition of small quantum corrections at low temperatures. If attention is further restricted to a gas composed of identical spherical molecules then the second virial coefficient is related to the potential-energy function $u_{12}(r_{12})$ between an isolated pair of molecules as follows:

$$B_{12} = -2\pi N_A \int_0^\infty f_{12} r_{12}^2 dr_{12}, \tag{6.6}$$

where

$$f_{ij} = \exp(-u_{ij}/k_B T) - 1, \tag{6.7}$$

r_{12} is the distance between the centres of mass, N_A is Avogadro's constant and k_B is Boltzmann's constant. The third virial coefficient of such a system is related to the intermolecular potential-energy function of a triplet of molecules, $u_{123}(r_{12}, r_{13}, r_{23})$, and this is often assumed to be given by the sum of the pair-interaction energies:

$$u_{123} = u_{12}(r_{12}) + u_{13}(r_{13}) + u_{23}(r_{23}). \tag{6.8}$$

With this pair additivity approximation, the third virial coefficient is given by

$$C_{123} = -(8^2 N_A^2 / 3) \int f_{12} f_{13} f_{23}\, r_{12} r_{13} r_{23} dr_{12} dr_{13} dr_{23}, \tag{6.9}$$

where the integral is over all possible values of r_{12}, r_{13} and r_{23} that form a triangle. Higher-order virial coefficients can be expressed in terms of similar, but more complicated, cluster integrals involving, for the mth virial coefficient, m molecules.

Both Equations (6.6) and (6.9) may be applied to unlike molecules to obtain the interaction virial coefficients appearing in Equations (6.2) and (6.3) for mixtures. In these cases, the pair potentials u_{ij} refer to the relevant unlike interaction.

Many of the approximations inherent in Equations (6.6) and (6.9) may be relaxed to accommodate a more realistic representation of the molecules and their interactions.[2] For example, non-spherical molecules may be treated with orientation-dependent intermolecular potentials and the corresponding expressions for the virial coefficients then involve an unweighted average over all possible relative orientations of the molecules. Low-temperature quantum effects can be accounted for through the inclusion of quantum corrections and, ultimately, a fully non-classical quantum treatment may be adopted. Thus the statistical mechanics of the virial equation and of the virial coefficients may be considered completely developed, lacking only exact knowledge of the true intermolecular potentials.

6.3 Virial Coefficients of Model Systems

For certain simple forms of the intermolecular pair potential, Equations (6.6) and (6.9) lead to closed-form analytical expressions for B and C. As an example, the hard-core square-well potential may be considered. This potential is defined by

$$\left.\begin{array}{ll} u_{12}=\infty, & r_{12}<\sigma \\ u_{12}=-\varepsilon, & \sigma\leq r_{12}\leq\lambda\sigma \\ u_{12}=0, & r_{12}>\lambda\sigma \end{array}\right\}, \tag{6.10}$$

where σ is the diameter of the hard spherical cores, ε is the depth of the potential-energy well surrounding the hard cores, and $\lambda\sigma$ is the range of the potential-energy well. In terms of this simple model, B and C are given by:[3]

$$B=b_0\{1-(\lambda^3-1)\Delta\} \tag{6.11}$$

$$C=\frac{1}{8}b_0^2\{5-(\lambda^6-18\lambda^4+32\lambda^3-15)\Delta-(2\lambda^6-36\lambda^4+32\lambda^3+18\lambda^2-16)\Delta^2$$
$$-(6\lambda^6-18\lambda^4+18\lambda^2-6)\Delta^3\} \tag{6.12}$$

where $b_0=(2\pi NA\sigma^3/3)$ and $\Delta=\{\exp(\varepsilon/k_BT)-1\}$. Although simple, this model incorporates much of the physics of the problem and yields the correct temperature dependence of the virial coefficients. From Equation (6.11) and Figure 6.1, one can observe that B approaches the positive constant b_0 at very high temperatures but diverges towards $-\infty$ as $T\rightarrow 0$. Similarly, from Equation (6.12) and Figure 6.1, we see that C approaches $5b_0^2/8$ as $T\rightarrow\infty$ and diverges towards $-\infty$ as $T\rightarrow 0$; in between these limits it passes through

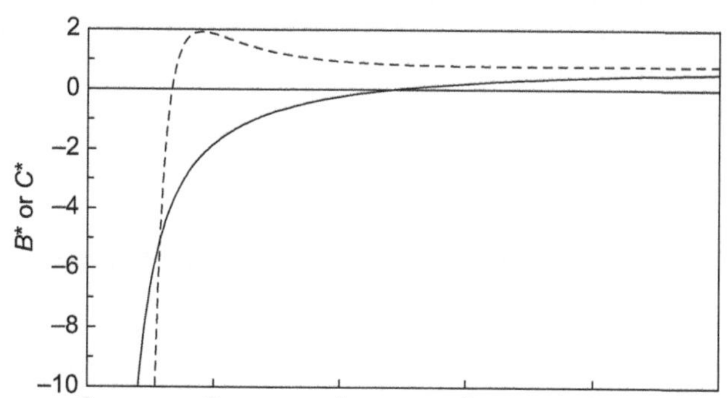

Figure 6.1 Reduced second virial coefficient $B^*=B/b_0$ and reduced third virial coefficient $C^*=8C/5b_0^2$ as functions of reduced temperature $T^*=k_BT/\varepsilon$ for the hard-core square-well potential with $\lambda=1.75$: ———, $B^*(T^*)$; - - - -, $C^*(T^*)$.

a single maximum. By treating σ, ε and λ as adjustable parameters, one can obtain an accurate description of the second virial coefficient over a wide range of temperatures; thus, Equation (6.11) provides a convenient means of correlating experimental second virial coefficients. It is also possible to represent third virial coefficients of real gases with Equation (6.12) provided that the square-well parameters are readjusted to fit that quantity.[3]

The virial coefficients of a number of other model intermolecular potentials have been evaluated, in most cases by numerical means. Evaluation of B and C by numerical quadrature is very straightforward, even for complex non-spherical and non-additive intermolecular potentials. However, the computation of higher-order virial coefficients becomes increasingly demanding in terms of both CPU time and memory. Nevertheless, extensive results are available for purely repulsive and additive potentials. A greatly improved algorithm has recently been developed by Wheatley,[4] permitting the computation of virial coefficients up to order 12 for hard spheres and up to order 10 for soft spheres that interact according to the potential $u(r)=\varepsilon(r/\sigma)^n$, with exponent $n=12$. For so-called Lennard-Jones molecules, having the intermolecular potential $u(r)=\varepsilon[(r/\sigma)^{12}-(r/\sigma)^6]$, virial coefficients up to order 8 have been reported.[5] Less simplistic intermolecular potentials have also been considered to high order. For example, Jager *et al.*[6] have determined *ab initio* both the pair potential and the non-additive three-body correction for argon and used these to determine the equation of state correct to the seventh virial coefficient. The availability of results up to quite high order sheds some light on the radius of convergence of the virial series. For the repulsive potentials, the series appears to converge satisfactorily for all fluid states that, according to computer simulations, are thermodynamically stable.[4] For Lennard-Jones molecules, and presumably other systems with an attractive branch in the intermolecular potential, the series converges up to roughly the critical density; however, it does not converge for liquid states, at least within the number of virial coefficients available hitherto.[5] For most real systems, virial coefficients above the third are not known with useful accuracy and, when truncated after the term in C, convergence to within order 10^{-2} is limited to roughly one half of the critical density.[6,7]

6.4 Measurement and Correlation of Virial Coefficients

The second virial coefficient has been studied extensively for a host of different substances, and there is also a smaller but still substantial database of experimental third virial coefficients. Little is known from experimental measurements about fourth and higher virial coefficients.

Much work on the second virial coefficient was motivated by a desire to gain a quantitative understanding of the intermolecular pair potentials of simple gases. In these cases, experimental techniques were developed that focused on the volumetric behaviour of gases at low densities, such that the terms in the third and higher-order virial coefficients were either very small or negligible. The second virial coefficient is not sensitive to fine details in the intermolecular potential (as evidenced by the success of the square-well model in correlating B data) and such sensitivity as it has diminishes with increasing temperature. Thus results at low temperatures are the most valuable. Unfortunately, it is at low temperatures that systematic errors arising from gas adsorption on the walls of the experimental apparatus become most troublesome. As a consequence, there are often quite large discrepancies between different measurements of B at the low temperatures.[8]

Traditional experimental techniques relied upon the determination of the pressure as a function of inverse volume for a fixed amount of gas.[1] Neglecting virial coefficients above the second, Equation (6.1) may be written in terms of the extensive variables volume, V, and amount of substance, n, as follows:

$$pV = nRT(1 + nB/V) \tag{6.13}$$

Thus by linear regression of isothermal p,V data one obtains both the amount of substance and the second virial coefficient. Numerous ingenious variations on this basic approach, many involving glass and mercury, have been devised but would not be the methods of choice in a modern laboratory.

Isothermal gas expansion techniques, especially that due to Burnett, have many advantages.[9,10] In this method, the gas is initially confined in a vessel of volume V_1, separated by a valve from a second evacuated vessel of volume V_2. The initial pressure p_i is measured and the gas is then allowed to expand to fill both vessels. After the establishment of equilibrium, the final pressure p_f is measured. Since the amount of substance is the same in both cases, n may be eliminated from the problem by considering the ratio p_i/p_f which, starting from Equation (6.4), may be developed by as a power series in p_f:[11]

$$\begin{aligned} p_i/p_f &= N[1 + (N-1)B'p_f + (N-1)\{(N+1)C' + (N-1)B'^2\}p_f^2 \cdots] \\ &= \sum_{i=0}^{\infty} a_i p_f^i , \end{aligned} \tag{6.14}$$

where $N = (V_1 + V_2)/V_1$. Usually, a sequence of Burnett expansions is carried out with decreasing initial pressures such that regression of the pressure ratios with Equation (6.14), truncated after an appropriate number of terms, yields both N and the coefficients B', C' $\cdots$ without the need to measure the extensive quantities V and n. However, the method requires very accurate pressure measurements and remains vulnerable to errors arising from gas adsorption on the walls of the apparatus. A differential Burnett method has been described that largely eliminates absolute pressure measurements in favour of the measurement of differences between the nearly equal pressures of a test gas and a reference gas.[11]

The direct determination of density by means of the two-sinker magnetically-coupled densimeter has revolutionised the measurement of the volumetric properties of fluids. These instruments operate on Archimedes' buoyancy principle and involve weighing a sinker of known mass and volume in the fluid under study. The modern instrument was developed by Wagner and co-workers in the 1980s and operates in a differential mode with two sinkers having almost identical mass, surface area and surface finish but different volumes.[12,13] The effect of this is to eliminate many sources of systematic error that accompany weighing of a single sinker. Such instruments are suitable for measuring the density of both gases and liquids, including saturated states. The close matching of the surface area and finish between the two sinkers also has the effect of compensating for the effects of adsorption on the surface of the sinkers, allowing reliable density measurements to be made in the gas phase, even at low temperatures, from which virial coefficients may be determined. The method has been used to determine the second and third virial coefficients of a number of gases including carbon dioxide,[14] methane[15] and other light hydrocarbons.[16]

Virial coefficients may also be determined indirectly, that is without recourse to pV_mT or equivalent measurements. Such indirect approaches have the advantage of being, in principle, independent of gas adsorption. One such approach exploits the isothermal Joule–Thomson coefficient φ which is defined by $\varphi = (\partial H_m/\partial p)_T = V_m - T(\partial V_m/\partial T)_p$, where H_m is molar enthalpy, and may be measured by means of isothermal flow calorimetry involving a throttle. The isothermal Joule–Thomson coefficient may be expressed as a power series in pressure as follows:

$$\varphi = -RT(\mathrm{d}B'/\mathrm{d}T) - RT^2(\mathrm{d}C'/\mathrm{d}T)p + \ldots. \tag{6.15}$$

Measurements of φ in the dilute gas therefore lead to the quantity $-RT(\mathrm{d}B'/\mathrm{d}T) = \{B - T(\mathrm{d}B/\mathrm{d}T)\}$ from which B may be obtained either by integration, starting from a known value of B, or by fitting with a model such as Equation (6.11) that prescribes a functional form for $B(T)$. In a similar way, flow-mixing calorimetry may be used to determine interaction second virial coefficient B_{12} in a binary mixture. Both methods have been applied extensively by Wormald and co-workers for pure substances[17–19] and for difficult-to-study mixtures such as steam with hydrocarbons,[20] or SO_2,[21] or H_2S[22] or HCl.[23]

Measurements of the speed of sound c in the gas phase offer a very favourable indirect route to the determination of second and third virial coefficients. The quantity c^2 may be developed as a power series in either inverse molar volume or pressure; in the latter case we have:

$$c^2 = A_0 + A_1 p + A_2 p^2 + \ldots, \tag{6.16}$$

where $A_0 = RT\gamma^{pg}/M$, γ^{pg} is the ratio of the perfect-gas heat capacity at constant pressure to that at constant volume and M is the molar mass. The higher-order coefficients in Equation (6.16) are related to the virial coefficients of the gas; in particular, the group $\beta_a = RTA_1/A_0$, known as the second acoustic virial coefficients, is related to the ordinary second virial coefficient by

$$\beta_a = 2B + 2(\gamma^{pg} - 1)T(dB/dT) + \{(\gamma^{pg} - 1)^2/\gamma^{pg}\}T^2(d^2B/dT^2). \tag{6.17}$$

The main advantage of acoustic measurements is that the speed of sound may be measured with extremely high precision, leading to second acoustic virial coefficients having very small uncertainties. The ordinary second virial coefficient may be obtained in principle by integration of Equation (6.17) starting from initial values of B and (dB/dT); however this is rarely carried out. Instead, a functional form for $B(T)$ is assumed, such as Equation (6.11), the parameters of which are fitted to the experimental second acoustic virial coefficients. A similar approach may be adopted for the third acoustic virial coefficient.[3] Sound speed measurements have been used to determine second virial coefficients for a wide variety of systems, including alkanes,[24–29] methanol[30] and refrigerant gases.[31–33]

6.5 Thermodynamic Properties from the Virial Equation of State

The virial equation of state provides a convenient means of determining the residual thermodynamic properties of a pure gas or mixture.[7] Starting from Equation (6.1), the residual Helmholtz energy corresponding to a state of specified temperature, pressure and composition may be obtained by integration with the following result:

$$A_m^{res} = RT\left\{B/V_m + \frac{1}{2}C/V_m^2 + \cdots\right\} - RT\ln Z, \tag{6.18}$$

where $Z = pV_m/RT$. Other residual thermodynamic properties follow by differentiation including residual molar entropy S_m^{res}, residual isochoric molar heat capacity $C_{V,m}^{res}$ and residual chemical potential μ_s^{res} of component s in a mixture:

$$S_m^{res} = -R\left\{(B + B_1)/V_m + \frac{1}{2}(C + C_1)/V_m^2 + \cdots\right\} + R\ln Z \tag{6.19}$$

$$C_{V,m}^{res} = -R\left\{(2B_1 + B_2)/V_m + \frac{1}{2}(2C_1 + C_2)/V_m^2 + \cdots\right\} \tag{6.20}$$

$$\mu_s^{res} = RT\left\{2\left(\sum_i x_i B_{is}\right)\Big/V_m + \frac{3}{2}\left(\sum_i\sum_j x_i x_j C_{ijs}\right)\Big/V_m^2 + \cdots\right\} - RT\ln Z. \tag{6.21}$$

In these equations, B_m and C_m denote the mth temperature derivatives of B and C, respectively. If one also knows, as functions of temperature, the perfect-gas heat capacities of all components present in the gas then a complete description of the observable thermodynamic properties may be readily obtained.

Alternatively, starting from the pressure series, Equation (6.4), residual thermodynamic properties (again at specified temperature, pressure and composition) may be determined directly. The residual molar Gibbs energy $G_{\mathrm{m}}^{\mathrm{res}}$ is first obtained by integration:

$$G_{\mathrm{m}}^{\mathrm{res}} = RT\left\{B'p + \frac{1}{2}C'p^2 + \cdots\right\}; \tag{6.22}$$

other properties follow from this by differentiation, including the residual molar entropy and the residual molar isobaric heat capacity:

$$S_{\mathrm{m}}^{\mathrm{res}} = -R\left\{(B' + B_1')p + \frac{1}{2}(C' + C_1')p^2 + \cdots\right\} \tag{6.23}$$

$$C_{p,\mathrm{m}}^{\mathrm{res}} = -R\left\{(2B_1' + B_2')p + \frac{1}{2}(2C_1' + C_2')p^2 + \cdots\right\}. \tag{6.24}$$

The expressions developed in this section are useful in thermodynamic calculations pertaining to the gas phase, and also in the computation of vapour–liquid equilibrium states for mixtures provided that a suitable thermodynamic model is available for the liquid phase.

6.6 Compendia and Correlations

There have been a number of reviews of the available experimental virial coefficient data of which perhaps the best known is that of Dymond and Smith,[8] dating from 1980. More recently, an extensive two-volume compendium of critically-assessed experimental pure-component and interaction virial coefficients has been published.[34,35] These reviews concentrate mainly on the second virial coefficient, although a fair amount of third-virial coefficient data is also reviewed. Recommended values of B are given based on a critical assessment of the data and, in the most recent reviews, these are correlated as $B = a - b\,\exp(c/T)$, which is a form of Equation (6.11) and thus related to the hard-core square-well potential.

Thermodynamic properties are often required for mixtures containing poorly-studied compounds for which little or no experimental data exist. In these cases, correlations can be applied and, in the present context, those based on the extended principle of corresponding states are the most useful. In the standard three-parameter corresponding-states approach, thermodynamic properties are expressed in reduced terms, using critical temperature T^{c} and critical pressure p^{c} as scaling parameters, as linear functions of Pitzer's acentric factor ω. The latter is given by

$$\omega = -1 - \log_{10}\{p^{\mathrm{g+l}}(T = 0.7T^{\mathrm{c}})/p^{\mathrm{c}}\}, \tag{6.25}$$

where $p^{\mathrm{g+l}}(T)$ denotes vapour pressure at given T. The second virial coefficient is then given by

$$(p^{\mathrm{c}}/RT^{\mathrm{c}})B = b_1 + \omega b_2 \tag{6.26}$$

where b_1 and b_2 are universal functions of T/T^{c}. Well-known correlations for b_1 and b_2 have been given by Tsonopoulos[36,37] and updated by Meng *et al.*[38]

Interaction second virial coefficients can also be estimated from correlations of this form by applying mixing rules to obtain the pseudo-critical constants and acentric factor characterising the unlike interaction.[37,39,40] The correlation in the form of Equation (6.26) can only be considered 'universal' among essentially non-polar compounds; however, additional terms involving the reduced dipole moment have been proposed to extend the method to polar compounds.[36,41] Similar methods have been proposed for the third virial coefficient.[38,41,42]

References

1. E. A. Mason and T. H. Spurling, *The Virial Equation of State*, Pergamon Press, Oxford, 1969.
2. J. P. M. Trusler, *The Virial Equation of State, in Equations of State for Fluids and Fluid Mixtures*, ed. R. F. Kayser, C. J. Peters, J. V. Sengers and H. J. White Jr., Elsevier, Amsterdam, 2000, vol. 1.
3. K. A. Gillis and M. R. Moldover, *Int. J. Thermophys.*, 1996, **17**, 1305.
4. R. J. Wheatley, *Phys. Rev. Lett.*, 2013, **110**, 200601.
5. A. J. Schultz and D. A. Kofke, *Mol. Phys.*, 2009, **107**, 2309.
6. B. Jager, R. Hellmann, E. Bich and E. Vogel, *J. Chem. Phys.*, 2011, **135**, 084308.
7. J. P. M. Trusler, *The Virial Equation of State, in Applied Thermodynamics of Fluids*, ed. A. R. H. Goodwin, J. V. Sengers and C. J. Peters, Royal Society of Chemistry, London, 2010.
8. E. B. Smith and J. H. Dymond, *The Virial Coefficients of Pure Gases and Mixtures: a Critical Compilation*, Clarendon Press, Oxford, 1980.
9. E. S. Burnett, *J. Appl. Mech.*, 1936, **3**, A136.
10. M. Waxman and J. R. Hastings, *J. Res. Natl. Bur. Stand., Sect. C*, 1971, **75C**, 165.
11. M. B. Ewing and K. N. Marsh, *J. Chem. Thermodyn.*, 1979, **11**, 793.
12. R. Kleinrahm and W. Wagner, *J. Chem. Thermodyn.*, 1986, **18**, 739.
13. W. Wagner and R. Kleinrahm, *Metrologia*, 2004, **41**, S24.
14. W. Duschek, R. Kleinrahm and W. Wagner, *J. Chem. Thermodyn.*, 1990, **22**, 827.
15. G. Handel, R. Kleinrahm and W. Wagner, *J. Chem. Thermodyn.*, 1992, **24**, 685.
16. S. Glos, R. Kleinrahm and W. Wagner, *J. Chem. Thermodyn.*, 2004, **36**, 1037.
17. N. Albizreh and C. J. Wormald, *J. Chem. Thermodyn.*, 1977, **9**, 749.
18. C. J. Wormald, *J. Chem. Thermodyn.*, 1979, **11**, 1127.
19. M. L. McGlashan and C. J. Wormald, *J. Chem. Thermodyn.*, 2000, **32**, 1489.
20. G. R. Smith, M. J. Fahy and C. J. Wormald, *J. Chem. Thermodyn.*, 1984, **16**, 825.
21. C. J. Wormald, *J. Chem. Thermodyn.*, 2003, **35**, 91.
22. C. J. Wormald, *J. Chem. Thermodyn.*, 2003, **35**, 1019.
23. C. J. Wormald, *J. Chem. Thermodyn.*, 2003, **35**, 417.

24. M. B. Ewing, A. R. H. Goodwin, M. L. Mcglashan and J. P. M. Trusler, *J. Chem. Thermodyn.*, 1987, **19**, 721.
25. M. B. Ewing, A. R. H. Goodwin, M. L. Mcglashan and J. P. M. Trusler, *J. Chem. Thermodyn.*, 1988, **20**, 243.
26. M. B. Ewing, A. R. H. Goodwin and J. P. M. Trusler, *J. Chem. Thermodyn.*, 1989, **21**, 867.
27. M. B. Ewing and A. R. H. Goodwin, *J. Chem. Thermodyn.*, 1991, **23**, 1107.
28. M. B. Ewing and A. R. H. Goodwin, *J. Chem. Thermodyn.*, 1992, **24**, 301.
29. J. P. M. Trusler and M. P. Zarari, *J. Chem. Thermodyn.*, 1995, **27**, 771.
30. S. J. Boyes, M. B. Ewing and A. R. H. Goodwin, *J. Chem. Thermodyn.*, 1992, **24**, 1151.
31. A. R. H. Goodwin and M. R. Moldover, *J. Chem. Phys.*, 1990, **93**, 2741.
32. A. R. H. Goodwin and M. R. Moldover, *J. Chem. Phys.*, 1991, **95**, 5230.
33. A. R. H. Goodwin and M. R. Moldover, *J. Chem. Phys.*, 1991, **95**, 5236.
34. J. H. Dymond, K. N. Marsh, R. C. Wilhoit and K. C. Wong, *Virial Coefficients of Pure Gases, Part A*, Springer-Verlag, Berlin, 2002.
35. J. H. Dymond, K. N. Marsh and R. C. Wilhoit, *Virial Coefficients of Pure Gases and Mixtures, Part B*, Springer-Verlag, Berlin, 2003.
36. C. Tsonopoulos, *AIChE J.*, 1974, **20**, 263.
37. C. Tsonopoulos, Second Virial Cross-Coefficients: Correlation and Prediction of k_{ij}, in *Equations of State in Engineering and Research*, ed. K. C. Chao and R. L. Robinson Jr., American Chemical Society, Washington, DC, 1979, vol. 182, pp. 143.
38. L. Meng, Y. Y. Duan and L. Li, *Fluid Phase Equilib*, 2004, **226**, 109.
39. C. Tsonopoulos, J. H. Dymond and A. M. Szafranski, *Pure Appl. Chem.*, 1989, **61**, 1387.
40. L. Meng and Y. Y. Duan, *Fluid Phase Equilib.*, 2005, **238**, 229.
41. L. A. Weber, *Int. J. Thermophys.*, 1994, **15**, 461.
42. H. Orbey and J. H. Vera, *AIChE J.*, 1983, **29**, 107.

CHAPTER 7

Excess Volumes of Liquid Nonelectrolyte Mixtures

EMMERICH WILHELM*[a] AND J.-P. E. GROLIER[b]

[a] Institute of Physical Chemistry, University of Wien, Währinger Straße 42, A-1090, Wien (Vienna), Austria; [b] Clermont University, Institute of Chemistry of Clermont-Ferrand, ICCF, CNRS UMR 6296, 24 Landais Av., F-63177, Aubière, France
*Email: emmerich.wilhelm@univie.ac.at

7.1 Introduction and Some Relevant Thermodynamics

Thermodynamics rests on an experiment-based axiomatic fundament. Experiments, together with theory and computer simulations, are the pillars of science, as indicated by Figure 1.1 and elaborated in Chapter 1 of this monograph:[1] the "knowledge triangle"[2–4] indicates what may be learned from a comparison of respective results under idealised conditions. The link between macroscopically observable physical properties of bulk phases in which the physical chemist or the chemical engineer is interested, and the properties of molecules and of their interactions in pairs, triplets, and so forth is commonly known as *molecular thermodynamics*. This term was coined by Prausnitz more than four decades ago.[5,6] It is a field of great academic fascination and of practical importance, *i.e.* it is an indispensable part of physical chemistry as well as of chemical engineering. The impressive growth of molecular thermodynamics has been stimulated by the continuously increasing need for thermodynamic property data and phase

Volume Properties: Liquids, Solutions and Vapours
Edited by Emmerich Wilhelm and Trevor M. Letcher

Published by the Royal Society of Chemistry, www.rsc.org

equilibrium data,[7–26] and has greatly profited from unprecedented advances in experimental techniques,[27–34] from advances in the theory of liquids in general and from advances in computer simulations of reasonably realistic model systems.[35–50]

More than 50 years ago, Hildebrand and Scott[51] remarked that "*Of the various thermodynamic functions for the mixing process, the volume change on mixing at constant pressure...is one of the most interesting, yet certainly still one of the least understood.*" This quotation was used by Battino in the opening paragraph of his benchmark review[52] of 1971 on volume changes on mixing for binary mixtures of liquids, and somewhat surprisingly, it is still *grosso modo* valid. Certainly, this field has not suffered from a lack of attention from the scientific community; and perusal of relevant journals reveals that, since the introduction of vibrating-tube densimetry,[53–58] routine application of this technique has resulted in a prolific outpouring on this topic. Unfortunately, the volume-change-on-mixing data are frequently reported at only one temperature (or over a very narrow temperature range) and/or only for a few compositions, it is frequently not clear how the mixtures/solutions were prepared and whether the liquid components were degassed (and how), and often hand-waving arguments replace rationally selected accompanying experiments, say enthalpies of mixing and heat capacities, and application of adequate model theories. In view of the large number of new experimental results published since Battino's review, it is not surprising at all that very few comprehensive surveys have been published since, the most important being the reviews of Handa and Benson,[59] and of Cibulka and Holub.[60] Whereas the latter authors in their compilation give excess molar volumes at equimolar composition only (similar to Battino[52]), Handa and Benson provide complete experimental information on each system *via* appropriate smoothing functions, the coefficients of which are listed. A bibliography of *PVT* properties of more than 350 pure liquids and 170 binary liquid mixtures has been prepared by Tekáč *et al.*[61] Currently, the main sources of data on volume changes on mixing two liquids are the Dortmund Data Bank[19] and Landolt–Börnstein.[23]

Reliable volumetric data of fluid systems in general, and excess molar volumes in particular, are of central importance in process design, in the storage of fluids and in their transport. From a theoretical point of view, volume changes on mixing can be caused by a variety of factors, hence heuristically a discussion of mixtures/solutions in general may be conducted in terms of differences between the components in terms of molecular size, molecular shape anisotropy, dispersion energy, polarity, molecular polarisability, flexibility, and so forth. Figure 7.1 presents an overview of the most important aspects at the *molecular* level as well as the *bulk* level.[4,62,63] In fact, in many liquid mixtures/solutions dipolar (and quadrupolar) interactions, or the breaking or forming of hydrogen bonds, contribute significantly to the thermodynamic properties and may result in cooperative phenomena. This aspect is of particular importance in *aqueous solutions*, and has for decades occupied the central stage of

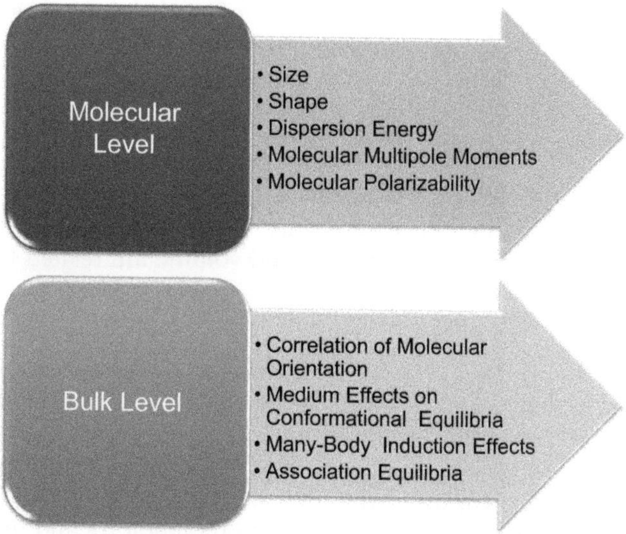

Figure 7.1 Heuristic summary of the most important physical aspects to be considered in practical descriptions of pure liquids and liquid mixtures/solutions at the molecular level as well as the bulk level (after Wilhelm[4,62,63]).

research in this area.[64–69] Generally volume changes on mixing dense molecular fluids are caused by a combination of several of the factors indicated in Figure 7.1, making it clear that it is a complex property. Only through judiciously planned experiments may one hope to learn more about their relative importance in intermolecular interactions and thus specifically improve our knowledge of the liquid state and associated macroscopic behaviour. These data provide information for the improvement and development of new, more detailed equations of state; not only for the widely used cubic and generalised van der Waals equations of state,[70] but also for modern fundamental equations explicit in the Helmholtz energy as a function of temperature and density.[71] Fundamental equations allow the description of the entire fluid region, and calculation of thermodynamic properties requires *only* differentiation of the equation. Such ambitious projects depend crucially on reliable, accurate data over wide ranges of temperature and pressure/density, which are provided, for instance, by the *International Thermodynamic Tables of the Fluid State*. This fundamentally important series started in 1972 with *argon* (Angus *et al.*[72]), and comprises also *carbon dioxide* (Angus *et al.*[73]), *methane* (Angus *et al.*,[74] Wagner and de Reuck[75]), *nitrogen* (Angus *et al.*[76]), *propene* (Angus *et al.*[77]), *chlorine* (Angus *et al.*[78]), *oxygen* (de Reuck and Wagner[79]), *ethene* (de Reuck[80]), *fluorine* (de Reuck[81]) and *methanol* (de Reuck and Craven[82]).

The molar volume V or, alternatively, the mass density ρ, characterises a single-phase fluid at temperature T and pressure P: a *pure* substance is indicated by a subscript i, while the absence of a subscript indicates a multicomponent mixture/solution with composition $\{x_i\}=\{x_1,x_2,x_3,...\}$, where

$$x_i = n_i \Big/ \sum_i n_i \tag{7.1}$$

denotes the mole fraction of component i, $\sum_i x_i = 1$, and $x_i = 1$ for a pure fluid; n_i denotes the amount of substance of component i, and $n = \sum_i n_i$ is the total amount of substance. Density and volume are related by

$$\rho(T,P,\{x_i\}) = m/nV(T,P\{x_i\}), \tag{7.2}$$

where $m = \sum_i m_i = \sum_i n_i m_{\mathrm{m},i}$ is the total mass of the phase, m_i is the mass of component i ($i = 1, 2, 3,\ldots$) with molar mass $m_{\mathrm{m},i}$, that is, $m_i = n_i m_{\mathrm{m},i}$, and $nV(T,P,\{x_i\})$ is the total (extensive) volume of the phase. The molar mass of the mixture is thus given by $m_\mathrm{m} = m/n = \sum_i x_i m_{\mathrm{m},i}$. As already pointed out by Rowlinson and Swinton,[35] the two material properties volume and density, and their derivatives with respect to temperature and pressure, are determined, to a high degree of accuracy, solely by the *intermolecular* forces. These derivatives are known as the mechanical coefficients: the *isobaric expansivity* is defined by

$$\alpha_P \equiv \frac{1}{V}\left(\frac{\partial V}{\partial T}\right)_P = -\frac{1}{\rho}\left(\frac{\partial \rho}{\partial T}\right)_P, \ \text{const}\,\{x_i\} \tag{7.3}$$

the *isothermal compressibility*† is defined by

$$\beta_T \equiv -\frac{1}{V}\left(\frac{\partial V}{\partial P}\right)_T = \frac{1}{\rho}\left(\frac{\partial \rho}{\partial P}\right)_T, \ \text{const}\,\{x_i\}, \tag{7.4}$$

and the *isochoric thermal pressure coefficient* is given by

$$\gamma_V \equiv \left(\frac{\partial P}{\partial T}\right)_V, \ \text{const}\,\{x_i\}, \tag{7.5}$$

and $\alpha_P/\beta_T = \gamma_V$. The *isentropic* coefficients, with which they are related through the thermal coefficients, *i.e.* the heat capacities, and the thermal coefficients themselves depend also on *internal molecular* properties.

The volume (or the density) and volume-related properties appear in numerous thermodynamic relations. For instance, for homogeneous constant-composition fluids we have

$$\left(\frac{\partial F}{\partial V}\right)_{T,\{x_i\}} = -P, \tag{7.6}$$

---

†In this chapter the isothermal compressibility is represented by the symbol β_T and not by κ_T, as was recently recommended by IUPAC.[83] Similarly, the isentropic compressibility is represented by the symbol β_S and not by κ_S. For their ratio, the symbol $\kappa \equiv \beta_T/\beta_S$ is used.

$$\left(\frac{\partial G}{\partial P}\right)_{T,\{x_i\}} = V, \tag{7.7}$$

$$\left(\frac{\partial S}{\partial V}\right)_{T,\{x_i\}} = \left(\frac{\partial P}{\partial T}\right)_{V,\{x_i\}}, \tag{7.8}$$

$$\left(\frac{\partial S}{\partial P}\right)_{T,\{x_i\}} = -\left(\frac{\partial V}{\partial T}\right)_{P,\{x_i\}}, \tag{7.9}$$

$$\left(\frac{\partial U}{\partial V}\right)_{T,\{x_i\}} = -P + T\left(\frac{\partial P}{\partial T}\right)_{V,\{x_i\}}, \tag{7.10}$$

$$\left(\frac{\partial H}{\partial P}\right)_{T,\{x_i\}} = V - T\left(\frac{\partial V}{\partial T}\right)_{P,\{x_i\}}, \tag{7.11}$$

$$\left(\frac{\partial C_V}{\partial V}\right)_{T,\{x_i\}} = T\left(\frac{\partial^2 P}{\partial T^2}\right)_{V,\{x_i\}}, \tag{7.12}$$

$$\left(\frac{\partial C_P}{\partial P}\right)_{T,\{x_i\}} = -T\left(\frac{\partial^2 V}{\partial T^2}\right)_{P,\{x_i\}}, \tag{7.13}$$

$$\left(\frac{\partial \alpha_P}{\partial P}\right)_{T,\{x_i\}} = -\left(\frac{\partial \beta_T}{\partial T}\right)_{P,\{x_i\}}, \tag{7.14}$$

$$C_P - C_V = -T\frac{(\partial V/\partial T)^2_{P,\{x_i\}}}{(\partial V/\partial P)_{T,\{x_i\}}}, \tag{7.15}$$

and

$$\left(\frac{\partial (V/T)}{\partial (1/T)}\right)_{P,\{x_i\}} = -\left(\frac{\partial T}{\partial P}\right)_{H,\{x_i\}} C_P. \tag{7.16}$$

Here, all the symbols have their usual meaning, that is F is the molar Helmholtz energy (noting that A is also an internationally accepted modern symbol), G is the molar Gibbs energy, S is the molar entropy, U is the molar internal energy, H is the molar enthalpy, C_V is the molar heat capacity at constant volume, and C_P is the molar heat capacity at constant pressure.

The *volumetric* behaviour of *real gases* or *vapours* (pure or mixed) at low to moderate densities is of considerable practical and theoretical importance.[84–89] Since Chapter 6 is devoted to this topic, only a few comments will be made here. From experiment we know that the behaviour of any real gas approaches that of the perfect gas[90] as $P \to 0$, or $\rho_n \to 0$, where $\rho_n \equiv 1/V = \rho/m_{\mathrm{m}}$ denotes the amount-of-substance density. Thus, expanding the compression factor

$$Z \equiv \frac{P}{\rho_n RT}, \tag{7.17}$$

at constant temperature and composition, as a power series in ρ_n or P, respectively, about the zero-density or zero-pressure limit of the real gas, respectively, yields the pressure-explicit virial equation in density in the former case,

$$Z = 1 + B\rho_n + C\rho_n^2 + D\rho_n^3 + \cdots, \tag{7.18}$$

or the volume-explicit virial equation in pressure in the latter case,

$$Z = 1 + B'P + C'P^2 + D'P^3 + \cdots \tag{7.19}$$

Here, B, C,... are the second, the third, *etc.* virial coefficient of the density series, and B',C',... are the second, the third, *etc.* virial coefficient of the pressure series, and R is the molar gas constant. Note that $B' = B/RT$, $C' = (C - B^2)/(RT)^2$ and so forth. According to their definition, the virial coefficients of *pure* fluids are *only* functions of T and, in the case of *mixtures*, they are *only* functions of T and $\{x_i\}$.

In principle, experimental determination of any property, the value of which depends on *gas imperfection*, may be used to obtain thermodynamic information on *virial coefficients*, in particular on second virial coefficients B.[91–94] For instance, flow calorimetry is the commonly used method for measuring the heat capacity at constant pressure of gases and vapours,[93] and isothermal determinations of C_P as a function of pressure are usually performed to obtain the molar heat capacity C_P^{pg} of the substance in the perfect-gas state, *i.e.* in the limit of zero pressure. We note that for a real *gas* or a real *vapour* in the limit of zero pressure, the pressure derivative of C_p is generally *non-zero* and *positive* up to *very* high temperatures:

$$\lim_{P \to 0}\left(\frac{\partial C_P}{\partial P}\right)_T = -T\frac{\mathrm{d}^2 B}{\mathrm{d}T^2}. \tag{7.20}$$

Flow calorimeters equipped with a throttle have been used to measure the *isenthalpic* Joule–Thomson coefficient $\mu_{\mathrm{JT}} \equiv (\partial T/\partial P)_{H,\{x_i\}}$ and the molar *isothermal* Joule–Thomson coefficient $\psi \equiv (\partial H/\partial P)_{T,\{x_i\}}$. They are related to the molar isobaric heat capacity through Equation (7.16), which may be rewritten as

$$\frac{\mu_{\mathrm{JT}} C_P}{\psi} = -1. \tag{7.21}$$

This equation is of considerable utility in *gas flow calorimetry*. For instance, a versatile recycle-flow calorimeter for the determination of all three quantities

μ_{JT},ψ and C_P has been described by Miyazaki *et al.*[95] The determination of deviations from perfect-gas behaviour *via* flow calorimetry has the advantage over conventional methods that errors due to *adsorption* are avoided, and that measurements can be made at much lower temperatures and pressures.

Definitely the most popular way of presenting and discussing the composition dependence of the molar volume of real *liquid* mixtures is based on the use of *excess molar volumes* $V^E(T, P, \{x_i\})$.[6,96–102] This quantity represents deviations from *ideal-solution* behaviour and is defined by

$$V^E(T, P, \{x_i\}) \equiv V(T, P, \{x_i\}) - V^{id}(T, P, \{x_i\}), \tag{7.22}$$

i.e. V^E is the difference between the molar volume of the real solution and the value calculated for an ideal solution at the same temperature, pressure, composition and phase. For an ideal solution, by definition

$$V^{id}(T, P, \{x_i\}) \equiv \sum_i x_i V_i^*(T, P), \tag{7.23}$$

for all temperatures, pressures and compositions, where $V_i^*(T, P)$ is the molar volume of *pure* component *i*, indicated by a superscript asterisk (*), in the same phase, usually the liquid phase. The use of excess volumes simply provides a practical way to better visualise and handle experimental data, since it magnifies the real-mixture deviations from the linear behaviour represented by Equation (7.23). Recalling the definition of a partial molar property,

$$V_i \equiv \left(\frac{\partial(nV)}{\partial n_i}\right)_{T,P,n_{j \neq i}} \tag{7.24}$$

is the *partial* molar volume associated with component *i* in the real solution, and

$$V_i^{id}(T, P, \{x_i\}) = V_i^*(T, P). \tag{7.25}$$

Since the total volume nV of a phase is a homogeneous function of the first degree in the amounts of substance $\{n_i\}$, Euler's theorem yields

$$nV = \sum_i n_i V_i, \tag{7.26}$$

and after division by the total amount of substance n

$$V = \sum_i x_i V_i. \tag{7.27}$$

In complete analogy, the corresponding *excess partial molar volume* for component *i* in solution is defined by

$$V_i^E \equiv \left(\frac{\partial(nV^E)}{\partial n_i}\right)_{T,P,n_{j \neq i}} = V_i - V_i^*, \tag{7.28}$$

and

$$V^{\mathrm{E}} = \sum_i x_i V_i^{\mathrm{E}} = \sum_i x_i \left(V_i - V_i^*\right). \tag{7.29}$$

Of practical and theoretical importance is the partial molar volume at *infinite dilution*,[103–106] which is defined by

$$V_i^{\infty} = \lim_{x_i \to 0} V_i, \text{ constant } T \text{ and } P, \tag{7.30}$$

and the excess partial molar volume at *infinite dilution*

$$V_i^{\mathrm{E}\infty} = \lim_{x_i \to 0} V_i^{\mathrm{E}}, \text{ constant } T \text{ and } P. \tag{7.31}$$

Chapter 8 is devoted to this topic.

Alternatively, one may characterise the volumetric behaviour of a real mixture by the *volume change per mole* accompanying the mixing process:

$$\Delta V(T, P, \{x_i\}) \equiv V(T, P, \{x_i\}) - \sum x_i V_i^*(T, P), \tag{7.32}$$

with a corresponding *partial molar volume change on mixing*

$$\Delta V_i(T, P, \{x_i\}) \equiv \left(\frac{\partial(n\Delta V)}{\partial n_i}\right)_{T, P, n_j \neq i} = V_i(T, P, \{x_i\}) - V_i^*(T, P). \tag{7.33}$$

Thus the excess molar volume is also the difference between the real molar change of volume on mixing and the ideal-solution change of volume on mixing, hence it is identical to the excess volume change on mixing,

$$\Delta V^{\mathrm{E}}(T, P, \{x_i\}) \equiv \Delta V(T, P, \{x_i\}) - \Delta V^{\mathrm{id}}(T, P, \{x_i\}) = V^{\mathrm{E}}(T, P, \{x_i\}). \tag{7.34}$$

An entirely similar relation holds for the partial molar quantities:

$$\Delta V_i^{\mathrm{E}}(T, P, \{x_i\}) \equiv \Delta V_i(T, P, \{x_i\}) - \Delta V_i^{\mathrm{id}}(T, P, \{x_i\}) = V_i^{\mathrm{E}}(T, P, \{x_i\}) \tag{7.35}$$

The terms excess molar volume and excess molar volume change on mixing may both be used interchangeably and are indeed found in the literature.[99,100,102] If the focus is on *mixture properties*, then $V^{\mathrm{E}}(T, P, \{x_i\})$ and $V_i^{\mathrm{E}}(T, P, \{x_i\})$ are preferred, while for *mixing processes* $\Delta V^{\mathrm{E}}(T, P, \{x_i\})$ and $\Delta V_i^{\mathrm{E}}(T, P, \{x_i\})$ may be regarded as more appropriate.

Since $\Delta V^{\mathrm{id}} = 0$ and $\Delta V_i^{\mathrm{id}} = 0$, we have

$$V^{\mathrm{E}} = \Delta V \tag{7.36}$$

and

$$V_i^{\mathrm{E}} = \Delta V_i, \tag{7.37}$$

for all temperatures, pressures and compositions. We emphasise that the definition in Equation (7.22) is not restricted to any phase, though excess molar properties in general, and excess molar volumes in particular, are predominantly used for *liquid* mixtures/solutions. In fact, the single-phase thermodynamic property most frequently determined is the excess volume/the volume of mixing in the liquid phase.

In solution thermodynamics, the *excess molar Gibbs energy* $G^{\mathrm{E}}(T,P,\{x_i\})$ is of central importance. Recalling that

$$G^{\mathrm{id}} \equiv \sum_i x_i G_i^* + RT\sum_i x_i \ln x_i \tag{7.38}$$

and

$$S^{\mathrm{id}} \equiv \sum_i x_i S_i^* - R\sum_i x_i \ln x_i, \tag{7.39}$$

we have then by definition

$$\begin{aligned} G^{\mathrm{E}} &\equiv G - G^{\mathrm{id}} = G - \sum_i x_i G_i^* - RT\sum_i x_i \ln x_i \\ &= \Delta G - RT\sum_i x_i \ln x_i, \end{aligned} \tag{7.40}$$

and for the excess mole entropy

$$\begin{aligned} S^{\mathrm{E}} &\equiv S - S^{\mathrm{id}} = S - \sum_i x_i S_i^* + R\sum_i x_i \ln x_i \\ &= \Delta S + R\sum_i x_i \ln x_i \end{aligned} \tag{7.41}$$

Here, ΔG and ΔS are the *molar Gibbs energy change of mixing* and the *molar entropy change of mixing*, respectively. Note that

$$G^{\mathrm{E}} = H^{\mathrm{E}} - TS^{\mathrm{E}} \tag{7.42a}$$

and

$$\Delta G = \Delta H - T\Delta S, \tag{7.42b}$$

where, depending on the point-of-view,

$$H^{\mathrm{E}} = \Delta H \tag{7.43}$$

is called either the excess molar enthalpy or the molar enthalpy change of mixing.

For convenience, the non-dimensional quantity G^{E}/RT is frequently used instead of G^{E}, and the corresponding *fundamental excess-property relation* is then given by

$$\mathrm{d}\left(\frac{nG^{\mathrm{E}}}{RT}\right) = -\frac{nH^{\mathrm{E}}}{RT^2}\mathrm{d}T + \frac{nV^{\mathrm{E}}}{RT}\mathrm{d}P + \sum_i \ln \gamma_i \mathrm{d}n_i. \tag{7.44}$$

The corresponding Gibbs–Duhem equation reads

$$-\frac{H^{\mathrm{E}}}{RT^2}\mathrm{d}T + \frac{V^{\mathrm{E}}}{RT}\mathrm{d}P - \sum_i x_i \mathrm{d}\ln \gamma_i = 0, \tag{7.45}$$

which at constant T and P becomes

$$\sum_i x_i \mathrm{d}\ln\gamma_i = 0. \tag{7.46}$$

Here, $\gamma_i(T, P, \{x_i\})$ is the *activity coefficient* of component i using the *symmetric* convention (Lewis–Randall convention).[5,6,96–106]

Inspection of Equation (7.44) leads directly to

$$\left(\frac{\partial(G^{\mathrm{E}}/RT)}{\partial T}\right)_{P,\{x_i\}} = -\frac{H^{\mathrm{E}}}{RT^2}, \tag{7.47}$$

$$\left(\frac{\partial(G^{\mathrm{E}}/RT)}{\partial P}\right)_{T,\{x_i\}} = \frac{V^{\mathrm{E}}}{RT}, \tag{7.48}$$

and to

$$\left(\frac{\partial(nG^{\mathrm{E}}/RT)}{\partial n_i}\right)_{T,P,n_{j\neq i}} = \frac{G_i^{\mathrm{E}}}{RT} = \ln\gamma_i. \tag{7.49}$$

Equation (7.49) demonstrates that In $\ln\gamma_i$ is a partial molar quantity with respect to G^{E}/RT, hence

$$\frac{G^{\mathrm{E}}}{RT} = \sum_i x_i \frac{G_i^{\mathrm{E}}}{RT} = \sum_i x_i \ln\gamma_i. \tag{7.50}$$

The partial molar analogue of Equation (7.47) is

$$\left(\frac{\partial \ln\gamma_i}{\partial T}\right)_{P,\{x_i\}} = -\frac{H_i^{\mathrm{E}}}{RT^2}, \tag{7.51}$$

and of Equation (7.48) it is

$$\left(\frac{\partial \ln\gamma_i}{\partial P}\right)_{T,\{x_i\}} = \frac{V_i^{\mathrm{E}}}{RT}. \tag{7.52}$$

Equations (7.48) and (7.52) may be used to calculate the influence of pressure on G^{E}, G_i^{E} or Inγ_i, respectively. Note that the partial molar Gibbs energy of component i is just the chemical potential of component i, that is,

$$G_i = \mu_i, \tag{7.53}$$

and for the excess partial molar Gibbs energy of component i

$$G_i^{\mathrm{E}} = \mu_i^{E} = RT\ln\gamma_i, \tag{7.54}$$

where μ_i^{E} is the excess chemical potential of component i. These properties hold key positions in solution thermodynamics.

Outside the critical region the influence of pressure on excess quantities is rather small. For instance, modern flow calorimeters[29,30,107–109] allow reliable measurements of enthalpy changes on mixing/excess enthalpies at elevated pressures (and elevated temperatures), and from the reciprocity relation the pressure dependence of the excess molar enthalpy is found to be

$$\left(\frac{\partial H^{\mathrm{E}}}{\partial P}\right)_{T,\{x_i\}} = V^{\mathrm{E}} - T\left(\frac{\partial V^{\mathrm{E}}}{\partial T}\right)_{P,\{x_i\}}. \tag{7.55}$$

As indicated by Equation (7.55), calorimetric results obtained at high pressures can be checked for consistency by determining excess molar volumes (and their temperature dependence) *via*, say, vibrating-tube densimetry (see below, Section 7.2), and *vice versa*.

7.2 Methods and Apparatus

This section deals with experimental methods for determining excess volumes, predominantly of *binary* liquid mixtures of nonelectrolytes, predominantly over the temperature range 273 K to about 350 K (this is the customary temperature range of water thermostats), *i.e.* well below the critical temperature, and at pressures up to a few bars (1 bar =0.1 MPa), and their correlation. For surveys see the detailed reviews of Battino,[52] Handa and Benson,[59] Weissberger,[110] and Marsh.[111,112]

The majority of volume changes of mixing/excess volumes for liquid mixtures have been determined *via* two methods: either *directly* by mixing the two components and measuring the volume change with the help of a dilatometer, or *indirectly* by calculating the excess volume from the measured density at known composition. The direct method is the most transparent one: it avoids any correction due to the volatility of the liquids and an accurate knowledge of the composition is not required. Many ingenious designs of apparatus have been reported in the literature, basically in two styles: *batch dilatometers* ("single shot" dilatometers), where known amounts of each component are initially separated (usually by mercury) and, after mixing them isothermally, the volume change is measured in a precision-bore capillary; and *dilution dilatometers* which allow consecutive additions of known amounts of one component to a mixing chamber which, at the start of the measurement, contains a known amount of the other component.

Perhaps the most popular designs of single shot dilatometers are based on that of Duncan *et al.*[113] This apparatus requires only 0.2 cm^3 to 0.8 cm^3 of each component, and yields excess molar volumes with an imprecision of less than about ± 0.002 $\mathrm{cm}^3\ \mathrm{mol}^{-1}$. Similarly, the apparatus of Keyes and Hildebrand[114] (essentially a double-arm pycnometer with mercury at the bottom to separate the two liquids) has been modified frequently, for instance by Brown and Smith.[115] A detailed analysis of possible errors associated with batch dilatometers was presented by Stookey *et al.*[116]

Modified versions of the dilution dilatometer of Geffcken *et al.*[117] have been presented, among others, by Watson *et al.*,[118] Pflug and Benson,[119] Beath *et al.*,[120] Stokes *et al.*,[121] Bottomley and Scott (see Figure 7.2),[122]

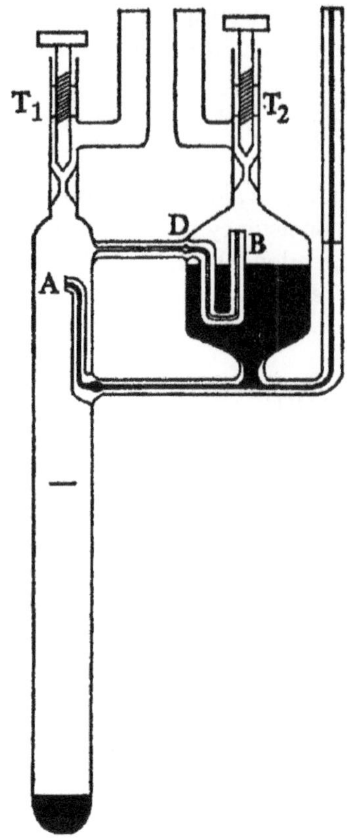

Figure 7.2 Grease-free tilting dilution dilatometer of Bottomley and Scott,[122] shown in diagrammatic form. The dilatometer uses mercury (in black) to displace the liquid components, designated 1 and 2, to be mixed. T_1 and T_2 are the 1.25 mm bore Teflon needle valves used to introduce the respective components 1 and 2 into the dilatometer. The apparatus can be rotated anticlockwise up to 45° without letting any volumetric part of it emerge from the thermostat fluid. The tilting of the apparatus allows some mercury to flow through the tip at A into the burette, then displacing an appropriate amount of component 1 into the bulb (capacity *ca.* 45 cm^3) to component 2 through the upper capillary D at B. Mixing of the two components causes the meniscus in C to change its position. The amount of component 1 added is obtained from the burette levels, and from the change of the meniscus position in C the volume change on mixing is determined. Stirring is carried out with a magnet operating a glass-encapsuled soft-iron bar inside the bulb. For additional details see reference 122.
(Reprinted with permission from: G. A. Bottomley and R. L. Scott, *J. Chem. Thermodyn.*, 1974, **6**, 973–981; copyright © 1974, Elsevier B. V.).

Tanaka *et al.*,[123] Dickinson *et al.*,[124] Kumaran and McGlashan (see Figure 7.3),[125] and Cibulka and Holub.[126] The precision attainable with such instruments has been impressively demonstrated by Marsh and Richards[127]

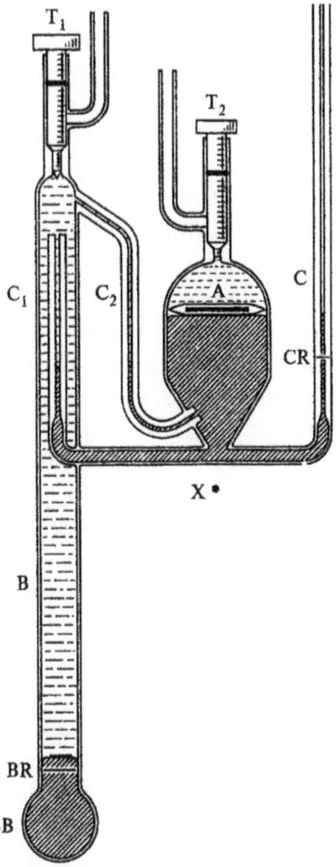

Figure 7.3 Improved grease-free tilting dilution dilatometer of Kumaran and McGlashan.[125] It is similar in design and operation to the Bottomley–Scott dilatometer[122] shown in Figure 7.2; it is, however, easier to load, easier to operate, and easier to calibrate. For details see reference 125. (Reprinted with permission from: M. K. Kumaran and M. L. McGlashan, *J. Chem. Thermodyn.*, 1977, **9**, 259–267; copyright © 1977, Elsevier B. V.).

who investigated the important system (ethanol + water) at 10 K intervals from 278.15 K to 338.15 K over the entire composition range, paying particular attention to the regions dilute in either ethanol or water. Modifying the dilatometer of Stokes *et al.*,[121] Bottomley and Scott[122] eliminated all grease taps in their design of a tilting dilution dilatometer. This dilatometer, shown in Figure 7.2, is the most precise instrument so far designed, and with adequate thermostating (temperature control to about 1 mK or better) a remarkably small imprecision in V^E of *ca.* ± 0.00015 cm^3 mol^{-1} is reported. Using this tilting-type continuous-dilution dilatometer, Bottomley and Scott measured the volume change on mixing D_2O with H_2O.[128] For this mixture the molar volumes are almost additive. The imprecision claimed for this experiment amounts to $\pm 0.2 \times 10^{-4}$ cm^3 mol^{-1}. Most interestingly, at 25 °C (t/°C = T/K − 273.15) the excess molar volume is positive with a maximum value of about 1.8×10^{-4}cm^3 mol^{-1} near $x_{D_2O} = 0.5$, while at 4 °C it is negative with a minimum of about -1.2×10^{-4} cm^3 mol^{-1} near $x_{D_2O} = 0.5$. Note that the temperature of maximum density for water is 4.0 °C, and for heavy water it is 11.6 °C. The Bottomley-Scott dilatometer is fairly easy to operate, yet the filling under vacuum is tedious, and it is limited to determining *small* volume changes on mixing. The grease-free tilting dilution dilatometer of Kumaran and McGlashan[125] is easier to handle and is not restricted by the size of the volume changes on mixing to be measured.

In almost all of the dilatometers presented, compressibility corrections caused by the changes of the mercury level during measurement must be considered.[112] In order to avoid corresponding corrections a few apparatus allowing operation at constant pressure have been designed. For instance, the dilatometer of Tanaka *et al.*[123] which is based on the earlier design of Chareyron and Clechet.[129]

Excess molar volumes can be calculated from experimentally determined mass densities $\rho(T,P,\{x_i\}) = m/nV(T,P,\{x_i\})$ at known composition. For a binary liquid mixture at constant temperature (and constant pressure),

$$V^{\mathrm{E}} = x_1 m_{\mathrm{m},1}\left(\frac{1}{\rho}-\frac{1}{\rho_1^*}\right) + x_2 m_{\mathrm{m},2}\left(\frac{1}{\rho}-\frac{1}{\rho_2^*}\right), \tag{7.56}$$

where $m_{\mathrm{m},1}$ and $m_{\mathrm{m},2}$ are the molar masses of the components with pure-substance densities ρ_1^* and ρ_2^*. For a multicomponent mixture with p components, the excess molar volume is calculated from

$$V^{\mathrm{E}} = \sum_{i=1}^{p} x_i m_{\mathrm{m},i}\left(\frac{1}{\rho}-\frac{1}{\rho_i^*}\right). \tag{7.57}$$

Classically, density measurements using pycnometers have been the main source of experimental data for the determination of excess volumes.[52,59] While simple in principle, great care must be exercised in thoroughly degassing the liquids, in filling the pycnometer, in applying corrections for buoyancy and for the amounts of evaporated mixture components in the vapour space: a rather comprehensive description of the measuring methodology and various types of pycnometers has been given in Weissberger's *The Physical Methods of Organic Chemisty.*[110] Of central importance is the determination of the composition of the liquid mixtures. Mixing bottles designed to circumvent uncertainties associated with directly filling the instrument with hypodermic syringes have been designed, for instance, by Wood and Brusie,[130] by Kohler and Rott,[131] and by Battino.[132] With proper care, this time-consuming classical method yields precise densities.

One of the most recent pycnometer designs is due to Westwood and Kabadi.[133] It is the first batch overflow-type pycnometer designed for measurements at elevated temperatures (up to 473.15 K) and moderately elevated pressures (up to 5 MPa). The experimental setup consists of four individual cells (stainless steel pycnometers) and of one cell for temperature measurements all connected through a network of tubings and fittings to a helium cylinder for pressurisation. The main feature of the design includes a bored-through expansion fitting, which allows the overflow due to thermal expansion from the cell *via* a small-bore tube to collect in a pressurised line. Temperature is controlled to about ± 0.2 K, and the pressure to about ± 69 kPa. The pycnometers were calibrated with degassed deionised water, leading to an estimated uncertainty in the density of the liquids of ± 0.6 kg m^{-3}. The instrument was used to measure the densities of 1-butanol and heptane, and their binary mixtures over the entire composition range[133] at 4.93 MPa and four temperatures: 316.85 K, 376.15 K, 427.55 K, and 458.15 K. Comparison with the excess molar volumes reported by Zielkiewicz[134] at 313.15 K and ambient pressure indicates reasonable agreement within the uncertainty limits of this new pycnometer.

Undoubtedly, the most frequently used method for measuring liquid densities to obtain V^{E} is *vibrating-tube densimetry*. More than 45 years have passed since the seminal publication of Kratky *et al.*,[53] and vibrating-tube densimeters have been extensively used for density measurements not only at ambient conditions but increasingly also at high temperatures and pressures. Attractive features of this precision method are simple operation, small sample volumes, the possible use in either the static (batch) mode or the flow mode,[54] in particular for measurements on aqueous solutions at high temperatures extending into the supercritical region (as pioneered by Albert and Wood[135] at the University of Delaware, Newark, USA), problem-oriented selection of the incremental pressure steps which may be kept quite small, and the possibility of computer control and automation of measurements. Vibrating-tube densimetry is particularly well suited for the determination of *small* density differences, a fact which has been used, for instance, in the determination of *partial molar volumes* $V_2^{\mathrm{L}\infty}$ *at infinite dilution* of gases in liquids (superscript L),[136–141] a topic covered in Chapter 9 of this monograph. For work at ordinary temperatures and low pressures, the oscillator is usually made of glass, while for elevated temperatures and pressures appropriate metal tubes are used (stainless steel, Pt-Rh (20 per cent), Hastelloy). During the last two decades, this versatile method has greatly profited from the introduction of many new, improved designs: a selection is presented in references 142 through 156.

Vibrating-tube densimeters are based on the principle that the density of a fluid contained in the U-shaped (or V-shaped) hollow oscillator of the instrument is related to the natural vibration frequency of the tube, that is to say for the usual range of interest, to an excellent approximation

$$\rho(T,P,\{x_i\}) = A(T,P) + B(T,P)\tau^2. \tag{7.58}$$

Here, τ is the period of vibration, and the temperature and pressure dependent parameters A and B are characteristic for a particular oscillator.[157,158] Usually, the densimeter is calibrated with pure degassed water and nitrogen at atmospheric pressure or, alternatively, with vacuum, *i.e.* zero density.[159] Once the constant B is determined, one may obtain density differences $\rho^{\mathrm{L}} - \rho_{\mathrm{ref}}^{\mathrm{L}*}$ of any solution with density ρ^{L} relative to that of pure water with density $\rho_{\mathrm{ref}}^{\mathrm{L}*}$ from

$$\rho - \rho_{\mathrm{ref}}^{\mathrm{L}*} = B\left(\tau^2 - \tau_{\mathrm{ref}}^2\right). \tag{7.59}$$

Thus, measurements with this instrument are preferably performed at the *same* temperature and pressure with the sample under investigation and a reference fluid (subscript ref) of precisely known density $\rho_{\mathrm{ref}}^{\mathrm{L}*}(T,P)$. In order to fully preserve the potential precision of the instrument, frequent calibration is obligatory.

The first measurements of densities of binary liquid mixtures were done in the dynamic mode, *i.e.* with liquids flowing through the oscillating tube at room temperature and atmospheric pressure.[54–58] This allows fast and precise measurements, and permits the simultaneous determination of densities *and* heat capacities by connecting in series a vibrating-tube densimeter (flow mode) and a flow heat capacity calorimeter (Picker type).[55] Extension of this experimental technique to high temperatures and high pressures originated at the University of Delaware, Newark, USA, with the work of Albert and Wood.[135] Their instrument was later improved at Blaise Pascal University, France, by Hynek *et al.*,[149] whose densimeter was developed for measuring corrosive solutions at temperatures up to 723 K and at pressures up to 40 MPa. Further improvements of this family of homemade vibrating-tube flow densimeters are due to Hnědkovský and Cibulka at the Prague Institute of Chemical Technology, Czech Republic. Lately, they developed a flow densimeter for measuring small density differences in dilute aqueous solutions of organic substances[154] over the temperature range 298 K to 573 K, and pressure range 0.1 MPa to 30 MPa. This instrument uses a photo-electric pick-up system to capture the periods of the mechanical oscillations of the vibrating tube instead of the classical magnetic detection system used in the original design of Stabinger *et al.*[53]

However, parallel to these research-laboratory-based improvements of flow densimetry, the current trend is to organise fully automated measuring installations, able to work over extended temperature and pressure ranges, using a commercial vibrating-tube densimeter (predominantly from Anton Paar, Graz, Austria) working in the static mode. Here, we cite just a few representative recent articles, and note quite similar temperature and pressure ranges: Troncoso *et al.*[160] ($278 < T/\mathrm{K} < 333$, $0.1 < P/\mathrm{MPa} < 40$; see Figure 7.4), Garcia *et al.*[161] ($278 < T/\mathrm{K} < 358$, $0.1 < P/\mathrm{MPa} < 60$), Zéberg-Mikkelsen *et al.*[162] ($298 < T/\mathrm{K} < 353$, $0.1 < P/\mathrm{MPa} < 45$), Outcalt and McLinden[163] ($270 < T/\mathrm{K} < 470$, $0.1 < P/\mathrm{MPa} < 50$), Segovia *et al.*[164] ($283 < T/\mathrm{K} < 383$, $0.1 < P/\mathrm{MPa} < 70$). The latter two contributions include measurements on industrial fluids.

In vibrating-tube densimetry, the reference liquid almost universally used for calibration measurements is pure deaerated water.[165] The molar volume

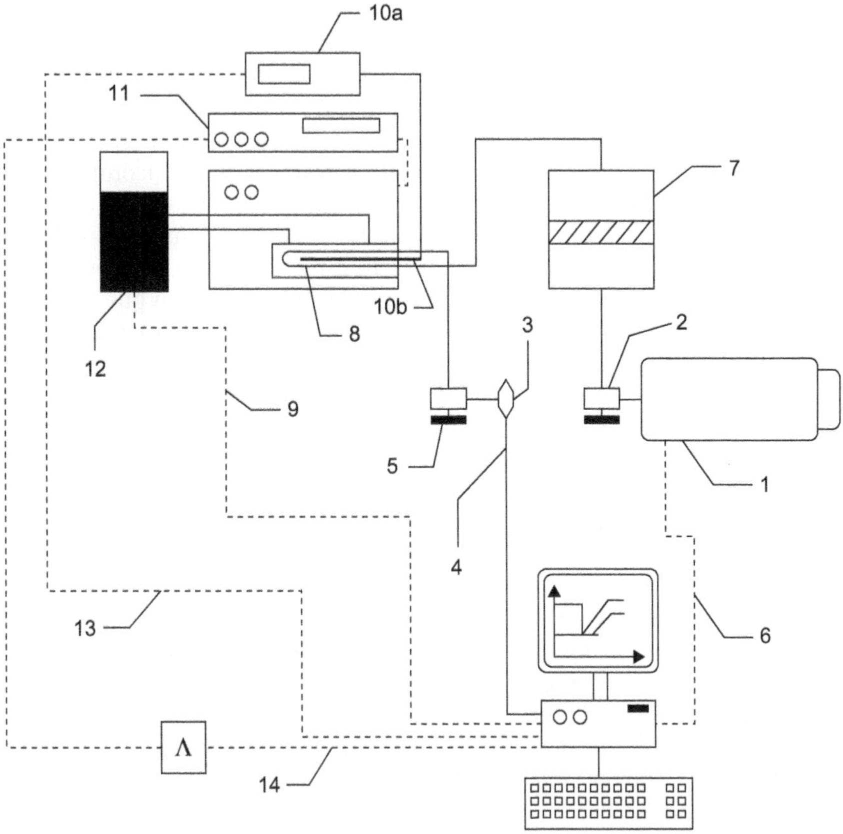

Figure 7.4 Fully automated measuring system for $P\rho T$-measurements on liquids as used by Troncoso *et al.*[160] It makes use of an Anton Paar high-pressure vibrating-tube densimeter (DMA 512P), and is thus illustrative of the present trend to measure densities of liquids very precisely by using vibrating bodies (tubes or wires), the oscillation frequency of which is directly related to the density of the liquid under investigation. Most of the recent systems presented in the literature use commercial Anton Paar instruments which allow density measurements over broad temperature and pressure ranges. In the figure, the high-pressure cell (8) is coupled with a seven-digit frequency counter (11). Temperature is controlled to ± 0.005 K by a thermostat (12) and measured with a Pt100 probe located close to the cell (10b). Pressure is regulated within ± 0.005 MPa by a Ruska pressure controller (1), transmitted *via* a pressure separator (7) that separates the sample from the pressurising oil, and measured upstream by a pressure transducer (3). The setup of Troncoso *et al.* permits data acquisition from 278.15 K to 333.15 K at pressures up to 40 MPa. For details see reference 160.
(Reprinted with permission from: J. Troncoso, D. Bessières, C. A. Cerdeiriña, E. Carballo and L. Romani, *Fluid Phase Equil.*, 2003, **208**, 141–154; copyright © 2003, Elsevier B. V.).

of pure liquid water may be obtained from the formulation of the International Association for the Properties of Water and Steam (IAPWS)[9] as implemented in a NIST database.[166] Alternatively one may use the property formulation adopted by the International Committee for Weights and Measures (CIPM) in 2001: Tanaka *et al.*[167] recommend a new standard for use in metrology for the density of gas-free water with the isotopic abundance of Vienna Standard Mean Ocean Water (VSMOW) within the temperature range 0 °C to 40 °C and at 101.325 kPa. This parallels the formulation endorsed by the IAPWS-95.[9] In a recent article, Harvey *et al.*[168] discussed the roles of IAPWS and CIPM standards and present guidelines for their appropriate use. We note that within the range of validity of the CIPM formulation, mutual agreement is within the respective quoted uncertainties. Both standards yield *very* similar values for the density, with the largest difference being observed at 40 °C (and 101.325 kPa): here, the IAPWS-95 density of pure water is higher by *ca.* 1.14×10^{-6} g cm^{-3}. The CIPM formulation for the density should *not* be extrapolated outside the range 0 °C to 40 °C. The selection of water property formulae for volume and flow calibration, including compressibility and viscosity, has been recently discussed by Batista and Paton.[169]

The influence of dissolved air on the density of water is of considerable interest in metrology. Quite a number of investigations have dealt with this problem,[170–176] though in part with inconsistent results. Using recently measured partial molar volumes at infinite dilution[136,137,141c,170c] of Ar, N_2, O_2 and CO_2 in conjunction with high-precision values of Henry fugacities,[177–180] Harvey *et al.*[181] calculated the molar volume of the solution of dry air[182] in water as

$$V^{\mathrm{L}} = x_{\mathrm{w}} V_{\mathrm{w}}^{\mathrm{L}*} + \sum_{i=2}^{n} x_i V_i^{\mathrm{L}\infty}, \tag{7.60}$$

where x_{w} is the mole fraction of water, $V_{\mathrm{W}}^{\mathrm{L}*}$ is the molar volume of pure water obtained from the formulation of the IAPWS-95, and $V_i^{\mathrm{L}\infty}$ is the partial molar volume of gaseous solute i (Ar, N_2, O_2, CO_2) at essentially infinite dilution, and the sum is over all solute species. For such highly dilute solutions far below the critical point of the solvent, Equation (7.60) is an excellent approximation. Aqueous carbon dioxide undergoes a weak ionisation reaction that was also taken into account[183,184] (the second ionisation step is negligible for this purpose). The density increment due to dissolved air in water is then computed as

$$\Delta\rho^{\mathrm{L}} \equiv \rho^{\mathrm{L}} - \rho_{\mathrm{w}}^{\mathrm{L}*} = \frac{m_{\mathrm{m}}}{V^{\mathrm{L}}} - \frac{m_{\mathrm{m,w}}}{V_{\mathrm{w}}^{\mathrm{L}*}}, \tag{7.61}$$

where ρ^{L} is the density of air-saturated water at a total pressure of 101.325 kPa, $\rho_{\mathrm{W}}^{\mathrm{L}*}$ is the density of pure, air-free water at 101.325 kPa, and m_{m} and $m_{\mathrm{m,w}}$ denote the molar mass of the solution and the molar mass of pure water, respectively. The calculated values $\Delta\rho^{\mathrm{L}}$ for air saturation of water at a total pressure of 101.325 kPa were expressed as a function of temperature from 0 °C to 50 °C by

$$\Delta\rho^{\mathrm{L}}/(\mu\mathrm{g\ cm^{-3}}) = 0.103 - 2.371\times10^{5}\theta^{-2.5} + 1.820\times10^{-7}\theta^{3}, \qquad (7.62)$$

where $\theta = t/^{\circ}\mathrm{C} + 75$.

Calibration issues of vibrating-tube densimetry and related topics have been carefully discussed by Majer and Pádua,[185] together with a general survey of instruments, including *vibrating-wire densimeters*. Vibrating-wire densimeters are widely used for measurements of both density *and* viscosity at high pressures;[186–191] they are essentially hydrostatic weighing densimeters where a solid sinker suspended in a fluid sets the wire under tension. The buoyancy force on the sinker, caused by the fluid, reduces the wire tension and thereby lowers its resonant frequency from that measured in vacuum.

Measuring techniques for fluid densities using instruments exploiting the *Archimedes principle* (buoyancy or hydrostatic methods) have been developed to unprecedented precision and accuracy (two-sinker densimeters), in particular by Wagner's group in Bochum, Germany.[192–195] For measurements at moderate and high densities they also developed single-sinker densimeters with a novel type of a magnetic suspension coupling, which significantly simplifies experiments.[196–200] For instance, Sommer *et al.*[200] recently reported comprehensive $P\rho T$ measurements on liquid cyclohexane, toluene and ethanol in the temperature range from 233.15 K to 473.15 K at pressures up to 30 MPa. The uncertainty of the measured densities was estimated to be $\pm 0.015\%$. This work complements previous efforts of Schilling *et al.*,[199] where the densities of liquid heptane, nonane, 2,4-dichlorotoluene and bromobenzene were determined over the same temperature and pressure ranges, though due to chemical instability a few of the measurements at the highest temperatures could not be used. The aim of their work was to provide comprehensive and accurate data sets of densities in this scientifically and technically important temperature and pressure range. There is a lack of density reference liquids at lower as well as higher densities than that of water (and water cannot be used at T <273.15 K!), and the investigated liquids qualify for use as *density reference liquids* for the calibration of densimeters. Their densities $\rho/\mathrm{kg\ m^{-3}}$ at $T = 293.15$ K and $P = 0.1$ MPa range from 684.0 (heptane) to 1493.6 (bromobenzene), covering a large range indeed. In order to facilitate their practical use, they also provide accurate correlating equations, the structure of which has been obtained through use of the optimisation technique of Setzmann and Wagner:[201]

$$\frac{\rho(T,P)}{\rho_{\mathrm{ref}}(T_{\mathrm{ref}},P_{\mathrm{ref}})} = \sum_{i=1}^{q} r_i \tau^{t_i} \pi^{p_i}. \qquad (7.63)$$

Here, $\tau \equiv (T/T_0 - 1)$, and $T_0 = 100\,\mathrm{K}$; $\pi \equiv (P/P_0 + 1)$, and $P_0 = 100\,\mathrm{MPa}$; and the coefficients r_i and the exponents t_i and p_i are substance-specific parameters. The number of terms varies between $q = 8$ for heptane and nonane,

$q=9$ for 2,4-dichlorotoluene and bromobenzene, $q=10$ for cyclohexane and toluene, and $q=20$ for ethanol. $\rho_{ref}(T_{ref},P_{ref})$ is a reference density measured with the single-sinker densimeter at $T_{ref}=293.15\,K$ and $P_{ref}=0.1\,MPa$: for instance, $\rho_{ref}(T_{ref},P_{ref})/kg\,m^{-3}$ amounts to 683.987 for heptane, to 778.632 for cyclohexane, to 789.433 for ethanol and to 1493.556 for bromobenzene. Though of great practical value, these correlating equations are in fact preliminary and should eventually be replaced by equations of state describing the *entire* thermodynamic surface in the spirit of the work contained in references 72 through 82. That is to say, by empirical equations explicit in the Helmholtz energy as a function of temperature and density.[71] An in-depth review of modern instruments developed in this field has been recently prepared by Wagner *et al.*[202] (see also reference 191).

The basic principle underlying density measurements *via* measuring the buoyant force on a sinker of known mass m_{sinker} and volume v_{sinker} is simple: after immersing the sinker in the thermostated fluid to be investigated, weighing, for instance, with a commercial analytic balance yields the fluid density as

$$\rho=\frac{m_{sinker}-m_{sinker}^{app}}{v_{sinker}}. \tag{7.64}$$

Here, m_{sinker}^{app} is the measured apparent mass of the sinker submerged in the sample fluid. A very thin platinum wire may be used to suspend the sinker, or an advanced type of magnetic suspension coupling (see above). Alternatively, one may use magnetic float densimeters which operate on the principle of exactly balancing gravity, buoyancy and magnetic field effects so that the hollow glass or quartz float containing a permanent magnet (or a soft-iron core) is levitated at a particular position in the liquid. This is achieved by passing a controlled electrical current through a solenoid positioned below the sample cell containing the liquid and the float. In turn, the magnetic force is determined from the coil current, usually by calibrating the instrument with liquids of accurately known density.[191,203–210]

An interesting instrument was developed by Keramati and Wolgemuth.[211] The density float consists of a disk on the periphery of which two cylindrical floats are permanently attached with a separation angle of 45° measured with respect to the horizontal pivot permanently mounted through the centre of the disk. The two cylinders possess nominally *equal* volumes but *different* densities. During an experiment the entire assembly, that is, disk, pivot and the two cylinders, is completely immersed in the fluid under study, which in turn is contained in a pressure vessel. The buoyancy forces acting on the two cylinders are different because their densities are different, hence the assembly will orient itself angularly (with respect to the pivot) into an equilibrium position where the net moment is equal to zero. For a given geometry of the float, the density of the fluid is simply related to the observed angle of rotation: thus the problem of measuring the density has been reduced to the task of measuring a *rotation angle*.

Excess molar volumes of $\{(1-x)H_2O+xCH_3OH\}$ from 0.1 MPa to 275 MPa at 278.15 K, 288.15 K, 298.15 K, 313.15 K, and 323.15 K were determined for six mixtures through use of a *bellows volumometer* by Easteal and Woolf.[212] In conjunction with literature data, they also determined partial molar volumes, isothermal compressibilities, isobaric expansivities, excess molar Gibbs energies G^E and excess molar enthalpies H^E over the entire ranges of pressure and temperature. Götze and Schneider[213] reported values of V^E for this system, measured dilatometrically with a stainless-steel instrument mounted in an autoclave, for $x=0.5$ up to 250 MPa and at T/K = 273.15, 298.15, 323.15, and 348.15; and Gibson,[214] using a different experimental method still (a variable-volume piezometer), obtained specific volumes over the entire composition range at 100 MPa and 298.15 K. Agreement is good throughout. Comparison with excess molar enthalpies at elevated pressures ($P \leq 60$MPa) determined by Heintz and Lichtenthaler[108] with an isothermal flow calorimeter, in the spirit of Equation (7.55), is also quite satisfactory. More recently, the liquid-phase isochoric heat capacity and the density of an aqueous methanol solution ($x=0.36$) were measured by Aliev *et al.*[215] with a twin-cell adiabatic calorimeter (for the density measurements this corresponds to an essentially direct weighing method). The temperature range extended from 333 K to 422 K, with pressures up to 20 MPa. From the measured densities excess molar volumes were calculated, which were in good agreement with results presented by Xiao *et al.*[148] (using a vibrating-tube densimeter at temperatures from 323 K through 573 K, and at pressures of 7.0 MPa and 13.3 MPa), and Yokoyama and Uematsu[216] (using a metal-bellows variable volumometer from 320 K to 400 K, and at pressures up to 200 MPa). Reasonable agreement was found between experimental values of V^E and values calculated with the model of Simonson *et al.*[217]

As pointed out in the Introduction, high-pressure density data of pure fluids and fluid mixtures, particularly at elevated temperatures, are of considerable importance for the chemical industry. Thus, during the last two decades, this field has received increased attention, and reviews have been recently presented by Woolf,[218] Holste,[219] and Palavra *et al.*[220,221]

The last sub-topic we shall touch upon concerns methods for density determinations at elevated pressures which are based on measuring the speed of ultrasound as a function of pressure. In this field, a quantity of central importance is the *isentropic compressibility* β_S, often loosely called the adiabatic compressibility:

$$\beta_S \equiv -V^{-1}(\partial V/\partial P)_S = \rho^{-1}(\partial \rho/\partial P)_S. \tag{7.65}$$

At *low sound frequencies* and *small sound amplitudes*, to an excellent approximation, this property is related to the *thermodynamic speed of ultrasound* v_0 by[222,223]

$$v_0^2 = \left(\frac{\partial P}{\partial \rho}\right)_S = \frac{1}{\rho\beta_S} = \frac{\kappa}{\rho\beta_T}, \tag{7.66}$$

where κ represents the *ratio* of the compressibilities, *i.e.* (see Chapter 1)

$$\kappa \equiv \beta_T/\beta_S = C_P/C_V \tag{7.67}$$

In passing, we note that β_S of liquids may also been determined by purely mechanical methods, that is, by directly measuring the volume increase on sudden decompression,[224] though this method is considerably less common and less accurate than that based on ultrasonics, Equation (7.66). The difference of the compressibilities may be expressed as

$$\beta_T - \beta_S = TV\alpha_p{}^2/C_P, \tag{7.68}$$

noting that this equation is of considerable importance in thermophysics (see below).

Within the constraints indicated above, ν_0 may be treated as a *thermodynamic equilibrium property*. However, at *higher* frequencies many fluids show sound speed dispersion due to relaxation effects,[222,223,225,226] *i.e.* the experimental sound speed becomes *larger* than ν_0 of Equation (7.66). Thus, care must always be exercised when deciding whether the measured speed of ultrasound is indeed the thermodynamic quantity to be used subsequently.[94] An attractive alternative to the experimental routes to *high-pressure PρT* data of liquids (superscript L) discussed so far, is to measure the thermodynamic speed of ultrasound ν_0 as a function of P and T, at *constant composition*, a field that was recently reviewed by Takagi and Wilhelm.[227] Combining these results on $\nu_0(T,P)$ at constant $\{x_i\}$, in the spirit of Equations (7.66) and (7.68), with data at ordinary pressure, say at $P_{\text{ref}} = 0.1$ MPa, *i.e.* with $\rho^{\text{L}}(T, P_{\text{ref}})$ and $C_{\text{P}}^{\text{L}}(T, P_{\text{ref}})$, the isothermal pressure dependence of the density may be expressed as

$$\left(\frac{\partial \rho^{\text{L}}}{\partial P}\right)_T = \frac{1}{\nu_0^2} + \frac{Tm_{\text{m}}\left(\alpha_P^{\text{L}}\right)^2}{C_P^{\text{L}}}. \tag{7.69}$$

Upon integration,[227–233] we obtain

$$\rho^{\text{L}}(T,P) = \rho^{\text{L}}(T,P_{\text{ref}}) + \int_{P_{\text{ref}}}^{P} \nu_0^{-2}\,\text{d}P + Tm_{\text{m}}\int_{P_{\text{ref}}}^{P} \left(\left(\alpha_P^{\text{L}}\right)^2 \Big/ C_P^{\text{L}}\right)\text{d}P. \tag{7.70}$$

The first integral is evaluated directly by fitting the isothermal ultrasonic speed data with suitably selected polynomials or Padé approximants, and for the second integral several successive integration algorithms have been devised. The simplicity, rapidity and precision of this method makes it highly attractive for the determination of the density (and hence for excess molar volumes), isobaric expansivity, isothermal compressibility, isochoric thermal pressure coefficient, isobaric heat capacity and isochoric heat capacity of liquid systems at high pressures. Its application to *Room Temperature Ionic Liquids* (RTILs)[234] has first been reported by Gomes de Azevedo and coworkers.[235] Details may be found in the original literature. Concerning the wide-temperature range/wide-pressure range results reported by Sun *et al.*,[230,231] for instance on the isobaric heat capacity and the isobaric expansivity of heptane and toluene,[231] the unequivocal proof of the existence of *minima* of the *isotherms* $C_P^L = C_P^L(P)$ at elevated pressures, and of a substance-specific *crossing "point"* (small crossing region?) of the *isotherms* of the isobaric expansivity $\alpha_P^L(P)$ at elevated pressures are particularly interesting (and intriguing). For heptane, this crossing "point" is found at *ca.* 120 MPa, at which pressure $\partial\alpha_P^L(P)$. Thus, for any given pressure *lower* than 120 MPa, α_P^L of heptane *increases* with temperature (at constant pressure), while at *higher* pressures α_P^L *decreases* with temperature (at constant pressure), *i.e.* $\partial\alpha_P^L/\partial T$ becomes *negative*. As evidenced by Equation (7.13), the pressure dependence of C_P^L,

$$\left(\frac{\partial C_P^L}{\partial P}\right)_{T,\{x_i\}} = -TV^L\left[\left(\alpha_P^L\right)^2 + \left(\partial\alpha_P^L/\partial T\right)_{P,\{x_i\}}\right], \tag{7.71}$$

is directly influenced by the temperature dependence of α_P^L. While $(\alpha_P^L)^2$ is always positive, depending on the pressure $(\partial\alpha_P^L/\partial T)_{P,\{x_i\}}$ can be positive or negative. That is, a minimum of the function C_P^L *vs.* P at constant T (and constant composition) is observed for that pressure where $(\alpha_P^L)^2 + (\partial\alpha_P^L/\partial T)_{P,\{x_i\}} = 0$.

The thermodynamic properties of pure liquid water are accessible as reference values through the IAPWS-95 formulation of the International Association for the Properties of Water and Steam[9] and, evidently, the speed of ultrasound in water is important for calibration purposes as well as for determining derived properties. Due to somewhat less precise experimental data of the speed of ultrasound in some pressure and temperature ranges, *i.e.* for $P > 200$ MPa and $T > 320$ K, new high-precision measurements are highly desirable. The recent publications of Benedetto *et al.*,[236] Meier and Kabelac,[237] Baltasar *et al.*,[238] and Lin and Trusler[239] are contributing to the improvement of the situation.

For older surveys of experimental techniques used for the measurement of *PVT* properties of liquids and liquid mixtures, in particular at elevated pressures, and their classification see Whalley[240] and Tekáč *et al.*[61] As already indicated above, recent reviews of topics of this field may be found in reference 30.

The binary mixture $\{x_1C_6H_6 + x_2c - C_6H_{12}\}$ has been suggested as a *test system* for measurements of excess volumes. In 1981, a critical evaluation of the extant literature at that time was presented by Cibulka and Holub.[126] They compared 32 published experimental values of V^E at 298.15 K and ambient pressure, determined with different measuring techniques (predominantly, dilatometers were used), with their own results obtained with a modified Kumaran–McGlashan dilution dilatometer.[125] Excess molar volumes were fitted to

$$V^E/(\mathrm{cm^3\ mol^{-1}}) = x_1x_2[2.59432 - 0.10198(x_1 - x_2) + 0.02509(x_1 - x_2)^2] \tag{7.72}$$

with a standard deviation of 0.0009 cm^3 mol^{-1} and a maximum deviation of $-0.0018\,cm^3\,mol^{-1}$. While *very* good agreement with the most reliable literature results was found, the deviations of some sets of literature data are considerable.[129,241,242] Cibulka and Holub[126] believe that the accuracy of the excess volumes correlated by Equation (7.72) is better than $\pm 0.002\,cm^3\,mol^{-1}$

7.3 Correlation of Experimental Data

Unfortunately, no general theory exists that adequately describes the composition dependence of excess molar quantities in general, and excess molar volumes in particular. The correlating equations commonly used are empirical or semiempirical at best.[6] Focusing on *binary* mixtures, definitely the most popular functional form is known as the Redlich–Kister equation.[243–245] Since by definition the excess molar volume is zero for the pure components, that is to say, for $x_1 = 0$ as well as $x_1 = 1$, V^E at constant temperature and pressure is usually expressed by

$$V^E = x_1 x_2 \left(A_0 + \sum_{m=1}^{a} A_m (x_1 - x_2)^m \right). \tag{7.73}$$

The parameters A_0 and A_m, $m = 1, 2, 3, \ldots$ are, of course, independent of composition but are, in general, functions of T and P.

The excess partial molar volume of component i at *infinite dilution* is defined by

$$V_i^{E\infty} = \lim_{x_i \to 0} V_i^E, \quad T \text{ and } P \text{ constant}, \tag{7.74}$$

and may thus be calculated for component 1 according to

$$V_1^{E\infty} = A_0 + \sum_{m=1}^{a} A_m (-1)^m, \tag{7.75a}$$

and for component 2 according to

$$V_2^{E\infty} = A_0 + \sum_{m=1}^{a} A_m. \tag{7.75b}$$

Over the years, quite a number of analytical representations of the composition dependence of thermodynamic excess properties of binary systems have been proposed. In particular, the use of power series in terms of mole fractions can be traced back to the seminal work of Margules[‡] of 1895[247] in which, in modern language, an empirical power series in the mole fraction

[‡]Max Margules was an Austrian Physicist. He was born in Brody, Galicia, a former crown-land of Austria, in 1856, and he died, because of malnutrition after World War I, in Perchtoldsdorf near Wien (Vienna), Austria, in 1920. He contributed significantly and lastingly to thermodynamics *and* meteorology.[246]

of one component was suggested for $\ln\gamma_i$ and thus for G^E/RT.[248] Using such an expansion also for the excess volume, the simple power series reads, for instance,

$$V^E = x_1x_2\left(a_0 + a_1x_2 + a_2x_2^2 + a_3x_2^3 + \cdots\right). \tag{7.76}$$

Alternatively, one could write an analogous expansion in x_1, though with *different* expansion parameters $a_0',a_1',a_2',\ldots$. Evidently, it is desirable to use a series with respect to a variable which reflects somehow the symmetry between the components.[243,244] Selecting $(x_1 - x_2)$ as the *independent* variable results in the *equivalent* Redlich–Kister expansion: here, an exchange of the components, *i.e.* $(x_2 - x_1)$, merely leads to a *change in sign* of the coefficients of the odd-power terms with $A_1,A_3\cdots$. Geometrically, the even-power terms are symmetrical in mole fraction and flatten or sharpen the parabola given by $V^E = A_0x_1x_2$, while the odd-power terms are asymmetrical in mole fraction and thus skew the parabola. Note that all terms with $(x_1 - x_2)^m$, $m \geq 1$, are zero for $x_1 = x_2 = 0.5$.

When the Redlich–Kister equation is truncated after the *linear* term under the summation sign, *i.e.* when $A_m = 0$ for $m \geq 2$, it may be rearranged into an equivalent polynomial form, often used to express the composition dependence of G^E/RTx_1x_2 (and thus that of $\ln\gamma_1$ and $\ln\gamma_2$) and known as the *three-suffix* Margules[§] equation:

$$\frac{V^E}{x_1x_2} = A_{21}x_1 + A_{12}x_2. \tag{7.77}$$

The excess partial molar volumes are thus given by

$$V_1^E = x_2^2[A_{12} + 2x_1(A_{21} - A_{12})] \tag{7.78a}$$

and

$$V_2^E = x_1^2[A_{21} + 2x_2(A_{12} - A_{21})], \tag{7.78b}$$

hence

$$A_{12} = A_0 - A_1 = V_1^{E\infty} \tag{7.79a}$$

and

$$A_{21} = A_0 + A_1 = V_2^{E\infty} \tag{7.79b}$$

For highly skewed data the use of more than four or five terms in the Redlich–Kister expansion may cause spurious oscillations in derived excess partial molar volumes and, in particular, may lead to unreliable limiting values $V_i^{E\infty}$. The flexibility required to fit strongly unsymmetrical curves

[§]The designation *three-suffix* indicates that the polynomial in x used to represent the excess molar volume is of order 3.

$V^E = V^E(x_1)$ at constant T and P is provided by *Padé approximants*[100,249,250] of order $[a/b]$

$$V^E = x_1 x_2 \frac{A_0 + \sum_{m=1}^{a} A_m (x_1 - x_2)^m}{1 + \sum_{n=1}^{b} B_n (x_1 - x_2)^n}, \tag{7.80}$$

where the denominater must never become zero. Note that $x_1 - x_2 = 2x_1 - 1 = 1 - 2x_2$.

As alternatives to Equations (7.73) and (7.80), expressions based on *orthogonal polynomials* have been suggested,[251–253] for instance expansions based on Legendre polynomials,[252,253] *i.e.*

$$V^E = x_1 x_2 \sum_{p=0}^{n} a_p L_p(x_1), \tag{7.81}$$

or alternatively by

$$V^E = x_1 x_2 \sum_{p=0}^{n} a_p L_p(z_{12}), \tag{7.82}$$

where $z_{12} = x_1 - x_2$. The first five polynomials are listed in Table 7.1. While $L_p(x_1)$ is identical to $L_p(z_{12})$ and the coefficients a_p of the two series are, of course, the same, as pointed out by Pelton and Bale[252] and by Hillert,[253] the use of Legendre expansions in terms of $L_p(z_{12})$ has certain advantages. Pelton and Bale also give expressions for the excess partial molar volumes, which are obtained from

$$\begin{aligned} V_1^E &= V^E - (z_{12} - 1)\frac{dV^E}{dz_{12}} \\ &= x_2^2 \sum_{p=0}^{n} a_p \left[L_p(z_{12}) + (z_{12} + 1)\frac{dL_p(z_{12})}{dz_{12}} \right], \end{aligned} \tag{7.83a}$$

and

$$\begin{aligned} V_2^E &= V^E - (z_{12} + 1)\frac{dV^E}{dz_{12}} \\ &= x_1^2 \sum_{p=0}^{n} a_p \left[L_p(z_{12}) + (z_{12} - 1)\frac{dL_p(z_{12})}{dz_{12}} \right], \end{aligned} \tag{7.83b}$$

with V^E being given by Equation (7.82).

Table 7.1 The first five Legendre polynomials $L_p(x_1)$ and $L_p(z_{12})$, where $z_{12} = x_1 - x_2$, to be used with Equation (7.81) or Equation (7.82), respectively.

p	Legendre polynomial $L_p(x_1)$	Legendre polynomial $L_p(z_{12})$
0	1	1
1	$2x_1 - 1$	z_{12}
2	$6x_1^2 - 6x_1 + 1$	$(3z_{12}^2 - 1)/2$
3	$20x_1^3 - 30x_1^2 + 12x_1 - 1$	$(5z_{12}^3 - 3z_{12})/2$
4	$70x_1^4 - 140x_1^3 + 90x_1^2 - 20x_1 + 1$	$(35z_{12}^4 - 30z_{12}^2 + 3)/8$
5	$25x_1^5 - 630x_1^4 + 560x_1^3 - 210x_1^2 + 30x_1 - 1$	$(63z_{12}^5 - 70z_{12}^3 + 15z_{12})/8$

The *recurrence relation* for calculating Legendre polynomials as used by Pelton and Bale[252b] is

$$L_p(z_{12}) = \frac{(2p-1)z_{12}}{p}L_{p-1}(z_{12}) - \frac{p-1}{p}L_{p-2}(z_{12}), \tag{7.84}$$

with $L_0(z_{12})=1$ and $L_1(z_{12})=z_{12}$; these two terms are identical with the corresponding Redlich–Kister terms. For the derivatives appearing in Equations (7.83a and b), the recurrence relation is thus[252b]

$$\frac{\mathrm{d}L_p(z_{12})}{\mathrm{d}z_{12}} = \frac{(2p-1)}{p}L_{p-1}(z_{12}) + \frac{(2p-1)z_{12}}{p}\frac{\mathrm{d}L_{p-1}(z_{12})}{\mathrm{d}z_{12}} - \frac{p-1}{p}\frac{\mathrm{d}L_{p-2}(z_{12})}{\mathrm{d}z_{12}}. \tag{7.85}$$

Legendre polynomials satisfy

$$\int_{-1}^{1} L_p(z_{12})L_q(z_{12})\mathrm{d}z_{12} = \begin{cases} 0 & \text{if } q \neq p \\ \dfrac{2}{2p+1} & \text{if } q = p \end{cases}. \tag{7.86}$$

This orthogonality property permits a polynomial approximation to any continuous function within its range [−1,1]. When used with necessarily *discrete* experimental results, Legendre polynomial expansions have the merit that increasing the number of terms to improve the fit with experimental data will only *slightly* influence the values of the lower order terms. Explicit conversion formulae to calculate Legendre coefficients from Redlich–Kister coefficients (or from simple power series coefficients) have been given by Pelton and Bale,[252b] and have also been described by Howald and Eliezer,[254] and by Tomiska.[255]

If the number of components in a mixture goes to three and beyond, the experimental effort for the determination of excess properties increases sharply. This explains the much smaller number of accurate excess property data for liquid multicomponent mixtures compared to the number of excess property data for liquid binary systems in general and of excess volume data in particular. This situation is paralleled by less reliable empirical or semi-empirical correlating functions to represent their composition dependence. The Redlich–Kister expansion can be extended without difficulty to multicomponent mixtures,[243] yielding, for instance, for a *ternary* system at constant T and P,

$$V^{\mathrm{E}} = V^{\mathrm{E}}_{12} + V^{\mathrm{E}}_{13} + V^{\mathrm{E}}_{23} + V^{\mathrm{E}}_{123}, \tag{7.87}$$

where the V^{E}_{ij} refer to the constituent binaries, *i.e.*

$$V^{\mathrm{E}}_{ij} = x_i x_j \left[A_{ij} + B_{ij}(x_i - x_j) + C_{ij}(x_i - x_j)^2 + \cdots\right], \tag{7.88}$$

with $i \neq j$ and $i < j$, and

$$V^{\mathrm{E}}_{123} = x_1x_2x_3[A_{123} + D_1(x_2 - x_3) + D_2(x_3 - x_1) + \ldots \tag{7.89}$$

represents the ternary term.

The frequently used prediction of multicomponent behaviour from data of the constituent binaries alone, *without* ternary (or higher) terms, is an approximate result arising from the model assumptions introduced. The most successful correlation of this type is Kohler's equation,[256] *i.e.*

$$V^{E} = (x_1 + x_2)^2 V_{12}^{E}(x_1{}', x_2{}') + (x_1 + x_3)^2 V_{13}^{E}(x_1{}', x_3{}') + (x_2 + x_3)^2 V_{23}^{E}(x_2{}', x_3{}'), \tag{7.90}$$

where the x_i, $i = 1$, 2 or 3, are the mole fractions in the ternary system. The quantities V_{ij}^{E} are assumed to depend *only* on the ratio x_i/x_j, $i \neq j$, *i.e.* on the ratio x_i'/x_j' of the corresponding binary system: the details of the molecular interaction between component i and component j are *not* influenced by the addition of the third component. Thus, V_{ij}^{E} denotes the excess molar volume of the binary mixture composed of substance i and substance j with corresponding superscripted mole fractions

$$x_i' = 1 - x_j' = x_i/(x_i + x_j). \tag{7.91}$$

Kohler's equation is *symmetrical*, that is, the three binary systems are treated identically.

Quite a few authors have used the Cibulka equation,[257] which has the same structure as the Redlich–Kister equation. However, the ternary contribution

$$V_{123}^{E} = x_1 x_2 x_3 \left[B_0 + B_1' x_1 + B_2' x_2\right] \tag{7.92}$$

differs only in form from the Redlich–Kister expression truncated after the D_2 term; they are mathematically equivalent.[258] Another three-term expression for the ternary contribution has been suggested by Singh *et al.*:[259]

$$V_{123}^{E} = x_1 x_2 x_3 \left[A + B x_1 (x_2 - x_3) + C x_1^2 (x_2 - x_3)^2\right]. \tag{7.93}$$

Brief reviews of the performance of several polynomial equations in correlating excess molar volumes (and some other properties) of rather complex ternary systems have been recently published by Djordjević *et al.*[260]

7.4 Selected Results

In the second half of the 20th century, experimental investigation of thermodynamic properties of mixtures of organic liquids intensified significantly, partly due to increased demand from the chemical industry, and partly due to the interplay between theoretical advances, progress in computer simulation techniques and significantly improved experimental methods including improved computer-supported data reduction techniques. Excess thermodynamic properties G^{E}, H^{E}, V^{E}, and $C_p{}^{E}$, as well as thermomechanical coefficients and ultrasonic speeds, are now measured at an ever increasing rate, and are used to either check/develop solution models or provide data needed for chemical engineering applications. The flow of new experimental results is reflected by the dramatic increase in the volume of primary research journals and of data compilations like Landolt-Börnstein (see Introduction 7.1). In the next three subsections, due to space limitations (and time limitations), we shall present results on densities/excess volumes for just some selected binary molecular non-electrolyte mixtures (and some ternary mixtures) belonging to one of the following three groups: (I) mixtures of aprotic liquids, (II) mixtures containing alcohols, and (III) aqueous mixtures. The selection is, of course, highly subjective and simply reflects our preferences and idiosyncrasies. For the omission of many important papers, as well as related research areas, we would like to offer our apologies in advance.

7.4.1 Mixtures of Aprotic Liquids

In recent years, just a few systematic studies of the volumetric properties of fairly simple hydrocarbon mixtures have been communicated. For instance, liquid densities of pentane, octane and nonane, and of their binary mixtures were measured by Ramos-Estrada *et al.*[261] from 273.15 K to 363.15 K at 0.1 MPa, using a vibrating-tube densimeter from Anton Paar (DMA 5000). Volumetric properties, *i.e.* $\rho = \rho(T, P, x)$, of the binary systems (hexane + octane) and (hexane + decane) within the temperature range 313 K to 363 K and up to 25 MPa in pressure, were determined by Quevedo-Nolasco *et al.*[262] (using a DMA HPM from Anton Paar). For each mixture, the experimental densities were correlated with an empirical six-parameter equation (a modified Toscani–Szwarc equation[263]):

$$\rho(T, P, x) = \frac{d_3 - d_4 T + d_5 T^{1/2} + d_6 P}{d_1 + d_2 P}. \tag{7.94}$$

Using optimized parameters d_i, V^E, α_P and β_T were determined over the entire composition range. Studies by Iloukhani and Rezaei-Sameti focus on binary mixtures of methylcyclohexane with *n*-alkanes (C_5 to C_{10}).[264] Excess molar volumes were determined from density measurements with an Anton Paar densimeter (DMA 4500) operating in the static mode at 293.15 K, 298.15 K and 303.15 K and ambient pressure, and isobaric expansivities were derived. The data were subsequently used to check the extent of applicability of the Prigogine–Flory–Patterson theory.[265–267] This work parallels theirs on mixtures of toluene with *n*-alkanes (C_5 to C_{10}).[268] For the mixture of tetralin (1,2,3,4-tetrahydronaphthalene) with hexadecane, Paredes *et al.*[269] measured densities and speeds of ultrasound (with the automatic DSA 5000 densimeter from Anton Paar), as well as refractive indexes (at $\lambda = 589.3$ nm) at 293.15 K to 343.15 K and atmospheric pressure. Agreement with the excess molar volumes at 298.15 K and 308.15 K reported by Letcher and Scoones[270] is good. Tetralin and hexadecane differ significantly in size, shape and chemical nature: while the Prigogine–Flory–Patterson model[265–267,271] predicts the sound speed and the small positive V^E reasonably well, it gives the *sign* of dV^E/dT incorrectly.

We conclude this subsection with presenting two studies involving mixtures containing silicones. Matteoli *et al.*[272] investigated (V^E and H^Eat 298.15 K) mixtures of heptane with several linear and cyclic silicone oligomers. All examined systems have (small) *positive* excess volumes with the exception of {heptane + hexamethyldisiloxane}, where $V^E(x=0.5) = -0.105$ cm^3 mol^{-1} is found. In the series with linear silicones, $V^E(x=0.5)$ is largest for {heptane + dodecamethylpentasiloxane}, where it amounts to 0.106 cm^3 mol^{-1}. Somewhat larger excess volumes are observed with the mixtures of heptane with cyclic silicones: for {heptane +octamethylcyclotetrasiloxane}, $V^E(x=0.5) = 0.468$ cm^3 mol^{-1}. These results will contribute to a better understanding of the weak interactions of silicones with hydrocarbons, as will the work of Dong *et al.*,[273] which focuses on the interaction of a cyclic silicone with an aromatic compound. Specifically, 2,4,6,8-tetramethylcyclotetrasiloxane (TMCTS, [-HSi(CH_3)O-]$_4$) was selected, which is an important chemical in the semiconductor industry. Excess molar volumes for binary mixtures of TMCTS with benzene, toluene, *o*-xylene, *m*-xylene and *p*-xylene, at 288.15 K, 298.15 K and 308.15 K, were determined from densities measured with a vibrating-tube densimeter. The composition dependence of V^E was represented by two correlating functions: Redlich–Kister, Equation (7.73), and Equation (7.81) with Legendre polynomials. Partial molar excess volumes were also calculated at the three temperatures, as were the isobaric expansivities. The excess volumes are rather symmetric: at 298.15 K, $V^E(x=0.5) = -0.427$ cm^3 mol^{-1} for the mixture with benzene, while for the mixture with *p*-xylene $V^E(x=0.5) = -1.120$ cm^3 mol^{-1} is observed. The excess volumes become more negative with increasing temperature.

In contradistinction to the situation outlined above, mixtures involving polar liquids, like nitriles, halocarbons, esters, ketones, amines and amides, have attracted much more attention. For instance, V^E, H^E, and C_P^E of the mixtures {$x_{BN}C_6H_5CN + (1 - x_{BN})C_6H_5CH_3$} and {$x_{BN}C_6H_5CN + (1 - x_{BN})C_6H_6$} were measured by Wilhelm *et al.*[274] between 293.15 K and 318.15 K and at atmospheric pressure. The former system was selected for the following reasons: (I) both benzonitrile (BN) and toluene (T) have roughly the same size and shape, and they are both aprotic "rigid" aromatic molecules; (II) while toluene is essentially non-polar with a very small dipole moment ($\mu_T = 1.20 \times 10^{-30}$ C m), benzonitrile has a very large dipole moment ($\mu_{BN} = 13.94 \times 10^{-30}$ C m). Similar comments apply to the mixture of benzonitrile with benzene, the main differences being the zero dipole moment of benzene and its smaller shape anisotropy compared to toluene. For both mixtures, V^E is negative (becoming more negative with increasing temperature), small and skewed "parabolic", with a minimum at 318.15 K amounting to *ca.* −0.37 cm^3 mol^{-1} at $x_{BN} \approx 0.4$ in the system with toluene, and with a minimum of about −0.22 cm^3 mol^{-1} at $x_{BN} \approx 0.35$ in the system with benzene. However, the composition dependence of H^E is remarkable: for {$C_6H_5CN + C_6H_5CH_3$} a *very* rare pronounced *M-shape* (double maxima) is observed, while {$C_6H_5CN + C_6H_6$} exhibits a distinct shoulder in the range $0.05 < x_{BN} < 0.25$. An M-shaped curve H^E has also been reported by Bjola *et al.*[275] for the mixture of γ-butyrolactone (oxolane-2-one) with benzene at 293.15 K, with the negative, essentially parabolic V^E (about -0.8 cm^3 mol^{-1} at $x_{\gamma B} = 0.5$) also becoming more negative with increasing temperature.

We note that double maxima are also shown by excess enthalpies of mixtures of tetrachloromethane (TCM) with aromatic hydrocarbons in certain temperature ranges: for instance, H^E for $\{CCl_4 + C_6H_5CH_3\}$ at around 313 K[276] and H^E for $\{CCl_4 + 1,3\text{-}C_6H_4(CH_3)\}$ at around 293 K.[277] Of particular note is the strongly skewed *M-shape* of V^E for $\{CCl_4 + C_6H_6\}$ observed in an extremely narrow temperature range, as shown by the careful investigation of Bottomley and Scott.[278] Below about 288 K, V^E is negative for all mole fractions. Above 288.2 K a maximum develops at the TCM-rich side, near $x_{TCM} \approx 0.9$, and with increasing temperature the maximum increases and moves to lower values of x_{TCM}; the minimum also moves towards the TCM-poor side. Around 295.2 K a second maximum appears at the TCM-poor side, and the resulting double-maximum-shape, or M-shape, persists to about 296.2 K. As stated by Bottomley and Scott, close to 295.65 K the *absolute* values of the V^Es of the low-x_{TCM} maximum and the minimum are about equal and each amounts to roughly 0.0005 cm^3 mol^{-1}, while for the high-x_{TCM} maximum $V^E \approx 0.004$ cm^3 mol^{-1}.

Studies of thermodynamic properties of mixtures of α,ω-dihaloalkanes with hydrocarbons have a long tradition.[58,279–281] It is based on interesting questions associated with the existence of rotamers in the α,ω-dihaloalkanes showing large differences in the corresponding dipole moments (and differences in the corresponding molecular volumes), and on interesting questions concerning proximity effects. Recently, Chorążewski *et al.*[282] presented experimental densities, heat capacities at constant pressure (Micro DSC III, from Setaram) and speeds of ultrasound at 4 MHz (pulse-echo-overlap apparatus) for mixtures of 1,6-dichlorohexane (DCH) with *n*-heptane at 293.15 K, 298.15 K, 303.15 K, 308.15 K and 313.15 K, covering the whole composition range. From these data, excess molar volumes, isobaric expansivities, isentropic and isothermal compressibilities, molar heat capacities at constant volume and internal pressures were calculated. The V^Es are negative throughout: they are somewhat skewed with minima at *ca.* $x_{DCH} = 0.6$, and they become less negative with decreasing temperature. Most interestingly, the C_P^Es are all negative and strongly asymmetric with minima at *ca.* $x_{DCH} = 0.3$ and with broad shoulders at larger values of mole fraction x_{DCH}. With decreasing temperature, the $C_P^E(x_{DCH})$ curves become somewhat less negative with more pronounced shoulders, behaviour which is reminiscent of the onset of a *W-shaped* composition dependence.[280d,e,f] Densities of binary mixtures of 1,2-dichloroethane with polyethers, such as ethylene glycol dimethyl ether, were recently measured by Valtz *et al.*[283] from 283.15 K to 333.15 K.

The temperature dependence of V^E of binary mixtures of tetrahydropyran with *n*-alkanes (C_7–C_9) was determined by Bravo *et al.*[284] by measuring densities at 288.15 K and 308. 15 K, and discussed in terms of the Prigogine–Flory–Patterson theory.[265–267] Sulfolane[285] [2,3,4,5-tetrahydrothiophene-1,1-dioxide, tetramethylene sulfone, $(CH_2)_4SO_2$], a stable, polar aprotic solvent having a high relative permittivity ($\varepsilon_r = 46.4$) and a large dipole moment ($\mu = 16.0 \times 10^{-30}$ C m), is miscible with water, and is widely used in the chemical industry for the extraction of aromatic hydrocarbons from hydrocarbon mixtures.[286] Excess molar volumes for binary mixtures of sulfolane with toluene, *o*-xylene, *m*-xylene, *p*-xylene, ethylbenzene and 1,2,4-trimethylbenzene were determined at 298.15 K by Yu and Li.[287] The V^Es are all negative with minima at approximately $x = 0.5$, and for equimolar composition they range from −0.566 cm^3 mol^{-1} for the mixture with 1,2,4-trimethylbenzene to −1.070 cm^3 mol^{-1} for the mixture with toluene.

Thermodynamic properties of mixtures containing alkyl alkanoates have been measured by several groups (frequently in connection with biofuel studies), for instance in Las Palmas de Gran Canaria,[288] and in Clermont-Ferrand.[289,290] Here, we just specifically quote the recent work of Ortega *et al.*[291] on excess volumes and excess enthalpies for mixtures of ethyl ethanoate with alkanes (C_5 to C_{10}) at temperatures ranging from 291.15 K to 328.15 K, and that of Rathnam *et al.*[292] on the excess volumes for 12 binary mixtures of alkyl esters with chlorobenzene, bromobenzene and nitrobenzene at 303.15 K, 308.15 K and 313.15 K. For all the systems, V^E is negative at all compositions; for instance, V^E ($x = 0.5$)/cm^3 mol^{-1} at 308.15 K amounts to −0.150 for {ethyl formate + chlorobenzene}, −0.206 for {ethyl formate + bromobenzene} and −0.271 for {ethyl formate + nitrobenzene}.

Excess molar volumes and speeds of ultrasound were determined, over the entire composition range at several temperatures between 293.15 K and 323.15 K, of binary mixtures of ketones with amines (such as triethylamine, aniline or pyridine) by Alonso *et al.*,[293] of ethers with arylamines (such as *N*-methylaniline) by Oswal *et al.*,[294] and of cyclic ethers (such as tetrahydrofuran) with *N*,*N*-dimethylformamide by Sinha *et al.*[295]

Densities (and thus excess volumes) of the ternary system {cyclohexane + tetrahydrofuran + chlorocyclohexane} and of the corresponding binaries were measured at 298.15 K and 313.15 K by Gascón *et al.*[296] In a detailed experimental investigation, Wisniak *et al.*[297] reported densities and excess volumes at 298.15 K of the ternary mixture {ethylbenzene + styrene + ethyl acrylate} and of the three constituent binaries. The binary data were correlated with Redlich–Kister equations and with series of Legendre polynomials; and for the ternary behaviour several empirical correlating methods were compared. References 298 and 299 contain related work, that is, excess volumes of binary mixtures of benzene with ethyl acrylate, butyl acrylate, methyl methacrylate, and styrene; and excess volumes of binary mixtures of dimethyl carbonate with butyl methacrylate, allyl methacrylate, vinyl acetate, and styrene, respectively. Dimethylcarbonate, $OC(OCH_3)_2$, has a rather small dipole moment[300] ($\mu = 3.04 \times 10^{-30}$ C m) and low solubility in water, but is miscible with many organic solvents, such as diethyl ether and methanol; it is a widely used non-toxic, biodegradable methylating agent, replacing carcinogenic iodomethane and dimethyl sulfate, as well as a carbonylating agent, replacing highly toxic phosgene.[301] Densities, excess molar volumes, isothermal compressibilities and isobaric expansivities of mixtures of dimethylcarbonate with cyclohexane were determined at five different temperatures from 293.15 K to 313.15 K and over the pressure range of 0.1 MPa to 40 MPa by Zhou *et al.*[302] The densities of the pure liquids and their mixtures at atmospheric pressure were measured with a vibrating-tube densimeter (model DA-505, KEM, Japan), while the densities at elevated pressures were measured with a variable-volume autoclave and a precise analytical balance. The excess molar volumes are rather symmetric: $V^E(x=0.5)$ at constant $P=0.1$ MPa increases from 1.346 cm^3 mol^{-1} at 293.15 K to 1.762 cm^3 mol^{-1} at 313.15 K, while at constant $T=298.13$ K $V^E(x=0.5)$ decreases from 1.440 cm^3 mol^{-1} at 0.1 MPa to 0.949 cm^3 mol^{-1} at 40 MPa. Densities of the mixtures of propylenecarbonate ($\mu=16.5\times10^{-30}$C m) with acetonitrile ($\mu=11.5\times10^{-30}$ C m) and with *N,N*-dimethylacetamide ($\mu=12.4\times10^{-30}$C m) were obtained at 308.15 K and 318.15 K by Piekarski *et al.*,[303] while heat capacities C_P at constant pressure and speeds of ultrasound were measured at 298.15 K only. From these experimental data, V^E, C_P^E, α_P, β_S, β_T and the internal pressure

$$\Pi=T\alpha_P/\beta_T-P \tag{7.95}$$

were derived and discussed.

7.4.2 Mixtures Containing Alcohols or Alkanoic Acids

Undoubtedly, the most common *chemical* phenomena observed in solution thermodynamics are those involving *hydrogen bonds*. In liquid alcohols, a class of substances of considerable technological interest, hydrogen bonding (self-association) is a prominent aspect, hence mixtures containing alcohols have been the subject of numerous experimental investigations to help design and improve solution theories. The models developed for solutions involving associating molecules are usually based on the idea of discussing excess properties in terms of physical and chemical, *i.e.* hydrogen bonding, contributions. Perhaps the most popular models used are those based on the Prigogine–Flory–Patterson (PFP) approach,[265–267] on the *dispersive-quasi-chemical* (DISQUAC) model due to Kehiaian,[304] the *extended-real-associated-solution* (ERAS) model proposed by Heintz,[305] and the *lattice-fluid-associated-solution* (LFAS) theory of Panayiotou.[306]

In the tradition of the studies of associated liquids, initiated by Henry V. Kehiaian almost 50 years ago at the Institute of Physical Chemistry of the Academy of Sciences in Warsaw, Poland, Andrej and Teresa Treszczanowicz and their co-workers[307] have published new measurements of excess volumes of mixtures of an 1-alkanol with 1-octene at 288.15 K, 298.15 K and 308.15 K. The evolution of the V^Es as a function of the alkanol chain length (C_3–C_{10}) is discussed on the basis of a modified version of the Treszczanowicz–Benson association model.[308] Troncoso *et al.*[160] have used the recommended reference mixture {1-hexanol + *n*-hexane} to check the reliability of their entirely automated density measurement device to obtain new (ρ,T,P) data over extended ranges of T and P. For a review of older experimental data for the excess molar volumes of (1-alkanol + n-alkane) mixtures, see reference 309; for an older survey of excess volume data on mixtures of methanol with either heptane, or cyclohexane, or benzene, or tolouene, see reference 310.

Incomplete combustion of motor fuels results in environmentally harmful emissions of carbon monoxide and hydrocarbons, hence oxygen-containing substances, such as *ethers* or *alcohols* are added. In order to find the optimal compositions for the gasoline blends, vapour–liquid phase equilibrium models for the mixtures of interest are needed.[311] Excess volumes have been determined for binary mixtures of *tert*-amylmethyl ether (TAME) with ethanol,[312] both components being gasoline additives, and TAME acts as octane booster. For binary mixtures of *tert*-amylethylether (TAEE, 2-ethoxy-2-methylbutane) with 1-butanol, or with 2-butanol, or with 2-methyl-2-butanol, Sundberg *et al.*[313] present isothermal vapour–liquid equilibrium (VLE) data at 358 K (deriving activity coefficients), and excess molar volumes and excess molar enthalpies at 298.15 K. The measured VLE data were well correlated with local composition models.

Excess volumes for six ternary mixtures formed by 2,2,4-trimethylpentane, diisopropyl ether or methyl *tert*-butyl ether, and methanol, ethanol or 1-propanol have been determined from measured densities at 298.15 K by Vela *et al.*[314] Densities of binary mixtures of dibutyl ether (DBE) with 1-propanol, 2-propanol, and 1-hexanol at mole fractions of *ca.* $x_{\mathrm{DBE}} = 0.15$, 0.325, 0.5, 0.675 and 0.85 between 293.15 K and 353.15 K, and for pressures up to 140 MPa, at 10 MPa intervals, have been measured by Alaoui *et al.*[315] In order to correlate the densities over the entire temperature and pressure ranges, a Tait-type equation,

$$\rho(T,P) = \frac{\rho_0(T)}{1 - C\ln[(B(T)+P)/(B(T)+0.1\,\mathrm{MPa})]} \tag{7.96}$$

was used, where

$$\rho_0(T) = A_0 + A_1 T + A_2 T^2 + A_3 T^3 \tag{7.97}$$

and

$$B(T) = B_0 + B_1 T + B_2 T^2. \tag{7.98}$$

Excess molar volumes, isobaric expansivities and isothermal compressibilities were derived.

Methyl isobutyl ketone [MIBK, 4-methylpentan-2-one, $(CH_3)_2CHCH_2C(O)CH_3$] is a polar aprotic solvent with a fairly large dipole moment, *i.e.* $\mu = 9.3\times10^{-30}$ C m. Unlike other common ketone solvents, its comparatively low solubility in water makes it well-suited for liquid–liquid extraction. New VLE data, excess molar enthalpies and excess molar volumes were obtained by Laavi *et al.*[316] for industrially interesting binary mixtures of MIBK with 2-butanol, *tert*-pentanol (*tert*-amyl alcohol, 2-methyl-2-butanol, $C_5H_{11}OH$) and 2-ethyl-1-hexanol. The mixture of MIBK with *tert*-pentanol exhibits azeotropic behaviour. Excess molar volumes for mixtures of cyclohexanone with 2-propanol, 2-butanol, 2-pentanol, 2-hexanol and 2-heptanol, at 298.15 K, 303.15 K, 308.15 K and 313.15 K, have been reported by Almasi and Sarkoohaki:[317] they are generally quite small and symmetrical, and increase with increasing temperature.

N,*N*-Dimethylformamide [DMF, $(CH_3)_2NC(O)H$] has a large dipole moment $\mu = 12.9 \times 10^{-30}$C m, and the liquid has a high relative permittivity (at 298.15 K, $\varepsilon_r = 36.7$); it is miscible with the majority of the common non-polar and polar solvents, including water. Bhuiyan and Uddin[318] report excess volumes for binary mixtures of DMF with methanol, with ethanol, and with 2-propanol at five temperatures from 303.15 K to 323.15 K. The V^Es are negative throughout and rather symmetrical; they become more negative with increasing temperature: for instance, for {DMF + methanol} at 303.15 K, $V^E(x_{DMF} = 0.5) = -0.387$ cm^3 mol^{-1} is found, while at 323.15 K $V^E(x_{DMF} = 0.5) = -0.451$ cm^3 mol^{-1}. Excess molar volumes for mixtures of formamide with 1-alkanols (C_1–C_3), at five temperatures from 303.15 K to 323.15 K, have been presented by Gómez Marigliano and Sólimo.[319] They are in good agreement with data reported by Garcia *et al.*[320] V^Es for the binary mixtures of DMF with benzyl alcohol ($C_6H_5CH_2OH$), and with acetophenone (methyl phenyl ketone, $C_6H_5COCH_3$), at 298.15 K, 303.15 K, 308.15 K and 313.15 K, have been reported by Nikam and Kharat;[321] for excess volumes for mixtures of 1,4-dioxane with 1-propanol and 2-propanol see reference 322, and for excess volumes for mixtures of 1,2-dibromoethane with 1-propanol, 1-butanol and 1-pentanol, see reference 323.

For the mixture of tetrahydrofuran with 2,2.2-trifluoroethanol (TFE), Pérez *et al.*[324] measured excess molar enthalpies (with a Thermometrics flow calorimeter 2277), excess molar volumes and speeds of ultrasound (determined with an Anton Paar densimeter and sound analyzer DSA 48) at 283.15 K, 298.15 K and 313.15 K. The system shows large *negative* H^Es, which become *more* negative with increasing temperature: at 283.15 K, $H^E(x_{TFE} = 0.5) = -1740$ J mol^{-1}, while at 313.15 K, $H^E(x_{TFE} = 0.5) = -2120$ J mol^{-1}. Most interestingly, the excess molar volumes are moderately *positive* throughout (and somewhat skewed towards the TFE side), thus showing a sign which is opposite to that usually associated with strongly negative excess enthalpies. They become *more* positive with increasing temperature: at 283.15 K, $V^E(x_{TFE} = 0.5) = 0.998$ cm^3 mol^{-1}, while at 313.15 K, $V^E(x_{TFE} = 0.5) = 1.103$ cm^3 mol^{-1}. All these results are very similar to those obtained for the mixture of tetrahydropyran with TFE.[325]

The excess volumes at 298.15 K of 10 binary systems formed from the normal alcohols methanol, ethanol, propanol, hexanol and decanol were measured directly by Benson and Pflug[326] using a dilatometer, described in detail in reference 119. All excess molar volumes are positive and small, though they increase with increasing difference in chain length. In particular, the excess molar volumes of {methanol (MeOH) + ethanol (EtOH)} and of {ethanol + propanol} are *very* small and thus definitely present a challenge for the determination *via* densities determined by vibrating-tube densimetry.[327] For equimolar composition, for the former system $V^E(x_{MeOH}=0.5)=0.0085$ cm^3 mol^{-1} with a standard deviation of the Redlich–Kister fit of 2.2×10^{-4} cm^3 mol^{-1}, while for the latter system $V^E(x_{EtOH}=0.5)=0.0084$ cm^3 mol^{-1} with a standard deviation of the Redlich–Kister fit of 1.3×10^{-4} cm^3 mol^{-1}. Empirical application of the Flory theory[266] provided a useful semiquantitative correlation between the excess volumes and the excess enthalpies of these mixtures with hydrogen bonds. Densities and excess volumes for mixtures of methanol with ethanol, 1-propanol, 2-propanol, 1-butanol, 2-methyl-1-propanol (*iso*-butanol), 2-butanol and 2-methyl-2-propanol (*tert*-butanol) over the entire composition range at six different temperatures from 288.15 K to 313.15 K (in steps of 5 K) were recently reported by Boruń *et al.*[328] Excess molar volumes for binary systems of 1,3-propanediol mixed with 1-alkanols (C_1 - C_5) at four temperatures from 293.15 K to 313.15 K were recently presented by Almasi and Khosravi.[329] The V^Es for 1,3-propanediol mixed with methanol are negative throughout, and with increasing 1-alkanol chain length the excess volumes become less negative; the excess volume for 1,3-propanediol mixed with 1-pentanol is already positive. For all systems dV^E/dT is positive. The Peng–Robinson–Stryjek–Vera cubic equation of state,[330,331] in conjunction with simple mixing rules,[332,333] was used to correlate/predict the excess volumes. For the key alkanols methanol and ethanol the most reliable fluid density data over wide ranges of temperature and pressure are provided by the *International Thermodynamic Tables of the Fluid State – 12: Methanol*,[82] and by the values computed using the fundamental equation for ethanol of Dillon and Penoncello.[334]

Excess molar volumes and speeds of ultrasound for binary systems of ethylenediamine (EDA, $NH_2CH_2CH_2NH_2$) mixed with 2-propanol, 1-butanol and 2-methyl-1-propanol, at five temperatures from 293.15 K to 313.15 K, were determined by Dubey and Kumar.[335] They are negative throughout, somewhat skewed towards the alcohol side, and show some indication of a *shoulder* at compositions rich in EDA; dV^E/dT is *very* small.

N-Methyl-2-pyrrolidone (NMP, C_5H_9NO) is a dipolar aprotic solvent with a five-membered ring structure (lactam of 4-methylaminobutanoic acid). It has a fairly high relative permittivity ($\varepsilon_r=32$, at 298.15 K) and a large dipole moment ($\mu=13.6\times10^{-30}$ C m). It is miscible with the majority of the common organic solvents, such as benzene, chloroform and ethyl acetate, and with water. In fact, NMP is one of the most widely used commercial solvents for the separation of hydrocarbons by extractive distillation.[336] Using a dual path pulse-echo apparatus operating at 5 MHz, Dávila and Trusler[337a] report high-precision measurements of the speed of ultrasound in mixtures of NMP and methanol at temperatures between 298.15 K and 343.15 K (in steps of 15 K) and at pressures up to 60 MPa. They also measured the isobaric heat capacity of each mixture as a function of temperature at 0.1 MPa (or orthobaric pressure at 338.15 K and 348.15 K) using a Setaram DSC III microcalorimeter. In conjunction with density data at ambient pressure obtained from earlier measurements,[337b] the experimental results were combined in the spirit of Equation (7.70), and yielded the density, the isothermal compressibility, the isobaric expansivity, the excess molar volume, and the *specific* isobaric heat capacity $c_P \equiv C_P/m_m$ at temperatures between 298.15 K and 343.15 K and at pressures up to 60 MPa.

The preferential solvation at 313.15 K in binary mixtures of NMP with ethanol,[338a] and with methanol,[338b] was investigated by Zielkiewicz using two complementary methods: (I) excess molar Gibbs energies (and thus activity coefficients), determined *via* total vapour pressure measurements, and excess molar volumes, measured with a Kumaran–McGlashan dilution dilatometer,[125] provided the basis for a discussion of the thermodynamic data in terms of the *Kirkwood–Buff theory*[339,340] (the isothermal compressibility for a liquid mixture was approximated by $\beta_T^L = \varphi_1 \beta_{T,1}^{L*} + \varphi_2 \beta_{T,2}^{L*}$, where φ_i denotes the volume fraction of component i, and $\beta_{T,i}^{L*}$ is the pure-component compressibility of i); (II) *molecular dynamics simulations* were performed and, from the calculated pair distribution functions, the solvation shell radius was estimated together with *local* mole fractions x_{ij} of component i around a molecule of component j. Additional data on volumetric properties of binary mixtures of NMP, DMF, sulfolane, *etc.* with 1-alkanols, 2-propoxyethanol, ethanolamine, *etc.* may be found in references 341–343.

Densities and speeds of ultrasound of binary systems of acetic acid (HAc) with water, methanol, ethanol, methyl acetate and ethyl acetate at 293.15 K, 298.15 K and 303.15 K were measured over the entire composition range by González *et al.*[344] At 298.15 K, $V^E(x_{HAc} = 0.5) = -1.133$ cm^3 mol^{-1} for the mixture with water, $V^E(x_{HAc} = 0.5) = -0.721$ cm^3 mol^{-1} for the mixture with methanol, and $V^E(x_{HAc} = 0.5) = -0.486$ cm^3 mol^{-1} for the mixture with ethanol. For the two alcohol systems dV^E/dT is negative and small, while for aqueous acetic acid dV^E/dT is positive and small. For excess volumes of six binary mixtures of acetic acid (ethanoic acid) and propionic acid (propanoic acid) with 2-butanol, methyl-2-propanol and 2-methyl-2-butanol at six temperatures from 293.15 K to 333.15 K, see Behroozi and Zarei.[345] For the binary system of trifluoroethanoic acid (TFA) mixed with propionic acid, Hnědkovský *et al.*[346] measured excess molar volumes with a tilting dilution dilatometer at 298.15 K and 318.15 K. The V^Es are negative throughout, showing a highly unusual composition dependence: at both temperatures, *W-shaped* curves (*i.e.* one maximum at, roughly, $x_{TFA} \approx 0.4$, and two minima at, roughly, $x_{TFA} \approx 0.25$ and $x_{TFA} \approx 0.8$, respectively) are observed, and dV^E/dT is negative. The density of pure trifluoroethanoic acid in the temperature range 278.15 K to 328.15 K has been reported in reference 347.

In a series of articles, an attempt was made by Kohler *et al.*[348–350] to present a unified treatment of systems consisting of a carboxylic acid and an amine. Substances considered were formic acid, acetic acid, trifluoroacetic acid, propionic acid, and trimethylacetic acid (pivalic acid); and triethylamine, tri-*n*-butylamine, di-*n*-butylamine and *n*-butylamine. Experiments included melting curves, excess volumes, expansivities, viscosities, conductometric measurements and NMR measurements. For instance, the system (acetic acid + triethylamine) is one of the rare examples where *phase separation* occurs at *negative* values of the excess Gibbs energy G^E, and the excess molar volumes are *negative* and *very* large. This unusual behaviour was interpreted in terms of a strong attractive interaction between a polar 1:1 complex of acetic acid with triethylamine and the acetic acid dimer, which causes the liquid mixture of equimolar composition to split into a phase rich in the 3:1 aggregate and a phase rich in triethylamine. For the mixture of propionic acid (HPr) with triethylamine, the excess molar volume at equimolar composition and 293.15 K amounts to $V^E(x_{HPr}=0.5)=-8.024\ \mathrm{cm^3\ mol^{-1}}$, and shows a strongly negative temperature dependence: $dV^E/dT=-25.6\times10^{-3}\ \mathrm{cm^3\ K^{-1}\ mol^{-1}}$. Trifluoroacetic acid mixed with triethylamine has an even stronger contraction.

The convenience, speed and reliability of vibrating-tube densimeters (provided predominantly by Anton Paar, Graz, Austria) has resulted in a marked increase in the number of publications devoted to the volumetric behaviour of ternary liquid systems of nonelectrolytes. Here, we just summarily present a selection of seven papers[351–357] dealing with mixtures containing an alcohol. Besides correlating the experimental results with empirical equations such as Redlich–Kister, Kohler, Cibulka, Singh, *etc.* (see above in Sub-Chapter 7.3), some authors have also used the Prigogine–Flory–Patterson approach or the ERAS model. For details we refer to the original work.

7.4.3 Aqueous Solutions of Nonelectrolytes

This field is huge and would certainly merit a monograph of its own, in particular when results obtained at high dilution are to be included (in fact, parts of Chapters 8 and 9 are devoted to this topic). It is also of great complexity, and a quantitative theory of aqueous solutions of nonelectrolytes is nowhere in sight. Water is frequently called the "unique solvent," and its unique properties[64,358,359] are all related to the ability of water molecules to form *three-dimensional hydrogen bond* networks. Life as we know it is sustained by liquid water. In fact, water is the principal constituent of all living organisms, making up about 70% by weight of the human body. In the biochemistry of the cell, water plays a central role for the structure, stability, dynamics and function of biomacromolecules,[360–364] to the extent that it may no longer be regarded as a mere medium for bioprocesses, but, as recently suggested by Ball,[362,363] rather as an indispensable *active matrix*, something like a "biomolecule" itself. Again, as in the preceding subsections, only relatively few papers will be presented and commented on, just enough to present to the reader the general flavour of this field, and to give an indication of current research activities.

Liquid substances that are miscible with water are usually composed of polar molecules and contain groups capable of forming hydrogen bonds.

Indeed, the thermophysical properties of aqueous solutions of hydroxy compounds are of interest in many areas of science and engineering, and the volumetric properties of mixtures of water with alkanols are particularly important. In contradistinction to water, the lower alcohols are unable to form three-dimensional networks, but hydrogen bonding leads to chains and rings. For many alcohols, the solute–water hydrogen bond is strong enough to make the two liquids miscible at all temperatures: this is the case for aqueous solutions of methanol, ethanol, and the two propanols. At ambient conditions, 1-butanol and the higher alcohols are only partially miscible with water; however, while {1-butanol + water} shows an upper critical solution temperature of about 398 K, tertiary butanol is completely miscible with water at all temperatures.

Mixtures of type {an alcohol + water} have been extensively investigated using a variety of experimental techniques and theoretical methods. Since alcohols are composed of a *hydrophilic* part (–OH group) and a *hydrophobic* part (alkyl group) interesting competitive interactions are to be expected and indeed observed. Thus, systematic measurements of the *PVT*-properties of systems of type {an alcohol + water} over extended ranges of temperature and pressure, *and* as a function of the size of the alkyl group (chain length, in the case of *n*-alkan-1-ols) as well as of the number of hydroxyl groups (diol, triols, *etc.*), play a pivotal role in improving our understanding of these complex mixtures. For the simplest system belonging to this series, *i.e.* {methanol + water}, densities, and thus excess volumes, have been repeatedly determined at ambient conditions,[365–369] and, as already presented above in subsection 7.2, also at elevated temperatures and pressures.[212–217] The excess molar volume is negative across the entire composition range, and becomes more negative with increasing temperature. In particular, Benson and Kiyohara[366] presented a detailed analysis of V^E and dV^E/dT for mixtures of water with methanol, and also of water with ethanol and with 1-propanol, obtained with a flow-type vibrating-tube densimeter (from Sodev, model 02D) at 5 degree intervals from 15 °C to 35 °C; special attention was directed towards the composition dependence of these quantities at small mole fractions of alcohol. Temperature wise, the most comprehensive contribution is that of Xiao *et al.*[148a] They determined V^E of {methanol + water} at 323 K, 373 K, 423 K, 473 K, 523 K and 573 K and at moderate pressures of 7.0 MPa and 13.5 MPa, using a vibrating-tube densimeter of the Albert–Wood type.[135] The data were successfully correlated with an improved corresponding-states model. *High-pressure* results up to 350 MPa and over the temperature range from 283.15 K to 348.15 K have been reported by Kubota *et al.*[370] A comprehensive compilation of thermodynamic data sets for this mixture covering the extant literature at that time was prepared by Aliev *et al.*[215] in 2003. Volumetric properties of mixtures of water (H_2O, D_2O) and methanol H/D-isotopomers (CH_3OH, CD_3OH, CH_3OD and CD_3OD) were measured by Ivanov and Abrossimov[371] over the entire composition range at 5 °C, 15 °C, 25 °C, 35 °C and 45 °C, and discussed in terms of structural changes caused by the isotope substitution.

Recently, it was found that many thermophysical properties of water can be modelled by using a Jagla pair-potential energy function:[372] it has a hard core and a linear repulsive ramp, and contains *two* characteristic length scales corresponding to the hard core and to the soft core. Methanol may be modelled as a "dimer" consisting of a hard sphere (CH_3) and a Jagla particle (OH), and this dimer is then dissolved in the Jagla water. Using discrete molecular dynamics,[373] the experimental results for the excess volumes of the system {methanol + water} are reproduced qualitatively.

In macroscopically homogeneous solutions of methanol in water, preferential solvation (*i.e.* the local composition differs from the bulk composition) may be deduced *via* the Kirkwood–Buff (KB)[339] integral method. In 1977, Ben-Naim[340] suggested an *inversion* procedure with respect to the KB theory,[65,66] in which the *thermodynamic* quantities $V_i, \beta_T, (\partial \ln \gamma_i / \partial x_i)_{T,P}$ are used to calculate *molecular* quantities, the so-called KB integrals, for a pair of species *i* and *j*:

$$G_{ij} = \int_0^\infty [g_{ij}(r) - 1] 4\pi r^2 \mathrm{d}r. \tag{7.99}$$

Here, $g_{ij}(r)$ is the pair distribution function for the pair *i*, *j* (defined in the grand canonical ensemble) and *r* is the intermolecular separation. The product $\rho_{N,i}G_{ij}$, where $\rho_{N,i}$ is the number density of species *i*, represents the total *average* excess, or deficiency, of molecules of species *i* in the entire space around a molecule *j*. Thus, G_{ij} provides a measure of the average tendency of molecules *i* to cluster around a molecule *j*. After Ben-Naim first reported G_{ij} for the mixture {ethanol + water },[340] KB integrals on a few alcohol–water systems were evaluated by Donkersloot,[374] by Patil[375] and, notably, by Matteoli and Lepori.[376] Using *judiciously* selected thermodynamic data for mixtures of methanol with water appropriate from subambient (−13 °C) to elevated temperatures (250 °C), *i.e.* $V^E(x)$, $\beta_T(x)$ and, in particular, $G^E(x)$ [because the composition dependence of the activity coefficient is required in a KB-based analysis, see Equation (7.49], Marcus[377] presented a comprehensive KB study of the effect of methanol on water structure. According to his analysis, the methanol self-preference in methanol-rich mixtures is small, but this hydrophobic effect becomes considerable at $x_{CH_3OH} = 0.15$, reaching a pronounced maximum near 150 °C. Extensive data on solvent mixtures describing their properties and structure (including preferential solvation) are presented and discussed in the recent monograph by Marcus.[378]

The composition and temperature dependence of the partial molar volumes of alcohols (ROH) dissolved in water are frequently used for discussions of the influence of alcohols on water structure.[379] Focusing, for instance, on the completely miscible 1-alkanols, the excess partial molar volume at infinite dilution in liquid water is negative in each case, reflecting the open water structure which facilitates the accommodation of solutes. At 298.15 K,[366] $V^{E\infty}_{ROH}$/cm^3 mol^{-1} amounts to −2.53 for methanol, to −3.52 for ethanol, and to −4.22 for 1-propanol. The composition dependence of the excess partial molar volumes of these *partially* hydrophobic solutes is characterised by an initially strong decrease of V^E_{ROH} with increasing x_{ROH} from its value at infinite dilution, $V^{E\infty}_{ROH}$, and eventually reaching a minimum at rather *small* alcohol mole fraction: the respective minima are situated, roughly, at $x_{ROH} \approx 0.1$ for methanol, at $x_{ROH} \approx 0.06$for ethanol, and at $x_{ROH} \approx 0.03$ for 1-propanol. These results are in satisfactory agreement with older literature.[380–382] However, Benson and Kiyohara[366] experienced some difficulty in finding algebraic representations to correlate and smooth the results on excess molar volumes as functions of composition and temperature. Cautionary remarks concerning the calculation of partial molar volumes from density measurements have been made by Marsh and Richards,[127] in particular when V_i changes rapidly with composition, as is the case in aqueous solutions of alcohols. Taking advantage of their extensive experience in developing dilution dilatometers,[121,383] Marsh and Richards,[127] in 1980, were the *first* to measure *directly* excess volumes of mixtures of ethanol with water from 278.15 K to 338.15 K (10 K intervals). Note that the dilatometer used by Marsh and Burfitt[383] for measuring excess volumes of dilute alcohol solutions in non-polar solvents allows the determination of volume changes to $\pm 2\times10^{-5}$ cm^3 mol^{-1} in a total volume of 120 cm^3, which corresponds to measuring the density to $\pm 2\times10^{-7}$ g cm^{-3}! Such a precision is at least better by an order of magnitude than that usually obtained with vibrating-tube densimeters. For the excess partial molar volume of ethanol at infinite dilution in liquid water at 298.15 K, Marsh and Richards,[127] obtain $V^{E\infty}_{C_2H_5OH}\big/\mathrm{cm^3\ mol^{-1}} = -3.488$.

The excess molar volume of the system {ethanol + water} at 298.15 K was determined by Grolier and Wilhelm[384] together with the excess molar heat capacity at constant pressure. Densities of solutions of H_2O in C_2H_5OH and D_2O in C_2H_5OD, in a composition range close to pure ethanol, *i.e.* $x_{water} \leq 0.037$, and for temperatures between 278.15 K and 318.15 K (in steps of 10 K) were measured by Ivanov[385] and discussed in terms of the solute H/D isotope effect. Volumetric properties of ethanol–water mixtures at 298.15 K and at pressures up to 5 MPa were reported by Petek *et al.*,[386] and Ott *et al.*[387] presented excess molar volumes for this system at 298.15 K, 323.15 K and 348.15 K at pressures up to 15 MPa; the same temperature range but a somewhat larger pressure range (0.1 MPa through 40 MPa) were covered by Pečar and Doleček.[388] Chauhdry and Lamb[389] measured excess volumes of {ethanol + water} at 298.15 K and 323.15 K by a direct method in a mixing dilatometer[390] at pressures up to 220 MPa, and a still larger pressure range (up to 350 MPa) and a temperature range extending from 283.15 K to 348.15 K was covered by Kubota *et al.*[370]

Critically evaluated $V^E(x)$, $G^E(x)$, $H^E(x)$ and viscosities of {methanol + water} and {ethanol + water} for temperatures between 0 °C and 100 °C (in steps of 10 °C) were compiled by Westmeier[391] in 1975 after a comprehensive survey of the extant literature at that time.

Compared to the situation with mixtures of methanol and ethanol with water, the number of scientific papers reporting densities or directly measured excess volumes for the mixtures {1-propanol + water} and/or {2-propanol + water} is much smaller. References 392–396 contain data obtained at ambient pressure and 298.15 K (or at nearby temperatures), and Benson and Kiyohara[366] determined excess volumes for {1-propanol + water} from 288.15 K to 308.15 K over the entire composition range (see above). The most comprehensive density measurements on aqueous 1-propanol and 2-propanol, pressure wise, are again those of Kubota *et al.*,[370] extending up to 350 MPa. The compressive behaviour of {water + 1-propanol, +1,2-ethanediol, +1,2-propanediol, +1,3-propanediol, and +glycerol} at 298.15 K and for pressures up to 200 MPa was studied by Miyamoto *et al.*[397] Isothermal compressibilities and excess molar volumes were derived, and for {1-propanol + water} agreement with the results of Kubota *et al.*[370] is good. The recently published experimental study of Abdulagatov and Azizov[398] on the density and derived volumetric properties (excess molar volumes, apparent molar volumes) of {1-propanol + water} provides data for pressures up to 40 MPa, and covers the temperature range 298 K to 582 K. Agreement with older literature results is good throughout.

Excess volumes for the ternary system {*N,N*-dimethylformamide + 1-propanol + water} at 313.15 K were measured with a dilution dilatometer by Zielkiewicz,[399] and Resa *et al.*[400] determined excess volumes for {ethanol + water + 1-propanol} over the range 288.15 K to 323.15 K. Density data for dilute aqueous solutions of methanol, ethanol, 1-propanol and 2-propanol were measured from 298.15 K to 573.15 K and at pressures up to 30 MPa by Hynčica *et al.*,[401] apparent molar volumes and isentropic compressibilities of *n*-alkanols and α,ω-alkane diols in dilute aqueous solutions were determined by Nakajima *et al.*,[402] and apparent molar volumes and apparent molar heat capacities of dilute aqueous solutions of ethanol, 1-propanol and 2-propanol were measured by Origlia-Luster and Woolley.[403]

The accuracy of prediction methods for excess molar volumes of liquid multicomponent mixtures affects, of course, the reliability of process design calculations. Densities of six binary mixtures, three ternary mixtures and the quaternary mixture formed by water, methanol, ethanol, and 1-propanol were measured by Mori *et al.*[404] over the entire composition range at 10 K intervals from 298.15 K to 338.15 K. An NRTL-type equation[405] was used to correlate the composition and temperature dependence of the excess molar volumes:

$$V^{\mathrm{E}} = \sum_{i=1}^{n} x_i \frac{\sum_{j=1}^{n} \tau_{ji} G_{ji} x_j}{\sum_{k=1}^{n} G_{ki} x_k}, \tag{7.100}$$

where

$$G_{ij} = \exp(-\alpha_{ij}\tau_{ij}), \tag{7.101}$$

with $\alpha_{ij} = \alpha_{ji}$. The temperature dependence of the τ_{ij} parameters was assumed to be of the form

$$\tau_{ij} = \frac{a_{ij}}{T} + b_{ij}, \tag{7.102}$$

with $a_{ij} \neq a_{ji}$ and $b_{ij} \neq b_{ji}$. For each *binary* system optimised parameters were determined, though for the three systems consisting of alcohols only, α_{ij} was set to 0.3.[6] Excess molar volumes for the ternary and quaternary mixtures were then calculated *via* Equations (7.100)– (7.102) using *binary* parameters *only*. Most of the experimental V^E data were estimated to within $\pm 20\%$.

As pointed out above, tertiary butyl alcohol [TBA, *tert*-butanol, 2-methyl-2-propanol, $(CH_3)_3COH$] is the only butanol isomer which is completely miscible with water at ambient pressure, as is methanol, ethanol, 1-propanol and 2-propanol. All these aqueous systems show quite unusual composition dependences of thermodynamic properties,[64,406] such as a large, *highly skewed* maximum of the excess heat capacity[384] C_P^E, which for {TBA + water} at 298.15 K amounts to about 16 J K^{-1} mol^{-1} at $x_{TBA} \approx 0.075$,[407] or a pronounced minimum of the excess partial molar volume of the alcohol at small alcohol mole fractions.[366,407,408] With increasing size of the alcohol, the minimum in V_{ROH}^E becomes deeper and more pronounced and shifts to *smaller* mole fractions of the alcohol (see above): the most significant effect is observed in {TBA + water}, which makes it a particularly attractive system to investigate. As an amphiphilic solute, TBA contains a hydrophilic as well as a hydrophobic group: the hydrophilic OH-group interacts with water by forming hydrogen bonds, and the hydrophobic $(CH_3)_3C$-group shows a tendency for self-aggregation, thereby influencing the adjacent water structure (hydrophobic hydration). In fact, *microscopic heterogeneity* in mixtures of alcohols with water has been suggested for quite some time and has found experimental support not only from unusual thermodynamic properties, but also from direct structural evidence obtained *via* neutron scattering studies,[409–412] small-angle X-ray scattering,[413] X-ray diffraction and mass spectrometric measurements,[414] low-frequency Raman spectroscopy[415] and light-scattering experiments.[416,417] This concept of micro-heterogeneity is supported by computer simulations.[379,418–422] We note that the polar interaction of water with the OH-group of an alcohol may contribute more to the thermodynamic properties of aqueous mixtures of alcohols than hydrophobic effects, a point-of-view which has recently been advanced by Ben-Naim.[423–425]

Volumetric properties of aqueous solutions of *tert*-butanol have been determined by many researchers, for instance, in chronological sequence, by Nakanishi,[408] Franks and Smith,[426] de Visser *et al.*,[407] Hvidt *et al.*,[427] Sakurai,[428] Kubota *et al.*,[370] Kim and Marsh,[429] Kipkemboi and Easteal,[430] and most recently, by Egorov and Makarov.[431] Again, as before with {ethanol + water},[127] Marsh was the first to measure *directly* the excess volumes of {TBA + water}, covering the entire composition range at 5 K intervals from 303.15 K to 323.15 K. The continuous-dilution dilatometer used was the one described by Stokes *et al.*[121] and Marsh and Richards,[127] and provides an accuracy of operation of better than ± 0.001 cm^3 mol^{-1}. At small mole fractions of TBA, *i.e.* for $x_{TBA} < 0.039$, V^E becomes more negative as the temperature increases, at $x_{TBA} = 0.039$, the excess molar volume $V^E = -0.0376$ cm^3 mol^{-1} is independent of temperature, and for $x_{TBA} > 0.039$, V^E increases with increasing temperature. A similar, unusual behaviour is shown by the excess partial molar volume of *tert*-butanol in water. At $x_{TBA} = 0.019$, $V^E_{TBA} = -9.83$ cm^3 mol^{-1} is independent of temperature; for $x_{TBA} < 0.019$, V^E_{TBA} becomes more negative with increasing temperature, while for $x_{TBA} > 0.019$, it increases with increasing temperature. Using a vibrating-tube densimeter, Egorov and Makarov[431] covered a larger temperature range, *i.e.* 274.15 K to 348.15 K, though with the usual *smaller* precision associated with this kind of instrument. Finally, it is perhaps worthwhile to point out that the freezing temperature of *tert*-butanol (2-methyl-2-propanol) is 298.87 K,[429] which clearly invites a critical attitude *vis-à-vis* some experimental results reported in the literature.

Volumetric properties of aqueous solutions of *tert*-butanol at high pressure have been determined by just a few researchers. For instance, Kubota *et al.*[370] presented specific volumes for five compositions ($x_{TBA} = 0.05$, 0.10, 0.25, 0.50 and 0.75) at 298.15 K, 323.15 K and 348.15 K at pressures up to 210 MPa. Harris *et al.*[432] studied the compressive behaviour of {TBA + water} at six compositions from $x_{TBA} = 0.025$ to $x_{TBA} = 0.45$ for temperatures from 5 °C to 75 °C at pressures up to 300 MPa. The measurements of Egorov *et al.*[433] extended up to 100 MPa, and were carried out at five temperatures from 278.15 K to 323.15 K for 17 mole fractions x_{TBA}. At equimolar composition and at 323.15 K, the increase of pressure causes a change from $V^E = -0.945$ cm^3 mol^{-1} at $P = 0.1$ MPa to $V^E = -0.285$ cm^3 mol^{-1} at $P = 100$ MPa.

Since Chapter 8 is devoted to partial molar volumes at high dilution (nonelectrolytes), we will present here only one recent contribution to this field to illustrate what may be achieved with modern instrumentation. Recently, Šimurka *et al.*[434] measured the densities of dilute aqueous solutions of four aliphatic alcohols and one ether alcohol using both an Anton Paar DSA 5000 vibrating-tube densimeter and a home-made flow densimeter. The densities were determined in the temperature range 278 K through 573 K and at pressures up to 30 MPa, and used to obtain partial molar volumes at infinite dilution $V^{E\infty}_{ROH}$, together with several derived quantities associated with the temperature and pressure dependence of $V^{E\infty}_{ROH}$. A comprehensive review and a critical evaluation of experimental data on the limiting activity coefficients γ^{∞}_{ROH}, excess partial molar enthalpies $H^{E\infty}_{ROH}$ and so forth of lower 1-alkanols (ROH) dissolved in water, from 0 °C to 100 °C, were prepared by Dohnal *et al.*[435]

Osmolytes are small intracellular organic molecules, ubiquitous in living organisms, which stabilise proteins against unfolding under adverse environmental stress.[436] Polyols are an important group of naturally occurring osmolytes, which includes glycerol (1,2,3-trihydroxypropane), trehalose, sorbitol, glucose *etc.* In particular, the cryoprotective properties of mixtures of glycerol with water have found widespread use in preserving the functionality of biological material during cooling and to protect it from damage due to intracellular ice formation,[437] and have thus stimulated increasing interest in thermodynamic properties of aqueous polyol systems.

Densities, viscosities and heat capacities at constant pressure of the binary mixture of ethylene glycol (1,2-ethanediol) with water were measured by Yang *et al.*[438] over the entire composition range from 273.15 K to 353.15 K. At lower temperatures, V^E is negative throughout, but with increasing temperature a sigmoid composition dependence develops, the small positive part being observed at low mole fractions of ethylene glycol. Densities, sound speeds and relative dielectric permittivities for the four binary systems {water + 1,2-ethanediol, + diethylene glycol [$HOCH_2CH_2OCH_2CH_2OH$], + triethylene glycol [$HOCH_2CH_2(OCH_2CH_2)_2OH$], and + 1,2-dimethoxyethane} were measured by George and Sastry[439] at several temperatures from 298.15 K to 348.15 K. Related work, including results on ternary systems, may be found in references 440–447.

Some thermodynamic and structural aspects of mixtures of glycerol (G) with water have been critically selected and reviewed by Marcus.[448] The excess molar Gibbs energy, the excess molar enthalpy and the excess molar volume at 298.15 K were obtained from published experimental data. Surprisingly few data on volumetric properties of aqueous glycerol solutions have been reported, and Marcus, using two sources at 298.15 K, suggested a Redlich–Kister correlation for V^E, which has been superseded by the recent results obtained by Xu *et al.*[449] *via* density measurements at 25 °C and 35 °C, using a vibrating-tube densimeter. For small weight fractions of solute, Neal and Goring[450] determined dilatometrically the change in apparent specific volume with temperature for 15 nonelectrolytes, including glycerol, in water between −4 °C and 60 °C.

High-resolution neutron diffraction coupled with hydrogen/deuterium isotopic substitution has been used by Towey and Dougan[451] to determine the structure of a dilute aqueous glycerol solution with $x_G = 0.05$. Interestingly, no change in the local structure of water at the first water–water neighbour level is observed. However, for the second water coordination shell the change of water structure is similar to that seen in water at high pressure and for dilute solutions of the cryoprotectants methanol[452] and trehalose.[453] In addition, their results indicate that the hydrogen bonded network is very effectively mixed between glycerol and water, which is consistent with molecular dynamics simulation results of Dashnau *et al.*[454] and Politi *et al.*[455] For neutron diffraction studies in a concentrated mixture of glycerol with water see Towey *et al.*[456]

Thermodynamic properties of mixtures of 2-*n*-alkoxyethanols, in particular molar heat capacities at constant pressure, have attracted attention because of the unusual composition dependence found with some mixtures, reminiscent of that associated with micellization. Molar heat capacities and molar volumes of several such systems have been determined by Roux *et al.*:[457] they studied mixtures of water with 2-methoxyethanol, 2-ethoxyethanol, 2-propoxyethanol and with 2-butoxyethanol (BE). The system {2-butoxyethanol + water} shows a liquid–liquid solubility loop[458–460] with a lower critical solution temperature (LCST) $T_{LC} = 48.8_5 \pm 0.1$ °C at (27.4 ± 0.4)% by mass of BE, and an upper critical solution temperature (UCST) $T_{UC} = 130.2 \pm 0.2$ °C at (29.7 ± 0.5)% by mass of BE. Aqueous solutions of ethoxyethanol (EE) have been studied Tamura *et al.*[461] Excess enthalpies, excess isobaric heat capacities, densities and speeds of ultrasound were measured covering the entire composition range. The excess molar enthalpy is negative and strongly skewed towards the water side, with the minimum, at 298.15 K, being $H^E \approx -1000$ J mol^{-1} at $x_{EE} \approx 0.2$. The excess molar isobaric heat capacity is positive and also strongly skewed towards the water side: at 298.15 K, the maximum is at $x_{EE} \approx 0.1$ with $C_P^E \approx 9.5$ J K^{-1} mol^{-1}, and for $x_{EE} > 0.2$ the C_P^E-curve shows an unusually broad shoulder. The value of C_P^E rises as temperature increases. The excess molar volume is negative throughout and also skewed towards the water side: at 298.15 K, the minimum is at $x_{EE} \approx 0.35$ with $V^E \approx -1.15$ cm^3 mol^{-1}. The temperature dependence of V^E is positive. From their experimental results, Tamura *et al.* conclude that the non-randomness in the mixture {ethoxyethanol + water} is very large. Similar comprehensive investigations were reported for {2-isopropoxyethanol + water}[462] and for {2-isobutoxyethanol (IBE) + water}[463] at 293.15 K, 298.15 K and 303.15 K, that is, below, near and above the lower critical solution temperature $T_{LC} = 299.61$ K at the critical composition $x_{LC} = 0.0664$ of IBE.[464] Excess volumes were measured directly by Reddy *et al.*,[465] using a batch dilatometer,[466] together with speeds of ultrasound for mixtures of water with 1,2-ethanediol, 2-methoxyethanol, 2-ethoxyethanol and 2-butoxyethanol at 308.15 K. A continuous-dilution dilatometer, described by Dickinson *et al.*,[124] was used by Pal *et al.*[467] to directly measure excess volumes of mixtures of water with 1-methoxy-2-propanol, and with 1-ethoxy-2-propanol, respectively, at 298.15 K. Densities[468] and speeds of ultrasound[469] for {1-propoxypropan-2-ol + water} were determined at 5 K intervals from 283.15 K to 323.15 K over the entire composition range, and compared with results obtained for mixtures of water with isomeric 2-butoxyethanols, and for mixtures of water with 2-isopropoxyethanol, to study the effect of branching (and chain length). The related topic of microemulsions has been treated exhaustively in the recent survey edited by Fanun.[470]

With the implemetation of the Kyoto Protocol in many countries, in particular Europe, carbon dioxide removal from gas streams of industrial plants, from exhaust air and from natural gas has stimulated the development of more economically viable processes. Triethylene glycol monomethyl ether [TEGMME, $CH_3(OCH_2CH_2)_3OH$], and related compounds, have received some interest from the gas industry because they absorb more CO_2[471,472] and less methane[473,474] than many solvents currently used.

Excess volumes of mixtures of water with ethylene glycol monomethyl ether [$CH_3O(CH_2)_2OH$], diethylene glycol monomethyl ether [$CH_3O(CH_2)_2O(CH_2)_2OH$] and triethylene glycol monomethyl ether were measured directly by Pal and Singh[475] with a continuous-dilution dilatometer, as described by Dickinson *et al.*,[124] at 308.15 K. A more comprehensive study of {TEGMME + water} has been presented by Henni *et al.*[476] They measured the density (vibrating-tube densimeter) at five temperatures in the range 25 °C to 80 °C covering the whole composition range. Frequently, the second derivative of the partial molar volume at infinite dilution,

$$\left(\frac{\partial^2 V_i^\infty}{\partial T^2}\right)_P = -\frac{1}{T}\left(\frac{\partial C_{P,i}^\infty}{\partial P}\right)_T, \tag{7.103}$$

is used in discussions of solution behaviour.[477] The partial molar volume of TEGMME at infinite dilution in water increases *linearly* with increasing temperature, hence the second derivative is zero: the solute may be regarded as having no net effect on the water structure. The partial molar volume of water at infinite dilution in TEGMME remains essentially constant in this temperature range. Related work has been communicated in references 478 and 479.

Polyethylene glycol [PEG, $H(OCH_2CH_2)_nOH$], also known as poly(ethylene oxide) [PEO], is a polyether of importance in many industrial, chemical, biological and medical applications. It is one of the most studied water-soluble synthetic polymers,[480–483] but PEG is also soluble in methanol, ethanol, acetonitrile and benzene. When dissolved in water, water forms a layer around PEG,[484,485] reminiscent of the hydration layer around proteins. In fact, PEG is frequently used as a model for biopolymers. In binary liquid mixtures with hydrogen bonding between the components, closed-loop immiscibility is frequently observed. Indeed, for PEG molar masses above, roughly, 2100 the temperature–composition phase diagram of {PEG + water} begins to exhibit closed-loop phase separation with an LCST and an UCST, and with increasing molar mass the loop expands.[486,487] For its pressure dependence see Bekiranov *et al.*,[480] Cook *et al.*[488] and Sun and King.[489]

In order to understand the physics associated with the interaction of PEG with water, Fenn *et al.*[490] investigated spectroscopically aqueous solutions of tetraethylene glycol dimethyl ether: this would yield information about the role played by the polyether backbone alone, since the two terminal hydrophilic hydroxyl groups are replaced by methyl groups. The experiments indicate two distinct subensembles of water molecules: those that are hydrogen bonded to other water molecules and those that are bound to ether oxygens. These findings are in accord with results from neutron scattering[491] and molecular dynamics simulation.[492]

As evidenced by this review, a huge amount of data on excess volumes of binary mixtures as a function of temperature is available at atmospheric pressure. The influence of pressure has been investigated to a much smaller extent, and for *ternary* systems very few studies have been reported. For the system {water (W) + diacetone alcohol (D) + 2-propanol (P)} and the constituent binaries, Boned *et al.*[493] measured densities between 0.1 MPa and 65 MPa (in steps of 5 MPa) at 303.15 K, 323.15 K and 343.15 K, covering the entire composition range (the IUPAC name of diacetone alcohol is 4-hydroxy-4-methylpentan-2-one, $CH_3C(O)CH_2C(OH)(CH_3)_2$). Altogether, 2772 densities were determined. Five frequently used correlating equations were tested (of which the Redlich–Kister equation, Equation (7.87), and the Kohler equation, Equation (7.90), were by far the most useful), but the best representation was obtained with this *new* correlating function

$$V^{\mathrm{E}} = \frac{a_1 x_{\mathrm{W}} x_{\mathrm{D}} + a_2 x_{\mathrm{W}} x_{\mathrm{P}} + a_3 x_{\mathrm{D}} x_{\mathrm{P}} + a_4 x_{\mathrm{W}} x_{\mathrm{D}} x_{\mathrm{P}}}{1 + a_5 x_{\mathrm{W}} + a_6 x_{\mathrm{D}} + a_7 x_{\mathrm{P}}} \left(1 + a_8 \sqrt{T} + a_9 P\right), \quad (7.104)$$

which requires a significantly *smaller* number of parameters compared to the other methods, *i.e.* 9. As stated by Boned *et al.*, comparable results are obtained when using $a_8 T$ or a_8/T instead of $a_8 T^{1/2}$.

Alkanolamines, with their combined physicochemical properties of both alcohols and amines, are valuable intermediates in organic synthesis. In addition, during the last 30 years, aqueous alkanolamine solutions and related mixtures have been intensively studied because of their potential usefulness in industrial gas-sweetening processes for the removal of carbon dioxide and hydrogen sulfide. In particular, the research group with Henni has been highly active. For instance, densities of aqueous diisopropanolamine (DIPA) solutions were measured from 25 °C to 70 °C over the whole composition range,[494] with the exception of measurements below 45.4 °C, the melting point of diisopropanolamine, where only a large, liquid part of the composition range was covered. The excess molar volumes are all negative and noticeably skewed towards the water side, and they become less negative with increasing temperature. At 50 °C, the minimum is near $x_{\mathrm{DIPA}} = 0.3$ with $V^{\mathrm{E}} \approx -1.2$ cm^3 mol^{-1}. The partial molar volume at infinite dilution of DIPA in water increases linearly with temperature. According to the Hepler criterion, Equation (7.103), diisopropanolamine has no net effect on the structure of water. This is in accord with conclusions of Maham *et al.*[495,496] and Henni *et al.*[497,498] Densities for aqueous 2-(methylamino)-ethanol (MAE) solutions at six temperatures in the range 298.15 K to 343.15 K were measured by Li *et al.*[499] V^{E} is noticeably skewed towards the water side: at 298.15 K, the minimum is near $x_{\mathrm{MAE}} = 0.35$ with $V^{\mathrm{E}} \approx -1.2$ cm^3 mol^{-1}. Densities for the system {2-((2-aminoethyl)amino)ethanol + water} were measured at six temperatures in the range 298.15 K to 343.15 K by Mundhwa *et al.*[500] Densities and speeds of ultrasound were determined for the binary system {3-ethoxypropan-1-amine (EPA) + water (W)} across the entire composition range at temperatures between 283.15 K and 303.15 K (at 5 K intervals) and ambient pressure by Moita *et al.*[501] and Pinheiro *et al.*[502] V^{E} is noticeably skewed towards the water side: at 283.15 K, the minimum is located near $x_{\mathrm{EPA}} = 0.4$ with $V^{\mathrm{E}} \approx -2.25$ cm^3 mol^{-1}. The excess molar volumes of the 51 mixtures at different mole fractions x_{EPA} were correlated by a Redlich–Kister equation with nine parameters for the 283.15 K isotherm and eight for the other isotherms. The excess molar volumes do not vary significantly with temperature. In agreement with previous observations, fitting thermodynamic excess quantities of aqueous solutions of amphiphilic substances over the entire composition range by a *single* Redlich–Kister equation appears to be not the best approach.[503] For instance, a better description of the composition dependence of V^E in the approximate range $x_{EPA} < 0.015$ (this concerns 12 to 15 experimental points, depending on the isotherm) is obtained with a Redlich–Kister equation of a much lower degree: a two-parameter equation, *i.e.* a linear correlating equation, suffices. This approach was used to obtain the excess partial molar volumes at infinite dilution. At 283.15 K, $V_{EPA}^{E\infty} = -9.22\ cm^3\ mol^{-1}$ and $V_W^{E\infty} = -6.02\ cm^3\ mol^{-1}$. Related work on aqueous solutions of amines, of various alkanolamines and diethanolamines [R-N($C_2H_4OH)_2$], and of diethylene glycol amine [2-(2-aminoethoxy)ethanol, $NH_2CH_2CH_2OCH_2CH_2OH$)] may be found in references 504 to 510.

Discussion of thermodynamic quantities of relevance for biochemical systems is generally based upon results obtained for relatively simple model compounds dissolved in liquid water. Excess molar volumes and excess molar isobaric heat capacities of mixtures of water with an alkanoic acid (formic acid, acetic acid, propionic acid or butyric acid) at 298.15 K were determined by Casanova *et al.*[511] The excess volumes are negative for all four systems, fairly large in magnitude and rather symmetrical about $x = 0.5$; they are in good agreement with literature data. For instance, for {acetic acid + water} $V^E(x_{HAc} = 0.5) = -1.146\ cm^3\ mol^{-1}$, while Campbell *et al.*[512a] give $-1.145\ cm^3\ mol^{-1}$. However, the measurements do not extend sufficiently into the high-dilution region to allow accurate evaluation of the partial molar volumes of the alkanoic acids at infinite dilution, though results agree within 4%, in the worst case, with literature values.[513,514] Densities, surface tensions and refractive indices of the same four systems were measured by Granados *et al.*[515] at 298.15 K, covering the entire composition range. The calculated excess molar volumes agree well with literature data. Excess molar volumes for the three binaries {formic acid + water}, {acetic acid + water} and {propionic acid + water} were determined by Apelblat and Manzurola[516] at 288.15 K, 298.15 K and 308.15 K. Combining their results with literature data on excess enthalpies and excess Gibbs energies, they discussed these systems in terms of various associated mixture models (*e.g.* Mecke–Kempter).[517] The most comprehensive study of the volumetric behaviour under hydrostatic compression of mixtures of water with formic acid, or acetic acid, or propionic acid, or isobutyric acid is due to Korpela.[518] It covers the pressure range 0.1 MPa up to 250 MPa at the three temperatures 298.15 K, 313.15 K and 328. 15 K and the compressions were correlated with the Tait equation. For all four systems, the excess volume and the isothermal compressibility at 0.1 MPa pressure have been tabulated.

Hydration of proteins is a crucial factor in determining structure and stability of proteins to which hydrophilic as well as hydrophobic interactions contribute.[519] Aqueous solutions of alkyl amides are commonly regarded as good model systems for peptides in water, allowing convenient variation of the hydrophobic part by varying the alkyl groups systematically. Because of this, and because *N,N*-dimethylformamide [DMF, $(CH_3)_2NC(O)H$] is a widely used dipolar aprotic solvent, with a low melting point (212.7 K) and a high normal boiling point (426 K), the mixture of DMF with water has been the subject of quite a few investigations. De Visser *et al.*[56] determined excess volumes and excess isobaric heat capacities of mixtures of *N,N*-dimethylformamide with water at 298.15 K; and Pal and Singh[520] measured excess volumes for mixtures of water with *N*-methylformamide (MF), DMF, *N,N*-dimethylacetamide (DMA), 2-pyrrolidone (PYR) and *N*-methyl-2-pyrrolidone (NMP) at 303.15 K and 308.15 K directly with a continuous-dilution dilatometer similar to that described by Dickinson *et al.*[124] For each system, the excess molar volumes are negative over the entire composition range. Excess molar volumes and viscosities at 298.15 K for mixtures of water with formamide (FA), MF, DMF, PYR and NMP, respectively, were reported by García *et al.*,[521] while for all six systems {FA, or MF, or DMF, or DMA, or PYR, or NMP + water} Papamatthaiakis *et al.*[522] measured the density and the speed of ultrasound at 298.15 K. Scharlin *et al.*[523] determined the densities of mixtures of DMF with water (H_2O), and with heavy water (D_2O) at nine temperatures from 277.134 K to 318.15 K, covering the entire composition range. They used a vibrating-tube densimeter from Sodev (model 03-D; Quebec, Canada) which was operated in the static mode. For each temperature, the calculated excess molar volumes were correlated with a Padé approximant of the Myers–Scott type,[249] see Equation (7.80). In the temperature range investigated, the excess volumes for both systems are negative, noticeably skewed towards the H_2O side or the D_2O side, respectively, and they become less negative with increasing temperature. For the mixture with water, at 298.15 K the minimum is located near $x_{H_2O} = 0.6$ and amounts to *ca.* $-1.12\ cm^3\ mol^{-1}$, and for the mixture with heavy water, the minimum is also located near $x_{D_2O} = 0.6$ and amounts to *ca.* $-1.16\ cm^3\ mol^{-1}$: at each temperature, the minimum is slightly more negative for $\{DMF + D_2O\}$ than for $\{DMF + H_2O\}$. The "isotope effect", that is, the difference $V^E(DMF + D_2O) - V^E(DMF + H_2O)$, is negative throughout with a minimum at *ca.* $x_{water} = 0.6$, which becomes less negative with increasing temperature. Agreement with reliable literature data, for instance, with those reported by Miyai *et al.*,[524] is excellent. A similar careful investigation of the systems $\{DMA + H_2O\}$ and $\{DMA + D_2O\}$ has been presented by Scharlin and Steinby.[525]

Densities of binary mixtures of water with formamide, or *N,N*-dimethylformamide, or *N,N*-dimethylacetamide, or 1,4-dioxane, or dimethyl sulfoxide [DMSO, $(CH_3)_2SO$], have been measured by Tôrres *et al.*[526] for the whole composition range at 288.15 K. 293.15 K, 298.15 K and 303.15 K. The excess molar volumes are negative throughout. At 298.15 K, $V^E(x_{H_2O} = 0.5)/cm^3 \cdot mol^{-1} = -0.169$ for {FA + water}, $= -0.595$ for {1,4-dioxane + water}, $= -0.933$ for {DMSO + water}, $= -1.067$ for

{DMF + water}, and −1.526 for {DMA + water}. Excess molar volumes for the ternary mixture {DMF + methanol + water} and for the three constituent binary mixtures,[527a] excess molar volumes for the ternary mixture {DMF + ethanol + water} and for the three constituent binary mixtures,[527b] and excess molar volumes for the ternary mixture {DMF + 1-propanol + water} and for the three constituent binary mixtures,[527c] all at 298.15 K, were determined by Bai *et al.* Several popular empirical expressions were used to correlate and predict the ternary excess molar volumes from experimentally determined excess molar volumes of the constituent binaries.

Densities and viscosities of the mixtures {DMF + H_2O} and {DMF + D_2O} were measured by Ueno *et al.*[528] at 5 °C, 25 °C and 45 °C under high pressure up to 196.1 MPa. From the densities, isothermal compressibilities β_T were calculated. For both systems, β_T exhibits a minimum in the water-rich region at all temperatures and pressures studied. The minimum is more pronounced at lower temperatures and lower pressures. This behaviour is in contradistinction to the composition dependence of β_T for {formamide + water} at 25 °C,[529] where the compressibility decreases monotonously with increasing x_{FA}. Interestingly, the viscosity η of {DMF + H_2O, or + D_2O} shows a remarkable maximum in the water-rich region at all temperatures and pressures studied, which becomes more pronounced at lower temperatures and higher pressures: for instance, at 25 °C and at $P = 0.1$ MPa, $\eta(\text{DMF}) = 0.717$ mPa s, $\eta(H_2O) = 0.890$ mPa s, and $\eta(x_{DMF} \approx 0.3) \approx 2.8$ mPa s; at 196.1 MPa , $\eta(x_{DMF} \approx 0.3) \approx 4.8$ mPa s. For the system {FA + H_2O} at 25 °C, $\eta(x_{FA})$ increases monotonously with increasing x_{FA}. These results were discussed in terms of hydrophobic hydration around the methyl groups of DMF.

Excess molar volumes for mixtures of the heterocycles morpholine [$O(CH_2CH_2)_2NH$] and *N*-methylmorpholine with water were determined by Maham *et al.*[530] from density measurements over the entire composition range between 298.15 K and 353.15 K (in steps of 5 K, 10 K and 20 K). The excess molar volumes are negative throughout and noticeably skewed towards the water side: at 298.15 K, the minimum for {morpholine + water} is at $x_{H_2O} \approx 0.58$ with $V^E(x_{H_2O} \approx 0.58) = -1.70$ cm^3 mol^{-1}, and the additional methyl group on the nitrogen atom leads to a more negative and stronger skewed excess volume for {*N*-methylmorpholine + water} with the minimum at $x_{H_2O} \approx 0.63$, and $V^E(x_{H_2O} \approx 0.63) = -2.25$ cm^3 mol^{-1}. Densities of aqueous solutions of morpholine at small mole fractions have been studied by Tremaine *et al.*[531]

Densities, speeds of ultrasound, relative permittivities ε_r and viscosities were measured for three binary mixtures of water with 2-pyrrolidone, with *N*-methyl-2-pyrrolidone and with *N*-vinyl-2-pyrrolidone by George and Sastry[532] at different temperatures in the range 298.15 K to 338.15 K. The excess molar volumes are negative throughout, noticeably skewed towards the water side, and they become less negative with increasing temperature. In order to obtain information on correlation among dipoles, the authors calculated the

Kirkwood correlation factor g_K using the Kirkwood–Fröhlich equation[533] in the form

$$g_K = \left[\frac{(\varepsilon_r - \varepsilon_\infty)(2\varepsilon_r + \varepsilon_\infty)}{\varepsilon_r(\varepsilon_r + 2)^2}\right]\left[\frac{9\varepsilon_0 k_B T}{\rho_N(x_1\mu_1 + x_2\mu_2)^2}\right], \quad (7.105)$$

where ε_0 denotes the permittivity of vacuum, ε_∞ was approximated by $1.1n_D^2$ with n_D being the refractive index, μ_1 and μ_2 are the dipole moments of water and of the pyrrolidone component, respectively, k_B is the Boltzmann constant, ρ_N is the number density, *i.e.*

$$\rho_N \equiv \frac{L}{V} = \frac{L\rho}{m_m} = L\rho_n, \quad (7.106)$$

where $\rho_n \equiv 1/V$ is the amount-of-substance density, and L is the Avogadro constant, *i.e.* $Lk_B = R$. The letter L is used in honour of Josef Loschmidt, pioneering Austrian physicist and chemist.[¶] For dielectric relaxation studies of these aqueous systems, see Dachwitz and Stockhausen.[535] Densities, excess molar volumes and viscosities for {NMP + water} in the temperature range from 298.15 K to 343.15 K were also reported by Henni's group.[536]

For the ternary system {NMP + methanol + water} and its constituent binary mixtures, Aparicio *et al.*[537] measured densities, speeds of ultrasound, isobaric heat capacities, viscosities and refractive indexes at 298.15 K and 0.1 MPa over the entire composition range. The binary excess molar volumes are all negative, the one for {NMP + water} being the most pronounced with $V^E = -1.2$ cm^3 mol^{-1}at $x_{H_2O} \approx 0.61$; the binary excess heat capacities are all positive, again the one for {NMP + water} being the most pronounced with $C_P^E = 12.6$ J K^{-1} mol^{-1} at $x_{H_2O} \approx 0.71$. The ternary excess/mixing properties were correlated with the Cibulka equation, Equation (7.92). The discussion was conducted in terms of competitive effects due to self-association and hetero-association, and was based on (I) density functional theory calculations, and (II) the perturbed chain statistical associating fluid theory (PC-SAFT) approach.[538]

The structure of {NMP + water} at 298.15 K was investigated by Usula *et al.*[539] by combining density measurements (vibrating-tube densitometer, model DMA 58, Anton Paar, Graz, Austria), wide-angle X-ray scattering

[¶]Josef Loschmidt was born near Karlsbad, a famous Austrian spa, in 1821, and he died in Wien (Vienna) in 1895. Starting in 1866, he was associated with the University of Wien (Privat-Dozent), where he became an Associated Professor in 1868 and a Full Professor of Physical Chemistry in 1872, to retire in 1891. In 1865, Loschmidt provided the first reasonable estimate for the number of particles in a given volume of gas at ambient conditions, *i.e.* for the number density. Perhaps less known, but highly important for the history of chemistry, is his publication in 1861 of the book entitled *Chemische Studien I. Constitutionsformeln der organischen Chemie in graphischer Darstellung*, in which he gave two-dimensional *graphical* representations for more than 300 molecules, including numerous *aromatic* molecules (including *benzene*), which are *similar to modern structural formulae.* In fact, Loschmidt indicated the relative positions of atoms in molecules, he introduced markings (short lines) for double and triple bonds, and he suggested the *cyclic* structure of benzene to consist of six connected CH groups, and the cyclic structure of the phenyl ring in benzene derivatives, *four years before* Kekulé (who, by the way, most likely knew Loschmidt's book).[534]

experiments and molecular dynamics simulations. In order to reproduce the experimentally observed density maximum at $x_{NMP} \approx 0.33$, it was necessary to incorporate the influence of varying solvation with varying composition on the charge polarisation. The MD simulations showed that water molecules can occupy a "cavity" existing in the pure NMP network, which can, however, accommodate not more than two: thus, adding water to NMP results in an increase of the system density up to a mole fraction $x_{NMP} \approx 0.33$; additional water changes the NMP structure and the density decreases. The X-ray scattering results support these findings.

Dimethyl sulfoxide (DMSO) is an important dipolar aprotic solvent with a large dipole moment, $\mu = 13.2 \times 10^{-30}$ C m, and relative permittivity, $\varepsilon_r = 48$ at 298.15 K; the melting point is at 292 K, the normal boiling point is at 462 K. It is miscible with water and with many organic solvents. DMSO can penetrate the skin readily without damaging it, and in medical research it is often used as a drug carrier across cell membranes. DMSO, glycerol and ethylene glycol are also the most common cryoprotectants to reduce cell damage during freezing and thawing. In fact, advances in the application of cryopreservation techniques to proteins, living cells, tissues and organs has had a profound effect in many areas of biotechnology, in plant and animal breeding programmes, in cryobanking for the conservation of biological diversity, and in medical applications.[437,540–543] DMSO is an amphiphilic molecule with two hydrophobic methyl groups and a highly polar S=O group (the partial negative charge is located on the oxygen atom). Thus, DMSO can only act as proton acceptor when forming *strong* hydrogen bonds with water molecules, and clear evidence has been provided by numerous spectroscopic studies[544] and neutron scattering studies,[545,546] dielectric relaxation spectroscopy,[547] ultrasonic absorption and sound speed measurements,[548,549] and further corroborated by molecular dynamics simulations.[550–554] Depending on the composition, different dominant structures are observed.[551,555] In particular, in water-rich mixtures the tetrahedral structure associated with water molecules results in local structures predominantly characterised by a molecular ratio of DMSO : $2H_2O$.

Compared to all these research activities, thermodynamic studies are surprisingly limited:[556] the majority were performed at or near ambient temperatures, and with very few exceptions, at atmospheric pressure. Of central importance is, of course, the solid–liquid phase diagram for {DMSO + water}:[557] the freezing point of DMSO is greatly depressed by the addition of water, the depression amounting to about 91 K at the eutectic at $x_{DMSO} = 0.33$ at 200 K. Densities and viscosities of mixtures {DMSO + water} have been measured by Schichman and Amey[558] over the temperature range $213 \leq T/\mathrm{K} \leq 293$. The maximum of the curve viscosity *vs.* composition is observed at *ca.* $x_{DMSO} = 0.33$ and is increasing considerably with decreasing temperature. Densities and viscosities of mixtures {DMSO + water} at 298.15 K, 318.15 K and 338.15 K, covering the entire composition range, were also reported in reference 559, densities and derived apparent molar volumes at several temperatures between 298.15 K and 338.15 K in reference 560, densities and isobaric heat capacities at 298.15 K in reference 561, and densities and speeds of ultrasound at 298.15 K in reference 562. Westh[563] measured specific volumes, isobaric expansivities and specific isobaric heat capacities of aqueous DMSO mixtures with *mass* fractions $0 \leq w_{DMSO} \leq 0.85$ in the range from about 248 K to room temperature. Apparent molar quantities of DMSO were calculated therefrom. For mixtures {DMSO + water} at 298.15 K, Lai *et al.*[564] measured the vapour pressures, the densities and the excess partial molar enthalpies, and calculated excess partial molar volumes (up to $x_{DMSO} = 0.44$), and excess partial molar Gibbs energies. The excess partial molar Gibbs energies for both DMSO and water are negative for the entire composition range. The excess partial molar volumes of DMSO are all negative. The excess molar enthalpy is strongly negative throughout, and for equimolar composition $H^E = -2540$ J mol^{-1}.

The volumes of mixing for DMSO with water were measured directly by Lau *et al.*[565] at 298.15 K and 308.15 K with a dilatometer.[566] For the equimolar mixture at 298.15 K, $V^E = -0.944$ cm^3 mol^{-1}; most of the older literature data on excess volume are in reasonable to good agreement with this directly obtained result. Kimura and Takagi[567] measured densities at 298.15 K and 318.15 K of binary mixtures containing dimethyl sulfoxide, and binary mixtures containing methyl methylthiomethyl sulfoxide [MMTSO, $CH_3(CH_3SCH_2)SO$], a potentially useful solvent. For their investigation they used bicapillary pycnometers described earlier.[568] At 298.15 K, near the minimum of the curve V^E *vs.* x_{DMSO}, that is, near $x_{DMSO} \approx 0.43$, their results give a slightly larger contraction than that obseved by Lau *et al.* Kimura and Takagi[567] fitted the excess molar volumes for $\{x_{DMSO}(CH_3)_2SO + (1 - x_{DMSO})H_2O\}$ to a Redlich–Kister polynomial

$$V^E/(\text{cm}^3 \cdot \text{mol}^{-1}) = x_{DMSO}(1 - x_{DMSO}) \sum_{i=1}^{k} A_i (2x_{DMSO} - 1)^{i-1}. \qquad (7.107)$$

with $k = 4$. At 298.15 K, the coefficients are: $A_1 = -3.8282$, $A_2 = 1.9507$, $A_3 = 0.9840$, and $A_4 = -1.8340$, and the standard deviation of the fit is 0.0028 cm^3 mol^{-1}. At 318.15 K, they report: $A_1 = -3.6574$, $A_2 = 1.9661$, $A_3 = 0.6920$, and $A_4 = -1.5432$, and the standard deviation of the fit is 0.0023 cm^3 mol^{-1}. Additional contributions, in particular directed towards determining reliable isobaric expansivities, are from Aminabhavi and Gopalakrishna,[442] from Tamura *et al.*,[569] from Saleh *et al.*,[570] from Markarian *et al.*,[571] from Tôrres *et al.*,[526] from del Carmen Grande *et al.*,[572] and from Egorov and Makarov.[573] Isotopic substitution effects were investigated by Sacco and Matteoli[574] by measuring densities and viscosities of {DMSO + H_2O} and {DMSO + D_2O} at 298.15 K. Isotope effects were also investigated by Miyai *et al.*,[524] who measured excess enthalpies, excess isobaric heat capacities, densities/excess volumes and speeds of ultrasound for mixtures of DMSO with water and heavy water at 298.15 K. Agreement between all these results ranges from fair to good. In particular, they show that (I) V^E becomes *less* negative with increasing temperature, and that (II) $\partial V^E/\partial T$ is rather small; in summary, the data strongly indicate that the excess molar volumes for {DMSO + water} as presented in Landolt–Börnstein, New Series IV/23, Subvolume A[23c] (pp. 2–407, file number LB0141), are *incorrect.*

While thermodynamic properties of {DMSO + water} have been studied repeatedly at ambient pressure, high-pressure investigations are scarce. The melting curve as well as *PVT* data of *pure* liquid DMSO were measured from 293 K to 322 K for pressures up to 150 MPa by Fuchs *et al.*[575] Raman spectroscopy was used by Czeslik and Jonas[576] to study intermolecular interactions, structure and dynamics of pure liquid DMSO between 303.15 K and 343.15 K and at pressures up to 220 MPa. In order to separate temperature effects from density effects,[577] the density $\rho^{L}(T,P)$ of DMSO was also measured over the same region and fitted with the Tait equation, Equation (7.96).

The compression

$$k \equiv \frac{V(T,P_{\mathrm{ref}}) - V(T,P)}{V(T,P_{\mathrm{ref}})} = 1 - \frac{\rho(T,P_{\mathrm{ref}})}{\rho(T,P)} \tag{7.108}$$

for aqueous mixtures of DMSO (up to $x_{\mathrm{DMSO}} = 0.5$) was measured at 298.15 K for pressures up to 150 MPa ($P_{\mathrm{ref}} = 0.1$ MPa) by Moriyoshi and Uosaki,[578] and also fitted by the Tait equation. Petitet *et al.*[579] determined the speed of ultrasound in the aqueous mixture with $x_{\mathrm{DMSO}} = 0.32$ on seven isotherms between 268 K and 313 K at pressures up to 500 MPa, and in pure DMSO on eight isotherms between 294 K and 338 K at pressures up to 250 MPa. Together with thermodynamic quantities at 0.1 MPa and introducing a quite rough approximation concerning the heat capacity of pure DMSO and the heat capacity of the mixture, they used an approach based on Equation (7.70) to calculate a number of thermophysical properties at high pressures. For instance, with increasing pressure the excess molar volume becomes less negative. Most recently, Egorov and Makarov[580] measured the compression k of liquid {DMSO + water} over the entire composition range at 278.15 K, 288.15 K, 298.15 K, 308.15 K and 323.15 K at pressures from atmospheric to 100 MPa. At 278.15 K and 288.15 K, measurements did not cover the whole composition range because some mixtures froze under pressure. Agreement with results of Moriyoshi and Uosaki[578] is excellent.

Using vibrating-tube densimetry, Markarian *et al.*[571] also determined excess molar volumes for mixtures of diethyl sulfoxide [DESO, $(CH_3CH_2)_2SO$] with water from 298.15 K to 343.15 K (in steps of 5 K), covering the entire composition range. The V^{E}s for {DESO + water} are all negative with a minimum at $x_{\mathrm{DESO}} \approx 0.4$. At each temperature this minimum is noticeably more negative (*e.g.* at 298.15 K, $V^{E} \approx -1.36$ cm^3 mol^{-1}) than the minimum for the corresponding {DMSO + water} system, and $\partial V^{E}/\partial T$ is much larger than for {DMSO + water}. Interaction between DESO and water is stronger than between DMSO and water, which is reflected by the excess molar Gibbs energy being more negative than for the aqueous DMSO system: at 298.15 K and equimolar composition, for {DESO + water}[581] $G^{E} = -1733$ J mol^{-1} which has to be compared with $G^{E} = -1218$ J mol^{-1} (Lam and Benoit[582]), $G^{E} = -1162$ J mol^{-1} (Chan and Van Hook,[583] extrapolated),

$G^E = -1263$ J mol^{-1} (Lai *et al.*[564]), and $G^E = -1527$ J mol^{-1} (Qiann *et al.*[584]). Clearly, these differences should be resolved. For a study of the cryoprotective properties of DESO, see reference 585.

7.5 Concluding Remarks

It was not the goal of this survey to present an exhaustive overview of every experimental technique used for the determination of excess molar volumes V^E, either directly or indirectly, and closely related quantities of *liquid* nonelectrolyte mixtures, but rather to provide an admittedly subjective selection of some of the more popular instruments as well as some novel or unusual ones. This will hopefully allow the interested reader to see how the latter compare with more conventional methods. We have also presented a concise overview of the relevant thermodynamic formalism with ramifications into neighbouring disciplines, and indicated fields of increased current research activities. With the focus on experiments, the field was divided into *three* subgroups comprising work on (I) mixtures of essentially aprotic liquids, (II) mixtures containing alcohols or alkanoic acids, and (III) aqueous mixtures of nonelectrolytes. Evidently, research activities in the general area of volumetric properties have *greatly* increased, and some of the accompanying problems for the scientific community concerning adequate, reasonable publication of experimental results should be resolved soon, essentially by the editors of scientific journals. Lastly, a few important aspects of data correlation, which plays such an important role in the practical application and dissemination of accurate thermodynamic data, have been touched upon.

Chemical thermodynamics in general, and *PVTx*-research in particular, are thriving fields combining theoretical importance with great practical utility. Although simple in principle, enormous effort and ingenuity had to go into scientific design to provide the vast array of instruments now at our disposal for the determination of volumetric properties of fluids over large ranges of temperature, pressure and composition. The major driving forces for progress in instrumentation are (I) the need to continuously expand the application ranges of the apparatus, (II) the desire to improve accuracy and speed of measurement, and (III) the practical demands to facilitate data management, data storage and data transfer. In combination with advances in the statistical–mechanical treatment of solutions and increasingly sophisticated computer simulations, ever more comprehensive and reliable experimental results provide new insights, and in turn initiate new scientific advances on microscopic, mesoscopic and macroscopic levels.

Excess molar volumes on mixing two (or more) liquids can arise from any one of the heuristic factors depicted in Figure 7.1, though, in general, a combination of several will take place, resulting in a particular, characteristic V^E. Thus we repeat our plea, already stated in the Introduction, for *more comprehensive* experimental investigations, including calorimetric measurements and/or vapour–liquid equilibria, and *adequate* molecular thermodynamical discussions.

Finally, we would like to point out that cross-fertilisation with neighbouring disciplines has always been a potent stimulus in science. This is also true for investigations of volumetric properties of aqueous solutions of proteins (and other large molecules), which are discussed in the last two chapters of this monograph: in conjunction with calorimetric and spectroscopic data, they will contribute greatly to our understanding of the role of hydrophobic and hydrophilic effects in biology, they will provide further insight into protein hydration and packing, and they will contribute towards solving the problems of protein folding and binding, with obvious biomedical implications. The increasing number of investigations with a strong biophysical flavour is thus not surprising.

References

1. E. Wilhelm, Volumetric Properties: Introduction, Concepts and Selected Applications, in *Volumetric Properties: Liquids, Solutions and Vapours*, ed. E. Wilhelm and T. M. Letcher, The Royal Society of Chemistry/IACT & IUPAC, Cambridge, UK, 2015, ch. 1, pp. 1–27.
2. E. Wilhelm, Thermodynamics of Solutions, Especially Dilute Solutions of Nonelectrolytes, in *Molecular Liquids: New Perspectives in Physics and Chemistry, NATO ASI Series C: Mathematical and Physical Sciences*, ed. J. J. C. Teixeira-Dias, Kluwer Academic Publishers, Dordrecht, The Netherlands, 1992, vol. 379, pp. 175–206.
3. E. Wilhelm, *Netsu Sokutei*, 2012, **39**(2), 61.
4. E. Wilhelm, *J. Solution Chem.*, 2014, **43**, 525.
5. J. M. Prausnitz, *Molecular Thermodynamics of Fluid-Phase Equilibria*, Prentice-Hall, Englewood Cliffs, NJ, USA, 1969.
6. J. M. Prausnitz, R. N. Lichtenthaler and E. G. de Azevedo, *Molecular Thermodynamics of Fluid-Phase Equilibria*, Prentice Hall PTR, Upper Saddle River, NJ, USA, 3rd edn, 1999.
7. V. Majer and V. Svoboda, *Enthalpies of Vaporization of Organic Compounds. A Critical Review and Data Compilation*, Blackwell Scientific Publications/IUPAC, Oxford, UK, 1985.
8. L. Haar, J. S. Gallagher and G. S. Kell, *NBS/NRC Steam Tables (NSRDS). Thermodynamic and Transport Properties and Computer Programs for Vapor and Liquid States of Water in SI Units*, Hemisphere Publishing Corporation, New York, USA, 1984.
9. W. Wagner and A. Pruß, The IAPWS formulation 1995 for the thermodynamic properties of ordinary water substance for general and scientific use, *J. Phys. Chem. Ref. Data*, 2002, **31**, 387.
10. M. W. Chase Jr., NIST-JANAF Thermochemical Tables, Parts I and II, *J. Phys. Chem. Ref. Data, Monograph No. 9*, American Chemical Society and American Institute of Physics, Woodbury, NY, USA, 4th edn, 1998.

11. *International DATA Series, SELECTED DATA ON MIXTURES, Series A*, published by the Thermodynamics Research Center, Texas A&M University, College Station, TX, from 1973 through 1994.
12. B. E. Poling, J. M. Prausnitz and J. P. O'Connell, *The Properties of Gases and Liquids*, McGraw-Hill, New York, USA, 5th edn., 2001.
13. A. Kazakov, C. D. Muzny, R. D. Chirico, V. V. Diky and M. Frenkel, Web Thermo Tables – an On-Line Version of the TRC Thermodynamic Tables, *J. Res. Natl. Inst. Stand. Technol.*, 2008, **113**, 209.
14. *NIST SDR 4. NIST Thermophysical Properties of Hydrocarbon Mixtures Database: Version 3.2*, NIST, Boulder, Colorado, USA: http://www.nist.gov./srd/nist4.cfm (Last updated: June 19, 2012).
15. *NIST SDR 10. NIST/ASME Steam Properties Database: Version 2.22*, NIST, Boulder, Colorado, USA: http://www.nist.gov./srd/nist10.cfm (Last updated: February 28, 2012).
16. *NIST SDR 23: NIST Reference Fluid Thermodynamic and Transport Properties Database (REFPROP): Version 9.0.* NIST, Boulder, Colorado, USA: http://www.nist.gov./srd/nist23.cfm (Last updated: February 22, 2013).
17. *NIST SDR 103b: NIST ThermoData Engine Version 7.0 – Pure Compounds, Binary Mixtures, Ternary Mixtures, and Chemical Reactions*, NIST, Boulder, Colorado, USA: http://www.nist.gov./srd/nist103b.cfm (Last updated: February 14, 2013).
18. *NIST SDR 203. NIST Web Thermo Tables (WTT) – Professional Edition*, NIST, Boulder, Colorado, USA: http://www.nist.gov./srd/nistwebsub3.cfm (Last updated: January 11, 2013).
19. *Dortmund Data Bank Software and Separation Technology:* http://www.ddbst.de.
20. (a) R. C. Wilhoit, K. N. Marsh, X. Hong, N. Gadalla and M. Frenkel, *Densities of Aliphatic Hydrocarbons: Alkanes, in Landolt–Börnstein, Numerical Data and Functional Relationships in Science and Technology, New Series; Group IV: Physical Chemistry, Vol. 8, Subvolume B*, ed. K. N. Marsh, Springer-Verlag, Berlin, Heidelberg, 1996; (b) R. C. Wilhoit, K. N. Marsh, X. Hong, N. Gadalla and M. Frenkel, *Densities of Aliphatic Hydrocarbons: Alkenes, Alkadienes, Alkynes, and Miscellaneous Compounds, ibid., Subvolume C*, ed. K. N. Marsh, Springer-Verlag, Berlin, Heidelberg, 1996; (c) R. C. Wilhoit, X. Hong, M. Frenkel and K. R. Hall, *Densities of Monocyclic Hydrocarbons, ibid., Subvolume D*, ed. K. R. Hall and K. N. Marsh, Springer-Verlag, Berlin, Heidelberg, 1997; (d) R. C. Wilhoit, X. Hong, M. Frenkel and K. R. Hall, *Densities of Aromatic Hydrocarbons, ibid., Subvolume E*, ed. K. R. Hall and K. N. Marsh, Springer-Verlag, Berlin, Heidelberg, 1998; (e) R. C. Wilhoit, X. Hong, M. Frenkel and K. R. Hall, *Densities of Polycyclic Hydrocarbons, ibid., Subvolume F*, ed. K. R. Hall and K. N. Marsh, Springer-Verlag, Berlin, Heidelberg, 1998; (f) M. Frenkel, X. Hong, R. C. Wilhoit and K. R. Hall, *Densities of Alcohols, ibid., Subvolume G*, ed. K. R. Hall and K. N. Marsh, Springer-Verlag, Berlin, Heidelberg, 1998; (g) M. Frenkel, X. Hong, R. C. Wilhoit and K. R. Hall, *Densities of Esters*

and Ethers, ibid., Subvolume H, ed. K. R. Hall and K. N. Marsh, Springer-Verlag, Berlin, Heidelberg, 2001; (h) M. Frenkel, X. Hong, Q. Dong, X. Yan and R. D. Chirico, *Densities of Phenols, Aldehydes, Ketones, Carboxylic Acids, Amines, Nitriles, and Nitrohydrocarbons, ibid., Subvolume I*, ed. M. Frenkel and K. N. Marsh, Springer-Verlag, Berlin, Heidelberg, 2002; (i) M. Frenkel, X. Hong, Q. Dong, X. Yan and R. D. Chirico, *Densities of Halohydrocarbons, ibid., Subvolume J*, ed. M. Frenkel and K. N. Marsh, Springer-Verlag, Berlin, Heidelberg, 2003.

21. J.-P. E. Grolier, C. J. Wormald, J.-C. Fontaine, K. Sosnkowska-Kehiaian and H. V. Kehiaian, Heats of Mixing and Solution, in *Landolt–Börnstein, Numerical Data and Functional Relationships in Science and Technology, New Series; Group IV: Physical Chemistry, Vol. 10A*, Springer-Verlag, Berlin, Heidelberg, 2004.
22. (a) I. Wichterle, J. Linek, Z. Wagner, J.-C. Fontaine, K. Sosnkowska-Kehiaian and H. V. Kehiaian, Vapor-Liquid Equilibrium in Mixtures and Solutions, in *Landolt–Börnstein, Numerical Data and Functional Relationships in Science and Technology, New Series; Group IV: Physical Chemistry, Vol. 13A1*, Springer-Verlag, Berlin, Heidelberg, 2007; (b) I. Wichterle, J. Linek, Z. Wagner, J.-C. Fontaine, K. Sosnkowska-Kehiaian and H. V. Kehiaian, Vapor-Liquid Equilibrium in Mixtures and Solutions, in *Landolt–Börnstein, Numerical Data and Functional Relationships in Science and Technology, New Series; Group IV: Physical Chemistry, Vol. 13A2*, Springer-Verlag, Berlin, Heidelberg, 2008.
23. (a) R. Lacmann and C. Synowietz, Nonaqueous Systems and Ternary Aqueous Systems, in *Landolt–Börnstein, Numerical Data and Functional Relationships in Science and Technology, New Series; Group IV: Macroscopic and Technical Properties of Matter, Vol. 1, Densities of Liquid Systems, Part a*, ed. Kl. Schäfer, Springer-Verlag, Berlin, Heidelberg, 1974; (b) J. D'Ans, H. Surawski and C. Synowietz, Densities of Binary Aqueous Systems and Hear Capacities of Liquid Systems, in *Landolt–Börnstein, Numerical Data and Functional Relationships in Science and Technology, New Series; Group IV: Macroscopic and Technical Properties of Matter, Vol. 1, Densities of Liquid Systems and their Heat Capacities, Part b*, ed. Kl. Schäfer, Springer-Verlag, Berlin, Heidelberg, 1977; (c) I. Cibulka, L. Hnědkovský, J.-C. Fontaine, K. Sosnkowska-Kehiaian and H. V. Kehiaian, Binary Liquid Systems of Nonelectrolytes, in *Landolt–Börnstein, Numerical Data and Functional Relationships in Science and Technology, New Series; Group IV: Physical Chemistry, Vol. 23, Volumetric Properties of Mixtures and Solutions, Subvolume A*, ed. H. V. Kehiaian, Springer-Verlag, Berlin, Heidelberg, 2009.
24. I. Cibulka, J.-C. Fontaine, K. Sosnkowska-Kehiaian and H. V. Kehiaian, Heats of Mixing, Vapor-Liquid Equilibrium, and Volumetric Properties of Mixtures and Solutions, in *Landolt–Börnstein, Numerical Data and Functional Relationships in Science and Technology, New Series; Group IV: Physical Chemistry, Vol. 26A, Binary Liqid Systems of Nonelectrolytes I*, Springer-Verlag, Berlin, Heidelberg, 2011.

25. I. Cibulka, J.-C. Fontaine, K. Sosnkowska-Kehiaian and H. V. Kehiaian, Heats of Mixing, Vapor-Liquid Equilibrium, and Volumetric Properties of Mixtures and Solutions, in *Landolt–Börnstein, Numerical Data and Functional Relationships in Science and Technology, New Series; Group IV: Physical Chemistry, Vol. 26B, Binary Liqid Systems of Nonelectrolytes II*, Springer-Verlag, Berlin, Heidelberg, 2012.
26. I. Cibulka, J.-C. Fontaine, K. Sosnkowska-Kehiaian and H. V. Kehiaian, Heats of Mixing, Vapor-Liquid Equilibrium, and Volumetric Properties of Mixtures and Solutions, in *Landolt–Börnstein, Numerical Data and Functional Relationships in Science and Technology, New Series; Group IV: Physical Chemistry, Vol. 26C, Binary Liqid Systems of Nonelectrolytes III*, Springer-Verlag, Berlin, Heidelberg, 2012.
27. *Experimental Thermodynamics, Vol. I: Calorimetry of Non-reacting Systems*, ed. J. P. McCullough and D. W. Scott, Butterworths/IUPAC, London, UK, 1968.
28. *Experimental Thermodynamics, Vol. II: Experimental Thermodynamics of Non-reacting Fluids*, ed. B. Le Neindre and B. Vodar, Butterworths/IUPAC, London, UK, 1975.
29. *Solution Calorimetry: Experimental Thermodynamics, Vol. IV*, ed. K. N. Marsh and P. A. G. O'Hare, Blackwell Scientific Publications/IUPAC, Oxford, UK, 1994.
30. *Measurement of the Thermodynamic Properties of Single Phases: Experimental Thermodynamics, Vol. VI*, ed. A. R. H. Goodwin, K. N. Marsh and W. A. Wakeham, Elsevier/IUPAC, Amsterdam, The Netherlands, 2003.
31. *Measurement of the Thermodynamic Properties of Multiple Phases: Experimental Thermodynamics, Vol. VII*, ed. R. D. Weir and Th. W. de Loos, Elsevier/IUPAC, Amsterdam, The Netherlands, 2005.
32. *The Experimental Determination of Solubilities*, ed. G. T. Hefter and R. P. T. Tomkins, *Wiley, Series in Solution Chemistry, Vol. 6*, Wiley, Chichester, UK, 2003.
33. *Development and Application in Solubility*, ed. T. M. Letcher, The Royal Society of Chemistry/IACT, Cambridge, UK, 2007.
34. *Heat Capacities: Liquids, Solutions and Vapours*, ed. E. Wilhelm and T. M. Letcher, The Royal Society of Chemistry/IUPAC & IACT, Cambridge, UK, 2010.
35. J. S. Rowlinson and F. L. Swinton, *Liquids and Liquid Mixtures*, Butterworth Scientific, London, UK, 3rd edn., 1982.
36. *Equations of State for Fluids and Fluid Mixtures. Experimental Thermodynamics, Vol. V*, ed. J. V. Sengers, R. F. Kayser, C. J. Peters and H. J. White, Elsevier Science/IUPAC, Amsterdam, The Netherlands, 2000.
37. F. J. Vesely, *Computational Physics: An Introduction*, Kluwer Academic/Plenum Publishers, New York, USA, 2nd edn., 2001.
38. D. Frenkel and B. Smit, *Understanding Molecular Simulation*, Academic Press, London, UK, 2nd edn., 2002.

39. J.-L. Barrat and J.-P. Hansen, *Basic Concepts for Simple and Complex Liquids*, Cambridge University Press, Cambridge, UK, 2003.
40. C. J. Cramer, *Essentials of Computational Chemistry: Theories and Models*, John Wiley & Sons, Chichester, UK, 2nd edn., 2004.
41. D. C. Rapaport, *The Art of Molecular Dynamics Simulation*, Cambridge University Press, Cambridge, UK, 2nd edn., 2004.
42. J.-P. Hansen and I. R. McDonald, *Theory of Simple Liquids*, Academic Press/Elsevier, London, UK, 3rd edn., 2006.
43. A. Ben-Naim, *Molecular Theory of Solutions*, Oxford University Press, Oxford, UK, 2006.
44. I. G. Kaplan, *Intermolecular Interactions: Physical Picture, Computational Methods and Model Potentials*, John Wiley & Sons, Chichester, UK, 2006.
45. K. Lucas, *Molecular Models for Fluids*, Cambridge University Press, New York, NY, USA, 2007.
46. M. E. Tuckerman, *Statistical Mechanics: Theory and Molecular Simulation*, Oxford University Press, New York, USA, 2010.
47. *Applied Thermodynamics of Fluids*, A. R. H. Goodwin, ed. J. V. Sengers and C. J. Peters, The Royal Society of Chemistry/IUPAC & IACT, Cambridge, UK, 2010.
48. C. G. Gray and K. E. Gubbins, *Theory of Molecular Fluids, Vol. 1: Fundamentals*, Clarendon Press, Oxford, UK, 1984.
49. C. G. Gray, K. E. Gubbins and C. G. Joslin, *Theory of Molecular Fluids. Vol. 2: Applications*, Oxford University Press, Oxford, UK, 2011.
50. U. K. Deiters and T. Kraska, *High-Pressure Fluid Phase Equilibria: Phenomenology and Computation*, Elsevier, Oxford, UK, 2012.
51. J. H. Hildebrand and R. L. Scott, *Regular Solutions*, Prentice-Hall, Englewood Cliffs, NJ, USA, 1962, ch. 8, p. 104.
52. R. Battino, *Chem. Rev.*, 1971, **71**, 5.
53. (a) H. Stabinger, O. Kratky and H. Leopold, *Monatsh. Chem.*, 1967, **98**, 436; (b) O. Kratky, H. Leopold and H. Stabinger, *Z. Angew. Phys.*, 1969, **27**, 273.
54. P. Picker, E. Tremblay and J. Jolicoeur, *J. Solution Chem.*, 1974, **3**, 377.
55. J.-P. E. Grolier, G. C. Benson and P. Picker, *J. Chem. Eng. Data*, 1975, **20**, 243.
56. C. de Visser, G. Perron, J. E. Desnoyers, W. J. M. Heuvelsland and G. Somsen, *J. Chem. Eng. Data*, 1977, **22**, 74.
57. J. R. Goates, J. B. Ott and J. F. Moellmer, *J. Chem. Thermodyn.*, 1977, **9**, 249.
58. E. Wilhelm, J.-P. E. Grolier and M. H. Karbalai Ghassemi, *Monatsh. Chem.*, 1978, **109**, 369.
59. Y. P. Handa and G. C. Benson, *Fluid Phase Equilib.*, 1979, **3**, 185.
60. I. Cibulka and R. Holub, *Chem. Listy*, 1978, **72**, 457.
61. V. Tekáč, I. Cibulka and R. Holub, *Fluid Phase Equilib.*, 1985, **19**, 33.
62. E. Wilhelm, *Thermochim. Acta*, 1990, **162**, 43.
63. E. Wilhelm, *Pure Appl. Chem.*, 2005, **77**, 1317.
64. *Water: A Comprehensive Treatise, Vols. I – VII*, ed. F. Franks, Plenum Press, New York, USA, 1972 through 1982.

65. A. Ben-Naim, *Water and Aqueous Solutions. Introduction to a Molecular Theory*, Plenum Press, New York, USA, 1974.
66. A. Ben-Naim, *Hydrophobic Interactions*, Plenum Press, New York, USA, 1980.
67. C. Tanford, *The Hydrophobic Effect: Formation of Micelles and Biological Membranes*, Wiley, New York, USA, 2nd edn., 1980.
68. A. Ben-Naim, *Molecular Theory of Water and Aqueous Solutions, Part I: Understanding Water*, World Scientific, Singapore, 2009.
69. A. Ben-Naim, *Molecular Theory of Water and Aqueous Solutions, Part II: The Role of Water in Protein Folding, Self-Assembly and Molecular Recognition*, World Scientific, Singapore, 2011.
70. I. G. Economou, Cubic and Generalized van der Waals Equations of State, in *Applied Thermodynamics of Fluids*, ed. A. R. H. Goodwin, J. V. Sengers and C. J. Peters, The Royal Society of Chemistry/IUPAC & IACT, Cambridge, UK, 2010, ch. 4, pp. 53–83.
71. E. W. Lemmon and R. Span, Multi-parameter Equations of State for Pure Fluids and Mixtures, in *Applied Thermodynamics of Fluids*, ed. A. R. H. Goodwin, J. V. Sengers and C. J. Peters, The Royal Society of Chemistry/IUPAC & IACT, Cambridge, UK, 2010, ch. 12, pp. 394–432.
72. S. Angus, B. Armstrong and K. M. de Reuck, *International Thermodynamic Tables of the Fluid State - 1: Argon: 1971*, IUPAC, Butterworths, London, UK, 1972.
73. S. Angus, B. Armstrong and K. M. de Reuck, *International Thermodynamic Tables of the Fluid State – 3: Carbon Dioxide*, IUPAC, Pergamon Press, Oxford, UK, 1976.
74. S. Angus, B. Armstrong and K. M. de Reuck, *International Thermodynamic Tables of the Fluid State – 5: Methane*, IUPAC, Pergamon Press, Oxford, UK, 1978.
75. W. Wagner and K. M. de Reuck, *International Thermodynamic Tables of the Fluid State – 13. Methane*, IUPAC, Blackwell Scientific, Oxford, UK, 1996.
76. S. Angus, K. M. de Reuck and B. Armstrong,*International Thermodynamic Tables of the Fluid State – 6: Nitrogen*, IUPAC, Pergamon Press, Oxford, UK, 1979.
77. S. Angus, B. Armstrong and K. M. de Reuck, *International Thermodynamic Tables of the Fluid State – 7: Propylene (Propene)*, IUPAC, Pergamon Press, Oxford, UK, 1980.
78. S. Angus, B. Armstrong and K. M. de Reuck, *International Thermodynamic Tables of the Fluid State – 8: Chlorine, Tentative Tables*, IUPAC, Pergamon Press, Oxford, UK, 1985.
79. W. Wagner and K. M. de Reuck, *International Thermodynamic Tables of the Fluid State – 9: Oxygen*, IUPAC, Blackwell Scientific, Oxford, UK, 1987.
80. K. M. de Reuck, *International Thermodynamic Tables of the Fluid State – 10: Ethylene (Ethene)*, IUPAC, Blackwell Scientific, Oxford, UK, 1988.
81. K. M. de Reuck, *International Thermodynamic Tables of the Fluid State – 11: Fluorine*, IUPAC, Blackwell Scientific, Oxford, UK, 1990.

82. K. M. de Reuck and R. J. B. Craven, *International Thermodynamic Tables of the Fluid State – 12: Methanol*, IUPAC, Blackwell Scientific, Oxford, UK, 1993.
83. *International Union of Pure and Applied Chemistry, Quantities, Units and Symbols in Physical Chemistry*, RSC Publishing, Cambridge, UK, 2007.
84. J. H. Dymond, K. N. Marsh, R. C. Wilhoit and K. C. Wong, Virial Coefficients of Pure Gases, in *Landolt–Börnstein, Numerical Data and Functional Relationships in Science and Technology, New Series; Group IV: Physical Chemistry, Virial Coefficients of Pure Gases and Mixtures, Vol. 21A*, ed. M. Frenkel and K. N. Marsh, Springer-Verlag, Heidelberg, Germany, 2002.
85. J. H. Dymond, K. N. Marsh and R. C. Wilhoit, Virial Coefficients of Mixtures, in *Landolt–Börnstein, Numerical Data and Functional Relationships in Science and Technology, New Series; Group IV: Physical Chemistry, Virial Coefficients of Pure Gases and Mixtures, Vol. 21B*, ed. M. Frenkel and K. N. Marsh, Springer-Verlag, Heidelberg, Germany, 2003.
86. E. A. Mason and T. H. Spurling, *The Virial Equation of State*, Pergamon Press, Oxford, UK, 1969.
87. D. A. McQuarrie, *Statistical Mechanics*, Harper & Row, New York, USA, 1976.
88. G. C. Maitland, M. Rigby, E. B. Smith and W. A. Wakeham, *Intermolecular Forces: Their Origin and Determination*, Clarendon Press, Oxford, UK, 1981.
89. J. P. M. Trusler, The Virial Equation of State, in *Applied Thermodynamics of Fluids*, ed. A. R. H. Goodwin, J. V. Sengers and C. J. Peters, The Royal Society of Chemistry/IUPAC & IACT, Cambridge, UK, 2010, ch. 3, pp. 33–52.
90. J. S. Rowlinson, *The Perfect Gas*, Pergamon Press, Oxford, UK, 1963.
91. E. Wilhelm, *Thermochim. Acta*, 1983, **69**, 1.
92. E. Wilhelm, Heat Capacities: Introduction, Concepts and Selected Applications, in *Heat Capacities: Liquids, Solutions and Vapours*, ed. E. Wilhelm and T. M. Letcher, The Royal Society of Chemistry/IUPAC & IACT, Cambridge, UK, 2010, ch. 1, pp. 1–27.
93. C. J. Wormald, Heat Capacities and Related Properties of Vapours and Vapour Mixtures, in *Heat Capacities: Liquids, Solutions and Vapours*, ed. E. Wilhelm and T. M. Letcher, The Royal Society of Chemistry/IUPAC & IACT, Cambridge, UK, 2010, ch. 6, pp. 112–131.
94. E. Wilhelm, *J. Solution Chem.*, 2010, **39**, 1777.
95. T. Miyazaki, A. V. Hejmadi and J. E. Powers, *J. Chem. Thermodyn.*, 1980, **12**, 105.
96. I. Prigogine and R. Defay, *Chemical Thermodynamics (translated and revised by D. H. Everett)*, Longmans, Green and Co, London, UK, 1954.
97. R. Haase, *Thermodynamik der Mischphasen*, Springer-Verlag, Berlin, Germany, 1956.
98. K. Denbigh, *The Principles of Chemical Equilibrium*, Cambridge University Press, Cambridge, UK, 4th edn., 1981.

99. G. Kortüm and H. Lachmann, *Einführung in die chemische Thermodynymik: Phänomenologische und statistische Behandlung*, Verlag Chemie, Weinheim and Vandenhoeck & Ruprecht, Göttingen, Germany, 7th edn., 1981.
100. H. C. Van Ness and M. M. Abbott, *Classical Thermodynamics of Nonelectrolyte Solutions: With Applications to Phase Equilibria*, McGraw-Hill Book Company, New York, USA, 1982.
101. S. E. Wood and R. Battino, *Thermodynamics of Chemical Systems*, Cambridge University Press, Cambridge, UK, 1990.
102. C. Lüdecke and D. Lüdecke, *Thermodynamik: Physikalisch-chemische Grundlagen der thermischen Verfahrenstechnik*, Springer-Verlag Berlin, Heidelberg, Germany, 2000.
103. E. Wilhelm, Precision Methods for the Determination of the Solubility of Gases in Liquids, in *CRC Critical Reviews in Analytical Chemistry*, CRC Press, Boca Raton, FL, USA, 1985, vol. 16 , pp. 129–175.
104. E. Wilhelm, Low-Pressure Solubility of Gases in Liquids, in *Experimental Thermodynamics, Vol. VII: Measurement of the Thermodynamic Properties of Multiple Phases*, R. D. Weir and Th. W. de Loos, eds., Elsevier/IUPAC, Amsterdam, The Netherlands, 2005, ch. 7, pp. 137–176.
105. E. Wilhelm, Thermodynamics of Nonelectrolyte Solubility, in *Development and Applications in Solubility*, ed. T. M. Letcher, The Royal Society of Chemistry/IUPAC & IACT, Cambridge, UK, 2007, ch. 1, pp. 3–18.
106. E. Wilhelm and R. Battino, Partial Molar Heat Capacity Changes of Gases Dissolved in Liquids, Chapter 21 in *Heat Capacities: Liquids, Solutions and Vapours*, ed. E. Wilhelm and T. M. Letcher, The Royal Society of Chemistry/IACT & IUPAC, Cambridge, UK, 2010, ch. 21, pp. 457–471.
107. J. J. Christensen, R. M. Izatt, D. J. Eatough and L. D. Hansen, *J. Chem. Thermodyn.*, 1978, **10**, 25.
108. A. Heintz and R. N. Lichtenthaler, *Ber. Bunsenges. Phys. Chem.*, 1979, **83**, 853.
109. M. A. Siddiqi and K. Lucas, *J. Chem. Thermodyn.*, 1982, **14**, 1183.
110. *Technique of Organic Chemistry. Vol. 1, Part 1: Physical Methods of Organic Chemistry*, ed. A. Weissberger, Interscience, New York, USA, 3rd edn., 1959.
111. K. N. Marsh, The Measurement of Thermodynamic Excess Functions of Binary Liquid Mixtures, in *Chemical Thermodynamics (Specialist Periodical Reports), Vol. 2*, ed. M. L. McGlashan, The Chemical Society, London, UK, 1978, ch. 1, pp. 1–45.
112. K. N. Marsh, Mixtures of Liquids, in *Measurement of the Thermodynamic Properties of Single Phases: Experimental Thermodynamics, Vol. VI*, ed. A. R. H. Goodwin, K. N. Marsh and W. A. Wakeham, Elsevier/IUPAC, Amsterdam, The Netherlands, 2003, ch. 8.2, pp. 404–407.
113. W. A. Duncan, J. P. Sheridan and F. L. Swinton, *Trans. Faraday Soc.*, 1966, **62**, 1090.

114. D. B. Keyes and J. H. Hildebrand, *J. Am. Chem. Soc.*, 1917, **39**, 2126.
115. I. Brown and F. Smith, *Aust. J. Chem.*, 1962, **15**, 1.
116. D. J. Stookey, H. M. Sallak and B. D. Smith, *J. Chem. Thermodyn.*, 1973, **5**, 741.
117. W. Geffcken, A. Kruis and L. Solana, *Z. Phys. Chem., Abt. B*, 1937, **35**, 317.
118. A. E. P. Watson, I. A. McLure, J. E. Bennett and G. C. Benson, *J. Phys. Chem.*, 1965, **69**, 2753.
119. H. D. Pflug and G. C. Benson, *Can. J. Chem.*, 1968, **46**, 287.
120. L. A. Beath, S. P. O'Neill and A. G. Williamson, *J. Chem. Thermodyn.*, 1969, **1**, 293.
121. R. H. Stokes, B. J. Levien and K. N. Marsh, *J. Chem. Thermodyn.*, 1970, **2**, 43.
122. G. A. Bottomley and R. L. Scott, *J. Chem. Thermodyn.*, 1974, **6**, 973.
123. R. Tanaka, O. Kiyohara, P. J. D'Arcy and G. C. Benson, *Can. J, Chem.*, 1975, **53**, 2262.
124. E. Dickinson, D. C. Hunt and I. A. McLure, *J. Chem. Thermodyn.*, 1975, **7**, 731.
125. M. K. Kumaran and M. L. McGlashan, *J. Chem. Thermodyn.*, 1977, **9**, 259.
126. I. Cibulka and R. Holub, *Collect. Czech. Chem. Commun.*, 1981, **46**, 2774.
127. K. N. Marsh and A. E. Richards, *Aust. J. Chem.*, 1980, **33**, 2121.
128. G. A. Bottomley and R. L. Scott, *Aust. J. Chem.*, 1976, **29**, 427.
129. R. Chareyron and P. Clechet, *Bull. Soc. Chim. Fr.*, 1971, 2853.
130. S. E. Wood and J. P. Brusie, *J. Am. Chem. Soc.*, 1943, **65**, 1891.
131. F. Kohler and E. Rott, *Monatsh. Chem.*, 1954, **85**, 703.
132. R. Battino, *J. Phys. Chem.*, 1966, **70**, 3408.
133. B. M. Westwood and V. N. Kabadi, *J. Chem. Thermodyn.*, 2003, **35**, 1965.
134. J. Zielkiewicz, *J. Chem. Thermodyn.*, 1994, **26**, 959.
135. H. A. Albert and R. H. Wood, *Rev. Sci. Instrum.*, 1984, **55**, 589.
136. J. C. Moore, R. Battino, T. R. Rettich, Y. P. Handa and E. Wilhelm, *J. Chem. Eng. Data*, 1982, **27**, 22.
137. T. Zhou and R. Battino, *J. Chem. Eng. Data*, 2001, **46**, 331.
138. I. Cibulka and A. Heintz, *Fluid Phase Equilib.*, 1995, **107**, 235.
139. P. Izák, I. Cibulka and A. Heintz, *Fluid Phase Equilib.*, 1995, **109**, 227.
140. D. R. Biggerstaff and R. H. Wood, *J. Phys. Chem.*, 1988, **92**, 1988.
141. (a) R. Crovetto, R. H. Wood and V. Majer, *J. Chem. Thermodyn.*, 1990, **22**, 231; (b) R. Crovetto and R. H. Wood, *Fluid Phase Equilib.*, 1992, **74**, 271; (c) L. Hnědkovský, R. H. Wood and V. Majer, *J. Chem. Thermodyn.*, 1996, **28**, 125.
142. A. H. Corti, R. F. Prini and F. Svarc, *J. Solution Chem.*, 1990, **19**, 793.
143. V. Majer, R. Crovetto and R. H. Wood, *J. Chem. Thermodyn.*, 1991, **23**, 333.
144. J. M. Simonson, C. S. Oakes and R. J. Bodnar, *J. Chem. Thermodyn.*, 1994, **26**, 345.

145. L. A. Galicia-Luna, D. Richon and H. Renon, *J. Chem. Eng. Data*, 1994, **39**, 424.
146. J. G. Blencoe, S. E. Drummond, J. C. Seitz and B. E. Nesbitt, *Int. J. Thermophys.*, 1996, **17**, 179.
147. R. F. Chang and M. R. Moldover, *Rev. Sci. Instrum.*, 1996, **67**, 251.
148. (a) C. Xiao, H. Bianchi and P. R. Tremaine, *J. Chem. Thermodyn.*, 1997, **29**, 261; (b) E. Bulemela, P. Tremaine and Shun-ichi Ikawa, *Fluid Phase Equilib.*, 2006, **245**, 125.
149. V. Hynek, M. Obšil, V. Majer, J. Quint and J.-P. E. Grolier, *Int. J. Thermophys.*, 1997, **18**, 719.
150. V. Hynek, L. Hnědkovský and I. Cibulka, *J. Chem. Thermodyn.*, 1997, **29**, 1237.
151. A. W. Hakin, D. C. Daisley, L. Delgago, J. L. Liu, R. A. Marriott, J. L. Marty and G. Tompkins, *J. Chem. Thermodyn.*, 1998, **30**, 583.
152. V. Hynek, S. Degrange, M. Polednicek, V. Majer, J. Quint and J.-P. E. Grolier, *J. Solution Chem.*, 1999, **28**, 631.
153. E. C. Ihmels and J. Gmehling, *Ind. Eng. Chem. Res.*, 2001, **40**, 4470.
154. L. Hnědkovský and I. Cibulka, *Int. J. Thermophys.*, 2004, **25**, 1135.
155. T. Aida, A. Yamazaki, M. Akutsu, T. Ono, A. Kanno, T. A. Hoshina, M. Ota, M. Watanabe, Y. Sato, R. L. Smith and H. Inomata, *Rev. Sci. Instrum.*, 2007, **78**, 115111–1.
156. J. Safarov, F. Millero, R. Feistel, A. Heintz and E. Hassel, *Ocean Sci.*, 2009, **5**, 235.
157. B. Lagourette, C. Boned, H. Saint-Guirons, P. Xans and H. Zhou, *Meas. Sci. Technol.*, 1992, **3**, 699.
158. R. Laznickova and H. Huemer, *Meas. Sci. Technol.*, 1998, **9**, 719.
159. J.-P. E. Grolier, E. Wilhelm and M. H. Hamedi, *Ber. Bunsenges. Phys. Chem.*, 1978, **82**, 1282.
160. J. Troncoso, D. Bessières, C. A. Cerdeiriña, E. Carballo and L. Romaní, *Fluid Phase Equilib.*, 2003, **208**, 141.
161. B. Garcia, S. Aparicio, R. Alcade, M. J. Dávila and J. M. Leal, *Ind. Eng. Chem. Res.*, 2004, **43**, 3205.
162. C. K. Zéberg-Mikkelsen, L. Lugo, J. Garcia and J. Fernández, *Fluid Phase Equilib.*, 2005, **235**, 139.
163. S. L. Outcalt and M. O. McLinden, *Ind. Eng. Chem. Res.*, 2007, **46**, 8264.
164. J. J. Segovia, O. Fandiño, E. R. López, L. Lugo, M. C. Martin and J. Fernández, *J. Chem. Thermodyn.*, 2009, **41**, 632.
165. R. Battino, M. Banzhof, M. Bogan and E. Wilhelm, *Anal. Chem.*, 1971, **43**, 806.
166. A. H. Harvey, A. P. Peskin and S. A. Klein, *NIST/ASME Steam Properties*, NIST Standard Reference Database 10, Version 2.2, National Institute of Standards and Technology, Gaithersburg, Maryland, USA, 2000.
167. M. Tanaka, G. Girard, R. Davis, A. Peuto and N. Bignell, *Metrologia*, 2001, **38**, 301.
168. A. H. Harvey, R. Span, K. Fujii, M. Tanaka and R. S. Davis, *Metrologia*, 2009, **46**, 196.

169. E. Batista and R. Paton, *Metrologia*, 2007, **44**, 453.
170. (a) N. Bignell, *J. Phys. E: Sci. Instrum.*, 1982, **15**, 378; (b) N. Bignell, *Metrologia*, 1983, **19**, 57; (c) N. Bignell, *J. Phys. Chem.*, 1984, **88**, 5409; (d) N. Bignell, *Metrologia*, 1986/87, **23**, 207; (e) N. Bignell, *J. Phys. Chem.*, 1987, **91**, 1687.
171. I. Lauder, *Aust. J. Chem.*, 1959, **12**, 40.
172. F. J. Millero and R. T. Emmet, *J. Mar. Res.*, 1976, **34**, 15.
173. G. S. Kell, *J. Phys. Chem. Ref. Data*, 1977, **6**, 1109.
174. H. Watanabe and K. Iizuka, *Jpn. J. Appl. Phys.*, 1981, **20**, 1979.
175. H. Watanabe and K. Iizuka, *Metrologia*, 1985, **21**, 19; Erratum: *ibid.*, 1986, **22**, 115.
176. G. Girard and M.-J. Coarasa, The Influence of Dissolved Air on the Density of Water, in *Precision Measurements and Fundamental Constants II*, Natl. Bur. Stand. (U.S.), Spec. Publ. 617, ed. B. N. Taylor and W. D. Phillips, U.S. Government Printing Office, Washington D.C., USA, 1984, pp. 453–456.
177. T. R. Rettich, R. Battino and E. Wilhelm, *J. Solution Chem.*, 1992, **21**, 987.
178. T. R. Rettich, R. Battino and E. Wilhelm, *J. Solution Chem.*, 1984, **13**, 335.
179. T. R. Rettich, R. Battino and E. Wilhelm, *J. Chem. Thermodyn.*, 2000, **32**, 1145.
180. J. J. Carroll, J. D. Slupsky and A. E. Mather, *J. Phys. Chem. Ref. Data*, 1991, **20**, 1201.
181. A. H. Harvey, S. G. Kaplan and J. H. Burnett, *Int. J. Thermophys.*, 2005, **26**, 1495.
182. P. Giacomo, *Metrologia*, 1982, **18**, 33.
183. H. S. Harned and R. Davis, Jr., *J. Am. Chem. Soc.*, 1943, **65**, 2030.
184. E. L. Shock and H. C. Helgeson, *Geochim. Cosmochim. Acta*, 1988, **52**, 2009.
185. V. Majer and A. A. H. Pádua, Measurement of Density with Vibrating Bodies, in *Measurement of the Thermodynamic Properties of Single Phases: Experimental Thermodynamics, Vol. VI*, ed. A. R. H. Goodwin, K. N. Marsh and W. A. Wakeham, Elsevier/IUPAC, Amsterdam, The Netherlands, 2003, ch. 5.2, pp. 149–168.
186. A. A. H. Pádua, J. M. N. A. Fareleira, J. C. G. Calado and W. A. Wakeham, *Int. J. Thermophys.*, 1996, **17**, 781.
187. F. Audonnet and A. A. H. Pádua, *Fluid Phase Equilib.*, 2001, **181**, 147.
188. H. M. T. Avelino, J. M. N. A. Fareleira and W. A. Wakeham, *Int. J. Thermophys.*, 2003, **24**, 323.
189. D. R. Caudwell, J. P. M. Trusler, V. Vesovic and W. A. Wakeham, *J. Chem. Eng. Data*, 2009, **54**, 359; Corrections: ibid., 2010, **55**, 5396.
190. F. Ciotta and J. P. M. Trusler, *J. Chem. Eng. Data*, 2010, **55**, 2195.
191. W. A. Wakeham, M. J. Assael, A. Marmur, J. De Coninck, T. D. Blake, S. A. Theron and E. Zussman, Material Properties: Measurement and Data, in *Springer Handbook of Experimental Fluid Mechanics*,

ed. C. Tropea, A. L. Yarin and J. F. Foss, Springer-Verlag, Berlin, Heidelberg, Germany, 2007, ch. 3, pp. 85–177.
192. R. Kleinrahm and W. Wagner, *J. Chem. Thermodyn.*, 1986, **18**, 739.
193. H. W. Lösch, R. Kleinrahm and W. Wagner, *Chem. Ing. Tech.*, 1994, **66**, 1055.
194. (a) R. Gilgen, R. Kleinrahm and W. Wagner, *J. Chem. Thermodyn.*, 1994, **26**, 383; (b) R. Gilgen, R. Kleinrahm and W. Wagner, *J. Chem. Thermodyn.*, 1994, **26**, 399.
195. (a) P. Nowak, R. Kleinrahm and W. Wagner, *J. Chem. Thermodyn.*, 1997, **29**, 1137; (b) P. Nowak, R. Kleinrahm and W. Wagner, *J. Chem. Thermodyn.*, 1997, **29**, 1157.
196. (a) M. Funke, R. Kleinrahm and W. Wagner, *J. Chem. Thermodyn.*, 2002, **34**, 717; (b) M. Funke, R. Kleinrahm and W. Wagner, *J. Chem. Thermodyn.*, 2002, **34**, 735.
197. W. Wagner, K. Brachthäuser, R. Kleinrahm and H. W. Lösch, *Int. J. Thermophys.*, 1995, **16**, 399.
198. J. Klimeck, R. Kleinrahm and W. Wagner, *J. Chem. Thermodyn.*, 2001, **33**, 251.
199. G. Schilling, R. Kleinrahm and W. Wagner, *J. Chem. Thermodyn.*, 2008, **40**, 1095.
200. D. Sommer, R. Kleinrahm, R. Span and W. Wagner, *J. Chem. Thermodyn.*, 2011, **43**, 117.
201. U. Setzmann and W. Wagner, *Int. J. Thermophys.*, 1989, **10**, 1103.
202. W. Wagner, R. Kleinrahm, H. W. Lösch and J. T. R. Watson, Hydrostatic Balance Densimeters with Magnetic Suspension Couplings, in *Measurement of the Thermodynamic Properties of Single Phases: Experimental Thermodynamics, Vol. VI*, ed. A. R. H. Goodwin, K. N. Marsh and W. A. Wakeham, Elsevier/IUPAC, Amsterdam, The Netherlands, 2003, ch. 5.1, pp. 127–149.
203. L. Benjamin, *J. Phys. Chem.*, 1966, **70**, 3790.
204. F. Franks and H. T. Smith, *Trans. Faraday Soc.*, 1967, **63**, 2586.
205. W. L. Masterton and H. K. Seiler, *J. Phys. Chem.*, 1968, **72**, 4257.
206. F. J. Millero, J. H. Knox and R. T. Emmet, *J. Solution Chem.*, 1972, **1**, 173.
207. I. A. Weeks and G. C. Benson, *J. Chem. Thermodyn.*, 1973, **5**, 107.
208. B. N. Barman and Z. Rahim, *Rev. Sci. Instrum.*, 1977, **48**, 1695.
209. R. Masui, W. M. Haynes, R. F. Chang, H. A. Davis and J. M. H. Levelt Sengers, *Rev. Sci. Instrum.*, 1984, **55**, 1132.
210. S. Toscani, P. Figuière and H. Szwarc, *J. Chem. Thermodyn.*, 1989, **21**, 1263.
211. (a) B. Keramati and C. H. Wolgemuth, *Rev. Sci. Instrum.*, 1975, **46**, 1573; (b) B. Keramati and C. H. Wolgemuth, *J. Chem. Eng. Data*, 1976, **21**, 423.
212. (a) A. J. Easteal and L. A. Woolf, *J. Chem. Thermodyn.*, 1985, **17**, 49; (b) A. J. Easteal and L. A. Woolf, *J. Chem. Thermodyn.*, 1985, **17**, 69.
213. G. Götze and G. M. Schneider, *J. Chem. Thermodyn.*, 1980, **12**, 661.
214. R. E. Gibson, *J. Am. Chem. Soc.*, 1935, **57**, 1551.

215. M. M. Aliev, J. W. Magee and I. M. Abdulgatov, *Int. J. Thermophys.*, 2003, **24**, 1551.
216. H. Yokoyama and M. Uematsu, *J. Chem. Thermodyn.*, 2003, **35**, 813.
217. J. M. Simonson, D. J. Bradley and R. H. Busey, *J. Chem. Thermodyn.*, 1987, **19**, 479.
218. L. A. Woolf, Bellows Volumetry, in *Measurement of the Thermodynamic Properties of Single Phases: Experimental Thermodynamics, Vol. VI*, ed. A. R. H. Goodwin, K. N. Marsh and W. A. Wakeham, Elsevier/IUPAC, Amsterdam, The Netherlands, 2003, ch. 5.3, pp. 168–174.
219. J. C. Holste, Piezometer, in *Measurement of the Thermodynamic Properties of Single Phases: Experimental Thermodynamics, Vol. VI*, ed. A. R. H. Goodwin, K. N. Marsh and W. A. Wakeham, Elsevier/IUPAC, Amsterdam, The Netherlands, 2003, ch. 5.4, pp. 174–184.
220. A. M. F. Palavra, Isochoric Methods, in *Measurement of the Thermodynamic Properties of Single Phases: Experimental Thermodynamics, Vol. VI*, ed. A. R. H. Goodwin, K. N. Marsh and W. A. Wakeham, Elsevier/IUPAC, Amsterdam, The Netherlands, 2003, ch. 5.5, pp. 185–191.
221. A. M. F. Palavra, M. A. Tavares Cardoso, J. A. P. Coelho and M. F. B. Mourato, *Chem. Eng. Technol.*, 2007, **30**, 689.
222. K. F. Herzfeld and T. A. Litovitz, *Absorption and Dispersion of Ultrasonic Waves*, Academic Press, New York, USA, and London, UK, 1959.
223. A. B. Bhatia, *Ultrasonic Absorption*, Oxford University Press, London, UK, 1967.
224. (a) L. A. K. Staveley, K. R. Hart and W. I. Tupman, *Discuss. Faraday Soc.*, 1953, **15**, 130; (b) L. A. K. Staveley, W. I. Tupman and K. R. Hart, *Trans. Faraday Soc.*, 1955, **51**, 323.
225. B. J. Berne and R. Pecora, *Dynamic Light Scattering*, Wiley, New York, USA, 1976.
226. E. Wilhelm and A. Asenbaum, Heat Capacities and Brillouin Scattering in Liquids, in *Heat Capacities: Liquids, Solutions and Vapours*, ed. E. Wilhelm and T. M. Letcher, The Royal Society of Chemistry/IACT & IUPAC, Cambridge, UK, 2010, ch. 11, pp. 238–263.
227. T. Takagi and E. Wilhelm, Speed-of-Sound Measurements and Heat Capacities of Liquid Systems at High Pressure, in *Heat Capacities: Liquids, Solutions and Vapours*, ed. E. Wilhelm and T. M. Letcher, The Royal Society of Chemistry/IUPAC & IACT, Cambridge, UK, 2010, ch. 10, pp. 218–237.
228. R. A. Fine and F. J. Millero, *J. Chem. Phys.*, 1973, **59**, 5529.
229. G. S. Kell and E. Whalley, *J. Chem. Phys.*, 1975, **62**, 3496.
230. T. F. Sun, C. A. ten Seldam, B. J. Kortbeek, N. J. Trappeniers and S. N. Biswas, *Phys. Chem. Liq.*, 1988, **18**, 107.
231. T. F. Sun, S. A. R. C. Bominaar, C. A. ten Seldam and S. N. Biswas, *Ber. Bunsenges. Phys. Chem.*, 1991, **95**, 696.
232. E. Zorębski and M. Dzida, *J. Chem. Eng. Data*, 2007, **52**, 1010.
233. M. Chorążewski and M. Skrzypek, *Int. J. Thermophys.*, 2010, **31**, 26.

234. S. Aparicio, M. Atilhan and F. Karadas, *Ind. Eng. Chem. Res.*, 2010, **49**, 9580.
235. R. Gomes de Azevedo, J. M. S. S. Esperança, V. Najdanovic-Visak, Z. P. Visak, H. J. R. Guedes, M. Nunes da Ponte and L. P. N. Rebelo, *J. Chem. Eng. Data*, 2005, **50**, 997.
236. G. Benedetto, R. M. Gavioso, P. A. G. Albo, S. Lago, D. M. Ripa and R. Spagnolo, *Int. J. Thermophys.*, 2005, **26**, 1667.
237. K. Meier and S. Kabelac, *Rev. Sci. Instrum.*, 2006, **77**, 123903.
238. E. H. Baltasar, M. Taravillo, V. G. Baonza, P. D. Sanz and B. Guignon, *J. Chem. Eng. Data*, 2011, **56**, 4800.
239. C.-W. Lin and J. P. M. Trusler, *J. Chem. Phys.*, 2012, **136**, 094511.
240. E. Whalley, The Compression of Liquids, in *Experimental Thermodynamics, Vol. II: Experimental Thermodynamics of Non-reacting Fluids*, ed. B. Le Neindre and B. Vodar, Butterworths/IUPAC, London, UK, ch. 9, 1975, pp. 421–500.
241. R. J. Powell and F. L. Swinton, *J. Chem. Eng. Data*, 1968, **13**, 260.
242. J. M. Janssens and M. Ruel, *Can. J. Chem. Eng.*, 1972, **50**, 591.
243. O. Redlich and A. T. Kister, *Ind. Eng. Chem.*, 1948, **40**, 345.
244. G. Scatchard, *Chem. Rev.*, 1949, **44**, 7.
245. O. Redlich, A. T. Kister and C. E. Turnquist, Thermodynamics of Solutions: Analysis of Vapor-Liquid Equilibria, *Chem. Eng. Prog., Symp. Ser. 2*, 1952, **48**, 49.
246. J. Wisniak, *J. Phase Equilib.*, 2003, **24**, 103.
247. M. Margules, *Sitzber. Akad. Wiss. Wien, Math.-Naturwiss. Kl. Abt. IIa*, 1895, **104**, 1243.
248. J. Tomiska and A. Neckel, *J. Phase Equilib.*, 1996, **17**, 11.
249. D. M. Myers and R. L. Scott, *Ind. Eng. Chem.*, 1963, **55**(7), 43.
250. K. N. Marsh, *J. Chem. Thermodyn.*, 1977, **9**, 719.
251. R. L. Klaus and H. C. Van Ness, The Orthogonal Polynomial Representation of Thermodynamic Excess Functions, in *Chem. Eng. Prog., Symp. Ser. No. 81*, 1967, **63**, 88.
252. (a) C. W. Bale and A. D. Pelton, *Metall. Trans.*, 1974, **5**, 2323; (b) A. D. Pelton and C. W. Bale, *Metall. Trans. A*, 1986, **17**, 1057.
253. M. Hillert, *CALPHAD: Comput. Coupling Phase Diagrams Thermochem.*, 1980, **4**, 1.
254. R. A. Howald and I. Eliezer, *Metall. Trans. B*, 1977, **8**, 190.
255. J. Tomiska, *CALPHAD: Comput. Coupling Phase Diagrams Thermochem.*, 1981, **5**, 93.
256. F. Kohler, *Monatsh. Chem.*, 1960, **91**, 738.
257. I. Cibulka, *Collect. Czech. Chem. Commun.*, 1982, **47**, 1414.
258. Z. P. Visak, A. G. M. Ferreira and I. M. A. Fonseca, *J. Chem. Eng. Data*, 2000, **45**, 926.
259. P. P. Singh, R. K. Nigam, S. P. Sharma and S. Aggarwal, *Fluid Phase Equilib.*, 1984, **18**, 333.
260. (a) B. D. Djordjević, S. P. Šerbanović, I. R. Radović, A. Ž. Tasić and M. L. Kijevčanin, *J. Serb. Chem. Soc.*, 2007, **72**, 1437; (b) B. D. Djordjević,

M. L. Kijevčanin, I. R. Radović, S. P. Šerbanović and A. Ž. Tasić, *J. Serb. Chem. Soc.*, 2013, **78**, 1079.
261. M. Ramos-Estrada, G. A. Iglesias-Silva, K. R. Hall and F. Castillo-Borja, *J. Chem. Eng. Data*, 2011, **56**, 4461.
262. R. Quevedo-Nolasco, L. A. de la Cruz-Hernández, L. A. Galicia-Luna and O. Elizalde-Solis, *J. Chem. Eng. Data*, 2011, **56**, 4226.
263. S. Toscani and H. Szwarc, *J. Chem. Eng. Data*, 2004, **49**, 163.
264. H. Iloukhani and M. Rezaei-Sameti, *J. Mol. Liq.*, 2006, **126**, 62.
265. I. Prigogine, A. Bellemans and V. Mathot, *The Molecular Theory of Solutions*, North Holland Publishing Company, Amsterdam, The Netherlands, 1957.
266. (a) P. J. Flory, R. A. Orwell and A. Vrij, *J. Am. Chem. Soc.*, 1964, **86**, 3507; (b) P. J. Flory, R. A. Orwell and A. Vrij, *J. Am. Chem. Soc.*, 1964, **86**, 3515; (c) P. J. Flory, *J. Am. Chem. Soc.*, 1965, **87**, 1833; (d) A. Abe and P. J. Flory, *J. Am. Chem. Soc.*, 1965, **87**, 1838.
267. (a) D. Patterson, *Pure Appl. Chem.*, 1976, **47**, 305; (b) M. Barbe and D. Patterson, *J. Solution Chem.*, 1980, **9**, 753; (c) H. T. Van and D. Patterson, *J. Solution Chem.*, 1982, **11**, 793; (d) M. Costas and D. Patterson, *J. Solution Chem.*, 1982, **11**, 807.
268. H. Iloukhani, M. Rezaei-Sameti and J. Basiri-Parsa, *J. Chem. Thermodyn.*, 2006, **38**, 975.
269. M. L. L. Paredes, R. A. Reis, A. A. Silva, R. N. G. Santos and G. J. Santos, *J. Chem. Eng. Data*, 2011, **56**, 4076.
270. T. M. Letcher and B. W. H. Scoones, *J. Solution Chem.*, 1981, **10**, 459.
271. M. Gepert, E. Zorebski and A. Leszczynska, *Fluid Phase Equilib.*, 2005, **233**, 157.
272. E. Matteoli, P. Gianni, L. Lepori and A. Spanedda, *J. Chem. Eng. Data*, 2011, **56**, 5019.
273. H. Dong, Y. Yue, C. Wu and G. Lai, *J. Chem. Eng. Data*, 2012, **57**, 1050.
274. E. Wilhelm, W. Egger, M. Vencour, A. H. Roux, M. Polednicek and J.-P. E. Grolier, *J. Chem. Thermodyn.*, 1998, **30**, 1509.
275. B. S. Bjola, M. A. Siddiqui and P. Svedja, *J. Chem. Eng. Data*, 2001, **46**, 1167.
276. (a) M. L. McGlashan, D. Stubley and H. Watts, *J. Chem. Soc. A*, 1969, 673; (b) P. J. Howell and D. Stubley, *J. Chem. Soc. A*, 1969, 2489; (c) P. J. Howell, B. J. Skillerne de Bristowe and D. Stubley, *J. Chem. Soc. A*, 1971, 397.
277. (a) G. Scatchard, S. E. Wood and J. M. Mochel, *J. Am. Chem. Soc.*, 1940, **62**, 712; (b) R. Kind, G. Kahnt, D Schmidt, J. Schumann and H.-J. Bittrich, *Z. Phys. Chem.*, 1968, **238**, 277.
278. G. A. Bottomley and R. L. Scott, *J. Chem. Thermodyn.*, 1974, **6**, 973.
279. E. Wilhelm, R. Schano, G. Becker, G. H. Findenegg and F. Kohler, *Trans. Faraday Soc.*, 1969, **65**, 1443.
280. (a) E. Wilhelm, J.-P. E. Grolier and M. H. Karbalai Ghassemi, *Ber. Bunsenges. Phys. Chem.*, 1977, **81**, 925; (b) E. Wilhelm, J.-P. E. Grolier

and M. H. Karbalai Ghassemi, *Thermochim. Acta*, 1979, **31**, 79; (c) E. Wilhelm, A. Faradjzadeh and J.-P. E. Grolier, *J. Chem. Thermodyn.*, 1979, **11**, 979; (d) A. Lainez, G. Roux-Desgranges, J.-P. E. Grolier and E. Wilhelm, *Fluid Phase Equilib.*, 1985, **20**, 47; (e) A. Lainez, J.-P. E. Grolier and E. Wilhelm, *Thermochim. Acta*, 1985, **91**, 243; (f) A. Lainez, E. Wilhelm, G. Roux-Desgranges and J.-P. E. Grolier, *J. Chem. Thermodyn.*, 1985, **17**, 1153; (g) E. Wilhelm, A. Lainez and J.-P. E. Grolier, *Fluid Phase Equilib.*, 1989, **49**, 233.

281. A. Lainez, M. R. Lopez, M. Caceres, J. Nuñez, R. G. Rubio, J.-P. E. Grolier and E. Wilhelm, *J. Chem. Soc., Faraday Trans.*, 1995, **91**, 1941.
282. M. Chorążewski, P. Góralski, M. Hrynko, J.-P. E. Grolier and E. Wilhelm, *J. Chem. Eng. Data*, 2010, **55**, 5471.
283. A. Valtz, C. Coquelet, G. Boukais-Belaribi, A. Dahmani and F. B. Belaribi, *J. Chem. Eng. Data*, 2011, **56**, 1629.
284. R. Bravo, M. Pintos and A. Amigo, *Can. J. Chem.*, 1995, **73**, 375.
285. J. F. Casteel and P. G. Sears, *J. Chem. Eng. Data*, 1974, **19**, 196.
286. U. Tilstam, *Org. Process Res. Dev.*, 2012, **16**, 1273.
287. Y.-X. Yu and Y.-G. Li, *Fluid Phase Equilib.*, 1998, **147**, 207.
288. (a) M. Chaar, J. Ortega, J. Plácido and E. Gonzalez, *ELDATA: Int. Electron. J. Phys.-Chem. Data*, 1995, **1**, 191; (b) M. Vidal, J. Ortega and J. Plácido, *J. Chem. Thermodyn.*, 1997, **29**, 47; (c) J. Ortega, G. Sabater, I. de la Nuez, J. J. Quintana and J. Wisniak, *J. Chem. Eng. Data*, 2007, **52**, 215.
289. (a) E. Jimenez, L. Romani, E. Wilhelm, G. Roux-Desgranges and J.-P. E. Grolier, *J. Chem. Thermodyn.*, 1994, **26**, 817; (b) M. Pintos-Barral, R. Bravo, G. Roux-Desgranges, J.-P. E. Grolier and E. Wilhelm, *J. Chem. Thermodyn.*, 1999, **31**, 1151.
290. E. Wilhelm, *High Temp. – High Pressures*, 1997, **29**, 613.
291. L. Fernández, E. Pérez, J. Ortega, J. Canosa and J. Wisniak, *J. Chem. Eng. Data*, 2010, **55**, 5519.
292. M. V. Rathnam, R. K. R. Singh and M. S. Kumar, *J. Chem. Eng. Data*, 2008, **53**, 265.
293. (a) I. Alonso, V. Alonso, I. Mozo, I. G. de la Fuente, J. A. González and J. C. Cobos, *J. Chem. Eng. Data*, 2010, **55**, 2505; (b) I. Alonso, I. Mozo, I. G. de la Fuente, J. A. González and J. C. Cobos, *J. Mol. Liq.*, 2010, **155**, 109; (c) I. Alonso, I. Mozo, I. G. de la Fuente, J. A. González and J. C. Cobos, *J. Chem. Eng. Data*, 2010, **55**, 5400.
294. (a) S. L. Oswal, V. Pandyian, B. Krishnakumar and P. Vasantharani, *Thermochim. Acta*, 2010, **507–508**, 27; (b) V. Pandyian, P. Vasantharani, S. L. Oswal and A. N. Kannappan, *J. Chem. Eng. Data*, 2011, **56**, 269.
295. B. Sinha, R. Pradhan, S. Saha, D. Brahman and A. Sarkar, *J. Serb. Chem. Soc.*, 2013, **78**, 1443.
296. L. Gascón, C. Lafuente, P. Cea, M. Dominguez and F. M. Royo, *Int. J. Thermophys.*, 2000, **21**, 1185.
297. J. Wisniak, L. E. Sandoval, R. D. Peralta, R. Infante, G. Cortez, L. E. Elizalde and H. Soto, *J. Solution Chem.*, 2007, **36**, 135.

298. R. D. Peralta, R. Infante, G. Cortez, A. Cisneros and J. Wisniak, *Thermochim. Acta*, 2003, **398**, 39.
299. J. Wisniak, G. Cortez, R. D. Peralta, R. Infante, L. E. Elizalde, T. A. Amaro, O. Garcia and H. Soto, *J. Chem. Thermodyn.*, 2008, **40**, 1671.
300. S. K. Reddy and S. Balasubramanian, *J. Phys. Chem. B*, 2012, **116**, 14892.
301. P. Tundo and M. Selva, *Acc. Chem. Res.*, 2002, **35**, 706.
302. J. Zhou, R. Zhu, H. Xu and Y. Tian, *J. Chem. Eng. Data*, 2010, **55**, 5569.
303. H. Piekarski, K. Kubalczyk and M. Wasiak, *J. Chem. Eng. Data*, 2010, **55**, 5435.
304. (a) H. V. Kehiaian, *Fluid Phase Equilib.*, 1983, **13**, 243; (b) H. V. Kehiaian, *Pure Appl. Chem.*, 1985, **57**, 15.
305. (a) A. Heintz, *Ber. Bunsenges. Phys. Chem.*, 1985, **89**, 172; (b) A. Heintz and D. Papaioannou, *Thermochim. Acta*, 1998, **310**, 69.
306. (a) C. G. Panayiotou, *J. Phys. Chem.*, 1988, **92**, 2960; (b) C. G. Panayiotou, *Pure Appl. Chem.*, 1989, **61**, 1453.
307. (a) A. J. Treszczanowicz, T. Treszczanowicz, T. Kasprzycka-Guttman and T. S. Pawłowski, *J. Solution Chem.*, 2002, **31**, 455; (b) T. Treszczanowicz, A. J. Treszczanowicz, T. Kasprzycka-Guttman and T. S. Pawłowski, *J. Solution Chem.*, 2004, **33**, 275; (c) A. J. Treszczanowicz, T. Treszczanowicz, T. S. Pawłowski and T. Kasprzycka-Guttman, *J. Solution Chem.*, 2004, **33**, 1049; (d) A. J. Treszczanowicz, T. S. Pawłowski, T. Treszczanowicz and A. M. Szafrański, *J. Chem. Eng. Data*, 2010, **55**, 5478.
308. A. J. Treszczanowicz and G. C. Benson, *Fluid Phase Equilib.*, 1985, **23**, 117.
309. A. J. Treszczanowicz, T. Treszczanowicz and G. C. Benson, *Fluid Phase Equilib.*, 1993, **89**, 31.
310. R. Srivastava and B. D. Smith, *J. Phys. Chem. Ref. Data*, 1987, **16**, 209.
311. K. N. Marsh, P. Niamskul, J. Gmehling and R. Bölts, *Fluid Phase Equilib.*, 1999, **156**, 207.
312. I.-C. Hwang, J-Y. Lee and S-J. Park, *Fluid Phase Equilib.*, 2009, **281**, 5.
313. A. T. Sundberg, H. Laavi, Y. Kim, P. Uusi-Kyyny and J.-P. Pokki, *J. Chem. Eng. Data*, 2012, **57**, 3502.
314. M. P. Vela, M. Artal, J. M. Embid, R. Bravo and S. Otín, *J. Chem. Eng. Data*, 2012, **57**, 1139.
315. F. E. M. Alaoui, E. A. Montero, J.-P. Bazile, F. Aguílar and C. Boned, *Fluid Phase Equilib.*, 2014, **363**, 131.
316. H. Laavi, A. Zaitseva, J.-P. Pokki, P. Uusi-Kyyny, Y. Kim and V. Alopaeus, *J. Chem. Eng. Data*, 2012, **57**, 3092.
317. M. Almasi and B. Sarkoohaki, *J. Chem. Eng. Data*, 2012, **57**, 309.
318. M. M. H. Bhuiyan and M. H. Uddin, *J. Mol. Liq.*, 2008, **138**, 139.
319. A. C. Gómez Marigliano and H. Sólimo, *J. Chem. Eng. Data*, 2002, **47**, 796.
320. B. Garcia, R. Alcalde, J. M. Leal and J. S. Matos, *J. Chem. Soc., Faraday Trans.*, 1996, **92**, 3347.

321. P. S. Nikam and S. J. Kharat, *J. Chem. Eng. Data*, 2003, **48**, 1291.
322. S. M. Contreras, *J. Chem. Eng. Data*, 2001, **46**, 1149.
323. T. Satyanarayana Rao, N. Veeraiah and C. Rambahu, *Indian J. Chem. Sect. A*, 2002, **41**, 2268.
324. E. Pérez, M. Cardoso, A. M. Mainar, J. I. Pardo and J. S. Urieta, *J. Chem. Eng. Data*, 2003, **48**, 1306.
325. E. Pérez, A. M. Mainar, J. Santafé and J. S. Urieta, *J. Chem. Eng. Data*, 2003, **48**, 723.
326. G. C. Benson and H. D. Pflug, *J. Chem. Eng. Data*, 1970, **15**, 382.
327. I. M. Abdulagatov, F. S. Aliyev, M. A. Talibov, J. T. Safarov, A. N. Shahverdiyev and E. P. Hassel, *Thermochim. Acta*, 2008, **476**, 51.
328. A. Boruń, M. Źurada and A. Bald, *J. Therm. Anal. Calorim.*, 2010, **100**, 707.
329. M. Almasi and L Khosravi, *J. Serb. Chem. Soc.*, 2012, **77**, 363.
330. D. Y. Peng and D. B. Robinson, *Ind. Eng. Chem. Fundam.*, 1976, **15**, 59.
331. R. Stryjek and J. H. Vera, *J. Chem. Eng.*, 1986, **64**, 820.
332. J. M. Resa, M. J. Iglesias, A. González and J. Lanz, *J. Chem. Thermodyn.*, 2001, **33**, 723.
333. I. R. Grgurić, S. P. Šerbanović, M. L. Kijevčanin, A. Ž. Tasić and B. D. Djordjević, *Thermochim. Acta*, 2004, **412**, 25.
334. H. E. Dillon and S. G. Penoncello, *Int. J. Thermophys.*, 2004, **25**, 321.
335. G. P. Dubey and K. Kumar, *J. Chem. Eng. Data*, 2011, **56**, 2995.
336. M. Krummen and J. Gmehling, *Fluid Phase Equilib.*, 2004, **215**, 283.
337. (a) M. J. Dávila and J. P. M. Trusler, *J. Chem. Thermodyn.*, 2009, **41**, 35; (b) S. Aparicio, B. Garcia, R. Alcalde, M. J. Dávila and J. M. Leal, *J. Phys. Chem. B*, 2006, **110**, 6933.
338. (a) J. Zielkiewicz, *Phys. Chem. Chem. Phys.*, 2003, **5**, 1619; (b) J. Zielkiewicz, *Phys. Chem. Chem. Phys.*, 2003, **5**, 3193.
339. J. G. Kirkwood and F. P. Buff, *J. Chem. Phys.*, 1951, **19**, 774.
340. A. Ben-Naim, *J. Chem. Phys.*, 1977, **67**, 4884.
341. B. García, R. Alcalde, S. Aparicio and J. M. Leal, *Phys. Chem. Chem. Phys.*, 2002, **4**, 1170.
342. A. Pal and R. K. Bhardwaj, *J. Chem. Eng. Data*, 2002, **47**, 1128.
343. J. Águila-Hernández, A. Trejo, B. E. Garcia-Flores and R. Molnar, *Fluid Phase Equilib.*, 2008, **267**, 172.
344. B. González, A. Domínguez and J. Tojo, *J. Chem. Eng. Data*, 2004, **49**, 1590.
345. M. Behroozi and H. Zarei, *J. Chem. Eng. Data*, 2012, **57**, 1089.
346. L. Hnědkovský, I. Cibulka and I. Malijevská, *J. Chem. Thermodyn.*, 1990, **22**, 135.
347. I. Malijevská and V. Mandlová, *J. Chem. Thermodyn.*, 2002, **34**, 569.
348. F. Kohler, E. Liebermann, G. Miksch and C Kainz, *J. Phys. Chem.*, 1972, **76**, 2764.
349. F. Kohler, H. Atrops, H. Kalali, E. Liebermann, E. Wilhelm, F. Ratkovics and T. Salamon, *J. Phys. Chem.*, 1981, **85**, 2520.
350. F. Kohler, R. Gopal, G. Götze, H. Atrops, M. A. Demiriz, E. Liebermann, E. Wilhelm, F. Ratkovics and B. Palagyi, *J. Phys. Chem.*, 1981, **85**, 2524.

351. M. Dominguez, P. Cea, C. Lafuente, F. M. Royo and J. S. Urieta, *J. Chem. Thermodyn.*, 1996, **28**, 779.
352. M. L. Kijevčanin, S. P. Šerbanović, I. R. Radović, B. D. Djordjević and A. Ž. Tasić, *Fluid Phase Equilib.*, 2007, **251**, 78.
353. M. M. Mato, S. M. Cebreiro, J. L. Legido and M. I. Paz Andrade, *J. Therm. Anal. Calorim.*, 2006, **84**, 279.
354. R. L. Gardas and S. L. Oswal, *J. Solution Chem.*, 2008, **37**, 1449.
355. H. Iloukhani and K. Khanlarzadeh, *Thermochim. Acta*, 2010, **502**, 77.
356. K. Khanlarzadeh and H. Iloukhani, *J. Chem. Thermodyn.*, 2011, **43**, 1583.
357. I. R. Radović, S. P. Šerbanović, B. D. Djordjević and M. L. Kijevčanin, *J. Chem. Eng. Data*, 2011, **56**, 344.
358. D. Eisenberg and W. Kauzmann, *The Structure and Properties of Water*, Clarendon Press, Oxford, UK, 1969; published in the Oxford Classics Series 2005, reprinted 2006.
359. F. Franks, *Water: 2nd Edition. A Matrix of Life*, The Royal Society of Chemistry, Cambridge, UK, 2000.
360. S. K. Pal and A. H. Zewail, *Chem. Rev.*, 2004, **104**, 2099.
361. Y. Levy and J. N. Onuchic, *Annu. Rev. Biophys. Biomol. Struct.*, 2006, **35**, 389.
362. P. Ball, *Chem. Rev.*, 2008, **108**, 74.
363. P. Ball, *ChemPhysChem*, 2008, **9**, 2677.
364. F. Mallamace, C. Corsaro, D. Mallamace, P. Baglioni, H. E. Stanley and S.-H. Chen, *J. Phys. Chem. B*, 2011, **115**, 14280.
365. M. L. McGlashan and A. G. Williamson, *J. Chem. Eng. Data*, 1976, **21**, 196.
366. G. C. Benson and O. Kiyohara, *J. Solution Chem.*, 1980, **9**, 791.
367. N. C. Patel and S. I. Sandler, *J. Chem. Eng. Data*, 1985, **30**, 218.
368. A. Arce, A. Blanco, Soto and I. Vidal, *J. Chem. Eng. Data*, 1993, **38**, 336.
369. G. Chen and H. Knapp, *J. Chem. Eng. Data*, 1995, **40**, 1001.
370. H. Kubota, Y. Tanaka and T. Makita, *Int. J. Thermophys.*, 1987, **8**, 47.
371. E. V. Ivanov and V. X. Abrossimov, *J. Solution Chem.*, 1996, **25**, 191.
372. E. A. Jagla, *Phys. Rev. E*, 2001, **63**, 061501.
373. Z. Su, S. V. Buldyrev, P. G. Debenedetti, P. J. Rossky and H. E. Stanley, *J. Chem. Phys.*, 2012, **136**, 044511.
374. M. C. A. Donkersloot, *J. Solution Chem.*, 1979, **8**, 293.
375. K. J. Patil, *J. Solution Chem.*, 1981, **10**, 315.
376. (a) E. Matteoli and L. Lepori, *J. Chem. Phys.*, 1984, **80**, 2856; (b) E. Matteoli and L. Lepori, *J. Chem. Soc., Faraday Trans.*, 1995, **91**, 431.
377. Y. Marcus, *Phys. Chem. Chem. Phys.*, 1999, **1**, 2975.
378. Y. Marcus, *Solvent Mixtures: Properties and Selective Solvation*, Marcel Dekker, New York, USA, 2002.
379. (a) M. Mijaković, B. Kežić, L. Zoranić, F. Sokolić, A. Asenbaum, C. Pruner, E. Wilhelm and A. Perera, *J. Mol. Liq.*, 2011, **164**, 66; (b) A. Asenbaum, C. Pruner, E. Wilhelm, M. Mijaković, L. Zoranić, F. Sokolić, B. Kežić and A. Perera, *Vib. Spectrosc.*, 2012, **60**, 102.

380. M. E. Friedman and H. A. Scheraga, *J. Phys. Chem.*, 1965, **69**, 3795.
381. T. Nakajima, T. Komatsu and T. Nakagawa, *Bull. Chem. Soc. Jpn.*, 1975, **48**, 783.
382. S. Cabani, G. Conti and E. Matteoli, *J. Solution Chem.*, 1976, **5**, 751.
383. K. N. Marsh and C. Burfitt, *J. Chem. Thermodyn.*, 1975, 7, 43.
384. J.-P. E. Grolier and E. Wilhelm, *Fluid Phase Equilib.*, 1981, **6**, 283.
385. E. V. Ivanov, *J. Chem. Thermodyn.*, 2012, **47**, 162.
386. A. Petek, D. Pečar and V. Doleček, *Acta Chim. Slov.*, 2001, **48**, 317.
387. J. B. Ott, J. T. Sipowska, M. S. Gruszkiewicz and A. T. Woolley, *J. Chem. Thermodyn.*, 1993, **25**, 307.
388. D. Pečar and V. Doleček, *Fluid Phase Equilib.*, 2005, **230**, 36; Erratum: Fluid Phase Equilib., 2005, **233**, 235.
389. M. S. Chauhdry and J. A. Lamb, *J. Chem. Eng. Data*, 1987, **32**, 431.
390. (a) O. Bozdag, M. S. Chauhdry and J. A. Lamb, *Rev. Sci. Instrum.*, 1984, **55**, 427; (b) M. S. Chauhdry and J. A. Lamb, *J. Chem. Thermodyn.*, 1986, **18**, 665.
391. (a) S. Westmeier, *Chem. Techn.*, 1976, **28**, 286; (b) S. Westmeier, *Chem. Technol.*, 1976, **28**, 350.
392. W. M. Langdon and D. B. Keyes, *Ind. Eng, Chem.*, 1943, **35**, 459.
393. K.-Y. Chu and A. R. Thompson, *J. Chem. Eng. Data*, 1962, **7**, 358.
394. S. Z. Mikhail and W. R. Kimel, *J. Chem. Eng. Data*, 1963, **8**, 323.
395. (a) S. G. Bruun and A. Hvidt, *Ber. Bunsenges. Phys. Chem.*, 1977, **81**, 930; (b) C. Dethlefsen, P. G. Sørensen and A. Hvidt, *J. Solution Chem.*, 1984, **13**, 191.
396. H. Yamamoto, K. Ichikawa and J. Tokunaga, *J. Chem. Eng. Data*, 1994, **39**, 155.
397. Y. Miyamoto, M. Takemoto, M. Hosokawa, Y. Uosaki and T. Moriyoshi, *J. Chem. Thermodyn.*, 1990, **22**, 1007.
398. I. M. Abdulagatov and N. D. Azizov, *J. Chem. Thermodyn.*, 2014, **71**, 155.
399. J. Zielkiewicz, *J. Chem. Thermodyn.*, 1995, **27**, 225.
400. J. M. Resa, J. M. Goenaga and M. Iglesias, *J. Therm. Anal. Calorim.*, 2007, **88**, 549.
401. P. Hynčica, L. Hnědkovsky and I. Cibulka, *J. Chem. Thermodyn.*, 2004, **36**, 1095.
402. (a) T. Nakajima, T. Komatsu and T. Nakagawa, *Bull. Chem. Soc. Jpn.*, 1975, **48**, 783; (b) T. Nakajima, T. Komatsu and T. Nakagawa, *Bull. Chem. Soc. Jpn.*, 1975, **48**, 788.
403. M. L. Origlia-Luster and E. M. Woolley, *J. Chem. Thermodyn.*, 2003, **35**, 1101.
404. H. Mori, S. Iwata, T. Kawachi, T. Matsubara, Y. Nobuoka and T. Aragaki, *J. Chem. Eng. Jpn.*, 2004, **37**, 850.
405. H. Renon and J. M. Prausnitz, *AIChE J.*, 1968, **14**, 135.
406. F. Franks and D. J. G. Ives, *Q. Rev., Chem. Soc.*, 1966, **20**, 1.
407. C. de Visser, G. Perron and J. E. Desnoyers, *Can. J. Chem.*, 1977, **55**, 856.
408. (a) K. Nakanishi, *Bull. Chem. Soc. Jpn.*, 1960, **33**, 793; (b) K. Nakanishi, N. Kato and M. Maruyama, *J. Phys. Chem.*, 1967, **71**, 814.

409. S. Dixit, J. Crain, W. C. K. Poon, J. L. Finney and A. K. Soper, *Nature*, 2002, **416**, 829.
410. D. T. Bowron, *Phil. Trans. R. Soc. London B*, 2004, **359**, 1167.
411. (a) M. Misawa, Y. Inamura, D. Hosaka and O. Yamamuro, *J. Chem. Phys.*, 2006, **125**, 074502; (b) M. Nakada, O. Yamamuro, K. Maruyama and M. Misawa, *J. Phys. Soc. Jpn.*, 2007, **76**, 054601.
412. D. T. Bowron and J. L. Finney, *J. Phys. Chem. B*, 2007, **111**, 9838.
413. (a) K. Nishikawa, Y. Kodera and T. Iijima, *J. Phys. Chem.*, 1987, **91**, 3694; (b) K. Nishikawa, H. Hayashi and T. Iijima, *J. Phys. Chem.*, 1989, **93**, 6559; (c) K. Nishikawa and T. Iijima, *J. Phys. Chem.*, 1990, **94**, 6227.
414. (a) N. Nishi, S. Takahashi, M. Matsumoto, A. Tanaka, K. Muraya, T. Takamuku and T. Yamaguchi, *J. Phys. Chem.*, 1995, **99**, 462; (b) M. Matsumoto, N. Nishi, T. Furusawa, M. Saita, T. Takamuku, M. Yamagami and T. Yamaguchi, *Bull. Chem. Soc. Jpn.*, 1995, **68**, 1775; (c) T. Takamuku, T. Yamaguchi, M. Asato, M. Matsumoto and N. Nishi, Z. *Naturforsch. A, Phys. Sci.*, 2000, **55**, 513.
415. K. Yoshida and T. Yamaguchi, *Z. Naturforsch., A: Phys. Sci.*, 2001, **56**, 529.
416. D. Subramanian, D. A. Ivanov, I. K. Yudin, M. A. Anisimov and J. V. Sengers, *J. Chem. Eng. Data*, 2011, **56**, 1238.
417. D. Subramanian and M. A. Anisimov, *J. Phys. Chem. B*, 2011, **115**, 9179.
418. A. Fornili, M. Civera, M. Sironi and S. L. Fornili, *Phys. Chem. Chem. Phys.*, 2003, **5**, 4905.
419. S. Paul and G. N. Patey, *J. Phys. Chem. B*, 2006, **110**, 10514.
420. S. K. Allison, J. P. Fox, R. Hargreaves and S. P. Bates, *Phys. Rev. B*, 2005, **71**, 024201.
421. M. D. Hands and L. V. Slipchenko, *J. Phys. Chem. B*, 2012, **116**, 2775.
422. B. Kežić and A. Perera, *J. Chem. Phys.*, 2012, **137**, 014501.
423. A. Ben-Naim, *Biophys. Chem.*, 2003, **105**, 183.
424. A. Ben-Naim, *J. Chem. Phys.*, 2006, **125**, 024901.
425. A. Ben-Naim, *Open J. Biophys.*, 2011, **1**, 1.
426. F. Franks and H. T. Smith, *J. Chem. Eng. Data*, 1968, **13**, 538.
427. A. Hvidt, R. Moss and G. Nielsen, *Acta Chem. Scand., Ser. B*, 1978, **32**, 274.
428. M. Sakurai, *Bull. Chem. Soc. Jpn.*, 1987, **60**, 1.
429. E. S. Kim and K. N. Marsh, *J. Chem. Eng. Data*, 1988, **33**, 288.
430. P. K. Kipkemboi and A. J. Easteal, *Can. J. Chem.*, 1994, **72**, 1937.
431. G. I. Egorov and D. M. Makarov, *J. Chem. Thermodyn.*, 2011, **43**, 430.
432. K. R. Harris, P. J. Newitt, P. J. Back and L. A. Woolf, *High Temp. - High Pressures*, 1998, **30**, 51.
433. (a) G. I. Egorov, D. M. Makarov and A. M. Kolker, *J. Chem. Thermodyn.*, 2013, **61**, 161; (b) G. I. Egorov, D. M. Makarov and A. M. Kolker, *J. Chem. Thermodyn.*, 2013, **61**, 169.
434. L. Šimurka, I. Cibulka and L. Hnědkovský, *J. Chem. Eng. Data*, 2011, **56**, 4564.

435. V. Dohnal, D. Fenclova and P. Vrbka, *J. Phys. Chem. Ref. Data*, 2006, **35**, 1621.
436. P. W. Hochachka and G. N. Somero, *Biochemical Adaption. Mechanism and Process in Physiological Evolution*, Oxford University Press, Oxford, UK, 2002.
437. *Life in the Frozen State*, ed. B. J. Fuller, N. Lane and E. E. Benson, CRC Press, Boca Raton, FL, USA, 2004.
438. C. Yang, P. Ma, F. Jing and D. Tang, *J. Chem. Eng. Data*, 2003, **48**, 836.
439. J. George and N. V. Sastry, *Fluid Phase Equilib.*, 2004, **216**, 307.
440. G. Czechowski, B. Żywucki and J. Jadżyn, *J. Chem. Eng. Data*, 1988, **33**, 55.
441. V. K. Reddy, K. S. Reddy and A. Krishnaiah, *J. Chem. Eng. Data*, 1994, **39**, 615.
442. T. M. Aminabhavi and B. Gopalakrishna, *J. Chem. Eng. Data*, 1995, **40**, 856.
443. A. Marchetti, A. Martignani and L. Tassi, *J. Chem. Thermodyn.*, 1998, **30**, 653.
444. N. G. Tsierkezos and I. E. Molinou, *J. Chem. Eng. Data*, 1998, **43**, 989.
445. P. Baraldi, G. C. Franchini, A. Marchetti, G. Sanna, L. Tassi, A. Ulrici and G. Vaccari, *J. Solution Chem.*, 2000, **29**, 489.
446. J. George and N. V. Sastry, *J. Chem. Eng. Data*, 2003, **48**, 1529.
447. B. B. Gurung and M. N. Roy, *Phys. Chem. Liq.*, 2007, **45**, 331.
448. Y. Marcus, *Phys. Chem. Chem. Phys.*, 2000, **2**, 4891.
449. L. Xu, X. Hu and R. Lin, *J. Solution Chem*, 2003, **32**, 363.
450. J. L. Neal and D. A. I. Goring, *J. Phys. Chem.*, 1970, **74**, 658.
451. J. J. Towey and L. Dougan, *J. Phys. Chem. B*, 2012, **116**, 1633.
452. S. Dixit, A. K. Soper, J. L. Finney and J. Crain, *Europhys. Lett.*, 2002, **59**, 377.
453. S. E. Pagnotta, M. A. Ricci, F. Bruni, S. McLain and S. Magazù, *Chem. Phys.*, 2008, **345**, 159.
454. J. L. Dashnau, N. V. Nucci, K. A. Sharp and J. M. Vanderkooi, *J. Phys. Chem. B*, 2006, **110**, 13670.
455. E. Politi, L. Sapir and D. Harries, *J. Phys. Chem. A*, 2009, **113**, 7548.
456. J. J. Towey, A. K. Soper and L. Dougan, *J. Phys. Chem. B*, 2011, **115**, 7799.
457. G. Roux, G. Perron and J. E. Desnoyers, *J. Solution Chem.*, 1978, **7**, 639.
458. G. Schneider and G. Wilhelm, *Z. Phys. Chem. Neue Folge*, 1959, **20**, 219.
459. C. M. Ellis, *J. Chem. Educ.*, 1967, **44**, 405.
460. R. M. Stephenson, *J. Chem. Eng. Data*, 1993, **38**, 134.
461. K. Tamura, S. Tabata and S. Murakami, *J. Chem. Thermodyn.*, 1998, **30**, 1319.
462. K. Tamura, T. Sonoda and S. Murakami, *J. Solution Chem.*, 1999, **28**, 777.
463. H. Doi, K. Tamura and S. Murakami, *J. Chem. Thermodyn.*, 2000, **32**, 729.
464. H.-J. Zimmer and D. Woermann, *Ber. Bunsenges. Phys. Chem.*, 1991, **95**, 533.

465. V. K. Reddy, K. S. Reddy and A. Krishnaiah, *J. Chem. Eng. Data*, 1994, **39**, 615.
466. K. S. Reddy and P. R. Naidu, *Can. J. Chem.*, 1977, **76**, 55.
467. A. Pal, S. Sharma and Y. P. Singh, *J. Chem. Eng. Data*, 1997, **42**, 1157.
468. Â. F. S. Santos, M.-L. C. J. Moita and I. M. S. Lampreia, *J. Chem. Thermodyn.*, 2009, **41**, 1387.
469. I. M. S. Lampreia, Â. F. S. Santos, M.-L. C. J. Moita, A. O. Figueiras and J. C. R. Reis, *J. Chem. Thermodyn.*, 2012, **45**, 75.
470. *Microemulsions: Properties and Applications*, ed. M. Fanun, CRC Press, Boca Raton, FL, USA, 2009.
471. A. Henni and A. Mather, *Can. J. Chem. Eng.*, 1995, **73**, 156.
472. A. Henni, P. Tontiwachwuthikul and A. Chakma, *Can. J. Chem. Eng.*, 2005, **83**, 358.
473. A. Henni and A. Mather, *Fluid Phase Equilib.*, 1995, **108**, 213.
474. A. Henni, P. Tontiwachwuthikul and A. Chakma, *J. Chem. Eng. Data*, 2006, **51**, 64.
475. A. Pal and Y. P. Singh, *J. Chem. Eng. Data*, 1996, **41**, 425.
476. A. Henni, P. Tontiwachwuthikul, A. Chakma and A. E. Mather, *J. Chem. Eng. Data*, 1999, **44**, 101.
477. L. G. Hepler, *Can. J. Chem.*, 1969, **47**, 4613.
478. X.-X. Li, G. Zhao, D.-S. Liu and W.-W. Cao, *J. Chem. Eng. Data*, 2009, **54**, 890.
479. X.-X. Li, G.-C. Fan, Y.-W. Wang, M. Zhang and Y.-Q. Lu, *J. Mol. Liq.*, 2010, **151**, 62.
480. S. Bekiranov, R. Bruinsma and P. Pincus, *Phys. Rev. E*, 1997, **55**, 577.
481. M. Polverari and T. G. M. van de Ven, *J. Phys. Chem.*, 1996, **100**, 13687.
482. B. Hammouda, D. L. Ho and S. Kline, *Macromolecules*, 2004, **37**, 6932.
483. H. Lee, A. H. de Vries, S.-J. Marrink and R. W. Pastor, *J. Phys. Chem. B*, 2009, **113**, 13186.
484. R. Kjellander and E. Florin, *J. Chem. Soc., Faraday Trans. 1*, 1981, **77**, 2053.
485. S. Lusse and K. Arnold, *Macromolecules*, 1996, **29**, 4251.
486. S. Saeki, N. Kuwahara, M. Nakata and M. Kaneko, *Polymer*, 1976, **17**, 685.
487. Y. C. Bae, J. J. Shim, D. S. Soane and J. M. Prausnitz, *J. Appl. Polym. Sci.*, 1993, **47**, 1193.
488. R. L. Cook, H. E. King, Jr. and D. G. Peiffer, *Phys. Rev. Lett.*, 1992, **69**, 3072.
489. T. Sun and H. E. King, Jr., *Macromolecules*, 1998, **31**, 6383.
490. E. E. Fenn, D. E. Moilanen, N. E. Levinger and M. D. Fayer, *J. Am. Chem. Soc.*, 2009, **131**, 5530.
491. F. R. Trouw, O. Borodin, J. C. Cook, J. R. D. Copley and G. D. Smith, *J. Phys. Chem. B*, 2003, **107**, 10446.
492. O. Borodin, D. Bedrov and G. D. Smith, *J. Phys. Chem. B*, 2002, **106**, 5194.

493. C. Boned, M. Moha-Ouchane and J. Jose, *Phys. Chem. Liq.*, 2000, **38**, 113.
494. A. Henni, J. J. Hromek, P. Tontiwachwuthikul and A. Chakma, *J. Chem. Eng. Data*, 2003, **48**, 1062.
495. Y. Maham, T. T. Teng, A. E. Mather and L. G. Hepler, *Can. J. Chem.*, 1995, **73**, 1514.
496. Y. Maham, T. T. Teng, L. G. Hepler and A. E. Mather, *J. Solution Chem.*, 1994, **23**, 195.
497. A. Henni, P. Tontiwachwuthikul, A. Chakma and A. E. Mather, *J. Chem. Eng. Data*, 2001, **46**, 56.
498. A. Henni, J. J. Hromek, P. Tontiwachwuthikul and A. Chakma, *J. Chem. Eng. Data*, 2003, **48**, 551.
499. J. Li, M. Mundhwa, P. Tontiwachwuthikul and A. Henni, *J. Chem. Eng. Data*, 2007, **52**, 560.
500. M. Mundhwa, R. Alam and A. Henni, *J. Chem. Eng. Data*, 2006, **51**, 1268.
501. M. L. C. J. Moita, L. M. V. Pinheiro, Â. F. S. Santos and I. M. S. Lampreia, *J. Chem. Eng. Data*, 2012, **57**, 2290.
502. L. M. V. Pinheiro, M. L. C. J. Moita, Â. F. S. Santos and I. M. S. Lampreia, *J. Chem. Thermodyn.*, 2013, **64**, 93.
503. Â. F. S. Santos and I. M. S. Lampreia, *Thermochim. Acta*, 2011, **512**, 268.
504. M. A. Saleh, S. Akhtar and A. R. Khan, *Phys. Chem. Liq.*, 2000, **38**, 137.
505. F. I. Chowdhury, M. A. R. Khan, M. A. Saleh and S. Akhtar, *J. Mol. Liq.*, 2013, **182**, 7.
506. Y. Maham, T. T. Teng, L. G. Hepler and A. E. Mather, *Thermochim. Acta*, 2002, **386**, 111.
507. L. Lebrette, Y. Maham, T. T. Teng, L. G. Hepler and A. E. Mather, *Thermochim. Acta*, 2002, **386**, 119.
508. F. I. Chowdhury, S. Akhtar and M. A. Saleh, *Phys. Chem. Liq.*, 2009, **47**, 638.
509. T. T. Teng, Y. Maham, L. G. Hepler and A. E. Mather, *Can. J. Chem.*, 1994, **72**, 125.
510. A. Valtz, M. Teodorescu, I. Wichterle and D. Richon, *Fluid Phase Equilib.*, 2004, **215**, 129.
511. C. Casanova, E. Wilhelm, J.-P. E. Grolier and H. V. Kehiaian, *J. Chem. Thermodyn.*, 1981, **13**, 241.
512. (a) A. N. Campbell, E. M. Kartzmark and J. M. T. M. Gieskes, *Can. J. Chem.*, 1963, **41**, 407; (b) J. M. T. M. Gieskes, *Can. J. Chem.*, 1965, **43**, 2448.
513. E. J. King, *J. Phys. Chem.*, 1969, **73**, 1220.
514. H. Høiland, *Acta Chem. Scand., Ser. A*, 1974, **28**, 699.
515. K. Granados, J. Gracia-Fadrique, A. Amigo and R. Bravo, *J. Chem. Eng. Data*, 2006, **51**, 1356.
516. A. Apelblat and E. Manzurola, *Fluid Phase Equilib.*, 1987, **32**, 163.
517. (a) H. Kehiaian and K. Sosnkowska-Kehiaian, *Bull. Acad. Pol. Sci., Ser. Sci. Chim.*, 1963, **11**, 591; (b) H. Kehiaian, *Bull. Acad. Pol. Sci., Ser. Sci.*

Chim., 1964, **12**, 479; (c) H. Kehiaian, *Bull. Acad. Pol. Sci., Ser. Sci. Chim.*, 1964, **12**, 567.
518. J. Korpela, *Acta Chem. Scand., Ser. A*, 1971, **25**, 2852.
519. T. H. Lilley, *Pure Appl. Chem.*, 1994, **66**, 429.
520. A. Pal and Y. P. Singh, *J. Chem. Eng. Data*, 1995, **40**, 818.
521. B. García, R. Alcalde, J. M. Leal and J. S. Matos, *J. Phys. Chem. B*, 1997, **101**, 7991.
522. D. Papamatthaiakis, F. Aroni and V. Havredaki, *J. Chem. Thermodyn.*, 2008, **40**, 107.
523. P. Scharlin, K. Steinby and U. Domańska, *J. Chem. Thermodyn.*, 2002, **34**, 927.
524. K. Miyai, M. Nakamura, K. Tamura and S. Murakami, *J. Solution Chem.*, 1997, **26**, 973.
525. P. Scharlin and K. Steinby, *J. Chem. Thermodyn.*, 2003, **35**, 279.
526. R. B. Tôrres, A. C. M. Marchiore and P. L. O. Volpe, *J. Chem. Thermodyn.*, 2006, **38**, 526.
527. (a) T.-C. Bai, J. Yao and S.-J. Han, *J. Chem. Eng. Data*, 1999, **44**, 491; (b) T.-C. Bai, J. Yao and S.-J. Han, *J. Chem. Thermodyn.*, 1998, **30**, 1347; (c) T.-C. Bai, J. Yao and S.-J. Han, *Fluid Phase Equilib.*, 1998, **152**, 283.
528. M. Ueno, R. Mitsui, H. Iwahashi, N. Tsuchihashi and K. Ibuki, *J. Phys.: Conf. Ser.*, 2010, **215**, 012074.
529. M. Ueno, Y. Mizumaki, N. Tsuchihashi and K. Ibuki, *Rev. High Pressure Sci. Technol.*, 2003, **13**, 134.
530. Y. Maham, M. Boivineau and A. E. Mather, *J. Chem. Thermodyn.*, 2001, **33**, 1725.
531. P. R. Tremaine, D. Shvedov and C. Xiao, *J. Phys. Chem.*, 1997, **101**, 409.
532. J. George and N. V. Sastry, *J. Chem. Eng. Data*, 2004, **49**, 235.
533. C. J. F. Böttcher, Theory of Electric Polarization, completely revised by O. C. van Belle, P. Bordewijk and A. Rip, Vol. 1: Dielectrics in Static Fields, Elsevier Scientific Publishing Company, Amsterdam, The Netherlands, 2nd edn., 1973.
534. *Pioneering Ideas for the Physical and Chemical Sciences*: Proceedings of the Josef Loschmidt Symposium held in Wien (Vienna), Austria, 25–27 June, 1995, ed. W. Fleischhacker and T. Schönfeld, Plenum Press, New York, USA, 1997.
535. E. Dachwitz and M. A. Stockhausen, *Ber. Bunsenges. Phys. Chem.*, 1987, **91**, 1347.
536. A. Henni, J. J. Hromek, P. Tontiwachwuthikul and A. Chakma, *J. Chem. Eng. Data*, 2004, **49**, 231.
537. S. Aparicio, R. Alcalde, M. J. Dávila, B. García and J. M. Leal, *J. Phys. Chem. B*, 2008, **112**, 11361.
538. (a) J. Gross and G. Sadowski, *Ind. Eng. Chem. Res.*, 2001, **40**, 1244; (b) J. Gross and G. Sadowski, *Ind. Eng. Chem. Res.*, 2002, **41**, 5510; (c) J. Gross, *AIChE J.*, 2005, **51**, 2556; (d) J. Gross and J. Vrabec, *AIChE J.*, 2006, **52**, 1194; (e) M. Kleiner and J. Gross, *AIChE J.*, 2006, **52**, 1951; (f) J. Vrabec and J. Gross, *J. Phys. Chem. B*, 2008, **112**, 51.

539. M. Usula, F. Mocci, F. C. Marincola, S. Porcedda, L. Gontrani and R. Caminiti, *J. Chem. Phys.*, 2014, **140**, 124503.
540. Z.-W. Yu and P. J. Quinn, *Biosci. Rep.*, 1994, **14**, 259.
541. Z. Hubálek, *Cryobiology*, 2003, **46**, 205.
542. A. I. Zhmakin, *Fundamentals of Cryobiology. Physical Phenomena and Mathematical Models*, Springer-Verlag, Berlin, Heidelberg, Germany, 2009.
543. B. Wowk, *Cryobiology*, 2010, **60**, 11.
544. D. B Wong, K. P. Sokolowsky, M. I. El-Barghouthi, E. E. Fenn, C. H. Giammanco, A. L. Sturlaugson and M. D. Fayer, *J. Phys. Chem. B*, 2012, **116**, 5479.
545. (a) A. K. Soper and A. Luzar, *J. Chem. Phys.*, 1992, **97**, 1320; (b) A. K. Soper and A. Luzar, *J. Phys. Chem.*, 1996, **100**, 1357; (c) J. T. Cabral, A. Luzar, J. Teixeira and M.-C. Bellissent-Funel, *J. Chem. Phys.*, 2000, **113**, 8736.
546. H. N. Bordallo, K. W. Herwig, B. M. Luther and N. E. Levinger, *J. Chem. Phys.*, 2004, **121**, 12457.
547. Z. Lu, E. Manias, D. D. Macdonald and M. Lanagan, *J. Phys. Chem. A*, 2009, **113**, 12207.
548. W. M. Madigosky and R. W. Warfield, *J. Chem. Phys.*, 1983, **78**, 1912.
549. U. Kaatze, M. Brai, F.-D. Scholle and R. Pottel, *J. Mol. Liq.*, 1990, **44**, 197.
550. A. Luzar and D. Chandler, *J. Chem. Phys.*, 1993, **98**, 8160.
551. I. A. Borin and M. S. Skaf, *J. Chem. Phys.*, 1999, **110**, 6412.
552. D. P. Geerke, C. Oostenbrink, N. F. A. van der Vegt and W. F. van Gunsteren, *J. Phys. Chem. B*, 2004, **108**, 1436.
553. N. Zhang, W. Li, C. Chen and J. Zuo, *Comput. Theor. Chem.*, 2013, **1017**, 126.
554. S. Banerjee and B. Bagchi, *J. Chem. Phys.*, 2013, **139**, 164301.
555. A. Wulf and R. Ludwig, *ChemPhysChem*, 2006, 7, 266.
556. D. Martin and H. G. Hauthal, *Dimethylsulfoxid*, Akademie-Verlag, Berlin, Germany, 1971.
557. D. H. Rasmussen and A. P. MacKenzie, *Nature*, 1968, **220**, 1315.
558. S. A. Schichman and R. L. Amey, *J. Phys. Chem.*, 1971, **75**, 98.
559. J. M. G. Cowie and P. M. Toporowski, *Can J. Chem.*, 1961, **39**, 2240.
560. D. J. Pruett and L. K. Felker, *J. Chem. Eng. Data*, 1985, **30**, 452.
561. C. de Visser, W. J. M. Heuvelsland, L. A. Dunn and G. Somsen, *J. Chem. Soc., Faraday Trans. 1*, 1978, **74**, 1159.
562. L. Werblan and J. Lesiński, *Pol. J. Chem.*, 1978, **52**, 1211.
563. P. Westh, *J. Phys. Chem.*, 1994, **98**, 3222.
564. J. T. W. Lai, F. W. Lau, D. Robb, P. Westh, G. Nielsen, C. Trandum, A. Hvidt and Y. Koga, *J. Solution Chem.*, 1995, **24**, 89.
565. C. F. Lau, P. T. Wilson and D. V. Fenby, *Aust. J. Chem.*, 1970, **23**, 1143.
566. R. W. Kershaw and G. N. Malcolm, *Trans. Faraday Soc.*, 1968, **64**, 323.
567. T. Kimura and S. Takagi, *J. Chem. Thermodyn.*, 1986, **18**, 447.
568. T. Kimura and S. Takagi, *J. Chem. Thermodyn.*, 1979, **11**, 119.

569. K. Tamura, M. Nakamura and S. Murakami, *J. Solution Chem.*, 1997, **26**, 1199.
570. M. A. Saleh, S. Akhtar, M. S. Ahmed and M. H. Uddin, *Phys. Chem. Liq.*, 2002, **40**, 621.
571. S. A. Markarian, A. M. Asatryan and A. L. Zatikyan, *J. Chem. Thermodyn.*, 2005, **37**, 768.
572. M. del Carmen Grande, J. A. Juliá, M. Garcia and C. M. Marschoff, *J. Chem. Thermodyn.*, 2007, **39**, 1049.
573. G. I. Egorov and D. M. Makarov, *Russ. J. Phys. Chem. A*, 2009, **83**, 693.
574. A. Sacco and E. Matteoli, *J. Solution Chem.*, 1997, **26**, 527.
575. A. H. Fuchs, M. Ghelfenstein and H. Szwarc, *J. Chem. Eng. Data*, 1980, **25**, 206.
576. C. Czeslik and J. Jonas, *J. Phys. Chem. A*, 1999, **103**, 3222.
577. J. Jonas and Y. T. Lee, *J. Phys.: Condens. Matter*, 1992, **4**, 305.
578. T. Moriyoshi and Y. Uosaki, *J. Soc. Mater. Sci. Jpn.*, 1984, **33**(365), 127.
579. J. P. Petitet, P. Bezot and C. Hesse-Bezot, *Phys. B (Amsterdam, Neth.)*, 1988, **153**, 181.
580. G. I. Egorov and D. M. Makarov, *Russ. J. Phys. Chem. A*, 2009, **83**, 2058.
581. S. A. Markarian, A. L. Zatikyan, V. V. Grigoryan and G. S. Grigoryan, *J. Chem. Eng. Data*, 2005, **50**, 23.
582. S. Y. Lam and R. L. Benoit, *Can. J. Chem.*, 1974, **52**, 718.
583. T. C. Chan and W. A. Van Hook, *J. Solution Chem.*, 1976, **5**, 107.
584. X. Qiann, B. Han, Y. Liu, H. Yan and R. Liu, *J. Solution Chem.*, 1995, **24**, 1183.
585. S. A. Markarian, S. Bonora, K. A. Bagramyan and V. B. Arakelyan, *Cryobiology*, 2004, **49**, 1.

CHAPTER 8

Partial Molar Volumes of Non-Ionic Solutes at Infinite Dilution

IVAN CIBULKA*[a] AND VLADIMIR MAJER[b,c]

[a] Department of Physical Chemistry, Institute of Chemical Technology, Technická 5, 166 28 Prague 6, Czech Republic; [b] Institute of Chemistry of Clermont-Ferrand, UMR 6296 CNRS - Blaise Pascal University, F-63171, Aubiere, France; [c] Department of Chemistry, Technical University of Liberec, Studentská 2, 460 01 Liberec, Czech Republic
*Email: ivan.cibulka@vscht.cz

8.1 Introduction

Partial molar volume of a component in a mixture is one of the important partial molar quantities providing information on how a component is accommodated in a mixture. In accordance with the general definition of partial molar quantities, the partial molar volume is defined as

$$V_{\mathrm{m},i} = \left(\frac{\partial V}{\partial n_i}\right)_{T,p,n_{j\neq i}}. \tag{8.1}$$

In other words, the partial molar volume of component i is the change of volume of the whole system by adding one mole of component i to a system of infinite size at constant temperature and pressure. The basic feature of partial molar quantities is the equality between their sum weighted by mole

Volume Properties: Liquids, Solutions and Vapours
Edited by Emmerich Wilhelm and Trevor M. Letcher

Published by the Royal Society of Chemistry, www.rsc.org

fractions over all components and molar quantity of the mixture, *i.e.*, for volume it holds:

$$V_\mathrm{m} = \sum_{i=1}^{N} x_i V_{\mathrm{m},i}. \tag{8.2}$$

Let us limit our considerations to a binary system consisting of a solute (numbered 2) dissolved in a solvent (numbered 1). If a small amount of the solute is dissolved in a large amount of solvent, then the molecules of the solute in the resulting solution are far away from each other and thus surrounded mainly by the molecules of the solvent. When the amount of the solute is negligible with respect to the amount of the solvent the standard state of infinite dilution is reached at which no solute–solute interactions occur. The corresponding volume change in such conditions is the partial molar volume at infinite dilution, *i.e.*, standard molar volume of the solute:

$$V^0_{\mathrm{m},2} = \lim_{n_2 \to 0} \left(\frac{\partial V}{\partial n_2}\right)_{T,p,n_1}. \tag{8.3}$$

Such a standard state convention is an asymmetrical one where the solute is referenced to the state of infinite dilution while the solvent is referenced to its pure state. This is in contrast to the symmetrical convention where all components are referenced to their pure states.

From the molecular point of view, the introduction of a solute molecule into an infinitely dilute solution can be divided into three steps: first, the point mass is placed in the solvent (the corresponding volume change is called co-volume), then a cavity to accommodate the solute molecule is created in the solvent around the point mass and the solute molecule is placed in it, and, at last, the interactions between solute and solvent molecules are "switched on". While the volumes relating to the first two steps are positive, the last one can be positive or negative depending on the character of the solute–solvent interactions.

The standard molar volume is related to other standard thermodynamic quantities through the well-known formulae of classical thermodynamics, *e.g.*, standard molar Gibbs energy (standard chemical potential):

$$\left(\frac{\partial G^0_{\mathrm{m},2}}{\partial p}\right)_{T,n_1,n_2} = \left(\frac{\partial \mu^0_2}{\partial p}\right)_{T,n_1,n_2} = V^0_{\mathrm{m},2}, \tag{8.4}$$

and the standard molar enthalpy, and standard molar isobaric heat capacity:

$$\begin{aligned} \left(\frac{\partial H^0_{\mathrm{m},2}}{\partial p}\right)_{T,n_1,n_2} &= V^0_{\mathrm{m},2} - T\left(\frac{\partial V^0_{\mathrm{m},2}}{\partial T}\right)_{p,n_1,n_2}, \\ \left(\frac{\partial C^0_{p,\mathrm{m},2}}{\partial P}\right)_{T,n_1,n_2} &= -T\left(\frac{\partial^2 V^0_{\mathrm{m},2}}{\partial T^2}\right)_{p,n_1,n_2}. \end{aligned} \tag{8.5}$$

Obviously, the standard molar volume enables us to evaluate variations of other standard thermodynamic and related properties with pressure.

On the other hand, the relations between volumetric and other thermodynamic properties may be employed for the indirect determination of standard molar volumes from other related properties (*e.g.* solubility, speed of sound).

The basic source of information on standard molar volumes is from experimental data and measurements. Several sections of this chapter are therefore devoted to this aspect: treatment of experimental data to evaluate standard molar volumes (Section 8.2), experimental techniques (Section 8.3), and sources of experimental data (Section 8.4). Among various solvents water has a dominant position, naturally due to its importance in various disciplines such as chemistry, biochemistry, biology, geology, environmental sciences, and industrial processes. Liquid systems containing solvents other than water have been studied to a much lesser extent and mostly in the close-to-ambient ranges of state parameters. On the other hand, data on standard molar volumes of both the non-ionic and electrolyte solutes in water are plentiful, not only for near ambient conditions but also over extended intervals of temperature and pressure. This is thanks to extensive density measurements with dilute solutions performed over the last 25 years or so using the non-commercial vibrating-tube instruments constructed in several laboratories. Such measurements are of great importance since the changes of the standard molar volume can be dramatic under super-ambient conditions due to divergence at the critical point of the solvent, as discussed below in Section 8.5. This section also includes a brief overview of predictive group contribution approaches, and equations of state for the standard molar volumes of non-ionic solutes in water.

8.2 Determination of Standard Molar Volumes from Experimental Data

Experimental data suitable for the evaluation of standard molar volumes are, basically, of two different origins. The prevailing amount of existing data on standard molar volumes comes from direct measurements of volumetric properties of solutions as a function of composition, temperature, and pressure and, in this chapter, we tend to focus on this approach. A small number of standard molar volumes have been evaluated from measurements of non-volumetric properties and these indirect methods are only briefly mentioned at the end of this section.

The core of direct methods for the evaluation of standard molar volumes is the extrapolation of properly selected volumetric quantities, resulting from an experiment, to infinite dilution. Direct volumetric experimental data can be reported in several forms, mostly as the solution density ρ, the density difference $\Delta\rho = \rho - \rho_1$ referenced to the density of a solvent ρ_1, or mixing volume V^M, measured for a set of mixture compositions. The selection of a particular method for evaluating standard molar volumes depends on both the input experimental quantity (ρ, $\Delta\rho$, V^M) and the character of the solute. In the following sections, for the sake of simplicity, only binary systems are considered.

One group of methods is employed in the case of density data measured for a set of dilute solutions of a solute dissolved in a liquid solvent. The concentration scale is usually molality. With this type of data, knowledge of any of the volumetric properties of the pure solute (density, molar volume) is not needed. The solute may not necessarily be a liquid in the pure state at the experimental conditions; pure solutes in all aggregate states can be considered. Since solutions of gaseous solutes in liquid solvents are discussed in Chapter 9, liquid and solid solutes are mainly considered here.

Another group of methods is applicable for mixtures where all components exist in the pure state as liquids at the conditions of experiment. Then the mixing volume can be evaluated from experimental data as:

$$V^{\mathrm{M}} = V_{\mathrm{m}} - [(1 - x_2)V^{\bullet}_{\mathrm{m},1} + x_2 V^{\bullet}_{\mathrm{m},2}], \tag{8.6}$$

where $V^{\bullet}_{\mathrm{m},i}$, $i = 1, 2$, are molar volumes of pure liquid components and x_2 is the mole fraction of solute. In this case the concentration scale is usually mole fraction. It should be noted that, contrary to the symmetrical standard state convention where mixing volume V^{M} is identical to excess volume V^{E}, the excess volume defined within the unsymmetrical convention is $V^{\mathrm{E}} = V_{\mathrm{m}} - [(1 - x_2)V^{\bullet}_{\mathrm{m},1} + x_2 V^{0}_{\mathrm{m},2}]$ and obeys the limits $\lim_{x_2 \to 0}(V^{\mathrm{E}}) = 0$ and $\lim_{x_2 \to 1}(V^{\mathrm{E}}) = V^{\bullet}_{\mathrm{m},2} - V^{0}_{\mathrm{m},2}$.

Let us explain in more detail how the standard molar volume of a solute can be obtained from solution density and mixing volume data.

(1) Solution density data. A data set of densities ρ or density differences $\Delta\rho$ *versus* molality is a typical form of input experimental data. They are related to the volumes of a solution containing 1 kg of the solvent V_{kg} as follows:

$$V_{\mathrm{kg}} = \frac{1 + m_2 M_2}{\rho}, \rho = \Delta\rho + \rho_1, \tag{8.7}$$

where m_2 and M_2 are the molality and molar mass (kg mol^{-1}) of the solute, respectively, and ρ is the density of the solution. The standard molar volume can be evaluated either *via* apparent molar volumes or from an analytical function of concentration fitted to measured density data.

(a) Apparent molar volume of the solute $V^{(\mathrm{app})}_{\mathrm{m},2}$ is defined as:

$$V^{(\mathrm{app})}_{\mathrm{m},2} = \frac{V - n_1 V^{\bullet}_{\mathrm{m},1}}{n_2} = \frac{V_{\mathrm{kg}} - V^{\bullet}_{\mathrm{m},1} / M_1}{m_2}, \tag{8.8}$$

where V is a total volume of the solution containing n_1 and n_2 moles of the solvent and solute, respectively, and $V^{\bullet}_{\mathrm{m},1} = M_1/\rho_1$ is the molar volume of pure solvent. Based on the definition in Equation (8.8), the relationship between partial and apparent molar volumes is:

$$V_{m,2} = \left(\frac{\partial V_{kg}}{\partial m_2}\right)_{T,p,n_1} = V_{m,2}^{(app)} + m_2\left(\frac{\partial V_{m,2}^{(app)}}{\partial m_2}\right)_{T,p}. \tag{8.9}$$

A combination of Equations (8.7) and (8.8) leads to the expression:

$$V_{m,2}^{(app)} = \frac{M_2}{\rho} - \frac{\rho - \rho_1}{m_2\rho\rho_1} = \frac{M_2}{\rho} - \frac{\Delta\rho}{m_2\rho\rho_1}. \tag{8.10}$$

The standard molar volume is the limiting value of the apparent molar volume since, employing the l'Hôpital rule to Equation (8.8), it holds:

$$\begin{aligned} \lim_{n_2\to 0}\left(V_{m,2}^{(app)}\right) &= \lim_{n_2\to 0}\left(\frac{V - n_1 V_{m,1}^{\bullet}}{n_2}\right) = \lim_{n_2\to 0}\left(\frac{\partial V}{\partial n_2}\right)_{T,p,n_1} \\ &= \lim_{n_2\to 0}\left(V_{m,2}\right) = V_{m,2}^{0}. \end{aligned} \tag{8.11}$$

The standard molar volume of the solute can then be obtained as the adjustable parameter $V_{m,2}^0$ of the fits of the apparent molar volumes:

$$V_{m,2}^{(app)} = V_{m,2}^0 + \sum_{i=1}^{N} S_{V,i} m_2^i \text{(non-ionic solutes)}$$

or (8.12)

$$V_{m,2}^{(app)} = V_{m,2}^0 + \sum_{i=1}^{N} S_{V,i} m_2^{i/2} \text{(electrolytes)}$$

calculated for each experimental $\rho(m_2)$ data point. For non-ionic solutes in dilute region the expansions in Equation (8.12) are usually truncated after the first (linear) polynomial term. For aqueous solutions of electrolytes, the first term of the expansion $S_{V,1}m_2^{1/2}$ must be consistent with the Debye–Hückel limiting law.

(b) If the density in Equation (8.7) can be expressed by an analytical function $\rho(m_2)$ then the partial molar volume can be obtained by differentiation as:

$$V_{m,2} = \left(\frac{\partial V_{kg}}{\partial m_2}\right)_{T,p,n_1} = \frac{M_2}{\rho} - \frac{1 + m_2 M_2}{\rho^2}\left(\frac{\partial\rho}{\partial m_2}\right)_{T,p}. \tag{8.13}$$

Instead of fitting the density itself, the fit of the expression $\Delta\rho/m_2$ is recommended when $\Delta\rho$ is an experimental input quantity. Such a fit assigns the higher weights to data points in the dilute region. The polynomial function:

$$\frac{\Delta\rho}{m_2} = \frac{\rho - \rho_1}{m_2} = a + bm_2 + cm_2^2 \tag{8.14}$$

in combination with Equation (8.13) leads to expressions for the partial molar volume and standard molar volume in the forms:

$$V_{\mathrm{m},2} = \frac{\rho_1 M_2 - a - 2bm_2 - (M_2 b + 3c)m_2^2 - 2\,M_2 c m_2^3}{(\rho_1 + am_2 + bm_2^2 + cm_2^3)^2} \quad \text{and}$$
$$V^0_{\mathrm{m},2} = \frac{M_2}{\rho_1} - \frac{a}{\rho_1^2}. \tag{8.15}$$

It can be easily verified that Equation (8.10) gives the identical limiting expression since:

$$\lim_{m_2\to 0}(V^{(\mathrm{app})}_{\mathrm{m},2}) = \lim_{m_2\to 0}\left(\frac{M_2}{\rho} - \frac{\rho-\rho_1}{m_2\rho\rho_1}\right)$$
$$= \frac{M_2}{\rho_1} - \frac{1}{\rho_1^2}\lim_{m_2\to 0}\left(\frac{\Delta\rho}{m_2}\right) = \frac{M_2}{\rho_1} - \frac{a}{\rho_1^2} = V^0_{\mathrm{m},2}. \tag{8.16}$$

The method was introduced and widely used by the group of Cibulka[1] and later extended[2] to the evaluation of the partial molar expansion $E_{\mathrm{m},2} = (\partial V_{\mathrm{m},2}/\partial T)_p$. It can be used for solutions of non-electrolytes only where extrapolation to infinite dilution is not constrained by the Debye–Hückel limiting law.

The number of polynomial terms on the right-hand side of Equation (8.14) depends predominantly on the concentration range of data, which is often limited by the solubility of the solute. For solutes with very low solubility it is advisable to evaluate parameter a simply as the average of $\Delta\rho/m_2$ values obtained from multiple measurements, while the other two parameters b, c are set to zero, *i.e.*, assuming a linear dependence $\rho(m_2)$.

(2) Mixing volume data. The usual form of experimental data is a data set of mixing volumes *versus* mole fractions. After one of the components is denoted as a solvent then the "dilute" concentration region is defined. Based on Equation (8.6) the relation between the partial molar volume of a solute 2 and mixing (molar) volume V^{M} is given by the expressions:

$$V_{\mathrm{m},2} = V^{\bullet}_{\mathrm{m},2} + V^{\mathrm{M}} + (1-x_2)\left(\frac{\partial V^{\mathrm{M}}}{\partial x_2}\right)_{T,p} \quad \text{and}$$
$$V^0_{\mathrm{m},2} = V^{\bullet}_{\mathrm{m},2} + \lim_{x_2\to 0}\left(\frac{\partial V^{\mathrm{M}}}{\partial x_2}\right)_{T,p}. \tag{8.17}$$

From the equality $V_{\mathrm{m}} = (1-x_2)V^{\bullet}_{\mathrm{m},1} + x_2 V^{(\mathrm{app})}_{\mathrm{m},2} = (1-x_2)V^{\bullet}_{\mathrm{m},1} + x_2 V^{\bullet}_{\mathrm{m},2} + V^{\mathrm{M}}$ it can be easily derived that:

$$V^{(\mathrm{app})}_{\mathrm{m},2} = V^{\bullet}_{\mathrm{m},2} + \frac{V^{\mathrm{M}}}{x_2} \quad \text{and} \quad V^0_{\mathrm{m},2} = V^{\bullet}_{\mathrm{m},2} + \lim_{x_2\to 0}\left(\frac{V^{\mathrm{M}}}{x_2}\right). \tag{8.18}$$

The standard molar volume can then be obtained either by a direct extrapolation of the apparent molar volumes using Equation (8.12) or by evaluating the limits in Equations (8.17) or (8.18). Experimental dependences of mixing volume on composition are usually correlated by empirical functions among which the most frequently used one is the Redlich–Kister expansion:[3]

$$V^{\mathrm{M}} = x_2(1 - x_2)\sum_{i=1}^{N} A_i(1 - 2x_2)^{(i-1)}. \tag{8.19}$$

Then, in combination with the above equations, the limits for $x_2 \rightarrow 0$ leads to the expression:

$$V^0_{\mathrm{m},2} = \lim_{x_2\rightarrow 0}(V_{\mathrm{m},2}) = \lim_{x_2\rightarrow 0}\left(V^{(\mathrm{app})}_{\mathrm{m},2}\right) = V^\bullet_{\mathrm{m},2} + \sum_{i=1}^{N} A_i. \tag{8.20}$$

For the reasons discussed by Desnoyers and Perron,[4] it is recommended to analyse (to plot or fit) experimental data on mixing volume using the quantity $V^{\mathrm{M}}/[x_2(1-x_2)]$, not the mixing volume itself, particularly over the entire concentration ranges. This procedure assigns to each V^{M} data point the weight $1/[x_2(1-x_2)]$ which asymptotically increases in the dilute region. Such fits can be employed for the evaluation of standard molar volumes since the limit of $V^{\mathrm{M}}/[x_2(1- x_2)]$ for $x_2 \rightarrow 0$ is identical to the limits in Equations (8.17) and (8.18).

Determination of standard molar volumes from experimental volumetric data may reach its limit for solutes with very low solubility in a particular solvent. Typical examples of such systems are aqueous aromatic hydrocarbons[5,6] where the density differences $\Delta\rho = \rho(\text{saturated solution}) - \rho(\text{water})$ at ambient conditions are as small as $-0.1\ \mathrm{kg\ m^{-3}}$ (benzene) and $-0.04\ \mathrm{kg\ m^{-3}}$ (toluene). Solubilities of hydrocarbons with larger molecules in water are much smaller and thus the use of direct volumetric data to evaluate standard molar volumes with a reasonable uncertainty becomes impossible. Then non-volumetric methods may find their use. Based on the relation in Equation (8.4) between volume and Gibbs energy, the variation of solubility $x_2^{(\mathrm{sat})}$ (mole fraction of the solute in saturated solution) with pressure at constant temperature is given by the relation:

$$\left[\frac{\partial \ln\left(x_2^{(\mathrm{sat})}\right)}{\partial p}\right]_T = -\frac{\Delta_{\mathrm{sol}}V}{RT} = -\frac{V^0_{\mathrm{m},2} - V^\bullet_{\mathrm{m},2}}{RT}, \tag{8.21}$$

where $\Delta_{\mathrm{sol}}V = V^0_{\mathrm{m},2} - V^\bullet_{\mathrm{m},2}$ is the standard volume of dissolution. Two assumptions are adopted in the derivation of Equation (8.21): (*i*) solubility is so low that the activity coefficient of the solute (asymmetrical standard state convention) approaches unity, and (*ii*) the solubility of the solvent in the solute is negligible. The method is widely used by Sawamura and coworkers[7,8] for solutions of hydrocarbons in water and in organic solvents.

The pressure range of the most used experimental devices (vibrating-tube densimeters, see Section 8.3) is limited to several tens of MPa and, for obvious reasons, can hardly be increased. The speed of sound, *c*, which is linked with volumetric properties, can however be measured up to much higher pressures. The relationship between the speed of sound and density is based on the well-known formula for the difference between isothermal and isentropic compressibilities:

$$\kappa_T - \kappa_S = \frac{1}{\rho}\left(\frac{\partial \rho}{\partial p}\right)_T - \frac{1}{\rho}\left(\frac{\partial \rho}{\partial p}\right)_S = \frac{TV_\mathrm{m}\alpha_p^2}{c_{p,\mathrm{m}}}, \tag{8.22}$$

where α_p is the isobaric thermal expansivity and $c_{p,\mathrm{m}}$ the molar isobaric heat capacity. After the introduction of the Newton–Laplace formula $\kappa_S = 1/(\rho c^2)$, Equation (8.22) turns into:

$$\left(\frac{\partial \rho}{\partial p}\right)_T = \frac{1}{c^2}\left(1 + c^2\frac{TM\alpha_p^2}{c_{p,\mathrm{m}}}\right) = \frac{c_{p,\mathrm{m}}}{c^2 c_{V,\mathrm{m}}} = \frac{\kappa_T}{c^2\kappa_S}. \tag{8.23}$$

Densities at elevated pressure can then be obtained by integration and consequently standard molar volumes evaluated from the densities using the procedures presented above. The method was recently applied to the determination of standard molar volumes of aqueous alkanols[9] and solutes of biological importance (peptides,[10] nucleosides[11]) at pressures up to 120 MPa.

8.3 Experimental Approaches

Since the properties of dilute solutions are close to those of the pure solvent, the requirements imposed on the accuracy and/or precision of the measurements are high. Thus the selection of a proper experimental technique is a substantial step in the design of the methodology.

As mentioned in the previous section, standard molar volumes can be obtained from results of both the volumetric and non-volumetric measurements and thus the experimental equipment may be based on various principles. Due to the fact that the majority of experimental values of standard molar volumes are obtained from volumetric measurements, we outline here the methods which provide direct experimental volumetric data. As mentioned above, such data may be in the form of solution density ρ, density difference $\Delta\rho = \rho - \rho_1$, or mixing volume V^M. The techniques that enable us to measure the solution density are basic ones and an extensive review has been published recently.[12] On the other hand, dilatometric methods, which directly provide the volume changes on mixing, can also be employed. Dilution dilatometers designed for measurements in the dilute region[13–15] are of interest. Dilatometers are capable of providing highly precise mixing volume data yet have a number of limitations (not feasible for applications at temperatures and pressures other than ambient, considerable sample consumption, rather complicated measurements, non-digital data output, *etc.*) and are, at present, rarely used for measurements of dilute solutions.

Nowadays, the vibrating-tube densimeters have replaced most "classical" methods for experimental investigation of volumetric properties of liquid systems (pycnometers, buoyancy methods, dilatometers) and are particularly adapted for investigating dilute solutions. Advantageous features of vibrating-tube densimeters are obvious: (*i*) small volume of sample, (*ii*) both the static (batch) and the flow regimes of measurements are possible, (*iii*) no contact of a sample with the surroundings and consequently ease of measurements at conditions remote from ambient, (*iv*) high precision, and (*v*) feasible automation of measurements and data acquisition.[16,17] Since Chapter 3 is devoted to vibrating-tube densitometry, only features relating to measurements in the dilute region are outlined here.

The working equation of vibrating-tube densimeters can be written as:

$$\rho = K\tau^2 + B, \qquad \Delta\rho = \rho - \rho_r = K(\tau^2 - \tau_r^2), \tag{8.24}$$

where τ is the period of oscillations of the vibrating tube, K and B are parameters of the tube (calibration constants), and the subscript r denotes a reference fluid, usually pure solvent, $\rho_r = \rho_1$. The values of parameters K and B are determined by calibration using at least two fluids with known density at a particular experimental condition. If each measurement of the sample (τ) is preceded and/or followed by measurements of the pure solvent (τ_1) then the drift of the oscillation period of the vibrating tube can be eliminated to a great extent. This procedure is equivalent to repeated recalibrations of the parameter B in Equation (8.24) ($B = \rho_1 - K\tau_r^2$) and is particularly advantageous for measurements in a flow regime if the solvent is also employed as a carrier liquid which pushes samples from a sampling loop into the densimeter. In this arrangement the vibrating-tube method is perfectly suitable for measurements of small density differences $\Delta\rho = \rho - \rho_1$, since, with appropriate stability of temperature and pressure, both the sample and the solvent are measured within a short period of time under almost identical conditions.

The determination of reliable values of the standard molar volume of a solute requires a large set of experimental data in the dilute region. This is not a problem for solutes soluble in a given solvent [usually an upper concentration limit in a study of dilute region is (0.5 to 1) mol kg^{-1}]. With less soluble solutes the upper concentration limit is conditioned by the solubility of the solute at the ambient temperature at which the solution is prepared. One must pay attention, however, to changes of solubility with temperature and pressure to avoid the formation of two phases at experimental conditions remote from the ambient ones.

The density difference $\Delta\rho = \rho - \rho_1$ of dilute solutions is less dependent on temperature and pressure than the density itself since the isobaric thermal expansivity and isothermal compressibility of dilute solutions are close to those of the pure solvent. Therefore the requirements concerning accuracy of temperature and pressure determination are not very stringent. Uncertainties of ± 0.05 K and ± 0.05 MPa are quite acceptable in regions sufficiently distant from the critical point of the solvent. Higher requirements are imposed upon short-time stability of both state variables. The measurement of the oscillation period for solution τ and preceding and/or subsequent measurements of the period for the solvent τ_1 should be performed, ideally at identical temperature and pressure. If the density difference should be determined to within 10^{-3} kg m^{-3} then the stability of the temperature of the measured liquid should be of the order of mK and pressure fluctuations should be within a few kPa at conditions remote from the critical point of the solvent. This can be achieved by precise temperature regulation (usually using several controlled systems of jackets that encase a vibrating-tube cell and by perfect pre-thermostatting of the incoming liquid). The use of elements that dampen pressure pulses in the pressure line and temperature control of the back-pressure regulator[17] are also advisable to minimise pressure fluctuations. Uncertainty in solute concentration also plays a role; here the accuracy of weighing the components during preparation of solutions and purity of solutes should be taken into account. Since impurities (except water) are usually of a similar nature to the main substance, purities of solute samples 99 mole percent and higher are acceptable. If a sample of the solute contains a significant amount of water, then the Fischer method is a good tool to get the data for evaluation of corrections in the case of aqueous solutions. Besides other benefits (minimization of routine work), automation[17] may also bring an opportunity for estimates of the concentration corrections due to vaporization of volatile solutes from solutions to a vapour space in a storage bottle directly connected to the sampling device. If the measurements are performed at elevated temperatures, one must pay attention to the thermal stability of the solution components. Distortion of experimental results indicates decomposition and analysis of the sample after passing through a hot zone of the apparatus may show its extent. A simple tool to reveal the decomposition is the "stop-flow" method[18,19] when the flow of the sample through the vibrating tube is stopped and the time dependence of the oscillation period is recorded. If there is any significant effect of the solute decomposition on solution density, then the drift of the oscillation period can be observed. In order to account for the inherent drift of the apparatus, the stop-flow experiment should be performed with a pure solvent regarded as thermally stable.

Most existing data of standard molar volumes have been obtained at close-to-ambient conditions using vibrating-tube densimeters which have been commercially available since the 1970s (Anton Paar GmbH, Austria; Sodev Inc., Canada). Some of the commercial models have been designed for measurements under extreme conditions and include the DMA HP, Anton Paar, (up to 473 K, 70 MPa) and the Sodev 003HP[20,21] (up to 423 K, 80 MPa). Many measurements, done over a wide range of temperature and pressure, have been performed using laboratory-made densimeters: the Albert and Wood design[22] and its modifications[23–27] and a few others.[1,28–31] As an example, a setup of an automated flow vibrating-tube densimeter[17] designed for measurements up to 573 K and 30 MPa is shown in Figure 8.1.

Other volumetric techniques capable of measurements under extreme conditions were employed occasionally for direct determination of

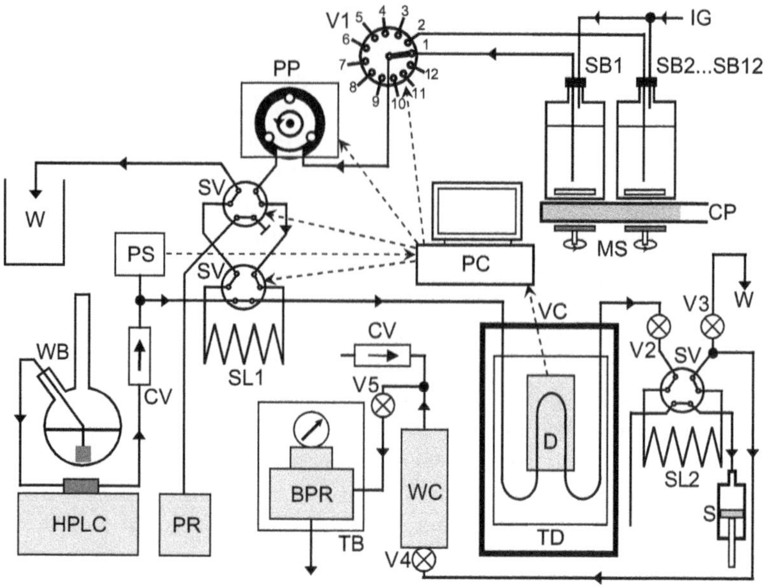

Figure 8.1 Simplified tubing diagram of automated flow vibrating-tube densimeter setup.[17] WB – water boiler, HPLC – high-pressure liquid chromatography pump, CV – check valves, PS – pressure sensor, SV – 3×2 sampling valves, SL1 – sampling loop for measured sample, PR – auxiliary pump used for pressurization of the sample loop SL1, PP – peristaltic pump, W – waste, V1 – twelve-to-one switch-over valve, PC – computer (control and data acquisition unit), SB1, SB2, ... SB12 – twelve sample storage bottles (two are shown), IG – input of inert gas (N_2, Ar), CP – cooled plate, MS – magnetic stirrers, VC – vacuum chamber, D – vibrating-tube densimeter cell,[1] TD – thermostats of the densimeter, V2 to V5 – closing valves, SL2 – sampling loop to withdraw samples passed through densimeter, S – syringe, WC – waste container filled entirely with nitrogen prior to automated run, BPR – back pressure regulator, TB – thermostat of BPR.

volumetric properties of dilute solutions. Constant-volume piezometers designed by Abdulagatov *et al.*,[32,33] a variable-volume piezometer,[34] and the piezometers of the Aimé's type[35,36] can be mentioned as examples.

8.4 Data Sources

The dominant significance of a solvent leads to the classification of solutions into separate groups, each with a common solvent. Thus two large groups of solutions can be distinguished: those with a non-aqueous (organic) solvent and those that are aqueous solutions. While the data for aqueous solutions are numerous, the amount of information for the first group is rather modest, despite the large number of liquid organic solvents. The above classification can be extended to multicomponent systems, typically solutions of a non-ionic solute in a mixed solvent made up of an organic solute in an aqueous electrolyte solution. The number of such studies has increased over the past few years; the overview below is, however, limited to binary solutions only.

As shown in Section 8.2, standard molar volumes can be, in principle, obtained from various volumetric data on binary systems. Compilations of such data are available either in the literature, *e.g.*, older data on mixing volume of binary liquid mixtures,[37,38] or as parts of Web-based multi-property databases (NIST/TRC, DIPPR, DETHERM, Landolt–Börnstein Database). However, the standard molar volumes should be derived only from data that properly cover the region of low concentrations of the solute. Therefore only exclusively selected data sources, which provide standard molar volumes based on experiments in dilute regions, are reviewed below.

8.4.1 Non-Aqueous Solutions

No comprehensive data compilation of standard molar volumes of non-ionic solutes in organic solvents is available. The following overview, which does not claim to be complete, summarises selected references that present sets of experimental data for groups of organic solutes in organic solvents along with, in some cases, a survey of data compiled from the literature.

The data, which are scattered over various literature sources, were obtained predominantly at near-ambient conditions, mostly at $T = 298.15$ K and atmospheric pressure. We begin with tetrachloromethane which, with its non-polar globular molecule seems to be the most "popular" non-aqueous solvent. A systematic study was performed by Edward and co-workers for several groups of solutes (hydrocarbons, alcohols, ethers, ketones, haloalkanes) in tetrachloromethane[39–43] and in benzene, ethanol, and cyclohexane.[44] Selected alkanes, cycloalkanes, aromatic hydrocarbons, haloalkanes, and ethers in tetrachloromethane were measured by Nishimura *et al.*[45] Standard molar volumes on both sides of the concentration interval were evaluated for mixtures of ethers with tetrachloromethane.[46] Values for numerous *n*-alkanes and cyclic hydrocarbons in tetrachloromethane were summarised by Matteoli *et al.*[47] Experimental standard molar volumes of derivatives of tetraphenylporphyrine along with the compiled extensive data for hydrocarbons and crown ethers in tetrachloromethane and for hydrocarbons, alcohols, and carboxylic acids in benzene were presented by Zielenkiewicz and Perlovich.[48] Data are available for amines dissolved in heptane,[49] benzene,[50,51] cyclohexane,[51] trichloromethane and tetrachloromethane,[52] and several other organic solvents.[53] Standard molar volumes of the derivatives of dibenzothiophene dissolved in alkanes and cyclohexane were measured by de Oliveira *et al.*[54] Experimental results for alkanes, haloalkanes, mono- and polyhydric aliphatic alcohols, ethers, and ether alcohols dissolved in ethanol were presented by Itsuki *et al.*[55] Standard molar volumes of water as a solute dissolved in organic solvent (aromatic hydrocarbons,[5,56] 1-alkanols,[56–59] cyclohexanol,[58] propylene carbonate,[60,61] and others[59,60]) have also been studied.

8.4.2 Aqueous Solutions

Several data compilations on standard thermodynamic properties (often standard molar volumes and standard molar heat capacities) of aqueous non-ionic solutes can be found in the literature. Experimental values at $T = 298.15$ K and $p = 0.1$ MPa were collected by Lepori and co-workers[62,63] and Høiland.[64] Development of experimental techniques since 1980 capable of providing data for dilute solutions at super-ambient conditions led to a growing number of publications of standard molar volumes of aqueous non-electrolytes over an extended range of temperatures and pressures. Such data are needed for designing models for calculations of standard molar properties and surveys on data bibliography are often available in this kind of publication.[65–67] A chapter in a recently published monograph[68] presents an overview of both the experimental techniques and the *pVTx* data for aqueous systems (the ionic and non-ionic solutes) at elevated parameters of state.

Two Web-based databases on aqueous systems are accessible by the internet: (*i*) the ORCHYD database[69,70] of the thermodynamic functions of hydration (the difference between the standard property of the aqueous solute and the property in an ideal gas state) at $T = 298.15$ K and $p = 0.1$ MPa and other related data, (*ii*) a database on the thermodynamic properties of aqueous systems which includes a section on standard molar volumes at ambient and super-ambient conditions, elaborated and continuously updated by Šedlbauer[71] at the Technical University of Liberec (Czech Republic).

8.5 Standard Molar Volumes of Non-Ionic Solutes in Water

Undoubtedly, water is a substance of basic importance for life, the environment, industry and of great interest from a scientific point of view. It is a universal solvent dissolving, at least to some extent, most substances. Water exhibits anomalous behaviour at near ambient conditions: (i) it has a density maximum at about 277 K and consequently a negative $(\partial V/\partial T)_p$ down to the freezing temperature, (ii) its volume change on freezing is positive (with the exception of very high pressures), (iii) isothermal compressibility as a function of temperature exhibits a minimum at about 320 K and its value at the freezing point is higher than that at the normal boiling temperature, and (iv) the static permittivity is high at near ambient conditions with a rapid decrease as temperature increases. These peculiar macroscopic properties of water are due to its ordered H-bonded structure, which changes with temperature. The balance between ordered low-density ice-like aggregates and free non-bonded high density water is affected by temperature; the order is gradually destroyed when temperature increases and water tends to behave as a "normal" liquid at high temperatures. All these changes in the structure of water are also reflected in the values of standard molar volumes.

8.5.1 General Features

Taking the Scaled Particle Theory[72,73] (SPT) as a basis, the standard molar volume is a sum of three terms: (*i*) the co-volume (partial molar volume of the non-interacting point mass, equal to $\kappa_{T,1}RT$ where $\kappa_{T,1}$ is the isothermal compressibility of the solvent), (*ii*) the cavity volume related to the intrinsic volume of the solute molecule, and (*iii*) the third contribution originating from intermolecular interactions of a solute molecule and surrounding molecules of solvent (the "hydration shell" in the case of aqueous solutions). The presence of the solute molecule causes structural changes in the surrounding water depending on the nature of the surface of the solute molecule. Non-polar surfaces enhance the network structure (hydrophobic hydration, structure making) while polar parts (charged groups, electron donor/acceptor atoms) cause a breaking of the H-bonded aggregates in favour of solute–water hydrophilic interactions (hydrophilic hydration, structure breaking). In the cases of groups such as –OH, $-NH_2$ and some others, the interactions may have a H-bonding character.

Abundant data are available in the literature for a variety of aqueous non-ionic organic solutes at $T = 298$ K and atmospheric pressure (see Section 8.4). Thus the effects of the solute molecular structure on the standard molar volume can be investigated. For example, one can observe that the replacement of a hydrogen atom bonded to the carbon atom of the molecular frame of a solute molecule with a much larger hydroxyl group leads to a very moderate change in the standard molar volume: its values at $T = 298$ K for all five mono- and polyhydric alcohols derived from propane[74–76] are within the interval ± 0.6 cm^3 mol^{-1}; standard molar volumes of four alcohols derived from 2,2-dimethylpropane (neopentane)[77,78] are even closer to each other, being within the interval ± 0.3 cm^3 mol^{-1}. This observation may lead to the conclusion that the increase in the intrinsic volume of the solute molecule due to introduction of the hydroxyl group is compensated by breaking low-density ice-like water aggregates due to strong interactions between the hydrophilic hydroxyl group and water molecules. Such a conclusion is, however, valid only for 298 K and low pressure.

Much more significant effects appear when standard molar volumes are investigated over wider ranges of temperature and pressure. Figure 8.2 presents smoothed experimental values[77,78] for the above mentioned aqueous hydroxyl derivatives of 2,2-dimethylpropane, $(H_3C)_{(4-n)}C(CH_2OH)_n$, $n = 1, 2, 3, 4$. Obviously the variation with state parameters is strongly affected by the number of hydroxyl groups n and even the inversion of the signs of the derivatives $(\partial V^0_{m,2}/\partial T)_p$ and $(\partial V^0_{m,2}/\partial p)_T$ is observed for the most hydrophilic alcohol ($n = 4$) at high temperatures. The derived quantities: the analogues of the isobaric thermal expansivity $\alpha^0_{p,2} = (1/V^0_{m,2})(\partial V^0_{m,2}/\partial T)_p$, and of the isothermal compressibility $\kappa^0_{T,2} = -(1/V^0_{m,2})(\partial V^0_{m,2}/\partial p)_T$, and the derivative of the standard molar heat capacity with respect to pressure $(\partial c^0_{p,m,2}/\partial p)_T = -T(\partial^2 V^0_{m,2}/\partial T^2)_p$, calculated using smoothing polynomials[77,78] and presented in Figure 8.3, show the differences between the alcohols in more detail.

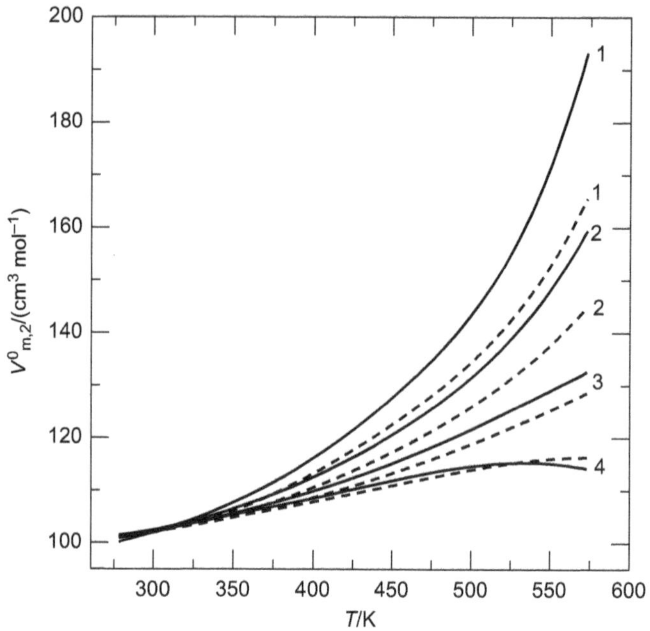

Figure 8.2 Plot of standard molar volumes of aqueous alcohols $(H_3C)_{(4-n)}C(CH_2OH)_n$, $n=1$, 2, 3, 4, against temperature. Full lines, at saturation pressure of water; dashed lines, at $p=30$ MPa. Integer numbers denote the number of hydroxyl groups n. $n=1$, 2,2-dimethylpropane-1-ol; $n=2$, 2,2-dimethylpropane-1,3-diol; $n=3$, 2,2-bis(hydroxymethyl)propane-1-ol; $n=4$, 2,2-bis(hydroxymethyl)propane-1,3-diol.

Interestingly, the effect of the number of hydrophilic hydroxyl groups is also observed in the low temperature range which can be attributed to the increase of the isothermal compressibility of water at low temperatures. According to Hepler's classification[79] based on the sign of the second derivative $(\partial^2 V^0_{m,2}/\partial T^2)_p$ and the shape of its temperature dependence, 2,2-dimethylpropane-1-ol ($n=1$) is the strongest structure maker (prevailing hydrophobic hydration) within this homologous series while 2,2-bis(hydroxymethyl)propane-1,3-diol ($n=4$) is a structure breaker (prevailing hydrophilic hydration), as can be seen in Figure 8.3(b). Obviously the behaviour shown in Figures 8.2 and 8.3 is predominantly governed by the ratio of the hydrophilic/hydrophobic parts of the solute molecule. The selected example illustrates general features that are observed, depending on the hydrophilic/hydrophobic ratio, in greater or lesser extent also for other series of solutes, such as alcohols derived from propane[76] and n-pentane,[80] cyclic ketones,[81] and cyclic ethers.[19]

The changes of $\alpha^0_{p,2}$ and $\kappa^0_{T,2}$ with temperature in Figure 8.3(a) are similar. Thus it seems likely that these two quantities are interrelated. The plot in Figure 8.4 was constructed using values calculated for: temperatures in the interval from 298 K to 573 K (or up to the highest experimental temperature

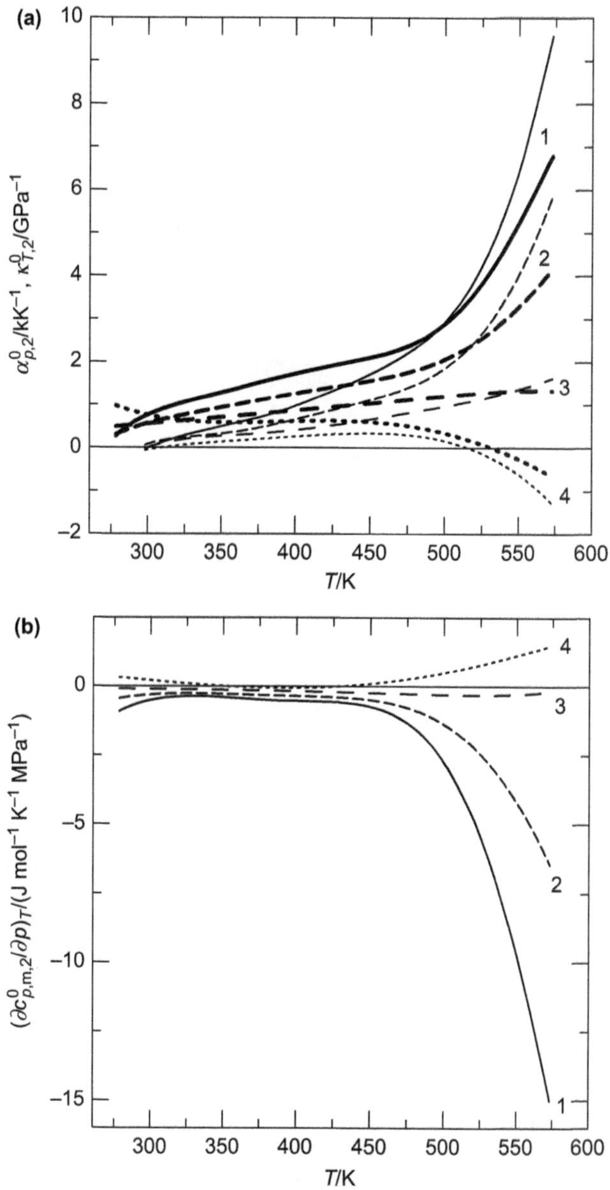

Figure 8.3 Plots of derived properties at saturation pressure of water of aqueous alcohols $(H_3C)_{(4-n)}C(CH_2OH)_n$, $n=1$, 2, 3, 4, against temperature. Integer numbers denote the number of hydroxyl groups n. (a) Isobaric thermal expansivity, $\alpha^0_{p,2}=(1/V^0_{m,2})(\partial V^0_{m,2}/\partial T)_p$ (thick curves), and isothermal compressibility, $\kappa^0_{T,2}=-(1/V^0_{m,2})(\partial V^0_{m,2}/\partial p)_T$(thin curves). (b) Derivative of standard heat capacity with respect to pressure, $(\partial c^0_{p,m,2}/\partial p)_T=-T(\partial^2 V^0_{m,2}/\partial T^2)_p$. $n=1$, 2,2-dimethylpropane-1-ol; $n=2$, 2,2-dimethylpropane-1,3-diol; $n=3$, 2,2-bis(hydroxymethyl)propane-1-ol; $n=4$, 2,2-bis(hydroxymethyl)propane-1,3-diol.

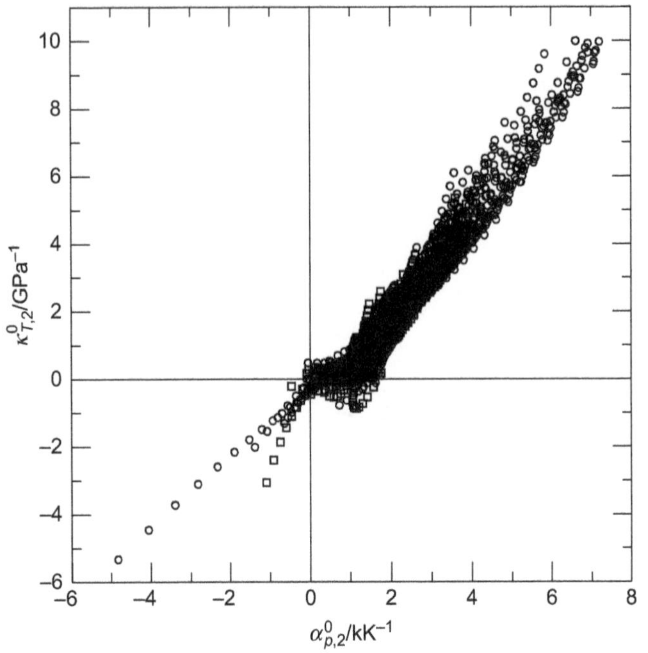

Figure 8.4 Plot of $\kappa^0_{T,2} = -(1/V^0_{m,2})(\partial V^0_{m,2}/\partial p)_T$ against $\alpha^0_{p,2} = (1/V^0_{m,2})(\partial V^0_{m,2}/\partial T)_p$. ○, at saturation pressure of water; □, at $p = 30$ MPa.

before thermally decomposing the solutes) in 5 K wide steps; for saturated pressure of water; and for $p = 30$ MPa from smoothing polynomials evaluated for various classes of aqueous solutes that included aliphatic alcohols, ketones, ethers,[77,78,82,83] cyclic ketones,[81] cyclic ethers,[19] cycloalkanols,[18] phenylalkanols,[84] as well as benzene and toluene[6] (a total of 53 solutes). The extremities of the plot correspond to the high-temperature range; the most hydrophobic solutes (benzene, toluene, monohydric alcohols, monoethers, monoketones) are on the right-hand side with both $\alpha^0_{p,2}$ and $\kappa^0_{T,2}$ positive, while strongly hydrophilic solutes (pentane-1,2,3,4,5-pentaol, 2,2-bis(hydroxymethyl)propane-1,3-diol, propane-1,2,3-triol) are located at the left-hand side with both $\alpha^0_{p,2}$ and $\kappa^0_{T,2}$ negative. Evidently there is a correlation between $\alpha^0_{p,2}$ and $\kappa^0_{T,2}$ which reflects the solute behaviour at high temperatures and particularly at the approach of the gas–liquid critical point of water.

It is of interest to examine the evolution of $V^0_{m,2}$ across the region close to the critical point of water ($T_c = 647.1$ K, $p_c = 22.06$ MPa, $\rho_c = 322$ kg m^{-3}). Applying the rules of partial differentiation, the standard molar volume of a solute can be expressed as:

$$V^0_{m,2} = \lim_{x_2 \to 0}\left[V_m + (1 - x_2)\left(\frac{\partial V_m}{\partial x_2}\right)_{T,p}\right] = V^{\bullet}_{m,1}\left[1 + \kappa_{T,1}\left(\frac{\partial p}{\partial x_2}\right)_{V,T,x_2 \to 0}\right]. \quad (8.25)$$

The derivative $(\partial p/\partial x_2)_{V,T,x_2\to 0}$ is a smooth function which is finite even at the critical point of water. On the other hand, the isothermal compressibility of water $\kappa_{T,1}$ diverges at its critical point and therefore the standard molar volume of a solute also diverges, scaling with $\kappa_{T,1}$. The sign of the divergence is governed by the sign of the initial slope of the critical line of a binary mixture called the Krichevskii parameter $A_{\mathrm{Kr}} = (\partial p\ /\ \partial x_2)^{\mathrm{c}}_{V,T,x_2\to 0}$. Its value is positive for volatile solutes (often, but not exclusively, hydrophobic) and negative for non-volatile ones (often, but not exclusively, hydrophilic). The Krichevskii parameter can be obtained from experimental data of various types,[85] including standard molar volumes using Equation (8.25).

Figure 8.5 shows experimental data for two different non-electrolyte solutes through the critical region of water: methane[86] as a representative of a hydrophobic volatile solute; and boric aid,[87] completely associated at elevated temperatures, as a hydrophilic non-volatile solute. The isobars at 28 MPa and 35 MPa exhibit extremes at supercritical temperature and pressure corresponding to the critical density of water, since the compressibility of water is at its maximum along the critical isochore and decreases on departure from the critical point.

8.5.2 Group Contribution Estimation Methods

When developing group contribution methods the molecule of a solute is split into segments: atoms in the 0th order methods, groups of atoms in the 1st order methods, and groups of atoms with their immediate neighbours in the 2nd order methods. Particular temperature and pressure dependent volumes, comprising both the cavity and interaction part, can be attributed to each kind of segment. Then the total standard molar volume is the sum of the mass point co-volume (the volume of a solute with no segments) and the sum of volumes of all relevant segments. This additivity principle implicitly assumes that the volume of a segment is independent of the presence of other segments and their configuration. This is partly true for the cavity component of the segment volume but rather questionable for the interaction part. This deficiency of the additivity principle can be compensated by the introduction of structural corrections reflecting the configuration of the segments. The number of segment kinds, and consequently the amount of experimental data necessary to evaluate segment volumes, increases substantially with the order while, as expected, departures from additivity decrease.

The most comprehensive 1st order group contribution approach was elaborated by Lepori and co-workers[62,63] employing compiled experimental data measured at $T = 298.15$ K and at atmospheric pressure for over four hundred aqueous non-ionic organic solutes. Various approaches to the co-volume term were discussed and structural correction parameters were introduced for bifunctional solutes. Group contributions (both 1st and 2nd order) to the standard properties of hydration at $T = 298.15$ K and $p = 0.1$ MPa were evaluated for several groups of aqueous solutes by Plyasunov and

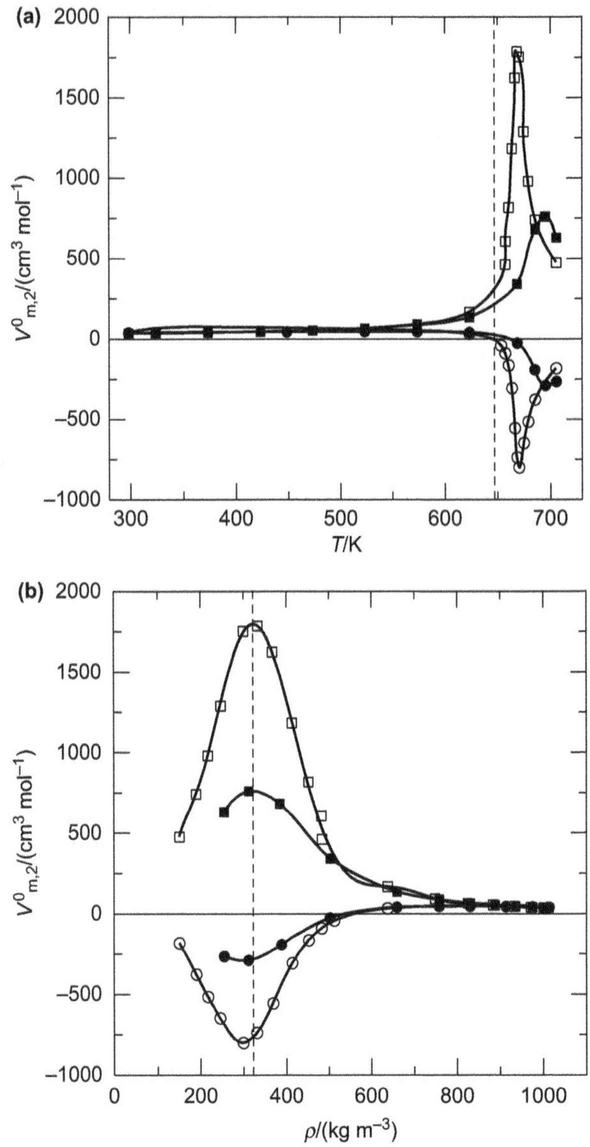

Figure 8.5 Plots of molar volumes against (a) temperature and (b) density of water. Open symbols, $p=28$ MPa; full symbols, $p=35$ MPa. Vertical dashed lines denote (a) critical temperature (647 K), (b) critical density (322 kg m^{-3}) of water. (□, ■, methane[86] (apparent molar volumes at $m_2=0.1$ mol kg^{-1}), ○, ●, boric acid[87] (standard molar volumes).

collaborators.[88–91] Extensive experimental data available for the $V^0_{m,2}(T,p)$ surfaces in the range from (298 K to 573 K) and up to 30 MPa were utilised for the evaluation of the 1st order group contributions in aqueous cyclic ethers,[19] cyclic ketones,[81] and open-chain aliphatic mono- and bifunctional alcohols,

well described by experimental data. Schulte *et al.*[103] proposed a simple but flexible four parameter relationship, having the correct behaviour near the critical point of water and successful in describing experimental data for organic aqueous solutes over an extended temperature and pressure range:

$$V^0_{\mathrm{m},2}(T,p) = a + a_T T - (\omega + \omega_T T)\kappa_{T,1}. \tag{8.28}$$

The models which result in practical use should have leading terms arising from some theoretical foundations. Though they always retain a considerable degree of empiricism such approaches allow a meaningful extrapolation on the basis of limited experimental data. The presence of the "standard-state term" $\kappa_{T,1}RT$, corresponding to the co-volume defined in Section 8.5.1, should be imbedded in any model with extrapolation capabilities. The magnitude of this standard-state term is small at ambient conditions, but increases with temperature and diverges at the solvent's critical point thus becoming the leading part in the near-critical region. At the same time it reduces to the ideal gas equation of state for molar volume when pressure approaches zero.

It was shown recently that, according to the Fluctuation Solution Theory[107] (FST), the spatial integral of the infinite-dilution solute–solvent direct correlation function C^0_{12} is simply related to a dimensionless parameter A^0_{12}:

$$1 - C^0_{12} = \frac{V^0_{\mathrm{m},2}}{\kappa_{T,1}RT} = A^0_{12}. \tag{8.29}$$

The parameter A^0_{12} is called the modified Krichevskii parameter[102] due to its direct connection with A_{Kr} defined in Section 8.5.1. Parameter A^0_{12} is well behaved at the solvent critical point and can be expressed in terms of a virial expansion in water density:

$$A^0_{12} = \lim_{n_2 \to 0}\left(\frac{\partial(pV/RT)}{\partial n_2}\right)_{T,V} = 1 + (2/M_1)\rho_1 B_{12} + \ldots, \tag{8.30}$$

where B_{12} is the second solute–water cross virial coefficient. A similar procedure for the pure solvent leads to a virial expansion for water where the A_{11} parameter is linked with the water–water direct correlation function. Subtracting this latter virial expansion from the former one, and introducing an empirical scaling parameter d accounting for the difference in the sizes of solute and solvent molecules, one gets:

$$A^0_{12} = 1 + d(A_{11} - 1) + \rho_1(2/M_1)(B_{12} - dB_{11}) + \ldots \tag{8.31}$$

This relationship is the basis for two thermodynamic models of practical importance that differ in approach to the expression of the second and higher virial terms.

Plyasunov *et al.*[108] (the model is often abbreviated as POCW) use directly experimental or estimated B_{12} and B_{11} values, and add additional empirical terms:

$$V^0_{\mathrm{m},2} = \kappa_{T,1}RT + d(V^\bullet_{\mathrm{m},1} - \kappa_{T,1}RT) + \kappa_{T,1}RT\rho_1\left\{2(B_{12} - dB_{11})\exp(-c_1\rho_1) + \left(\frac{a}{T^5} + b\right)[\exp(c_2\rho_1) - 1]\right\}, \tag{8.32}$$

where $c_1 = 0.0033\ \mathrm{m^3\ kg^{-1}}$ and $c_2 = 0.002\ \mathrm{m^3\ kg^{-1}}$ are predetermined constants and a, b, d are adjustable parameters. Direct use of second virial coefficients limits the use of this equation to non-ionic solutes only. This problem is circumvented in the approach of Šedlbauer *et al.*[65] (the model is often abbreviated as SOCW) which does not operate explicitly with virial coefficients. The effect of solute–solvent interaction is here modelled by a linear combination of exponential terms in temperature and density:

$$V^0_{\mathrm{m},2} = (1 - z)\kappa_{T,1}RT + d(V^\bullet_{\mathrm{m},1} - \kappa_{T,1}RT) + \kappa_{T,1}RT\rho_1\left\{a + b[\exp(\vartheta\rho_1) - 1] + c\exp\left(\frac{\theta}{T}\right) + \delta[\exp(\lambda\rho_1) - 1]\right\}, \tag{8.33}$$

where $\vartheta = 0.005\ \mathrm{m^3\ kg^{-1}}$, $\lambda = -0.01\ \mathrm{m^3\ kg^{-1}}$, $\theta = 1500$ K, and a, b, c, and d are four adjustable parameters. Parameter z is the particle charge, being $z = 0$ for neutral molecules. Parameter δ for non-electrolytes is calculated as $\delta = 0.35\ a$. This somewhat more empirical equation, compared to Equation (8.32), can be applied to all types of solutes at the cost of one additional adjustable parameter. Both equations diverge at the critical point of water and reduce to an ideal gas equation of state when pressure approaches zero. Both relationships can be integrated in Equation (8.26) to obtain the Gibbs energy of hydration.

In relation to Equation (8.29), Plyasunov *et al.*[109] used the corresponding states principle introducing the reduced spatial integral of the infinite-dilution solute–solvent direct correlation function $C^0_{12,\mathrm{r}} = C^0_{12}(T,\rho)\,/\,C^0_{12}(T_{\mathrm{ref}},\rho_{\mathrm{ref}})$. This reduced integral proved to be a general function of water density within particular classes of solutes. The deviations between solute classes apparently depend on the strength of solute–water interactions and can be correlated with the Gibbs energy of hydration.

Acknowledgements

One of the authors (V.M.) was supported by the "National Programme for sustainability I", project CZ.1.05/2.1.00/01.0005.

References

1. V. Hynek, L. Hnědkovský and I. Cibulka, *J. Chem. Thermodyn.*, 1997, **29**, 1237.
2. M. J. Blandamer and H. Høiland, *Phys. Chem. Chem. Phys.*, 1999, **1**, 1873.

3. O. Redlich and A. T. Kister, *Ind. Eng. Chem.*, 1948, **40**, 345.
4. J. E. Desnoyers and G. Perron, *J. Solution Chem.*, 1997, **26**, 749.
5. M. Sakurai, *Bull. Chem. Soc. Jpn.*, 1990, **63**, 1695.
6. P. Hynčica, L. Hnědkovský and I. Cibulka, *J. Chem. Thermodyn.*, 2003, **35**, 1905.
7. S. Sawamura and N. Fujita, *Carbon*, 2007, **45**, 965.
8. S. Sawamura and H. Ise, *J. Solution Chem.*, 2011, **40**, 1899.
9. A. W. Hakin and H. Høiland, *Phys. Chem. Chem. Phys.*, 2005, **7**, 2199.
10. G. R. Hedwig, E. Høgseth and H. Høiland, *Phys. Chem. Chem. Phys.*, 2008, **10**, 884.
11. G. R. Hedwig, E. Høgseth and H. Høiland, *J. Chem. Thermodyn.*, 2013, **61**, 117.
12. W. Wagner, R. Kleinrahm, H. W. Lösch, J. T. R. Watson, V. Majer, A. A. H. Pádua, L. A. Woolf, J. C. Holste, A. M. De Figueiredo Palavra, K. Fuji and J. W. Stansfeld, Density, in *Measurement of the Thermodynamic Properties of Single Phases*, ed. A. R. H. Goodwin, K. N. Marsh and W. A. Wakeham, Elsevier, Amsterdam, 1st edn, 2003, Experimental Thermodynamics Series vol. VI, ch. 5, pp. 125–235.
13. G. A. Bottomley, L. G. Glossop and W. P. Staunton, *Aust. J. Chem.*, 1979, **32**, 699.
14. R. Tanaka, O. Kiyohara, P. J. D'Arcy and G. C. Benson, *Can. J. Chem.*, 1975, **53**, 2262.
15. M. K. Kumaran, C. J. Halpin and G. C. Benson, *J. Chem. Thermodyn.*, 1982, **14**, 1099.
16. E. C. Ihmels, C. Aufderhaar, J. Rarey and J. Gmehling, *Chem. Eng. Technol.*, 2000, **23**, 409.
17. L. Hnědkovský and I. Cibulka, *Int. J. Thermophys.*, 2004, **25**, 1135.
18. I. Cibulka and L. Hnědkovský, *J. Chem. Thermodyn.*, 2009, **41**, 489.
19. I. Cibulka, *J. Chem. Thermodyn.*, 2010, **42**, 502.
20. A. Inglese, P. Robert, R. De Lisi and S. Milioto, *J. Chem. Thermodyn.*, 1996, **28**, 873.
21. R. De Lisi, S. Milioto and N. Muratore, *Langmuir*, 2001, **17**, 8078.
22. H. J. Albert and R. H. Wood, *Rev. Sci. Instrum.*, 1984, **55**, 589.
23. H. R. Corti, R. Fernández-Prini and F. Svarc, *J. Solution Chem.*, 1990, **19**, 793.
24. V. Majer, R. Crovetto and R. H. Wood, *J. Chem. Thermodyn.*, 1991, **23**, 333.
25. V. Hynek, M. Obšil, V. Majer, J. Quint and J.-P. E. Grolier, *Int. J. Thermophys.*, 1997, **18**, 719.
26. V. Hynek, S. Degrange, M. Poledníček, V. Majer, J. Quint and J.-P. E. Grolier, *J. Solution Chem.*, 1999, **28**, 631.
27. C. Xiao, H. Bianchi and P. R. Tremaine, *J. Chem. Thermodyn.*, 1997, **29**, 261.
28. J. M. Simmonson, C. S. Oakes and R. J. Bodnar, *J. Chem. Thermodyn.*, 1994, **26**, 345.
29. J. G. Blencoe, S. E. Drummond, J. C. Seitz and B. E. Nesbit, *Int. J. Thermophys.*, 1996, **17**, 179.

30. R. F. Chang and M. R. Moldover, *Rev. Sci. Instrum.*, 1996, **67**, 251.
31. A. W. Hakin, D. C. Daisley, L. Delgado, J. L. Liu, R. A. Marriott, J. L. Marty and G. Tompkins, *J. Chem. Thermodyn.*, 1998, **30**, 583.
32. I. M. Abdulagatov, A. R. Bazaev and A. E. Ramazanova, *Ber. Bunsen-Ges. Phys. Chem.*, 1994, **98**, 1596.
33. I. M. Abdulagatov and N. D. Azizov, *J. Chem. Thermodyn.*, 2004, **36**, 17.
34. C. Yokoyama and S. Takahasi, *Int. J. Thermophys.*, 1989, **10**, 35.
35. T. Moriyoshi, Y. Morishita and H. Inubushi, *J. Chem. Thermodyn.*, 1977, **9**, 577.
36. T. Moriyoshi, T. Tsubota and K. Hamaguchi, *J. Chem. Thermodyn.*, 1991, **23**, 155.
37. Y. P. Handa and G. C. Benson, *Fluid Phase Equilib.*, 1979, **3**, 185.
38. A. J. Treszczanowicz and T. Treszczanowicz, *Fluid Phase Equilib.*, 1993, **89**, 31.
39. J. T. Edward, P. G. Farrell and F. Shahidi, *J. Phys. Chem.*, 1978, **82**, 2310.
40. F. Shahidi, P. G. Farrell and J. T. Edward, *J. Phys. Chem.*, 1979, **83**, 419.
41. F. Shahidi, P. G. Farrell, J. T. Edward and P. Canonne, *J. Org. Chem.*, 1979, **44**, 950.
42. J. T. Edward, P. G. Farrell and F. Shahidi, *Can. J. Chem.*, 1979, **57**, 2585.
43. J. T. Edward, P. G. Farrell and F. Shahidi, *Can. J. Chem.*, 1979, **57**, 2892.
44. J. T. Edward, P. G. Farrell and F. Shahidi, *Can. J. Chem.*, 1979, **57**, 2887.
45. N. Nishimura, T. Tanaka and T. Motoyama, *Can. J. Chem.*, 1987, **65**, 2248.
46. A. Spanedda, L. Lepori and E. Matteoli, *Fluid Phase Equilib.*, 1991, **69**, 209.
47. E. Matteoli, L. Lepori and A. Spanedda, *J. Solution Chem.*, 1994, **23**, 619.
48. W. Zielenkiewicz and G. L. Perlovich, *J. Mol. Liq.*, 2005, **121**, 27.
49. L. Lepori, P. Gianni, A. Spanedda and E. Matteoli, *J. Chem. Thermodyn.*, 2011, **43**, 1453.
50. E. F. G. Barbosa and I. M. S. Lampreia, *Can. J. Chem.*, 1986, **64**, 387.
51. S. L. Oswal, J. S. Desai and S. P. Ijardar, *Thermochim. Acta*, 2006, **449**, 73.
52. S. L. Oswal, J. S. Desai, S. P. Ijardar and D. M. Jain, *J. Mol. Liq.*, 2009, **144**, 108.
53. S. P. Ijardar, N. I. Malek and S. L. Oswal, *Indian J. Chem., Sect. A*, 2011, **50**, 1709.
54. L. H. de Oliveira, J. L. da Silva, Jr. and M. Aznar, *J. Chem. Eng. Data*, 2011, **56**, 3955.
55. H. Itsuki, H. Yamamoto, H. Okazaki and S. Terasawa, *J. Chem. Soc., Perkin Trans. 2*, 1990, 1545.
56. M. Sakurai and T. Nakagawa, *J. Chem. Thermodyn.*, 1982, **14**, 269.
57. M. Sakurai and T. Nakagawa, *J. Chem. Thermodyn.*, 1984, **16**, 171.
58. N. Šegatin and C. Klofutar, *Monatsh. Chem.*, 2004, **135**, 161.
59. S. K. Kushare, R. R. Kolhapurkar, D. H. Dagade and K. J. Patil, *J. Chem. Eng. Data*, 2006, **51**, 1617.
60. S. K. Kushare, D. H. Dagade and K. J. Patil, *J. Chem. Thermodyn.*, 2008, **40**, 78.

61. E. V. Ivanov and E. Yu. Lebedeva, *J. Mol. Liq.*, 2011, **159**, 124.
62. S. Cabani, P. Gianni, V. Mollica and L. Lepori, *J. Solution Chem.*, 1981, **10**, 563.
63. L. Lepori and P. Gianni, *J. Solution Chem.*, 2000, **29**, 405.
64. H. Høiland, Partial Molar Volumes of Biochemical Model Compounds in Aqueous Solutions, in *Thermodynamic Data for Biochemistry and Biotechnology*, ed. H.-J. Hinz, Springer-Verlag, Berlin, 1986, ch. 2, pp. 17–44.
65. J. Šedlbauer, J. P. O'Connell and R. H. Wood, *Chem. Geol.*, 2000, **163**, 43.
66. E. M. Yezdimer, J. Šedlbauer and R. H. Wood, *Chem. Geol.*, 2000, **164**, 259.
67. J. Šedlbauer and V. Majer, *Eur. J. Mineral.*, 2000, **12**, 1109.
68. H. R. Corti and I. M. Abdulagatov, pVTx Properties of Hydrothermal Systems, in *Hydrothermal Properties of Materials: Experimental Data on Aqueous Phase Equilibria and Solution Properties at Elevated Temperatures and Pressures*, ed. V. Valyashko, Wiley, Chichester, UK, 1st edn, 2008, ch. 2, pp. 135–193.
69. N. V. Plyasunova, A. V. Plyasunov and E. L. Shock, *Int. J. Thermophys.*, 2004, **25**, 351.
70. ORganic Compounds HYDration properties database, URL: geopig.asu.edu/?q=tools, orchyd.asu.edu.
71. J. Šedlbauer, *Database on the infinite dilution partial molar volumes of aqueous organic nonelectrolytes and electrolytes*, URL: www.kch.tul.cz/texty/sedlbauer/vorg.nb. Updated 2012. Part of the database *Thermodynamic properties of aqueous solutions at geochemical conditions*, URL: www.kch.tul.cz/texty/sedlbauer/download.htm.
72. R. A. Pierotti, *Chem. Rev.*, 1976, **76**, 717.
73. R. N. French and C. M. Criss, *J. Solution Chem.*, 1981, **10**, 713.
74. P. Hynčica, L. Hnědkovský and I. Cibulka, *J. Chem. Thermodyn.*, 2004, **36**, 1095.
75. P. Hynčica, L. Hnědkovský and I. Cibulka, *J. Chem. Thermodyn.*, 2006, **38**, 801.
76. T. Katriňák, L. Hnědkovský and I. Cibulka, *J. Chem. Eng. Data*, 2012, **57**, 1152.
77. L. Šimurka, I. Cibulka and L. Hnědkovský, *J. Chem. Eng. Data*, 2011, **56**, 4564.
78. L. Šimurka, I. Cibulka and L. Hnědkovský, *J. Chem. Eng. Data*, 2012, **57**, 1570.
79. L. G. Hepler, *Can. J. Chem.*, 1969, **47**, 4613.
80. L. Hnědkovský and I. Cibulka, *J. Chem. Thermodyn.*, 2007, **39**, 833.
81. I. Cibulka, L. Šimurka, L. Hnědkovský and A. Bolotov, *J. Chem. Thermodyn.*, 2011, **43**, 1028.
82. I. Cibulka and L. Hnědkovský, *J. Chem. Thermodyn.*, 2011, **43**, 1215.
83. I. Cibulka and L. Hnědkovský, *J. Chem. Thermodyn.*, 2013, **64**, 231.
84. L. Stříteská, L. Hnědkovský and I. Cibulka, *J. Chem. Thermodyn.*, 2004, **36**, 401.

85. A. V. Plyasunov, *J. Phys. Chem. Ref. Data*, 2012, **41**, 033104–1.
86. L. Hnědkovský, R. H. Wood and V. Majer, *J. Chem. Thermodyn.*, 1996, **28**, 125.
87. L. Hnědkovský, V. Majer and R. H. Wood, *J. Chem. Thermodyn.*, 1995, **27**, 801.
88. A. V. Plyasunov and E. L. Shock, *J. Chem. Eng. Data*, 2001, **46**, 1016.
89. A. V. Plyasunov, N. V. Plyasunova and E. Shock, *J. Chem. Eng. Data*, 2004, **49**, 1152.
90. A. V. Plyasunov, N. V. Plyasunova and E. L. Shock, *J. Chem. Eng. Data*, 2006, **51**, 276.
91. A. V. Plyasunov, N. V. Plyasunova and E. L. Shock, *J. Chem. Eng. Data*, 2006, **51**, 1481.
92. J. Šedlbauer and P. Jakubu, *Ind. Eng. Chem. Res.*, 2008, **47**, 5048.
93. G. R. Hedwig and H.-J. Hinz, *Biophys. Chem.*, 2003, **100**, 239.
94. J. P. Amend and H. C. Helgeson, *Geochim. Cosmochim. Acta*, 1997, **61**, 11.
95. J. Šedlbauer, G. Bergin and V. Majer, *AIChE J.*, 2002, **48**, 2936.
96. M. Čenský, J. Šedlbauer, V. Majer and V. Růžička, *Geochim. Cosmochim. Acta*, 2007, **71**, 580.
97. H. C. Helgeson, D. H. Kirkham and G. C. Flowers, *Am. J. Sci.*, 1981, **281**, 1249.
98. J. C. Tanger and H. C. Helgeson, *Am. J. Sci.*, 1988, **288**, 19.
99. E. L. Shock, E. H. Oelkers, J. W. Johnson and D. A. Sverjensky, *J. Chem. Soc., Faraday Trans.*, 1992, **88**, 803.
100. E. L. Shock, H. C. Helgeson and D. A. Sverjensky, *Geochim. Cosmochim. Acta*, 1989, **53**, 2157.
101. E. L. Shock and H. C. Helgeson, *Geochim. Cosmochim. Acta*, 1988, **52**, 2009.
102. J. P. O'Connell, A. V. Sharygin and R. H. Wood, *Ind. Eng. Chem. Res.*, 1996, **35**, 2808.
103. M. D. Schulte, E. L. Shock, M. Obšil and V. Majer, *J. Chem. Thermodyn.*, 1999, **31**, 1195.
104. R. G. Clarke, L. Hnědkovský, P. R. Tremaine and V. Majer, *J. Phys Chem. B*, 2000, **104**, 11 781.
105. A. V. Plyasunov and E. L. Shock, *Geochim. Cosmochim. Acta*, 2000, **64**, 439.
106. P. R. Tremaine, D. Shvedov and C. Xiao, *J. Phys. Chem. B*, 1997, **101**, 409.
107. J. G. Kirkwood and F. P. Buff, *J. Chem. Phys.*, 1951, **19**, 774.
108. A. V. Plyasunov, J. P. O'Connell and R. H. Wood, *Geochim. Cosmochim. Acta*, 2000, **64**, 495.
109. A. V. Plyasunov, E. L. Shock and J. P. O'Connell, *Fluid Phase Equilib.*, 2006, **247**, 18.

CHAPTER 9

Partial Molar Volumes of Gases Dissolved in Liquids

EMMERICH WILHELM*[a] AND RUBIN BATTINO[b]

[a] Institute of Physical Chemistry, University of Wien, Währinger Straße 42, A-1090, Wien (Vienna), Austria; [b] Department of Chemistry, Wright State University, Dayton OH 45435, USA
*Email: emmerich.wilhelm@univie.ac.at

9.1 Introduction

For more than a century physico-chemical investigations of solubility and related phenomena have belonged to the most important topics in chemistry, a fact concisely summarised in 1950 by the introductory statement in Hildebrand and Scott's monograph:[1] "*The entire history of chemistry bears witness to the extraordinary importance of the phenomena of solubility.*" Indeed, the scientific insights gained in the study of nonelectrolyte solutions can hardly be overrated and have been of immense value for the development of the general discipline of solution thermodynamics,[2–9] for instance by providing idealised solution models, such as the one based on the Lewis–Randall (LR) rule, or the one based on Henry's law (HL). An important subdiscipline of this field concerns the *solubility of gases in liquids*, which has a long and distinguished tradition by itself. A number of seminal experimental investigations were made quite early. For instance, in the early 1800s William Henry (1774–1836) described investigations which showed that the amount of gas dissolved in a liquid is proportional to its pressure, a relation which bears his name. In the mid- and late 1800s, we note the influential work of Robert Bunsen (1811–1899), of Ivan M. Sechenov (1829–1905), and of Wilhelm Ostwald (1853–1932).

Volume Properties: Liquids, Solutions and Vapours
Edited by Emmerich Wilhelm and Trevor M. Letcher

Published by the Royal Society of Chemistry, www.rsc.org

The contributions of Joel H. Hildebrand (1881–1983) and collaborators in the mid-20th century promoted our modern knowledge of the thermodynamics of solutions of gases in liquids,[10] and inspired a generation of scientists. A few selected representative reviews are provided by references 11 through 23. They document the impressive number of vapour–liquid equilibrium (VLE) measurements involving a supercritical solute (usually at high dilution), that is, experiments at conditions where the experimental temperature exceeds the critical temperature of the solute.

Gas solubility data are of importance in many areas of the applied sciences, ranging from chemical engineering, to geochemistry, to the environmental sciences, and to biomedical technology.[9,10,24–32] In addition, during the last few decades major advances in experimental techniques have permitted the determination of gas solubilities and associated caloric properties (*partial molar enthalpy changes on solution* and *partial molar isobaric heat capacity changes on solution*) and volumetric properties (*partial molar volumes*) with unprecedented accuracy over increasing ranges of temperature and pressure. These data revealed new and interesting effects which in turn stimulate advances in theory. While some caloric properties have been reviewed quite recently,[33–35] the review of Handa and Benson on partial molar volumes of gases at infinite dilution in liquids[36,37] dates back to 1982; and their survey of experimental data is limited to *non-aqueous* polar and non-polar solvents.

Essentially, all of our contribution will focus on *highly dilute binary* solutions of a *gaseous solute* (component 2) in a *liquid solvent* (component 1). When considering only interactions with nearest neighbours, a rough estimate of the dilute region is provided by $x_2 < 0.01$, where x_2 denotes the solute mole fraction. Of particular interest are properties in the limit of *infinite dilution*: they will be indicated by a superscript ∞ attached to the property symbol, for instance to identify the *partial molar volume* $V_2^{L\infty}(T,P)$ of solute 2 at infinite dilution in the liquid (L) solution phase at the indicated temperature T and pressure P (for a rigorous definition see below, Section 9.2, and the introductory Chapter 1). Given the wide scope of solubility-related fields indicated above, it is not surprising that the subject has a vast literature, and that, in particular, studies of the solubility of gases in water (for instance of the rare gases and of simple hydrocarbons) hold such a prominent position in biophysics. They are a major source of information on *hydrophobic effects*[38–58] that are thought to be of pivotal importance for the formation and stability of higher order structures of biological substances, such as proteins, nucleic acids, and cell membranes. In summary, the preponderance of scientific papers dealing with aqueous solutions is not surprising.

Whenever the focus is on water[59,60] as solvent, the following three aspects should be kept in mind, reminding us that water is indeed unique:

- Water is the most abundant substance on the surface of the earth.[61] The approximate water supply amounts to roughly 13.4×10^{20} kg, of which, however, only a small fraction, *i.e.* roughly 0.001×10^{20} kg, is found as freshwater in rivers and lakes.

- Water is the only substance on earth that occurs naturally in all three states of matter, which fact is closely related to the size of the earth and that the earth's orbit is located in the so-called habitable zone of the solar system,[62] a prerequisite for the emergence of life on earth.[63–71]
- Liquid water (and to a lesser extent water vapour and ice) sustains life on earth as we know it. In fact, water is the principal constituent of all living organisms, making up about 70% by weight of the human body. In the biochemistry of the cell, water plays an essential role in the structure, stability, dynamics and function of biomacromolecules,[72–76] to the extent that it may no longer be regarded as being a mere medium in which bioprocesses occur, but, as recently suggested by Ball,[74,76] rather as an indispensable *active* matrix, something like a "biomolecule" itself.

In a classical paper, Frank and Evans[77] discussed some unusual partial molar thermodynamic properties of non-polar solutes in water (compared to those of the same solute in other solvents) in terms of an unspecified ordering of the water molecules around the solute—the famous "iceberg" formation, though this term was just conceived as a convenient descriptor and not intended to suggest a structure similar to that of ice. While this concept should not be taken literally, it has nevertheless provided a platform for discussing semi-quantitatively the large entropy losses and the large heat capacity changes[19,33–35] observed when dissolving a non-polar solute in water. What has been indicated by computer simulations and model calculations has been confirmed by structural studies using the method of neutron diffraction and isotopic substitution (NDIS) on aqueous solutions of argon[78] and methane.[79] The picture that emerges from this work is one in which the solute, say methane, is surrounded by a relatively *strong first* coordination shell containing about 19 water molecules which are oriented tangentially to the CH_4 molecule, with the pair distribution functions for methane in water showing *no* evidence of a second coordination shell. A similar behaviour was observed by Broadbent and Neilson in their NDIS study of the hydration structure around an argon atom.[78] However, the interaction between the apolar molecules and water is much shorter-ranged than suggested by the models used for computer simulations. The results also suggest that water molecules point their OH bonds towards the bulk. The recent femtosecond mid-infrared pump-probe spectroscopical study of Rezus and Bakker[80] to investigate the orientational mobility of water molecules in the hydration shells of hydrophobic groups showed that the molecular picture of Frank and Evans has to be clarified as follows: only about 20% of the water OH groups in the hydration shell of a methyl group represent strongly immobilised OH groups, while the remaining 80% maintain essentially the same reorientational mobility of bulk water. Their results are in line with Stillinger's suggestion[81] that *small* hydrophobic solutes can be accommodated by the hydrogen-bond network of water without breaking hydrogen bonds. At elevated temperatures, these *pseudo*-clathrate entities should gradually disappear, and concomitantly the partial molar heat capacity $C_{P,2}^{L\infty}$ at constant pressure of solute 2 at infinite dilution in the liquid solvent 1 (water) should also diminish, though it will increase again when the critical temperature $T_{c,1}$ of water is approached: $C_{P,2}^{L\infty}$ will diverge to $+\infty$ as $T_{c,1}$ is approached from *lower* temperatures (at $P = P_{c,1}$, the critical pressure of water). In fact, the partial molar volume $V_2^{L\infty}$ of a gas at infinite dilution in a liquid solvent also *diverges* to $+\infty$ at the critical point of the solvent. The effects of this divergence are felt relatively far from the critical point.[82] Using a lattice gas model, Wheeler[83] showed that for such a system $V_2^{L\infty}$ will tend to $+\infty$, proportional to the *isothermal compressibility* $\beta_{T,1}^{L*}$ of the pure orthobaric liquid solvent. The important experiments of Wood *et al.*[84–88] confirm these expectations.

The partial molar volume of a gas in a solution is also an indispensable property for the thermodynamically correct description of the corresponding vapour–liquid equilibrium and associated data reduction. However, whereas a large body of gas solubility data has been reported in the literature, there have been only relatively few measurements of the corresponding partial molar volumes of dissolved gases. Fortunately, the situation improved noticeably when vibrating-tube densimeters[89] were applied to this problem at ambient temperatures and low pressures,[90] as well as at temperatures up to the critical temperature of the solvent (and beyond) and elevated pressures.[87,88]

9.2 Thermodynamics

The partial molar volume of component i in solution at temperature T and pressure P is defined as:

$$V_i(T, P, \{x_i\}) \equiv \left(\frac{\partial(nV(T, P, \{x_i\}))}{\partial n_i}\right)_{T,P,n_{j \neq i}} \tag{9.1}$$

where $V(T,P,\{x_i\})$ denotes the molar volume of the solution, n_i is the amount of substance of component i, $n = \sum_i n_i$ is the total amount of substance of the solution phase, nV is the total volume of the solution phase, $x_i = n_i / \sum_i n_i$ is the mole fraction of i, and $\{x_i\}$ is the set of compositional variables with $\sum_i x_i = 1$; for a pure fluid $x_i = 1$. Since the total volume nV of a solution phase is a homogeneous function of the first degree in the amounts of substance $\{n_i\}$, Euler's theorem yields:

$$nV = \sum_i n_i V_i, \tag{9.2}$$

or after division by the total amount of substance n:

$$V = \sum_i x_i V_i. \tag{9.3}$$

It is important to keep in mind that *the partial molar volume V_i has no physical reality per se*. In general, *a partial molar property is an intensive property of the mixture/solution and not of a particular component*. It has to be evaluated for each mixture/solution, and usually $V_i \neq V_i^*$, where $V_i^*(T,P)$ is the molar volume of pure component *i*. Here, the superscript asterisk (*) denotes, as always, a *pure-substance* property. However, a partial molar property defined in analogy to Equation (9.1) can *always* be used to provide a systematic *formal* subdivision of the corresponding extensive property, say, *nV*, into a sum of contributions of the individual species in solution constrained by Equation (9.2), or a systematic *formal* subdivision of the intensive property *V* into a sum of contributions of the individual species *i* in solution constrained by Equation (9.3). Thus one may use partial molar properties as though they possess property values referring to the individual constituent species in solution. Such a formal subdivision may also be based on *mass* instead of amount of substance, in which case partial *specific* properties are obtained with similar physical significance.

For the special case of a *binary* system, the total volume is given by:

$$nV = n_1 V_1 + n_2 V_2 \tag{9.4}$$

and the partial molar volume V_2 of the solute is obtained as the slope of a plot of nV against n_2 at constant *T*, *P*, and n_1. However, an alternative method may be more convenient for calculating the partial molar volumes from any *analytical* expression for the molar volume as a function of composition, say, mole fraction x_2, at constant temperature and pressure. Because of the constraint $x_1 + x_2 = 1$, one gets:

$$V_1 = V - x_2 \frac{dV}{dx_2}, \tag{9.5}$$

and

$$V_2 = V + (1 - x_2) \frac{dV}{dx_2}. \tag{9.6}$$

Application of the Gibbs–Duhem equation yields:

$$\frac{dV_1}{dx_2} = -\frac{x_2}{x_1} \frac{dV_2}{dx_2}, \tag{9.7}$$

which shows that the slopes of the curves V_i *vs.* x_2, $i = 1, 2$, are of opposite sign. Experiment shows that the *limiting slopes* dV_i/dx_i at infinite dilution, *i.e.* at $x_i = 0$, remain finite, hence for $x_2 = 0$ we obtain $dV_1/dx_2 = 0$, and for $x_2 = 1$ we obtain $dV_2/dx_2 = 0$. In other words, the partial molar volume curves V_i *vs.* x_i are horizontal at $x_i = 1$ when $V_i \rightarrow V_i^*$. As indicated, for a given system partial molar volumes vary with temperature, pressure and composition. The partial molar volume at *infinite dilution* is defined by:

$$V_2^{\infty}(T, P) = \lim_{x_2 \to 0} V_2(T, P, x_2), \tag{9.8}$$

Non-reacting gases are much less soluble in water than in non-polar or weakly polar organic solvents. In parallel, as pointed out by Hildebrand and Scott,[1] if we take the normal boiling point $T_{b,2}$ as a reasonable reference state, large increases of $V_2^{L\infty}$ in non-polar solvents over the molar volumes $V_2^{L*}(T_{b,2})$ of the pure liquid solutes are observed, whereas *no* such increases occur in solutions of gases in water. This is illustrated in Table 8.6 of Hildebrand, Prausnitz, and Scott's monograph[10] and is reproduced here in Table 9.1: the partial molar volumes at infinite dilution $V_2^{L\infty}(T,P)$ at $T=298.15$ K of the four gases hydrogen, nitrogen, methane and ethane dissolved in water and benzene, respectively, are compared with the molar volumes of the pure gases at their normal boiling points. Observations of this kind have frequently been used in discussions of aqueous solution structures.

The partial molar volume at infinite dilution is of considerable importance in solution thermodynamics in general, and in the discussion of *gas solubilities* in particular. In this field of research, the main property of interest is the Henry fugacity $h_{2,1}(T,P_{\sigma,1}(T))$ (also known as Henry's law (HL) constant) of component 2 dissolved in the liquid solvent 1 at T and the vapour pressure of the solvent $P_{\sigma,1}(T)$. Experimentally, it is obtained from VLE measurements according to:[9,24,31,34]

$$h_{2,1}(T,P_{\sigma,1}(T)) = \lim_{x_2\to 0}\left(\frac{f_2^{L}(T,P,x_2)}{x_2}\right) = \lim_{\substack{x_2\to 0\\ y_2P\to 0}}\left(\frac{\phi_2^{V}(T,P,y_2)y_2P}{x_2}\right),\ \text{constant } T. \qquad (9.9)$$

Here, $f_2^L(T,P,x_2)$ is the liquid-phase fugacity of the gaseous solute 2 at total pressure P, x_2 denotes the mole fraction of dissolved gas in the liquid phase, y_2 denotes the mole fraction of gas in the coexisting vapour phase, and $\phi_2^V(T,P,y_2)$ is the fugacity coefficient of component 2 in the vapour phase. Equation (9.9) represents the classical experimental basis for the determination of the Henry fugacity from *isothermal* VLE measurements

Table 9.1 Molar volumes of the pure *liquid* solutes V_2^{L*} at their normal boiling points $T_{b,2}$ compared with their partial molar volumes at infinite dilution $V_2^{L\infty}$ in water and in benzene at 298.15 K and atmospheric pressure (from ref. 10). Note the significantly different isothermal compressibilities of the two solvents at 298.15 K and atmospheric pressure: $\beta_{T,H_2O}^{L*} = 4.525\times10^{-10}\ \text{Pa}^{-1}$,[a] and $\beta_{T,C_6H_6}^{L*} = 9.71\times10^{-10}\ \text{Pa}^{-1}$.[b]

	H_2	N_2	CH_4	C_2H_6
$V_2^{L*}(T_{b,2})/\text{cm}^3\text{mol}^{-1}$	28	35	34	55
$V_2^{L\infty}(298.15\ \text{K})/\text{cm}^3\text{mol}^{-1}$ Gas dissolved in liquid H_2O	<28	33	37	51
$V_2^{L\infty}(298.15\ \text{K})/\text{cm}^3\text{mol}^{-1}$ Gas dissolved in liquid C_6H_6	35	58	57	73

[a]Reference 175.
[b]Reference 208.

(determination of P, x_2 and y_2) at decreasing pressures $P \to P_{\sigma,1}(T)$ and concomitantly decreasing $x_2 \to 0$ and $y_2 \to 0$: the Henry fugacity referring to the *liquid phase* is obtained as the intercept of an *isothermal* plot of $\phi_2^{\mathrm{V}}(T,P,y_2)y_2P/x_2$ against x_2. The second part of Equation (9.9) is a consequence of the vapour–liquid equilibrium condition expressed as:

$$\begin{aligned} f_2^{\mathrm{L}}(T,P,x_2) &= f_2^{\mathrm{V}}(T,P,y_2) \\ &= \phi_2^{\mathrm{V}}(T,P,y_2)y_2P. \end{aligned} \tag{9.10}$$

The value of the Henry fugacity depends not only on T and P but also on the identities of solute 2 *and* solvent 1, hence the *double* subscript has been added. The vapour-phase fugacity coefficient ϕ_2^{V} *must be calculated*, and perfectly general equations allow the calculation of the component fugacity coefficients at any desired pressure or amount-of-substance density $\rho_n \equiv 1/V$ from suitable *PVT* equations of state (EOS) for the solution (see below).

Once $h_{2,1}(T,P_{\sigma,1}(T))$ has been obtained by this extrapolation method, the VLE measurements at $P > P_{\sigma,1}(T)$ allow, in principle, extraction of the liquid-phase activity coefficients based on ideal-solution behaviour in the sense of Henry's law. Ideality based on Henry's law is characterised by the assumption of the validity of the linear relation (at constant T and P)

$$f_2^{\mathrm{HL}}(T,P,x_2) = x_2h_{2,1}(T,P) \tag{9.11}$$

over the entire composition range $0 \le x_2 \le 1$. The corresponding activity coefficient of the solute in the real solution is then given by:

$$\begin{aligned} \gamma_2^{\mathrm{HL}}(T,P,x_2) &\equiv f_2^{\mathrm{L}}(T,P,x_2)/f_2^{\mathrm{HL}}(T,P,x_2) \\ &= f_2^{\mathrm{L}}(T,P,x_2)/x_2h_{2,1}(T,P), \end{aligned} \tag{9.12}$$

and thus for finite compositions:

$$f_2^{\mathrm{L}}(T,P,x_2) = \gamma_2^{\mathrm{HL}}(T,P,x_2)x_2h_{2,1}(T,P). \tag{9.13}$$

The activity coefficients are said to be normalised *unsymmetrically* when for solute 2 the definition Equation (9.12) applies, and for solvent 1 the usual Lewis–Randall (LR) definition:

$$\begin{aligned} \gamma_1^{\mathrm{LR}}(T,P,x_1) &\equiv f_1^{\mathrm{L}}(T,P,x_1)/f_1^{\mathrm{LR}}(T,P,x_1) \\ &= f_1^{\mathrm{L}}(T,P,x_1)/x_1f_1^{\mathrm{L}*}(T,P), \end{aligned} \tag{9.14}$$

is used, where $f_1^{\mathrm{L}*}(T,P)$ is the fugacity of pure component 1 at T and P of the liquid solution. Their limiting behaviour is thus given by:

$$\gamma_1^{\mathrm{LR}} \to 1 \text{ as } x_1 \to 1, \quad \text{constant } T,P, \tag{9.15a}$$

$$\gamma_2^{\mathrm{HL}} \to 1 \text{ as } x_2 \to 0, \quad \text{constant } T,P. \tag{9.15b}$$

For isothermal conditions, at *each* experimental composition x_2 Equation (9.13) applies, necessarily with a different corresponding equilibrium pressure. Thus, for each VLE experiment the respective Henry fugacities $h_{2,1}(T,P)$, as well as the respective HL-based activity coefficients $\gamma_2^{\mathrm{HL}}(T,P,x_2)$, refer to *different* pressures. For the reduction, correlation and further use of high-precision gas solubility data, it is advantageous to select, at each temperature, the vapour pressure $P_{\sigma,1}(T)$ of the solvent as the *reference pressure*. With this convention and taking into account the pressure dependences of the various quantities involved: *i.e.*

$$\left(\frac{\partial \ln h_{2,1}(T,P)}{\partial P}\right)_T = \frac{V_2^{\mathrm{L}\infty}(T,P)}{RT}, \tag{9.16}$$

and

$$\left(\frac{\partial \ln \gamma_2^{\mathrm{HL}}(T,P,x_2)}{\partial P}\right)_{T,x_2} = \frac{V_2^{\mathrm{L}}(T,P,x_2) - V_2^{\mathrm{L}\infty}(T,P)}{RT}, \tag{9.17}$$

we obtain:

$$h_{2,1}(T,P) = h_{2,1}(T,P_{\sigma,1}) \exp\left\{\int_{P_{\sigma,1}}^{P} \frac{V_2^{\mathrm{L}\infty}(T,P)}{RT}\,\mathrm{d}P\right\} \tag{9.18}$$

and

$$\gamma_2^{\mathrm{HL}}(T,P,x_2) = \gamma_2^{\mathrm{HL}}(T,P_{\sigma,1},x_2) \exp\left\{\int_{P_{\sigma,1}}^{P} \frac{V_2^{\mathrm{L}}(T,P,x_2) - V_2^{\mathrm{L}\infty}(T,P)}{RT}\,\mathrm{d}P\right\}. \tag{9.19}$$

Use of Equation (9.13) and insertion into the VLE equilibrium condition, Equation (9.10), yields the pressure-corrected isothermal–isobaric liquid-phase activity coefficient based on Henry's law at the *reference pressure* $P_{\sigma,1}(T)$ at mole fraction x_2 *via*

$$\gamma_2^{\mathrm{HL}}(T,P_{\sigma,1},x_2) = \frac{\phi_2^{\mathrm{V}}(T,P,y_2)y_2P}{x_2 h_{2,1}(T,P_{\sigma,1})} \exp\left\{-\int_{P_{\sigma,1}}^{P} \frac{V_2^{\mathrm{L}}(T,P,x_2)}{RT}\,\mathrm{d}P\right\}. \tag{9.20}$$

Here, $V_2^{\mathrm{L}}(T,P,x_2)$ is the partial molar volume of the solute at mole fraction x_2 in the liquid phase, $V_2^{\mathrm{L}\infty}(T,P)$ is the partial molar volume of the solute at infinite dilution in the liquid phase, and R is the gas constant. The exponentials in Equations (9.18)–(9.20) are known as *Poynting correction factors*; and the integrals as *Poynting integrals*. The preexponential factor on the right-hand side of Equation (9.20) is a dimensionless group containing the experimental data, x_2 and y_2 at T and P, the Henry fugacity $h_{2,1}(T,P_{\sigma,1})$ already determined by extrapolation, and the vapour-phase fugacity coefficient $\phi_2^{\mathrm{V}}(T,P,y_2)$ of the gas. In order to evaluate the Poynting integral in Equation (9.20), information is needed on the composition dependence of the partial molar volume of the solute as well as on its pressure dependence, information which is usually not readily available.

Equations (9.9) and (9.20) are the *key equations* in the *classical sequential* approach to gas-solubility *data reduction* (supercritical solute) and most frequently adopted. They simply reflect the focusing of *interest on the solute* in a composition range *close* to pure solvent, though, of course, "close" varies from system to system. As a matter of fact, the majority of low-pressure gas solubility studies focus on a composition range *very close* to pure solvent, that is the measurements are made to determine *only* the Henry fugacity, since the experimental imprecision as well as the lack of volumetric data for the reliable evaluation of the Poynting integral and the inevitable experimental scatter preclude the determination of the activity coefficients γ_2^{HL} at high dilution. This is even the case for the most precise gas solubility measurements to date due to Benson and Krause,[91–94] and Battino and collaborators.[95–99] Gas solubility experiments are usually conducted at rather low pressures, say, below a few bar (bar = 100 kPa), and the Poynting correction factors typically differ from unity by only a few parts per thousand.

At *high* dilution $V_2^{\mathrm{L}}(T,P,x_2) \cong V_2^{\mathrm{L}\infty}(T,P_{\sigma,1})$, and for gaseous solutes well below the critical temperature $T_{c,1}$ of the solvent, the pressure dependence of the partial molar volumes is rather small. If one now assumes that within experimental precision $\gamma_2^{\mathrm{HL}} \cong 1$, independent of composition, the Krichevsky–Kasarnovsky equation is obtained:[100]

$$\ln\left(\frac{\phi_2^{\mathrm{V}}(T,P,y_2)y_2P}{x_2}\right) = \ln h_{2,1}(T,P_{\sigma,1}) + \frac{(P-P_{\sigma,1})V_2^{\mathrm{L}\infty}(T,P_{\sigma,1})}{RT}. \tag{9.21}$$

From the condensed thermodynamic formalism presented above, applicable to precise measurements of the solubility of gases in liquids, it is clear that in addition to the experimental VLE data, *i.e.* P, x_2 and y_2 at temperature T, *three* auxilliary properties are indispensable: the vapour pressure $P_{\sigma,1}(T)$ of the solvent, the fugacity coefficient $\phi_2^{\mathrm{V}}(T,P,y_2)$ of the gas in the vapour phase, and the partial molar volume $V_2^{\mathrm{L}\infty}(T,P_{\sigma,1})$ at infinite dilution in the liquid phase.

A large amount of experimental data for vapour pressures of liquids, together with tabulated correlation parameters, may be found, for instance, in the monograph by Poling *et al.*,[101] in the Dortmund Data Bank,[102] or in the VDI Heat Atlas.[103] The Antoine equation is the most popular of all the empirical vapour-pressure equations:

$$\ln\frac{P_{\sigma,1}}{P_0} = A - \frac{B}{T+C}, \tag{9.22}$$

with A, B and C being substance specific parameters, and P_0 a convenient pressure unit that ensures a dimensionless argument of the logarithm. The Antoine equation should never be used outside the specified temperature range.

A much more accurate representation of vapour-pressure data over very large temperature ranges has been developed by Wagner.[104] The Wagner equation expresses the reduced vapour pressure as a function of reduced temperature $T_{r,1} \equiv T/T_{c,1}$:

$$\ln\frac{P_{\sigma,1}}{P_{c,1}} = \frac{1}{T_{r,1}}\left(a\tau_1 + b\tau_1^{1.5} + c\tau_1^3 + d\tau_1^6\right), \tag{9.23}$$

where $P_{c,1}$ is the critical pressure of the solvent, $T_{c,1}$ is the critical temperature of the solvent, $\tau_1 \equiv 1 - T_{r,1}$, and a, b, c and d are substance-specific constants. Its application requires accurate critical data. Note that due to the structure of the equation the critical point condition is automatically met. Various other forms of Wagner-type equations have been presented.

Perfectly general equations are at our disposal for calculating $\phi_2^V(T,P,y_2)$. With a *volume-explicit* EOS

$$Z^V \equiv PV^V/RT = Z^V(T,P,\{y_i\}), \tag{9.24}$$

where Z^V is the compression factor of the vapour-phase solution and V^V is the vapour-phase molar volume:

$$\ln\phi_i^V = \int_0^P \left[\left(\partial(n^V Z^V)/\partial n_i^V\right)_{T,P,n_{j\neq i}^V} - 1\right]\frac{dP}{P}, \quad \text{constant } T \text{ and } \{y_i\}, \tag{9.25}$$

with $n^V = \sum_i n_i^V$. When a *pressure-explicit* EOS

$$Z^V = Z^V(T,V^V,\{y_i\}) \tag{9.26}$$

is used, we obtain:

$$\ln\phi_i^V = \int_{V^V}^{\infty} \left[\left(\partial(n^V Z^V)/\partial n_i^V\right)_{T,n^V V^V,n_{j\neq i}^V} - 1\right]\frac{dV^V}{V^V} - \ln Z^V, \quad \text{constant } T \text{ and } \{y_i\}. \tag{9.27}$$

Most of the EOS in use are pressure-explicit rather than volume-explicit, the most important exception being the *volume-explicit virial EOS*. Since gas solubility measurements are usually performed in the low to moderate pressure domain, virial equations are the equations of choice. They are superior to cubic EOS and computationally convenient. Using a two-term volume-explicit virial equation:

$$Z^V(T,P,y_2) = 1 + B(T,y_2)P/RT \tag{9.28}$$

in Equation (9.25) leads to the widely used expression for the fugacity coefficient of component i at mole fraction y_i in a binary vapour mixture:

$$\ln\phi_i^V = \frac{P}{RT}\left(B_{ii} + y_j^2\delta_{12}\right), \quad i,j = 1,2, i \neq j, \tag{9.29}$$

with

$$\delta_{12} \equiv 2B_{12} - (B_{11} + B_{22}). \tag{9.30}$$

For the second virial coefficient of the mixture we obtain:

$$B = y_1 B_{11} + y_2 B_{22} + y_1 y_2 \delta_{12}. \tag{9.31}$$

Here, B_{11} and B_{22} designate the second virial coefficients of the pure components, and B_{12} is a composition-independent interaction virial coefficient (cross coefficient). The fugacity coefficient of the solute at infinite dilution in the vapor phase is thus given by:

$$\ln\phi_2^{V\infty} = \frac{P}{RT}(2B_{12} - B_{11}), \tag{9.32}$$

and the fugacity coefficient of pure component 2 by:

$$\ln\phi_2^{V*} = \frac{P}{RT}B_{22}. \tag{9.33}$$

The quite popular rule-of-thumb $\phi_2^{V}(T,P,y_2) \approx \phi_2^{V*}(T,P)$ may often be rather unsatisfactory: *e.g.* for the evaluation of $\phi_2^{V\infty}$ it only holds if $B_{12} = (B_{11} + B_{22})/2$.

Frequently, experimental results on second virial coefficients,[105a,b] in particular for mixtures, are not available. Even for pure water vapour the situation below about 400 K is not entirely satisfactory and subject to intensive research.[106] Thus, for VLE data reduction and VLE calculation one has to depend heavily on semi-empirical estimation methods, which are predominantly based on the *extended corresponding states theorem.* One of the most popular and reliable methods is due to Tsonopoulos which, since its inception in 1974, has been revised and modified several times.[107–110] Specifically, the reduced pure-substance second virial coefficient $B_{ii}(T_{r,i})$ at a reduced temperature $T_{r,i} \equiv T/T_{c,i}$ of a substance i with acentric factor ω_i and a *reduced* dipole moment:[28,34,111,112]

$$p_{r,i} \equiv \left(\frac{Lp_i^2}{4\pi\varepsilon_0 V_{c,i} k_B T_{c,i}}\right)^{1/2} \tag{9.34}$$

is given by[101]

$$\frac{P_{c,i}B_{ii}(T_{r,i})}{RT_{c,i}} = \sum_{l=0}^{3} b_{l,i}B^{(l)}(T_{r,i}). \tag{9.35}$$

Here, $T_{c,i}$ is the critical temperature of substance i, $P_{c,i}$ is its critical pressure, and $V_{c,i}$ its critical molar volume. The numerical value of the permanent molecular dipole moment of i is denoted by p_i, ε_0 is the permittivity of vacuum, L is the Avogadro constant, k_B is the Boltzmann constant, and $R = Lk_B$ is the gas constant. As indicated, the quantities $B^{(l)}(T_{r,i})$ are *universal* functions of the reduced temperature: $B^{(0)}$ with $b_{0,i} = 1$ describes second virial coefficients of simple fluids, $B^{(1)}$ with $b_{1,i} = \omega_i$ corrects for *PVT* behaviour of normal fluids, $B^{(2)}$ with $b_{2,i} = b_{2,i}(p_{r,i})$ must be included for substances with sizeable reduced dipole moment $p_{r,i}$, and $b_{3,i}B^{(3)}$ is to be added for hydrogen-bonded substances.

Estimation of the second interaction virial coefficients (second cross virial coefficients) B_{ij} follows routes based on the extended corresponding states principle and is essentially similar to the methods used for pure fluids. The basis is, for instance, the assumption that Equation (9.35) with the same universal functions $B^{(l)}(T_{r,i})$ may also be used for the calculation of second cross virial coefficients[108,113,114] B_{ij} if the pure substance quantities are replaced by characteristic interaction parameters: for instance, $T_{c,i} \Rightarrow T_{c,ij}$, $P_{c,i} \Rightarrow P_{c,ij}$, $V_{c,i} \Rightarrow V_{c,ij}$, $\omega_i \Rightarrow \omega_{ij}$ and so forth. These interaction parameters are obtained through use of appropriate *combining rules*,[9,101,108,113,114] thereby linking them to pure-substance properties (see also Chapter 1).

In the key relations for the classical, sequential approach to gas-solubility data reduction, the influence of composition upon the liquid-phase fugacity of the solute has been separated formally from the influence of pressure. As already outlined above, detailed knowledge of the composition dependence, as well as the pressure dependence of the partial molar volume at each temperature of interest, is rarely available, hence for the great majority of gas-solubility measurements approximations at various levels of sophistication must be introduced to allow adequate VLE data reduction. The situation becomes particularly demanding at elevated pressures and/or when the *solvent critical region* is approached, where Poynting corrections become significant. At temperatures well below the critical temperature $T_{c,1}$ and at pressures reasonably close to $P_{\sigma,1}$ (and thus at *very* small mole fractions x_2), to an excellent approximation Equation (9.21) applies and $V_2^{L\infty}(T,P_{\sigma,1})$ must be known in order to obtain $\ln h_{2,1}(T,P_{\sigma,1})$ from the measured data.

The preferred experimental methods for determining $V_2^{L\infty}(T,P_{\sigma,1})$ are either high-precision dilatometry or high-precision density determinations of the equilibrated solutions. Both methods will be discussed in the next section, following a brief literature survey based on the more extensive one in references 36 and 37.

9.3 Experimental Determination of the Partial Molar Volumes of Gases Dissolved in Liquids at Infinite Dilution

Summaries of the experimental methods used to determine partial molar volumes have been presented quite early by Horiuti in 1931,[115] by Battino and Clever in 1966,[16] by Clever and Battino in 1975,[116] and most recently by Handa and Benson in 1982.[36,37] The partial molar volume of a gas dissolved in a liquid can be determined either directly from the change in volume when the gas dissolves, that is *dilatometrically*, or from the change in density, that is *via* precision *density measurements*. The calculation of the partial molar volume from density measurements requires knowledge of the density of the pure solvent and of the solution of *known* composition, that is, this measuring technique is limited to systems where the solubility of the gas is known. Mole fraction gas solubilities x_2 are usually taken from the literature:[18,19] for most organic liquids at, say, 100 kPa gas pressure, they are in the range 10^{-2} to 10^{-4}, and for aqueous systems they are still smaller. Thus the density difference between pure solvent and a dilute solution is generally quite small and the precision of the density measurements must be correspondingly high for obtaining reliable partial molar volumes. To an excellent approximation, at high dilution the *apparent* molar volume (see Chapter 1) is derived from the measured mass density $\rho_1^{\mathrm{L}*}$ of the pure degassed solvent and the mass density ρ^{L} of a solution with mole fraction solubility x_2 at the actual experimental conditions according to:

$$V_2^{\mathrm{L,app}}(T,P,x_2)=\frac{1}{\rho^{\mathrm{L}}\rho_1^{\mathrm{L}*}}\left[m_{\mathrm{m},1}\left(\rho^{\mathrm{L}}-\rho_1^{\mathrm{L}*}\right)(1-1/x_2)+m_{\mathrm{m},2}\rho_1^{\mathrm{L}*}\right]\cong V_2^{\mathrm{L}\infty}(T,P), \tag{9.36}$$

where $m_{\mathrm{m},1}$ and $m_{\mathrm{m},2}$ are the molar masses of the solvent and the solute, respectively. For very small mole fractions $V_2^{\mathrm{L,app}}$ becomes essentially the *partial* molar volume $V_2^{\mathrm{L}\infty}$ at *infinite dilution* of component 2. If the densities are measured with a precision of 2×10^{-6} and the mole fraction solubility is known to 1%, then the partial molar volume at infinite dilution can be determined with a precision of about 1% to 5%, depending on the absolute value of the solubility.

It is, however, easier to determine partial molar volumes directly using a dilatometric technique; the precision is frequently better than that of densimetric methods, and the measurements can be made without any prior knowledge of the actual gas solubility. Thus, it is not surprising that before Moore *et al.*[90] introduced *vibrating-tube densimetry* for the measurement of partial molar volumes of gases dissolved in liquids, most data on $V_2^{\mathrm{L}\infty}$ have been obtained by dilatometric techniques. Vibrating-tube densimetry is less precise than the best dilatometric methods, and it is also less precise than the magnetic float technique for measuring densities developed by Bignell;[117–121] but is faster and, most importantly, flow vibrating-tube densimeters have been successfully developed for measuring apparent molar volumes of gases at high dilution in water at elevated pressures (20 MPa to 35 MPa) from about 300 K to 716 K, that is into supercritical conditions.[87,88,122] For water, the most important and interesting solvent, Poling *et al.*[101] recommend $T_\mathrm{c}=647.14$ K, $P_\mathrm{c}=22.064$ MPa, and $V_\mathrm{c}=55.95\ \mathrm{cm}^3\ \mathrm{mol}^{-1}$.

A few techniques for measuring *gas solubilities* in liquids have been reported which allow the *simultaneous* determination of partial molar volumes, though their precision is generally less than that of *direct* experimental methods for obtaining $V_2^{\mathrm{L}\infty}$ (the two most important methodological groups will be discussed below in Sections 9.3.1 and 9.3.2, respectively). For instance, Kennan and Pollack[123] used a modified Van Slyke method which allows the simultaneous determination of the Ostwald coefficient, the mole fraction solubility and the partial molar volume of the solute gas in the solvent. Specifically, measurements are reported for nitrogen, argon, krypton and xenon in water at $T=298.15$ K and pressures up to 11.6 MPa, and the precision of the partial molar volumes is estimated to be 2% to 6%. *Ionic liquids* (IL) exhibit a huge diversity in chemical and physical properties as well as in phase behaviour with other substances, and this tunability makes them highly attractive as solvents for chemical reaction systems involving gases, gas storage applications, gas separation applications, *etc.*, hence there is a growing interest in the experimental determination of the solubility of gases in ILs.[124–126] At elevated pressures and increasing solubility, the thermodynamically correct separation of the influence of the Poynting terms (containing partial molar volumes) from the influence of intermolecular interaction between solute molecules (quantified by activity coefficients) can only be achieved through use of *independently* obtained data on partial molar volumes.[24,31,34] The gas solubility apparatus designed by Maurer and collaborators[125,127,128] also allows the determination of partial molar volumes of gases dissolved in ILs, and recently this group has reevaluated already published experimental solubilities of Xe, H_2, O_2, CO, CO_2, CH_4, and CF_4 in three imidazolium-based ILs ([bmim][PF_6], [bmim][CH_3SO_4] and [hmim][Tf_2N]), thereby obtaining reliable information on the partial molar volumes of those gases dissolved in the three ILs at temperatures from about 293 K to 413 K.[129]

The approximation represented by Equation (9.21), the Krichevsky–Kasarnovsky equation, is extremely useful in the reduction of precise gas-solubility data at low to moderate pressures. However, it was also frequently used for the extraction of $V_2^{L\infty}$ from gas-solubility data obtained in measurements at *elevated* pressures. Under these conditions, the solubility x_2 may be large enough for liquid-phase non-idealities to become noticeable and the assumption $V_2^{L} \cong V_2^{L\infty}$ and essentially incompressible partial molar volumes, may become too severe. Thus, partial molar volumes at infinite dilution obtained through use of the Krichevsky–Kasarnovsky equation in gas-solubility studies at elevated pressures may be unreliable.[24,31,34,130–132] Concerning the modelling of the composition dependence of $\gamma_2^{HL}(T,P_{\sigma,1},x_2)$, the correlating equation has to be compatible with the number and the precision of the experimental data points. In the simplest case one may assume the validity of the two-suffix Margules equations: *i.e.*

$$\ln \gamma_1^{LR} = Ax_2, \tag{9.37a}$$

$$\ln \gamma_2^{HL} = A(x_1^2 - 1), \tag{9.37b}$$

hence the Krichevsky–Ilinskaya equation is obtained:[133,134]

$$\ln\left(\frac{\phi_2^{V}(T,P,y_2)y_2P}{x_2h_{2,1}(T,P_{\sigma,1})}\right) - \frac{(P - P_{\sigma,1})V_2^{L\infty}(T,P_{\sigma,1})}{RT} = A(x_1^2 - 1). \tag{9.38}$$

Here, the partial molar volume of the solute is still approximated by a pressure-independent and composition-independent $V_2^{L\infty}$. Because of the additional parameter $A = A(T,P_{\sigma,1})$, it has a wider application range than

Equation (9.21), and has been especially useful for solutions of light gases (He, H_2) in solvents where the solubility is appreciable.[135]

9.3.1 Dilatometric Methods

As reported by Horiuti,[115] attempts to determine partial molar volumes using a dilatometric method were already made as early as in 1878! However, the first detailed description of a dilatometer suitable for measuring $V_2^{L\infty}$ of gases dissolved in liquids appears to be that of Horiuti himself. Subsequently, modified versions of his dilatometer have been used by many researchers, for instance by Kritchevsky and Ilinskaya,[133] Schumm and Brown,[136] Jolley and Hildebrand,[137] Walkley and Hildebrand,[138] Walkley and Jenkins,[139] Ng and Walkley,[140] and Tiepel and Gubbins.[141] Handa *et al.*[37] have used the principle of Horiuti's apparatus and, drawing on their expertise in determining volume changes on mixing two liquids, constructed a dilatometer which virtually eliminates any need for the compressibility correction, maintains exactly the same pressure in the two capillaries, and allows repeated measurements at the same pressure. Only a very brief description of the instrument will be presented below (see Figure 9.1); for details the original article should be consulted.

The apparatus was mounted in a water bath controlled to ± 0.0002 K with no temperature drift over several days. The large mixing vessel (M) of about 150 cm^3 was fitted with two magnetic stirrers (K). The two gas burettes S ranged in volume from 3 cm^3 to 20 cm^3 and were separated by calibrated precision capillary bore segments where the mercury levels could be read with a cathetometer. Temperature was measured with a quartz thermometer, and pressure measurements were made using a Texas Instruments precision pressure gauge. All volumes were calibrated using pure mercury. Volume changes were measured with a 5 cm-stroke micrometer (A) readable to 10^{-4} cm. After loading with pure degassed solvent,[142] *PVT* measurements were made of the gas in the gas burettes applying corrections for non-ideality. After each injection of gas into the water and equilibrium was reached, mercury manometer levels were adjusted to their initial levels using the micrometer A to determine the volume change. Approximately 2×10^{-4} mol of gas were loaded into the cell during each charge, and after ten loadings the mole fraction of the gas in the solution was *ca.* 0.002. The total pressure was adjusted and kept at 1 atm (atm = 101.325 kPa). The authors estimated that the imprecision of the partial molar volumes at infinite dilution was 0.4%, regardless of the solubility of the gas. Results for $V_2^{L\infty}$ were presented for methane dissolved in tetrachloromethane, and for a number of gases in several *n*-alkanes ranging from hexane to hexadecane, and several alkan-1-ols ranging from ethanol to decanol: 40 liquid–gas systems altogether. All measurements were made at $T = 298.15$ K. $V_2^{L\infty}$ for some solutions could be compared with previously reported experimental results, and in most cases agreement was good.

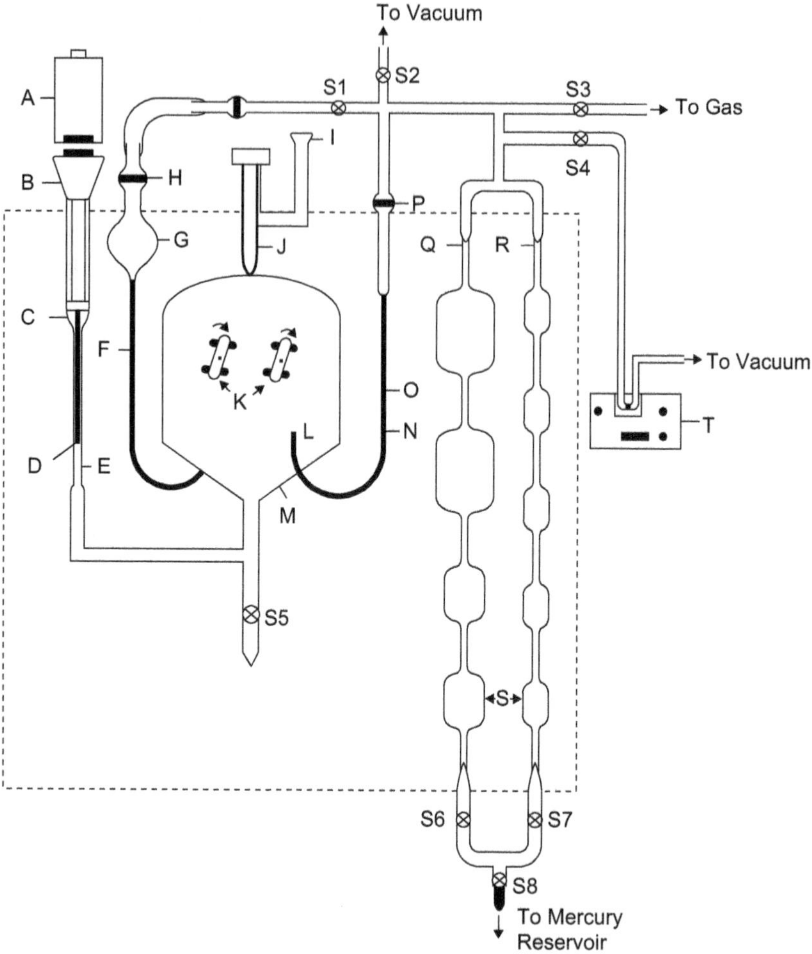

Figure 9.1 Schematic diagram of the dilatometer and burette for determining infinite-dilution partial molar volumes of gases dissolved in liquids as designed by Handa *et al.*:[37] (A) micrometer; (B) ground-glass socket, 14/20 standard taper; (C) stainless steel extension of the micrometer shaft (6 cm) connected to a Teflon tip (D) which fits snugly into (E) Trubore tube, 3 mm id; (F) Trubore capillary, 1 mm id; (G) buffer bulb for mercury; (H and P) O-ring joints for connection to vacuum line; (I) O-ring joint for connection to liquid-degassing apparatus; (J) right-angled stopcock (Fischer and Porter) with 4 mm-bore Teflon plug; (K) twin magnetic stirrers; (L) tip of gas delivery tube; (M) mixing cell with volume of *ca.* 150 cm^3; (N, Q and R) reference marks; (O) Trubore capillary, 1.5 mm id; (S) gas burette; (S1, S2, ..., S8) valves; (T) quartz-spiral pressure gauge (Texas Instruments). For details concerning the operation of the instrument see reference 37.
(Reprinted with permission from: Y. P. Handa, P. J. D'Arcy and G. C. Benson, *Fluid Phase Equil.*, 1982, **8**, 181–196; copyright © 1982, Elsevier B. V.)

The most extensive dilatometric determinations of partial molar volumes of gases dissolved in hydrocarbon solvents up to about 15 MPa and over large temperature ranges, that is from ambient temperatures up to 533 K, are due to Connolly and Kandalic.[143–147] Though they focused on hydrogen dissolved in alkanes, cyclohexane and aromatics, they also presented partial molar volumes for the systems *n*-octane + carbon monoxide and benzene + carbon monoxide. At gas mole fractions smaller than 0.1 or 0.15, the dilatometrically measured volume expansion:

$$\Delta v^{\mathrm{L}} \equiv n_1^{\mathrm{L}} V_1^{\mathrm{L}} + n_2^{\mathrm{L}} V_2^{\mathrm{L}} - n_1^{\mathrm{L}} V_1^{\mathrm{L}*} \tag{9.39}$$

of an initially gas-free liquid divided by the amount of substance n_2 of gas dissolved varies essentially linearly with composition x_2 at constant temperature and pressure:

$$\Delta v^{\mathrm{L}} / n_2^{\mathrm{L}} = V_2^{\mathrm{L}\infty} + c x_2, \tag{9.40}$$

hence

$$\lim_{n_2^{\mathrm{L}} \to 0} \left(\Delta v^{\mathrm{L}} / n_2^{\mathrm{L}}\right)_{T,P,n_1^{\mathrm{L}}} = V_2^{\mathrm{L}\infty}. \tag{9.41}$$

$V_2^{\mathrm{L}\infty}$ and c are functions of T and P only. Equation (9.40) is a consequence of the fact that at each temperature and pressure the molar volume of the *dilute* solution ($x_2 < 0.15$) can be represented by:

$$V^{\mathrm{L}}(T, P, x_2) = V_1^{\mathrm{L}*}(T, P) + b(T, P) x_2 + c(T, P) x_2^2 + \cdots, \tag{9.42}$$

where cubic and higher terms have been neglected. Using the definition of partial molar volume, *i.e.* Equation (9.1), one obtains:

$$V_1^{\mathrm{L}} = V_1^{\mathrm{L}*} - c x_2^2, \tag{9.43}$$

and

$$V_2^{\mathrm{L}} = V_2^{\mathrm{L}\infty} + c\left(1 - x_1^2\right) = V_2^{\mathrm{L}\infty} + c x_2 (2 - x_2). \tag{9.44}$$

By using Equations (9.43) and (9.44) in conjunction with Equation (9.39), Equation 9.40 is obtained. Thus, values of the parameter c of Equation (9.40) are *approximately* related to the composition variation of the partial molar volumes V_1^{L} and V_2^{L}. Surprisingly, at the lower pressures and higher temperatures, the partial molar volumes of hydrogen at infinite dilution in hydrocarbons often exceed the molar volumes of the hydrocarbons. Since $c > 0$, V_1^{L} decreases as hydrogen is added and V_2^{L} increases. Correlating equations for $V_1^{\mathrm{L}*}$, $V_2^{\mathrm{L}\infty}$ and c over the indicated large temperature and pressure ranges have been presented by Connolly and Kandalic,[143–148] and the huge amount of experimental values for the solution dilation $\Delta v^{\mathrm{L}}/n_2^{\mathrm{L}}$ (several thousand of these values) are available as supplementary material, as indicated in the original articles.

9.3.2 Densimetric Methods

Most densimetric techniques are used at low gas pressures (typically 100 kPa) and hence low gas solubilities. As already pointed out, under conditions where Equation (9.36) applies, and in conjunction with solubility data (x_2), to an excellent approximation the apparent molar volume $V_2^{L,app}(T,P,x_2)$ so obtained may essentially be regarded as the infinite dilution property $V_2^{L\infty}(T,P)$. Larger temperature and pressure ranges were covered by Connolly and Kandalic who also measured *densities* of liquid solutions of hydrogen in liquid benzene and *n*-octane[144] at temperatures from 403.15 K to 533.15 K at pressures up to 15 MPa and up to a hydrogen content of about $x_2 = 0.15$. At each T and P the corresponding molar volumes of the solutions were fitted by equations of type Equation (9.42), which in turn yields the composition dependence of the partial molar volumes of the solute V_2^L as well as the solvent V_1^L *via* Equations (9.43) and (9.44), respectively. The resulting values of $1/V_2^{L\infty}$ were found to be almost linear with pressure at constant temperature. Agreement with the dilatometric results was satisfactory throughout. Another early determination of densities of gases in liquids at elevated pressures (P <10 MPa) to derive apparent molar volumes of nitrogen and argon in several organic solvents is due to Masterton *et al.*[149] The work of Wood *et al.*[87,88,122] will be separately discussed below.

A micropycnometer method was described by Lauder[150] which is capable of measuring density differences of the order of a few ppm (10^{-6}), and he used this technique to determine the partial molar volumes of a few atmospheric gases in water.[151] Other excellent designs for measuring *small* density differences between two liquids have been reported by Masterton,[152] and by Blair and Quinn.[153] As indicated above, Bignell[117–121] developed a magnetic float technique for measuring densities with high precision, *i.e.*, to 0.15×10^{-6}. At this level, adequate temperature control is critical and the thermostat maintained it to ± 1 mK. Partial molar volumes of nitrogen, oxygen and argon dissolved in water were determined from 0 °C to 21 °C with reported inaccuracies of ± 0.18 cm^3 mol^{-1} for N_2, of ± 0.10 cm^3 mol^{-1} for O_2, and of ± 0.08 cm^3 mol^{-1} for Ar; this is quite exceptional.

As already pointed out, measuring the density of aqueous solutions of gases *via* vibrating-tube densimetry under *flow* conditions is perhaps not the most precise technique, but definitely the most convenient. Moore *et al.*[90] used this technique to determine the partial molar volumes at infinite dilution of 20 "gases" (about half of them were supercritical) in water at 298.15 K and a total pressure of about 100 kPa. The instrument used was from Sodev, and is shown in Figure 9.2. The principle of operation is that the density ρ of a fluid contained in a U-shaped hollow oscillator is related to the natural vibration frequency of the tube, in the density range of interest, by

$$\rho = a + b\tau^2. \tag{9.45}$$

Here, τ is the period of vibration, and a and b are temperature-dependent constants characteristic of a particular oscillator. The densimeter and the auxilliary apparatus (see Figure 9.2) were housed in an air bath controlled to

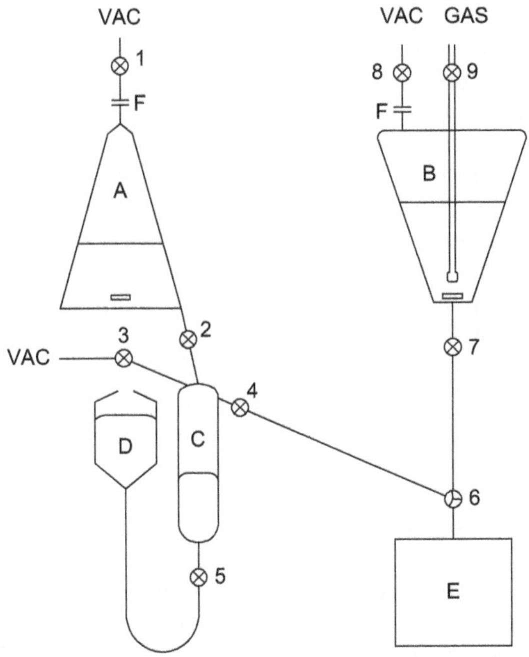

Figure 9.2 Schematic diagram of the instrumental set-up for the determination of apparent molar volumes at high dilution/partial molar volumes at infinite dilution of gases dissolved in liquids based on vibrating-tube densitometry under flow conditions as designed by Moore *et al.*:[90] (A) degassing vessel for pure water; (B) degassing vessel for water to be saturated with a gas; (C) pressurising vessel; (D) mercury reservoir; (E) vibrating-tube densimeter; (F) O-ring joints. Stopcocks are indicated by ⊗ and numbered.
(Reprinted with permission from: J. C. Moore, R. Battino, T. R. Rettich, Y. P. Handa and E. Wilhelm, *J. Chem. Eng. Data*, 1982, **27**, 22–24; copyright © 1982, American Chemical Society.)

± 0.1 K. Upon entering the densimeter the fluid in the U-tube was thermostated to ± 0.0005 K by water circulated from a temperature controlled water bath. The period of the filled tube was measured with a high-resolution digital frequency meter. Ten successive readings were averaged for each experiment. The densimeter was calibrated with pure degassed water and nitrogen at atmospheric pressure. Once the constant b is determined, one may obtain density differences $\Delta\rho^{\mathrm{L}} = \rho^{\mathrm{L}} - \rho_1^{\mathrm{L}*}$ of any solution with density ρ^{L} relative to that of pure water with density $\rho_1^{\mathrm{L}*}$ from:

$$\Delta\rho^{\mathrm{L}} = b(\tau^2 - \tau_1^2). \tag{9.46}$$

The constancy of b was frequently checked. The estimated maximum imprecision of the results is ± 1.5 cm^3 mol^{-1}, or about $\pm$(1.5% to 4%), and they compare well with literature values determined conventionally. Figure 9.2 shows the degassing vessel[142] for pure water and the vessel for saturating the water with a gas, along with the flow patterns. From the measured densities $V_2^{L\infty}$ is obtained *via* Equation (9.36). Using the same instrument, Zhou and Battino[154] determined partial molar volumes at infinite dilution of 13 gases in water at temperatures 298.15 K and 303.15 K (at atmospheric pressure).

Vibrating-tube densimetry has also been used by Cibulka and co-workers[155,156] to determine $V_2^{L\infty}$ of nitrogen, oxygen, argon, carbon dioxide and air dissolved in several liquid *n*-alkanes and alkan-1-ols at 298.15 K and 313.15 K and ambient pressure. During an earlier investigation of the shrinkage observed when crude oils are spiked with light alkanes, Ashcroft *et al.*[157] noted the paucity of experimental data on the effect of dissolved gases on the densities of pure liquid hydrocarbons. Thus, measurements with an Anton Paar DMA 602 vibrating-tube densimeter of the density changes caused by dissolved air, nitrogen, oxygen, hydrogen, methane and carbon dioxide in *n*-alkanes (heptane, octane, nonane, decane, dodecane, tetradecane and hexadecane), cyclohexane, methylcyclohexane and toluene were undertaken[158] at 298.15 K and a total pressure of 101 kPa to determine $V_2^{L\infty}$. There is some overlap between this work and the data set of the research group of Cibulka and Heintz,[155,156] with reasonably good agreement. Ashcroft *et al.* found that the partial molar volumes of all the gases decrease with increasing carbon number of the alkane series, and that at the same experimental conditions nitrogen and methane have almost the same partial molar volume in the *n*-alkanes.

As already indicated, high precision and rapid measurement time make vibrating-tube flow densimetry ideally suited for investigations of volumetric properties of liquids/solutions at temperatures up to about 720 K and pressures up to about 40 MPa.[87,88,122] In fact, Wood *et al.* used it in their pioneering measurements of the apparent molar volumes of argon, xenon, ethylene, methane and carbon dioxide in water at pressures a few MPa above the critical pressure of water up to temperatures above the critical temperature of water. The apparent molar volumes of these solutes exhibit quite a spectacular temperature dependence which is correlated with the temperature dependence of the product $\rho_1^* \beta_{T,1}^*$ of pure water, where $\beta_{T,1}^*$ denotes the isothermal compressibility. For instance, $V_2^{L,\mathrm{app}}$ of argon, xenon and ethene (ethylene) in water at about 26.7 MPa increases slowly up to 600 K. Above 600 K the temperature dependence increases dramatically (so does the pressure dependence) and the apparent molar volumes go through maxima. For the three solutes, the maxima at approximately 26.7 MPa are observed at (669 ± 2) K, they are *greater* than 2200 $\mathrm{cm^3\ mol^{-1}}$ and sharper than the maxima at approximately 34 MPa, which occur at (690 ± 2) K and are about 1000 $\mathrm{cm^3\ mol^{-1}}$.

9.4 Estimation Methods

By now it should be evident that the partial molar volume of a gas in solution is a quantity of importance in the reduction and correlation of precise gas-solubility measurements. Poynting-type corrections are rather ubiquitous in this field: whenever VLE data are discussed at "constant T and P", it is always tacitly assumed that the pressure corrections have been performed properly.[159] Fortunately, for the reasonably small pressure ranges $P - P_{\sigma,1}$ generally used in gas-solubility studies, and at temperatures well below the critical temperature $T_{c,1}$, uncertainties associated with commonly used approximations (see above) are frequently small, perhaps even negligible when compared to other uncertainties. However, compared to the large body of data on gas *solubilities*, experimental results on $V_2^{L\infty}$ (and on V_2^L in general) of gases dissolved in liquids are not plentiful. Hence the continuing interest in the experimental determination of partial molar volumes on the one hand, and the development of reliable semi-empirical estimation methods on the other. Clearly, more experimentally determined partial molar volumes, covering wide temperature ranges including the critical region, and wide pressure ranges are highly desirable.

For $V_2^{L\infty}$ for a *single* gas in *different* non-polar liquids, Schumm and Brown[136] suggested a correlation with the solubility parameter δ_1 of the solvent[9,160,161], *i.e.*

$$V_2^{L\infty}/V_1^{L*} = f(\delta_1). \tag{9.46}$$

This correlation was tested by Handa and Benson[36] using more recent data: in many cases they found a *linear* dependence on δ_1.

Based on earlier work of Smith and Walkley,[162] Lyckman *et al.*[163] suggested the functional form

$$V_2^{L\infty} P_{c,2}/RT_{c,2} = f\left(TP_{c,2}/T_{c,2}\delta_1^2\right), \tag{9.47}$$

where $T_{c,2}$ and $P_{c,2}$ are the critical temperature and the critical pressure, respectively, of the solute. For systems well below the critical temperature $T_{c,1}$ of the solvent, an approximately linear correlation is obtained yielding reasonable estimates of $V_2^{L\infty}$. However, for expanded solvents, *i.e.* for solvents at temperatures close to $T_{c,1}$, Lyckman *et al.*[163] found that their correlation predicts $V_2^{L\infty}$ values which are much smaller than the experimental values.

For gases dissolved in water Moore *et al.*[90] represented their results to within *ca.* ±10% by

$$V_2^{L\infty} = 10.74\ \text{cm}^3/\text{mol} + 0.2683 V_{c,2}. \tag{9.48}$$

The Handa–Benson correlation[36,37] allows the prediction of partial molar volumes of gases at infinite dilution in non-aqueous solvents (non-polar, polar, hydrogen-bonded) at atmospheric pressure to within about 10%, as long as the system is well below the solvent's critical temperature:

$$\frac{P_{c,2} V_2^{L\infty}}{RT_{c,2}} = 0.088 + 2.763 \frac{TP_{c,2}}{T_{c,2}\Pi_1^{L*}}. \tag{9.49}$$

Here,

$$\Pi_1^{L*} = \frac{T\alpha_{P,1}^{L*}}{\beta_{T,1}^{L*}} - P_{\sigma,1} \tag{9.50}$$

is the internal pressure, $\alpha_{P,1}^{L*}$ is the isobaric expansivity, and $\beta_{T,1}^{L*}$ is the isothermal compressibility of the pure liquid solvent at saturation. On the basis of their measurements of partial molar volumes of air-component gases in *n*-alkanes and alkan-1-ols, Cibulka and coworkers[155,156] arrive at an equation which is very close to Equation (9.49).

Brelvi and O'Connell[164] correlated $V_2^{L\infty}$ with the dimensionless infinite dilution solvent–solute direct correlation function integral $C_{12}^{L\infty}$:

$$V_2^{L\infty} \Big/ RT\beta_{T,1}^{L*} = 1 - C_{12}^{L\infty}. \tag{9.51}$$

In turn, $C_{12}^{L\infty}$ may be obtained from

$$\ln\left[-C_{12}^{L\infty}(\hat{v}_1/\hat{v}_2)^{0.62}\right] = -2.4467 + 2.12074\tilde{\rho}_1, \quad \text{for } 2.0 \le \tilde{\rho}_1 \le 2.785 \tag{9.52}$$

$$\ln\left[-C_{12}^{L\infty}(\hat{v}_1/\hat{v}_2)^{0.62}\right] = 3.02214 - 1.87085\tilde{\rho}_1 + 0.71955\tilde{\rho}_1^2, \quad \text{for } 2.785 \le \tilde{\rho}_1 \le 3.2. \tag{9.53}$$

Here, $\tilde{\rho}_1 = \hat{v}_1/V_1^{L*}$ is the reduced amount-of-substance density of the solvent, and the $\hat{v}_i$ ($i = 1, 2$) are characteristic molar reducing volumes. However, it was pointed out that this correlation should not be used to describe the partial molar volumes of liquids dissolved in liquids; that is to say, it should not be used at temperatures substantially below the critical temperature $T_{c,2}$ of the solute. For more recent work along these lines, correlating and estimating partial molar volumes of nonelectrolytes at infinite dilution in water over large temperature and pressure ranges (up to 700 K and 50 MPa), see O'Connell *et al.*[165] and Plyasunov *et al.*[166,167] In particular, the corresponding-states correlation of Plyasunov *et al.*,[167] developed for obtaining partial molar volumes at infinite dilution of aqueous solutes at elevated temperatures and pressures, will serve as a reliable empirical method to *predict* V_2^{∞} of neutral solutes. It is based on the use of the *reduced* direct correlation function integral defined as

$$C_{12,r}^{\infty} \equiv \frac{C_{12}^{\infty}(T, \rho_1^*)}{C_{12}^{\infty}\left(T_{ref}, \rho_{1,ref}^{L*}\right)}, \tag{9.54}$$

where $T_{ref} = 298.15$ K and $\rho_{1,ref}^{L*} = 997$ kg m^{-3} (density of pure water at 298.15 K and 0.1 MPa). The corresponding-states correlation covers a density range of $500 \le \rho_1^*/\text{kg m}^{-3} \le 1010$ and a temperature range of $273 \le T/\text{K} \le 670$, with separate parameters a_0, a_1, a_2, a_3 for *different* classes of solutes (simple fluids, non-polar substances, alcohols, amines, amides, nitriles, *etc.*):

$$C_{12,r}^{\infty} = \rho_1^*\left\{a_0 + \left[a_1 + a_2(T/\text{K} - 228)^{-2}\right]\exp(a_3\rho_1^*)\right\}. \tag{9.55}$$

This relation is similar in form to that used in reference 165.

Scaled particle theory (SPT) has been used by Pierotti[168–170] and Wilhelm *et al.*,[19] among others, to calculate $V_2^{L\infty}$ for non-polar and polar gases in both non-polar and polar solvents and water. If a two-step dissolution process is considered, that is (I) creation of a cavity in the solvent large enough to accommodate a solute molecule, and (II) introduction of a solute molecule into the cavity which interacts with the surrounding solvent molecules, such a heuristic application of SPT yields[98]

$$V_2^{L\infty} = V_{cav} + \beta_{T,1}^{L*} G_{int} + \beta_{T,1}^{L*} RT. \tag{9.56}$$

Here, $V_{cav} = (\partial G_{cav}/\partial P)_T$, G_{cav} is the partial molar Gibbs energy of cavity formation, and G_{int} is the partial molar Gibbs energy of interaction between a solute molecule and the solvent. The SPT expressions for these quantities are well-known and may easily be found in the literature.[19,24,26a,168–174] The isothermal compressibility of water at 101.325 kPa (=1 atm) exhibits a *minimum* value[175] at 46.5 °C. Together with Equation (9.56), this quite unusual temperature dependence suggests that, at least for some solutes in water, a plot $V_2^{L\infty}$ *vs.* T may also show a shallow minimum, behaviour which has indeed been reported for a few solutions.[98,119,121] Additional high-precision work in this area would be desirable. For many liquids, a *self-consistent* set of *effective* Lennard-Jones (6,12) interaction parameters has been given by Wilhelm and Battino.[173] The correlational and predictive powers of SPT-based methods can be substantially improved by introducing the concept of *temperature-dependent size parameters* as suggested by Wilhelm[176] and more recently by Montfort and Perez,[177] Schaffer and Prausnitz[178] and Schulze and Prausnitz.[179] The popularity of SPT is partly due to the fact that its predictions can be easily connected with experiment, and partly because it is based on an intuitively appealing interpolative view of the connection between known microscopic, molecular and macroscopic limits.[40,180,181]

Recently, Klähn *et al.*[182] determined *computationally* partial molar volumes at infinite dilution of H_2, CO and CO_2 dissolved in liquid heptane ($\beta_{T,1}^{L*} = 14.6 \times 10^{-10}$ Pa^{-1}), diethylether ($\beta_{T,1}^{L*} = 19.3 \times 10^{-10}$ Pa^{-1}), acetone (2-propanone; $\beta_{T,1}^{L*} = 12.9 \times 10^{-10}$ Pa^{-1}) and methanol ($\beta_{T,1}^{L*} = 12.5 \times 10^{-10}$ Pa^{-1}), where all data refer to 298.15 K and ambient pressure. Using molecular dynamics simulations[183–185] in combination with solvent OPLS-AA (Optimized Potentials for Liquid Simulations-All Atom) force field parameters[186,187] and customised force field parameters for the solutes, $V_2^{L\infty}$ is computed by the *direct* method, that is, the volume difference between the pure liquid solvent and a solution containing a single solute molecule is calculated.[188–190] In addition, the simulation results may also be used to derive the solute (2)–solvent (1) pair distribution function $g_{21}(r)$,[189,191] where r denotes the solute–solvent centre-of-mass distance. At infinite dilution, in the grand canonical ensemble, $V_2^{L\infty}$ is given by the Kirkwood–Buff expression[192,193] as

$$V_2^{L\infty} = -4\pi \int_0^{\infty} [g_{21}(r) - 1] r^2 dr + \beta_{T,1}^{L*} RT. \tag{9.57}$$

Equation (9.57) is exact and explicitly shows that $V_2^{L\infty}$ is determined only by the spatial solvent structure around the solute and the isothermal compressibility of the *pure* solvent. Klähn *et al.*[182] now suggest decomposition of the integral into physically reasonable and intuitively appealing contributions that will facilitate the discussion of the observed variations of the partial molar volume in different solutions. Specifically, Equation (9.57) may be written as follows:

$$V_2^{L\infty} = V_{ex} + (V_{cs1} + V_{cs2} + V_{lr}) + \beta_{T,1}^{L*}RT. \tag{9.58}$$

The first term on the right-hand side of Equation (9.58), V_{ex}, represents the contribution of the solvent cage around the solute, that is, the contribution of the integral between $r=0$ and $r=R_0$. For hard spheres, R_0 is the distance of closest approach, $g_{21}(r < R_0)=0$, and $V_{ex}=4\pi R_0^3/3$ is the solvent *excluded* volume. For real solutions, with a r^{-12} repulsive potential-energy function, R_0 becomes an *effective* distance parameter. The second term V_{cs1} represents the contribution due to integration over the region $R_0 \leq r < R_1$, where R_1 denotes the *first* minimum of the pair distribution function; that is, the integration interval corresponds to the *first* solvent *coordination sphere* around the solute. Analogously, V_{cs2} results from integration over the region $R_1 \leq r < R_2$, where R_2 denotes the *second* minimum of the pair distribution function; that is, the integration interval corresponds to the *second* solvent coordination sphere around the solute. While V_{cs1} and V_{cs2} describe solute-induced volume changes of the solvent in the close solute vicinity, V_{lr} represents *long-range* contributions beyond the second solvation shell resulting from integration over the region between $r=R_2$ and $r=\infty$. Due to appropriate averaging, deviations from a spherical solvent cage shape do not influence the solute–solvent pair distribution function and thus do not influence $V_2^{L\infty}$. Of particular interest is the suggestive similarity between Equation (9.58) and the SPT expression Equation (9.56).[98] It indicates that the compressibility influences $V_2^{L\infty}$ not only *directly* through the term $\beta_{T,1}^{L*}RT$, a large compressibility would also facilitate the displacement of solvent molecules around the solute towards the first and second solvation shells, thereby contributing to an increase of $V_2^{L\infty}$. As pointed out by Klähn *et al.*,[182] this would help explain the observed large partial molar volumes of H_2 and CO in diethylether: the isothermal compressibility of diethylether is significantly larger than the compressibility of the other three solvents. A similar behaviour is evidenced by Table 9.1: the isothermal compressibility of liquid benzene is more than twice as large as that of water.

Calculated[182] $V_2^{L\infty}$ deviate only by about 7% from experimental results, and the trends observed across the 12 investigated solutions are reproduced quantitatively in most cases. The *combined* approach of Klähn *et al.*[182] (that is, direct molecular dynamics simulations and decomposition of Kirkwood–Buff integrals obtained *via* pair distribution functions) indeed provides valuable insights into the physical–chemical factors that influence $V_2^{L\infty}$.

9.5 Concluding Remarks, Outlook and Acknowledgements

By common consent, the liquid state of matter still houses by far the largest number of unsolved/crudely solved problems in modern physical chemistry. When studying liquid mixtures/solutions, the appearance of *new* phenomena not present in the pure components constitutes a most stimulating intellectual challenge and is also of enormous practical importance in neighbouring scientific disciplines, including chemical engineering. These aspects represent the principal reasons for investing so much experimental and theoretical work in the study of mixture/solution properties. As a prominent example we quote here the recent careful work on the effect of dissolved air on the density (and refractive index) of water by Harvey *et al.*[194] Water, in particular at ambient temperature and pressure, *i.e.* at 298.15 K and 0.1 MPa, serves as a reference standard *par excellence* for thermophysical property measurements. The influence of dissolved air on the density of water is thus of considerable interest in metrology. Quite a number of investigations have dealt with this problem,[118–121,151,195–199] though in part with inconsistent results. Using recently measured partial molar volumes at infinite dilution[88b,90,119,154] of N_2, O_2, Ar and CO_2 in conjunction with high-precision values of Henry fugacities,[97–99,200] Harvey *et al.*[194] calculated the molar volume of the solution of dry air[201] in water as:

$$V^{\mathrm{L}} = x_{\mathrm{w}} V_{\mathrm{w}}^{\mathrm{L}*} + \sum_{i=2}^{n} x_i V_i^{\mathrm{L}\infty}, \tag{9.59}$$

where x_{w} is the mole fraction of water, $V_{\mathrm{w}}^{\mathrm{L}*}$ is the molar volume of pure water obtained from the formulation of the IAPWS-95 (International Association for the Properties of Water and Steam, 1995),[202] and $V_i^{\mathrm{L}\infty}$ is the partial molar volume of solute i at infinite dilution, and the sum is over all solute species. For such highly dilute solutions, far below the critical point of the solvent, Equation (9.59) is an excellent approximation. Aqueous carbon dioxide undergoes a weak ionisation reaction,

$$CO_2(aq) + H_2O \rightleftharpoons H^+ + HCO_3^-, \tag{9.60}$$

which was also taken into account[203,204] (the second ionisation step is negligible for this purpose). The density increment due to dissolved air in water is then computed as

$$\Delta\rho^{\mathrm{L}} \equiv \rho^{\mathrm{L}} - \rho_{\mathrm{w}}^{\mathrm{L}*} = \frac{m_{\mathrm{m}}}{V^{\mathrm{L}}} - \frac{m_{\mathrm{m,w}}}{V_{\mathrm{w}}^{\mathrm{L}*}}, \tag{9.61}$$

where ρ^{L} is the density of air-saturated water at a total pressure of 101.325 kPa, $\rho_{\mathrm{w}}^{\mathrm{L}*}$ is the density of pure, air-free water at 101.325 kPa, and m_{m} and $m_{\mathrm{m,w}}$ denote the molar mass of the solution and the molar mass of pure water, respectively. The calculated values $\Delta\rho^{\mathrm{L}}$ for air saturation of water at a total pressure of 101.325 kPa were expressed as a function of temperature from 0 °C to 50 °C by:

$$\Delta\rho^{L}/(\mu g\ cm^{-3}) = 0.103 - 2.371\times10^{5}\,\tau^{-2.5} + 1.820\times10^{-7}\,\tau^{3}, \qquad (9.62)$$

where $\tau = t/°C + 75$.

Because of the role of accurately known physical properties of water, it is important to also indicate the formulation for the density of liquid water endorsed by the International Committee for Weights and Measures (CIPM) in 2001: Tanaka *et al.*[205] recommend a new standard for use in metrology for the density of deaerated water with the isotopic abundance of Vienna Standard Mean Ocean Water (VSMOW) in the temperature range 0 °C to 40 °C and at 101.325 kPa. In a recent article, Harvey *et al.*[206] discussed the role of IAPWS and CIPM standards and present guidelines for their appropriate use. We note that, within the range of validity of the CIPM formulation, mutual agreement is within the respective quoted uncertainties. Both standards yield *very* similar values for the density, with the largest difference being observed at 40 °C (and 101.325 kPa): here, the IAPWS-95 density is higher by *ca.* 1.14×10^{-6} g cm^{-3}. The CIPM formulation for the density should *not* be extrapolated outside the range 0 °C to 40 °C. The selection of water property formulas for volume and flow calibration, including compressibility and viscosity, has recently been discussed by Batista and Paton.[207]

PVT measurements, calorimetry and vapour–liquid equilibrium determinations (and liquid–liquid equilibrium determinations and solid–liquid equilibrium determinations as well) are the oldest and most fundamental experimental disciplines in physical chemistry. They provide quantitative information on properties to be used for theoretical advances on the one hand, and to improve the practical application of science (chemical engineering) on the other. Although simple in principle, enormous effort and ingenuity have gone into designing the vast array of apparatuses now at our disposal for the determination of *PVT* properties of pure and mixed fluids over large ranges of temperature and pressure. The major driving forces for progress in instrumentation are the desire to increase the area of applicability, to improve precision and accuracy, to increase the speed of measurement and to improve data management. Undoubtedly, automation of instruments will progress, and so will miniaturisation. In summary, chemical thermodynamics of solutions in general, and the experimental study of volumetric properties of dilute (aqueous) solutions in particular, continue to be exciting, developing fields. When combined with advances in the *statistical–mechanical treatment* of mixed systems and increasingly sophisticated *computer simulations*, new insights and important connections at the microscopic, mesoscopic and macroscopic level are continuously communicated. Cross-fertilisation with other neighbouring disciplines, notably with biophysics, is becoming increasingly important.

During the last two decades, the field treated in this chapter has grown enormously, hence a few areas have only been touched upon, such as the one associated with *critical* phenomena. This topic definitively deserves the special coverage which is provided by Chapter 11. Here, we just stress the fact that the difficulties of measuring thermodynamic properties of fluids in the critical region arise from their unusual mechanical, thermal and optical properties in this region.

We trust that the various topics treated in this chapter will provide a feeling for the scope of the field, its position in science, and its future potential. In this connection, it is a pleasure to acknowledge the many years of fruitful scientific collaboration (since 1969!) between us, the *Dayton–Wien connection*, so to speak, and the good fortune to have dedicated coworkers such as T. R. Rettich, Y. P. Handa, T. Tominaga and R. J. Wilcock, to name but a few.

References

1. J. H. Hildebrand and R. L. Scott, *The Solubility of Nonelectrolytes*, Reinhold Publishing Corporation, New York, USA, 3rd edn, 1950.
2. E. A. Guggenheim, *Mixtures*, Oxford at the Clarendon Press, London, UK, 1952.
3. I. Prigogine and R. Defay, *Chemical Thermodynamics*, translated and revised by D. H. Everett, Longmans, Green and Co, London, UK, 1954.
4. R. Haase, *Thermodynamik der Mischphasen*, Springer-Verlag, Berlin, Germany, 1956.
5. K. Denbigh, *The Principles of Chemical Equilibrium*, Cambridge University Press, Cambridge, UK, 4th edn, 1981.
6. J. S. Rowlinson and F. L. Swinton, *Liquids and Liquid Mixtures*, Butterworth Scientific, London, UK, 3rd edn, 1982.
7. H. C. Van Ness and M. M. Abbott, *Classical Thermodynamics of Nonelectrolyte Solutions*, McGraw-Hill, New York, USA, 1982.
8. S. E. Wood and R. Battino, *Thermodynamics of Chemical Systems*, Cambridge University Press, Cambridge, UK, 1990.
9. J. M. Prausnitz, R. N. Lichtenthaler and E. G. de Azevedo, *Molecular Thermodynamics of Fluid-Phase Equilibria*, Prentice Hall PTR, Upper Saddle River, NJ, USA, 3rd edn, 1999.
10. J. H. Hildebrand, J. M. Prausnitz and R. L. Scott, *Regular and Related Solutions. The Solubility of Gases, Liquids, and Solids*, Van Nostrand Reinhold Company, New York, USA, 1970.
11. A. E. Markham and K. A. Kobe, *Chem. Rev.*, 1941, **28**, 519.
12. T. J. Morrison and N. B. Johnstone, *J. Chem. Soc.*, 1954, 3441.
13. D. M. Himmelblau, *J. Phys. Chem.*, 1959, **63**, 1803.
14. D. M. Himmelblau and E. Arends, *Chem. Ing. Tech.*, 1959, **31**, 791.
15. D. M. Himmelblau, *J. Chem. Eng. Data*, 1960, **5**, 10.
16. R. Battino and H. L. Clever, *Chem. Rev.*, 1966, **66**, 395.
17. K. W. Miller and J. H. Hildebrand, *J. Am. Chem. Soc.*, 1968, **90**, 3001.
18. E. Wilhelm and R. Battino, *Chem. Rev.*, 1973, **73**, 1.
19. E. Wilhelm, R. Battino and R. J. Wilcock, *Chem. Rev.*, 1977, **77**, 219.

20. R. Battino, *Rev. Anal. Chem.*, 1989, **9**, 131.
21. R. Fernández-Prini, J. L. Alvarez and A. H. Harvey, *J. Phys. Chem. Ref. Data*, 2003, **32**, 903.
22. (a) H. L. Clever and R. Battino, Solubility of Gases in Liquids, in *The Experimental Determination of Solubilities*, ed. G. T. Hefter and R. P. T. Tomkins, Wiley Series in Solution Chemistry, Vol. 6, John Wiley & Sons/IUPAC, Chichester, UK, 2003, ch. 2.1, pp. 101–150; (b) R. Battino and H. L. Clever, The Solubility of Gases in Water and Seawater, in *Development and Applications in Solubility*, ed. T. M. Letcher, The Royal Society of Chemistry/IUPAC & IACT, Cambridge, UK, 2007, ch. 6, pp. 66–77.
23. J. L. Alvarez and R. Fernández-Prini, *J. Solution Chem*, 2008, **37**, 1379.
24. E. Wilhelm, *CRC Crit. Rev. Anal. Chem.*, 1985, 16 , 129.
25. E. Wilhelm, *Pure Appl. Chem.*, 1985, **57**, 303.
26. (a) E. Wilhelm, *Fluid Phase Equilib.*, 1986, **27**, 233; (b) E. Wilhelm, Dilute Aqueous Solutions of Nonelectrolytes: Recent Advances, in *Interactions of Water in Ionic and Nonionic Hydrates*, ed. H. Kleeberg, Springer-Verlag, Berlin, BRD, 1987, pp. 117–123; (c) J.-P. E. Grolier and E. Wilhelm, *Pure Appl. Chem.*, 1991, **63** , 1427.
27. E. Wilhelm, Thermodynamics of Solutions, Especially Dilute Solutions of Nonelectrolytes, in *Molecular Liquids: New Perspectives in Physics and Chemistry*, ed. J. J. C. Teixeira-Dias, Kluwer Academic Publishers, Amsterdam, The Netherlands, 1992, pp. 175–206.
28. E. Wilhelm, *High Temp. – High Pressures*, 1997, **29**, 613.
29. E. Wilhelm, Thermodynamics of Nonelectrolyte Solubility, in *Development and Applications in Solubility*, ed. T. M. Letcher, The Royal Society of Chemistry/IUPAC & IACT, Cambridge, UK, 2007, ch. 1, pp. 3–18.
30. E. Wilhelm, *J. Therm. Anal. Calorim.*, 2012, **108**, 547.
31. E. Wilhelm, *Netsu Sokutei*, 2012, **39**(2), 61.
32. E. Wilhelm, *J. Solution Chem.*, 2014, **43**, 525.
33. E. Wilhelm, *Thermochim. Acta*, 1997, **300**, 159.
34. E. Wilhelm, Low-Pressure Solubility of Gases in Liquids, in *Experimental Thermodynamics, Vol. VII: Measurement of the Thermodynamic Properties of Multiple Phases*, ed. R. D. Weir and Th. W. de Loos, Elsevier/IUPAC, Amsterdam, The Netherlands, 2005, ch. 7, pp. 137–176.
35. E. Wilhelm and R. Battino, Partial Molar Heat Capacity Changes of Gases Dissolved in Liquids, in *Heat Capacities: Liquids, Solutions and Vapours*, ed. E. Wilhelm and T. M. Letcher, The Royal Society of Chemistry/IACT & IUPAC, Cambridge, UK, 2010, ch. 21, pp. 457–471.
36. Y. P. Handa and G. C. Benson, *Fluid Phase Equilib.*, 1982, **8**, 161.
37. Y. P. Handa, P. J. D'Arcy and G. C. Benson, *Fluid Phase Equilib.*, 1982, **8**, 181.
38. W. Kauzmann, *Adv. Protein Chem.*, 1959, **14**, 1.
39. C. Tanford, *The Hydrophobic Effect: Formation of Micelles and Biological Membranes*, Wiley, New York, 1973.
40. F. H. Stillinger, *J. Solution Chem.*, 1973, **2**, 141.

41. L. R. Pratt and D. Chandler, *J. Chem. Phys.*, 1977, **67**, 3683.
42. A. Ben-Naim, *Hydrophobic Interactions*, Plenum Press, New York, 1980.
43. P. L. Privalov and S. J. Gill, *Adv. Protein Chem.*, 1988, **39**, 191.
44. K. P. Murphy, P. L. Privalov and S. J. Gill, *Science*, 1990, **247**, 559.
45. K. A. Dill, *Biochemistry*, 1990, **29**, 7133.
46. W. Blokzijl and J. B. F. N. Engberts, *Angew. Chem., Int. Ed. Engl.*, 1993, **32**, 1545.
47. G. Hummer, S. Garde, A. E. Garcia, M. E. Paulaitis and L. R. Pratt, *J. Phys. Chem. B*, 1998, **102**, 10469.
48. K. A. T. Silverstein, A. D. J. Haymet and K. A. Dill, *J. Am. Chem. Soc.*, 1998, **120**, 3166.
49. G. Hummer, S. Garde, A. E. Garcia and L. R. Pratt, *Chem. Phys.*, 2000, **258**, 349.
50. L. R. Pratt, *Ann. Rev. Phys. Chem.*, 2002, **53**, 409.
51. N. T. Southall, K. A. Dill and A. D. J. Haymet, *J. Phys Chem. B*, 2002, **106**, 521.
52. B. Widom, P. Bhimalapuram and K. Koga, *Phys. Chem. Chem. Phys.*, 2003, **5**, 3085.
53. D. Chandler, *Nature*, 2005, **437**, 640.
54. D. Ben-Amotz, *J. Chem. Phys.*, 2005, **123**, 184504.
55. D. Ben-Amotz and R. Underwood, *Acc. Chem. Res.*, 2008, **41**, 957.
56. D. Paschek, R. Ludwig and J. Holzmann, Computer Simulation Studies of Heat Capacity Effects Associated with Hydrophobic Effects, in *Heat Capacities: Liquids, Solutions and Vapours*, ed. E. Wilhelm and T. M. Letcher, The Royal Society of Chemistry/IACT & IUPAC, Cambridge, UK, 2010, ch. 20, pp. 436–456.
57. A. J. Patel, P. Varilly and D. Chandler, *J. Phys. Chem. B*, 2010, **114**, 1632.
58. A. Ben-Naim, *Open J. Biophys.*, 2011, **1**, 1.
59. *Water: A Comprehensive Treatise*, Vols. I–VII, ed. F. Franks, Plenum Press, New York, USA, 1972 through 1982.
60. D. Eisenberg and W. Kauzmann, *The Structure and Properties of Water*, Oxford Classics Series, Clarendon Press, Oxford, UK, 2005.
61. F. Franks, *Water: A Matrix of Life: Edition 2*, The Royal Society of Chemistry, Cambridge, UK, 2000.
62. J. F. Kasting, D. P. Whitmire and R. T. Reynolds, *Icarus*, 1993, **101**, 108.
63. *Geochemistry and the Origin of Life*, ed. S. Nakashima, S. Maruyama, A. Brack and B. F. Windley, Universal Academy Press, Tokyo, Japan, 2001.
64. W. Martin and M. J. Russell, *Philos. Trans. R. Soc. B*, 2003, **358**, 59.
65. R. M. Hazen: *Genesis: The Scientific Quest for Life's Origin*, Joseph Henry Press, Washington, DC, USA, 2005.
66. J. I. Lunine, *Philos. Trans. R. Soc. B*, 2006, **361**, 1721.
67. C. S. Cockell, *Philos. Trans. R. Soc. B*, 2006, **361**, 1845.
68. J. Jortner, *Philos. Trans. R. Soc. B*, 2006, **361**, 1877.
69. A. Ricardo and J. W. Szostak, *Sci. Am.*, 2009, **301**(3), 54.
70. I. Budin and J. W. Szostak, *Annu. Rev. Biophys.*, 2010, **39**, 245.

71. K. Michaelian, *Earth Syst. Dynam.*, 2011, **2**, 37.
72. S. K. Pal and A. H. Zewail, *Chem. Rev.*, 2004, **104**, 2099.
73. Y. Levy and J. N. Onuchic, *Annu. Rev. Biophys. Biomol. Struct.*, 2006, **35**, 389.
74. P. Ball, *Chem. Rev.*, 2008, **108**, 74.
75. F. Mallamace, C. Corsaro, D. Mallamace, P. Baglioni, H. E. Stanley and S.-H. Chen, *J. Phys. Chem. B*, 2011, **115**, 14280.
76. P. Ball, *ChemPhysChem*, 2008, **9**, 2677.
77. H. S. Frank and M. W. Evans, *J. Chem. Phys.*, 1945, **13**, 507.
78. R. D. Broadbent and G. W. Neilson, *J. Chem. Phys.*, 1994, **100**, 7543.
79. P. H. K. De Jong, J. E. Wilson, G. W. Neilson and A. D. Buckingham, *Mol. Phys.*, 1997, **91**, 99.
80. Y. L. A. Rezus and H. J. Bakker, *Phys. Rev. Lett.*, 2007, **99**, 148301.
81. F. H. Stillinger, *J. Solution Chem.*, 1973, **2**, 141.
82. J. M. H. Levelt Sengers, Thermodynamics of Solutions Near the Solvent's Critical Point, in *Supercritical Fluid Technology: Reviews in Modern Theory and Applications*, ed. T. J. Bruno and J. F. Ely, CRC Press, Boca Raton, FL, USA, 1991, ch. 1, pp. 1–56.
83. J. C. Wheeler, *Ber. Bunsenges. Phys. Chem.*, 1972, **76**, 308.
84. D. R. Biggerstaff, D. E. White and R. H. Wood, *J. Phys Chem.*, 1985, **89**, 4378.
85. D. R. Biggerstaff and R. H. Wood, *J. Phys Chem.*, 1988, **92**, 1994.
86. L. Hnedkovský and R. H. Wood, *J. Chem. Thermodyn.*, 1997, **29**, 731.
87. D. R. Biggerstaff and R. H. Wood, *J. Phys. Chem.*, 1988, **92**, 1988.
88. (a) R. Crovetto, R. H. Wood and V. Majer, *J. Chem. Thermodyn.*, 1990, **22**, 231; (b) R. Crovetto and R. H. Wood, *Fluid Phase Equilib.*, 1992, **74**, 271; (c) L. Hnedkovský, R. H. Wood and V. Majer, *J. Chem. Thermodyn.*, 1996, **28**, 125.
89. (a) H. Stabinger, O. Kratky and H. Leopold, *Monatsh. Chem.*, 1967, **98**, 436; (b) O. Kratky, H. Leopold and H. Stabinger, *Z. Angew. Phys.*, 1969, **27**, 273.
90. J. C. Moore, R. Battino, T. R. Rettich, Y. P. Handa and E. Wilhelm, *J. Chem. Eng. Data*, 1982, **27**, 22.
91. B. B. Benson and D. Krause Jr., *J. Chem. Phys.*, 1976, **64**, 689.
92. B. B. Benson, D. Krause Jr. and M. A. Peterson, *J. Solution Chem.*, 1979, **8**, 655.
93. B. B. Benson and D. Krause Jr., *J. Solution Chem.*, 1980, **9**, 895.
94. D. Krause Jr. and B. B. Benson, *J. Solution Chem.*, 1989, **18**, 823.
95. T. R. Rettich, Y. P. Handa, R. Battino and E. Wilhelm, *J. Phys. Chem.*, 1981, **85**, 3230.
96. T. R. Rettich, R. Battino and E. Wilhelm, *Ber. Bunsenges. Phys. Chem.*, 1982, **86**, 1128.
97. T. R. Rettich, R. Battino and E. Wilhelm, *J. Solution Chem.*, 1984, **13**, 335.
98. T. R. Rettich, R. Battino and E. Wilhelm, *J. Solution Chem.*, 1992, **21**, 987.

99. T. R. Rettich, R. Battino and E. Wilhelm, *J. Chem. Thermodyn.*, 2000, **32**, 1145.
100. I. R. Krichevsky and J. S. Kasarnovsky, *J. Am. Chem. Soc.*, 1935, **57**, 2168.
101. B. E. Poling, J. M. Prausnitz and J. P. O'Connell, *The Properties of Gases and Liquids*, McGraw-Hill, New York, USA, 5th edn, 2001.
102. *Dortmund Data Bank Software and Separation Technology*: http://www.ddbst.de.
103. M. Kleiber and R. Joh, *VDI Heat Atlas*, Springer, Berlin, Heidelberg, Germany, 2nd edn, 2010.
104. W. Wagner, *Cryogenics*, 1973, **13**, 470.
105. (a) J. H. Dymond, K. N. Marsh, R. C. Wilhoit and K. C. Wong, Virial Coefficients of Pure Gases, in *Landolt–Börnstein, Numerical Data and Functional Relationships in Science and Technology, New Series; Group IV: Physical Chemistry, Virial Coefficients of Pure Gases and Mixtures, Vol. 21A*, ed. M. Frenkel and K. N. Marsh, Springer-Verlag, Heidelberg, Germany, 2002; (b) J. H. Dymond, K. N. Marsh and R. C. Wilhoit, *Virial Coefficients of Mixtures, in Landolt–Börnstein, Numerical Data and Functional Relationships in Science and Technology, New Series; Group IV: Physical Chemistry, Virial Coefficients of Pure Gases and Mixtures, Vol. 21B*, ed. M. Frenkel and K. N. Marsh, Springer-Verlag, Heidelberg, Germany, 2003.
106. A. H. Harvey and E. W. Lemmon, *J. Phys. Chem. Ref. Data*, 2004, **33**, 369.
107. C. Tsonopoulos, *AIChE J.*, 1974, **20**, 263.
108. C. Tsonopoulos and J. H. Dymond, *Fluid Phase Equilib.*, 1997, **133**, 11.
109. L. Meng, Y.-Y. Duan and L. Li, *Fluid Phase Equilib.*, 2004, **228**, 109.
110. L. Meng and Y.-Y. Duan, *Fluid Phase Equilib.*, 2007, **258**, 29.
111. E. Wilhelm, *Thermochim. Acta*, 1990, **162**, 43.
112. S. Bo, R. Battino and E. Wilhelm, *J. Chem. Eng. Data*, 1993, **38**, 611; Correction: J. Chem. Eng. Data, 1996, **41**, 644.
113. C. Tsonopoulos, Second Virial Cross-Coefficients: Correlation and Prediction of kij, in *Equations of State in Engineering and Research, Advances in Chemistry Series, Vol. 182*, ed. K. C. Chao and R. L. Robinson, Jr., American Chemical Society, Washington, DC, USA, 1979, ch. 8, pp. 143–162.
114. L. Meng and Y.-Y. Duan, *Fluid Phase Equilib.*, 2005, **238**, 229.
115. J. Horiuti, On the Solubility of Gas and Coefficient of Dilatation by Absorption, *Sci. Pap. Inst. Phys. Chem. Res. (Jpn)*, 1931, **17**, 125.
116. H. L. Clever and R. Battino, The Solubility of Gases in Liquids, in *Solutions and Solubilities, Part 1. Techniques of Chemistry, Vol. VIII*, ed. M. R. J. Dack, Wiley, New York, USA, 1975, pp. 379–441.
117. N. Bignell, *J. Phys. E: Sci. Instrum.*, 1982, **15**, 378.
118. N. Bignell, *Metrologia*, 1983, **19**, 57.
119. N. Bignell, *J. Phys. Chem.*, 1984, **88**, 5409.
120. N. Bignell, *Metrologia*, 1986/87, **23**, 207.
121. N. Bignell, *J. Phys. Chem.*, 1987, **91**, 1687.
122. H. A. Albert and R. H. Wood, *Rev. Sci. Instrum.*, 1984, **55**, 589.

123. R. P. Kennan and G. L. Pollack, *J. Chem. Phys.*, 1990, **93**, 2724.
124. J. L. Anderson, J. K. Dixon and J. F. Brennecke, *Acc. Chem. Res.*, 2007, **40**, 1208.
125. G. Maurer and Á. Pérez-Salado Kamps, Solubility of Gases in Ionic Liquids, Aqueous Solutions, and Mixed Solvents, in *Development and Applications in Solubility*, ed. T. M. Letcher, The Royal Society of Chemistry/IUPAC & IACT, Cambridge, UK, 2007, ch. 4, pp. 41–58.
126. J. L. Anderson, J. L. Anthony, J. F. Brennecke and E. J. Maginn, Gas Solubilities in Ionic Liquids, in *Ionic Liquids in Synthesis*, ed. P. Wasserscheid and T. Welton, Wiley-VCH, Weinheim, Germany, 2008, 2nd edn, pp. 103–129.
127. J. Kumelan, Á. Pérez-Salado Kamps, D. Tuma and G. Maurer, *J. Chem. Eng. Data*, 2006, **51**, 1364.
128. J. Kumelan, D. Tuma, Á. Pérez-Salado Kamps and G. Maurer, *J. Chem. Eng. Data*, 2010, **55**, 165.
129. J. Kumelan, D. Tuma and G. Maurer, *Fluid Phase Equilib.*, 2009, **275**, 132.
130. R. E. Gibbs and H. C. Van Ness, *Ind. Eng. Chem. Fundam.*, 1971, **10**, 312.
131. P. M. Mathias and J. P. O'Connell, *Chem. Eng. Sci.*, 1981, **36**, 1123.
132. J. J Carroll and A. E. Mather, *J. Solution Chem.*, 1992, **21**, 607.
133. I. Kritchevsky and A. Iliinskaya, *Acta Physicochim.* URSS, 1945, 20 , 327.
134. E. Bender, U. Klein, W. Ph. Schmitt and J. M. Prausnitz, *Fluid Phase Equilib.*, 1984, **15**, 241.
135. M. Orentlicher and J. M. Prausnitz, *Chem. Eng. Sci.*, 1964, **19**, 775.
136. R. H. Schumm and O. L. I. Brown, *J. Am. Chem. Soc.*, 1953, **75**, 2520.
137. J. E. Jolley and J. H. Hildebrand, *J. Am. Chem. Soc.*, 1958, **80**, 1050.
138. J. Walkley and J. H. Hildebrand, *J. Am. Chem. Soc.*, 1959, **81**, 4439.
139. J. Walkley and W. I. Jenkins, *Trans. Faraday Soc.*, 1968, **64**, 19.
140. W. Y. Ng and J. Walkley, *J. Phys. Chem.*, 1969, **73**, 2274.
141. E. W. Tiepel and K. E. Gubbins, *J. Phys. Chem.*, 1972, **76**, 3044.
142. R. Battino, M. Banzhof, M. Bogan and E. Wilhelm, *Anal. Chem.*, 1971, **43**, 806.
143. J. F. Connolly and G. A. Kandalic, *J. Chem. Phys.*, 1962, **36**, 2897.
144. J. F. Connolly and G. A. Kandalic, *Chem. Eng. Prog., Symp. Ser.*, 1963, **44**(59), 8.
145. J. F. Connolly and G. A. Kandalic, *J. Chem. Thermodyn.*, 1984, **16**, 1129.
146. J. F. Connolly and G. A. Kandalic, *J. Chem. Eng. Data*, 1986, **31**, 396.
147. J. F. Connolly and G. A. Kandalic, *J. Chem. Thermodyn.*, 1989, **21**, 851.
148. J. F. Connolly and G. A. Kandalic, *J. Chem. Eng. Data*, 1962, **7**, 137.
149. W. L. Masterton, D. A. Robins and E. J. Slowinski Jr., *J. Chem. Eng. Data*, 1961, **6**, 531.
150. I. Lauder, *Aust. J. Chem.*, 1959, **12**, 32.
151. I. Lauder, *Aust. J. Chem.*, 1959, **12**, 40.
152. W. L. Masterton, *J. Chem. Phys.*, 1954, **22**, 1830.
153. L. M. Blair and J. A. Quinn, *Rev. Sci. Instrum.*, 1968, **39**, 75.
154. T. Zhou and R. Battino, *J. Chem. Eng. Data*, 2001, **46**, 331.

155. I. Cibulka and A. Heintz, *Fluid Phase Equilib.*, 1995, **107**, 235.
156. P. Izák, I. Cibulka and A. Heintz, *Fluid Phase Equilib.*, 1995, **109**, 227.
157. S. J. Ashcroft, D. R. Booker and J. C. R. Turner, *J. Inst. Energy*, 1992, **65**, 131.
158. S. J. Ashcroft and M. Ben Isa, *J. Chem. Eng. Data*, 1997, **42**, 1244.
159. H. C. Van Ness, *Ind. Eng. Chem. Fundam.*, 1979, **18**, 431.
160. A. F. M. Barton, *CRC Handbook of Solubility Parameters and Other Cohesion Parameters*, CRC Press, Boca Raton, USA, 2nd edn, 1991.
161. C. M. Hansen, *Hansen Solubility Parameters: A User's Handbook*, CRC Press, Boca Raton, USA, 2nd edn, 2007.
162. E. B. Smith and J. Walkley, *J. Phys. Chem.*, 1962, **66**, 597.
163. E. W. Lyckman, C. A. Eckert and J. M. Prausnitz, *Chem. Eng. Sci.*, 1965, **20**, 685.
164. S. W. Brelvi and J. P. O'Connell, *AIChE J.*, 1972, **18**, 1239.
165. J. P. O'Connell, A. V. Sharygin and R. H. Wood, *Ind. Eng. Chem. Res.*, 1996, **35**, 2808.
166. A. V. Plyasunov, J. P. O'Connell and R. H. Wood, *Geochim. Cosmochim. Acta*, 2000, **64**, 495.
167. A. V. Plyasunov, E. L. Shock and J. P. O'Connell, *Fluid Phase Equilib.*, 2006, **247**, 18.
168. R. A. Pierotti, *J. Phys. Chem.*, 1963, **67**, 1840.
169. R. A. Pierotti, *J. Phys. Chem.*, 1965, **69**, 281.
170. R. A. Pierotti, *Chem. Rev.*, 1976, **76**, 717.
171. H. Reiss, *Adv. Chem. Phys.*, 1965, **9**, 1.
172. E. Wilhelm and R. Battino, *J. Chem. Thermodyn.*, 1971, **3**, 379.
173. E. Wilhelm and R. Battino, *J. Chem. Phys.*, 1971, **55**, 4012.
174. E. Wilhelm and R. Battino, *J. Chem. Phys.*, 1972, **56**, 563.
175. G. S. Kell, *J. Chem. Eng. Data*, 1975, **20**, 97.
176. E. Wilhelm, *J. Chem. Phys.*, 1973, **58**, 3558.
177. J.-P. Montfort and J. L. Perez, *Chem. Eng. J.*, 1978, **16**, 205.
178. S. K. Schaffer and J. M. Prausnitz, *AIChE J.*, 1981, **27**, 844.
179. G. Schulze and J. M. Prausnitz, *Ind. Eng. Chem. Fundam.*, 1981, **20**, 175.
180. H. S. Ashbaugh and L. R. Pratt, *Rev. Mod. Phys.*, 2006, **78**, 159.
181. H. S. Ashbaugh and L. R. Pratt, *J. Phys. Chem. B*, 2007, **111**, 9330.
182. M. Klähn, A. Martin, D. W. Cheong and M. V. Garland, *J. Chem. Phys.*, 2013, **139**, 244506.
183. H. J. C. Berendsen, J. P. M. Postma, W. F. van Gunsteren, A. DiNola and J. Haak, *J. Chem. Phys.*, 1984, **81**, 3684.
184. D. van der Spoel, E. Lindahl, B. Hess, G. Groenhof, A. E. Mark and H. J. C. Berendsen, *J. Comput. Chem.*, 2005, **26**, 1701.
185. B. Hess, C. Kutzner, D. van der Spoel and E. Lindahl, *J. Chem. Theory Comput.*, 2008, **4**, 435.
186. W. L. Jorgensen, D. S. Maxwell and J. Tirado-Rives, *J. Am. Chem. Soc.*, 1996, **118**, 11225.
187. G. A. Kaminski, R. A. Friesner, J. Tirado-Rives and W. L. Jorgensen, *J. Phys. Chem. B*, 2001, **105**, 6474.

188. M. S. Moghaddam and H. S. Chan, *J. Chem. Phys.*, 2007, **126**, 114507.
189. N. Patel, D. N. Dubins, R. Pomès and T. V. Chalikian, *J. Phys. Chem. B*, 2011, 4856.
190. J. Spooner, H. Wiebe, N. Boon, E. Deglint, E. Edwards, B. Yanciw, B. Patton, L. Thiele, P. Dance and N. Weinberg, *Phys. Chem. Chem. Phys.*, 2012, **14**, 2264.
191. A. V. Sangwai and H. S. Ashbaugh, *Ind. Eng. Chem. Res.*, 2008, **47**, 5169.
192. J. G. Kirkwood and F. P. Buff, *J. Chem. Phys.*, 1951, **19**, 774.
193. A. Ben-Naim, *Molecular Theory of Solutions*, Oxford University Press, Oxford, UK, 2006.
194. A. H. Harvey, S. G. Kaplan and J. H. Burnett, *Int. J. Thermophys.*, 2005, **26**, 1495.
195. F. J. Millero and R. T. Emmet, *J. Mar. Res.*, 1976, **34**, 15.
196. G. S. Kell, *J. Phys. Chem. Ref. Data*, 1977, **6**, 1109.
197. H. Watanabe and K. Iizuka, *Jpn. J. Appl. Phys.*, 1981, **20**, 1979.
198. H. Watanabe and K. Iizuka, *Metrologia*, 1985, **21**, 19; Erratum: *ibid.*, 1986, **22**, 115.
199. G. Girard and M.-J. Coarasa, The Influence of Dissolved Air on the Density of Water, in *Precision Measurements and Fundamental Constants II, Natl. Bur. Stand. (U.S.), Spec. Publ. 617*, ed. B. N. Taylor and W. D. Phillips, US Government Printing Office, Washington DC, USA, 1984, pp. 453–456.
200. J. J. Carroll, J. D. Slupsky and A. E. Mather, *J. Phys. Chem. Ref. Data*, 1991, **20**, 1201.
201. P. Giacomo, *Metrologia*, 1982, **18**, 33.
202. W. Wagner and A. Pruß, *J. Phys. Chem. Ref. Data*, 2002, **31**, 387.
203. H. S. Harned and R. Davis, Jr., *J. Am. Chem. Soc.*, 1943, **65**, 2030.
204. E. L. Shock and H. C. Helgeson, *Geochim. Cosmochim. Acta*, 1988, **52**, 2009.
205. M. Tanaka, G. Girard, R. Davis, A. Peuto and N. Bignell, *Metrologia*, 2001, **38**, 301.
206. A. H. Harvey, R. Span, K. Fujii, M. Tanaka and R. S. Davis, *Metrologia*, 2009, **46**, 196.
207. E. Batista and R. Paton, *Metrologia*, 2007, **44**, 453.
208. A. Asenbaum and E. Wilhelm, *Adv. Mol. Relax. Interact. Processes*, 1982, **22**, 187.

CHAPTER 10

Saturated Liquid Density of Pure Liquids and of Mixtures

TOSHIHARU TAKAGI*[a] AND TOMOYA TSUJI*[b]

[a] Department of Chemistry and Materials Technology, Kyoto Institute of Technology, Kyoto 606-8585, Japan; [b] College of Industrial Technology, Nihon University, Narashino 275-8575, Japan
*Email: taka-301@mbox.kyoto-inet.or.jp; tsuji.tomoya@nihon-u.ac.jp

10.1 Introduction

The vapour–liquid equilibrium of pure liquids and of their mixtures, together with the saturated liquid densities, constitute important thermodynamic properties in the understanding of matter. As a consequence, research into saturated liquid densities of operating fluids is very active and many papers have reported on pure saturated liquid densities and on the densities of mixtures. These experimental results have been fitted by empirical equations like the Tait equation,[1] and various kinds of equations of state. A summary of the research on the equilibrium properties of fluids is described by Poling *et al.*[2] In our previous studies, the saturated liquid density and related properties were widely investigated for pure liquids and mixtures and the data were utilised for the estimation of other physical properties and process design like heat pumps and refrigerators.[2–16]

In this chapter, we describe the latest experimental research and trends in saturated liquid densities of pure liquids and of liquid mixtures. Furthermore, the equations of state used to estimate saturated liquid density estimations are reviewed.

Volume Properties: Liquids, Solutions and Vapours
Edited by Emmerich Wilhelm and Trevor M. Letcher

Published by the Royal Society of Chemistry, www.rsc.org

10.2 Latest Experimental Data

Most of the saturated liquid density experimental data are reported in the following journals: *Journal of Chemical & Engineering Data; Journal of Chemical Thermodynamics;* and *Fluid Phase Equilibria*. Furthermore, most of the data for natural products, such as fatty oils, tend to be reported in *Industrial & Engineering Chemistry Research* and *The Journal of Supercritical Fluids.* The saturated density is defined at a boundary between a compressed liquid phase and the coexisting phase. As such, the saturated density can refer to the boundary among two phases, *viz.* a vapour and a liquid; a solid and liquid; or a liquid and liquid. However, most measurements refer to the boundary between a vapour and a liquid and these are discussed in this chapter. The experimental data for pure substances are often measured together with high pressure liquid densities, to determine pressure–volume–temperature (*P–V–T*) relationships. Similarly, the data for mixtures usually involve vapour–liquid equilibrium measurements.

10.2.1 Pure Substances

Over the past decade, most of the reported data related to light hydrocarbons, dimethylether and fluorohydrocarbons, these being alternative and environmentally friendly refrigerants. The data have been used to calculate various thermodynamic properties and to construct equations of state. Miyamoto and Uematsu[17] measured the saturated liquid density for propane[17] and butane[18] over the temperature range from 280 to 365 K. Miyamoto *et al.* also reported the density of isobutene (2-methylpropane),[19] as indicated graphically in Figure 10.1. The measurement was carried out at temperatures from 280 to 407 K, near the critical temperature, ($T_C = 407.84$ K), and pressures up to 10 MPa

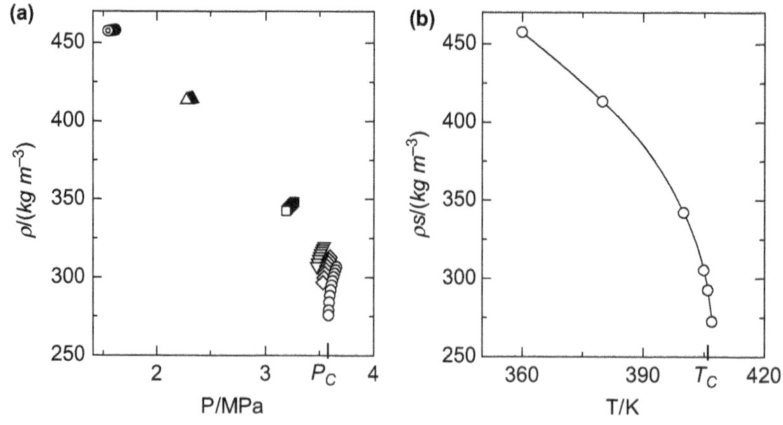

Figure 10.1 Pressure and temperature dependences of (a) density, ρ, and (b) saturated liquid density, ρ_S, for isobutene. ◎, $T = 360$ K; △, 380 K; □, 400 K; ▽, 405 K; ◇, 406 K; ○, 407 K.

with intervals of about 12 kPa around the saturation boundary at each temperature [Figure 10.1(a)]. The saturated liquid density, ρ_S, [Figure 10.1(b)] was obtained by extrapolation to the vapour pressure.

Although Ihmels *et al.*[20] measured and reported the liquid densities of 1-butene, isobutene, *cis*-2-butene and *trans*-2-butene over a wide range of temperature and pressure, there is no datum at the saturation point. They did however list the extrapolated values of the saturated density for the four substances. Similar data can be seen for dimethylether.[21] Dimethyether has versatile properties which makes it: a potential new refrigerant; a possible alternative fuel for petroleum gas; and a new type of spray propellant. This has attracted many researchers to measure its thermodynamic properties. Wu *et al.*[22] reported the saturated liquid density for dimethylether over the temperature range from 301.8 K to the critical point. They also reported the density and the viscosity within the temperature range from 233 to 363 K.[23] The saturated density data for dimethylether have been reported by Bobbo *et al.*,[24] and by Tanaka and Higashi[25] and their results have been used to produce equations of state. The saturated liquid density for other potential refrigerants, such as chlorohydrocarbons and fluorohydrocarbons, have also been reported. Goncalves *et al.*[26] measured the *P–V–T* relationship for dichloromethane, and the saturated liquid density was estimated by extrapolation. However, most reports relate to fluorohydrocarbons. Hu and Chen[27] reported the saturated liquid density of 1,1,1,2,3,3,3-heptafluoropropane within the temperature range from 326 to 374 K. Fedele *et al.*,[28] reported compressed liquid density measurements on the same compound (see NIST standard reference database 23 REFPROP ver. 9.1[29]).

Other reported data include work by Bobbo *et al.*[30] for 1,1,1,3,3-pentafluorobutane from 283 to 343 K; Fedele *et al.*[31] for 2,3,3,3-tetrafluoroprop-1-ene from 283 to 333 K; and Brown *et al.*[32] for *trans*-1,3,3,3-tetrafluoroprop-1-ene from 283 to 353 K. Data on systems other than refrigerants are few and far between. Wu *et al.*[33] reported *P–V–T* relationships for three hydrocarbons: hexadecane, octadecane and eicosane over the temperature range 323 to 523 K and for pressures up to 265 MPa. Abdulagatov *et al.*[34] determined the saturated liquid density for 1-propanol over the temperature range 298 to 423 K from compressed liquid density and saturated vapour pressure data. In the restricted region near the critical point, Rasulov *et al.*[35] measured the saturated liquid density for 1-butanol. Similar data were obtained for toluene by Abdulagatov *et al.*[36] Of interest, Anderson *et al.*[37] reported the saturated liquid density for tetramethylorthosilicate over the temperature range from 334 to 532 K. The significance of this data is that this substance has potential as a useful industrial reagent.

10.2.2 Mixtures

Most of the reported experimental saturated liquid density data are for the binary mixtures, together with a few sets of data for ternary and quaternary mixtures. The publications involving binary liquid density data tend to include vapour–liquid equilibrium data. The binary systems reported included mixtures involving carbon dioxide with a hydrocarbon or an alkanol, methane or ethane or ethylene with a hydrocarbon or an alkanol at pressures of up to the critical point. Table 10.1 lists typical saturated liquid density data for mixtures reported over the past decade.

Figure 10.2 shows the experimental results containing the saturated liquid density for (carbon dioxide + methylacetate) at $T = 313.15$ K.[49] The liquid density shows a convex shape composition dependency [Figure 10.2(a)], as is common with mixtures containing a supercritical fluid. The end points of the isobar line lower than 6 MPa correspond to the saturated liquid density [Figure 10.2(b)]. Binary experimental data have also been reported for supercritical carbon dioxide mixed with industrial materials such as canola oil[68] and also with cocoa butter.[69] Other mixtures, for which saturated liquid densities have been reported, usually include a refrigerant. For example, Akasaka *et al.*[70] reported the saturated liquid density of (2,3,3,3-tetrafluoropropane + difluoromethane) in the vicinity of the critical point. Similar data for (difluoromethane + propane) were reported by Higashi.[71] Some of the systems reported contained water or an inorganic salt or even a polymer. For example, Zhou *et al.*[72] measured the liquid density for (water + glycerol containing NaCl or KCl or RbCl or CsCl at $T = 298.15$ K. Pawar *et al.*[73] reported liquid density measurements for (ethanol + water + KI) at 298.15, 303.15, 308.15 and 313.15 K. Taboada *et al.*[74] measured the liquid density for {water + polyethylene glycol 4000 (PEG4000) + NaCl or KCl} at 333.15 K. Strictly speaking, these are not saturated systems but, as the pressure dependence is negligible, the liquid density values may be regarded as saturated ones. These liquid density data are often measured together with refractive indices or conductivities or dielectric constants. These systems are generally not measured in the high pressure region, so the procedure and apparatus are generally simple.

10.2.3 Measuring Devices

In this section, interesting and well-designed measuring apparatuses are discussed. Density is measured by a direct or an indirect method and an oscillation type density meter, such as the Anton Paar 512HPM or 512P, has often been used. The saturated liquid density tends to be measured using a gas solubility method or a vapour–liquid equilibrium method.

Figure 10.3 shows a schematic diagram of the experimental apparatus of Nourozieh *et al.*[53] The apparatus is based on a variable volume method, and the pressure and the temperature in the equilibration cell can be maintained and controlled. After attaining equilibrium, the saturated vapour phase and the saturated liquid move to sampling cells and the density is measured using an oscillating U-tube density meter. The liquid composition is measured by the expanded gas volume method in a gas meter.

Table 10.1 Typical saturated liquid density data for mixtures reported recently.

System	Temperature T /[K]	Pressure range P/ [MPa]	Researcher	Ref.
CO_2 + decane	323.2	1.02–5.96	Kariznovi *et al.*	38
CO_2 + tetradecane	323.2	0.98–6.00	Kariznovi *et al.*	38
CO_2 + pentadecane	313.15	2.092–6.800	Kodama *et al.*	39
CO_2 + methanol	303.2	1.02–5.05		
	323.2	1.11–5.95		
CO_2 + ethanol	303.2	1.14–5.94	Kariznovi *et al.*	40
	323.2	1.02–6.08		
CO_2 + ethanol	313.2	0.1–8.51	Seifried and Temelli	41
	328.2	0.11–9.90		
	343.2	0.16–11.22		
CO_2 + ethanol	313.2	0.62–4.64	Tsivintzelis *et al.*	42
	328.2	1.65–4.37		
CO_2 + 1-propanol	303.2	0.90–4.99	Kariznovi *et al.*	40
	323.2	0.77–6.00		
CO_2 + 1-propenol	313.24	3.133–7.883	Galicia-Luna and Elizalde-Solis	43
	363.17	3.924–14.004		
CO_2 + 2-propanol	313.22	2.402–7.788	Galicia-Luna and Elizalde-Solis	43
	333.03	2.577–10.145		
	348.25	1.801–11.570		
	362.92	2.580–12.423		
CO_2 + 2-propanol	302.8	0.99–3.97	Nourozieh *et al.*	44
	323.1	0.98–5.94		
CO_2 + 1-butanol	313.22	3.144–8.063	Elizalde-Solis and Galicia-Luna	45
	363.17	3.093–14.895		
CO_2 + 1-butanol	303.2	1.05–4.97	Kariznovi *et al.*	46
	323.1	0.92–6.25		
CO_2 + 2-butanol	313.21	2.564–7.759	Kariznovi *et al.*	46
	333.12	2.656–10.067		
	363.06	2.776–13.073		
CO_2 + 2-methyl-2-propanol	313.15	3.435–7.709	Kodama *et al.*	47
CO_2 + diglymes	313.15	1.195–7.126	Kodama *et al.*	39
CO_2 + triglymes	313.15	1.503–7.202	Kodama *et al.*	39
CO_2 + tetraglymes	313.15	1.459–7.316	Kodama *et al.*	39
CO_2 + tetrahydrofuran	298.15	2.137–6.147	Kodama *et al.*	48
	313.15	2.060–7.629		
CO_2 + methyl acetate	313.15	1.313–7.114	Kato *et al.*	49
CO_2 + dichloromethane	308.2	1.92–5.08	Tsivintzelis *et al.*	42
	318.2	1.49–4.43		
	328.2	2.05–6.49		
CO_2 + ethoxyethanol	313.2	1.40–5.12	Missopolinou *et al.*	50
	320.2	2.06–5.68		
	328.2	1.53–6.24		

Table 10.1 (*Continued*)

System	Temperature T /[K]	Pressure range P/ [MPa]	Researcher	Ref.
Methane + tetradecane	294.4–294.8	2.05–9.49	Nourozieh *et al.*	51
	324.0–324.1	2.07–9.54		
	373.3–373.5	2.10–9.50		
	447.5–447.6	1.99–9.47		
Methane + octadecane	323.0–323.1	2.08–9.49	Nourozieh *et al.*	52
	347.9	2.09–9.47		
	397.2–398.2	2.03–9.59		
	447.5–447.7	1.92–9.51		
Methane + methanol	294.5–294.8	1.05–7.99	Nourozieh *et al.*	53
Methane + ethanol	294.6–295.2	1.02–7.99	Nourozieh *et al.*	53
Methane + 1-propanol	295.1–294.7	0.99–8.02	Nourozieh *et al.*	53
Methane + 2-propanol	302.8	1.07–6.06	Nourozieh *et al.*	44
	323.1	0.98–6.04		
Methane + 1-butanol	303.2	1.06–6.07	Kariznovi *et al.*	46
	323.1	1.05–6.07		
Ethane + tetradecane	323.2–323.3	1.05–5.94	Kariznovi *et al.*	54
	373.4–373.6	1.01–7.07		
	422.5–422.7	1.08–8.00		
Ethane + octadecane	323.1–323.3	1.15–5.89	Nourozieh *et al.*	55
	372.8	0.94–7.98		
	422.5–422.6	0.95–8.17		
Ethane + 2-propanol	308.15	2.191–4.990	Kodama *et al.*	56
	313.15	3.169–5.357		
Ethane + 1-butanol	303.2	1.05–3.07	Kariznovi *et al.*	46
	323.1	1.14–6.02		
Ethylene + methanol	278.15	2.050–5.769	Haneda *et al.*	57
	283.65	2.222–5.906		
Ethylene + ethanol	283.65	1.243–5.267	Kodama *et al.*	58
Ethylene + 1-propanol	283.65	2.262–5.438	Kodama *et al.*	59
Ethylene + 2-propanol	283.65	2.227–5.029	Kodama *et al.*	60
Ethylene + 1-butanol	283.65	1.583–5.750	Kodama *et al.*	61
CO_2 + decane + tetradecane	303.2	0.97–6.03	Kariznovi *et al.*	38
CO_2 + decane + octadecane	323.2	0.93–6.01	Nourozieh *et al.*	62
CO_2 + ethanol + decane	303.2	0.02–5.05	Nourozieh *et al.*	63
	323.2	1.07–5.95		
CO_2 + ethanol + decane	303.2	1.02–5.05	Nourozieh *et al.*	64
Methane + decane + tetradecane	294.6–295.8	0.99–8.02	Kariznovi *et al.*	65
Methane + decane + hexadecane	295.0–295.7	0.93–8.01	Kariznovi *et al.*	66
Methane + decane + octadecane	294.4–295.7	0.94–8.05	Nourozieh *et al.*	67

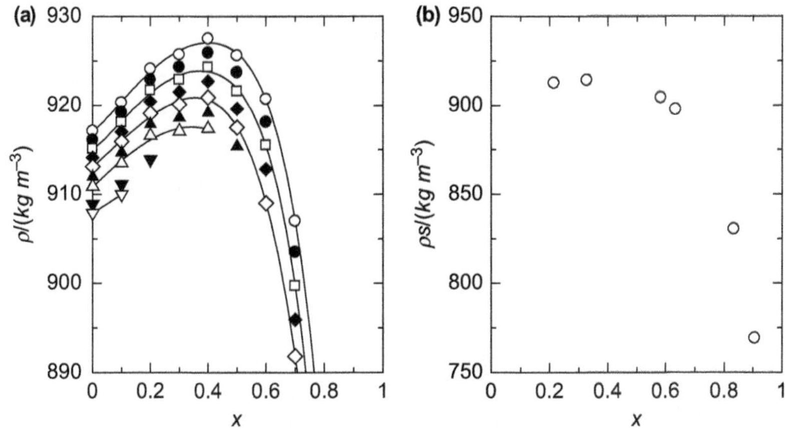

Figure 10.2 Experimental liquid density (a) and saturated density (b) for {x carbon dioxide $+(1-x)$ methylacetate} at 313.15 K.[49] ●, 10 MPa; ▲, 9 MPa; ■, 8 MPa; ◆, 7 MPa; ▼, 6 MPa; ○, 5 MPa; △, 4 MPa; □, 3 MPa; ◇, 2 MPa; ▽, 1 MPa.

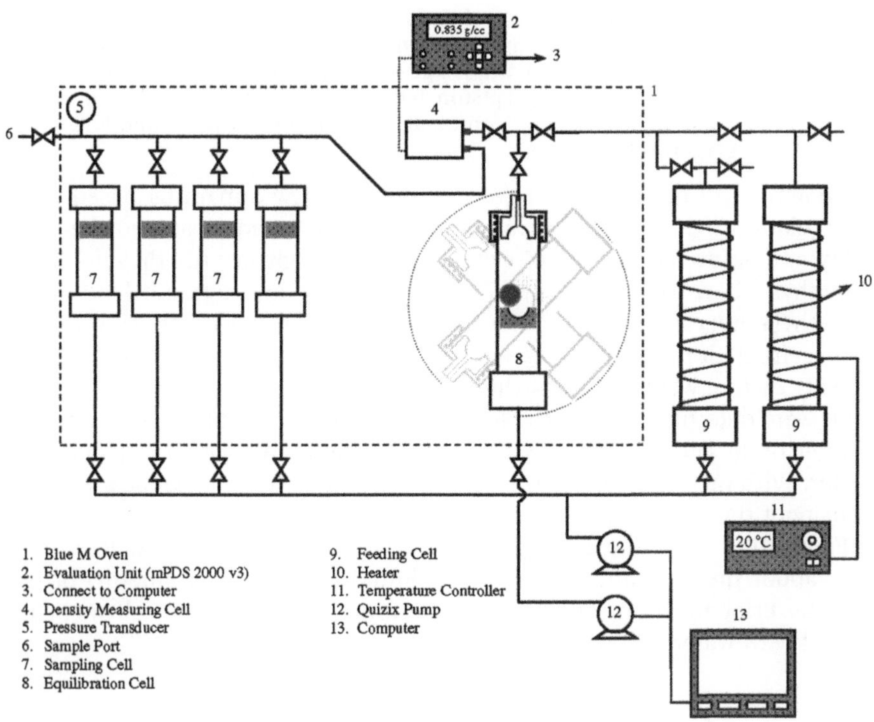

Figure 10.3 Experimental apparatus of Nourozieh *et al.* Reproduced with permission from ref. 53.

Using this apparatus, they have measured the vapour–liquid equilibrium for methane–alcohols at $T=295$ K and pressure, every 1 MPa, up to 10 MPa, with uncertainty within ± 1 kg m^{-3} in the saturated liquid density.

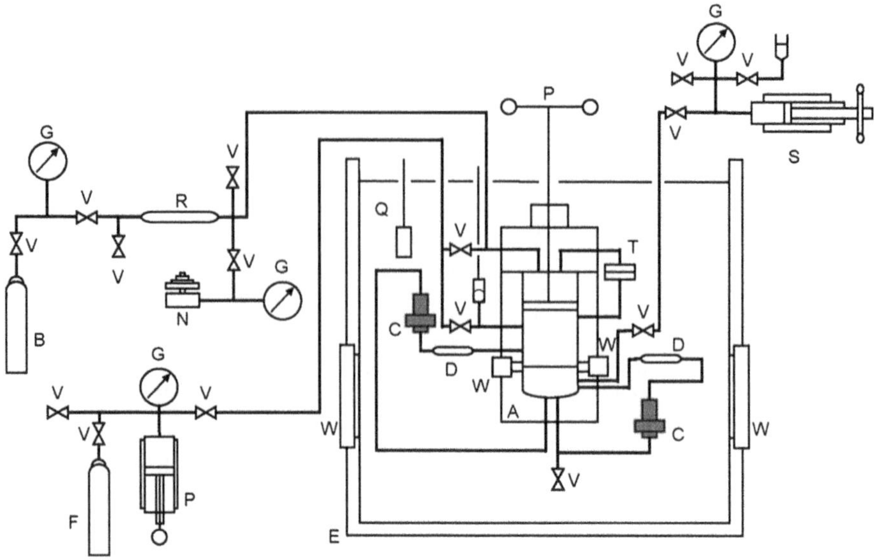

Figure 10.4 Experimental apparatus of Kodama *et al.*[39] (A) Variable volume cell, (B) nitrogen reservoir, (C) circulation pump, (D) density meter, (E) constant temperature liquid bath, (F) gas reservoir, (G) pressure gauge, (N) air dead weight gauge, (P) piston, (Q) quartz thermometer, (R) reservoir, (S) hand syringe pump, (T) pressure transducer, (V) valve, and (W) visual glass window.

Kodama and co-workers[39] have developed a new apparatus based on a recirculation method, as illustrated in Figure 10.4, and measured the solubility and saturated density of carbon dioxide in glymes (diethylene glycol dimethyl ether, triethylene glycol dimethyl ether, and tetraethylene glycol dimethyl ether) at 313.15 K without any analysis of phase compositions. The two recirculation pumps were equipped with cells and oscillating U-tube density meters were placed in the lines. The composition of vapour and liquid were determined from the loaded mass, the measured density and the volume of the cell. The volume of the cell was calibrated with a displacement meter. With increasing pressure, the vapour phase should disappear and the density of the compressed liquid phase can be determined.

Figure 10.5 shows the density meter used by Anderson *et al.*[37] to determine the vapour pressure and the saturated liquid density of tetramethyl orthosilicate. Prior to the measurements, the inner volume of the cell was calibrated with water and methanol *via*

$$\rho = m_{liq}/V_{cell}. \tag{10.1}$$

The saturated vapour pressure is given by:

$$P^s = P_{total} - \frac{n_{air}RT}{V_{call} - m_{liq}\rho} \tag{10.2}$$

where the amount of air, n_{air}, was determined from the initial condition.

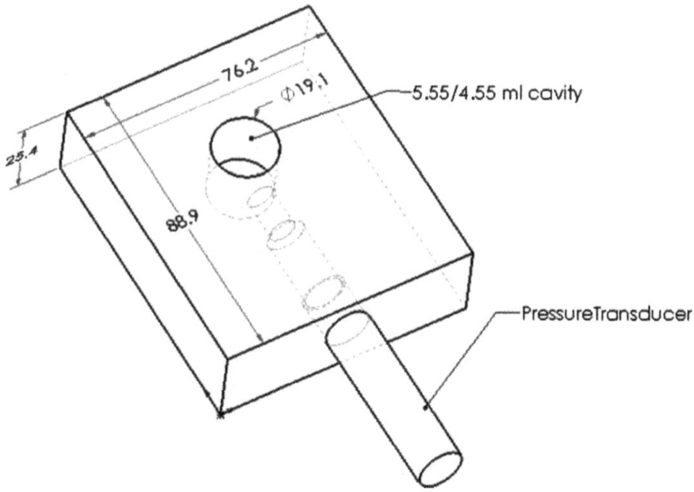

Figure 10.5 Experimental apparatus of Anderson *et al.*
Reproduced with permission from ref. 37. Copyright 2008 American Chemical Society.

10.3 Thermodynamic Model

To evaluate the saturated liquid density, a thermodynamic model must be used. Some models are not just for evaluating the saturated liquid density but also for determining the *P–V–T* relationship. In this section, typical calculations are described.

10.3.1 Corresponding State Theory

Generally the saturated liquid density for a pure substance can be correlated using the following empirical equation:

$$\rho = \rho_c(a\tau^{0.3x} + b\tau^{1/2} + c^{10/6} + d\tau^{11/6}) \tag{10.3}$$

where ρ_c is the critical density, τ is non-dimensional temperature given by:

$$\tau = 1 - T/T_c \tag{10.4}$$

and a,b,c,d are constants which depend on the substance. A similar equation is used for the saturated vapour pressure. The constants should be optimised using precise experimental data. The equation involves a non-dimensional polynomial, and the constants depend on the substance under investigation. The equation has an upper limit of temperature but the lower limit is not indicated. Yonglove *et al.*[75] proposed the following equation, which can be applied from the triple point to the critical point:

$$\rho = \rho_c + (\rho_t - \rho_c)\exp(h(x)) \tag{10.5}$$

where $h(x)$ is given by:

$$x = \frac{T_c - T}{T_c - T_t} \tag{10.6}$$

$$h(x) = a_7 + \sum_{n=9}^{10} a_n \left[1 - x^{\frac{n-11}{3}}\right] + \sum_{n=11}^{13} a_n \left[1 - x^{\frac{n-10}{2}}\right] \tag{10.7}$$

and the constants, a_7–a_{13}, are optimised by experimental values.

The Hankinson–Brobst–Thomson (HBT) equation[76,77] has been widely used for the estimation of the saturated liquid density. The HBT equation is given by:

$$\frac{v_s}{v^*} = v_R^{(0)} \left[1 - \omega_{SRK} v_R^{(\delta)}\right] \tag{10.8}$$

where

$$v_R^{(0)} = 1 + a(1-T_r)^{1/3} + b(1-T_r)^{2/3} + c(1-T_r) + d(1-T_r)^{4/3} \quad (0.25 < T_r < 1.0).$$

$$v_R^{(\delta)} = \left[e + fT_r + gT_r^2 + hT_r^3\right](T_r - 1.00001) \qquad (0.25 < T_r < 1.0)$$

where v_s is the saturated molar volume, v^* and ω_{SRK} constants depending on the substance. The a–h are comprehensive constants listed elsewhere.[76,77] If the constants are not available, the acentric factor ω_{SRK} is evaluated from the saturated vapour pressure and Soave–Redlich–Kwong (SRK) equation.[78] Then v^* is given by:

$$v^* = \frac{RT_c}{P_c}(a' + b'\omega_{SRK} + c'\omega_{SRK}^2) \tag{10.9}$$

where a'–c' are obtained from the literature. The values correspond to the classification of the molecular structure.

The modified Rackett equation[77,79,80] is one of the most popular equations for estimating saturated liquid density. The Rackett equation is given by:

$$v_s = \frac{RT_c}{P_c} z_{RA}^{\left[1+(1-T_r)^{2/7}\right]} \tag{10.10}$$

where z_{RA} is the Racket compressibility factor, and its value is given elsewhere.[77,79,80] If the parameters are not available, the following equation may be used:

$$z_{RA} = 0.29056 - 0.08775\omega \tag{10.11}$$

Instead of Equation (10.11), the following equation involving a reference temperature and the density, may be used:

$$v_s = v_s^{ref} z_{RA}^{\left[(1-T_r)^{2/7} - (1-T_r^{ref})^{2/7}\right]} \tag{10.12}$$

Yet another corresponding state equation is the Bhirud equation:[77,81]

$$\ln\frac{P_c v_s}{RT} = \ln v^{(0)} + \omega \ln v^{(1)} \tag{10.13}$$

where

$$\ln v^{(0)} = 1.3964 - 24.076T_r + 102.615T_r^2 - 255.719T_r^3 + 355.805T_r^4 - 256.617T_r^5 + 75.1088T_r^6$$

and

$$\ln v^{(1)} = 13.4412 - 135.7437T_r + 533.380T_r^2 - 1091.453T_r^3 + 1231.43T_r^4 - 728.227T_r^5 + 176.737T_r^6$$

In the Bhirud equation, the acentric factor and the critical point are important. These three equations make the calculation of the saturated liquid density of a pure substance a relatively easy operation. Many modified versions and comparisons of the Bhirud equation have been reported. Recently, Patel and Joshipura[82] reviewed this type of equation. These equations have now been used to construct volume-translated cubic equations. The Bhirud equation is not perfect as it depends on the corresponding state theory. As a way of improving the calculation of density, Kato *et al.*[83] proposed the following empirical equation for binaries:

$$\rho = \frac{x_1\rho_1^{0*} + x_2\rho_2^{0*} + \alpha x_1 x_2[(x_1 + \beta x_2)/(x_1 + \kappa x_1 x_2 + \lambda x_2)]}{1 - \ln\left[1 - \{1 - (P/P^*)\}\Big/\left\{1 + \exp\left(A - \dfrac{B}{x_1 + 6x_2}\right)\right\}\right]} \tag{10.14}$$

where ρ_i^{0*} is the density of pure components i at the arbitrary reference pressure P^* and α, β, κ, λ, A, B are adjustable parameters.

10.3.2 Helmholtz Type Equation

The Helmholtz equation is based on thermodynamics principles and is generally given by:

$$a(\tau,\delta) = a^0(\tau,\delta) + a^r(\tau,\delta) \tag{10.15}$$

where a^o, a^r is Helmholtz energy for the ideal and residual parts, respectively, and τ and δ are given by:

$$\tau = T_c/T \tag{10.16}$$

and

$$\delta = \rho/\rho_c \tag{10.17}$$

Examples involving the application of saturated liquid densities include work by Ihmels *et al.*[20,21,84,85] who applied the Helmholtz equation to saturated liquid density data of 1-butene, isobutene, *cis*-2-butene, *trans*-2-butene, dimethyl ether and sulfur dioxide. The Helmholtz equation has recently been applied to the saturated liquid density of new refrigerants such as 1,1,1,3,3,3-hexafluoropropane,[86] 1,1,1,2,3,3-hexafluoropropane.[87]

Furthermore, some attempt has been made by Estela-Uribe[88–91] to estimate the *P–V–T* relationship for mixtures involving air, natural gas, *etc.*

10.3.3 Volume Translated Cubic Equation

The conventional cubic equation has been widely used for estimating phase equilibrium properties. The SRK[78] and Peng–Robinson (PR)[92] equations are generally considered to be two of the most standard ones. Though the equations were generalised with C_1–C_{10} alkane, nitrogen, carbon dioxide, hydrogen sulfide and so on, the good reproducibilities are obtained even for fluorocarbons, especially for the vapour pressure and *P–V–T* relationship in the compressibility factor range larger than $z = 0.4$. However the saturated liquid density and the compressibility factor near the critical point do not always show good reproducibility. Recently, volume translated SRK and PR equations have been developed, as seen in the publications of Magoulas and Tassios,[93] de Hemptinne and Ungerer,[94] Ji and Lempe.[95] and Wang and Gmehling.[96] In this section, a recently reported cubic equation, based on the volume translated theory, is discussed.

The PR equation is given by:

$$P = \frac{RT}{v - b} - \frac{a}{v^2 + 2bv - b^2} \tag{10.18}$$

$$\begin{aligned} a &= 0.45724\frac{R^2T_c^2}{P_c}\alpha(T),\ b = 0.0778\frac{RT_c}{P_c} \text{ and} \\ \alpha(T) &= [1 + m(\omega)(1 - T_r^{0.5})]^2 \end{aligned} \tag{10.19}$$

Hinojosa-Gómez *et al.*[97] proposed the following usage of P_c', ω', T_c', instead of P_c, ω, T_c . The values of P_c', ω', T_c', were optimised with the vapour pressure and the saturated density data as follows:

$$P_c' = \eta_p + \mu_p T_c' \quad \omega' = \eta_w + \mu_w T_c' \quad T_c' = j_1 + j_2 T + j_3 T^2 \tag{10.20}$$

where the parameters, η_p, η_w, μ_p, μ_w, j_1, j_2, j_3 are constants which depend on the substance under investigation. This method involves a temperature dependence correction. Furthermore, the temperature dependence of the a and b terms has also been taken into account:

$$a = 0.45724\frac{R^2T_c^{*2}}{P_c^*}\alpha(T) \quad b = 0.077796\frac{RT_c^*}{P_c^*}\xi(T) \tag{10.21}$$

$$\alpha(T) = 1 + [c_1 + c_2(1 - T/T_c^*)](1 - T/T_c^*) \tag{10.22}$$

$$\xi(T) = 1 + [c_3 + c_4(1 - T/T_c^*)](1 - T/T_c^*) \tag{10.23}$$

where T_c^*, P_c^*, c_1, c_2, c_3, c_4 are constant, depending on the substance. Parameters are available for 71 compounds, mainly containing C_1–C_{36} alkanes, C_1–C_{20} alcohols, and C_2–C_9 ketenes.[97] According to the method, the average absolute deviations are from 0.1 to 2.4% for the saturated liquid densities.

Another method, using the PR or SRK equation, involves a correction for the molar volume and is one of the most popular methods related to volume translated cubic equations of state. According to Lin and Duan[98] the molar volume is given by:

$$v = v_{PR} - c \tag{10.24}$$

The PR equation then converts to:

$$P = \frac{RT}{v + c - b} - \frac{a}{(v-c)^2 + 2b(v-c) - b^2} \tag{10.25}$$

The following temperature dependence has been used:

$$c = (0.3074 - z_c)\frac{RT_c}{P_c} f(T_r)$$

and

$$f(T_r) = \beta + (1-\beta)\exp(-\gamma|1 - T_r|) \tag{10.26}$$

A generalised method has also been proposed by Lin and Duan[98] and β and γ are given by a function using the critical compressibility factor, z_c. They calculated the saturated liquid density for C_1–C_{10} alkanes, C_2–C_9 alkenes, aromatics, fluorocarbons, inorganic compounds, and demonstrated that the average absolute deviations are within from 0.32% to 4.36%. Using a different function of β and γ, α similar method was also proposed by Nazarzadeh and Moshfeghian.[99]

In the equation proposed by Baled *et al.*,[100] the volume translated parameters are given by:

$$c = A + B\left(\frac{T}{T_c}\right) \tag{10.27}$$

$$A = k_0 + k_1 \exp\left(\frac{-1}{k_2 M\omega}\right) + k_3 \exp\left(\frac{-1}{k_4 M\omega}\right) + k_5 \exp\left(\frac{-1}{k_5 M\omega}\right) \tag{10.28}$$

$$B = k_0' + k_1' \exp\left(\frac{-1}{k_2' M\omega}\right) + k_3' \exp\left(\frac{-1}{k_4' M\omega}\right) + k_5' \exp\left(\frac{-1}{k_5' M\omega}\right) \tag{10.29}$$

In the equation proposed by Abudour *et al.*,[101] the parameters are defined as:

$$v = v_{PR} + c - \delta_c\left(\frac{0.35}{0.35 + d}\right) \tag{10.30}$$

$$d = \frac{1}{RT_c}\left(\frac{\partial P}{\partial \rho}\right)_T \tag{10.31}$$

$$c = \left(\frac{RT_c}{P_c}\right)[c_1 - (0.004 + c_1)\exp(-2d)] \tag{10.32}$$

$$\delta_c = \left(\frac{RT_c}{P_c}\right)(0.3074 - z_c) \tag{10.33}$$

According to the method of Baled *et al.*, the average absolute deviation of the saturated liquid density for 15 compounds, including normal, blanched, and cyclic alkanes, is estimated to be within a precision of 0.70 to 2.63% for SRK-, and of 0.6 to 4.36% for PR-based equations in average absolute deviations including the supercritical region. Abudour *et al.* also demonstrated the average absolute deviations for the saturated and supercritical density of 65 compounds with a precision of 0.39 to 2.04%.

For mixtures, the following mixing rule is generally employed, because the parameter c has the same meaning as the excluded parameter, b:

$$a = \sum_i \sum_j x_i x_j (1 - k_{ij})(a_i a_j)^{0.5} \quad b = \sum_i x_i b_i \quad c = \sum_i x_i c_i \tag{10.34}$$

The methods developed by Lin and Duan[98] and Nazarzadeh *et al.*[99] for binary mixtures showed good reproducibility for saturated liquid densities. For example, using the method of Lin and Duan, the average absolute deviation of the saturated liquid density yields a precision within from 0.83 to 4.58% for the five fluorocarbon binaries. The method of Nazarzadeh *et al.* was used to apply to the mixtures with fewer than 8 components, mainly containing light hydrocarbon and nitrogen. The average absolute deviations were reported to be from 0.18 to 5.5%.

In recent years, with advances in the chemical industry, petroleum plants, heat pumps, biosciences and others, knowledge of thermophysical properties with high accuracy for the fluid have become increasingly important. The performance of the saturated liquid density measurement device has improved significantly. In particular, the high precision vibration densitometer is used in combination so that the compressed and saturated liquid density for pure and mixture fluids can now be measured with high precision over wide ranges of temperature and pressure including the critical region, as described in Figures 10.1 and 10.2, On the other hand, to reproduce the experimental results of P–V–T, an optimal equation of state has been devised to obtain the saturated liquid density over a wide range of temperature and pressure. In particular, in order to accurately fit the density in the case of systems containing a supercritical region or mixed systems, the far more complex formulae are used.

10.4 Conclusions

In this chapter, data on saturated liquid densities of pure liquids and of mixtures have been obtained from papers published over the past decade. The discussion of the data for pure substances has been largely classified into light hydrocarbons, refrigerants, fluorocarbons and ethers. Some data have been extrapolated using values of the saturated vapour pressure and compressed liquid density. Most of the data on mixtures relate to carbon dioxide, hydrocarbons and alcohols. The mixtures are mainly binary systems and the measured properties are largely vapour–liquid equilibrium or gas solubility. Also reported in this chapter are interesting types of apparatus and devices used to obtain volumetric data. Finally, an evaluation of theoretical approaches involving theories such as corresponding state theory, the Helmholtz equation and the volume translated cubic equation has been included. Many equations have been developed to describe saturated liquid properties of pure substances, but unfortunately the compatibility is generally poor for both binary and ternary systems. In the field of physical chemistry and chemical engineering, knowledge of the saturated liquid density of the pure liquid and/or mixtures will become increasingly important.

References

1. J. H. Dymond and R. Malhotra, *Int. J. Thermophys.*, 1988, **9**, 491.
2. *The Properties of Gases and Liquids*, ed. B. E. Poling, J. M. Prausnitz and J. P. O'Connell, McGraw-Hill, New York, 5th edn, 2001.
3. T. Takagi, H. Teranishi, C. Yokoyama and S Takahashi, *Thermochim. Acta*, 1989, **141**.
4. M. Hongo, M. Kusunoki, H. Matsuyama, T. Takagi, K. Mishima and Y. Arai, *J. Chem. Eng. Data*, 1990, **35**, 414.
5. T. Takagi, H. Teranishi, C. Yokoyama and S. Takahashi, *Thermochim. Acta*, 1992, **195**, 239.
6. K. Mishima, M. Hongo, T. Takagi and Y. Arai, *J. Chem. Eng. Data*, 1993, **38**, 49.
7. M. Hongo, T. Tsuji, K. Fukuchi and Y. Arai, *J. Chem. Eng. Data*, 1994, **39**, 688.
8. T. Tsuji, T. Hiaki and M. Hongo, *Ind. Eng. Chem. Res.*, 1998, **37**, 1685.
9. T. Takagi, T. Sakura, T. Tsuji and M. Hongo, *Fluid Phase Equilib.*, 1999, **162**, 171.
10. K. Otake, T. Tsuji, I. Sato, T. Akiya, T. Sako and M. Hongo, *Fluid Phase Equilib.*, 2000, **171**, 175.
11. T. Takagi and K. Shiba, *High Temp - High Pressures*, 2001, **33**, 303.
12. T. Takagi, K. Fujita, D. Furuta and T. Tsuji, *Fluid Phase Equil.*, 2003, **212**, 279.
13. T. Takagi, K. Sawada, H. Urakawa, T. Tsuji and I. Cibulka, *J. Chem. Eng. Data*, 2004, **49**, 1657.

14. T. Tsuji, S. Tanaka, T. Hiaki and R. Saito, *Fluid Phase Equilib.*, 2004, **219**, 87.
15. *DME Handbook*, ed. Japan DME Forum, Ohmsha, Tokyo, 2007.
16. T. Tsuji, K. Ohya, T. Hoshina, T. Hiaki, K. Maeda, H. Kuramochi and M. Osako, *Fluid Phase Equilib.*, 2014, **362**, 383.
17. H. Miyamoto and M. Uematsu, *J. Chem. Thermodyn.*, 2007, **39**, 225.
18. H. Miyamoto and M. Uematsu, *J. Chem. Thermodyn.*, 2007, **39**, 827.
19. H. Miyamoto, T. Koshi and M. Uematsu, *J. Chem. Thermodyn.*, 2008, **40**, 1222.
20. E. C. Ihmels, K. Fisher and J. Gmehling, *Fluid Phase Equilib.*, 2005, **228–229**, 155.
21. E. C. Ihmels and E. W. Lemmon, *Fluid Phase Equilib.*, 2007, **260**, 36.
22. J. Wu, Z. Liu, B. Wang and J. Pan, *J. Chem. Eng. Data*, 2005, **50**, 956.
23. J. Wu, Z. Xu, Z. Liu and B. Wang, *J. Chem. Eng. Data*, 2005, **50**, 966.
24. S. Bobbo, M. Scattolini, L. Fedele, R. Camporese and V. De Stefani, *J. Chem. Eng. Data*, 2005, **50**, 1667.
25. K. Tanaka and Y. Higashi, *J. Chem. Eng. Data*, 2010, **55**, 2658.
26. F. A. M. M. Goncalves, C. S. M. F. Costa, J. C. S. Bernardo, I. Johnson, I. M. A. Fonseca and A. G. M. Ferreira, *J. Chem. Thermodyn.*, 2011, **43**, 105.
27. P. Hu and Z.-S Chen, *Fluid Phase Equilib.*, 204, **221**, 7.
28. L. Fedele, F. Pernechele, S. Bobbo and M. Scattolini, *J. Chem. Eng. Data*, 2007, **52**, 1955.
29. NIST Standard Reference Database 23, REFPROP, Ver. 9.1, 2013, (http://www.nist.gov/srd/nist23.cfm).
30. S. Bobbo, M. Scattolini, L. Fedele and R. Camporese, *Fluid Phase Equilib.*, 2006, **222–223**, 291.
31. L. Fedele, J. S. Brown, L. Colla, A. Ferron, S. Bobbo and C. Zilio, *J. Chem. Eng. Data*, 2012, **57**, 482.
32. J. B. Brown, G. Nicola, C. Zillio, L. Fedele, S. Bobbo and F. Palonara, *J. Chem. Eng. Data*, 2012, **57**, 3710.
33. Y. Wu, B. Bamgbade, K. Liu, M. A. McHugh, H. Baled, R. M. Enick, W. A. Burgess, D. Tapriyal and B. D. Morreale, *Fluid Phase Equilib.*, 2011, **311**, 17.
34. I. M. Abdulagatov, J. T. Safarov, F. Sh. Aliyev, M. A. Talibov, A. N. Shahverdiyev and E. P. Hassel, *Fluid Phase Equilib.*, 2008, **268**, 21.
35. S. M. Rasulov, L. M. Radzhabova, I. M. Abdulagatov and G. V. Stepanov, *Fluid Phase Equilib.*, 2013, **337**, 323.
36. I. M. Abdulagatov, N. G. Polikhronidi, T. J. Bruno, R. G. Batyrova and G. V. Stepanov, *Fluid Phase Equilib.*, 2008, **263**, 71.
37. A. M. Anderson, T. R. Ross, M. R. Ernst and M. K. Carroll, *J. Chem. Eng. Data*, 2008, **53**, 1015.
38. M. Kariznovi, H. Nourozieh and J. Adedi, *J. Chem. Thermodyn.*, 2013, **57**, 189.
39. D. Kodama, M. Kanakubo, M. Kokubo, S. Hashimoto, H. Nanjo and M. Kato, *Fluid Phase Equilib.*, 2011, **302**, 103.

40. M. Kariznovi, H. Nourozieh and J. Adedi, *J. Chem. Thermodyn.*, 2013, **57**, 408.
41. R. Seifried and F. Temelli, *J. Chem. Eng. Data*, 2013, **55**, 2410.
42. I. Tsivintzelis, D. Missopolinou, K. Kalogiannis and C. Panayiotou, *Fluid Phase Equilib.*, 2004, **224**, 89.
43. L. Galicia-Luna and O. Elizalde-Solis, *Fluid Phase Equilib.*, 2010, **296**, 46.
44. H. Nourozieh, M. Kariznovi and J. Abedi, *J. Chem. Thermodyn.*, 2013, **65**, 191.
45. O. Elizalde-Solis and L. Galicia-Luna, *Fluid Phase Equilib.*, 2010, **296**, 66.
46. M. Kariznovi, H. Nourozieh and J. Abedi, *J. Chem. Thermodyn.*, 2013, **67**, 227.
47. D. Kodama, M. Mato and T. Kaneko, *Fluid Phase Equilib.*, 2013, **357**, 57.
48. D. Kodama, T. Yagihashi, T. Hosoya and M. Kato, *Fluid Phase Equilib.*, 2010, **297**, 168.
49. M. Kato, K. Sugiyama, M. Sato and D. Kodama, *Fluid Phase Equilib.*, 2007, **257**, 207.
50. D. Missopolinou, I. Tsivintzelis and C. Panayiotou, *Fluid Phase Equilib.*, 2005, **238**, 204.
51. H. Nourozieh, M. Kariznovi and J. Abedi, *Fluid Phase Equilib.*, 2012, **318**, 96.
52. H. Nourozieh, M. Kariznovi and J. Abedi, *Fluid Phase Equilib.*, 2012, **314**, 102.
53. H. Nourozieh, M. Kariznovi and J. Abedi, *J. Chem. Thermodyn.*, 2012, **54**, 165.
54. M. Kariznovi, H. Nourozieh and J. Abedi, *J. Chem. Eng. Data*, 2011, **56**, 3669.
55. M. Kariznovi, H. Nourozieh and J. Abedi, *J. Chem. Eng. Data*, 2012, **57**, 137.
56. D. Kodama, M. Ogawa, T. Kimura, H. Tanaka and M. Kato, *J. Chem. Eng. Data*, 2002, **47**, 916.
57. A. Haneda, T. Seki, D. Kodama and M. Kato, *J. Chem. Eng. Data*, 2006, **51**, 268.
58. D. Kodama, R. Sato, A. Haneda and M. Kato, *J. Chem. Eng. Data*, 2005, **50**, 122.
59. D. Kodama, J. Miyazaki, M. Kato and T. Sako, *Fluid Phase Equilib.*, 2004, **219**, 19.
60. D. Kodama, R. Sato, A. Haneda and M. Kato, *J. Chem. Eng. Data*, 2005, **50**, 1902.
61. D. Kodama, T. Seki and M. Kato, *Fluid Phase Equilib.*, 2007, **261**, 99.
62. H. Nourozieh, B. Bayestehparvin, M. Kariznovi and J. Abedi, *J. Chem. Eng. Data*, 2013, **58**, 1236.
63. H. Nourozieh, M. Kariznovi and J. Abedi, *J. Chem. Thermodyn.*, 2013, **58**, 377.
64. H. Nourozieh, M. Kariznovi and J. Abedi, *Fluid Phase Equilib.*, 2013, **337**, 246.

65. M. Kariznovi, H. Nourozieh and J. Abedi, *Fluid Phase Equilib.*, 2012, **334**, 30.
66. M. Kariznovi, H. Nourozieh and J. Abedi, *J. Chem. Eng. Data*, 2012, **57**, 2535.
67. H. Nourozieh, M. Kariznovi and J. Abedi, *J. Chem. Eng. Data*, 2012, **57**, 2513.
68. E. Jenab and F. Temelli, *J. Supercrit. Fluids*, 2012, **70**, 57.
69. M. J. Venter, P. Willems, S. Kareth, E. Weidner, N. J. M. Kuipers and A. B. de Haan, *J. Supercrit. Fluids*, 2007, **41**, 195.
70. R. Akasaka, K. Tanaka and Y. Higashi, *Int. J. Refrig.*, 2013, **36**, 1341.
71. Y. Higashi, *Fluid Phase Equilib.*, 2004, **219**, 99.
72. Y. Zhou, S. Li, Q. Zhai, Y. Jin and M. Hu, *J. Chem. Eng. Data*, 2010, **55**, 1289.
73. R. R. Pawar, S. B. Nahire and M. Husan, *J. Chem. Eng. Data*, 2009, **54**, 1935.
74. M. E. Taboada, H. R. Gallguillos, T. A. Graber and S. Balado, *J. Chem. Eng. Data*, 2005, **50**, 264.
75. B. A. Younglove and J. F. Ely, *J. Phys. Chem. Ref. Data*, 1987, **16**, 557.
76. R. W. Hankinson and G. H. Thomson, *AIChE J.*, 1979, **25**, 653.
77. R. C. Reid, J. M. Prausnitz and B. E. Poling, *The Properties of Gases & Liquids*, McGraw-Hill, New York, 4th edn, 1987.
78. G. Soave, *Chem. Eng. Sci*, 1972, **27**, 1197.
79. H. G. Rackett, *J. Chem. Eng. Data*, 1970, **15**, 514.
80. C. F. Spencer and R. P. Danner, *J. Chem. Eng. Data*, 1972, **17**, 236.
81. V. L. Bhirud, *AIChE J.*, 1978, **24**, 880.
82. N. K. Patel and M. H. Joshipura, *Procedia Eng.*, 2013, **51**, 386.
83. M. Kato, M. Kokubo, K. Ohashi, A. Sato and D. Kodama, *Open Thermodyn. J.*, 2009, **3**, 1.
84. E. C. Ihmels and E. W. Lemmon, *Fluid Phase Equilib.*, 2005, **228–229**, 173.
85. E. C. Ihmels, E. W. Lemmon and J. Gmehling, *Fluid Phase Equilib.*, 2003, **207**, 111.
86. J. Pan, X. Rui, X. Zhao and L. Qju, *Fluid Phase Equilib.*, 2012, **321**, 10.
87. X. Rui, J. Pan and Y. Wang, *Fluid Phase Equilib.*, 2013, **341**, 78.
88. J. F. Estela-Uribe, *Fluid Phase Equilib.*, 2006, **246**, 64.
89. J. F. Estela-Uribe, *Fluid Phase Equilib.*, 2010, **287**, 95.
90. J. F. Estela-Uribe, *Fluid Phase Equilib.*, 2013, **354**, 326.
91. J. F. Estela-Uribe, *Fluid Phase Equilib.*, 2006, **356**, 229.
92. D. Peng and D. B. Robinson, *Eng. Chem. Fundam.*, 1976, **15**, 59.
93. K. Magoulas and D. Tassios, *Fluid Phase Equilib.*, 1990, **56**, 119.
94. J. C. de Hemptinne and P. Ungerer, *Fluid Phase Equilib.*, 1995, **106**, 81.
95. W. Ji and D. A. Lempe, *Fluid Phase Equilib.*, 1997, **130**, 49.
96. L. Wang and J. Gmehling, *J. Chem. Eng. Sci.*, 1999, **54**, 3885.
97. H. Hinojosa-Gómez, J. F. Barragán-Aroche and E. R. Bazúa-Rueda, *Fluid Phase Equilib.*, 2010, **298**, 12.
98. H. Lin and Y. Duan, *Fluid Phase Equilib.*, 2005, **233**, 194.

99. M. Nazarzadeh and M. Moshfeghian, *Fluid Phase Equilib.*, 2012, **337**, 214.
100. H. Baled, R. M. Enick, Y. Wu, M. A. McHugh, W. Burgess, D. Tapriyal and B. D. Morreale, *Fluid Phase Equilib.*, 2012, **317**, 65.
101. A. M. Abudour, S. A. Mohammad, R. L. Robinson and K. A. M. Gasem, *Fluid Phase Equilib.*, 2012, **335**, 74.

CHAPTER 11

Critical Behaviour: Pure Fluids and Mixtures

CLAUDIO A. CERDEIRIÑA,*[a] PATRICIA LOSADA-PÉREZ,[b] GERMÁN PÉREZ-SÁNCHEZ[c] AND JACOBO TRONCOSO*[a]

[a] Departamento de Física Aplicada, Universidad de Vigo - Campus del Agua, 32004 Ourense, Spain; [b] Institute for Materials Research IMO, Hasselt University, Wetenschapspark 1, B-3590, Diepenbeek, Belgium; [c] Department of Chemistry and Biochemistry, Faculty of Sciences, University of Porto, Rúa Campo Alegre 687 / 4169 - 007, Porto, Portugal
*Email: calvarez@uvigo.es; jacobotc@uvigo.es

11.1 Introduction

Thermodynamic properties display anomalies near critical points that are described by power laws which are based on the renormalisation group theory.[1,2] Asymptotically close to the critical point, a property y behaves as a function of a state variable x according to $y \approx \Gamma|x-x_c|^{\varepsilon}$, where x_c is the value of x at criticality while Γ and ε denote the critical amplitude and critical exponent, respectively.[3,4]

Liquid–vapour criticality of pure fluids belongs to the universality class of the three-dimensional Ising model (3D-Ising). The concept of universality is a fascinating one in that it implies that systems of markedly different physical nature, like fluids and magnets, exhibit common, universal features near their respective critical points. Thus, upon establishing appropriate analogies between thermodynamic variables, the power laws are the same,[5] *i.e.*, they have the same mathematical form and the critical exponents take on the same values for all systems of a given universality class. Critical

Volume Properties: Liquids, Solutions and Vapours
Edited by Emmerich Wilhelm and Trevor M. Letcher

Published by the Royal Society of Chemistry, www.rsc.org

amplitudes are, however, nonuniversal (*i.e.*, system-dependent), but some ratios among them are universal.

Perhaps the best illustration of fluid–magnet analogies has to do with the spontaneous magnetisation M of a ferromagnet, which goes to zero as the Curie point is approached. This can be expressed as: $M \approx M_0|t|^{\beta}$, where $t \equiv (T - T_c)/T_c$ is usually referred to as the temperature critical deviation and $\beta \approx 0.326$ is the corresponding critical exponent. The fluid analogue of the spontaneous magnetisation curve is the liquid–gas coexistence curve in the density–temperature $\rho - T$ plane (see Figure 11.1);[6] in fact, $\rho_\sigma^{\text{liq}} - \rho_\sigma^{\text{vap}} \approx B|t|^{\beta}$, where σ denotes coexistence. It was first noted by Guggenheim in 1945 that experimental data for fluids reveal $\beta \approx 1/3$.[7] That result was in contradiction with the predicted value from mean-field theory, $\beta \approx 1/2$. Nowadays, there is no doubt that both theory and experiment yield $\beta \approx 1/3$.

The density of coexisting liquid and vapour phases is represented by an expansion involving higher-order terms beyond the leading one, $|t|^{\beta}$.[8] In the first instance, one encounters correction-to-scaling terms, the first of which is $|t|^{\beta+\theta}$, with $\theta \approx 0.52$. There are also higher-order contributions that reflect the asymmetry of the liquid–gas coexistence curve. They are absent in ferromagnetic–paramagnetic transitions: the spontaneous magnetisation curve displays a full, trivial symmetry upon magnetic field reversal.

The nature of such asymmetry-related corrections has remained unclear for many decades. This is one of the reasons why liquid–gas criticality has received a great deal of attention. In particular, the critical behaviour of the diameter of the coexistence curve in the $\rho - T$ plane, namely the mid-points of the phase boundary $\rho_d(T) \equiv (\rho_\sigma^{\text{liq}} + \rho_\sigma^{\text{vap}})/2$, has played a pivotal role. In the early 70s, various statistical mechanical models supported a $|t|^{1-\alpha}$

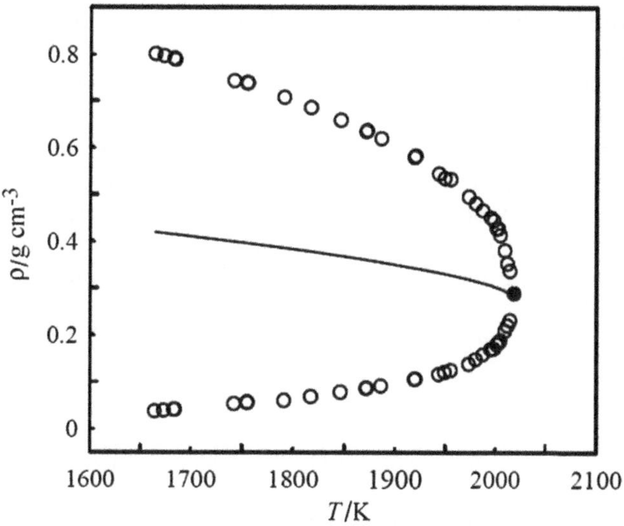

Figure 11.1 Coexistence curve of rubidium in the density–temperature plane.[6] The solid line represents the coexistence-curve diameter.

(with $\alpha \approx 0.109$) singularity, thereby suggesting the breakdown of the classical law of the rectilinear diameters. Deviation from the law became evident shortly afterwards, from experiments, but it has been a matter of debate until recently whether the $|t|^{1-\alpha}$ anomaly is accompanied by a $|t|^{2\beta}$ term which is dominating at near-criticality since $2\beta < 1-\alpha$.[8] To accommodate the $|t|^{1-\alpha}$ anomaly, the original scaling formulation of fluids had to be modified leading to what has been traditionally termed revised scaling.[9] A complete scaling ansatz that augments the phenomenology predicted by revised scaling needs to be introduced in order to account for the $|t|^{2\beta}$ term.[8] It is important to note that the behaviour of $\rho_d(T)$ is related to the so-called Yang–Yang dilemma: as long as the $|t|^{2\beta}$ anomaly exists, the chemical potential is nonanalytical.[10] Therefore, accurately determining the critical behaviour of ρ_d at near-criticality may reveal what is the correct thermodynamic scaling formulation of asymmetric fluid criticality. Despite difficulties inherent in experimentation in the liquid–gas critical region that make the detailed study of ρ_d a complicated task,[11–13] the validity of complete scaling is now accepted. Thus:

$$\rho_\sigma^{\text{liq,vap}} = \rho_c\left\{ 1 + A_{2\beta}|t|^{2\beta} + A_{1-\alpha}|t|^{1-\alpha} + A_1|t| + \ldots \pm B|t|^{\beta}\left[1 + b_\theta|t|^{\theta} + \ldots\right]\right\}, \tag{11.1}$$

where + and – in ± apply to liquid and vapour, respectively.

The isolated liquid–gas critical point of a pure fluid develops a critical line upon the addition of a second component without changing the universality class. In fact, experiments have confirmed that $\beta \approx 0.326$.[14] As regards asymmetry-related corrections, we are not aware of any detailed study. For binary mixtures, much work has been done for liquid–liquid phase transitions, for which gravity effects are of little importance. As regards liquid–liquid coexistence, much attention has been focused on the behaviour of the diameter of the coexistence curve in the mole fraction–temperature plane $x_2 - T$.[15,16] This is because x_2 now plays the role that density has in pure-fluid criticality. Rather than the $\rho - T$ plane, the ideal volume fraction $\varphi_2 \equiv x_1 v_1^\circ / (x_1 v_1^\circ + x_2 v_2^\circ)$, where $\overset{\circ}{v}_i$ is the molar volume of pure component i, is the volumetric variable of interest. This is closely related to the partial molar density ρx_2 *via* $\rho x_2 = \varphi_2 / \overset{\circ}{v}_2$. For these two composition variables, the extension of complete scaling to liquid–liquid criticality predicts:[17]

$$x_{2,\sigma}^{\pm} = x_c + A_{2\beta}^{x_2}|t|^{2\beta} + A_{1-\alpha}^{x_2}|t|^{1-\alpha} + A_1^{x_2}|t| + \ldots \pm B^{x_2}|t|^{\beta}\left[1 + b_\theta^{x_2}|t|^{\theta} + \ldots\right], \tag{11.2}$$

$$\begin{aligned}(\rho x_2)_\sigma^{\pm} = (\rho x_2)_c\Big\{ & 1 + A_{2\beta}^{\rho x}|t|^{2\beta} + A_{1-\alpha}^{\rho x}|t|^{1-\alpha} + A_1^{\rho x}|t| + \ldots \\ & \pm B^{\rho x}|t|^{\beta}\left[1 + b_\theta^{\rho x}|t|^{\theta} + \ldots\right]\Big\},\end{aligned} \tag{11.3}$$

where + and – apply to the two coexisting liquid phases.

Following this, the universality class of ionic criticality has been elucidated in the light of experiments on liquid–liquid phase transitions. Coexistence curves have been the subject of many discussions during the 1990s. Singh and Pitzer reported that $\beta \approx 0.5$, the mean-field value, for a solution of triethyl-*n*-hexylammonium triethyl-*n*-hexylboride in diphenyl ether.[18,19] A controversy arose whether the critical behaviour of systems containing charged particles is classical or Ising-like.[20] Plenty of theoretical and experimental work led to the conclusion that ionic criticality belongs to the 3D-Ising universal class.[21,22] This has now been widely accepted for mixtures, simple or complex.[4] In addition, it was noted that liquid–liquid phase transitions in ionic solutions display particular non-universal features so that systems are now classified as Coulombic or solvophobic.

Liquid–liquid phase transitions allow the behaviour of the density in the one-phase region to be evaluated. Along a path of approach to criticality of constant composition, asymptotically close to the critical point: $\rho \approx R_1^+ |t|^{1-\alpha}$. Detecting such a weak singularity, as well as checking the thermodynamic consistency between the density and the isobaric heat capacity, has been the subject of a number of works.[23]

Here we review all these issues. Section 11.2 describes all relevant experimental information regarding ρ_d for pure fluids, while the situation of coexistence-curve diameters in liquid–liquid phase equilibria is described in Section 11.3. As will be shown, all this information points towards the validity of the concept of 'complete scaling' as the correct theory of fluid–fluid criticality. Furthermore, Section 11.3 contains a description of ionic criticality. The behaviour of density in the one-phase region and its thermodynamic consistency with the heat capacity are treated in Section 11.4. Some concluding remarks are made in Section 11.5.

11.2 Coexistence Curves in Pure-Fluid Criticality

Experimental reports in the early 1970s indicated that a deviation from the law of the rectilinear diameters should not be ruled out. The first clear manifestation of such a phenomenon was a downward shift in the diameter of sulphur hexafluoride (SF_6), encountered in 1974 by Weiner *et al.*[24] Further studies for insulating fluids[25–28] were discussed and summarised in 1987.[29] They concluded that all available evidence was consistent with a $|t|^{1-\alpha}$ singularity, in accord with the accepted scaling formulation of that time. In 1998, Shimanskaya and Shimanskii[30] pointed out that, in contradiction with previous analyses, the data produced by Weiner *et al.* for SF_6 suggest a stronger anomaly than the $|t|^{1-\alpha}$ case, and, as a result, did not support the theory. Accurate measurements in 1985 for some alkali metals also showed remarkably high critical anomalies.[31] With the advent of complete scaling, the data for SF_6, Rb and Cs were reviewed. It was noted that the diameters of such fluids compare favourably with those of the hard-core square-well fluid and the restrictive primitive model electrolyte (as obtained from careful simulations).[32] A systematic study of all those data pointed towards the existence of $|t|^{2\beta}$ terms.[33]

Experimental data for SF_6 were obtained from ε measurements and converted to ρ by means of a separate experiment that was never published.[24] Data for N_2, Ne and HD[28] were also derived from ε measurements with the aid of the Clausius–Mossotti (CM) equation:

$$\frac{\varepsilon - 1}{(\varepsilon + 2)\rho} = \frac{4\pi}{3} < \alpha > , \tag{11.4}$$

where $<\alpha>$ denotes the average molecular polarisability. Similarly, for C_2H_4 and C_2H_6, original measurements of the refractive index n[25–27] were converted to ρ using the Lorentz–Lorenz (LL) equation:

$$\frac{n^2 - 1}{(n^2 + 2)\rho} = \frac{4\pi}{3} < \alpha > . \tag{11.5}$$

In turn, Wagner and coworkers reported data for all the above fluids[34–37] from direct ρpT measurements. Thus, results from two sources were available for those four fluids. Data for C_3H_8 were also determined from calorimetric measurements.[38] From the combination of refractive index and magnetic float densitometry measurements, Shimanskii and coworkers obtained data for Freon-113 and C_7H_{16}.[39,40] Finally, data for Rb[31] were obtained using ρpT measurements.

A common feature is that, when not linear, the diameter bends towards the gas branch of the coexistence curve. However, for N_2, Ne, HD, C_2H_4 and C_2H_6, ρ_d apparently does not show any significant curvature, but rather a slight deviation from linearity near the critical point that can be accounted for without a $|t|^{2\beta}$ term. A (slightly) different picture is encountered for SF_6, C_3H_8, C_7H_{16}, Freon-113 and Rb, for they display higher curvature and so the $|t|^{2\beta}$ singularity is needed for an appropriate description. Figures 11.2 and 11.3 show those data.[6] Values for the effective exponents, as determined from log–log plots, embody all the above-discussed features (see Figures 11.4 and 11.5).[6]

Data from Wagner and collaborators suggest a linear diameter. It is important to mention that, as explicitly reported, their studies yielded the classical β value, *i.e.*, 0.5. Specifically, those authors state that critical exponents in fluid criticality may be classical when obtained from earth-bounded experiments, while they are Ising-like in the absence of gravity. This suggests that gravity effects, which are dramatic for this sort of experiment could not have been properly accounted for. More detailed comments on the work of Wagner and coworkers have been provided by Sengers and Shanks.[4]

In spite of what seems to be compelling evidence for the deviation from the law of the rectilinear diameters and even for the existence of a stronger singularity asymptotically close to T_c than a $|t|^{1-\alpha}$ one, care must be taken with various issues that cast doubts on the reliability of experiments. First,

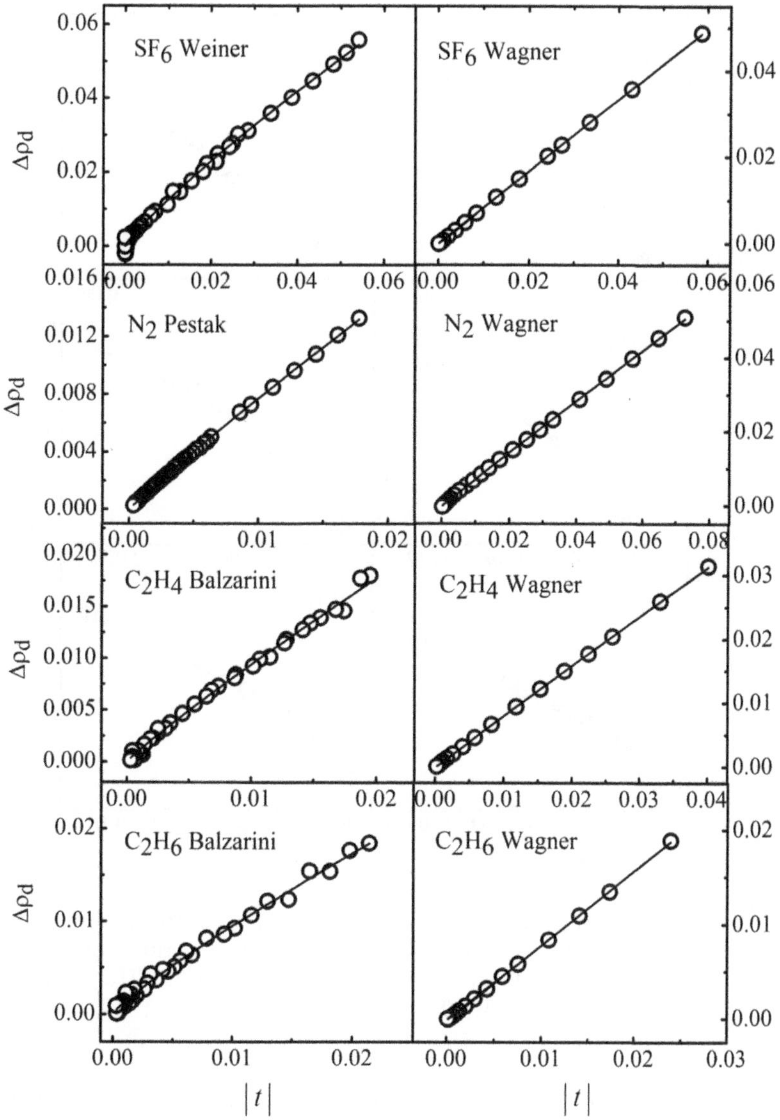

Figure 11.2 Dimensionless reduced diameter of the coexistence curve $\Delta\rho_d \equiv (\rho^{liq} + \rho^{vap})/(2\rho_c - 1)$ as a function of the temperature critical deviation t for SF_6, N_2, C_2H_4 and C_2H_6 from two different experimental sources (see text).[6] The solid line is included to guide the eye.

gravity effects in the pure-fluid critical region are severe,[11–13] and so a density profile along the sample is created as a result. Errors are reduced by keeping the height of the measuring cell to a minimum. Figure 11.6 illustrates how gravity can change the shape of the ρ_d curve of C_7H_{18}.[6] A second matter of concern has to do with the use of non-linear equations to

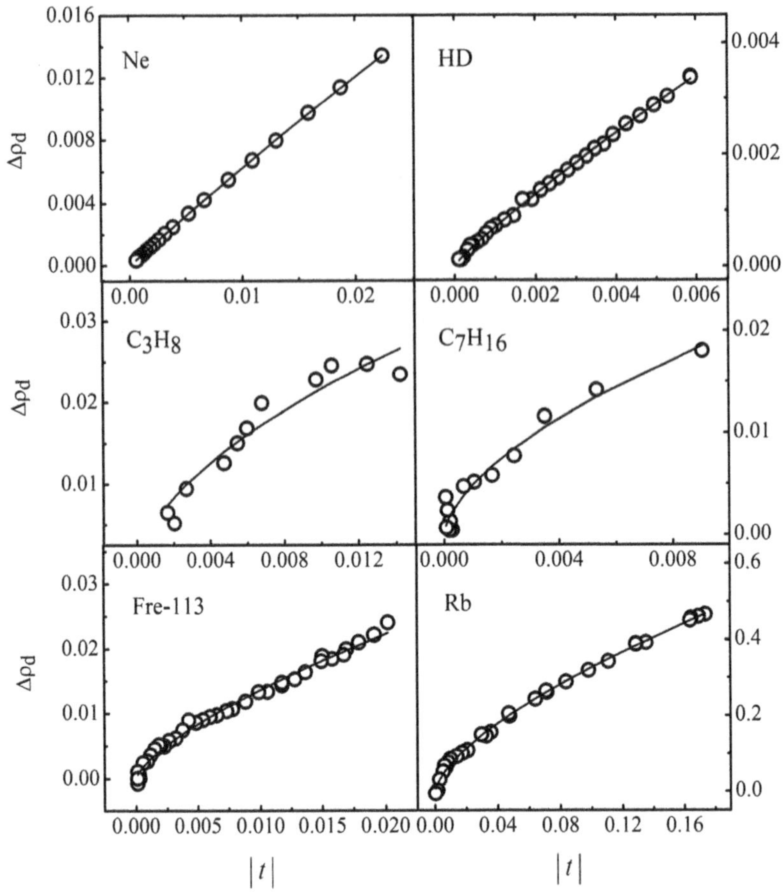

Figure 11.3 Dimensionless reduced diameter $\Delta\rho_d \equiv (\rho^{liq} + \rho^{vap})/(2\rho_c - 1)$ of the coexistence curve as a function of the temperature critical deviation t for Ne, HD, C_3H_8, C_7H_{16}, Fre-113 and Rb (see text).[6] The solid line is included to guide the eye.

transform primary, raw data to density. This is the case in ε and n measurements, for which, as commented, the non-linear CM and LL equations were employed. CM has been used under the assumption that the average molecular polarisability is constant. Such a condition holds for the case of dilute gases; however, in the critical region we have a condensed state. Therefore, one is inevitably introducing some error, the magnitude of which is difficult to quantify, when using CM with $\alpha \equiv$ constant. Understandably, the situation for n measurements is the same. Even in the case where the above difficulties are overcome, problems in detecting $|t|^{2\beta}$ anomalies remain. They are inherent to realistically separating such contributions from the $|t|^{1-\alpha}$ and $|t|$ ones, for they are characterized by critical-exponent values that are closely spaced numerically.

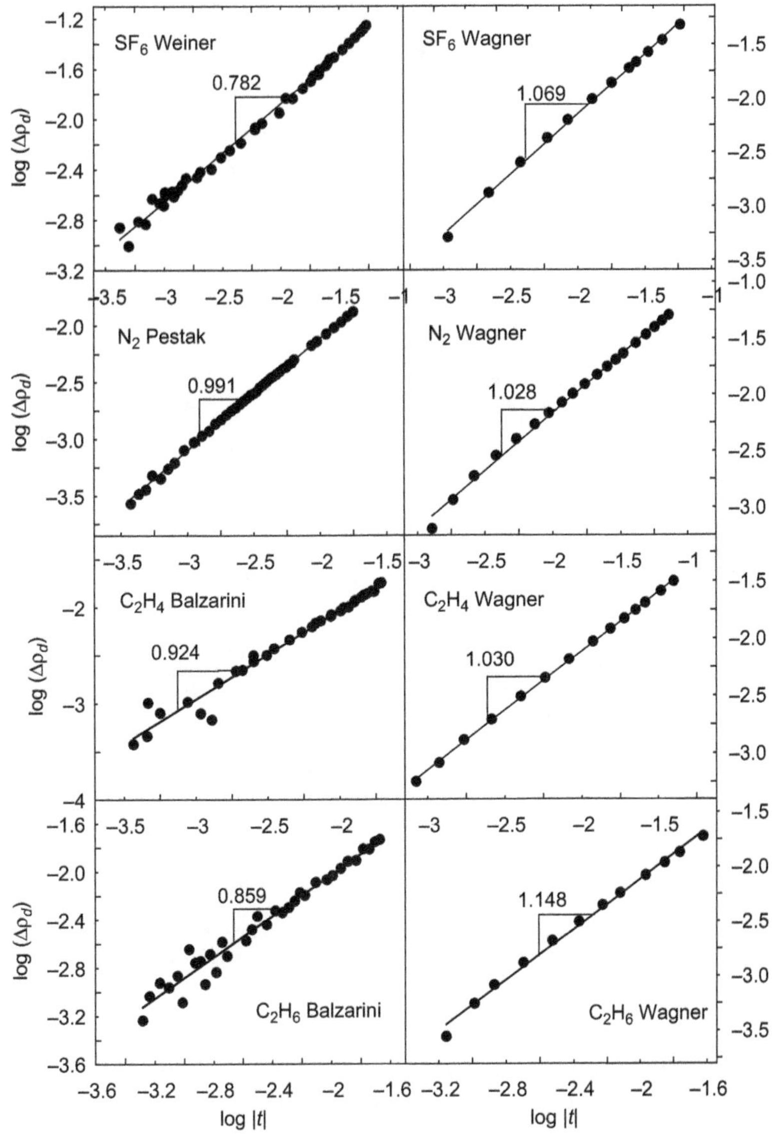

Figure 11.4 Log–log plot for SF_6, N_2, C_2H_4 and C_2H_6 from two different experimental sources.[6] Points are experimental data; solid lines are linear fits to experimental data.

11.3 Coexistence Curves in Liquid–Liquid Criticality

11.3.1 Experimental Methods

The determination of liquid–liquid coexistence curves is usually carried out *via* two main approaches. The first one (I) consists of the visual observation

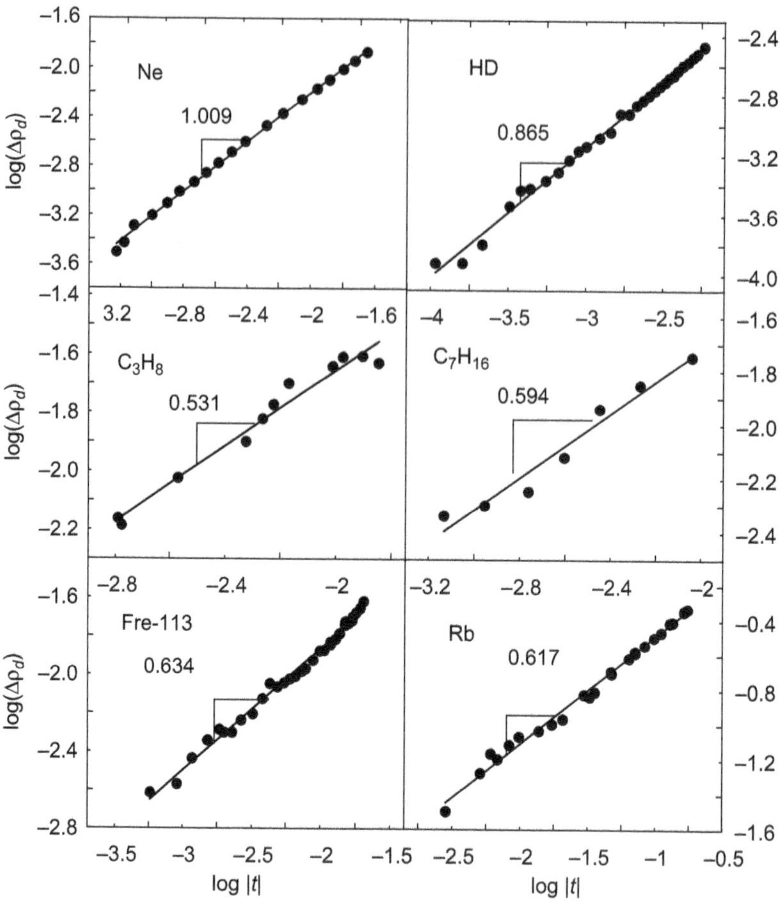

Figure 11.5 Log–log plot for Ne, HD, C_3H_8, C_7H_{16}, Fre-113 and Rb. Points are experimental data;[6] solid lines are linear fits to experimental data.

of the phase separation temperature for a homogeneous mixture with a fixed composition: a temperature scan is made in the one-phase region until phase separation occurs. To obtain the $x_2 - T$ coexistence curve, a number of mixtures of varying mole fraction must be studied [see Figure 11.7(a)]. In the second method (II) a mixture with near-critical composition is prepared. Then, at a fixed temperature, the refractive index of the two coexisting phases is determined, and the $n - T$ coexistence curve is completed from a collection of measurements at various temperatures [see Figure 11.7(b)].

There are two ways for converting n data to x_2 and ρx_2. Method (IIa) uses the LL equation as extended to mixtures:

$$\frac{n^2-1}{(n^2+2)} = \frac{4\pi}{3}\left(<\alpha_1> \frac{\phi_1}{v_1^\circ} + <\alpha_2> \frac{\phi_2}{v_2^\circ} \right), \tag{11.6}$$

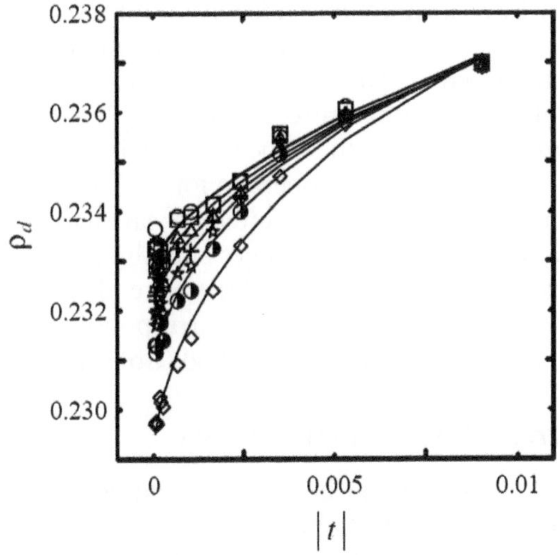

Figure 11.6 Coexistence-curve diameters ρ_d for C_7H_{16} at different cell heights.

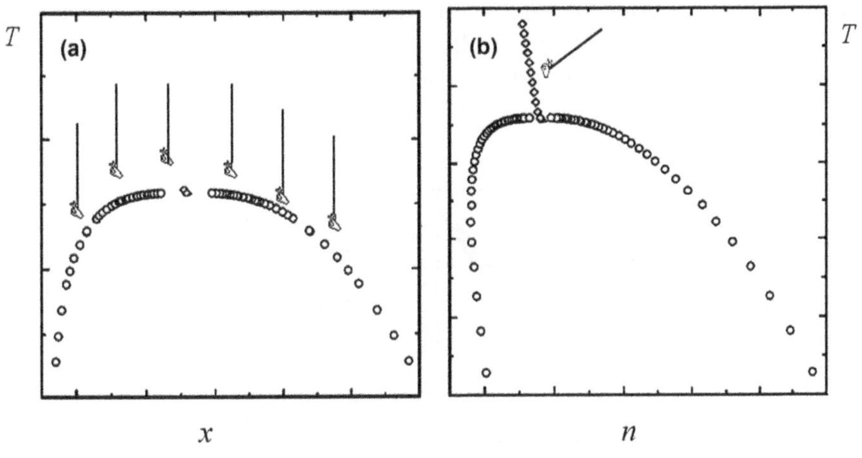

Figure 11.7 Illustration of experimental procedures for determining liquid–liquid coexistence curves: (a) method (I); (b) method (II).

where $<\alpha_i>$, φ_i, and $\overset{\circ}{v}_2$ are, respectively, the mean polarisability, volume fraction and molar volume of component *i*. Then, the LL equation provides $(\varphi - T)$ data, which are transformed to $\rho x_2 - T$ and $x_2 - T$ *via* the definition of volume fraction and its close connection with the partial molar density. In a second procedure (IIb), *n* is transformed to x_2 with the aid of a previously determined $n(x_2,T)$ function which describes the behaviour close to the phase boundary in the homogeneous region.[41]

Method (I) straightforwardly yields $(x_2 - T)$ data. However, a strict control of impurity levels is not feasible because of the need to manage many different samples. Such a shortcoming is absent for method (II), since only a near-critical mixture is studied. Variant (IIa) presents problems inherent in using an approximate equation, since it is assumed that $<\alpha_i>$ is constant or varies smoothly with temperature. Such a problem is absent in (IIb).[41] A general advantage of (II) with respect to (I) is that it allows the determination of the values of the properties for each coexisting phase at the same temperature; this is very well suited to calculating the coexistence-curve diameter at any temperature. To get such information from data obtained using method (I), manipulations must be made: given a mole fraction of one branch of the coexistence curve, the corresponding datum in the other branch is interpolated. In conclusion, method (IIb) seems the most reliable.

11.3.2 Mixtures of Molecular Liquids

Historically, the study of critical behaviour at liquid–liquid coexistence revealed the existence of non-vanishing $|t|^{2\beta}$ contributions to $x_{2,d}$ not supported by theory before the generalisation of complete scaling to liquid–liquid phase transitions.[17,42,43] Such a $|t|^{2\beta}$ contribution was considered spurious because of a "wrong choice" of the composition variable. In fact, when the mole fraction is converted to the volume fraction or, equivalently, the partial molar density ρx_2, the $|t|^{2\beta}$ singularity is frequently eliminated. Figure 11.8 displays this picture for a particular mixture, which is

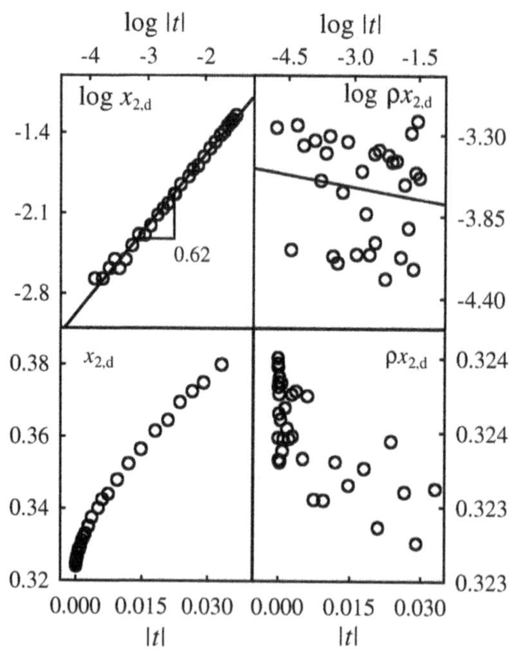

Figure 11.8 Log–log plots for the coexistence-curve diameter of the nitrobenzene + n-tetradecane system in the $x_2 - T$ and $\rho x_2 - T$ planes. Lower panels show diameters as a function of t.[44]

representative of many systems. It is then clear why ρx_2 has been claimed as a symmetrising variable, often referred to as the right composition variable or best order parameter. On the other hand, the absence of the $|t|^{2\beta}$ singularity in the $(\rho x_2 - T)$ plane is not a universal feature since some solutions exhibit large $|t|^{2\beta}$ contributions.[15,16]

Within the concept of complete scaling for liquid–liquid criticality, such $|t|^{2\beta}$ anomalies appear, as Equations (11.2) and (11.3) indicate, in their own right, so they may no longer be considered unrealistic.[17,42,43] Complete scaling then implies that there is no need to find a composition variable for which the $|t|^{2\beta}$ vanishes. It is now the basis for analysing data[41] and its validity has been checked against literature experimental data in systematic studies.[17,41] Under certain reasonable assumptions, a thermodynamic analysis predicts that the magnitude of the $|t|^{2\beta}$ term in the x_d diameter is driven by the solute/solvent molecular ratio. Such a size effect has been confirmed with great accuracy by experiments, which have also shown that the non-linear, singular diameter curves towards the phase rich in the component of lower molecular volume.[17]

11.3.3 Ionic Criticality

Liquid–liquid phase behaviour of ionic solutions has received a great deal of attention over the last two decades both from experimental and theoretical points of view. It was not until early this century that the nature of ionic criticality was demonstrated to be Ising-like. Almost simultaneously, it was realised that systems are classified as Coulombic or solvophobic.

To explain such an oversimplified classification, we appeal to the restrictive primitive model (RPM), the most basic picture of electrolytes.[20] It consists of hard spheres of diameter σ in a volume V that carry charges $+q$ and $-q$ in a vacuum. To convert it to a model for ionic solutions, one assumes that charged spheres are immersed in a medium of dielectric constant, ε, which is the solvent's dielectric constant. The usual temperature and composition variables in the model are:

$$T^* \equiv \frac{k_B T \varepsilon \sigma}{q^2}\ ;\ \rho^* \equiv \rho' \sigma^3, \qquad (11.7)$$

where k_B denotes the Boltzmann constant and ρ' the ionic number density, that is, the total number of ions per unit volume. Since the RPM displays a critical point, the associated critical phenomena are, obviously, dominated by Coulombic interactions. Then, to the extent that the observed behaviour for ionic solutions is close to or departed from that of the RPM, Coulombic (*i.e.*, driven by Coulombic interactions) or solvophobic (*i.e.*, driven by van der Waals and other interactions typical of molecular liquids) criticality is invoked.

A principal result is that ε plays a crucial role. Thus, experimental ρ_c^* values for solvents with low ε are small and close to that of the RPM, while solvents with high ε lead to an increase in ρ_c^*. As regard critical temperatures, T_c^* increases linearly as the dielectric constant does, from $T_c^* \approx T_c^*$(RPM) for low ε

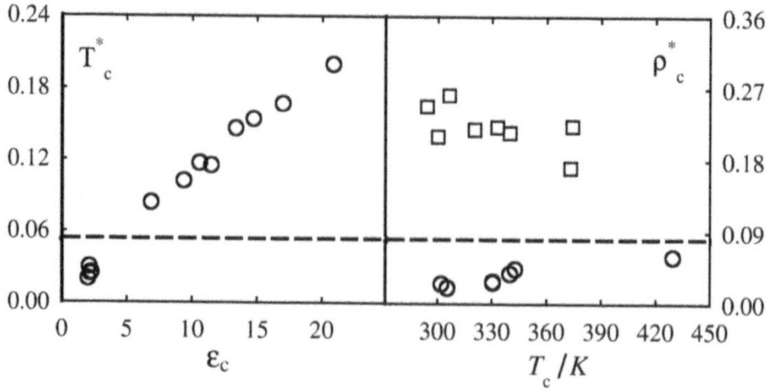

Figure 11.9 Critical temperatures T_c^* and critical densities ρ_c^* for ionic solutions. The critical temperature is represented as a function of solvent's dielectric constant. Circles and squares correspond to Coulombic and solvophobic systems, respectively.[44] Dashed lines are the RPM values.

values (*i.e.*, close to 1). On the basis that aprotic solvents (*e.g.*, alkanes) have low ε values, they must display Coulombic criticality. On the other hand, protic solvents like water and alcohols tend to exhibit solvophobic criticality. Water, with a high ε value, is a prototype of a solvophobic solvent, whereas an intermediate behaviour is expected for solutions of alcohols, given their intermediate ε values. Figure 11.9 illustrates these facts.[41]

The experimental values of variables like the critical mole fraction $x_{2,c}$ and the critical amplitudes of the width of the coexistence curve, B^{x_2} and $B^{\rho x_2}$ in Equations (11.2) and (11.3), also reflect the striking difference between Coulombic and solvophobic criticality (see Figure 11.10).[41] Thus, $x_{2,c}$ is significantly smaller for Coulombic systems. And the same holds for B^{x_2} and for $B^{\rho x_2}$. Data for ionic solutions with water and alcohols (recall, solvophobic) or alkanes (Coulombic) as solvents support these statements. It is worth undertaking an analysis of the behaviour of coexistence-curve diameters for Coulombic and solvophobic systems.

11.4 Thermodynamic Consistency Between the Density and the Heat Capacity

Close to the liquid–liquid critical point in the homogeneous region, theory predicts that the density shows a $|t|^{1-\alpha}$ anomaly, so that:

$$\rho = \rho_c + R^+_{1-\alpha}\,|\,t\,|^{1-\alpha} + R_1\,|\,t\,| + \dots \qquad (11.8)$$

This is a weak singularity since the $|t|^{1-\alpha}$ and linear terms are characterised by critical exponents which are numerically closely spaced. Furthermore, typical values of the critical amplitude $R^+_{1-\alpha}$ are small. This fact is illustrated in Figure 11.11,[6] which shows that the magnitude of the density anomaly is very small compared to that of the dielectric constant, also with a leading

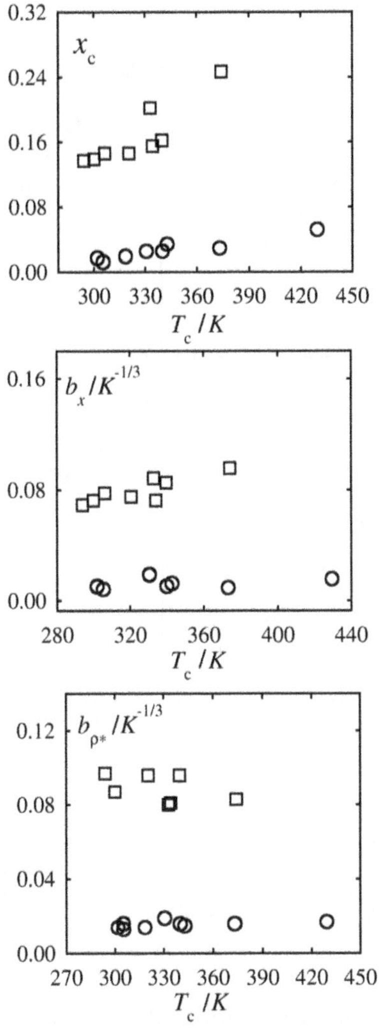

Figure 11.10 Critical mole fractions $x_{2,c}$ and critical amplitudes of the coexistence curve width in the $x_2 - T$ and $\rho^* - T$ planes for Coulombic (circles) and solvophobic (squares) systems as characterised by their T_c^* values.[44]

$|t|^{1-\alpha}$ singularity: in contrast to the former, the anomaly in ε can be observed by a direct observation of plotted data. Therefore, characterisation of density anomalies requires highly precise measurements and a careful data analysis for accurately separating the term $|t|^{1-\alpha}$ and the linear contributions. Figure 11.12 shows systematic deviations of experimental data from equation (11.1) with $R_{1-\alpha}^{+}=0$. As can be seen, such deviations become increasingly large as the critical point is approached. When the $|t|^{1-\alpha}$ term is included, no systematic deviation is found, thereby confirming the presence of the previously mentioned anomaly.

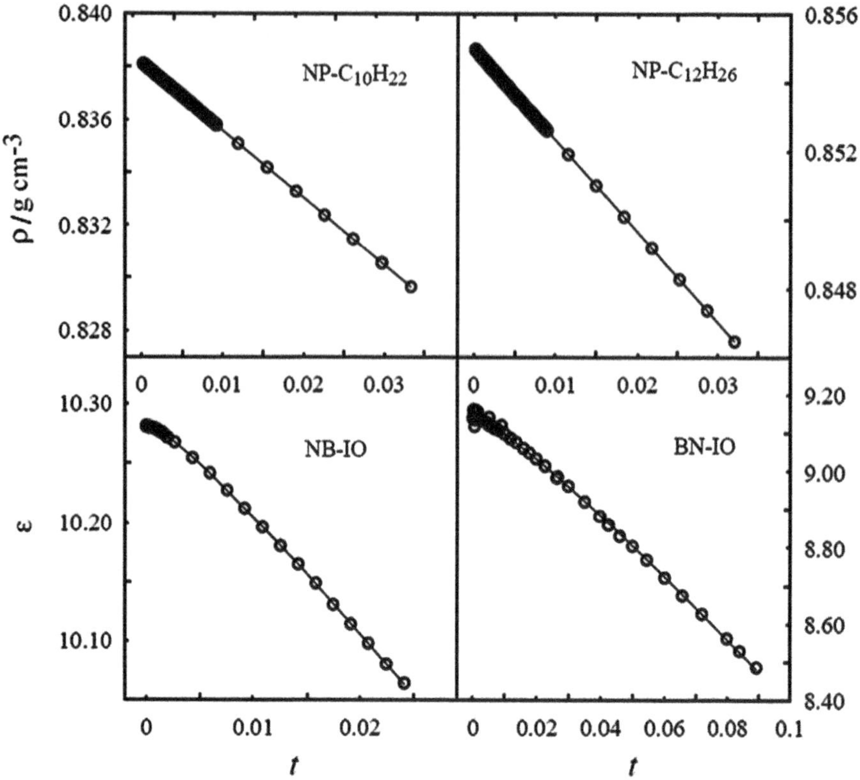

Figure 11.11 The temperature dependence of the density ρ for {1-nitropropane + n-decane ($NP\text{-}C_{10}H_{22}$)} and {1-nitropropane + n-dodecane ($NP\text{-}C_{12}H_{26}$)} and of the dielectric constant ε for {nitrobenzene + isooctane (NB-IO)} and {benzonitrile + isooctane (BN-IO)} in the one-phase critical region. Open circles are experimental data;[6] (—) lines to guide the eye.

It has been shown that an accurate determination of $R^{+}_{1-\alpha}$ allows one to check the thermodynamic consistency between the density and the heat capacity near the liquid–liquid critical point. The problem is stated as follows. The isobaric thermal expansivity $\alpha_p \equiv (\partial \ln\rho/\partial\rho)_T$ diverges to infinity at criticality so that in the neighborhood of the liquid–liquid critical point:[16]

$$\alpha_p \approx \frac{R_1(1-\alpha)}{\rho_c T_c}\,|t|^{-\alpha}. \tag{11.9}$$

The critical behaviour of the isobaric heat capacity per unit volume is given by:[16]

$$C_p \approx \frac{A^{+}}{\alpha}\,|t|^{-\alpha}. \tag{11.10}$$

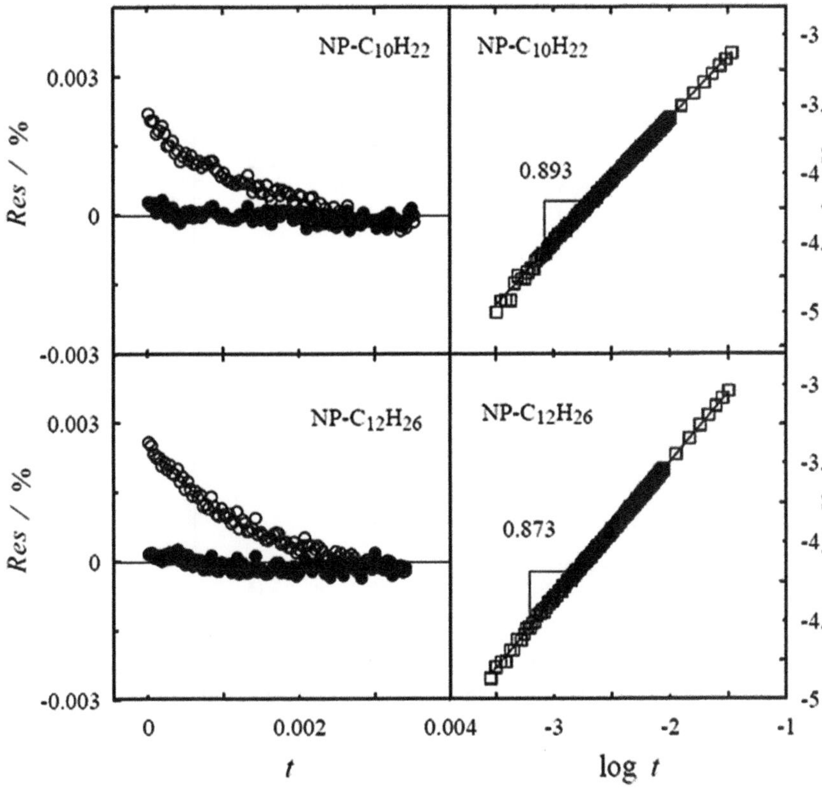

Figure 11.12 (a) Density residuals Res $\equiv 100(\rho - \rho_{calc})$ as a function of t for {1-nitropropane + n-decane ($NP-C_{10}H_{22}$)} and {1-nitropropane + n-dodecane ($NP-C_{12}H_{26}$)}; ρ^{calc} reads for the fitted values to Equation (11.8) with $R^+_{1-\alpha} = 0$ (open circles) and $R^+_{1-\alpha} \neq 0$ (filled circles). (b) Log–log plots for $\rho - \rho_c - R^+_{1-\alpha}t$. Solid lines are linear fits to experimental data.

Thermodynamic relations show that: $C_p = C_V + T\alpha_p(\partial p/\partial T)_V$, which when asymptotically close to the critical point converts to:

$$\left(\frac{dT}{dp}\right)_c = \pm\frac{R^+_{1-\alpha}\,\alpha(1-\alpha)}{\rho_c A^+}, \tag{11.11}$$

where + and − correspond to lower and upper consolute points, respectively. Here we have used the fact that the isochoric heat capacity remains finite at the critical point whereas $(\partial p/\partial T)_V \rightarrow (dp/dT)_c$ as $T \rightarrow T_c$.[45] Actual experimental devices allow one to determine A^+ and the slope of the critical line $(dT/dp)_c$ with great accuracy. On combining this information with $R^+_{1-\alpha}$ values, a consistency test is then feasible.

An analysis for 1-nitropropane-alkane systems[23] has clarified matters.[46] Figure 11.13 shows $(dT/dp)_c$ calculated from experimental A^+ and $R^+_{1-\alpha}$ values using Equation (11.11). As can be seen, the results compare favourably with directly measured data, and the agreement is very good within a scale of

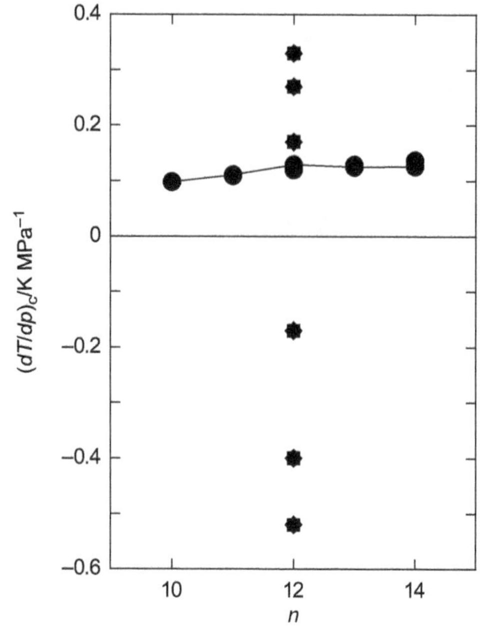

Figure 11.13 Slope of the critical line $(dT/dp)_c$ for {nitrobenzene + alkane (NB-C_nH_{2n+2})} [open circles] and other systems (stars); filled circles are the calculated values from A^+ and $R^+_{1-\alpha}$ *via* Equation (11.11).

typical variations in $(dT/dp)_c$. A careful, detailed analysis on how to accurately determine $R^+_{1-\alpha}$ has proved the key point for resolving this issue.[23]

Some final comments are in order. Dilatometry has been successfully employed in the past with a view to directly obtain α_p.[16] On the other hand, because of the very small compressibility of liquids, the (liquid–liquid) $|t|^{-\alpha}$ anomaly of the isothermal compressibility κ_T is almost undetectable using the current experimental methodologies.[16]

11.5 Concluding Remarks

This review is by no means exhaustive but it provides a survey of some topics dealing with the critical behaviour of the density and volumetric properties that have been studied experimentally over the past few years. One major issue is the complete scaling formulation of asymmetric fluid–fluid criticality. Actual experimental data lack the consistency required for accurately characterising $|t|^{2\beta}$ terms in the coexistence-curve diameters. Since heat capacity measurements have confirmed the existence of the Yang–Yang anomaly, there is no doubt that complete scaling is correct. The analogous effects for liquid–liquid criticality are far easier to detect. It is then fair to conclude that the predicted behaviour of coexistence curves in liquid–liquid phase transitions from complete scaling is a more settled issue than for liquid–vapour criticality in pure fluids.

Liquid–liquid phase transitions have proved very useful with a view to elucidating the universality class of ionic criticality. Quite interestingly, they reveal two types of critical behaviour, Coulombic and solvophobic. Thermodynamic consistency between the density and heat capacity near the liquid–liquid critical point is another issue that has recently found a satisfactory outcome.

It seems that, from an experimental viewpoint, there is little that remains to be done. Insights as to what are the microscopic mechanisms responsible for the effects predicted from complete scaling is a route to be analysed. Certainly, exploring exactly soluble statistical mechanical models may be a powerful tool for these purposes.

References

1. M. E. Fisher, *Rev. Mod. Phys.*, 1974, **46**, 597.
2. K. G. Wilson, *Rev. Mod. Phys.*, 1975, **47**, 773.
3. J. V. Sengers and J. M. Levelt-Sengers, in *Progress in Liquid Physics* ed. C. A. Croxton, Wiley, New York, 1978.
4. J. V. Sengers and J. G. Shanks, *J. Stat. Phys.*, 2009, **137**, 857.
5. H. E. Stanley, *Introduction to Phase Transitions and Critical Phenomena*, Oxford University Press, New York, 1971.
6. Plots taken from P. Losada-Pérez, PhD Thesis, University of Vigo, 2009.
7. E. A. Guggenheim, *J. Chem. Phys.*, 1945, **13**, 253.
8. Y. C. Kim, G. Orkoulas and M. E. Fisher, *Phys. Rev. E*, 2003, **67**, 061506.
9. J. J. Rehr and N. D. Mermin, *Phys. Rev. A*, 1973, **8**, 472.
10. M. E. Fisher and G. Orkoulas, *Phys. Rev. Lett.*, 2000, **85**, 696.
11. P. C. Hohenberg and M. Barmatz, *Phys. Rev. A*, 1972, **6**, 289.
12. M. R. Moldover, J. V. Sengers, R. W. Gammon and R. J. Hocken, *Rev. Mod. Phys.*, 1979, **51**, 79.
13. M. Barmatz, I. Hahn, J. A. Lipa and R. V. Duncan, *Rev. Mod. Phys.*, 2007, **79**, 1.
14. J. V. Sengers and J. M. H. Levelt-Sengers, *Annu. Rev. Phys. Chem.*, 1986, **37**, 189.
15. S. C. Greer and M. R. Moldover, *Annu. Rev. Phys. Chem.*, 1981, **32**, 233.
16. A. Kumar, H. R. Krishnamurthy and E. S. R. Gopal, *Phys. Rep.*, 1983, **98**, 57.
17. G. Pérez-Sánchez, P. Losada-Pérez, C. A. Cerdeiriña, J. V. Sengers and M. A. Anisimov, *J. Chem. Phys.*, 2010, **132**, 154502.
18. R. R. Singh and K. S. Pitzer, *J. Am. Chem. Soc.*, 1988, **110**, 8723.
19. R. R. Singh and K. S. Pitzer, *J. Chem. Phys.*, 1990, **92**, 6775.
20. M. E. Fisher, *J. Stat. Phys.*, 1994, **75**, 1.
21. H. Weingärtner and W. Schröer, *Adv. Chem. Phys.*, 2001, **116**, 1.
22. E. Luijten, M. E. Fisher and A. Z. Panagiotopoulos, *Phys. Rev. Lett.*, 2002, **88**, 185701.

23. P. Losada-Pérez, G. Pérez-Sánchez, C. A. Cerdeiriña, J. Troncoso and L. Romaní, *J. Chem. Phys.*, 2009, **130**, 044506 and references therein.
24. J. Weiner, K. H. Langley and N. C. Ford Jr., *Phys. Rev. Lett.*, 1974, **32**, 879.
25. D. A. Balzarini and M. Burton, *Can. J. Phys.*, 1974, **52**, 2011.
26. D. A. Balzarini and M. Burton, *Can. J. Phys.*, 1979, **57**, 1516.
27. J. R. de Bruyn and D. A. Balzarini, *Phys. Rev. A*, 1987, **36**, 5677.
28. M. W. Pestak and M. H. W. Chan, *Phys. Rev. B*, 1984, **30**, 274.
29. M. W. Pestak, R. E. Goldstein, M. H. W. Chan, J. R. de Bruyn, D. A. Balzarini and N. W. Ashcroft, *Phys. Rev. B*, 1987, **36**, 599.
30. E. T. Shimanskaya and Y. I Shimanskii, *High Temp. - High Pressures*, 1998, **30**, 635.
31. S. Jüngst, B. Knuth and F. Hensel, *Phys. Rev. Lett.*, 1985, **55**, 2160.
32. Y. C. Kim and M. E. Fisher, *Chem. Phys. Lett.*, 2005, **414**, 185.
33. J. T. Wang and M. A. Anisimov, *Phys. Rev. E*, 2007, **75**, 051107.
34. P. Novak, R. Kleirahm and W. Wagner, *J. Chem. Thermodyn.*, 1996, **28**, 1441.
35. P. Novak, R. Kleirahm and W. Wagner, *J. Chem. Thermodyn.*, 1997, **29**, 1157.
36. M. Funke, R. Kleirahm and W. Wagner, *J. Chem. Thermodyn.*, 2001, **33**, 735.
37. M. Funke, R. Kleirahm and W. Wagner, *J. Chem. Thermodyn.*, 2001, **34**, 2017.
38. I. M. Abdulagatov, L. N. Levina, Z. R. Zakatyaev and O. N. Mamchenkova, *J. Chem. Thermodyn.*, 1995, **27**, 1385.
39. L. M. Arykhovskaya, E. T. Shimanskaya and Y. I. Shimanskii, *Sov. Phys. JETP*, 1973, **37**, 848.
40. E. T. Shimanskaya, I. V. Bezruchko, V. I. Basok and Y. I. Shimanskii, *Sov. Phys. JETP*, 1981, **53**, 139.
41. Look for extensive work by Shen and coworkers at Lanzhou University, China.
42. C. A. Cerdeiriña, M. A. Anisimov and J. V. Sengers, *Chem. Phys. Lett.*, 2006, **424**, 414.
43. J. T. Wang, C. A. Cerdeiriña, M. A. Anisimov and J. V. Sengers, *Phys. Rev. E*, 2008, **77**, 031127.
44. Plots taken from G. Pérez-Sánchez, PhD Thesis, University of Vigo 2010.
45. R. B. Griffiths and J. C. Wheeler, *Phys. Rev. A*, 1970, **2**, 1047.
46. D. T. Jacobs and S. C. Greer, *Phys. Rev. E*, 1996, **54**, 5358.

CHAPTER 12

Ultrasonics 1: Speed of Ultrasound, Isentropic Compressibility and Related Properties of Liquids

AUGUSTINUS ASENBAUM,*[a] CHRISTIAN PRUNER*[a] AND EMMERICH WILHELM*[b]

[a] Department of Materials Research and Physics, Section for Experimental Physics, University of Salzburg, A-5020, Salzburg, Austria; [b] Institute of Physical Chemistry, University of Wien (Vienna), Waehringer Strasse 42, A-1090, Wien (Vienna), Austria
*Email: augustinus.asenbaum@sbg.ac.at; christian.pruner@sbg.ac.at; emmerich.wilhelm@univie.ac.at

12.1 Introduction

Ultrasound, that is, sound in the frequency range of roughly 20 kHz to 1 GHz, is widely used in industrial applications, such as cleaning, emulsification and non-destructive material testing, and in medical imaging where ultrasonography serves as a diagnostic technique used for the visualisation of internal body structures and organs, and for obstetric sonography. On a fundamental physico–chemical level, measurement of the speed of sound as a function of frequency, and thus of sound dispersion, and of the absorption of acoustic energy provide a wealth of information not only on equilibrium thermodynamic properties but also on transport coefficients and relaxation processes.[1–5]

Volume Properties: Liquids, Solutions and Vapours
Edited by Emmerich Wilhelm and Trevor M. Letcher

Published by the Royal Society of Chemistry, www.rsc.org

At sufficiently *low frequencies* f and *small amplitude* of the sound wave, to an excellent approximation (*i.e.* neglecting dissipative processes, such as those due to shear viscosity, volume viscosity (also known as bulk viscosity) and thermal conductivity), the *thermodynamic* speed of ultrasound ν_0 of a fluid at constant composition $\{x_i\}$ is related to the isentropic compressibility β_S (often loosely called the adiabatic compressibility) by

$$\nu_0^2 = \left(\frac{\partial P}{\partial \rho}\right)_S = \frac{1}{\rho \beta_S}, \tag{12.1}$$

where†

$$\beta_S \equiv -V^{-1}(\partial V/\partial P)_S = \rho^{-1}(\partial \rho/\partial P)_S. \tag{12.2}$$

Here, V is the molar volume, P is the pressure, $\rho \equiv m_\mathrm{m}/V = m/(nV)$ is the mass density, m is the mass, $m_\mathrm{m} = m/n$ denotes the molar mass, $n = \sum_i n_i$ is the total amount of substance while n_i is the amount of substance of component i, $\{x_i\}$ is the set of compositional variables with $x_i = n_i/\sum_i n_i$ denoting the mole fraction of i, and S is the molar entropy. While the requirement of a small amplitude of the sound wave is easily realised, the second assumption of low enough frequency is more delicate: the thermodynamic formalism may only be applied if the sound frequency is so low that the system is always in local equilibrium. However, when the sound period $1/f$ becomes comparable to a characteristic relaxation time of the liquid, say, associated with thermal relaxation, that is, with processes responsible for the exchange of energy between translational and internal degrees of freedom of the molecules, the sound speed becomes *frequency dependent*, that is $\nu(f) > \nu_0$. This point will be discussed below in some detail. Fortunately, in *many* organic liquids relaxation processes occur only at frequencies well *above* the ultrasonic range usually used in experiments, see for instance reference 6. Thus the speed of ultrasound determined with commercially available instruments is mostly, but *not* always, essentially the thermodynamic speed ν_0. However, particular care must be exercised when investigating liquids with molecules exhibiting *rotational isomerism*. For instance, ultrasonic absorption studies on liquid esters (alkyl alkanoates) have been used to study rotation about the C–O bond. The relaxation found is attributed to the perturbation of the equilibrium between the two planar rotamers formed by rotation about the ester C–O bond, and the relaxation frequencies determined at 296 K amount to[7] 0.6 MHz for ethyl formate, 10.5 MHz for ethyl acetate, and 12.2 MHz for ethyl propionate. Perhaps the most interesting class of liquids showing rotational isomerism about C–C bonds is that of halogen substituted alkanes.[8] For instance, in 1-chloro-2-methylpropane

---

†In this chapter the isentropic compressibility is represented by the symbol β_S and not by κ_S, as was recently recommended by IUPAC. Similarly, the isothermal compressibility is represented by the symbol β_T and not by κ_T. For their ratio, the symbol $\kappa \equiv \beta_T/\beta_S$ is used.

[$(CH_3)_2CH–CH_2Cl$, isobutyl chloride], three staggered rotamers may be identified by rotation about the central C–C bond: two degenerate enantiomeric states and a higher energy state where, in a Newman projection formula, Cl is between the terminal two CH_3 groups, *i.e.* in an *anti* position to H. At 273 K, the experimentally determined relaxation frequency is 36 MHz. Ultrasonic relaxation is also observed in 1-fluoro-1,1,2,2-tetrachloroethane: two degenerate enantiomeric rotamers and a state where, in a Newman projection formula, F is between the two chlorines, *i.e.* in an *anti* position to H. At 328 K, the relaxation frequency is 1.7 MHz. In summary, we reiterate that only speed of sound measurements at frequencies *well below* the dispersive region are a highly accurate means for obtaining thermodynamic speeds of ultrasound v_0 and other properties not so easily measured directly.

As shown in Chapter 1, for the *ratio* κ of the compressibilities of a constant composition fluid we have

$$\kappa \equiv \frac{\beta_T}{\beta_S} = \frac{C_P}{C_V}, \tag{12.3}$$

where

$$\beta_T \equiv -\frac{1}{V}\left(\frac{\partial V}{\partial P}\right)_T = \frac{1}{\rho}\left(\frac{\partial \rho}{\partial P}\right)_T \tag{12.4}$$

is the *isothermal compressibility*, C_P denotes the molar heat capacity at constant pressure (molar isobaric heat capacity), and C_V is the molar heat capacity at constant volume (molar isochoric heat capacity). For the *difference* between β_T and β_S we obtain

$$\beta_T - \beta_S = TV\alpha_P^2/C_P, \tag{12.5}$$

while

$$\beta_S^{-1} - \beta_T^{-1} = TV\gamma_V^2/C_V. \tag{12.6}$$

Here, at constant composition $\{x_i\}$,

$$\alpha_P \equiv V^{-1}(\partial V/\partial T)_P = -\rho^{-1}(\partial \rho/\partial T)_P \tag{12.7}$$

denotes the *isobaric expansivity*, and

$$\gamma_V \equiv (\partial P/\partial T)_V \tag{12.8}$$

is the *isochoric thermal pressure coefficient*. The quantities α_P, β_T and γ_V are commonly called the *mechanical coefficients* of a fluid, in contradistinction to the *isentropic coefficients* of which β_S is the most interesting one. The three *mechanical coefficients* are related as follows:

$$\gamma_V = \frac{\alpha_P}{\beta_T}, \tag{12.9}$$

and

$$\left(\frac{\partial \alpha_P}{\partial P}\right)_T = -\left(\frac{\partial \beta_T}{\partial T}\right)_P. \tag{12.10}$$

Using Equation (12.5) in conjunction with Equation (12.2) leads to

$$\kappa = 1 + \frac{T m_{\mathrm{m}} \alpha_P^2 v_0^2}{C_P}, \tag{12.11}$$

which is one of the *most important* equations in thermophysics. The majority of the isochoric heat capacity data for liquids reported in the literature has been obtained *indirectly* through use of Equation (12.11) *via*

$$C_V = C_P/\kappa, \tag{12.12}$$

that is to say, from experimentally determined molar isobaric heat capacities, isobaric expansivities and ultrasonic speeds[5,9–15] at sufficiently low frequencies. With modern equipment, these three quantities can be measured accurately and speedily, thereby making the indirect method for determining C_V of liquids quite attractive. In addition, a valuable *alternative* to the direct method for determining β_T, where hydrostatic pressure is applied to the fluid and the resulting volume change is measured, is also provided by Equation (12.11), since

$$\beta_T = \kappa \beta_S. \tag{12.13}$$

This approach is known as the *indirect* method for determining isothermal compressibilities and usually yields highly accurate results. Evidently, the most important use of β_S data obtained *via* speed of sound measurements is to calculate β_T and/or C_V using the appropriate equations given above. As pointed out by Rowlinson and Swinton,[16] the so-called mechanical coefficients α_P, β_T and γ_V are determined, to a high degree of accuracy, solely by the intermolecular forces, whilst the isentropic coefficients, to which they are related through the thermal coefficients, *i.e.* the heat capacities, and the thermal coefficients themselves depend also on *internal molecular* properties.

Alternative relations to Equation (12.1) are easily found: for instance,

$$v_0^2 = \frac{\kappa}{\rho \beta_T}, \tag{12.14}$$

and with Equation (1.126) of Chapter 1

$$v_0^2 = \frac{\kappa V^2}{m_{\mathrm{m}}} \left(\frac{\partial^2 F}{\partial V^2} \right)_T. \tag{12.15}$$

Equation (12.15) clearly indicates why high-precision speed of sound measurements are of such great value for the development of fundamental equations of state based on the molar Helmholtz energy F.

At frequencies where dispersion is not important, according to Equation (12.5) the isothermal pressure dependence of the density may be expressed as

$$\left(\frac{\partial\rho}{\partial P}\right)_T = \frac{1}{v_0^2} + \frac{Tm_\mathrm{m}\alpha_P^2}{C_P}. \tag{12.16}$$

This equation offers an alternative to the direct experimental route to high-pressure $P\rho T$ data, heat capacities and related quantities. Upon integration,[17–22] for the density at T and P we obtain

$$\rho(T,P) = \rho(T,P_\mathrm{ref}) + \int_{P_\mathrm{ref}}^{P} v_0^{-2}\mathrm{d}P + Tm_\mathrm{m}\int_{P_\mathrm{ref}}^{P}(\alpha_P^2/C_P)\mathrm{d}P, \tag{12.17}$$

where $\rho(T,P_\mathrm{ref})$ is the density at T and at a conveniently selected reference pressure P_ref. At low temperatures, *i.e.* below the normal boiling point temperature, typically the isobar at $P_\mathrm{ref} = 0.1$ MPa is selected, and the required reliable initial values are $\rho(T,P_\mathrm{ref})$ and thus $\alpha_P(T,P_\mathrm{ref})$, and $C_P(T,P_\mathrm{ref})$. The first integral is evaluated directly by fitting the isothermal high-pressure ultrasonic speed data $v_0(T,P)$ with suitably selected polynomials or Padé approximants (the pressure dependence of v_0 in organic liquids has recently been reviewed by Oakley *et al.*[23]). For the evaluation of the second integral one may use

$$\left(\frac{\partial C_P}{\partial P}\right)_T = -\frac{Tm_\mathrm{m}}{\rho}\left[\alpha_P^2 + \left(\frac{\partial\alpha_P}{\partial T}\right)_P\right] \tag{12.18}$$

and Equation (12.10) in conjunction with a successive integration algorithm. Thus starting at temperature T from pressure P_ref, the liquid density, isobaric expansivity, isobaric and isentropic compressibilities, isochoric thermal pressure coefficient, and isobaric and isochoric heat capacities at elevated pressures are obtained. The simplicity, rapidity and precision of this method makes it highly attractive for the determination of these thermodynamic quantities.

We note that the difference between the isobaric heat capacity and the isochoric heat capacity depends on *volume properties only* and is given by

$$C_P - C_V = TV\alpha_P^2/\beta_T, \tag{12.19}$$

Hence

$$\frac{C_V}{C_P - C_V} = \frac{\beta_S}{\beta_T - \beta_S}. \tag{12.20}$$

Equation (12.20) establishes a connection with *Rayleigh–Brillouin light scattering*.[6] For simple liquids (liquid noble gases), the ratio of the integrated intensity of the central, unshifted Rayleigh peak, I_R, and the integrated intensity of the *two* Brillouin peaks, $2I_B$, is given by the *Landau–Placzek* ratio, *i.e.*

$$\frac{I_R}{2I_B}=\frac{C_P-C_V}{C_V}=\kappa-1. \tag{12.21}$$

For *molecular* (normal) liquids, the ratio of the integrated intensity of the central, unshifted components of the scattered light (Rayleigh and Mountain) to the integrated intensity of the Brillouin peaks is a rather complicated expression,[24,25] and the ratio is *greater*[6] than $(\kappa-1)$. Evidently, if $C_P\approx C_V$ it will be rather difficult to observe the central Rayleigh peak. Liquid water, at temperatures around that of the density maximum, is such an interesting case and will be discussed below in Section 12.4. For glass-forming liquids, such as toluene, α-picoline, ethanol or glycerol, good agreement with theory is only observed at elevated temperatures, while at low temperatures the Landau–Placzek ratio becomes *significantly* larger.[26]

Relaxation processes not only cause absorption but they are also responsible for dispersion (for details we refer to the appropriate sections of the monograph by Herzfeld and Litovitz[1]). Note that few molecular collisions are sufficient to reach energy equipartition among external degrees of freedom (*i.e.* molecular translation and rotation), but *many* collisions are needed to change the energy distribution in the internal (vibrational) degrees.

Thermal relaxation in nonassociated liquids occurs *via* an interchange of energy between the sound wave and the internal (vibrational) molecular modes. The net effect is to introduce a frequency-dependent *volume* viscosity[6] η_v (also known as bulk viscosity) with relaxation time τ_v. In nonassociated liquids, where translation–vibration relaxation is the dominant feature, it can be shown[1] that for a *single* relaxation process the corresponding *dispersion of sound* may be described by

$$\frac{\nu(f)^2}{\nu_0^2}=1+\frac{(C_P-C_V)C_{\text{int}}}{(C_V-C_{\text{int}})C_P}\frac{\omega^2\tau_v^2}{1+\omega^2\tau_v^2}. \tag{12.22}$$

Here, $\nu(f)$ is the sound speed at frequency f, $\omega=2\pi f$ is the circular frequency, and C_{int} is the relaxing molar heat capacity due to internal (vibrational) degrees of freedom involved in the relaxation process. If sufficiently high frequencies (or in gases/vapours, sufficiently low pressures) can be applied to measure the entire dispersion curve, one obtains

$$\frac{\nu_\infty^2}{\nu_0^2}=1+\frac{(C_P-C_V)C_{\text{int}}}{(C_V-C_{\text{int}})C_P}, \tag{12.23}$$

where ν_∞ is the *high-frequency limit* of the speed of ultrasound. The relaxation frequency f'' associated with sound dispersion may be obtained from the halfpoint of the dispersion curve, *i.e.* from $\left[\left(\nu(f)^2/\nu_0^2\right)-1\right]$ graphed against $\log f$ (or $\log(f/P)$ for gases/vapours). The relaxation frequency f' associated with sound absorption is obtained from the halfpoint of the absorption curve, *i.e.* from a plot of the quantity $(\alpha/f^2)(v_0/v)$ against log f (or log (f/P) for gases/vapours), where α is the absorption coefficient of amplitude which depends on frequency. Note that f' is always somewhat smaller than f'', since $f'' = (v_\infty^2/v_0^2)f'$, and that the dispersion and absorption of sound due to thermal relaxation extend over a significant frequency range on *both* sides of the relaxation frequency.

The formalism indicated above has been modified to include *multiple* vibrational relaxation processes. It was recently used in the discussion of the clearly resolved *double* vibrational relaxation[27,28] observed in *liquid* SO_2. At ambient temperature (293.75 K) and orthobaric pressure the first relaxation step with a relaxation frequency $f''_1 = 22.8$ MHz is associated with the two stretching modes (1152 cm^{-1} and 1362 cm^{-1}) of this triatomic nonlinear molecule. The second relaxation step is associated with the bending mode (518 cm^{-1}), and the corresponding relaxation frequency is $f''_2 = 1.6$ GHz. The total sound dispersion is given by $v_0 = 914$ m s^{-1} and $v_\infty = 939$ m s^{-1}, *i.e.* $v_\infty^2/v_0^2 = 1.055$. For *gaseous* SO_2, ultrasonic speed and absorption measurements also indicate two relaxation steps. Apart from sulfur dioxide, gaseous dichloromethane, CH_2Cl_2, also exhibits a marked double-dispersion curve.[29] From a plot of $v(f)^2$ *vs.* log(f/P) at 303.15 K, the relaxation frequencies $f''_1 = 2.27$ MHz atm^{-1} and $f''_2 = 105$ MHz atm^{-1}, respectively, were obtained (atm = 101.325 kPa). The total dispersion of the speed of ultrasound in gaseous CH_2Cl_2 is given by $v_0 = 188.1$ m s^{-1} and $v_\infty = 198.9$ m s^{-1}, *i.e.* $v_\infty^2/v_0^2 = 1.118$. CS_2 is another nonassociated liquid where the relaxation frequency lies well within the usual frequency range of ultrasonic measurements. Experimental results suggest that the absorption above the classical value is due to vibrational relaxation only, and that the total vibrational heat capacity relaxes at *one* frequency. For liquid carbon disulfide at 298.15 K, the observed relaxation is centred at about 78 MHz,[30] and the dispersion of the speed of sound obtained therefrom is about $v_\infty^2/v_0^2 \approx 1.08$. For such a liquid, the primary effect of increasing the hydrostatic pressure is a decrease in the free volume, which in turn leads to an increase in the rate of molecular collisions. Thus, any process depending on this rate should become more rapid with increasing pressure, and this is indeed observed with liquid CS_2.[31] In passing we note that in *associated* liquids, such as glycerol, the relaxation frequency *decreases* with increasing pressure. This result is consistent with the view that different mechanisms are operative in these two types of liquids. A high-precision study of gaseous CF_4 between 175 K and 300 K, using a spherical resonator, is due to Ewing and Trusler,[32] see also Trusler.[5] Here, the dispersion is also dominated by vibrational relaxation, and the entire vibrational contribution to the heat capacity evidently relaxes as one. At 300 K, for the relaxation frequency $f''/P = 0.455$ MHz atm^{-1} is obtained, $v_0 = 180.9$ m s^{-1} and $v_\infty = 194.4$ m s^{-1}, *i.e.* $v_\infty^2/v_0^2 = 1.155$. For older results on sound absorption and sound dispersion see Landolt–Börnstein[33] and Lambert's monograph on vibrational and rotational relaxation in gases.[34]

As already indicated above, when sound waves propagate through a polyatomic liquid, there are a number of mechanisms by which the acoustic energy is dissipated and the sound wave weakened. Besides the *classical* mechanisms causing absorption, *i.e.* those due to shear viscosity and heat conduction (Kirchhoff-Stokes equation), thermal molecular relaxation and structural relaxation may contribute to make the experimentally observed absorption coefficient significantly larger than that predicted classically. The low-frequency sound attenuation α/f^2 in a liquid is given by[1]

$$\begin{aligned}\alpha/f^2 &= (\alpha/f^2)_s + (\alpha/f^2)_v + (\alpha/f^2)_{hc} \\ &= \frac{2\pi^2}{\rho v_0^3}\left[\frac{4}{3}\eta_s + \eta_v + lm_m\left(C_V^{-1} - C_P^{-1}\right)\right],\end{aligned} \quad (12.24)$$

where the subscripts s, v and hc indicate the contributions due to shear viscosity η_s, volume viscosity η_v(also known as bulk viscosity), and heat conductivity l to the total sound absorption. The *volume viscosity* is thus accessible *via* experimental *sound absorption* data. Note, that for most liquids the thermal conductivity term is much smaller than the terms due to shear or volume viscosity.

For single relaxation behaviour, the frequency dependence of the volume viscosity is given by

$$\eta_v(f) = \eta_{v,nr} + \eta_{v,r}(f), \quad (12.25)$$

where $\eta_{v,nr}$ is the *nonrelaxing* contribution to $\eta_v(f)$, and

$$\eta_{v,r}(f) = \eta_{v,r}(0)/(1 + \omega^2\tau_v^2), \quad (12.26)$$

is the relaxing contribution. Here,

$$\eta_{v,r}(0) = \lim_{f\to 0}\eta_{v,r}(f) = \eta_v(0) - \eta_{v,nr} \quad (12.27)$$

denotes the low-frequency value of $\eta_{v,r}(f)$. In turn, this quantity is related to the sound dispersion, and hence, *via* Equation (12.23), to the relaxing molar heat capacity:

$$\eta_{v,r}(0) = \rho v_0^2\tau_v(v_\infty^2\,/\,v_0^2 - 1) \quad (12.28)$$

$$= \rho v_0^2\tau_v\frac{(C_P - C_V)C_{int}}{(C_V - C_{int})C_P}. \quad (12.29)$$

Since $\eta_{v,nr}$ and τ_v are obtained by fitting experimental Brillouin spectra, and since the low-frequency value of the volume viscosity, $\eta_v(0)$, is accessible, *via* Equation (12.24), from ultrasound absorption measurements, $\eta_{v,r}(0)$and hence the sound dispersion $v_\infty^2\,/\,v_0^2$ may be determined *via* Equation (12.28), and the relaxing heat capacity C_{int} *via* Equation (12.29). In turn, C_{int} so obtained may be compared with vibrational contributions calculated with the help of the Planck–Einstein equation,

$$C_{\text{int,PE}}/R = \sum_i \frac{g_i(hf_i/k_BT)^2\exp(-hf_i/k_BT)}{[1-\exp(-hf_i/k_BT)]^2}, \tag{12.30}$$

where $R = Lk_B$ is the gas constant, L is the Avogadro constant, k_B is the Boltzmann constant, h is the Planck constant, and f_i is the i-th fundamental vibrational frequency of the molecule with degeneracy g_i.

When identifying $(2\pi^2/\rho v_0^3)\eta_{\text{v,r}}(0)$ with the low-frequency vibrational contribution to sound absorption, that is when assuming

$$(\alpha/f^2)_{\text{v}} = (\alpha/f^2)_{\text{v,nr}} + (\alpha/f^2)_{\text{vib}}, \tag{12.31}$$

the sound absorption due to a *single* relaxation process characterised by an *energy relaxation time* τ_e(and $2\pi f\tau_e \ll 1$) is given by

$$(\alpha/f^2)_{\text{vib}} = 2\pi^2\tau_e v_0^{-1}\frac{(C_P - C_V)C_{\text{int}}}{C_P C_V} \tag{12.32}$$

with

$$\tau_e = \tau_v/(1 - C_{\text{int}}/C_V). \tag{12.33}$$

The formalism presented above may be extended to include *multiple* vibrational relaxation processes.[27,28,35–38] In this case, however, for the calculation of the appropriate $C_{\text{int,PE}}$ only a subset of the normal modes of the molecule has to be used. Unfortunately, the different physical meaning of the various relaxation times, *i.e.* the difference between the τ_e and τ_v, and $\tau' = 1/2\pi f'$ as obtained from sound absorption measurement, see above, is *not* always fully appreciated.[39]

12.2 Experimental Ultrasonics

For measuring sound speed and sound absorption at ultrasonic frequencies, the following three main methods have been used: resonator techniques, pulse-modulated propagating wave transmission techniques and continuous wave techniques with variable path length. They have been discussed in two recent reviews on ultrasonic instrumentation for measuring sound speeds and sound absorption in liquids,[11,14] in the monograph by Trusler in 1991 on *Physical Acoustics and Metrology of Fluids*,[5] and the valuable review by Goodwin and Trusler[10] in 2003.

Eggers and Kaatze[11] reviewed various methods for broad-band ultrasonic absorption and sound speed measurements in liquids, using continuous wave and pulsed signals. They also discussed several different ultrasonic spectrometric techniques to cover the wide range of ultrasonic frequencies from kHz to GHz. Kaatze *et al.*[14] presented an introduction into the principles of sound speed measurements. The fundamentals of acoustical methods using continuous waves, pulse modulated signals and sharp pulses were also summarised. Eggers and Funck[40] described and discussed a resonator and a pulse method for sound speeds and relative ultrasonic absorption measurements in liquids from about 0.5 to 100 MHz. Sedlacek and Asenbaum[41] developed an ultrasonic spectrometer which used the correlation between the input and output signals of an ultrasonic transmitter–receiver system with two quartz transducers. The space-dependent cross-correlation function of the signals, which depicts the sound field in the liquid, was recorded on a $x-t$ recorder by continuous variation of the transmitter–receiver distance. Figure 12.1 shows the experimental set-up. The precision of the measurements of the sound speed was $\pm 0.1\%$, that of the sound attenuation was within $\pm 3\%$. Figure 12.2 shows a direct printout of the sound field in the test sample, a mixture of water with *tert*-butanol (TB) at $x_{TB} = 0.1$ at 298.15 K and ambient pressure.

By using a Hewlett-Packard network analyser and an oven-stabilised Hewlett-Packard frequency generator, the experimental set-up was later

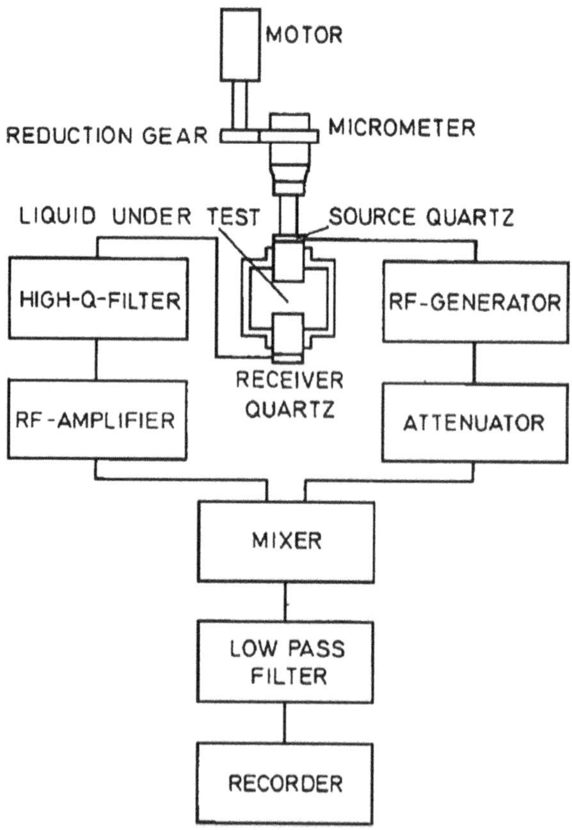

Figure 12.1 Experimental set-up for the measurement of the speed of ultrasound and sound absorption, as introduced by Sedlacek and Asenbaum[41] in 1977.

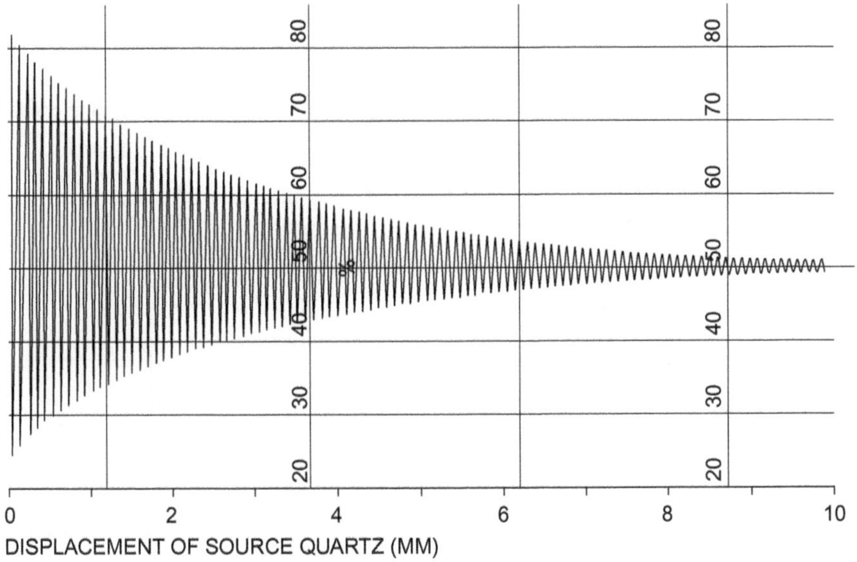

Figure 12.2 Original recording of the correlation between the pressure in a liquid and the signal driving the source quartz, for a mixture of *tert*-butanol (TB) and water with $x_{TB} = 0.1$. The temperature is 298.15 K, the sound frequency $f = 15.69$ MHz, and the pressure is about 0.1 MPa.

improved by the Sedlacek Group at the Institute for Experimental Physics, University of Wien (Vienna), Austria (PhD Dissertation of Müllner,[42] 1984), see Figure 12.3. Additional improvements, including full personal computer control, are due to Wagner[43] (PhD dissertation at the Department of Materials Research and Physics, University of Salzburg, Asenbaum Group, 2008), see Figures 12.4 and 12.5.

Takagi and Negishi[44] introduced an UHF ultrasonic technique which can be used for simultaneous measurement of ultrasonic speed and sound absorption in liquids. A transducer excites a continuous ultrasonic wave in the test liquid which is illuminated by a highly collimated beam from a He–Ne laser. The light scattered by the sound is detected by an optical heterodyne technique. The accuracy is better than 0.1% for the sound speed and 5% for the sound damping. Experiments were performed on carbon disulfide and bromoform covering the frequency range from 50 MHz to 1.5 GHz. The results were combined with the data obtained by a pulse method and by spontaneous Brillouin scattering to obtain ultrasonic spectra over the frequency range of 3 MHz to 6 GHz.

Uhlendorf *et al.*[45] designed an ultrasonic spectrometer, using odd harmonics up to 500 MHz of a transducer with peak power below 5 mW, a 0.01 mm to 7.5 mm path length cell with linear positioner driven by a stepping-motor, sensitive superheterodyne detection, and signal averaging. The ultrasonic speed and the sound absorption in liquids were determined within $\pm 1\%$.

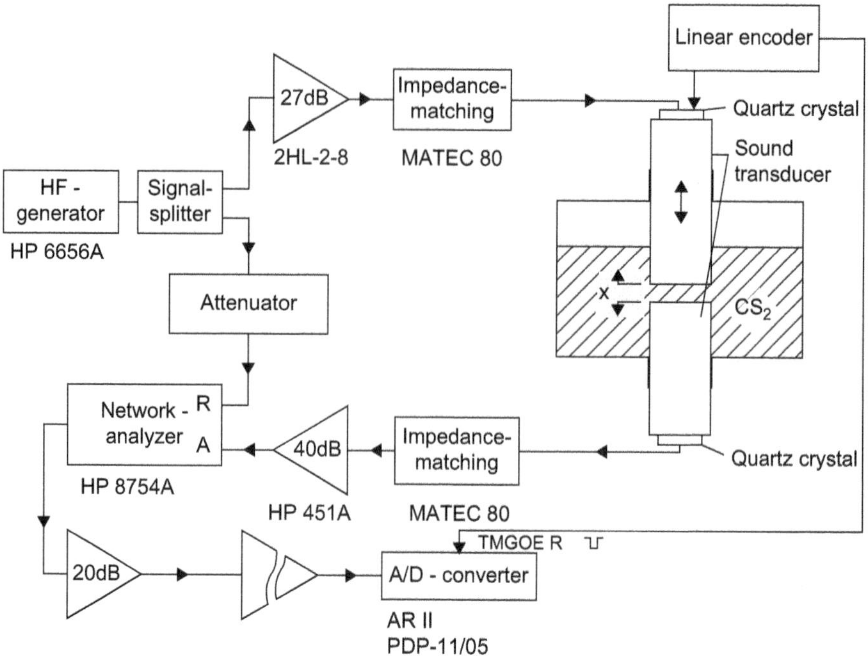

Figure 12.3 Experimental set-up taken from the PhD Dissertation of Müllner,[42] University of Wien (Vienna), Austria, 1984.

McClements and Fairley[46] described an ultrasonic pulse-echo technique which can be used to measure the ultrasonic speed and the attenuation coefficient of liquids. A year later, in 1992, these authors[47] developed a frequency scanning ultrasonic pulse-echo reflectometer which can be used to measure the ultrasonic speed and the attenuation in liquid samples as a function of frequency in the range 0.3 MHz to 6 MHz.

Tardajos *et al.*[48] described an automatic high-precision ultrasonic system with the corresponding ultrasonic cell to measure simultaneously the speed of sound and the attenuation as a function of frequency. The technique is based on the pulse-echo method. Eggers *et al.*[49] described ultrasonic resonator cells for sound speed measurements and sound attenuation measurements in liquids below 1 MHz with one planar and one concave piezoelectric transducer. Such cells with a limited sample volume are needed for broad-band ultrasonic relaxation spectrometry and for studies in chemical reaction kinetics, particularly for frequencies below 200 kHz.

Another method for measuring the speed of sound and sound attenuation in liquids between 300 MHz and 5 GHz is due to Kaatze *et al.*[50] One cell, using surface excitation of small rod-shaped piezoelectric lithium niobate crystals, allows for broad-band measurements up to about 3 GHz. The other cell utilises thickness vibrations of thin ZnO films. It is appropriate for measurements above 2 GHz.

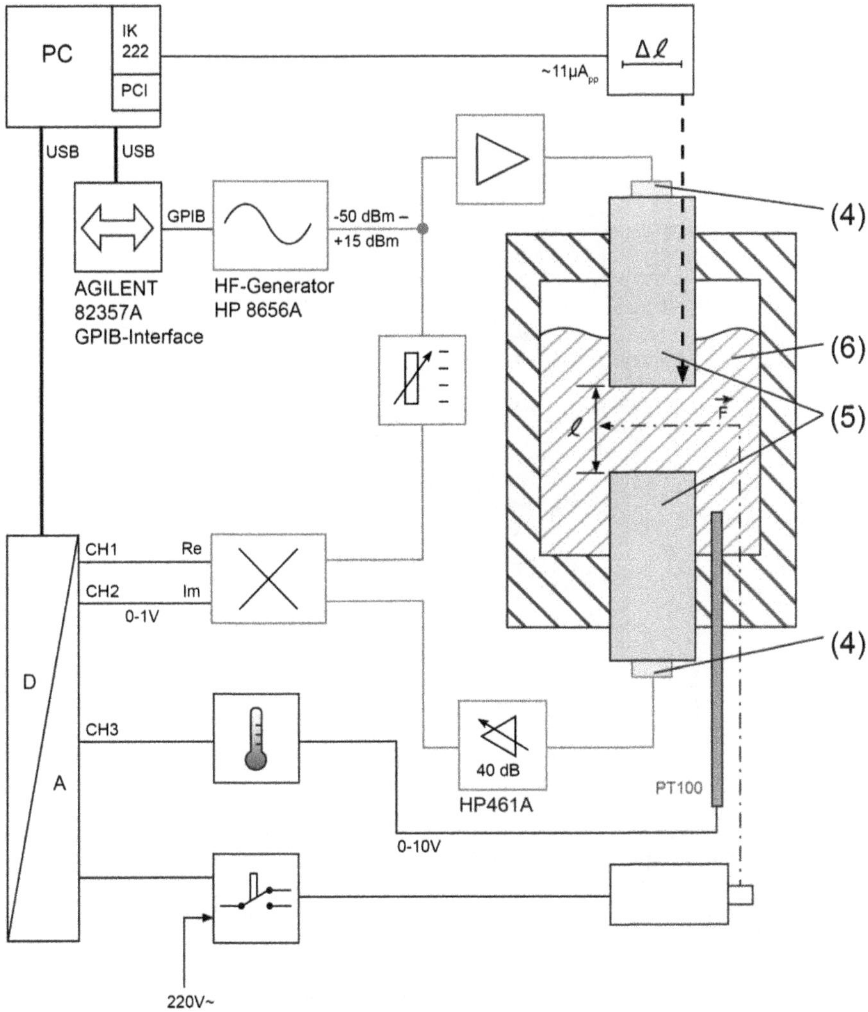

Figure 12.4 Experimental set-up taken from the PhD Dissertation of Wagner[43] at the University of Salzburg, Austria, 2008: 4... quartz crystals, 5... sound transducers, 6... liquid sample.

Schultz and Kaatze[51] presented a variable-path length method that allows the measurement of ultrasound speeds and ultrasound attenuation of liquids as a function of frequency. Focusing on the effect the sample liquid has on the transfer function for continuous acoustical waves, this automated method is particularly useful because of its simple and easy-to-handle electronic set-up, essentially consisting of a computer-controlled network analyser only.

A pulsed phase-sensitive technique for measuring sound speed and attenuation in fluids was developed in 1999 by Kozhevnikov *et al.*[52] The time

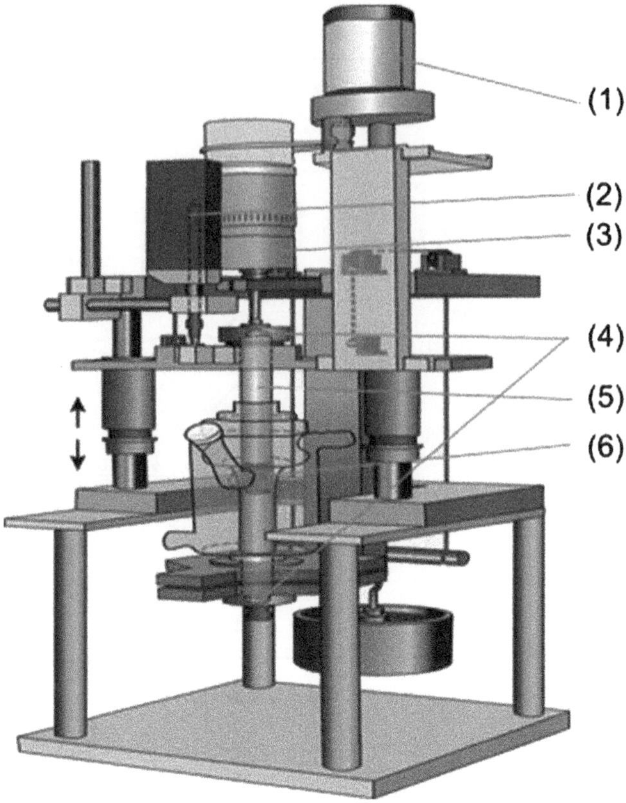

Figure 12.5 Apparatus for ultrasonic measurements described in detail in the PhD Dissertation of Wagner[43] (University of Salzburg, Austria, 2008): 1… motor drive, 2… linear encoder, 3… micrometer screw, 4… quartz crystals, 5… sound transducers, 6… liquid sample.

delay is measured between any two acoustical pulses transmitted through a sample or reflected from its boundaries. A current realisation of the technique allows the resolution of the time-delay variation down to 0.1 ns. Precise sound speed data can be obtained for samples of small thickness of about 1 mm.

Recently, Zak *et al.*[53] described a new measuring device for the determination of the speed of sound in liquids under high pressures up to 300 MPa. The apparatus operates on the principle of the pulse-echo-overlap method. A single transmitting–receiving piezoelectric ceramic transducer operating at 4 MHz and an acoustic reflector are used.

Lautscham *et al.*[12] described a method for precise measurement of the sound speed in liquids. They also presented a special construction of a resonator cell so that the sample liquid and a reference liquid could be investigated simultaneously. In addition, they presented a low-priced electronic set-up designed for the computer-controlled determination of the complex transfer functions of the cavity resonators and also for automatic temperature monitoring and control. Possible sources of error are discussed and some representative results are presented in order to illustrate the reliability of the method and the accuracy of the sound speed data relative to a reference liquid. Ultrasonic resonator cells for sound speed and sound attenuation measurements in liquids down to 70 kHz were described in 2003 by Polacek and Kaatze.[54] The resonators are provided with easy-to-obtain concavely shaped shaving mirrors as acoustic reflectors and separated piezoelectric devices for the coupling of the cell to the electronic set-up. The mode spectrum of the resonators is discussed and measurement and evaluation procedures, also considering higher-order satellite peaks, are presented. In 2005, Benedetto *et al.*[55] developed a prototype cylindrical ultrasonic cell designed to apply a microwave resonance technique. Since the apparatus is intended for speed of sound measurements in pressurised liquid-phase media, the cell design is such that a double reflector pulse-echo technique can be used for time-of-flight measurements. The main absorption and dispersion effects that influence the acoustic measurement in this interferometer-like configuration and the values of the corresponding corrections are considered and discussed. The performance of the apparatus and the method in terms of achievable precision and accuracy was tested by measuring the speed of sound in water on a single isotherm at 325 K between 0.1 MPa and 90 MPa.

Gedanitz *et al.*[56] described an apparatus for accurate measurements of the sound speed in fluids which is based on the pulse-echo technique, and operates at pressures up to 30 MPa in the temperature range between 250 K and 350 K. In order to validate the new instrument, measurements of the speed of sound for nitrogen from 250 K to 350 K and for water between 303.15 K and 323.15 K were presented for pressures up to 30 MPa. The overall uncertainties of the measurements on nitrogen and water were estimated to be 0.011% and 0.006%, respectively. Fortin *et al.*[57] demonstrated that commercial benchtop measurement systems can yield results of high quality. The importance of sample purity, reproducibility, expanded calibration and adjustment protocols, and rigorous uncertainty estimates are emphasised. They reported ultrasonic speed measured at atmospheric pressure and at temperatures from 278 K to 343 K (including uncertainty estimates) for isooctane (2,2,4-trimethylpentane) and for toluene. These data are useful for validating the performance of such instruments, which provide a repeatability of ± 0.1 m s^{-1} and an uncertainty of ± 0.5 m s^{-1}. Specifically, this is the case for the Density and Sound Velocity Meter DSA 5000 M by A. Paar, Graz, Austria: a speed of sound chamber with two piezoelectric transducers is the core of the apparatus. The propagation time of short acoustic pulses, which are repeatedly transmitted through the liquid sample, is measured. Using a tunable voltage controlled oscillator this technique was invented by Stabinger *et al.*[58a] in 1967 and Kratky *et al.*[58b] in 1969.

12.3 Brillouin Scattering

12.3.1 Introduction

In a Brillouin light scattering experiment, laser light is scattered by thermally driven density fluctuations in the liquid. These density fluctuations can be thermodynamically separated into pressure fluctuations at constant entropy, causing two Brillouin peaks, and entropy fluctuations at constant pressure, causing the Rayleigh peak. The two Brillouin peaks are symmetrically located because scattering occurs from thermally driven sound waves of frequency f travelling in opposite directions at the same hypersonic speed $\nu_H(f)$, and light scattered from them experiences a Doppler shift $\omega_B = 2\pi f$. The anti-Stokes Brillouin peak is located at the frequency $c/\lambda_0 + |\omega_B/2\pi|$, and the Stokes Brillouin peak is at $c/\lambda_0 - |\omega_B/2\pi|$.

Light scattered by the entropy fluctuations at constant pressure shows no frequency shift, and this *Rayleigh peak* is located at the frequency of incident light c/λ_0, where c is the speed of light and λ_0 is the wavelength of the incident laser light. The whole spectrum is frequently called the *Rayleigh–Brillouin triplet.* The frequency shift ω_B of the two Brillouin peaks is given by

$$\omega_B = \pm k \cdot \nu_H(f), \tag{12.34}$$

where

$$k = \frac{4\pi n}{\lambda_0}\sin(\theta/2) \tag{12.35}$$

denotes the absolute value of the wave vector transfer, that is transfer from incident to scattered light, due to the scattering angle θ, $f = \omega_B/2\pi$ is the frequency of the hypersound, and n is the refractive index of the liquid. The half width of these Brillouin peaks is due to various dissipative processes.[35,36,59–64] The above discussion of Rayleigh–Brillouin scattering is exact only for simple (*monoatomic*) liquids. In liquids consisting of *polyatomic* molecules, there is at least one additional fourth peak, a broad, unshifted central peak, the so-called *Mountain peak.*[60,63] This peak is due to relaxation processes associated with the weak coupling of internal vibrational modes to the translational modes. The Mountain peak was first observed experimentally by Gornall *et al.*[65] and later studied extensively by Nichols *et al.*[35,36] Figure 12.6 shows a typical Rayleigh–Brillouin spectrum of a dense normal liquid, *i.e.* of tetrachloromethane,[66] where the broad Mountain peak, extending out to the two Brillouin peaks, is clearly visible.

The separation ω_B of the Brillouin doublet from the Rayleigh peak is determined by the hypersound speed, *i.e.* the high-frequency speed of sound (GHz region)

$$\nu_H(f) = \omega_B/k. \tag{12.36}$$

The hypersonic frequency region accessible *via* Brillouin scattering exceeds the *typical* ultrasonic region by two to three decades. Thus, sound dispersion

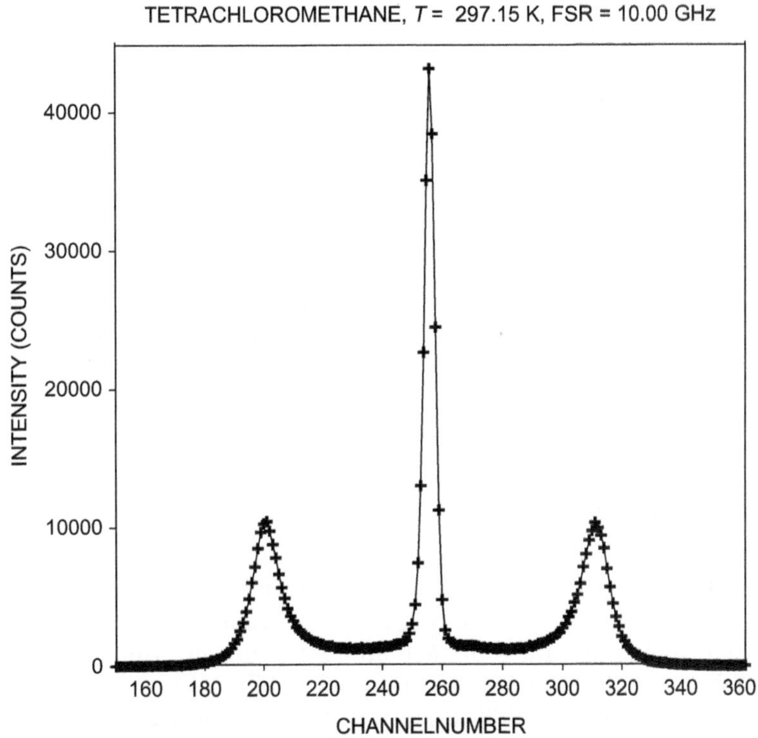

Figure 12.6 Rayleigh–Brillouin spectrum of liquid tetrachloromethane at 297.15 K and atmospheric pressure.[66] The free spectral range (FSR) was 10.00 GHz. Each channel number corresponds to a frequency interval of 97.66 MHz. The broad Mountain line is clearly visible.

and sound absorption studies become feasible over very large frequency ranges. At high pressures, Brillouin scattering is one of the most useful and frequently used spectroscopic methods for measuring sound speed and sound attenuation in liquids and solids, thereby yielding the pressure dependence of important thermophysical properties.

We conclude this section by showing the explicit expressions for the full width at half height (FWHH) of the two Brillouin peaks, $\Delta\omega_B$(FWHH), and of the unshifted Rayleigh peak, $\Delta\omega_R$(FWHH):[6]

Brillouin Components

The width (FWHH) of each of the two peaks positioned at frequencies $(c/\lambda_0 + \omega_B/2\pi)$ and $(c/\lambda_0 - \omega_B/2\pi)$, respectively, is given by

$$\Delta\omega_B(\text{FWHH}) = 2\Gamma_B k^2, \tag{12.37}$$

$$\Gamma_B = \frac{1}{2\rho}\left[\frac{4}{3}\eta_s + \eta_v + \frac{l m_m}{C_P}(\kappa - 1)\right]. \tag{12.38}$$

The quantity $D_v = \frac{1}{\rho}\left(\frac{4}{3}\eta_s + \eta_v\right)$ is frequently called the *longitudinal kinematic viscosity*, and $D_t = \frac{l m_m}{\rho C_P}$ is known as the *thermal diffusivity*. Thus, Equation (12.38) may be written in a more compact form as

$$\Gamma_B = \frac{1}{2}[D_v + D_t(\kappa - 1)]. \tag{12.39}$$

Rayleigh Component

The Rayleigh peak is positioned at frequency c/λ_0, and its width is given by

$$\Delta\omega_R(\text{FWHH}) = 2D_t k^2. \tag{12.40}$$

Experimental determination of the width of the central peak requires very high resolution, since $\Delta\omega_R$(FWHH) is usually of the order of 10 MHz.

12.3.2 Experimental Brillouin Spectroscopy

Experimental Brillouin spectra are usually recorded using a Fabry–Perot interferometer as a high resolution spectrometer. To obtain an adequate signal-to-noise ratio, long measuring times are required when recording low intensity Brillouin spectra with such an interferometer, which should thus have high optical, mechanical and thermal stability (Asenbaum[67]). In addition, laser frequency instabilities, as well as interferometer misalignments, should be eliminated as well as possible (Aschauer *et al.*[68]). To achieve an acceptable signal-to-noise ratio, a large number of individual scans are added up with the unshifted laser frequency c/λ_0 as reference. Due to the drift of the laser frequency, as well as the drift of the interferometer pass frequency, this referencing procedure prevents broadening of the spectral components of the Rayleigh–Brillouin spectra. Since piezoelectric transducers are somewhat nonlinear, each piezo stack behaves characteristically when applying a voltage ramp. The three piezoelectric scanning stacks necessary for scanning the Fabry–Perot may lead to minor mirror tilts during the scan and the almost perfect parallel alignment of the mirrors will be lost. To avoid such a nonlinear scanning and tilting of the mirrors during the scan, Sandercock[69,70] introduced an *almost perfect linear* scan by using a parallel scanning stage and by measuring the capacity of a plate condenser as a measure for the separation of two Fabry–Perot mirrors. Furthermore, Sandercock[71] developed an electronically stabilised *six-pass tandem multipass Fabry–Perot interferometer*. This technique represents the *state of the art* for high-resolution, high-contrast Brillouin light scattering measurements. We are now using this spectrometer system with a finesse of about 80. The scattered light is detected by a photodetector and stored in the memory of a personal computer. The time for each scan is about 0.5 s corresponding to 1024 channels. After each scan the parallelism of the Fabry–Perot mirrors is checked and the mirrors are afterwards realigned. Figure 12.7 shows a schematic representation of our experimental Brillouin scattering set-up.

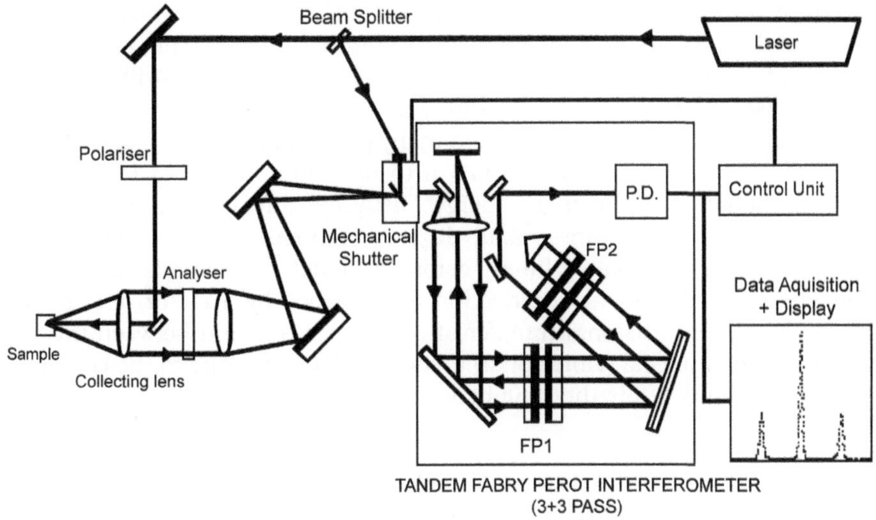

Figure 12.7 Experimental Brillouin scattering set-up with a six-pass tandem Fabry–Perot interferometer (FP1 and FP2), Sandercock type.[69–71] P.D. denotes the photon detector.

Mountain[59] has derived the spectral distribution of light scattered by density fluctuations in a dense, monatomic, one-component fluid from the time dependence of the density fluctuations predicted by the linearised hydrodynamic equations of irreversible thermodynamics. In additional papers, for liquids consisting of polyatomic molecules Mountain[60–63] showed that internal molecular vibrational modes generally couple weakly to the translational modes, thereby leading to a broad, unshifted central peak, the so-called Mounain peak (see above). Nichols and coworkers[35,36] derived formulas for the roots of the dispersion equation for temporally damped hydrodynamic waves in a viscous, thermally conducting liquid also having a frequency-dependent volume viscosity. They went beyond Mountain's treatment of Brillouin scattering in relaxing liquids by (I) deriving more accurate mathematical expressions for the frequency spectrum of the scattered light intensity, and (II) extending the formalism to multiply relaxing liquids. This formalism due to Nichols and coworkers, in conjunction with the translational hydrodynamics approach of Desai and Kapral,[64] has supplied the basis for analysing the Brillouin spectra recorded in our laboratory. Essentially, the spectral distribution is the sum of *four* Lorentzians representing the two unshifted central lines (Rayleigh and Mountain) and the Brillouin doublet, and a non-Lorentzian correction which shifts the apparent Brillouin peaks *slightly* towards the centre and at the same time renders them *slightly* asymmetric. However, the total spectrum is still symmetric about $\omega = 0$. The Brillouin shift ω_B, the relaxation time τ_v of the relaxing volume viscosity $\eta_v(f)$ and the nonrelaxing volume viscosity $\eta_{v,nr}$ are obtained *simultaneously* by least-squares fitting each experimental Brillouin spectrum with a theoretical Brillouin spectrum[35,36] convoluted with the instrumental function of the Fabry–Perot interferometer. The hypersonic speed $\nu_H(f)$ at sound frequency $f = \omega_B/2\pi$ is obtained from the Brillouin shift through use of Equations (12.35) and (12.36).

Contrary to liquified noble gases, in many polyatomic liquids, such as CS_2, CCl_4, and C_6H_6, the *main* contribution to ultrasonic damping is due to a delayed exchange of acoustic energy[1] between translational degrees of freedom and internal (vibrational) degrees of freedom of the molecules. Using the heuristic isolated binary collision model,[72,73] the relaxation time characterising vibrational deactivation is described by a factor due to the probability p for collisional transfer of energy per collision between the external and the internal (vibrational) modes, and a factor due to the collision frequency Z (thus $1/Z$ is the time between collisions):

$$\tau_e = \frac{1}{pZ}. \tag{12.41}$$

This *ansatz* separates the relaxation dynamics contained in p (and assumed to be binary) and the translational dynamics contained in Z, yielding the *relaxation frequency* as $1/\tau_e$. Depending on the value of this relaxation frequency, two scenarios are possible:

(I) This relaxation frequency is *higher* than the ultrasound frequency (about 3 MHz) but *lower* than the accessible Brillouin frequency (about 5 GHz); as a consequence, sound absorption is decreasing with increasing sound frequency, while the sound speed is increasing (sound dispersion). Well investigated liquids showing such a behaviour are, for example, tetrachloromethane, benzene (see Figure 12.8), and toluene.

(II) This relaxation frequency is *higher* than the ultrasound frequency (about 3 MHz) and also *higher* than the accessible Brillouin frequency (about 5 GHz); as a consequence, sound absorption remains constant with increasing sound frequency and also the sound speed remains constant with increasing frequency. Well investigated liquids showing no dispersion in the currently accessible frequency range are, for example, water, *n*-hexane, methanol and ethanol. Figure 12.9 presents nine Brillouin spectra of liquid *n*-hexane[75] in the pressure range from 0.1 MPa to 135.8 MPa at 303.15 K, and Figure 12.10 shows the corresponding ultrasound speed ν_0 and hypersound speed ν_H of *n*-hexane as a function of pressure.[75] Within the experimental uncertainty, *no* difference is discernible between ultrasound speed and hypersound speed, hence no relaxation takes place in *n*-hexane in the frequency region between 31.5 MHz and about 5 GHz.

In summary, for most liquids consisting of polyatomic molecules the hypersonic speed ν_H is *not* equal to the *thermodynamic* low-frequency ultrasonic speed ν_0 which is related to the isentropic compressibility β_S, as given by Equation (12.1). Unfortunately, this fact is frequently overlooked in the literature.

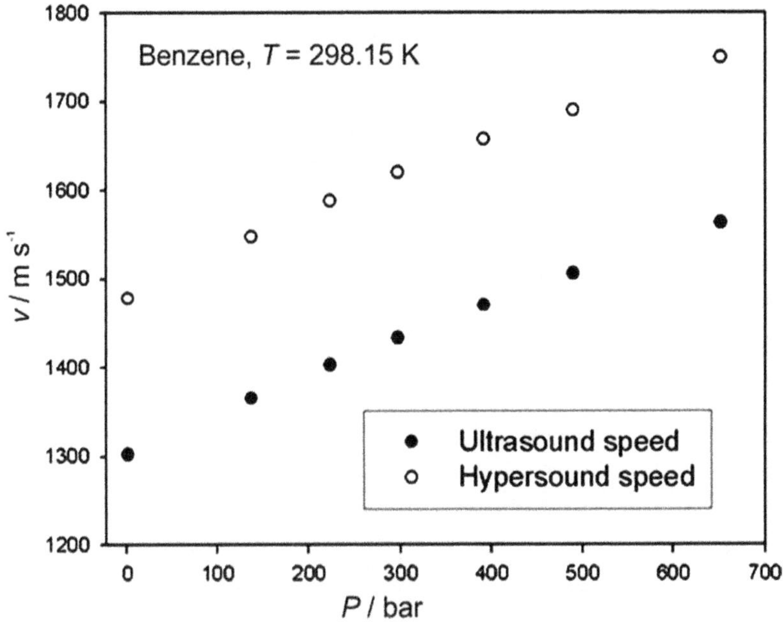

Figure 12.8 Ultrasound speed and hypersound speed, globally denoted by v, of benzene as a function of pressure at 298.15 K. Data were taken by Asenbaum and Hochheimer:[74] the considerable difference between hypersound speed and ultrasound speed can clearly be seen.

12.4 Selected Experimental Results

We do not endeavour to present a complete review of published work in ultrasonics and hypersonics; only a few selected, representative pure liquids will be considered (as already indicated, a comprehensive survey of the pressure dependence of the sound speed in organic liquids was presented by Oakley *et al.*[23]). Our presentation here will start with inorganic liquids and proceed to organic liquids of increasing complexity.

12.4.1 Ultrasonic Data

12.4.1.1 Inorganic Liquids

12.4.1.1.1 Argon. Van Itterbeek *et al.*[76] measured the sound speed in liquid argon under pressure of up to 7.4 MPa and in the temperature range 84 K to 90.3 K, using an acoustical interferometer. At 90.3 K they calculated the compressibility and the ratio of heat capacities C_P/C_V as a function of pressure. Naugle *et al.*[77] measured the sound speed in liquid argon at temperatures between 85 K and 145 K and pressures up to 15.7 MPa. Thoen *et al.*[78] reported measurements of the speed of ultrasonic pulses in liquid argon along seven isotherms at temperatures ranging from 100 K to

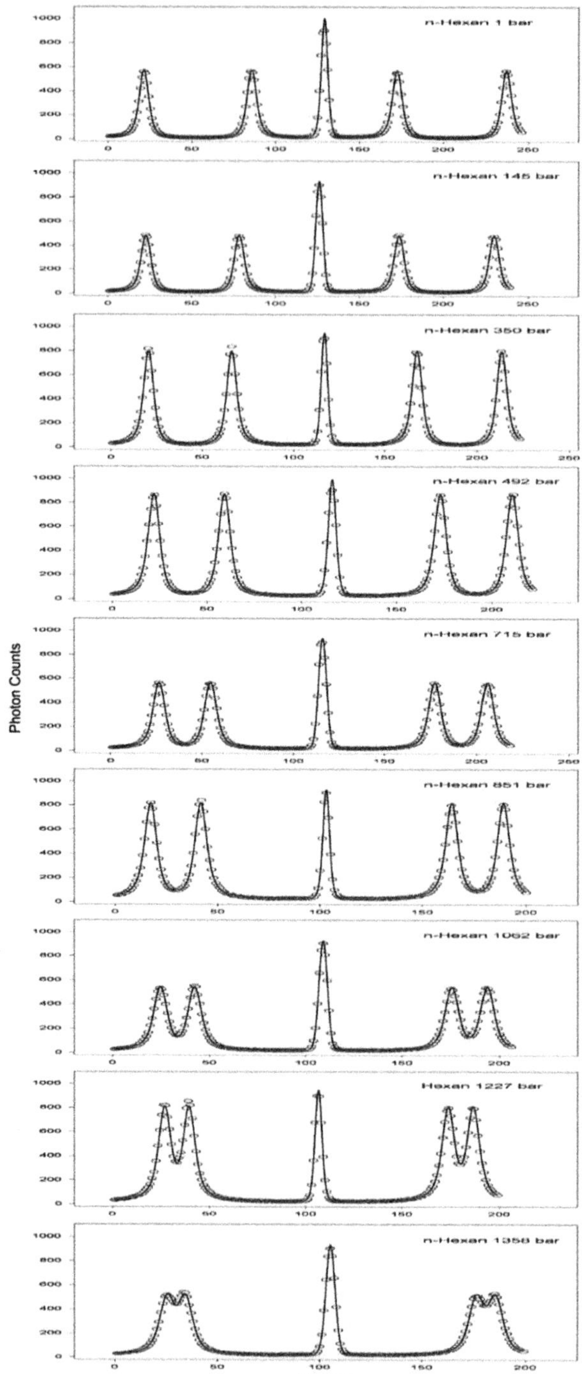

Figure 12.9 Brillouin spectra of *n*-hexane at 303.15 K and for pressures between 0.1 MPa and 135.8 MPa show the increase of the Brillouin shift with increasing pressure,[75] and thus the increase of the hypersound speed with increasing pressure, see Equation (12.36).

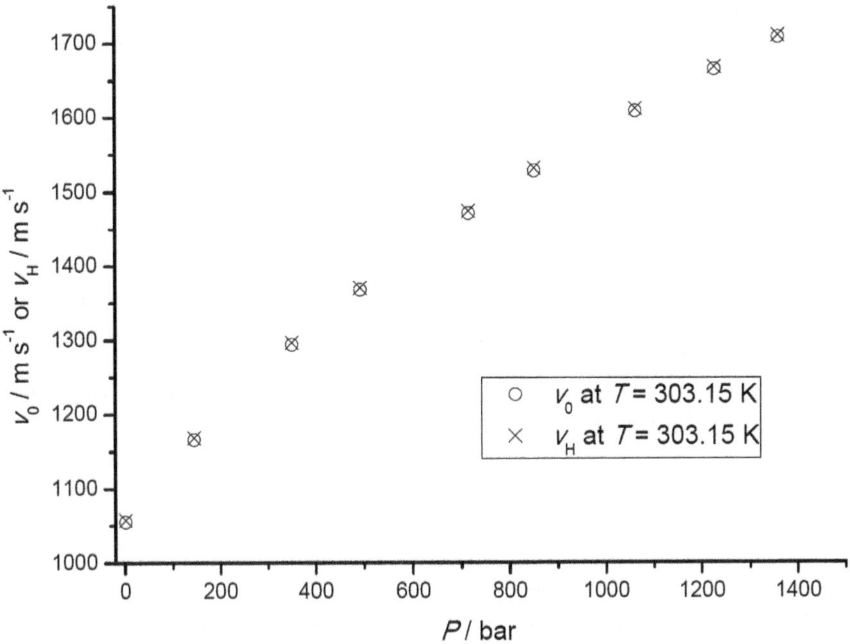

Figure 12.10 Ultrasound speed v_0 and hypersound speed v_H of *n*-hexane as a function of pressure at 303.15 K. The data were taken from Asenbaum *et al.*[75] Evidently, within the experimental uncertainty there is *no* difference between ultrasound speed and hypersound speed, hence no relaxation takes place in *n*-hexane in the frequency region between 31.5 MHz and about 5 GHz.

150 K and pressures up to 50 MPa. Combining these results with available density data, the following thermodynamic quantities were calculated at integer values of P and T: the isentropic and isothermal compressibility, the isobaric expansivity, the heat capacities C_P and C_V, and their ratio κ.

Liebenberg *et al.*[79] developed a piston-cylinder apparatus in which ultrasonic waves were transmitted into a high-pressure sample through the unsupported area of a Bridgman seal. The sound-wave transmission could be divided so that one portion returned from a fixed reflector in the sample, giving sound speed measurements at 295 K and at pressures up to 1.3 GPa, and the other portion returned from the piston itself, giving the relative volume of the sample. In this way, isentropic and isothermal compressibilities could be measured simultaneously.

12.4.1.1.2 Nitrogen. Mills *et al.*[80] measured the sound speed in fluid nitrogen along five isotherms from 247.5 K to 320.8 K at pressures up to the freezing pressure or 2.2 GPa, whichever is applicable. Simultaneously, the relative volume was measured in a piston-cylinder apparatus. The data were combined to yield a consistent equation of state from which thermodynamic quantities can be derived. Specifically, the isobaric expansivity, the isentropic compressibility, the isothermal compressibility and the heat capacity ratio κ were determined for the indicated ranges of temperature and pressure.

12.4.1.1.3 Oxygen. Van Dael *et al.*[81] measured the sound speed in saturated liquid oxygen covering the range between 61.14 K and 153.88 K. In a continuing effort to generate accurate thermodynamic and transport properties of cryogenic fluids, Straty and Younglove[82] measured the ultrasound speed in saturated liquid oxygen from 58 K to 150 K and in compressed fluid oxygen along isotherms from 70 K to 300 K at pressures up to 34 MPa. Using previously measured $P\rho T$ data, isentropic compressibilities and heat capacity ratios were calculated.

Abramson *et al.*[83] measured the speed of sound in supercritical fluid oxygen up to the freezing points of 6.27 GPa at 303.15 K and 10.74 GPa at 473.15 K. The oxygen was contained in a diamond anvil cell and pressure was measured on the ruby scale. The measurements were used to establish an equation of state.

12.4.1.1.4 Water. Since water is the most important liquid on earth, water is the best investigated liquid of all. Hawley *et al.*[84] measured the sound speed and ultrasonic absorption in water at 273.0 K, 283.3 K and 303.0 K. and at pressures up to 463 MPa, see Figure 12.11.

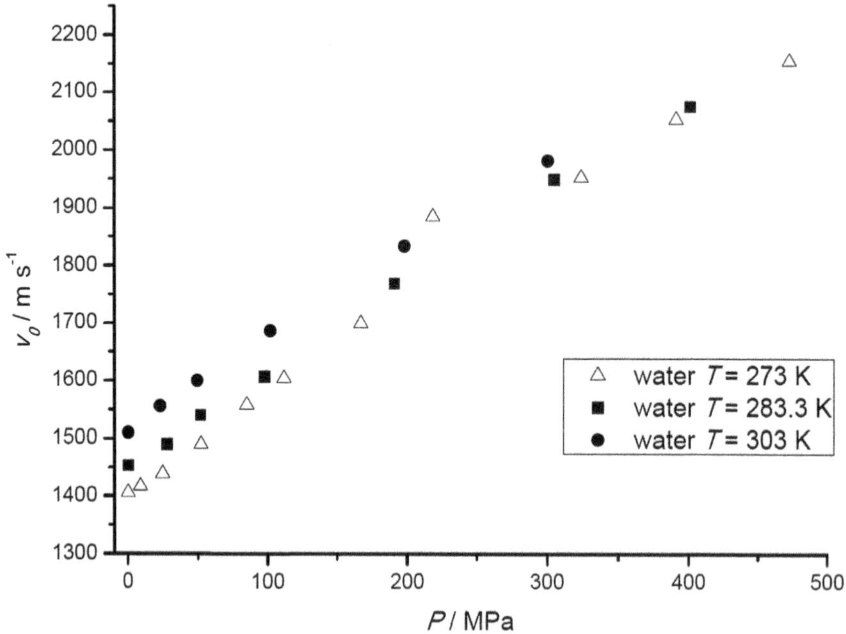

Figure 12.11 Ultrasound speed of water as a function of pressure at 273 K, 283.3 K, and 303 K and at a sound frequency of about 30 MHz; the data were measured by Hawley *et al.*[84]

Del Grosso and Mader[85] made 148 measurements of the sound speed of pure water between 0.001 °C and 95.126 °C on the T_{68} scale (t/°C = T/K − 273.15). The accuracy was believed to be 0.015 m sec^{-1}, and the reproducibility was 0.005 m sec^{-1}. Trinh and Apfel[86] used a standard optical Schlieren method for measuring the speed of sound in moderately supercooled and in superheated water under atmospheric pressure. The estimated accuracy was ± 3 m s^{-1}. Results in supercooled water down to −16.75 °C and in superheated water up to 176.5 °C were reported. Belogolskii *et al.*[87] have measured the sound speed in distilled water at pressures up to 60 MPa and in the temperature range 0 °C to 40 °C. The results are given in the form of tables and an equation derived by least-squares processing of the experimental data. Benedetto *et al.*[88] measured the speed of sound in high-purity water on nine isotherms between 274 K and 394 K and at pressures up to 90 MPa. The speeds of sound have an overall estimated uncertainty of 0.05%. The acoustic data were combined with available values of density and isobaric heat capacity along one isobar at atmospheric pressure (0.1 MPa) to calculate the same quantities over the whole temperature and pressure range by means of a numerical integration technique [see Equation (12.17)]. Gedanitz *et al.*[56] have measured the speed of sound for water at temperatures between 303.15 K and 323.15 K and at pressures up to 30 MPa. The overall uncertainty was estimated to be 0.006%. In 2011, Baltasar *et al.*[89] provided a complete set of accurate speed of sound results in water at pressures up to 700 MPa and at temperatures between 253.15 K and 348.15 K. The relative uncertainty of the sound speeds was estimated to be below 0.32%.

12.4.1.1.5 Carbon Dioxide. Bass and Lamb[90] measured the ultrasonic absorption and the sound speed of liquid carbon dioxide at temperatures from 273.15 K to 308.15 K and at pressures between 3.75 MPa and 10.3 MPa, covering the frequency range 1 MHz to 50 MHz. The observations cover the relaxation region centred at about 10 MHz. The results are adequately described in terms of a relaxation of the total vibrational heat capacity associated with a single relaxation time. Pecceu and Van Dael[91] measured the speed of sound pulses in liquid carbon dioxide between 217 K and 293 K. Numerical data of the isentropic compressibility, the isobaric and isochoric heat capacities, and the isobaric expansivity, as well as the thermal pressure coefficient, have been derived.

12.4.1.1.6 Carbon Disulfide. Mifsud and Nolle[92] measured the ultrasonic speed and sound absorption in carbon disulfide at temperatures between 273.15 K and 313.15 K and at pressures between 0.1 MPa and 138 MPa. The data were presented graphically only. Takagi[93] measured the ultrasonic speed and sound absorption in liquid carbon disulfide at 10 °C, 20 °C and 30 °C over a frequency range from 3 MHz to 5 GHz. Three experimental techniques were used: pulse-echo overlap at 3 MHz, high-resolution Bragg reflection from 50 MHz to 1.5 GHz, and Brillouin scattering in the hypersonic region. The results obtained on sound dispersion and absorption suggest two relaxation processes, one at 100 MHz and the other at 6 MHz (see also references 30 and 31).

12.4.1.1.7 Sulfur Dioxide. Bass and Lamb[94] reported sound absorption and sound speed data in liquid sulfur dioxide at 273.15 K, 298.15 K and 323.15 K, covering the frequency range from 3 MHz to 45 MHz. A pronounced relaxation process, centred at about 23 MHz, was associated with the time delay in deactivation of the two high wave-number modes of the three vibrational modes. The mode which does not participate is that of lowest wave-number which was presumed to be responsible for a relaxation at higher frequency of the order of 1500 MHz.

Fenner *et al.*[95] measured sound absorption and sound speed in liquid sulfur dioxide over a wide range of temperature, thereby greatly extending the range of the measurements of Bass and Lamb on the same system. The data now covered the temperature range 203 K to 323 K and the frequency range 6 MHz to 90 MHz. All the data were fitted satisfactorily to a single-relaxation equation, and gave a temperature-independent relaxation frequency of 22.5 MHz. The speed of sound, as measured, is a linear function of the temperature.

12.4.1.1.8 Sulfur Hexafluoride. Bass and Lamb[90] measured the ultrasonic absorption and the sound speed in liquid sulfur hexafluoride at temperatures from 273.15 K to 313.15 K covering the frequency range 1 MHz to 50 MHz. It was not possible to cover a substantial part of the relaxation region. However, the results are consistent with the assumption that the observed non-classical absorption is entirely due to vibrational relaxation and that the total vibrational specific heat relaxes with a single relaxation time. The corresponding characteristic frequency at 298.15 K is about 75 MHz. The speed of sound in liquid sulfur hexafluoride has been measured by Vacek and Zollweg[96] along isotherms at temperatures from 230 K, that is, close to the triple point, to 333 K which is above the critical temperature. Pressures ranged from near saturation to 60 MPa. The measurements were carried out using a pulse-echo-overlap technique at a frequency of 1 MHz. Experimental results were also used to locate the freezing transition between 230 K and 253 K.

12.4.1.2 Organic Liquids

12.4.1.2.1 *n*-Alkanes

12.4.1.2.1.1 Methane. Van Dael *et al.*[97] measured the sound speed in liquid methane between 94 K and 189 K. In conjunction with density data, heat capacities, compressibilities, isobaric expansivities and isochoric pressure coefficients were derived. In 1967 Van Itterbeek *et al.*[98] determined the speed of sound in liquid methane as a function of pressure along seven isotherms between 111 K and 190 K with pressures up to

20 MPa. Straty[99] measured sound speeds in saturated and compressed fluid methane. Specifically, measurements were made on the saturated liquid from 91 K to 186 K and on the compressed fluid along selected isotherms from 100 K to 300 K at pressures up to about 35 MPa. Kortbeek *et al.*[100] determined the sound speed at temperatures between 148.15 K and 298.15 K in steps of 25 K, and pressures up to 1000 MPa, by a phase comparison pulse-echo technique. The experimental data at 298.15 K have been used to calculate the isothermal and isentropic compressibilities and the ratio of the specific heat capacities.

In 1990, Kortbeek and Schouten[101] reported new density measurements on methane at 298.15 K up to 1 GPa, and speeds of sound were remeasured between 148.15 K and 298.15 K at intervals of 25 K and at pressures up to 1 GPa. The isothermal and the isentropic compressibility and the ratio of heat capacities κ have been calculated for 298.15 K.

12.4.1.2.1.2 Ethane. Tsumura and Straty[102] measured the speed of sound in saturated and compressed fluid ethane at temperatures between 91 K and 323.15 K and at pressures up to 35 MPa. These data were combined with newly available $P\rho T$ data to obtain the isentropic compressibility and the ratio of the heat capacities. Bücker and Wagner[103] presented high-precision sound speeds of ethane in the gaseous and supercritical region, generally within less than 0.015% uncertainty.

12.4.1.2.1.3 Propane. Younglove[104] reported sound speed measurements on liquid propane from 90 K to 300 K and for pressures up to 34 MPa. Also included are saturated liquid sound speeds from 90 K to 290 K. The data were combined with $P\rho T$ data to compute compressibilities and heat capacity ratios. Niepmann[105] obtained speeds of sound of liquid propane between 200 K and 340 K at pressures up to 60 MPa. In 2012, Meier and Kabelac[106] carried out measurements of the speed of sound in pure propane in the liquid region by a double-path length pulse-echo technique. The measured data cover the temperature range from 240 K to 420 K with pressures up to 100 MPa.

12.4.1.2.1.4 n-*Butane.* Using the pulse-echo technique, Niepmann[105] obtained speeds of sound of liquid *n*-butane between 200 K and 375 K at pressures up to 60 MPa. The estimated total uncertainty is approximately $\pm 0.5\%$.

12.4.1.2.1.5 n-*Pentane.* Richardson and Tait[107] reported measurements of sound speed and attenuation in liquid *n*-pentane at pressures up to 55 MPa and at temperatures between 15 °C and 44 °C. Lainez *et al.*[108] obtained sound speeds for *n*-pentane measured on isotherms between 263 K and 433 K and at pressures up to 210 MPa.

12.4.1.2.1.6 n-*Hexane.* Eden and Richardson[109] presented sound speed data in *n*-hexane graphically at 20 °C and 37 °C for pressures up to about 69 MPa; for 20 °C, sound speeds were also given in tabular form up to 98 MPa. Boelhouwer[110] measured sound speeds of liquid *n*-hexane in the temperature range from 253.15 K to 333.15 K and at pressures up to 140 MPa. Hawley *et al.*[84] reported the sound speed and sound damping of *n*-hexane at 303 K and at pressures up to 392 MPa; and Allegra *et al.*[111] measured sound speed and sound damping in *n*-hexane at 303.15 K and at pressures up to 981 MPa.

In 1998, Daridon *et al.*,[112] performed speed of sound measurements in liquid *n*-hexane at pressures up to 150 MPa between 293.15 K and 373.15 K. These data were used to evaluate the isentropic and isothermal compressibility in the same ranges of pressure and temperature. Khasanshin and Shchemelev[113] measured the speed of sound in liquid *n*-hexane at pressures up to 50 MPa at temperatures between 298.15 K and 433.15 K. Ball and Trusler[114] reported measurements of the speed of sound of *n*-hexane at 298.3 K, 323.15 K, 348.15 K, and 373.15 K and at pressures up to 100 MPa. The technique was based on a pulse-echo method with a single transducer placed between two plane parallel reflectors. The speed of sound is obtained from the difference between the round-trip transit times in the two paths. Khasanshin *et al.*[115] measured the sound speed of *n*-hexane at temperatures between 298 K and 433 K and pressures up to 100 MPa. Clearly, this makes liquid *n*-hexane one of the best investigated organic solvents.

12.4.1.2.1.7 n-*Heptane.* Boelhouwer[110] measured sound speeds in liquid *n*-heptane in the temperature range 253.15 K to 453.15 K and at pressures up to 140 MPa. Muringer *et al.*[116] measured the sound speed in *n*-heptane at elevated pressures between 185.6 K and 310.6 K using increments of about 12.5 K. At lower temperatures, the maximum pressure was limited by the freezing of *n*-heptane, while for 223.3 K and higher temperatures the pressure range extended up to 263.4 MPa. Based on the approach represented by Equation (12.17), a number of thermophysical properties of liquid *n*-heptane were calculated. Using an *improved* computational method,[21] these speed of sound data were reevaluated by Sun *et al.*,[22] who determined densities, isobaric expansivities, isothermal compressibilities, isobaric heat capacities and isochoric heat capacities for the indicated temperature and pressure ranges. Of particular note is the *negative* temperature dependence of α_P at higher pressures, and the confirmation of shallow minima in the C_P *vs.* P curves at higher temperatures, which is a consequence of Equation (12.18).

12.4.1.2.1.8 n-*Octane.* Boelhouwer[110] measured sound speeds of liquid *n*-octane in the temperature range from 253.15 K to 393.15 K and at pressures up to 140 MPa. Khasanshin and Shchemelev[113] measured the speed of sound in liquid *n*-octane at pressures up to 50 MPa at temperatures between 303.15 K and 433.15 K. Subsequently Khasanshin *et al.*[115] measured the sound speeds in a slightly enlarged temperature range for pressures up to 100 MPa.

12.4.1.2.1.9 n-*Nonane.* Boelhouwer[110] measured sound speeds of liquid *n*-nonane in the temperature range from 253.15 K to 413.15 K and at pressures up to 140 MPa.

12.4.1.2.1.10 n-*Decane.* Khasanshin and Shchemelev[113] measured the speed of sound in liquid *n*-decane at pressures up to 50 MPa at temperatures between 298.15 K and 433.15 K. Again in 2008, Khasanshin *et al.*[115] reported sound speeds for a slightly enlarged temperature range and for pressures up to 100 MPa.

12.4.1.2.1.11 n-*Dodecane,* n-*Tridecane,* n-*Tetradecane and* n-*Pentadecane.* Khasanshin and Shchemelev[113] measured the speed of sound in these liquid *n*-alkanes at pressures up to 50 MPa and at temperatures between 303 K and 433 K.

12.4.1.2.1.12 n-*Hexadecane.* Boelhouwer[110] measured sound speeds of liquid *n*-hexadecane in the temperature range from 293.15 K to 473.15 K and at pressures up to 140 MPa. Ball and Trusler[114] reported measurements of the speed of sound of *n*-hexadecane at 298.3 K for pressures up to 25 MPa, at 323.15 K for pressures up to 80 MPa, and at 348.15 K and 373.15 K for pressures up to 101 MPa. Khasanshin *et al.*[115] measured the sound speed of *n*-hexadecane at temperatures between 298 K and 433 K and pressures up to 100 MPa.

12.4.1.2.2 Branched Alkanes, Cycloalkanes and Propene

12.4.1.2.2.1 2,2-Dimethylpropane. Lainez *et al.*[108] obtained sound speeds for 2,2-dimethylpropane measured on isotherms between 263 K and 433 K and at pressures up to 54 MPa.

12.4.1.2.2.2 Isopentane (Methylbutane). Eden and Richardson[109] presented sound speed data for liquid isopentane graphically at five isotherms from 0 °C to 10.5 °C, and for pressures up to about 69 MPa; for 0 °C, sound speeds were also given in tabular form up to 98 MPa.

12.4.1.2.2.3 2-Methylpentane, 2,3-Dimethylpentane and 2,2,4-Trimethylpentane (Isooctane). Plantier and Daridon[117] measured ultrasonic speeds (at 3 MHz) in the compressed liquid phase of the three branched alkanes, 2-methylpentane, 2,3-dimethylpentane and 2,2,4-trimethylpentane (isooctane), from 293.15 K to 373.15 K and for pressures up to 150 MPa. Using Equation (12.17) together with low-pressure values for densities and isobaric heat capacities, densities, isentropic and isothermal compressibilities were calculated for the entire ranges of temperature and pressure.

For the most volatile liquid, *i.e.* 2-methylpentane, the saturation pressure was considered as the reference pressure, while for the other two liquids the initial values for ρ and C_P were at $P_{ref} = 0.1$ MPa. Agreement, for instance, with the directly determined density of 2,2,4-trimethylpentane from 198 K to 348 K and for pressures up to 100 MPa using a vibrating-wire instrument (Pádua *et al.*[118]), was very good.

12.4.1.2.2.4 Cyclohexane and Methylcyclohexane. Eden and Richardson[109] presented sound speed data for cyclohexane graphically at 19 °C and for pressures up to *ca.* 28 MPa, *i.e.* close to the freezing pressure. Sun *et al.*[20] determined the speed of sound in cyclohexane in the temperature range from 288 K to 323 K with increments of about 5 K. The isothermal experimental data were taken from 0.1 MPa upwards with pressure increments of 5 MPa, and each isotherm was terminated at a pressure close to the freezing pressure of cyclohexane. Thus, for instance, the highest experimental pressure was 15 MPa at 288.15 K, 35 MPa at 298.13 K, 55 MPa at 308.19 K and 85 MPa at 323.06 K. For these ranges of temperature and pressure, the density, the isobaric expansivity, the isothermal compressibility and the isobaric heat capacity were evaluated from the measured sound speeds, using a computational method based on Equation (12.17). Using a high-resolution Bragg reflection technique, Takagi and Negishi[119] determined the onset of ultrasonic speed dispersion in liquid cyclohexane and methylcyclohexane at 293.15 K.

12.4.1.2.2.5 Propene. Meier and Kabelac[120] reported measurements of the speed of sound in pure liquid propene. The data have been measured by a double-path length pulse-echo technique and cover the temperature range from 240 K to 420 K with pressures up to 100 MPa.

12.4.1.2.3 Aromatics

12.4.1.2.3.1 Benzene. Mifsud and Nolle[92] measured the ultrasonic speed and the sound absorption of liquid benzene at 298.15 K, 323.15 K and 343.15 K, and at pressures between 0.1 MPa and 69 MPa at the lowest temperature, and between 0.1 MPa and 138 MPa at the higher two temperatures; the data were presented graphically only. Richardson and Tait[107] reported sound speed in benzene at temperatures between 19 °C and 40 °C and at pressures ranging up to 56.5 MPa (close to the freezing pressure) at 19 °C, and up to 67.4 MPa at 40 °C. Makita and Takagi[121] carried out sound speed measurements in liquid benzene at temperatures from 283.15 K to 343.15 K at pressures up to 210 MPa. The isentropic compressibility and the heat capacity ratio κ were derived for the entire ranges of temperature and pressure. Bobik[122] published sound speeds for benzene at temperatures between 290 K and 461 K at pressures up to 62 MPa. Sun *et al.*[20] determined the speed of sound in liquid benzene in the temperature range from 283 K to 323 K with increments of about 5 K.

The isothermal experimental data were taken from 0.1 MPa upwards with pressure increments of 10 MPa, and each isotherm was terminated at a pressure close to the freezing pressure of cyclohexane. Thus, for instance, the highest experimental pressure was 50 MPa at 293.15 K, 90 MPa at 303.13 K, 130 MPa at 313.16 K, and 170 MPa at 323.13 K. For these ranges of temperature and pressure, the density, the isobaric expansivity, the isothermal compressibility and the isobaric heat capacity were evaluated from the measured sound speeds, using a computational method based on Equation (12.17). Takagi *et al.*[123] measured speeds of sound in liquid benzene at temperatures between 283.15 K and 333.15 K and at pressures up to about 30 MPa. The method used was a sing-around technique employing a fixed path acoustic interferometer operated at a frequency of 2 MHz.

12.4.1.2.3.2 Toluene. Hawley *et al.*[84] measured the sound speed and damping in toluene at 303 K and at pressures up to 522 MPa, and at 348 K and at pressures up to 426 MPa, see Figure 12.12. Allegra *et al.*[111] measured the sound speed and sound damping in toluene at 303.15 K and at pressures up to 981 MPa. Muringer *et al.*[116] measured the sound speed in toluene at elevated pressures between 173.2 K and 320.3 K using small temperature increments of about 5 K at lower temperatures, and about 25 K increments at higher temperatures. At lower temperatures, the

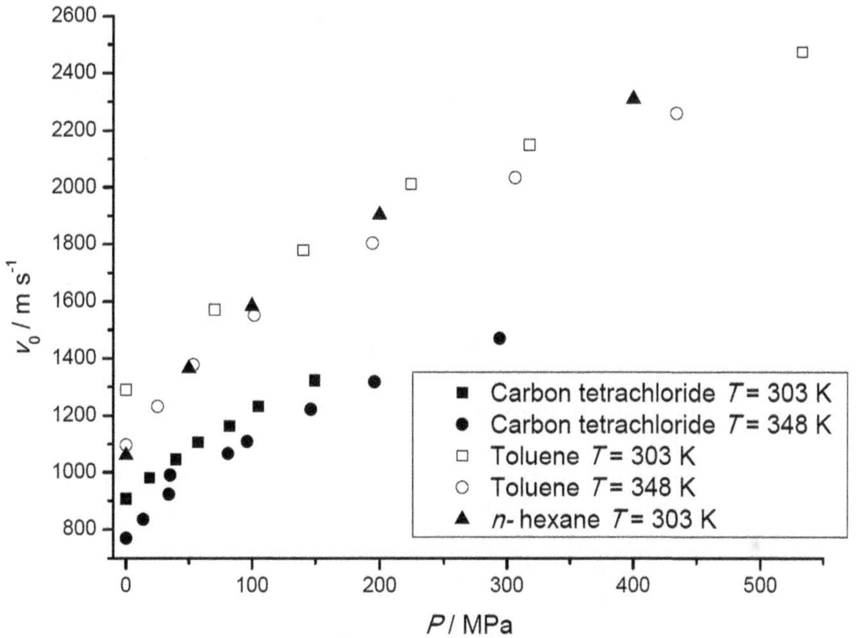

Figure 12.12 Sound speed of tetrachloromethane, toluene, and *n*-hexane as a function of pressure; data were taken from Hawley *et al.*[84]

maximum pressure was limited by the freezing of toluene, while for 223.1 K and higher temperatures the pressure range extended up to 263.4 MPa. Based on the approach represented by Equation (12.17), a number of thermophysical properties of liquid toluene were calculated. Using an *improved* computational method,[21] these speed of sound data were reevaluated by Sun *et al.*,[22] who determined densities, isobaric expansivities, isothermal compressibilities, isobaric heat capacities and isochoric heat capacities for the indicated temperature and pressure ranges. Of particular note is the *negative* temperature dependence of α_P at higher pressures, and the confirmation of shallow minima in the C_P *vs.* P curves at most temperatures, which is a consequence of Equation (12.18). Meier and Kabelac[124] carried out measurements of the speed of sound in liquid toluene by a double-path length pulse-echo technique. The data were obtained for the temperature range 240 K to 420 K with pressures up to 100 MPa.

12.4.1.2.4 Halogen Derivatives

12.4.1.2.4.1 Halomethanes. Using high-resolution Bragg reflection, Takagi *et al.*[125] measured ultrasonic speed and absorption in liquid dichloromethane (at 283.15 K, 293.15 K, and 303.15 K) and liquid dibromomethane (at 293.15 and 303.15 K) over the frequency range 60 MHz to 700 MHz. In both liquids, considerable speed dispersion was observed and described with single relaxation frequencies that is, 192 MHz for dichloromethane and 393 MHz for dibromoethane at 293.15 K. The observed relaxation strengths were roughly consistent with the theoretical values calculated from the vibrational relaxation associated with all but the lowest vibrational mode of the molecules. Takagi and Negishi,[119] again using Bragg reflection technique, report simultaneous measurements over the frequency range from 100 MHz to 1000 MHz for several organic liquids at 293.15 K. Specifically, the onset of sound speed dispersion for trichloromethane (chloroform) was presented. Since the Bragg reflection technique allows the simultaneous measurement of $\nu(f)$ and α/f^2, the experimental results may be discussed in analogy to the analysis commonly used in the discussion of dielectric properties, that is, similar to the Cole–Cole plot: for a *single* relaxation process a graph of $(\alpha\lambda_{us}/\pi)(\nu_0/\nu)^2$ *vs.* $(\nu_0/\nu)^2$ forms a semicircle centred on the abscissa with a diameter of $(\nu_\infty^2 - \nu_0^2)/\nu_\infty^2$; here, λ_{us} is the wavelength of ultrasound at each frequency f, that is, $\nu(f) = f\lambda_{us}$.

Takagi[126] studied ultrasonic speeds in liquid difluoromethane at temperatures from 243.15 K to 373.15 K and at pressures between the vapour pressure and about 35 MPa. Niepmann *et al.*[127] measured speeds of sound (at 2 MHz) in liquid chlorodifluoromethane (R22) between 200 K and 300 K, and in liquid dichloromethane between 200 K and 420 K. Starting at the coexistence line, pressures were varied up to 60 MPa. The inaccuracy of the experimental results is less than 0.15%. Two equations for each liquid were presented: one to report the speed of sound as a function of pressure and temperature over the entire experimental range and another for the variation of the sound speed as a function of temperature along the coexistence line only.

A considerably more comprehensive study of the temperature and pressure dependence of trichlorofluoromethane was presented by Lainez *et al.*[128] Sound speeds in liquid trichlorofluoromethane were measured at four temperatures between 353 K and 413 K and at pressures up to 210 MPa.

Mifsud and Nolle[92] measured the ultrasonic speed and the sound absorption in tetrachloromethane at 298.15 K and at 323.15 K and at pressures between 0.1 MPa and *ca.* 138 MPa (the data were presented graphically only). Richardson and Tait[107] determined sound speed of tetrachloromethane at 15.5 °C, 25 °C and 40 °C, and at pressures up to 68.9 MPa. McSkimin[129] measured sound speed and sound attenuation for tetrachloromethane as a function of temperature and pressure: at 5.5 °C up to about 77 MPa, at 25 °C up to about 132 MPa, and at 50 °C up to about 214 MPa. The results were presented graphically only. Hawley *et al.*[84] measured the sound speed and sound damping in tetrachloromethane at 303 K at pressures up to 147 MPa, and at 348 K at pressures up to 289 MPa. For a selection of their experimental data, see Figure 12.12. Bobik *et al.*[130] measured sound speeds for tetrachloromethane between 265 K and 435 K and up to 62 MPa, and in 1987 Lainez *et al.*[131] performed sound speed measurements at temperatures between 283 K and 455 K and at pressures up to 130 MPa.

12.4.1.2.4.2 Haloaromatics. Eden and Richardson[109] published sound speeds and sound absorption data in halogenated aromatics. Specifically, fluorobenzene, chlorobenzene, bromobenzene and iodobenzene were investigated at 22 °C and at pressures up to 68 MPa. These results were only presented graphically. At 50 °C, the pressure dependence of the sound speed in chlorobenzene and bromobenzene was determined up to 98 MPa and shown in tabular form. In addition, sound speeds of the four monohalogenated benzenes at atmospheric pressure were measured in the temperature range from 20 °C to *ca.* 60 °C and presented graphically.

12.4.1.2.5 Alcohols

12.4.1.2.5.1 Methanol. As part of a study of the pressure dependence of sound propagation in liquid primary alcohols, Carnevale and Litovitz[132] presented measurements of sound speed and sound absorption in methanol at 303.15 K and at pressures up to 196 MPa. The sound absorption decreases with increasing pressure, whereas the sound speed increases with increasing pressure. Eden and Richardson[109] presented their speed of sound data in methanol graphically at 20 °C, 30 °C and 40 °C, and for pressures up to about 68 MPa (an additional value for the sound speed at 20 °C and 98 MPa was given in tabular form). Wilson and Bradley[133] measured sound speeds of methanol at temperatures between of 273.15 K and 323.15 K and pressures between 0.1 MPa and 96.5 MPa. Hawley *et al.*[84] determined the sound speed and sound damping in methanol at 303 K and at pressures up to about 413 MPa (see Figure 12.13).

Sun *et al.*[134] measured the speed of sound in liquid methanol at temperatures from 274.7 K to 332.9 K and at pressures up to 276 MPa, using a phase comparison pulse-echo technique operating at 2 MHz. Combining their results with data at $P_{ref} = 0.1$ MPa, that is, $\rho(T,P_{ref})$ and $C_P(T,P_{ref})$, a simple and rapid successive integration algorithm (see Equation (12.17) and Sun *et al.*[21] for details) was used to derive precise values for the liquid state density, the isobaric expansivity, the isothermal compressibility, the isobaric heat capacity and the isochoric heat capacity as functions of pressure for the entire temperature range. In 1990, Sun *et al.*[135] extended the temperature range to lower temperatures, *i.e.* 203.0 K to 261.2 K, and measuring the sound speed in liquid methanol up to 276 MPa, they also determined the corresponding high-pressure thermodynamic properties listed above; that is, densities, isobaric expansivities, isothermal compressibilities *etc.* for $203.15 \leq T/\mathrm{K} \leq 263.15$ and $0.1 \leq P/\mathrm{MPa} \leq 280$. The derived density data have been used to examine the validity of several empirical equations of state. More recently, Plantier *et al.*[136] measured the sound speed of methanol at pressures up to 50 MPa and at temperatures between 303.15 K and 373.15 K.

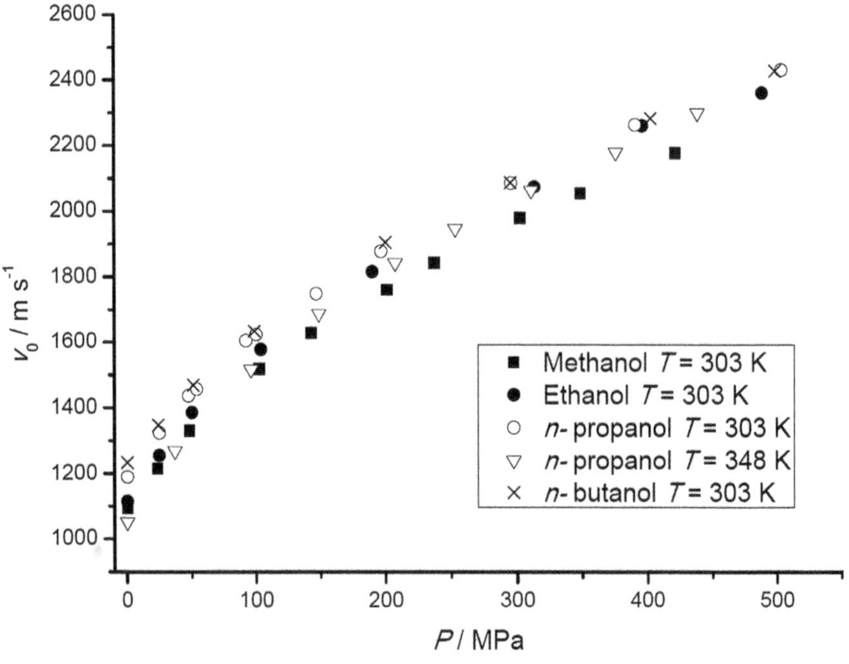

Figure 12.13 Sound speed of methanol, ethanol, propanol and *n*-butanol as a function of pressure; data were taken from Hawley *et al.*[84]

12.4.1.2.5.2 Ethanol. Ethanol is certainly the best investigated alcohol. Carnevale and Litovitz[132] determined sound speed and sound absorption in liquid ethanol at 303.15 K for pressures up to 196 MPa. With increasing pressure, the sound speed increases, while the sound absorption decreases. Eden and Richardson[109] published data of sound speed in ethanol at 20 °C for pressures up to *ca.* 68 MPa in graphical form (an additional value for the sound speed at 20 °C and 98 MPa is given in tabular form). Wilson and Bradley[133] measured sound speeds of ethanol at temperatures between 273.15 K and 323.15 K and pressures between 0.1 MPa and 96.5 MPa. Hawley *et al.*[84] measured the sound speed and sound damping in ethanol at 303 K and at pressures up to 479 MPa (see Figure 12.13). Similar to their work on methanol, Sun *et al.*[21] measured the sound speed in liquid ethanol from 273.9 K to 333.0 K and at pressures up to 276 MPa. Subsequently, Equation (12.17), in conjunction with an improved integration algorithm, was used to determine the density, the isobaric expansivity, the isothermal compressibility and the isobaric heat capacity for the temperature range 273.15 K to 333.15 K and for the pressure range 0.1 MPa to 280 MPa.

12.4.1.2.5.3 Propan-1-ol. Carnevale and Litovitz[132] measured sound speed and sound absorption in liquid *n*-propanol at 303.15 K and at pressures up to 196 MPa. With increasing pressure, the speed of ultrasound increases, while sound absorption decreases. Wilson and Bradley[133] measured sound speeds of *n*-propanol at temperatures between 273.15 K and 323.15 K and at pressures between 0.1 MPa and 96.5 MPa. Hawley *et al.*[84] measured the sound speed and the sound damping in *n*-propanol at temperatures between 296.2 K and 348.0 K and at pressures up to about 493 MPa (see Figure 12.13). With increasing pressure, the sound speed increases, while the sound absorption decreases.

12.4.1.2.5.4 Butanols. Carnevale and Litovitz[132] reported sound speed and sound absorption in liquid in *n*-butanol at 303.15 K and at pressures up to 196 MPa. Again, with increasing pressure the sound speed increases, while the sound absorption decreases. In addition, sound speed and sound absorption were measured at 196 MPa at 273.15 K, 288.15 K, 303.15 K and 318.15 K. These data allow one to determine the influence, if any, of pressure on the activation energies for shear and compressive flow: within experimental accuracy, no change in activation energy was detected in *n*-butanol. Wilson and Bradley[133] measured the sound speed in *n*-butanol at temperatures between 273.15 K and 323.15 K and at pressures between 0.1 MPa and 96.5 MPa. Hawley *et al.*[84] measured the speed of sound and sound damping in *n*-butanol at 303 K and at pressures up to about 489 MPa (see Figure 12.13).

Plantier *et al.*[136] have measured the sound speed of butan-1-ol in the temperature range 303.15 K to 373.15 K and at pressures up to 50 MPa. A DSA 5000 analyzer from Anton Paar, Graz, Austria, was used by

Outcalt *et al.*[137a] to measure density and speed of sound of butan-1-ol and butan-2-ol at ambient pressure (83 kPa) from 278.15 K to 343.15 K in 5 K increments, and corresponding isentropic compressibilities were calculated. Compressed liquid density measurements in the temperature range 270 K to 470 K and for pressures up to 50 MPa were carried out in an automated densimeter described by Outcalt and McLinden in a previous publication:[137b] the heart of this apparatus is a commercial vibrating-tube densimeter from Anton Paar (DMA-HPM).

12.4.1.2.5.5 Pentan-3-ol. Gonzalez-Salgado *et al.*[138] have measured the speed of sound for liquid pentan-3-ol at temperatures between 303.15 K and 373.15 K and at pressures up to 100 MPa. These results were combined with density and isobaric heat capacity data at atmospheric pressure obtained from the literature to calculate, in the spirit of Equation (12.17), the density, the isentropic compressibility and the isothermal compressibility for the entire range of temperature and pressure.

12.4.1.2.5.6 Hexanols. Chorazewski *et al.*[139] measured the speed of sound in hexan-1-ol and 2-ethyl-1-butanol over the temperature range 293.15 K to 318.15 K and at pressures up to 101 MPa. Densities were measured in the temperature range 283.15 K to 343.15 K or 353.15 K, respectively, at atmospheric pressure. For the measurements, a pulse-echo-overlap method and a vibrating-tube densitometer have been used. For liquid 3-methyl-pentan-3-ol, Gonzalez-Salgado *et al.*[138] measured the speed of sound between 303.15 K and 373.15 K and at pressures up to 100 MPa. Using Equation (12.17), these results, combined with densities and isobaric heat capacities at atmospheric pressure obtained from the literature, were used to calculate the density, the isentropic compressibility and the isothermal compressibility for the entire ranges of temperature and pressure.

12.4.1.2.6 Higher Alkanols. Dzida[140] measured the speed of sound in heptan-1-ol, octan-1-ol, nonan-1-ol and decan-1-ol at pressures up to 101 MPa within the temperature range 293 K to 318 K. The densities, isobaric heat capacities, isentropic and isothermal compressibilities, isobaric expansivities and internal pressures as functions of temperature and pressure were calculated *via* Equation (12.17), using experimental results on densities and isobaric heat capacities at ambient pressure taken from the literature.

For liquid 3-ethyl-pentan-3-ol, Gonzalez-Salgado *et al.*[138] measured the speed of sound at temperatures between 303.15 K and 373.15 K and at pressures up to 100 MPa. Using Equation (12.17), these results, combined with the densities and isobaric heat capacities at atmospheric pressure obtained from the literature, were used to calculate the density, the isentropic compressibility and the isothermal compressibility in the same range of pressure and temperature. Plantier *et al.*[136] determined the sound speed of octan-1-ol at pressures up to 50 MPa and at temperatures between 303.15 K and 373.15 K.

12.4.1.2.7 Polyols

12.4.1.2.7.1 Propanediols. Zorebski *et al.*[141] measured the speeds of sound in 1,2- and 1,3-propanediol at temperatures from 293 K to 318 K and at pressures up to 101 MPa by the pulse-echo-overlap method. The densities at atmospheric pressure of the propanediol isomers were measured in the temperature range from 283.15 K to 363.15 K with a vibrating-tube densitometer. From the experimental results, the densities, isobaric heat capacities, isobaric expansivities, isentropic compressibilities and isothermal compressibilities were calculated.

12.4.1.2.7.2 Butanediols. Zorebski and Dzida[142] measured the speeds of sound in 1,2- and 1,3-butanediol in the temperature range from 293 K to 318 K and at pressures up to 101 MPa by the pulse-echo-overlap method. The densities at atmospheric pressure of both butanediol isomers were measured in the temperature range from 288.15 K to 363.15 K with a vibrating-tube densitometer. From the experimental results the densities, isobaric heat capacities, isobaric expansivities, isentropic compressibilities and isothermal compressibilities were calculated.

12.4.1.2.7.3 Glycerol. Richardson and Tait[107] measured the sound speed in glycerol at temperatures between 23 °C and 79 °C and at pressures up to about 68 MPa (the data were presented in graphical form only). Litovitz and Lyon[143] reported the sound speed (at 22 MHz) in glycerol at temperatures between −100 °C and −72 °C, that is, below and above the glass transition temperature (the data were presented in graphical form only). The absolute value of the temperature coefficient of the sound speed in the glassy state ($-1.5\ \mathrm{ms^{-1}K^{-1}}$) is significantly less than in the liquid ($-5.0\ \mathrm{ms^{-1}K^{-1}}$). No discontinuity of the sound speed was observed at the "transition temperature" between glassy and liquid state at about −90 °C.

12.4.1.2.8 Miscellaneous Compounds

12.4.1.2.8.1 Ethyl Ether. Richardson and Tait[107] measured the sound speed in ethyl ether between 16 °C, 25 °C, 30 °C and 44 °C at pressures up to 62 MPa.

12.4.1.2.8.2 Acetone. Eden and Richardson[109] published data of sound speed in acetone at 20 °C, 30.5 °C and 41 °C and at pressures up to 98 MPa (graphical presentation only). An additional value for the sound speed at 20 °C and 98 MPa was given in tabular form. Recently, a novel high-pressure cell for the determination of the speed of ultrasound in liquids was designed by Azevedo *et al.*[144] For 2-propanone (acetone) the sound speed was measured at temperatures between 265 K and 340 K and at pressures up to 160 MPa. Szydlowski *et al.*[145] used a new high-pressure ultrasonic microcell to carry out sound speed measurements in deuterated 2-propanone (acetone-d6) at temperatures between 288 K and 338 K and at pressures up to 160 MPa. Comparison with the data for acetone-h6 enabled the authors to establish the magnitude and sign of deuterium isotope effects. Lago and Albo[146] described speed of sound measurements in liquid acetone along eleven isotherms at temperatures between 248.15 K and 298.15 K and at pressures up to 100 MPa.

12.4.1.2.8.3 Acetaldehyde. Eden and Richardson[109] published sound speeds in acetaldehyde at 8 °C and at 15 °C for pressures up to about 62 MPa (graphical presentation only).

12.4.1.2.8.4 Acetic Acid. Litovitz and Carnevale[147] measured the speed of sound and ultrasonic absorption in acetic acid at 323.15 K and at 155 MPa. Becker and Kohler[148] measured the sound speed in acetic acid at ambient pressure between 293.15 K and 313.15 K.

12.4.2 Brillouin Scattering Data

Thorough discussions of the theoretical background of Brillouin scattering in liquids are provided by Berne and Pecora,[24] Boon and Yip,[25] and Chu.[149] From Equations (12.34) and (12.35), the *angular* frequency shift of the two Brillouin peaks is related to the hypersonic sound speed in the medium by

$$\omega_{\mathrm{B}} = \pm \frac{4\pi n \nu_{\mathrm{H}}}{\lambda_0} \sin\frac{\theta}{2}. \tag{12.42}$$

Thus, determination of $\nu_{\mathrm{H}}(T,P)$ as a function of temperature and pressure requires knowledge of the temperature dependence as well as the pressure dependence of the refractive index $n(T,P)$ at the respective laser light frequency. In particular the pressure dependence is a formidable problem by itself, as indicated by the paucity of reliable data. A brief list of selected articles presenting experimental results for the isothermal piezo-optic coefficient $(\partial n/\partial P)_T$ of liquids was presented by Wilhelm and Asenbaum[6] in 2010.

12.4.2.1 Inorganic Liquids

12.4.2.1.1 Argon. High-pressure Brillouin scattering studies on *simple* fluids, such as argon, are quite scarce. As pointed out by Fleury and Boon[150] this is rather surprising, since such studies may disclose, perhaps, nonclassical behaviour (*i.e.* departure from the predictions of the Navier–Stokes equations) for very-high-frequency sound waves[151–153] due to the onset of frequency dependence of the liquid's transport coefficients. The dispersion phenomena should be most important at frequencies of the order of the reciprocal collision time, *i.e.* at 10^{12} Hz to 10^{13} Hz for dense simple fluids, a frequency range which is not yet accessible for Brillouin experiments. However, as yet no model for estimating the frequency above which these new effects are expected to become more easily measurable is known.

The speed of hypersound (between 2.5 GHz and 3.1 GHz) in liquid argon was measured by Fleury and Boon[150] along the vapour–liquid equilibrium curve between 85 K and 100 K (that is at pressures smaller than 0.33 MPa): ν_H decreases linearly from 850 m s^{-1} at 85 K to 742 m s^{-1} at 100 K. Carraresi *et al.*[154] determined the hypersound speed for frequencies between 2 GHz and 4 GHz at 298 K and at pressures between 70 MPa and 260 MPa, and did *not* find any dispersion. Because of the importance of the issue, Hochheimer *et al.*[155] undertook a high-pressure Brillouin scattering study of dense fluid argon (and nitrogen) between 157.7 K and 296.8 K and pressures up to 500 MPa. Their results indicate negative speed dispersion, thus supporting the older results of Fleury and Boon,[150] but contradicting the findings of Carraresi *et al.*[154] Hypersound speeds in dense fluid argon have been reported by Jia[156] between 293 K and 503 K for pressures up to the respective solidification pressure (at 293 K the solidification pressure of argon is about 1300 MPa: for recent experimental work on the melting curve of argon see Abramson[157]). Since only (very small) graphical representations of their results were provided, no quantitative conclusions can be drawn. Some results presented in this article were subsequently discussed by Datchi[158] and Li and Zhou.[159]

12.4.2.1.2 Water. Sedlacek and Asenbaum[160] measured hypersound speed and sound damping in water at 298.15 K at pressures up to 175 MPa. No sound dispersion was detected. Cunsolo and Nardone[161] determined Brillouin scattering spectra of water from 30 °C down to supercooling temperatures of about −26 °C at different scattering angles and for different laser wavelengths. The authors found relevant dispersion in the wave vector dependence of the hypersonic sound speed in the supercooled regime. For the observed dispersion, they suggest a single relaxation process, involving shear and volume viscosities. Decremps *et al.*[162] determined the hypersonic sound speed in water at temperatures from 300 K up to 723 K and pressures up to 9 GPa using a Brillouin scattering technique.

12.4.2.1.3 Carbon Dioxide. Giordano *et al.*[163] determined the melting curve and the fluid equation of state of carbon dioxide under high pressure in a resistively heated diamond anvil cell. The melting line was determined from room temperature up to 800 K and at pressures up to 11.1 GPa by visual observation of the solid–fluid equilibrium and *in situ* measurements of pressure and temperature. Interferometric and Brillouin scattering experiments were conducted to determine the refractive index and the hypersound speed in the fluid phase. Sound dispersion was detected and modelled by postulating a thermal relaxation process.

12.4.2.1.4 Carbon Disulfide. Stith *et al.*[164] measured hypersonic speeds in carbon disulfide at 300.15 K and at pressures up to 100 MPa. In carbon disulfide the dispersion between ultrasonic and hypersonic speeds

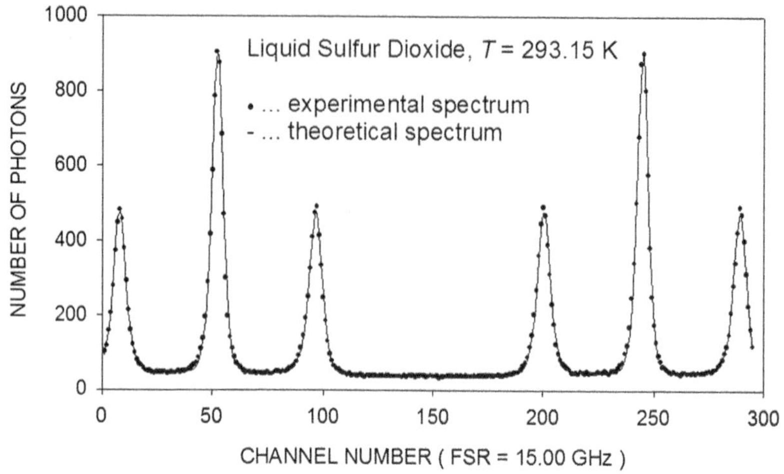

Figure 12.14 Brillouin spectrum of liquid sulfur dioxide at 293.15 K and at saturation pressure. The dots are the experimental points (photon counts at the detector), the solid curve represents the theoretical Brillouin spectrum according to Mountain's theory convoluted with the instrumental function of the Fabry–Perot interferometer; data were taken from Asenbaum *et al.*[28]

remains essentially constant over the pressure range studied. Shimizu *et al.*[165] measured Brillouin spectra at 300 K in liquid carbon disulfide in a diamond anvil cell up to 1.3 GPa.

12.4.2.1.5 Sulfur Dioxide. As already indicated above in the Introduction 12.1, liquid sulfur dioxide shows a clearly resolved double vibrational relaxation process. The first relaxation step is associated with the two stretching modes (at 1152 cm^{-1} and at 1362 cm^{-1}), with a relaxation frequency of 22.5 MHz, and the second relaxation step is associated with the bending mode (at 518 cm^{-1}), with a relaxation frequency in the GHz region. By using scattering angles of 45°, 60°, 90°, 120° and 135°, Asenbaum *et al.*[28] have determined the frequency dependence (1.9 GHz to 4.9 GHz) of the Brillouin shift, see Figure 12.14, and the half-width of the Brillouin peaks in orthobaric liquid sulfur dioxide at 293.15 K. The relaxation frequency of the second relaxation step was determined to be 1.6 GHz, and the corresponding nonrelaxing volume viscosity amounted to 5.0×10^{-4} Pa s $\approx 2\eta_s$.

12.4.2.2 *Organic Liquids*

12.4.2.2.1 Methane. For supercritical fluid CH_4 at 300 K, Shimizu *et al.*[166] studied the sound speed and the refractive index at pressures up to the fluid–solid phase transition at about 1.7 GPa. Li *et al.*[167] performed Brillouin scattering measurements on fluid methane along five isotherms (from 298 K to 539 K) and at pressures up to solidification (for recent experimental work on the melting curve of methane see Abramson[157]). A modified Merill–Bassett type four-screw diamond anvil cell was used to generate high pressure. The pressure was determined by the ruby fluorescence shift at room temperature, and by the YAG scale (Sm-doped $Y_3Al_5O_{12}$) at higher temperatures. Along each isotherm, sound speed and refractive index were determined: they both increase monotonously with increasing pressure; along an isobar they decrease slightly with increasing temperature. The maximum pressure was 5.12 GPa at 539 K.

12.4.2.2.2 Cyclohexane. Asenbaum[168] measured the temperature dependence of the ultrasonic speed and ultrasonic damping as well as the hypersonic speed and relaxation time τ_{v1} in liquid cyclohexane by Brillouin scattering. These data suggest a two-step relaxation process for the vibrational relaxing heat capacity. The relaxation time τ_{v2} of the second step was calculated based on the volume viscosity derived from the Brillouin line width. For the non-relaxing part of the volume viscosity the corresponding values of argon and krypton, using the theorem of corresponding states, were applied. Asenbaum and Wilhelm[169] measured Brillouin spectra of cyclohexane between 283.15 K and 343.15 K, and reevaluated data for the heat capacities at constant pressure and at constant volume. The vibrational contribution to C_V was determined and compared with calculated values obtained from the Planck–Einstein relation, Equation (12.30). Utilising perfect gas state heat capacity data, the residual isochoric heat capacity was determined and subsequently used to obtain an estimate for the residual rotational heat capacity (external, overall molecular rotation).

12.4.2.2.3 Benzene. Stith *et al.*[164] measured hypersonic speed in benzene at 300.15 K and at pressures up to 100 MPa. Asenbaum and Wilhelm[169] measured Brillouin spectra of benzene between 283.15 K and 343.15 K, and reevaluated data for the heat capacities at constant pressure and at constant volume. The vibrational contribution to C_V was determined and compared with calculated values obtained from the Planck–Einstein relation, Equation (12.30). Utilising perfect gas state heat capacity data, the residual isochoric heat capacity was determined and subsequently used to obtain an estimate for the residual rotational heat capacity (external, overall molecular rotation).

Asenbaum and Hochheimer[74] measured Brillouin spectra of liquid benzene at pressures up to 130 MPa and at temperatures from 298.15 K to 343.15 K. The results for the hypersound speeds were in good agreement with results obtained by Medina and O'Shea.[170] Figure 12.8 shows the ultrasonic speed and the hypersonic speed as a function of pressure at 298.15 K.

12.4.2.2.4 Toluene. Asenbaum *et al.*[171] recorded Brillouin spectra of liquid toluene at 0.1 MPa pressure and between 293.15 K and 313.15 K at

45°, 60°, 90°, 120°, 136° and 177° scattering geometry. In Figure 12.15, toluene Brillouin spectra at 278.15 K, 333.15 K and 363.15 K at 0.1 MPa and at 90° scattering angle are shown. The intensity of both the Brillouin peaks and the Rayleigh peak is increasing with increasing temperature, whereas the Brillouin shift decreases with increasing temperature. Between 273.15 K and 333.15 K, ultrasound speed and sound attenuation were also measured at frequencies around 109 MHz. In addition, Asenbaum *et al.*[172,173] measured Brillouin spectra of liquid toluene at

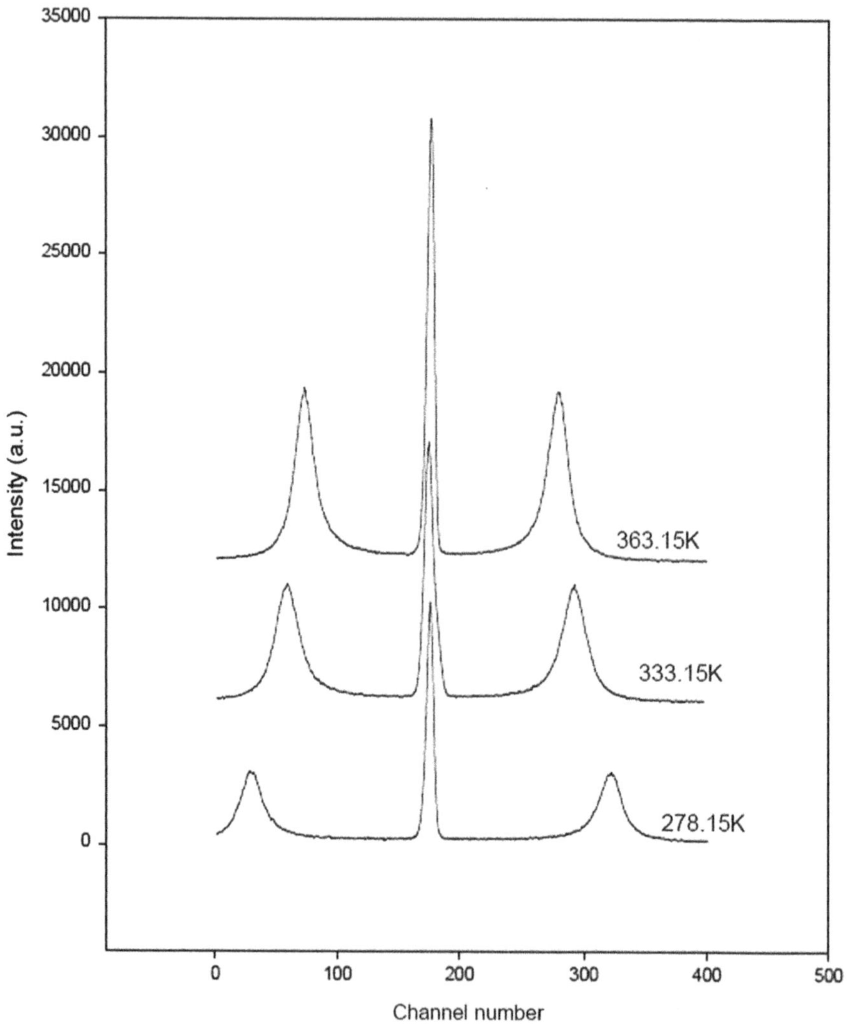

Figure 12.15 Brillouin spectra of toluene at 278.15 K, 333.15 K and 363.15 K at 0.1 MPa. The intensity of both the Brillouin peaks and the Rayleigh peak is increasing with increasing temperature, whereas the Brillouin shift decreases with increasing temperature.[171]

303.15 K and at pressures up to 162.5 MPa. From the experimental spectra the hypersound speed ν_H was determined as a function of pressure, as were the relaxing heat capacity C_{int}, the nonrelaxing $\eta_{v,nr}$ and the relaxing volume viscosities $\eta_{v,r}(f)$, the corresponding relaxation time τ_v, and the energy relaxation time τ_e. In contradistinction to former experimental results on tetrachloromethane and benzene, for toluene only a very slight decrease of τ_v and τ_e with increasing pressure was found.

12.4.2.2.5 Tetrachloromethane. Asenbaum and Wilhelm[169] measured Brillouin spectra of tetrachloromethane and reevaluated data for the heat capacities at constant pressure and at constant volume; the vibrational contribution to C_V was determined and compared with calculated values obtained from the Planck–Einstein relation, Equation (12.30). Utilising perfect gas state heat capacity data, the residual isochoric heat capacity was determined and subsequently used to present an estimate for the residual rotational heat capacity (external, overall molecular rotation).

Asenbaum and Hochheimer[174] measured Brillouin spectra at 298.15 K, 323.15 K and 348.15 K at pressures up to 150 MPa. The hypersound speeds were determined by fitting experimental Brillouin spectra to theoretical spectra according to Mountain's theory with consideration of the instrumental half-width of the Fabry–Perot interferometer. The dispersion remains constant as a function of pressure for all three temperatures.

12.4.2.2.6 Methanol and Ethanol. Brown *et al.*[175] determined hypersound speeds of methanol and ethanol in a diamond anvil cell at pressures from 0.0001 GPa up to 6.82 GPa and up to 3.19 GPa, respectively. A Brillouin scattering study of methanol at pressures up to 8.4 GPa was reported by Lee *et al.*[176] Ko and Kojima[177] measured Brillouin spectra of ethanol at temperatures between 100 K and 320 K. From the temperature dependence of the sound speed and the damping factor, it was found that the sound dispersion was negligible in the high-temperature range above 300 K.

12.4.2.2.7 Acetone. Stith *et al.*[164] measured the speed of hypersound in acetone at 300.15 K and at pressures up to 100 MPa. The hypersound speed increases almost linearly with increasing pressure.

12.4.2.2.8 Acetic Acid. Eastman *et al.*[178] determined the hypersound speed, at various frequencies, in ten liquids, including water, as function of temperature at ambient pressure. For acetic acid, the temperature range extended from 290.25 K to 368.05 K. Asenbaum[179] measured the hypersound speed in acetic acid by Brillouin scattering at temperatures between 293.15 K and 353.15 K. A Brillouin scattering study by Bohidar[180] on ten liquids provided hypersound speeds at 293.15 K and at pressures up to 82.5 MPa (in steps of, roughly, 10 MPa). For acetic acid, the pressure range extended from 0.1 MPa to 72 MPa.

12.5 Concluding Remarks

Ultrasonic speed and sound attenuation provide valuable insights into molecular processes occurring in dense liquids/fluids. In addition, sound speeds well below any dispersion region, that is, the thermodynamic speeds v_0, are key quantities for the high-precision determination of thermodynamic properties at elevated pressures, such as densities, isothermal compressibilities, isobaric expansivities, heat capacities at constant pressure and heat capacities at constant volume [see Equations (12.11)–(12.13) and (12.17)]. With modern instruments yielding reliable high-precision thermodynamic speeds of sound, v_0 data have become indispensable ingredients for the development of fundamental equations of state based on the Helmholtz energy [see Equation (12.15)]. Chapter 5, authored by Span and Lemmon, is devoted to this fundamental topic.

The invention of the laser (1960) has greatly stimulated interest in optics in general, and in the scattering of laser light from liquids in particular. It was Raymond Mountain with his seminal article, *Spectral Distribution of Scattered Light in a Simple Fluid*,[59] who really opened up the field for the experimentalists. Rayleigh–Brillouin spectra contain a wealth of information on important thermophysical quantities, such as the speed of hypersound, the volume (or bulk) viscosity, the thermal conductivity, the relaxing heat capacity and the corresponding relaxation time; Rayleigh–Brillouin spectroscopy thus extends and complements conventional ultrasonics. Of course, the study of the influence of temperature is important and interesting, but so is the influence of pressure: more work in this area would be highly desirable.

References

1. K. F. Herzfeld and T. A. Litovitz, *Absorption and Dispersion of Ultrasonic Waves*, Academic Press, New York, USA, 1959.
2. W. Schaafs, *Molekularakustik*, Springer-Verlag, Berlin, Germany, 1963.
3. A. B. Bhatia, *Ultrasonic Absorption*, Oxford University Press, London, UK, 1967.
4. M. J. Blandamer, *Introduction to Chemical Ultrasonics*, Academic Press, London, UK, 1973.
5. J. P. M. Trusler, *Physical Acoustics and Metrology of Fluids*, Adam Hilger, Bristol, UK, 1991.
6. E. Wilhelm and A. Asenbaum, Heat Capacities and Brillouin Scattering in Liquids, in *Heat Capacities: Liquids, Solutions and Vapours*, ed. E. Wilhelm and T. M. Letcher, The Royal Society of Chemistry/IUPAC & IACT, Cambridge, UK, 2010, ch. 11, pp. 238–263.
7. E. Wyn-Jones and W. J. Orville-Thomas, *Trans. Faraday Soc.*, 1968, **64**, 2907.
8. R. A. Pethrick and E. Wyn-Jones, *J. Chem. Phys.*, 1968, **49**, 5349.

9. W. Van Dael, Thermodynamic Properties and the Velocity of Sound, in *Experimental Thermodynamics, Vol. II: Experimental Thermodynamics of Non-reacting Fluids*, ed. B. Le Neindre and B. Vodar, Butterworths/IUPAC, London, UK, 1975, ch. 11, pp. 527–577.
10. A. R. H. Goodwin and J. P. M. Trusler, Sound Speed, in *Measurement of the Thermodynamic Properties of Single Phases: Experimental Thermodynamics, Vol. VI*, ed. A. R. H. Goodwin, K. N. Marsh and W. A. Wakeham, Elsevier/IUPAC, Amsterdam, The Netherlands, 2003, ch. 6, pp. 237–323.
11. F. Eggers and U. Kaatze, *Meas. Sci. Technol.*, 1996, **7**, 1.
12. K. Lautscham, F. Wente, W. Schrader and U. Kaatze, *Meas. Sci. Technol.*, 2000, **11**, 1432.
13. U. Kaatze, T. O. Hushcha and F. Eggers, *J. Solution Chem.*, 2000, **29**, 299.
14. U. Kaatze, F. Eggers and K. Lautscham, *Meas. Sci. Technol.*, 2008, **19**, 062001.
15. T. Takagi and E. Wilhelm, Speed-of-Sound Measurements and Heat Capacities of Liquid Systems at High Pressure, in *Heat Capacities: Liquids, Solutions and Vapours*, ed. E. Wilhelm and T. M. Letcher, The Royal Society of Chemistry/IUPAC & IACT, Cambridge, UK, 2010, ch. 10, pp. 218–237.
16. J. S. Rowlinson and F. L. Swinton, *Liquids and Liquid Mixtures*, Butterworth Scientific, London, UK, 3rd edn, 1982.
17. L. A. Davies and R. B. Gordon, *J. Chem. Phys.*, 1967, **46**, 2650.
18. R. Vedam and G. Holton, *J. Acoust. Soc. Am.*, 1968, **43**, 108.
19. G. S. Kell and E. Whalley, *J. Chem. Phys.*, 1975, **62**, 3496.
20. T. F. Sun, B. J. Kortbeek, N. J. Trappeniers and S. N. Biswas, *Phys. Chem. Liq.*, 1987, **16**, 163.
21. T. F. Sun, C. A. ten Seldam, B. J. Kortbeek, N. J. Trappeniers and S. N. Biswas, *Phys. Chem. Liq.*, 1988, **18**, 107.
22. T. F. Sun, S. A. R. C. Bominaar, C. A. ten Seldam and S. N. Biswas, *Ber. Bunsenges. Phys. Chem.*, 1991, **95**, 696.
23. (a) B. A. Oakley, G. Barber, T. Worden and D. Hanna, *J. Phys. Chem. Ref. Data*, 2003, **32**, 1501; (b) B. A. Oakley, D. Hanna, M. Shillor and G. Barber, *J. Phys. Chem. Ref. Data*, 2003, **32**, 1535.
24. B. J. Berne and R. Pecora, *Dynamic Light Scattering*, Wiley, New York, USA, 1976.
25. J. B. Boon and S. Yip, *Molecular Hydrodynamics*, McGraw-Hill, New York, USA, 1980.
26. V. A. Popova and N. V. Surovtsev, *J. Chem. Phys.*, 2011, **135**, 134510.
27. M. Musso, F. Aliotta, C. Vasi, R. Aschauer, A. Asenbaum and E. Wilhelm, *J. Mol. Liq.*, 2004, **110**, 33.
28. A. Asenbaum, R. Aschauer, C. Theisen, T. Fritsch and E. Wilhelm, *J. Mol. Liq.*, 2007, **134**, 55.
29. D. Sette, A. Busala and J. C. Hubbard, *J. Chem. Phys.*, 1955, **23**, 787.
30. J. H. Andreae, E. L. Heasell and J. Lamb, *Proc. Phys. Soc., London, Sect. B*, 1956, **69**, 625.

31. T. A. Litovitz, E. H. Carnevale and P. A. Kendall, *J. Chem. Phys.*, 1957, **26**, 465.
32. M. B. Ewing and J. P. M. Trusler, *J. Chem. Phys.*, 1989, **90**, 1106.
33. *Landolt-Börnstein, Group II: Atomic and Molecular Physics, Volume 5: Molecular Acoustics*, ed. W. Schaaffs, Springer-Verlag, Berlin, Germany, 1967.
34. J. D. Lambert, *Vibrational and Rotational Relaxation in Gases*, Clarendon, Oxford, UK, 1977.
35. W. H. Nichols and E. F. Carome, *J. Chem. Phys.*, 1968, **49**, 1000.
36. (a) E. F. Carome, W. H. Nichols, C. R. Kunsitis-Swyt and S. P. Singal, *J. Chem. Phys.*, 1968, **49**, 1013; (b) W. H. Nichols, C. R. Kunsitis-Swyt and S. P. Singal, *J. Chem. Phys.*, 1969, **51**, 5659.
37. H. C. Lucas, D. A. Jackson and H. T. Pentecost, *Opt. Commun.*, 1970, **2**, 239.
38. K. Takagi, P.-K. Choi and K. Negishi, *Acustica*, 1976, **34**, 336.
39. J. Chesnoy and G. M. Gale, *Ann. Phys. Fr.*, 1984, **9**, 893.
40. F. Eggers and T. Funck, *Rev. Sci. Instrum.*, 1973, **44**, 969.
41. M. Sedlacek and A. Asenbaum, *J. Acoust. Soc. Am.*, 1977, **62**, 1420.
42. M. Müllner, PhD Dissertation, *Vibrationsrelaxation in flüssigem Schwefelkohlenstoff*, University of Wien (Vienna), Institute for Experimental Physics, 1984.
43. U. Wagner, PhD Dissertation, *Ultraschall Sensor Systeme zur Bestimmung von Schallgeschwindigkeit, Schallimpedanz und Absorption*, University of Salzburg, Department of Materials Research and Physics, Section for Experimental Physics, 2008.
44. K. Takagi and K. Negishi, *J. Phys. D: Appl. Phys.*, 1982, **15**, 757.
45. V. Uhlendorf, K.-H. Richmann and W. Berger, *J. Phys. E: Sci. Instrum.*, 1985, **18**, 151.
46. D. J. McClements and P. Fairley, *Ultrasonics*, 1991, **29**, 58.
47. D. J. McClements and P. Fairley, *Ultrasonics*, 1992, **30**, 403.
48. G. Tardajos, G. G. Gaitano and F. R. Montero de Espinosa, *Rev. Sci. Instrum.*, 1994, **65**, 2933.
49. F. Eggers, U. Kaatze, K. H. Richmann and T. Telgmann, *Meas. Sci. Technol.*, 1994, **5**, 1131.
50. U. Kaatze, V. Kühnel and G. Weiss, *Ultrasonics*, 1996, **34**, 51.
51. J. Schultz and U. Kaatze, *Meas. Sci. Technol.*, 1998, **9**, 1266.
52. V. Kozhevnikov, D. I. Arnold, M. E. Briggs, S. P. Naurzakov, J. M. Viner and P. C. Taylor, *J. Acoust. Soc. Am.*, 1999, **106**, 3424.
53. A. Zak, M. Dzida, M. Zorebski and S. Ernst, *Rev. Sci. Instrum.*, 2000, **71**, 1756.
54. R. Polacek and U. Kaatze, *Meas. Sci. Technol.*, 2003, **14**, 1068.
55. G. Benedetto, R. M. Gavioso, P. A. G. Albo, S. Lago, D. M. Ripa and R. Spagnolo, *Int. J. Thermophys.*, 2005, **26**, 1651.
56. H. Gedanitz, M. J. Dávila, E. Baumhögger and R. Span, *J. Chem. Thermodyn.*, 2010, **42**, 478.
57. T. J. Fortin, A. Laesecke, M. Freund and S. Outcalt, *J. Chem. Thermodyn.*, 2013, **57**, 276.

58. (a) H. Stabinger, O. Kratky and H. Leopold, *Monatsh. Chem.*, 1967, **98**, 436; (b) O. Kratky, H. Leopold and H. Stabinger, Z. Angew. Phys., 1969, **27**, 273.
59. R. D. Mountain, *Rev. Mod. Phys.*, 1966, **38**, 205.
60. R. D. Mountain, *J. Res. Natl. Bur. Stand., Sect. A*, 1966, **70A**, 207.
61. R. D. Mountain, *J. Res. Natl. Bur. Stand., Sect. A*, 1968, **72A**, 95.
62. R. D. Mountain, *CRC Crit. Rev. Solid State Sci.*, 1970, **1**, 5.
63. R. D. Mountain, *Adv. Mol. Relax. Processes*, 1976, **9**, 225.
64. R. C. Desai and R. Kapral, *Phys. Rev. A*, 1972, **6**, 2377.
65. W. S. Gornall, G. I. Stegeman, B. P. Stoicheff, R. H. Stolen and V. Volterra, *Phys. Rev. Lett.*, 1966, **17**, 297.
66. A. Asenbaum, E. Wilhelm and C. Pruner, private communication.
67. A. Asenbaum, *Appl. Opt*, 1979, **18**, 540.
68. R. Aschauer, A. Asenbaum and H. Gerl, *Appl. Opt.*, 1990, **29**, 953.
69. S. M. Lindsay, M. W. Anderson and J. R. Sandercock, *Rev. Sci. Instrum.*, 1981, **52**, 1478.
70. R. Mock, B. Hillebrands and J. R. Sandercock, *J. Phys. E: Sci. Instrum.*, 1987, **20**, 656.
71. (a) J. R. Sandercock, *J. Sci. Instrum.*, 1976, **9**, 566; (b) J. R. Sandercock in *Light Scattering in Solids III*, ed. M. Cardona and G. Güntherodt, Vol. 51 of *Topics in Applied Physics*, Springer, Berlin, 1982.
72. T. A. Litovitz, *J. Chem. Phys.*, 1957, **26**, 469.
73. W. M. Madigosky and T. A. Litovitz, *J. Chem. Phys.*, 1961, **34**, 489.
74. A. Asenbaum and H. D. Hochheimer, *Z. Naturforsch.*, 1983, **38a**, 980.
75. A. Asenbaum, J. Laubereau, C. Pruner and E. Wilhelm, to be published.
76. A. Van Itterbeek, W. Grevendonk, W. Van Dael and G. Forrez, *Physica*, 1959, **25**, 1255.
77. D. G. Naugle, J. H. Lunsford and J. R. Singer, *J. Chem. Phys.*, 1966, **45**, 4669.
78. J. Thoen, E. Van Geel and W. Van Dael, *Physica*, 1969, **45**, 339.
79. D. H. Liebenberg, R. L. Mills and J. C. Bronson, *J. Appl. Phys.*, 1974, **45**, 741.
80. R. L. Mills, D. H. Liebenberg and J. C. Bronson, *J. Chem. Phys.*, 1975, **63**, 1198.
81. W. Van Dael, A. Van Itterbeek, A. Cops and J. Thoen, *Physica*, 1966, **32**, 611.
82. G. C. Straty and B. A. Younglove, *J. Chem. Thermodyn.*, 1973, **5**, 305.
83. E. H. Abramson, L. J. Slutsky, M. D. Harrell and J. M. Brown, *J. Chem. Phys.*, 1999, **110**, 10493.
84. S. Hawley, J. Allegra and G. Holton, *J. Acoust. Soc. Am.*, 1970, **47**, 137.
85. V. A. Del Grosso and C. W. Mader, *J. Acoust. Soc. Am.*, 1972, **52**, 1442.
86. E. Trinh and R. E. Apfel, *J. Acoust. Soc. Am.*, 1978, **63**, 777.
87. V. A. Belogolskii, S. S. Sekoyan, L. M. Samorukova, S. R. Stefanov and V. I. Levtsov, *Meas. Tech.*, 1999, **42**, 406.
88. G. Benedetto, R. M. Gavioso, P. A. G. Albo, S. Lago, D. M. Ripa and R. Spagnolo, *Int. J. Thermophys.*, 2005, **26**, 1667.

89. E. H. Baltasar, M. Taravillo, V. G. Baonza, P. D. Sanz and B. Guignon, *J. Chem. Eng. Data*, 2011, **56**, 4800.
90. R. Bass and J. Lamb, *Proc. R. Soc. London, Ser. A*, 1958, **247**, 168.
91. W. Pecceu and W. Van Dael, *Physica*, 1973, **63**, 154.
92. J. F. Mifsud and A. W. Nolle, *J. Acoust. Soc. Am.*, 1956, **28**, 469.
93. T. Takagi, *J. Acoust. Soc. Am.*, 1982, **71**, 74.
94. R. Bass and J. Lamb, *Proc. R. Soc. London, Ser. A*, 1957, **243**, 94.
95. D. B. Fenner, D. E. Bowen and M. P. Eastman, *J. Chem. Phys.*, 1979, **71**, 4849.
96. V. Vacek and J. A. Zollweg, *Fluid Phase Equilib.*, 1993, **88**, 219.
97. W. Van Dael, A. Van Itterbeek, J. Thoen and A. Cops, *Physica*, 1965, **31**, 1643.
98. A. Van Itterbeek, J. Theon, A. Cops and W. Van Dael, *Physica*, 1967, **35**, 162.
99. G. C. Straty, *Cryogenics*, 1974, **14**, 367.
100. P. J. Kortbeek, S. N. Biswas and N. J. Trappeniers, *Physica B + C*, 1986, **139–140**, 109.
101. P. J. Kortbeek and J. A. Schouten, *Int. J. Thermophys.*, 1990, **11**, 455.
102. R. Tsumura and G. C. Straty, *Cryogenics*, 1977, **17**, 195.
103. D. Bücker and W. Wagner, *J. Phys. Chem. Ref. Data*, 2006, **35**, 205.
104. B. A. Younglove, *J. Res. Natl. Bur. Stand.*, 1981, **86**, 165.
105. R. Niepmann, *J. Chem. Thermodyn.*, 1984, **16**, 851.
106. K. Meier and S. Kabelac, *J. Chem. Eng. Data*, 2012, **57**, 3391.
107. E. Richardson and R. Tait, *Philos. Mag.*, 1957, **2**, 441.
108. A. Lainez, J. Zollweg and W. Streett, *J. Chem. Thermodyn.*, 1990, **22**, 937.
109. H. Eden and E. Richardson, *Acustica*, 1960, **10**, 309.
110. J. Boelhouwer, *Physica*, 1967, **34**, 484.
111. J. Allegra, S. Hawley and G. Holton, *J. Acoust. Soc. Am.*, 1970, **47**, 144.
112. J. Daridon, B. Lagourette and J. Grolier, *Int. J. Thermophys.*, 1998, **19**, 145.
113. T. S. Khasanshin and A. P. Shchemelev, *High Temp.*, 2001, **39**, 60.
114. S. J. Ball and J. P. M. Trusler, *Int. J. Thermophys.*, 2001, **22**, 427.
115. T. S. Khasanshin, V. S. Samuilov and A. P. Shchemelev, *J. Eng. Phys. Thermophys.*, 2008, **81**, 760.
116. M. J. P. Muringer, N. J. Trappeniers and S. N. Biswas, *Phys. Chem. Liq.*, 1985, **14**, 273.
117. F. Plantier and J. L. Daridon, *J. Chem. Eng. Data*, 2005, **50**, 2077.
118. A. A. H. Pádua, J. M. N. A. Fareleira, J. C. G. Calado and W. A. Wakeham, *J. Chem. Eng. Data*, 1996, **41**, 1488.
119. K. Takagi and K. Negishi, *Ultrasonics*, 1978, **16**, 259.
120. K. Meier and S. Kabelac, *J. Chem. Eng. Data*, 2013, **58**, 1621.
121. T. Makita and T. Takagi, *Rev. Phys. Chem. Jpn.*, 1968, **38**, 41.
122. M. Bobik, *J. Chem. Thermodyn.*, 1978, **10**, 1137.
123. T. Takagi, K. Sawada, H. Urakawa, M. Ueda and I. Cibulka, *J. Chem. Thermodyn.*, 2004, **36**, 659.
124. K. Meier and S. Kabelac, *J. Chem. Eng. Data*, 2013, **58**, 1398.

125. K. Takagi, P. K. Choi and K. Negishi, *J. Acoust. Soc. Am.*, 1977, **62**, 354.
126. T. Takagi, *High Temp. – High Pressures*, 1993, **25**, 685.
127. R. Niepmann, G. J. Esper and K. A. Riemann, *J. Chem. Thermodyn.*, 1987, **19**, 741.
128. A. Lainez, P. Gopal, J. Zollweg and W. Streett, *J. Chem. Thermodyn.*, 1989, **21**, 773.
129. H. McSkimin, *J. Acoust. Soc. Am.*, 1957, **29**, 1185.
130. M. Bobik, R. Niepmann and W. Marius, *J. Chem. Thermodyn.*, 1979, **11**, 351.
131. A. Lainez, J. Miller, J. Zollweg and W. Streett, *J. Chem. Thermodyn.*, 1987, **19**, 1251.
132. E. H. Carnevale and T. A. Litovitz, *J. Acoust. Soc. Am.*, 1955, **27**, 547.
133. W. Wilson and D. Bradley, *J. Acoust. Soc. Am.*, 1964, **36**, 333.
134. T. F. Sun, S. N. Biswas, N. J. Trappeniers and C. A. ten Seldam, *J. Chem. Eng. Data*, 1988, **33**, 395.
135. T. F. Sun, J. A. Schouten and S. N. Biswas, *Ber. Bunsenges. Phys. Chem.*, 1990, **94**, 528.
136. F. Plantier, J. L. Daridon and B. Lagourette, *J. Phys. D: Appl. Phys.*, 2002, **35**, 1063.
137. (a) S. L. Outcalt, A. Laesecke and T. J. Fortin, *J. Mol. Liq.*, 2010, **151**, 50; (b) S. L. Outcalt and M. O. McLinden, *Ind. Eng. Chem. Res.*, 2007, **46**, 8264.
138. D. Gonzalez-Salgado, J. Troncoso, F. Plantier, J.-L. Daridon and D. Bessières, *J. Chem. Thermodyn.*, 2006, **38**, 893.
139. M. Chorazewski, M. Dzida, E. Zorebski and M. Zorebski, *J. Chem. Thermodyn.*, 2013, **58**, 389.
140. M. Dzida, *J. Chem. Eng. Data*, 2007, **52**, 521.
141. E. Zorebski, M. Dzida and M. Piotrowska, *J. Chem. Eng. Data*, 2008, **53**, 136.
142. E. Zorebski and M. Dzida, *J. Chem. Eng. Data*, 2007, **52**, 1010.
143. T. A. Litovitz and T. Lyon, *J. Acoust. Soc. Am.*, 1958, **30**, 856.
144. R. G. de Azevedo, J. Szydlowski, P. F. Pires, J. M. S. S. Esperança, H. J. R. Guedes and L. P. N. Rebelo, *J. Chem. Thermodyn.*, 2004, **36**, 211.
145. J. Szydlowski, R. G. de Azevedo, L. P. N. Rebelo, J. M. S. S. Esperança and H. J. R. Guedes, *J. Chem. Thermodyn.*, 2005, **37**, 671.
146. S. Lago and P. A. G. Albo, *J. Chem. Thermodyn.*, 2009, **41**, 506.
147. T. A. Litovitz and E. H. Carnevale, *J. Acoust. Soc. Am.*, 1958, **30**, 134.
148. G. Becker and F. Kohler, *Monatsh. Chem.*, 1972, **103**, 556.
149. B. Chu, *Laser Light Scattering: Basic Principles and Practice*, Academic Press, San Diego, USA, 2nd edn, 1991.
150. P. A. Fleury and J. P. Boon, *Phys. Rev.*, 1969, **186**, 244.
151. R. Zwanzig and R. D. Mountain, *J. Chem. Phys.*, 1965, **43**, 4464.
152. N. S. Gillis and P. D. Puff, *Phys. Rev. Lett.*, 1966, **16**, 606.
153. B. J. Berne, J. B. Boon and S. A. Rice, *J. Chem. Phys.*, 1967, **47**, 2283, Erratum: *J. Chem. Phys.*, 1968, **48**, 2883.

154. L. Carraresi, M. Celli and F. Barocchi, *Phys. Chem. Liq.*, 1993, **25**, 91.
155. H. D. Hochheimer, K. Weishaupt and M. Takesada, *J. Chem. Phys.*, 1996, **105**, 374.
156. R. Jia, F. Li, M. Li, Q. Cui, Z. He, L. Wang, Q. Zhou, T. Cui, G. Zou, Y. Bi, S. Hong and F. Jing, *J. Chem. Phys.*, 2008, **129**, 154503.
157. E. H. Abramson, *High Pressure Res.*, 2011, **31**, 549.
158. F. Datchi, *J. Chem. Phys.*, 2010, **132**, 017101.
159. F. Li and Q. Zhou, *J. Chem. Phys.*, 2010, **132**, 017102.
160. M. Sedlacek and A. Asenbaum, *Phys. Lett. A*, 1974, **50**, 245.
161. A. Cunsolo and M. Nardone, *J. Chem. Phys.*, 1996, **105**, 3911.
162. F. Decremps, F. Datchi and A. Polian, *Ultrasonics*, 2006, **44**, 1495.
163. V. M. Giordano, F. Datchi and A. Dewaele, *J. Chem. Phys.*, 2006, **125**, 054504.
164. J. H. Stith, L. M. Peterson, D. H. Rank and T. A. Wiggins, *J. Acoust. Soc. Am.*, 1974, **55**, 785.
165. H. Shimizu, S. Sasaki and T. Ishidate, *J. Chem. Phys.*, 1987, **86**, 7189.
166. H. Shimizu, N. Nakashima and S. Sasaki, *Phys. Rev. B*, 1996, **53**, 111.
167. M. Li, F. Li, W. Gao, C. Ma, L. Huang, Q. Zhou and Q. Cui, *J. Chem. Phys.*, 2010, **133**, 044503.
168. A. Asenbaum, *Adv. Mol. Relax. Interact. Processes*, 1982, **23**, 223.
169. A. Asenbaum and E. Wilhelm, *Adv. Mol. Relax. Interact. Processes*, 1982, **22**, 187.
170. F. D. Medina and D. C. O'Shea, *J. Chem. Phys.*, 1977, **66**, 1940.
171. A. Asenbaum, P. Soufi-Siavoch and E. Wilhelm, *Acustica*, 1989, **67**, 284.
172. A. Asenbaum, E. Wilhelm and P. Soufi-Siavoch, *Acustica*, 1989, **68**, 131.
173. A. Asenbaum, P. Soufi-Siavoch, R. Aschauer and E. Wilhelm in *Laser Materials and Laser Spectroscopy*, ed. W. Zhijiang and Z. Zhiming, World Scientific, Singapore, 1989, pp. 277–279.
174. A. Asenbaum and H. D. Hochheimer, *J. Chem. Phys.*, 1981, **74**, 1.
175. J. M. Brown, L. J. Slutsky, K. A. Nelson and L.-T. Cheng, *Science*, 1988, **241**, 65.
176. S. A. Lee, A. Anderson, S. M. Lindsay and R. C. Hanson, *High Pressure Res.*, 1990, **3**, 230.
177. J.-H. Ko and S. Kojima, *J. Non-Cryst. Solids*, 2002, **307–310**, 154.
178. D. P. Eastman, A. Hollinger, J. Kenemuth and D. H. Rank, *J. Chem. Phys.*, 1969, **50**, 1567.
179. A. Asenbaum, *Z. Naturforsch. A*, 1979, **34**, 207.
180. H. Bohidar, *J. Appl. Phys.*, 1988, **64**, 1810.

CHAPTER 13

Ultrasonics 2: High Pressure Speed of Sound, Isentropic Compressibility

TOSHIHARU TAKAGI

Department of Chemistry and Materials Technology, Kyoto Institute of Technology, Kyoto 606-8585, Japan
Email: taka-301@mbox.kyoto-inet.or.jp

13.1 Introduction

The pressure–volume (density)–temperature (P–ρ–T) surface of a fluid presents the basic thermodynamic properties for investigating the behaviour of fluids. The isothermal compressibility, κ_T, and isentropic (adiabatic) compressibility, κ_S, of a fluid are defined as:

$$\kappa_T = -\frac{1}{\rho}\left(\frac{\partial \rho}{\partial p}\right)_T, \tag{13.1}$$

$$\kappa_S = \frac{1}{\rho}\left(\frac{\partial \rho}{\partial P}\right)_S \tag{13.2}$$

where ρ refers to the specific molar volume, P to pressure, T to temperature and S to entropy. The pioneering studies on the P–ρ–T relationships for various compounds were conducted by Bridgman,[1,2] Gibson and Kincaid[3,4] and many colleagues in the early 20th century. Today, the experimental P–ρ–T data of a large number of liquids and/or mixtures have been stored or published in a variety of styles. Examples include the databases of

Volume Properties: Liquids, Solutions and Vapours
Edited by Emmerich Wilhelm and Trevor M. Letcher

Published by the Royal Society of Chemistry, www.rsc.org

NIST-TRC[5] and DECHEMA,[6] and the work by Cibulka and Takagi.[7] The latter group have arranged, analysed and selected the most reliable P–ρ–T experimental data of approximately 350 organic liquid compounds. The direct measurement of isentropic compressibility, κ_S, is extremely difficult and there are only a few reported measurements. These have been measured by Tyrer[8,9] at atmospheric pressure.

When low power and low frequency sound waves are generated in a sample, it propagates an adiabatic compression wave. The speed of sound, u in a fluid is a purely thermodynamic property which is linked to the isentropic compressibility, κ_S, through the Newton–Laplace equation:[10]

$$\kappa_S = \frac{1}{\rho}\left(\frac{\partial \rho}{\partial P}\right)_S = \frac{1}{\rho u^2}. \tag{13.3}$$

The study of the speed of sound in liquids at high pressure was initiated by Swanson[11] in 1934. Since then the measurement of the speed of sound as a function of pressure and temperature, $u(T,P)$, in organic and related compounds has been reported by many authors. Details of up to 2000 of these publications have been correlated and reviewed by many researchers including Oakley *et al.*[12,13] (2003), Neruchev *et al.*[14] (2005), Lemmon,[15] (2006) and Takagi and Wilhelm[16] (2010). In addition, the reviews of $u(T,P)$ data of specialised substances such as alcohols or alkanes have been correlated by Queimada *et al.*[17] (2006), Liang *et al.*[18] (2012) and by Padilla-Victoria *et al.*[19] (2013).

The speed of sound measurements in fluids can be observed with high accuracy even under severe conditions. These measurements offer a powerful method for deriving various thermodynamic quantities indirectly. In this chapter, we describe the effect of temperature, pressure and concentration on the speed of sound in water, organic compounds and in recently reported binary mixtures (2000 or later).

13.2 Experimental Method for Elevated Pressure Speed of Sound in Liquids

The speed of sound, u $(=L/t)$, in a fluid is obtained by measuring the travelling time, t, required for low power and low frequency sound waves to travel a distance L in the fluid. Up to about 1970, the pulse echo method was the main technique used to measure the speed of sound in a fluid. In 1985, Kortbeek *et al.*[20,21] developed a new phase-comparison-pulse echo technique with double refractors for measuring u at high pressures. As a result the accuracy of the measurement improved by leaps and bounds. This method, together with the more recent pulse echo-overlap and sing-around techniques[22–32] are the methods used today. Recent developments in electronic equipment, such as pulse generators, digital storage oscilloscopes and digital quartz thermometers, digital quartz pressure transducers together with amended designs of high pressure vessels have improved the accuracy and made measurements easier. For example, in the study carried out by Meier and Kabelac[104] on the speed of sound, $u(T,P)$ for liquid propane, the pressure was measured to within 0.005% up to 100 MPa, using a pressure balance operating through a transmitting fluid. Furthermore, the uncertainty in temperature measurement, using a Pt25 sensor was only 3 mK over a wide temperature range of 250 to 420 K. The overall uncertainty in the speed of sound measurement was only 0.02%. Considering the harsh conditions this is a remarkable result and one that would be difficult to attain in isobaric heat capacity or density of liquid measurements.

13.3 High Pressure Speed of Sound in Liquid and Thermodynamic Properties

Knowledge of the speed of sound at elevated pressures, measured with precision, can be used to determine, indirectly, thermodynamic properties of liquids and liquid mixtures. The speed of sound, u is related to the isothermal compressibility, κ_T, through the following equations:

$$\kappa_T = \gamma\kappa_S = \left(1 + \frac{T\alpha u^2}{C_P}\right)\frac{1}{\rho u^2}, \tag{13.4}$$

or

$$\left(\frac{\partial\rho}{\partial P}\right)_T = \left(\frac{\partial\rho}{\partial P}\right)_S + \frac{T}{\rho^2 C_P}\left(\frac{\partial\rho}{\partial T}\right)_P^2, \tag{13.5}$$

where $\alpha = -1/\rho(\partial\rho/\partial T)_P$ is the thermal expansion and $\gamma = C_p/C_v$ the ratio of molar isobaric, C_p, and molar isochoric, C_v, heat capacity. Thus, the density, ρ_P, at pressure, P, can be estimated by:

$$\rho_P = \rho_0 + \int_{P_0}^{P} u^{-2}dP + T\int_{P_0}^{P}\left(\frac{\alpha^2}{C_P}\right)dP, \tag{13.6}$$

where ρ_0 is the density at 0.1 MPa or at the saturated pressure. The values of C_p at pressure, P, required in this calculation can be estimated by:

$$\left(\frac{\partial C_P}{\partial P}\right)_T = -\frac{T}{\rho}\left[\alpha_P^2 + \left(\frac{\partial\alpha_P}{\partial T}\right)_P\right]. \tag{13.7}$$

From these relations, together with the initial values of ρ_0 and $C_{p(0)}$ (determined experimentally) and the speed of sound at elevated pressures, the density of a fluid at arbitrary conditions of T and P can readily be determined.

13.3.1 Speed of Sound in Liquid Water

The thermophysical properties of water related to temperature and pressure effects are of great importance in many industrial areas including electric power plants and food- and marine-science. Since the work of Greenspan and Tschiegg (1957),[33] Wilson (1959),[34] Fine and Millero (1973),[35] the investigations have continued with the aim of determining the thermophysical properties of water to higher degrees of accuracy and over wider ranges of temperature and pressure. Recently, Benedetto *et al.*[36] (2005) measured the $u(T,P)$ in pure water over the temperature range of 274 to 394 K and at pressures of up to 90 MPa with a total uncertainty of less than 0.05%. Vance and Brown[37] (2010) measured the $u(T,P)$ in water from 263.15 to 373.15 K and up to 700 MPa with an uncertainty of less than 0.2%. More recently Lin and Trusler[38] (2012) measured $u(T,P)$ in water from 253.15 to 473.15 K and up to 400 MPa with the incredible uncertainty of 0.03%. This deviation corresponds to an uncertainty of 0.5 m s^{-1} for a speed of sound value of 1680 m s^{-1} measured at $T = 300$ K and 100 MPa.

The speed of sound in pure water, reported by Benedetto *et al.*,[36] is given in Figures 13.1(a) and (b). As shown in Figure 13.1(a), the speed of sound increases at first with increasing temperature, and then reaches a maximum, u_{max}, at about 347 K, followed by a decrease. The temperature at which this maximum occurs shifts with a rise of pressure. This is anomalous behaviour when compared to organic liquids. In Figure 13.1(b), the speed of sound at constant temperature is seen to increase linearly with increasing pressure. This is similar to that found in organic liquids. The pressure dependence, $(\partial u/\partial P)_T$ at 364 K ($>u_{max}$) and that at 273.98 K in the boundary solid–liquid, are very little different from those found at 334 K and 303.99 K. These changes in the thermodynamic properties of water with temperature or pressure have been discussed elsewhere.[16]

For water, the contribution of the right hand side of Equation (13.6) is shown graphically in Figure 13.2(a). The third term, $T\int_{P_0}^{P}(\alpha^2 / C_P)dP$, for water is very small compared with the second one. The densities for water

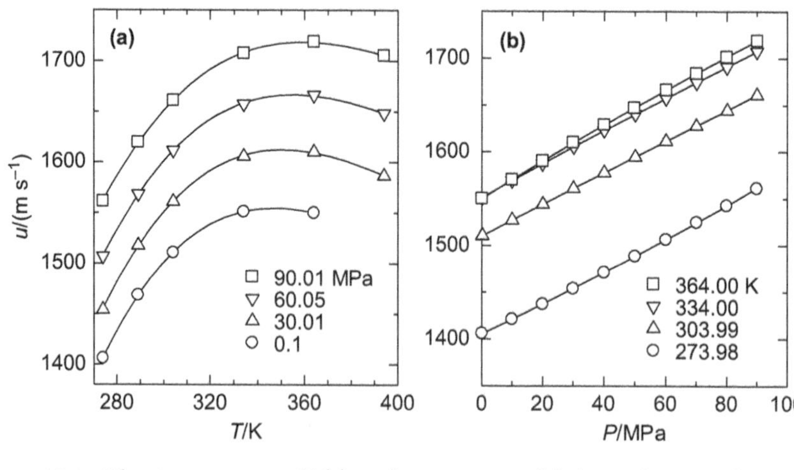

Figure 13.1 The temperature, T (a) and pressure, P (b) dependences of speed of sound, u, in water.[36]

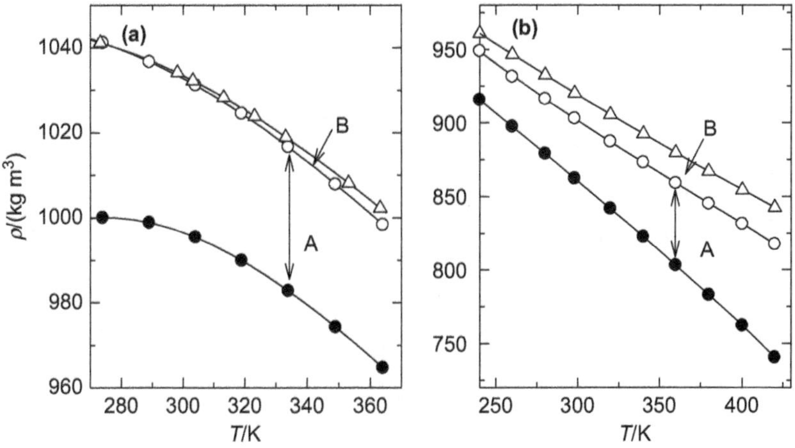

Figure 13.2 The temperature, T, dependence of densities, ρ, for: (a) water; ● densities, ρ_0, at 0.1 MPa;[39] ○ densities, ρ_{calc}, derived by the integration of speed of sound, u,[36] in pressures ranging from 0.1 to 90 MPa; △ densities, ρ, at 90 MPa[39] and (b) toluene; ● densities, ρ_0,[121] at 0.1 MPa; ○ densities, ρ_{calc}, derived by the integration of u[36] from 0.1 to 100 MPa; △, ρ [118] at 100 MPa. The A and B contributed to terms $\int_{P_0}^{P} u^{-2} dP$ and $T \int_{P_0}^{P} (\partial\alpha^2 / \partial C_P) dP$, respectively, in Equation (13.6).

thus estimated by the integration of $u(T,P)$ agree within 0.1%[27] within the temperature range 274–394 K and pressures up to 90 MPa, and within 0.03%[38] over the wide range 253–473 K and up to 400 MPa, when compared with the IAPWS-95 Formulation.[124] On the other hand, for toluene [see Figure 13.2(b)], the third term is large, about 20–30%. Meier and Kabelac[103] have measured the $u(T,P)$ for toluene at temperatures 248–298 K and pressures up to 100 MPa within 0.1%. The density was not derived in this paper.[103] For acetone, Lago and Albo[95] reported the $u(T,P)$ at temperatures 248.15–298.15 K and at pressures of up to 100 MPa with an uncertainty of 0.1%, [see Figure 13.3(b)]. The estimated density, for example, 869.38 kg m^{-3} at $T = 278.15$ K and 100 MPa is in excellent agreement with 869.99 kg m^{-3} evaluated by Cibulka *et al.*[118]

In 1975, Kell and Whalley[39] used Equation (13.6) to determine the density, (P–ρ–T) of water over the temperature range 273.15 to 373.15 K and at pressures up to 100 MPa to within ±0.02% from the elevated pressure speed of sound data of Millero.[35] This was a valuable achievement and has greatly contributed to the development of this field of research.

13.3.2 Speed of Sound in Liquid Organic Compounds

Because of its importance in indirectly determining thermodynamic properties, the measurement of speed of sound as a function of pressure in liquids and liquid mixtures continues to flourish. Khasanshin *et al.*[40–44]

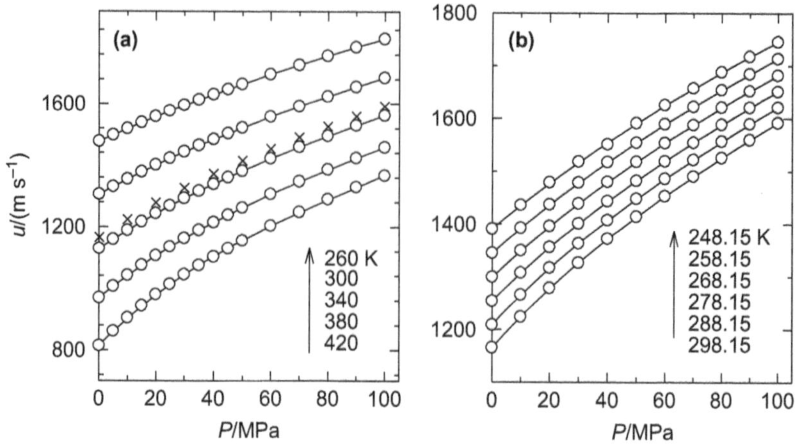

Figure 13.3 The pressure, P, dependence of speed of sound, u, in liquids (a) toluene[103] and (b) acetone.[95] Here, × in (a) is $u(P)$ for acetone at $T = 298.15$ K.

Bolotnikov *et al.*[45–48] and Hasanov *et al.*[49,50] reported the $u(T,P)$ for mainly liquid *n*-alkanes. Daridon and colleagues[51–68] have measured $u(T,P)$, for ethers, esters and *n*-alkanes. Their work on long chain alkanes is of particular importance in the petrochemical field. Dzida and co-workers[69–90] have studied $u(T,P)$ for *n*-alkanes, alcohols and their mixtures, and have discussed their results from a physical chemistry point of view. The research groups of Lago *et al.*,[91–100] Kabelac *et al.*,[101–104] Guedes *et al.*,[106–109] Trusler *et al.*[110–112] and others[105,113,114] have reported measurements of $u(T,P)$ on compounds such as: water, hydrocarbons, refrigerants, oils over a wide range of temperature and pressure. Research on the $u(T,P)$ for biomass-derived fuels and related compounds has recently been reported.[65,78,113,114]

High purity compounds such as toluene and acetone, which are now available, are very important in this field as they serve as standard liquids and/or solvents for the determination of accurate thermodynamic properties. The sound speed in liquid toluene is graphed in Figure 13.3(a). The measurements were done by Meier and Kabelac[103] over a wide range of temperature from 240 to 420 K and pressures of up to 100 MPa with an uncertainty of less than 0.03%. The speed of sound measurements in acetone from 248.15 to 298.15 K and at pressures of up to 100 MPa to within 0.1%, by Lago and Albo[95] are graphed in Figure 13.3(b). For these liquids, the speed of sound, u, at constant pressure decreases with increasing temperature, and the pressure dependence of $u(P)$ is positive and varies smoothly with pressure. In order to compare the $u(T,P)$ in toluene at $T = 300$ K, the sound speed for acetone at 298.15 K is plotted (×) in Figure 13.3(a). At the critical temperature, T_c, the u value reaches a minimum. The absolute speed of sound corresponding to T_c in toluene ($T_c = 591.80$ K) is higher than it is for acetone ($T_c = 508.10$ K). The pressure effects showed the same tendency.

High purity liquids as standard substances are very important in testing the reliability of experimental measuring devices such as speed of sound equipment, densimeters, viscometers and calorimeters. In the case of high pressure speed of sound devices, the physical measurement of the acoustic path length, L, is very difficult and can seriously affect the accuracy of experimental results. This is largely because the distance L is dependent on temperature, pressure and/or the electric connections between the transducer and the instrument. In general, the reliability is determined by a comparison of results in standard compounds such as water, toluene, acetone, and carbon tetrachloride for which acoustic velocities are known to a high degree of accuracy.

In the 1980s, global warming and the destruction of the ozone layer by chlorine-based fluorocarbons (largely used as refrigerants) became a major issue. In order to find new refrigerants and to understand the issues at stake, many research groups, some linked to international projects, have measured the thermodynamic properties of various hydrofluorocarbons at high pressure.

This work has resulted in difluoromethane (CH_2F_2, critical temperature $T_c = 351.25$ K, pressure $P_c = 5.783$ MPa) being accepted as a new refrigerant for household air conditioners in Japan as from 2013. The global warming potential of CH_2F_2 is one third of that of previously used refrigerants.

The measurements of $u(T,P)$ for CH_2F_2 within the temperature range 243–373 K and near the saturation line up to 35 MPa were done by Takagi[115] and the values within the temperature range 248–343 K and at pressure up to 65 MP were done by Pires and Guedes.[106] These values are graphed in Figure 13.4(a), together with values for 1,1,1,2,3,3,3-heptafluoropropane

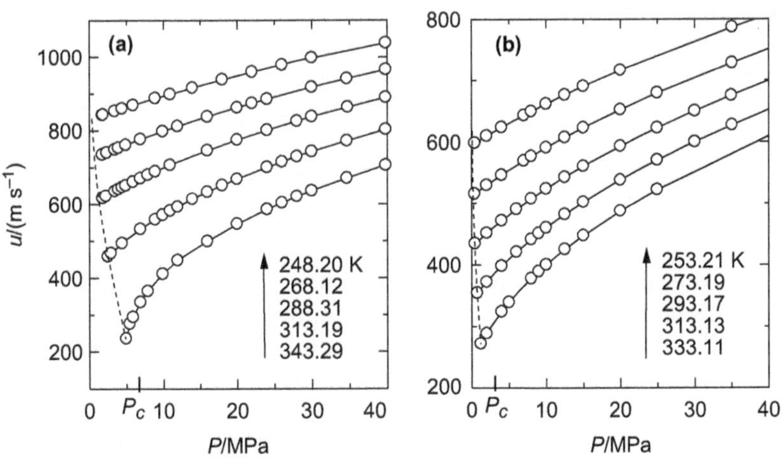

Figure 13.4 The pressure, P, dependence of speed of sound, u, in liquid (a) difluoromethane[106,115] and (b) 1,1,1,2,3,3,3-heptafluoropropane.[102] Here, p_c is the critical pressure and the broken line (---) refers to the saturated liquid.

(CF_3CHFCF_3, T_c: 351.56 K, P_c: 5.83 MPa) in Figure 13.4(b). This compound has recently been accepted as a fire extinguishing agent. The pressure dependence of $u(T,P)$ for this latter compound was measured by Meier and Kabelac.[102] Near the critical point, the speed of sound depends strongly on temperature or pressure changes. At the critical point, the ratio of molar isobaric, C_P, and isochoric heat capacity, C_v, ($\gamma = C_p/C_V$) becomes infinite, and the C_p measurement becomes impossible. The speed of sound, u_c, is given by:

$$u_C^2 = \frac{T}{C_{V_C}\rho_C^2}\left(\frac{\partial P}{\partial T}\right)^2_{\rho_C} \tag{13.8}$$

The experimental uncertainty of $u(T,P)$ is large near the critical region due to acoustic absorption. Leipertz and co-workers have measured the sound speed for several compounds such as toluene,[125] 1,1-difluoroethane (CH_3CHF_2)[126] and 1,1,1,2,3,3,3-heptafluoropropane (CF_3CHFCF_3)[127] along the saturation line, including the critical region, by using the dynamic light scattering method. The $u(T,P)$ values measured over a wide set of temperatures and pressures offer an effective way to check the reliability of the relevant equation of state.

Research involving speed of sound measurements for alkanes at elevated temperatures and pressures has been continued with the aim of widening the temperature and pressure conditions and also attempting to produce data with higher and higher degrees of accuracy. Khasanshin and Shchemelev[40] measured the speed of sound, $u(T,P)$, in liquid alkanes of carbon number 6 to 16 at temperatures between 303 and 433 K and pressures of up to 50 MPa. The relationship between the speed of sound, u, and pressure, P, are shown in Figure 13.5(a), and the relationship between the speed of sound and carbon number, N, at $T = 303.15$ K is given in Figure 13.5(b).

For *n*-hexane, since the $u(P)$ data involves only four points, the speed of sound values measured by Daridon *et al.*[51] were supplemented [× in (a)] and the $u(P)$ at 9.91 MPa and 29.52 MPa [• in (b)] were estimated by interpolation.[40,51]

These results[40] are summarised by the equation:

$$u = u_0 \exp\left(AN^{-1/m}\right) \tag{13.9}$$

where u is the speed of sound, m is an exponent, u_0 is the speed of sound, A is a coefficient which depends on temperature and pressure and N is the number of carbon atoms in an alkane molecule. With $m = 1$, Equation (13.9) reproduced the experimental values in the range up to 40 MPa with a maximum error of less than 0.1%.

In many cases, the measured results of speed of sound, u, in liquid have been fitted by polynomial equations as a function of pressure, P, and temperature, T. As an example, Figure 13.6 shows the relationship between the speed of sound and density for (a) toluene and (b) 1,1,1,2,3,3,3-heptafluoropropane, CF_3CHFCF_3 (HFC227ea). The densities at pressure, ρ_P, for

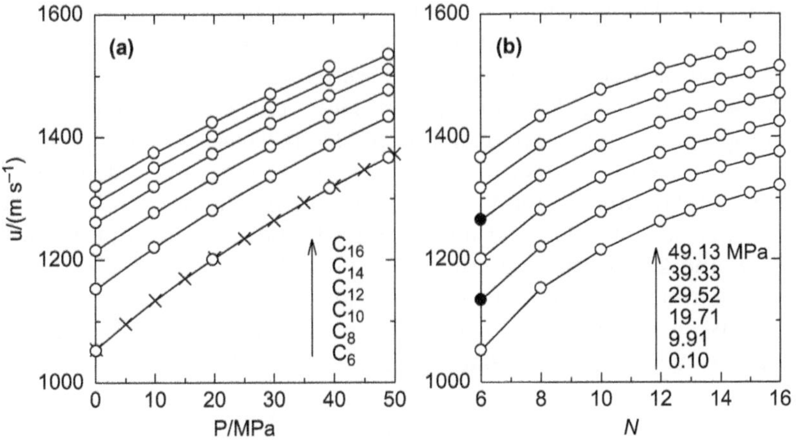

Figure 13.5 The relationship of speed of sound, u, with pressure, P, for n-alkanes[40] (a), and with N (number of carbon atoms in molecule) (b) at 303.15 K. For n-hexane, the data[51] were supplemented by the data points [× in (a)] which were used to obtain the u *vs.* P smooth curve; and the $u(P)$ values at 9.91 MPa and 29.52 MPa [● in (b)] were estimated by interpolation.

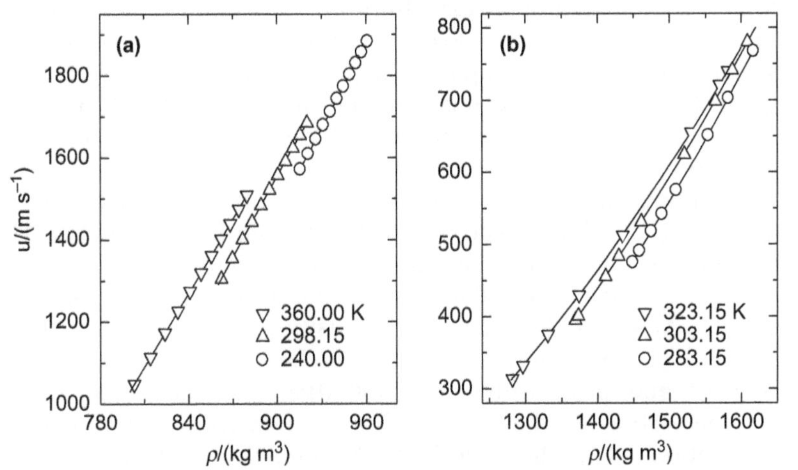

Figure 13.6 The relationship between the sound speed, u, and density, ρ. (a) For toluene; u[103] and ρ[119], corresponding to pressure from 0.1 to 100 MPa, and (b) for HFC227ea; u[102] and ρ[123] corresponding to pressures from saturated pressure to 65 MPa.

toluene, were correlated by Cibulka and Takagi[119] and for HFC227ea by Fedele *et al.*[123]

The speed of sound for toluene increases linearly with increasing density, while that for 1,1,1,2,3,3,3-heptafluoropropane (critical temperature, T_c, 374.9 K, pressure, P_C, 2.925 MPa) deviated slightly from linearity near the critical temperature.

Takagi and Fukushima[116] measured $u(T,P)$ at temperatures between 283 and 343 K and at pressures of up to 30 MPa for two other liquids having nearly the same physical properties as HFC227ea, namely: 1,1,1,2,2-pentafluoro-3,3-dichloropropane, $CF_3CF_2CHCl_2$ (HCFC225ca) and 1,1,2,2,3-pentafluoro-1,3-dichloropropane, $CClF_2CF_2CHClF$ (HCFC225cb). These compounds are used as cleaning agents in microelectronics and for precision machinery. The critical temperature, T_c, for the former liquid is 476.71 K and the critical pressure, P_c, is 3.06 MPa, and for the latter liquid the critical properties are 484.91 K and 2.98 MPa, respectively. For both these liquids the relationship between sound speed and density was found to be a linear relation (apart from the results at 343.15 K) and was well represented by the Tait equation,[120] which is usually used to correlate compressed liquid densities:

$$\frac{1/u_P}{1/u_0} = 1 - C\ln\frac{(B+P)}{(B+P_0)} \tag{13.10}$$

where u_P and u_0 are, respectively, the sound speed at pressure P and atmospheric pressure or saturated pressure P_0, C is a constant, and B a parameter which depends on temperature. The speed of sound in liquids $CF_3CF_2CHCl_2$ and $CClF_2CF_2CHClF$, calculated by above equation, reproduced the experimental data (apart from that measured at 343.15 K) with a maximum deviation $\pm 0.25\%$. Recently, Padilla-Victoria *et al.*[19] correlated the speed of sound for the liquid *n*-alkanes C_5 to C_{40} [see Figure 13.5 (b)] using the Tait equation and included the carbon number, N, in the parameters C and B. The equation reproduced the experimental values with an average absolute deviation of only 0.17%.

By combining the density, ρ, and sound velocity, u, the isentropic compressibility, κ_S, of the fluid can be calculated from Equation (13.3). The κ_S (T,P) for toluene and acetone calculated in this way have been plotted against pressure, P, in Figure 13.7. For toluene, the speed of sound, $u(T,P)$ was measured by Meier and Kabelac,[103] and density, ρ, was correlated by Cibulka and Takagi.[119] For acetone, the speed of sound over the temperature range of 248 to 298 K and for pressures up to 100 MPa, was measured by Lago and Albo.[95] They estimated the density $\rho(T,P)$ by numerical integration of u^{-2} using Equation (13.5). The critical temperature, T_c, for toluene is 591.80 K, and for acetone it is 508.10 K. In order to compare the κ_S (T,P) values of the two chemicals, the values for acetone at $T = 298.15$ K were plotted in Figure 13.8(a). This curve fits well on the line for toluene at $T = 348.15$ K.

The relationship of isentropic compressibility, κ_S, with pressure, P (a), and with N (carbon number) (b), for alkanes at $T = 303.15$ K is graphed in Figure 13.8. The u data used were measured by Khasanshin and Shchemelev,[40] and the ρ data used were correlated by Cibulka and Hnedkovsky.[117]

The pressure, P, dependence of isentropic compressibility, κ_S, is shown in Figure 13.9 for refrigerants difluoromethane, HFC32 (a) and 1,1,1,2,3,3,3-heptafluoropropane, HFC227ea (b). In this figure, the pressure dependence of κ_S is particularly pronounced near the critical region for both substances.

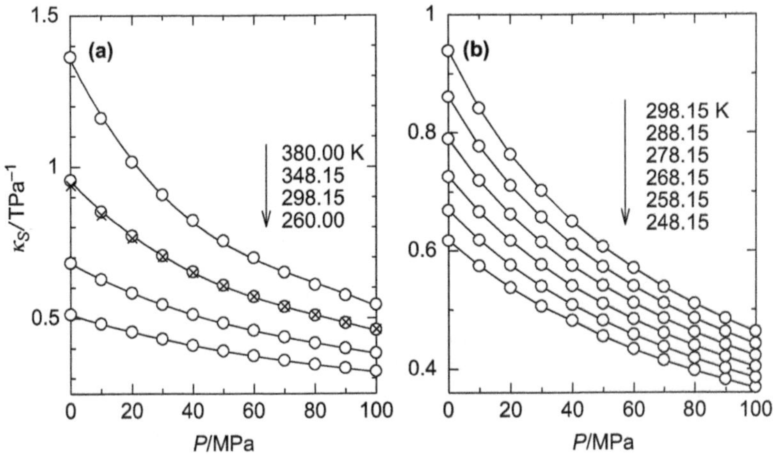

Figure 13.7 The pressure, P, dependence of isentropic compressibility, κ_S, for (a) toluene and (b) acetone. The values of u and ρ required for the calculation κ_S for toluene were obtained from references 103 and 119, respectively, and for acetone from reference 95. The × in (a) is the value of κ_S for acetone at $T = 298.15$ K.

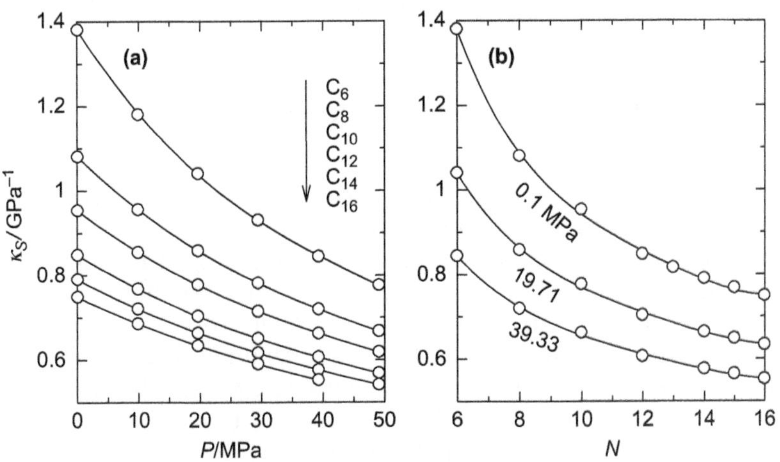

Figure 13.8 The relationship of isentropic compressibility, κ_S, with pressure, P (a), and with carbon number, N (b), at $T = 303.15$ K for alkanes. The values required for the κ_S calculation were obtained from references 40 (for u) and 118 (for ρ). For *n*-hexane, $u(P)$ is included in reference 51.

The critical temperature, T_c, and pressure, P_c, are 351.25 K and 5.783 MPa for HFC32, and 374.80 K and 2.925 MPa for HFC227ea, respectively. The curve for HFC32 at $T = 353.15$ K (×) in Figure 13.9(a) shown here is at a value slightly higher than the T_c. These curves show the behaviour around the critical point more clearly than was seen with the pressure effect of the speed of sound in Figure 13.4.

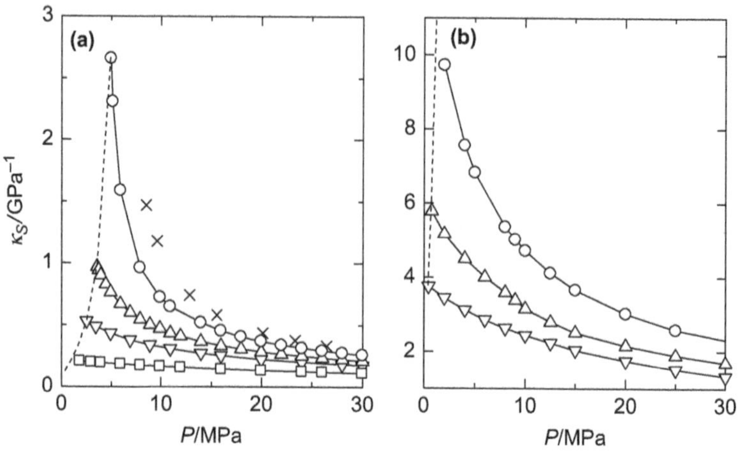

Figure 13.9 The pressure, P, dependence of isentropic compressibility, κ_S, for refrigerants: difluoromethane, HFC32 (a) and 1,1,1,2,3,3,3-heptafluoropropane, HFC227ea (b). (a) ○, $T = 343.29$ K; △, 328.25 K; ▽, 313.19 K; □, 278.12 K; ×, 353.15 K.[115] (b) ○, 333.11 K; △, 313.13 K; ▽, 293.17 K. Dotted line is the saturated liquid. The values of u and ρ required for κ_S calculation for HFC32 were obtained from references 106 and 122, respectively; for HFC227ea, the values were obtained from references 102 and 123, respectively.

13.3.3 Speed of Sound in Liquid Mixtures

Many industrial liquids are indeed mixtures or solutions and, as a result, there is a strong need to understand how this influences thermodynamic properties. The influence of composition on the speed of sound, for two simple cases, is summarised graphically in Figures 13.10 and 13.11(a). Here the speed of sound, u, has been plotted against the mole fraction x for {x *n*-heptane + (1 − x) *n*-dodecane} and {x propan-1-ol + (1 − x) *n*-heptane}. These measurements, done at 298.15 K, were reported by Dzida *et al.*[69,70,77]

They were also responsible for the isentropic compressibility, κ_S, for the two binary mixtures indicated in Figures 13.10 and 13.11. These two systems, one involving a mixture of non-polar liquids and the other a weakly polar liquid mixed with a non-polar liquid, show weak curves at 0.1 MPa. At higher pressures, the relationships become more linear and ideal. From these results, $u(T, P, x)$, Dzida *et al.*[69,70,77] calculated the excess quantities: molar volume, V^E, and molar heat capacity, $C_p{}^E$ (where excess properties refer to the difference between ideal solution properties and derived properties for the mixtures). In their reports, the derived thermodynamic properties were used to discuss the molecular interaction and structural behaviour as a function of temperature and pressure. In particular, the effect of pressure on the excess functions provided compelling information on the behaviour of the mixtures. It should be noted that the measurements on solutions at high pressures add a further degree of complication to the experimental

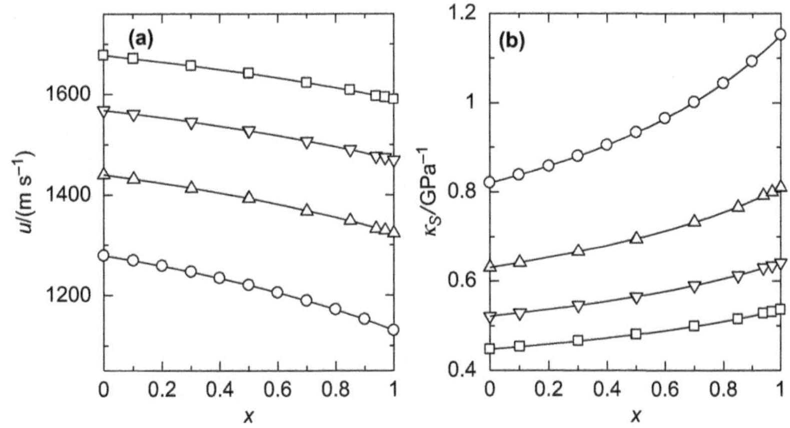

Figure 13.10 The composition, x, dependence of speed of sound, u (a) and isentropic compressibility, κ_S (b) at $T = 298.15$ K for binary mixtures of $\{x$ n-heptane $+ (1-x)$ n-dodecane$\}$[77]: ○, 0.1 MPa; △, 30.39 MPa; ▽, 60.79 MPa; □, 91.18 MPa. The $u(T,P)$ and $\rho(T,P)$, required for κ_S determination, are obtained from the experimental u (reference 77) and density (in same reference) derived by integration of $u(T,P)$, respectively.

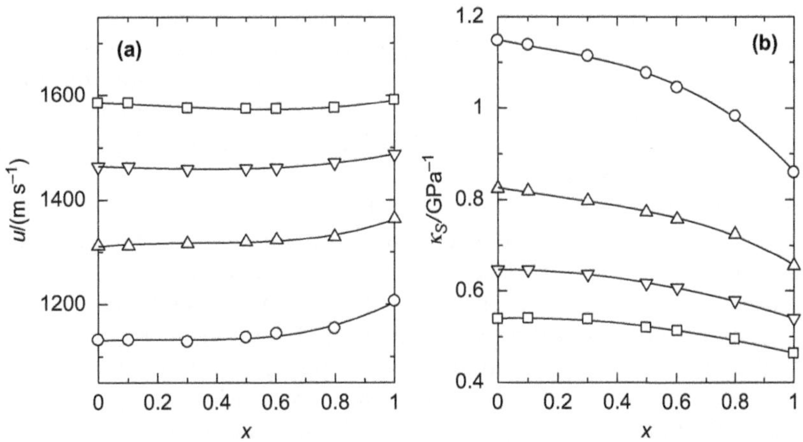

Figure 13.11 The composition, x, dependence of speed of sound, u (a) and isentropic compressibility, κ_S (b) at $T = 298.15$ K for binary mixtures of $\{x$ propan-1-ol $+ (1-x)$ n-heptane$\}$[69,70]: ○, 0.1 MPa; △, 30 MPa; ▽, 60 MPa; □, 90 MPa. The density $\rho(T,P)$ was given in reference 70 and estimated using the $u(T,P)$ in reference 69.

procedure. The excess isentropic compressibility, κ_S^{E}, at atmospheric pressure, an intensive variable, has frequently been used to investigate the composition behaviour of liquid mixtures. Such work was reported by Benson *et al.*[128,129] and many other workers.[130–132]

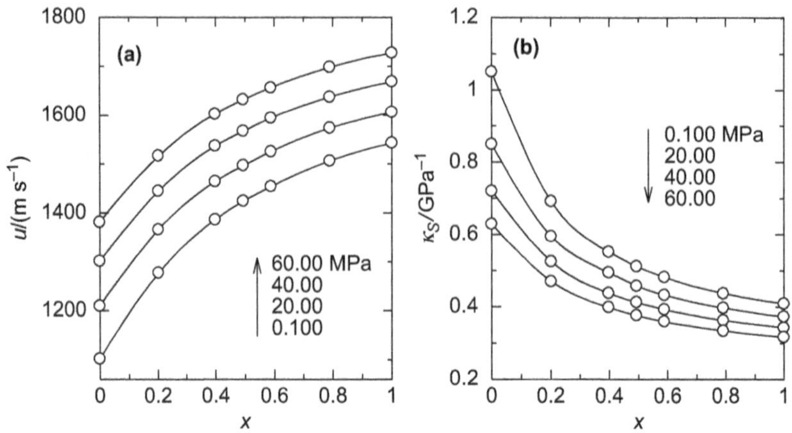

Figure 13.12 The composition x dependence: (a) of speed of sound, u, and (b) isentropic compressibility, κ_S, for {x N-methyl-2-pyrrolidone + (1 − x) methanol} at T = 298.15 K.[110]

Davila and Trusler[110] have measured the speed of sound, u, for the binary mixture of N-methyl-2-pyrrolidone (NMP) and methanol in the temperature range from 298.15 to 343.15 K and at pressures of up to 60 MPa with a relative uncertainty of 0.1%. These results were used to derive the density, ρ, thermal expansion, α, isothermal compressibility, κ_T, and isobaric heat capacity, C_p. The composition, x, dependence of $u(T,P)$ and $\kappa_S(T,P)$ calculated from $u(T,P,x)$ at T = 298.15 K are shown in Figure 13.12. For this binary mixture, the composition dependence of u and κ_S show the characteristic behaviour of a polar–polar mixture. The dipole moment for NMP is 4.1 D and for methanol it is 1.69 D.

The concentration dependence, x of $u(T,P)$ and $\kappa_S(T,P)$ for the two mixtures: n-heptane and n-dodecane, and propan-1-ol and n-heptane have been graphed in Figures 13.10 and 13.11. The curves are almost linear and become more so with increasing pressure. On the other hand, the x dependence of u and κ_S in the (NMP + methanol) system, in Figure 13.12, is very different. This suggests, qualitatively, that the molecular interactions are strongly involved. In their report, the composition, temperature and pressure effects of excess molar volume, V^E and isobaric heat capacity, $C_p{}^E$, derived from experimental $u(T,P)$, were discussed in terms of hydrogen bonding and other effects.

13.4 Conclusions

A knowledge of accurate thermodynamic properties of liquid and liquid mixtures as a function of pressure and temperature is important in the field of chemical science and technology. The speed of sound in a fluid under high pressure is an effective way to derive, indirectly, thermodynamic properties such as density and isobaric specific heat capacity. Today, the techniques used to measure the speed of sound at high pressure have improved significantly and relevant studies have become increasingly popular. In this chapter, the recently reported temperature, pressure and/or concentration dependences of the speed of sound, u, and of isentropic compressibility, κ_S, derived by combining u and density, ρ, for liquid water, hydrocarbons, *n*-alkanes, refrigerants and their binary mixtures have been described and discussed. For some refrigerants, the behaviour of the pressure dependence of isentropic compressibility, κ_S, around the critical point is clearly shown. Also in this chapter, the composition dependence of κ_S for a mixture of *N*-methyl-2-pyrrolidinone and methanol was discussed. This is of particular interest as the values of κ_S and density had been derived only from experimental speed of sound measurements at high pressure. With new chemical science and engineering challenges, such as the development of shale gas and of biodiesel, facing society in the future, it is expected that the study of the thermodynamic properties of novel compounds under high pressure will play an important part.

References

1. P. W. Brigman, *Proc. Am. Acad. Arts. Sci.*, 1911, **47**, 441.
2. P. W. Bridgman, *Proc. Am. Acad. Arts Sci.*, 1931, **66**, 185.
3. R. E. Gibson and J. F. Kincaid, *J. Am. Chem. Soc.*, 1937, **59**, 25.
4. R. E. Gibson and J. F. Kincaid, *J. Am. Chem. Soc.*, 1938, **60**, 511.
5. NIST-TRC database (http://trc.nist.gov/).
6. DECHEMA database (http://www.dechema.de/).
7. I. Cibulka and T. Takagi, *Int. J. Thermophys.*, 2004, **25**, 361.
8. T. Tyrer, *J. Chem. Soc. Trans.*, 1913, **103**, 1675.
9. T. Tyrer, *J. Chem. Soc. Trans.*, 1914, **105**, 2534.
10. A. J. Matheson, *Molecular Acoustics*, John Wiley, & Sons, Ltd., London, 1971.
11. J. C. Swanson, *J. Chem. Phys.*, 1934, **2**, 689.
12. B. A. Oakley, G. Barber, T. Worden and D. Hanna, *J. Phys. Chem. Ref. Data*, 2003, **32**, 1501.
13. B. A. Oakley, D. Hanna, M. Shillor and G. Barber, *J. Phys. Chem. Ref. Data*, 2003, **32**, 1535.
14. Y. A. Neruchev, M. F. Bolotnikov and V. V. Zotov, *High Temp.*, 2005, **43**, 266.
15. E. W. Lemmon and R. Span, *J. Chem. Eng. Data*, 2006, **51**, 785.
16. E. Wilhelm and T. Takagi, in *Heat Capacities: Liquids, Solutions and Vapours*, ed. E. Wilhelm and T. M. Letcher, RSC Publishing, Cambridge 2010.
17. A. J. Queimada, J. A. P. Coutinho, I. M. Marrucho and J.-L. Daridon, *Int. J. Thermophys.*, 2006, **27**, 1095.
18. X. Liang, B. Maribo-Mogensen, K. Thomsen, W. Yan and G. M. Kontogeorgis, *Ind. Eng. Chem. Res.*, 2012, **51**, 14903.

19. H. Padilla-Victoria, G. A. Iglesias-Silva, M. Ramos-Estrada and K. R. Hall, *Fluid Phase Equilib.*, 2013, **338**, 119.
20. P. J. Kortbeek, M. J. P. Muringer, N. J. Trappeniers and S. N. Biswas, *Rev. Sci. Instrum.*, 1985, **56**, 1269.
21. T. F. Sun, C. A. Ten Seldam, P. J. Kortbeek, N. J. Trappeniers and S. N. Biswas, *Phys. Chem. Liq.*, 1988, **18**, 107.
22. T. Takagi and H. Teranishi, *J. Chem. Thermodyn.*, 1987, **19**, 1299.
23. J. L. Daridon, A. Lagrabette and B. Lagourette, *J. Chem. Thermodyn.*, 1998, **30**, 607.
24. A. Zak, M. Dzida, M. Zorebski and S. Ernst, *Rev. Sci. Instrum.*, 2000, **71**, 1756.
25. *Measurement of the Thermodynamics Properties of Single Phases*, ed. A. R. H. Goodwin, K. N. Marsh and W. A. Wakeham, Elsevier, 2003.
26. Y. A. Neruchev, M. F. Bolotnikov and V. V. Zotov, *High Temp.*, 2005, **43**, 266.
27. G. Benedetto, R. M. Gavioso, P. A. G. Albo, S. Lago, D. M. Ripa and R. Spagnolo, *Int. J. Thermophys.*, 2005, **26**, 1651.
28. K. Meier and S. Kabelac, *Rev. Sci. Instrum.*, 2006, **77**, 123903.
29. M. Dzida, M. Chorążewski, M. Zorębski and R. Mańka, *J. Phys. IV*, 2006, **137**, 203.
30. H. Gedanitz, M. J. Davila, E. Baumhogger and R. Span, *J. Chem. Thermodyn.*, 2010, **42**, 478.
31. E. H. I. Ndiaye, D. Nasri and J.-L. Daridon, *J. Chem. Eng. Data*, 2012, **57**, 2667.
32. H. P. Victoria, G. A. I. Silva, M. R. Estrada and K. R. Hall, *Fluid Phase Equilib.*, 2013, **338**, 119.
33. M. Greenspan and C. E. Tschiegg, *J. Res. Natl. Bur. Stand.*, 1957, **59**, 249.
34. W. D. Wilson, *J. Acoust. Soc. Am.*, 1959, **31**, 1067.
35. R. A. Fine and F. J. Millero, *J. Chem. Phys.*, 1973, **59**, 5529.
36. G. Benedetto, R. M. Gavioso, P. A. Giuliano Albo, S. Lago, D. M. Ripa and R. Spagnlo, *Int. J. Thermophys.*, 2005, **26**, 1667.
37. S. Vance and J. M. Brown, *J. Acoust. Soc. Am.*, 2010, **127**, 174.
38. C.-W. Lin and J. P. M. Trusler, *J. Chem. Phys.*, 2012, **136**, 094511.
39. G. S. Kell and E. Whalley, *J. Chem. Phys.*, 1975, **62**, 3496.
40. T. S. Khasanshin and A. P. Shchemelev, *High Temp.*, 2001, **39**, 60.
41. T. S. Khasanshin, O. G. Poddubskii and A. P. Shchemelev, *High Temp.*, 2005, **43**, 530.
42. T. S. Khasanshin, O. G. Poddubskij, A. P. Shchamialiou and V. S. Samuilov, *Fluid Phase Equilib.*, 2006, **245**, 26.
43. T. S. Khasanshin, O. G. Poddubskij, A. P. Shchamialiou and V. S. Samuilov, *Int. J. Thermophys.*, 2006, **27**, 1746.
44. T. S. Khasanshin, V. S. Samuilov and A. P. Shchemelev, *J. Eng. Phys. Thermophys.*, 2008, **81**, 760.
45. M. F. Bolotnikov and Y. A. Neruchev, *J. Chem. Eng. Data*, 2004, **49**, 202.

46. M. F. Bolotnikov, Y. A. Neruchev, Y. F. Melikhov, V. N. Verveyko and M. V. Verveyko, *J. Chem. Eng. Data*, 2005, **50**, 1095.
47. V. V. Melent'ev, M. F. Bolotnikov and Y. A. Neruchev, *J. Chem. Eng. Data*, 2005, **50**, 1357.
48. V. V. Melent'ev, M. F. Bolotnikov and Y. A. Neruchev, *J. Chem. Eng. Data*, 2006, **51**, 181.
49. V. H. Hasanov, *High Temp.*, 2012, **50**, 44.
50. Y. M. Naziyev, A. N. Shahverdiyev and V. H. Hasanov, *J. Chem. Thermodyn.*, 2005, **37**, 1268.
51. J.-L. Daridon, B. Lagourette and J.-P. E. Grolier, *Int. J. Thermophys.*, 1998, **19**, 145.
52. F. Plantier, A. Danesh, M. Sohrabi, J.-L. Daridon, F. Gozalpour and A. Todd, *J. Chem. Eng. Data*, 2005, **50**, 673.
53. D. Gonzalez-Salgado, J. Troncoso, F. Plantier, J.-L. Daridon and D. Bessières, *J. Chem. Thermodyn.*, 2006, **38**, 893.
54. F. Plantier and J.-L. Daridon, *J. Chem. Eng. Data*, 2005, **50**, 2077.
55. T. Lafitte, D. Bessieres, M. M. Pineiro and J.-L. Daridon, *J. Chem. Phys.*, 2006, **124**, 024509.
56. E. R. Lopez, J.-L. Daridon, F. Plantier, C. Boned and J. Fernandez, *Int. J. Thermophys.*, 2006, **27**, 1354.
57. F. Plantier, D. Bessières, J.-L. Daridon and F. Montel, *Fuel*, 2007, **87**, 196.
58. T. Lafitte, F. Plantier, M. M. Pineiro, J.-L. Daridon and D. Bessieres, *Ind. Eng. Chem. Res.*, 2007, **46**, 6998.
59. H. Khelladi, J.-L. Daridon and F. Plantier, *J. Acous. Soc. Am.*, 2010, **128**, 672.
60. J. A. P. Coutinho, M. Goncalves, M. J. Pratas, M. L. S. Batista, V. F. S. Fernandes, J. Pauly and J.-L. Daridon, *Energy Fuels*, 2010, **24**, 2667.
61. E. H. I. Ndiaye, J.-P. Bazile, D. Nasri, C. Boned and J.-L. Daridon, *Fuel*, 2012, **98**, 288.
62. J. Pauly, J. A. P. Coutinho and J.-L. Daridon, *Fluid Phase Equilib.*, 2012, **313**, 32.
63. E. H. I. Ndiaye, D. Nasri and J.-L. Daridon, *J. Chem. Eng. Data*, 2012, **57**, 2667.
64. J.-L. Daridon, J. A. P. Coutinho, E. H. I. Ndiaye and M. L. L. Paredes, *Fuel*, 2013, **105**, 466.
65. S. V. D. Freitas, M. L. L. Paredes, J.-L. Daridon, A. S. Lima and J. A. P. Coutinho, *Fuel*, 2013, **103**, 1018.
66. E. H. I. Ndiaye, M. Habrioux, J. A. P. Coutinho, M. L. L. Paredes and J.-L. Daridon, *J. Chem. Eng. Data*, 2013, **58**, 1371.
67. E. H. I. Ndiaye, M. Habrioux, J. A. P. Coutinho, L. L. Paredes and J.-L. Daridon, *J. Chem. Eng. Data*, 2013, **58**, 2345.
68. J.-L. Daridon, J. A. P. Coutinho, E. H. I. Ndiaye and M. L. L. Paredes, *Fuel*, 2013, **105**, 466.
69. M. Dzida and S. Ernst, *J. Chem. Eng. Data*, 2003, **48**, 1453.
70. M. Dzida, *J. Solution Chem.*, 2004, **33**, 529.

71. M. Dzida, A. Zak and S. Ernst, *J. Chem. Thermodyn.*, 2005, **37**, 405.
72. M. Dzida and W. Marczak, *J. Chem. Thermodyn.*, 2005, **37**, 826.
73. M. Dzida, *J. Chem. Eng. Data*, 2007, **52**, 521.
74. E. Zorebski and M. Dzida, *J. Chem. Eng. Data*, 2007, **52**, 1010.
75. E. Zorebski, M. Dzida and M. Piotrowska, *J. Chem. Eng. Data*, 2008, **53**, 136.
76. E. Zorebski, M. Dzida and M. Cempa, *J. Chem. Eng. Data*, 2008, **53**, 1950.
77. M. Dzida and M. Cempa, *J. Chem. Thermodyn.*, 2008, **40**, 1531.
78. M. Dzida and P. Prusakiewicz, *Fuel*, 2008, **87**, 1941.
79. M. Dzida and P. Goralski, *J. Chem. Thermodyn.*, 2009, **41**, 402.
80. M. Dzida, *J. Chem. Eng. Data*, 2009, **54**, 1034.
81. M. Dzida, *J. Phys. Chem. B*, 2009, **113**, 11649.
82. M. Dzida and L. Waleczek, *J. Chem. Thermodyn.*, 2010, **42**, 312.
83. M. Dzida, *Int. J. Thermophys.*, 2010, **31**, 55.
84. M. Chorazewski and M. Skrzypek, *Int. J. Thermophys.*, 2010, **31**, 26.
85. E. Zorebski, M. Dzida and E. Wysocka, *J. Chem. Eng. Data*, 2011, **56**, 2680.
86. M. Dzida and A. Kaczmarczyk, *Int. J. Thermophys.*, 2012, **33**, 583.
87. E. Zorebski and M. Dzida, *J. Chem. Thermodyn.*, 2012, **54**, 100.
88. M. Chorazewski, M. Dzida, E. Zorębski and M. Zorębski, *J. Chem. Thermodyn.*, 2013, **58**, 389.
89. M. Dzida, *Int. J. Thermophys.*, 2013, **33**, 583.
90. M. Dzida, S. Jezak, J. Sumara, M. Zarska and P. Goralski, *Fuel*, 2013, **111**, 165.
91. G. Benedetto, R. M. Gavioso, P. A. G. Albo, S. Lago, D. M. Ripa and R. Spagnolo, *Int. J. Thermophys.*, 2005, **26**, 1651.
92. S. Lago, P. A. G. Albo and D. M. Ripa, *Int. J. Thermophys.*, 2006, **27**, 1083.
93. G. Scalabrin, M. G. Concion, P. Marchi and S. Lago, *Exp. Therm. Fluid Sci.*, 2007, **31**, 539.
94. S. Lago and P. A. G. Albo, *J. Chem. Thermodyn.*, 2008, **40**, 1558.
95. S. Lago and P. A. G. Albo, *J. Chem. Thermodyn.*, 2009, **41**, 506.
96. S. Lago and P. A. G. Albo, *J. Chem. Thermodyn.*, 2010, **42**, 462.
97. S. Lago, P. A. G. Albo and S. Brignolo, *J. Chem. Eng. Data*, 2011, **56**, 161.
98. P. A. G. Albo, S. Lago, R. Romeo and S. Lorefice, *J. Chem. Thermodyn.*, 2013, **58**, 95.
99. P. A. G. Albo and S. Lago, *Fuel*, 2014, **115**, 740.
100. S. Lago, S. Brignolo, R. Cuccaro, C. Musacchio, P. A. G. Albo and P. Tarizzo, *Appl. Acoust.*, 2014, **75**, 10.
101. K. Meier and S. Kabelac, *J. Chem. Eng. Data*, 2012, **57**, 3391.
102. K. Meier and S. Kabelac, *J. Chem. Eng. Data*, 2013, **58**, 446.
103. K. Meier and S. Kabelac, *J. Chem. Eng. Data*, 2013, **58**, 1398.
104. K. Meier and S. Kabelac, *J. Chem. Eng. Data*, 2013, **58**, 1621.
105. H. Gedanitz, M. J. Davila, E. Baumhogger and R. Span, *J. Chem. Thermodyn.*, 2010, **42**, 478.
106. P. F. Pires and H. J. R. Guedes, *J. Chem. Thermodyn.*, 1999, **31**, 55.

107. J. M. S. S. Esperanca, P. F. Pires, H. J. R. Guedes, N. Ribeiro, T. Costa and A. Aguiar-Ricardo, *J. Chem. Eng. Data*, 2006, **51**, 1148.
108. N. Ribeiro, T. Costa, A. Aguiar-Ricardo, J. M. S. S. Esperanca, P. F. Pires and H. J. R. Guedes, *J. Chem. Eng. Data*, 2006, **51**, 1906.
109. J. M. S. S. Espreanca, P. F. Pires, H. J. R. Guedes, N. Ribeiro, T. Costa and A. Aguiar-Ricardo, *J. Chem. Eng. Data*, 2006, **51**, 2161.
110. M. J. Davila and J. P M. Trusler, *J. Chem. Thermodyn.*, 2009, **41**, 35.
111. F. Peleties, J. J. Segovia, J. P. M. Trusler and D. Vega-Maza, *J. Chem. Thermodyn.*, 2010, **42**, 631.
112. C.-W. Lin, D. Ramjugernath, P. Reddy and J. P. M. Trusler, *J. Chem. Eng. Data*, 2012, **57**, 2568.
113. M. E. Tat and J. H. V. Gerpen, *J. Am. Oil Chem. Soc.*, 2003, **80**, 1127.
114. S. L. Outcalt and T. J. Fortin, *J. Chem. Eng. Data*, 2012, **57**, 2869.
115. T. Takagi, *High Temp. – High Pressures*, 1993, **25**, 685.
116. T. Takagi and M. Fukushima, *Thermochim. Acta*, 1998, **317**, 65.
117. I. Cibulka and L. Hnedkovsky, *J. Chem. Eng. Data*, 1996, **41**, 657.
118. I. Cibulka, L. Hnedkovsky and T. Takagi, *J. Chem. Eng. Data*, 1997, **42**, 2.
119. I. Cibulka and T. Takagi, *J. Chem. Eng. Data*, 1999, **44**, 411.
120. J. H. Dymond and R. Malhotra, *Int. J. Thermophys.*, 1988, **6**, 941.
121. J. W. Magee and T. J. Bruno, *J. Chem. Eng. Data*, 1996, **41**, 900.
122. J. W. Magee, *Int. J. Thermophys.*, 1996, **17**, 803.
123. L. Fedele, F. Pernechele, S. Bobbo and M. Scattolini, *J. Chem. Eng. Data*, 2007, **52**, 1955.
124. W. Wagner and A. Pruβ, *J. Phys. Chem. Ref. Data*, 2002, **31**, 387.
125. S. Will, A. P. Froba and A. Leipertz, *Int. J. Thermophys.*, 1998, **19**, 403.
126. K. Kraft and A. Leipertz, *Int. J. Thermophys.*, 1994, **15**, 791.
127. A. P. Froba, C. Botero and A. Leipertz, *Int. J. Thermophys.*, 2006, **27**, 1609.
128. O. Kiyohara, P. J. D'Arcy and G. C. Benson, *Can. J. Chem.*, 1979, **57**, 1006.
129. O. Kiyohara and G. C. Benson, *J. Chem. Thermodyn.*, 1979, **11**, 861.
130. H. Houkhani and Z. Rostami, *J. Chem. Eng. Data*, 2007, **52**, 921.
131. S. M. García-Abarrio, L. Viloria, L. Haya, J. S. Urieta and A. M. Mainar, *Fluid Phase Equilib.*, 2011, **308**, 78.
132. B. K. Sarkar, A. Choudhury and B. Sinha, *J. Solution Chem.*, 2012, **41**, 53.

CHAPTER 14

High-Pressure "Maxwell Relations" Measurements

STANISLAW L. RANDZIO,*[a] JEAN PIERRE E. GROLIER[b] AND MIROSLAW CHORAZEWSKI[c]

[a] Polish Academy of Sciences, Institute of Physical Chemistry, Kasprzaka 44/52, Pl-01-224, Warszawa, Poland; [b] Clermont University, ENSCCF, Institute of Chemistry of Clermont-Ferrand, ICCF, CNRS UMR 6296, 24 Landais Av., 63171, Aubière, France; [c] Institute of Chemistry, University of Silesia, Szkolna 9, Pl-40-006 Katowice, Poland
*Email: stanislaw.randzio@ichf.edu.pl

14.1 Introduction

The thermodynamic potentials, Gibbs energy G and Helmholtz energy A, when expressed in terms of common independent variables p, V and T yield the following "Maxwell relations":

$$\frac{\partial^2 G}{\partial p \partial T} = \frac{\partial^2 G}{\partial T \partial p} = \left(\frac{\partial V}{\partial T}\right)_P = -\left(\frac{\partial S}{\partial p}\right)_T = -\frac{1}{T}\left(\frac{dQ}{dp}\right)_T \tag{14.1}$$

$$-\frac{\partial^2 A}{\partial V \partial T} = -\frac{\partial^2 A}{\partial T \partial V} = \left(\frac{\partial p}{\partial T}\right)_V = \left(\frac{\partial S}{\partial V}\right)_T = \frac{1}{T}\left(\frac{dQ}{dV}\right)_T \tag{14.2}$$

The experimental importance of the above relations comes mainly from the fact that the thermomechanical coefficients $\left(\frac{\partial V}{\partial T}\right)_P$ and $\left(\frac{\partial P}{\partial T}\right)_V$ can be determined directly from calorimetric measurements of the heat Q developed by variations of pressure or volume, respectively, performed

Volume Properties: Liquids, Solutions and Vapours
Edited by Emmerich Wilhelm and Trevor M. Letcher

Published by the Royal Society of Chemistry, www.rsc.org

under isothermal conditions.[1] The respective techniques are known as the piezothermal technique,[2,3] as pressure-controlled[1,4–6] or volume controlled scanning calorimeters,[1,7] and as pressure perturbation calorimetry.[8,9] These techniques are especially important for investigation at extreme conditions of pressure and temperature, where the volumetric or densitometric techniques are often not precise enough to derive proper differential thermodynamic quantities or for investigation of liquid solutions of biological importance, where the isothermal conditions are advantageous and the measurements require small samples.[8]

The third coefficient $\left(\frac{\partial V}{\partial P}\right)_T$ is not related directly to the heat, but can be determined from the calorimetrically measured quantities with the use of known thermodynamic relations

$$\left(\frac{\partial S}{\partial V}\right)_T = \left(\frac{\partial p}{\partial T}\right)_V = -\frac{\left(\frac{\partial V}{\partial T}\right)_P}{\left(\frac{\partial V}{\partial p}\right)_T} = \frac{\alpha}{\kappa_T} \tag{14.3}$$

or

$$\left(\frac{\partial \alpha}{\partial p}\right)_T = -\left(\frac{\partial \kappa_T}{\partial T}\right)_P \tag{14.4}$$

where α is the coefficient of thermal expansion and κ_T is the isothermal coefficient of compressibility. From the above short introduction, one can see that a calorimeter properly controlled by an independent mechanical variable can be very useful in the investigation of volumetric properties of various substances. This chapter will be focused on the use of such techniques for the investigation of liquids of various natures over wide ranges of pressures and temperatures.

14.2 Pressure as the Controlled Variable

Pressure is an intensive thermodynamic variable and thus its precise control is possible from an external source without important corrections being made. A schematic diagram of an example of a calorimetric installation[10] adapted to measurements of the coefficient of thermal expansion of liquids on the basis of Maxwell relation, Equation 14.1 is presented in Figure 14.1. The installation consists of a calorimeter equipped with high-pressure vessels, a $(p;V;T)$ system and software. Two heat-flux calorimetric detectors, made from 672 thermocouples each, are mounted differentially and connected to a nanovolt amplifier. The calorimetric detectors are placed in a calorimetric metallic block, the temperature of which is directly controlled, with an uncertainty of near 10^{-4} K, with an entirely digital feedback loop of 22 bit resolution that is part of the controlling software. The calorimetric block is surrounded by a heating–cooling shield. The temperature difference

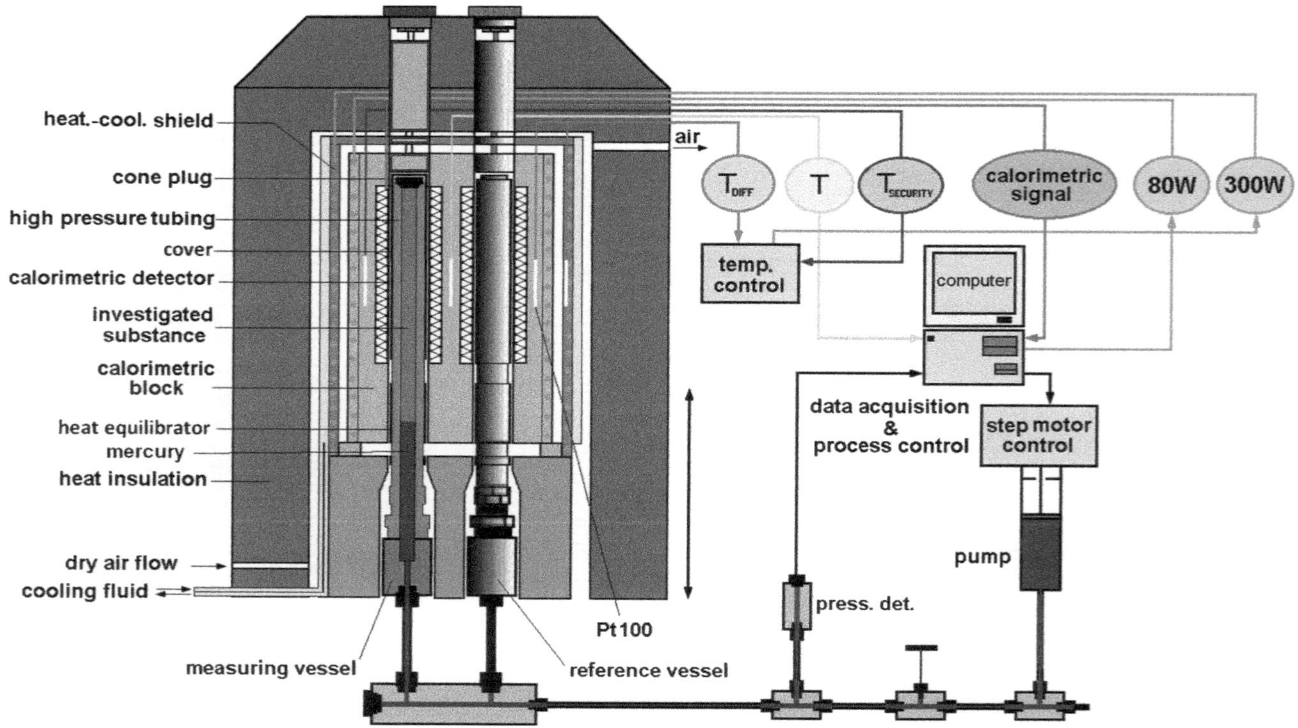

Figure 14.1 A calorimetric installation adapted to measurements of the coefficient of thermal expansion of liquids over wide temperature and pressure ranges.

between the block and the heating–cooling shield is set at a constant value and is controlled by an additional analogue controller.

The temperature measurements, both absolute and differential, are performed with calibrated platinum resistance 100 Ω sensors. The heaters are homogeneously embedded on the outer surfaces of both the calorimetric block and the heating–cooling shield. The whole assembly is placed in thermal insulation embedded in a stainless steel body and placed on a stand which permits the calorimeter to move up and down over the calorimetric vessels. When performing measurements at temperatures near 273 K or below, dry air is pumped through the apparatus. The hydraulic line is filled completely with mercury and is composed of a pump, pressure detector, manifold and other high-pressure connecting elements. The piston pump with a 9 cm^3 displacement is driven by a stepper motor controlled either manually or by software. Each motor step corresponds to a volume displacement of $(5.24 \pm 0.04) \times 10^{-6}$ cm^3. The pressure detector operates up to 400 MPa with a precision of 0.15%.

The pressure detector, the output of the calorimetric amplifier and the stepping motor are connected to a NI PCI-MIO-16XE-50 multifunction board through a NI SCB-68 shielded connector block. The temperature measurements and the digital control of the calorimetric block are performed through a serial port. The software elaborated with the use of LabView language performs as a Virtual Instrument (VI). It consists of ninety subVIs, each responsible for a particular function: pressure measurement, temperature measurement, counting the motor steps and recording the volume variations, measuring the calorimetric signal, *etc.* and each performs independently. However, all the subVIs form a hierarchical structure with a top window where the experimenter can see, simultaneously, all the four variables $(p; V; T; Q)$ of the process under investigation and the current functioning of the temperature and pressure control loops. For this reason, the complete technique has been called scanning transitiometry.[11]

The calorimetric vessels, made from 316 SS tubing with 0.48 cm internal diameter, are fixed on a mounting table attached to the stand. Only the measuring vessel is connected to the hydraulic line. The reference vessel serves only as a thermal reference. The tubing of both measuring and reference vessels are connected to reducers which are inside the calorimeter body. The connections from the reducers to the manifold are made with thin stainless steel capillaries in order to reduce the heat losses to the external environment. A liquid sample under investigation is placed directly on the mercury, the level of which is significantly below the lower part of the calorimetric detector, but inside the calorimetric thermostat. The vessels are closed with a cone plug and fixed in place by an internally threaded cover which also serves as a heat exchanger between the calorimetric vessel tubing and the calorimetric detector. Sleeves are also fixed on the calorimetric vessel tubing below the cover in order to help the heat exchange between the tubing and both the calorimetric block and the heating–cooling shield. Closing with the cone plug should be performed in such a way as to avoid any air space above the liquid sample. An overflow of the liquid during closing is permitted, because the mass of the investigated liquid sample is irrelevant to the determination of thermal expansion in a pressure-controlled calorimeter. Calorimetric vessels for determination of thermal expansion coefficients of liquids and solutions on the basis of the Maxwell relation in Equation (14.1), similar to the vessels described above but designed to work with other calorimeters and different hydraulic systems, have been described in the literature.[2,3,8,11–14]

The main requirements for the correct performance of measurements and treatment of the data are the conceptual definition of an active volume V_E of the calorimetric vessel and the transmission of pressure to it. The active volume V_E is defined as that part of the calorimetric vessel tubing which is entirely covered by the calorimetric detector; it implies that the active volume contains that part of the liquid sample which exchanges heat with the calorimetric detector when the pressure is varied. As is shown in Figure 14.1, the pressure variation is transmitted to the active volume through the investigated liquid itself. Thus, the mass m of the investigated sample contained in the active volume V_E is inversely proportional to the molar volume V_m of the investigated liquid (which is a function of pressure):

$$m = \frac{V_E}{V_m} \tag{14.5}$$

The pressure variation can be performed as a linear function of time:

$$p = p_0 \pm at, \quad \mathrm{d}p = \pm a\mathrm{d}t \tag{14.6}$$

where a is a rate of linear pressure variation. When introducing Equations (14.5) and (14.6) into Maxwell relation, Equation (14.1), one obtains:

$$P_T(p) = -\left[\pm \frac{1}{V_m}\left(\frac{\partial V_m}{\partial T}\right)_p V_E T a\right] = -(\pm \alpha V_E T a) \tag{14.7}$$

where $P_T(p) = dQ/\mathrm{d}t$ is the power (in units of J s^{-1} or W) developed as a function of pressure at constant temperature in the liquid sample contained in the active volume V_E. When pressure is applied inside a cylinder of internal volume V_E, the volume of the wall of the cylinder V_w will increase by an amount that can be approximated by:[6]

$$(dV_w) = V_E \kappa_{SS} dp \tag{14.8}$$

where κ_{SS} is the isothermal coefficient of compressibility of stainless steel, the material from which the cylinder is made. When introducing Equation (14.8) into Equation (14.3), Equation (14.9) is obtained, which defines the rate at which the heat is produced in the wall of the calorimetric cylinder during linear pressure variations:

$$P_{T,w}(p) = \pm \alpha_{SS} V_E T a \tag{14.9}$$

Thus, for a linear pressure scan the power generated in a calorimetric cylindrical vessel of internal active volume V_E filled with the substance under investigation is given by the sum of Equations (14.7) and (14.9):

$$\Delta P_T(p) = -[\pm(\alpha - \alpha_{SS})V_E Ta] \tag{14.10}$$

Taking into consideration that the calorimetric signal U is usually recorded in volts and the rate of pressure scan is low, the power recorded in the stationary state of the heat exchange can be expressed by Equation (14.11):

$$\Delta P_T(p) = k_C U \tag{14.11}$$

where k_c is the static gain (sensitivity) of the calorimetric detector expressed in units of W V^{-1}. Then Equation (14.10) can be rewritten in the following form, suitable for the determination of the coefficient of thermal expansion of the liquid under investigation:

$$\alpha = \alpha_{SS} - \left(\pm \frac{k_C}{V_E}\frac{U}{Ta}\right) \tag{14.12}$$

or in the form suitable for calibration purposes:

$$\frac{k_C}{V_E} = -\left[\pm(\alpha_{cl} - \alpha_{SS})\frac{Ta}{U}\right] \tag{14.13}$$

where α_{cl} is the coefficient of thermal expansion of the calibrating liquid.

In the case of pressure variations as a series of quasi stepwise changes Δp_i and recording respective responses of the calorimetric signal, Equations (14.12) and (14.13) take the forms of Equations (14.14) and (14.15) respectively:

$$\alpha = \alpha_{SS} - \left(\pm \frac{k_C}{V_E}\frac{I_i}{T\Delta p_i}\right) \tag{14.14}$$

$$\frac{k_C}{V_E} = -\left[\pm(\alpha_{cl} - \alpha_{SS})\frac{T\Delta p_i}{I_i}\right] \tag{14.15}$$

where I_i is the integral (in units of V s) of the calorimetric signal response to the pressure step Δp_i.

An example of the two modes of pressure variation and of respective calorimetric signal responses is presented in Figure 14.2. In the case of the linear pressure variation the calorimetric signal deviation is taken from the beginning of the stationary state at the respective pressure to the end of pressure scanning [from point A to point B in Figure 14.2(b)]. The pressure scanning rate should be sufficiently low in order to cope with the thermal inertia of the calorimetric system[15] and maintain equilibrium in the investigated sample. Typical pressure scanning rates are 0.02–0.002 MPa s^{-1}. In the case of the stepwise mode the integral of the complete calorimetric

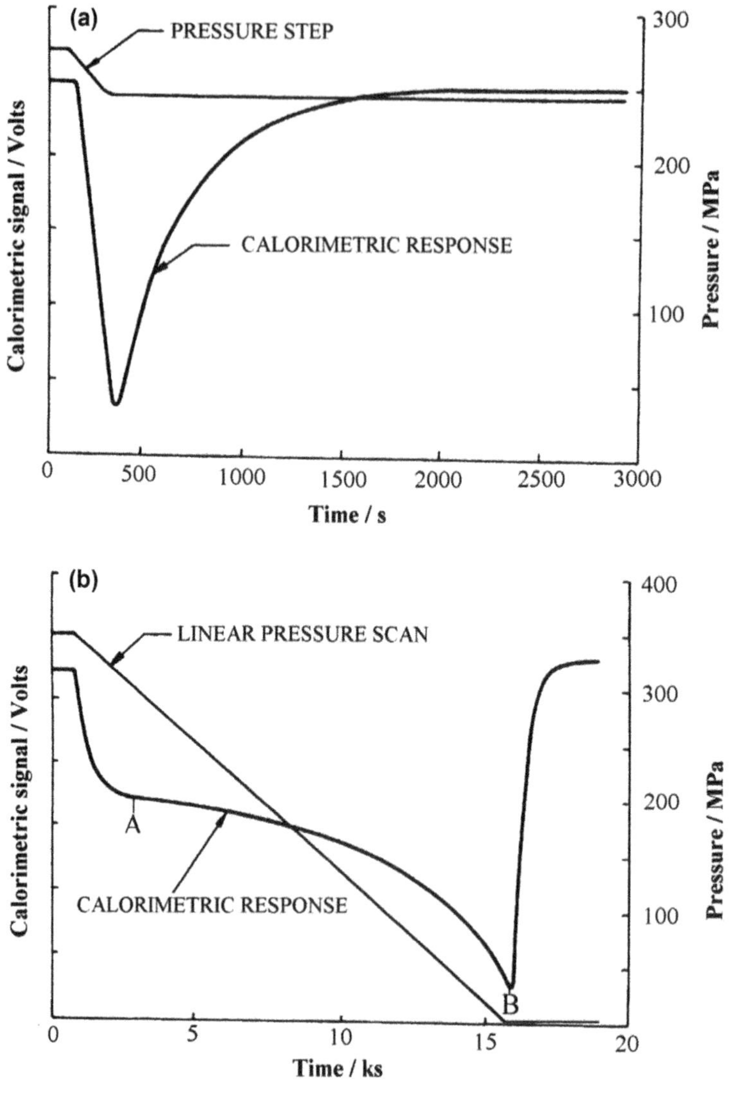

Figure 14.2 Examples of input and output signals for stepwise (a) and linear (b) pressure scans.

response is taken into consideration. It is also worth noting that the liquid under investigation must be at thermodynamic equilibrium (keeping in mind that Maxwell relations apply at thermodynamic equilibrium). Thus, it is recommended to pressurise the investigated liquid sample to the desired level and leave it for few hours to reach thermal and mechanical equilibrium. Only then can the experiments with pressure variations be started in the decompression mode; this gives a better reproducibility of measurements. The compression mode is also possible, but in this case there is a bigger risk of performing measurements in metastable states.

The calibration is performed by filling the calorimetric vessel with a liquid of known coefficient of thermal expansion, such as *n*-hexane[16] or water,[12] and varying pressure either by steps or linearly at selected temperatures. If the experiments are performed correctly the agreement between the two modes is satisfactory and the calibration constant k_c/V_E (in units of MPa (V s)$^{-1}$ is independent of pressure,[6] but depends slightly on temperature.[13] It is however interesting that the same calibration constant k_c/V_E can also be obtained from measurements of constant pressure heat capacity by scanning temperature using the same vessels.[17] Thus, taking into consideration that there is an important number of liquids with a known heat capacity as a function of both pressure and temperature,[18] the temperature-scanning calibration can at least be used as a verification of the calibration performed with the pressure-scanning mode.

A certain problem of accuracy of the methods described above is the contribution to the measured values coming from the vessel itself. According to the present approach, this contribution is defined as the coefficient of thermal expansion of the material from which the vessel is made. For stainless steel α_{ss} =5.51×10^{-5} K^{-1} and is considered as practically independent of both temperature and pressure. For most liquids at low pressures the coefficient of thermal expansion is near 10^{-3} K^{-1} or higher, thus an error of even 10% in α_{ss} can have only a small influence on the precision of the measurements. However, at high pressures the coefficient of thermal expansion of liquids decreases significantly and the imprecision in α_{ss} can influence both the calibration and the measurements. To avoid this inconvenience and to reduce other experimental problems, such as heat leaks from the vessels, an attempt has been made by Navia *et al.*[19] to perform a calibration procedure with two calibrating liquids and thus eliminate the contribution from the vessel itself. Taking into consideration the approach and experience of various laboratories, one can summarise that the accuracy of determination of the coefficient of thermal expansion of liquids by calorimetric methods based on the Maxwell relation, Equation (14.1), is about 2–3% depending on the pressure and temperature ranges.

14.3 Volume as the Controlled Variable

Similar to pressure, the volume variation can be performed as a linear function of time:

$$V = V_0 \pm at, \quad dV = \pm c dt \tag{14.16}$$

where c is a rate of linear volume variation. When introducing Equation (14.16) into Maxwell relation Equation (14.2) one obtains:

$$P_T(V) = \pm \left(\frac{\partial p}{\partial T}\right)_V cT \tag{14.17}$$

where $P_T(V) = dQ/dt$ is the power (in units of J s^{-1} or W) developed as a function of volume at constant temperature. Integration of Equation (14.17) gives:

$$\beta = \left(\frac{\partial p}{\partial T}\right)_V = \left(\frac{\partial S}{\partial V}\right)_T = \frac{Q}{T\Delta V} \tag{14.18}$$

Further similarities with pressure are limited, because volume is an extensive variable and its use as a controlled variable is more complicated.[7] For this reason, the simple relation Equation (14.18) holds only when the measured heat Q corresponds to the same amount of substance under investigation as the volume change, ΔV. In practice, two other situations are encountered.

(i) The whole active volume, V_E, of the measuring vessel is filled with the investigated liquid and a part of the liquid sample is not covered by the calorimetric detector, such as presented in Figure 14.1. In this case the external volume variations achieved by the pump must be multiplied by a factor V_E/V_s, before being introduced into Equation (14.18), V_s is the volume of the whole sample under investigation, which now must be accurately known, contrary to the pressure-scanning measurements.
(ii) It can happen, especially at high compressions, that the whole investigated sample occupies only a part of the active volume, V_E, the other part being occupied by the hydraulic fluid (mercury).[20] In this case, the measured heat effect contains a contribution from mercury which must subtracted.

Thus, in the case where the investigated sample is not completely covered by the calorimetric detector, $V_s > V_E$, Equation (14.18) takes the following form:

$$\beta = \frac{Q}{\Delta V_{e,cor} T} \frac{V_s}{V_E} \tag{14.19}$$

where $\Delta V_{e,cor}$ is the external volume variations, corrected for compressibility of the hydraulic fluid. The external volume variations V_e are determined from the recorded number of motor steps multiplied by the calibration constant of the hydraulic system, k_s. The compressibility of the hydraulic fluid is determined from compression without any other substance present in the system and fitted to a correlation equation.[20] In the case where the volume of the investigated liquid sample, V_s, at given pressure and temperature conditions is smaller than the active volume, V_E, covered by the calorimetric detector, $V_s < V_E$, Equation (14.18) takes the following form:

$$\beta = \frac{Q}{\Delta V_e T} \frac{V_E}{V_s} - \beta_{Hg} \frac{V_E - V_s}{V_s} \tag{14.20}$$

where β_{Hg} is the temperature coefficient of pressure for mercury.

As will be evident from subsequent considerations, the volume controlled technique is especially helpful in measurements near critical points where pressure as the controlled variable has limited possibilities. The following example will show details of such an investigation.[7] A *n*-hexane sample of 0.5246 g was placed in the calorimetric vessel with a syringe. The vessel was then immediately closed with the cone plug and the cover. Care was taken to avoid vapour or air space in the vessel. After introduction of the vessels into the calorimeter it was heated to $T = 507.2$ K, near the critical temperature of *n*-hexane ($T_{cr} = 507.8$ K, $p_{cr} = 3.09$ MPa) and left for a few hours to attain thermal and mechanical equilibrium at about 25 MPa. The measurements were performed by increasing the volume using stepwise variations. Data recorded during selected experiments are presented in Figure 14.3. One can see that, over the whole pressure range, all the volume variations are accompanied by respective heat effects. On the contrary, the respective pressure variations are smaller and smaller, and when approaching the critical point they become negligible and, finally, practically not measurable.

The details of calculations are given in Table 14.1. The parameters of the instrument needed for the calculation have been taken from the previous calibrations of the instrument [$k_s = (5.228 \pm 0.026) \times 10^{-6}$ cm^3 per step, $k_c = 38.10$ W$\times$V^{-1}, $V_E = 1.11$ cm^3].[20] The temperature coefficient of

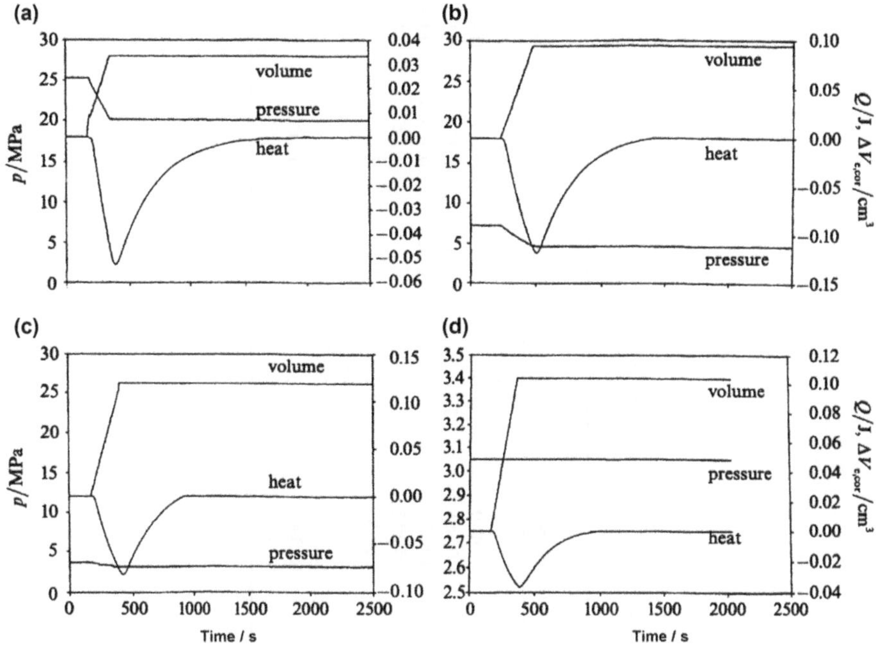

Figure 14.3 Examples of experimental data of volume controlled scanning transitiometry of *n*-hexane at $T = 507.2$ K: (a) decompression 1 (see Table 14.1); (b) decompression 5; (c) decompression 7; (d) decompression 9.[7]

Table 14.1 Results of volume controlled scanning transitiometric measurements of thermomechanical coefficients of *n*-hexane at $T = 507.2$ K.[7] The symbols are defined in the text.

Decompression number	Number of motor steps	p_{mean}/MPa	Δp/MPa	ΔV_e/mL	ΔV_{hydr}/mL	$\Delta V_{e,cor}$/mL	Q/J	V_s/mL	$(\partial p/\partial T)_V$/MPa K^{-1}	$10^3\ \alpha$/1/K	$10^3 \kappa_T$/1/MPa
1	83426	22.7	5.30	0.09353	0.06085	0.03268	3.592	1.5503	0.3027	1.205	3.979
2	83880	17.8	4.89	0.09590	0.06029	0.03562	3.686	1.5860	0.2910	1.339	4.601
3	86641	13.0	4.60	0.11034	0.05873	0.05161	4.119	1.6186	0.2294	1.592	6.940
4	89699	9.01	5.42	0.11064	0.04513	0.06551	4.122	1.6702	0.1867	2.142	11.47
5	89984	6.00	2.59	0.12781	0.03512	0.09269	4.770	1.7357	0.1587	3.266	20.59
6	86926	4.23	1.18	0.11182	0.01621	0.09561	3.703	1.8284	0.1258	5.579	44.36
7	55267	3.41	0.47	0.111884	0.00662	0.11222	3.110	1.9240	0.0947	11.71	123.6
8	85968	3.11	0.12	0.10682	0.00157	0.10525	1.634	2.0362	0.0562	24.60	438.0
9	85392	3.05	0.00	0.10381	0.0	0.10381	1.149	2.1415	0.0421	∞	∞
10	89835	3.05	0.00	0.12704	0.0	0.12707	1.293	2.2453	0.0406	∞	∞

pressure, β, for the investigated sample of *n*-hexane was derived with Equation (14.19). From both the pressure variations (being the results of the controlled volume variations) and the respective heat effects, Q, the values of the coefficient of thermal expansion have also been derived:

$$\alpha = \frac{Q}{\Delta p T V_E} \tag{14.21}$$

The isothermal coefficients of compressibility, κ_T, have also been derived directly from the volume and pressure variations:

$$\kappa_T = \frac{\Delta V_{e,cor}}{V_s \Delta p} \tag{14.22}$$

In the above derivations, it was assumed that the contribution from the calorimetric vessel itself was negligible with respect to the derived values. Detailed discussion of the results will be presented in the next section.

14.4 Results

14.4.1 Simple Liquids

Simple liquids, where the predominant intermolecular interactions are based on dispersive forces, can serve as references in the studies of more complex liquids. For a long time, *n*-hexane has been proposed as a model for simple dense liquids.[21–23] Figure 14.4 presents thermal expansion coefficient

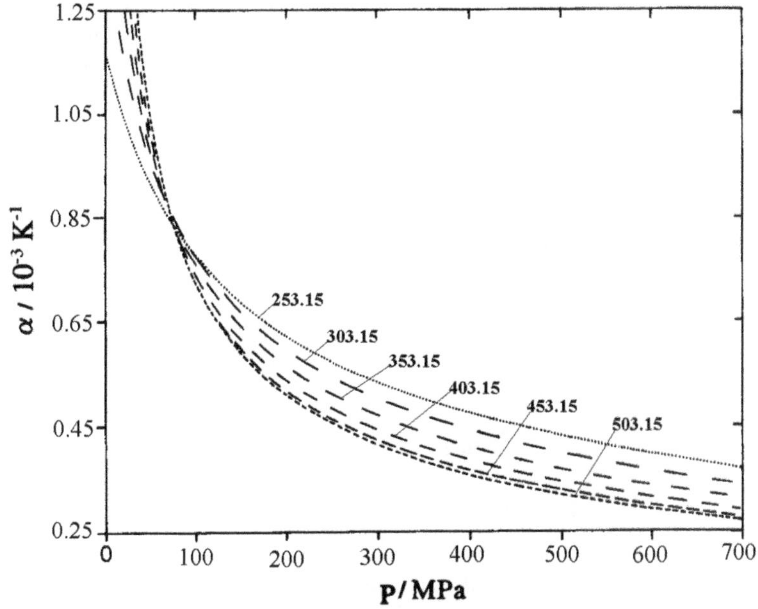

Figure 14.4 Isotherms of thermal expansion coefficient α for liquid *n*-hexane.[16]

α for n-hexane over a large temperature range from 243.15 K to 503.15 K under pressures of up to 700 MPa, established on the basis of Maxwell relation measurements performed in various laboratories.[16] The most interesting feature, which can be seen in this figure, is the existence of a pressure 62 ± 2 MPa, below which the thermal expansion coefficient α increases with temperature and above which it decreases with temperature (crossing point of α isotherms).

Similar behaviour of α has also been observed for liquid n-butane and CO_2. The crossing of α isotherms is observed for n-butane at higher pressures, near 100 MPa and near 120 MPa for CO_2.[24] Furthermore, the higher alkanes, such as nonane, decane, undecane, dodecane, tridecane, tetradecane and pentadecane, also exhibit a similar behaviour.[25] From data presented in Figure 14.5, one can see that the pressure at the crossing of α isotherms for n-alkanes decreases with increasing carbon atoms in the alkane chain, down to about 20 MPa for pentadecane.

Other organic liquids, such as benzene[25,26] and toluene,[10,27] also demonstrate crossing points of the α isotherms. It appears to be a characteristic feature of simple liquids. It is worth noting that a reliable investigation of

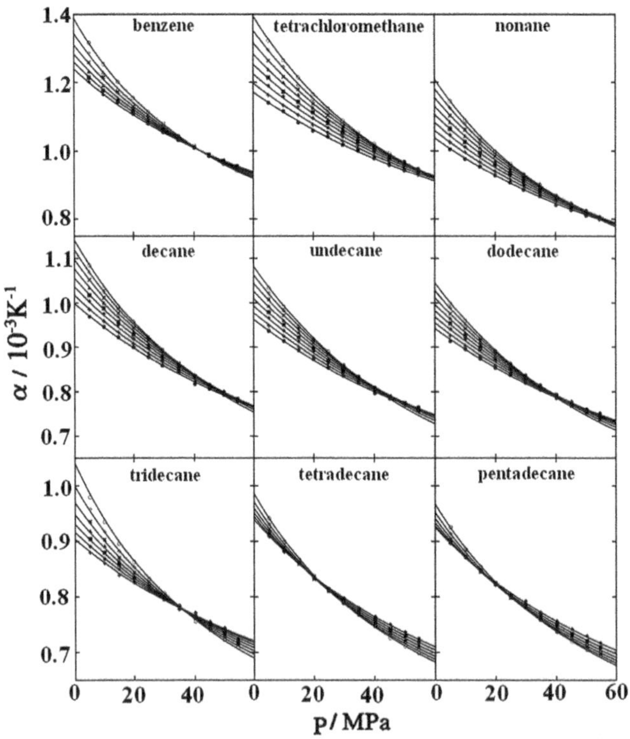

Figure 14.5 Isotherms of thermal expansion coefficient α for selected liquid hydrocarbons.[25] ● 278.15 K, + 288.15 K, ◆ 298.15 K, ■ 308.15 K, ▲ 318.15 K, X 328.15 K, ✙ 338.15 K, ○ 348.15 K.

this property requires sufficiently wide ranges of pressure and temperature of calorimetric measurements. For example, results obtained for tetrachloromethane,[25] pentane, heptane and nonane[19] cannot be conclusive in this respect, because the applied pressure and temperature ranges were too small.

The unique property of the crossing point of thermal expansion isotherms can be used as a criterion for a verification of equations of state for simple liquids.[28,29] It was found that a reproduction of the crossing point of thermal expansion isotherms for *n*-hexane could be done only with an equation of state with a modified repulsive contribution of the Carnahan–Starling type.[30] From the molecular point of view the change of sign of $d\alpha/dT$ at the crossing point of α isotherms can be interpreted as a result of a change in the shape of effective intermolecular potential in dense liquids caused by high pressure, where the main contribution to the thermal expansion comes from oscillations.[31] Those observations have led to the development of an equation of state with shifted Lennard-Jones pair potentials.[32] It was found that the best reproduction of the crossing point of thermal expansion isotherms for liquid methane can be done with a soft sphere van der Waals equation of state for an 8/4 pair potential.[32]

It is not the aim of the present chapter to go more deeply into the theory of the liquid state. It can only be concluded that for simple liquids there are two regions of significantly different properties: at low pressures below the crossing point of α isotherms, where $d\alpha/dT > 0$, the dominant contribution is fluctuations and at high pressures above the crossing point of α isotherms, where $d\alpha/dT < 0$, the system is dominated by short-range repulsive forces, with negligible fluctuations.[25,33] This property can be used as a reference in the interpretation of experimental data for complex liquids over wide pressure and temperature ranges.

14.4.2 Complex Liquids

Hexan-1-ol has been investigated with a pressure-controlled scanning calorimeter over wide pressure and temperature ranges.[34] Due to a self-association effect there is no unique crossing point of thermal expansion isotherms and the isotherms start to cross only at high pressures and at high temperatures. To distinguish this effect, Figure 14.6 presents isotherms of $d\alpha/dT$ for hexan-1-ol as a function of pressure. One can see that only at 403.15 K does $d\alpha/dT$ become negative under a pressure of close to 300 MPa. At higher temperatures the change of sign appears at lower pressures, because the self-association is weaker. Figure 14.7 presents a comparison of α isotherms for hexan-1-ol, *n*-hexane and binary mixtures of hexan-1-ol and *n*-hexane at selected concentrations.[35] One can see that by diluting hexan-1-ol with *n*-hexane the effect of self-association progressively disappears and the shape of the crossing of thermal expansion isotherms approaches the form typical of simple liquids. Other alcohols such as methanol, ethanol, propan-1-ol, butan-1-ol, pentan-1-ol, pentan-2-ol, pentan-3-ol, heptan-1-ol, octan-1-ol,

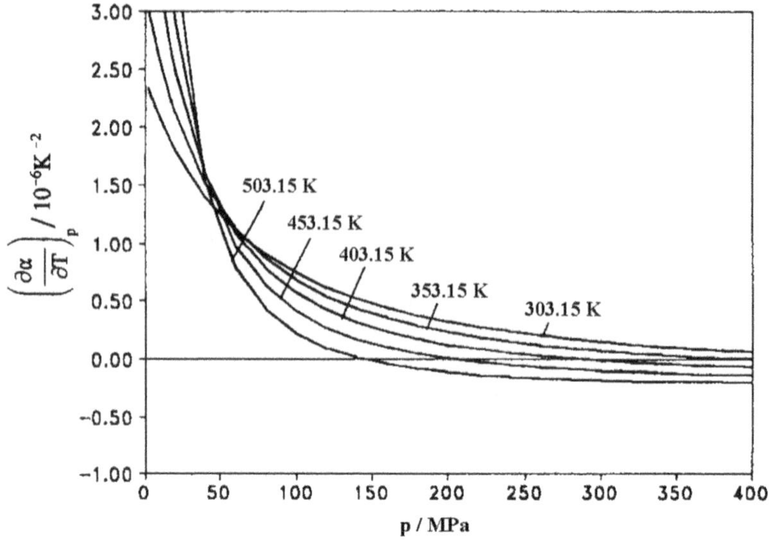

Figure 14.6 Selected isotherms of $d\alpha/dT$ for liquid hexan-1-ol.[34]

nonan-1-ol, and decan-1-ol have also been investigated with a pressure-controlled scanning calorimeter,[36] but the results cannot be compared to the above results for hexan-1-ol, because the ranges of temperature 278.15–348.15 K and of pressure 5–55 MPa were too small to see the crossing of α isotherms (change of sign in $d\alpha/dT$).

Water is another self-associated liquid where wide pressure and temperature ranges are required to investigate its properties. The unusual behaviour of thermal expansion isotherms for water obtained by piezothermal techniques is presented in Figure 14.8[12] One can see that, depending on temperature, the thermal expansion coefficient can decrease or even increase with pressure. Near $T=323$ K, the thermal expansion coefficient for water hardly changes with pressure.[12] The change of sign of $d\alpha/dT$ for liquid water occurs at very high pressures.

Quinoline and *m*-cresol are self-associated liquids and their binary mixtures form strong intermolecular complexes. Quinoline and *m*-cresol are also components of coal-derived liquids and thus the determination of their properties over wide pressure and temperature ranges is considered important. Figure 14.9 presents selected α isotherms for liquid quinoline.[37]

Surprisingly, the isotherms cross at a unique pressure of 60 ± 0.4 MPa, where $\alpha = (0.65 \pm 0.02) \times 10^{-3}$ K over the whole temperature range under investigation. However, the temperature dependence of α at high pressures is greater than for *n*-hexane. This has an influence on the shapes of isotherms of pressure effects on heat capacity, which for quinoline exhibit minima which move to higher pressures as the temperature decreases.[37] This is the opposite behaviour to that found for *n*-hexane.[16] The difference between quinoline and *n*-hexane can be explained by a weak self-association

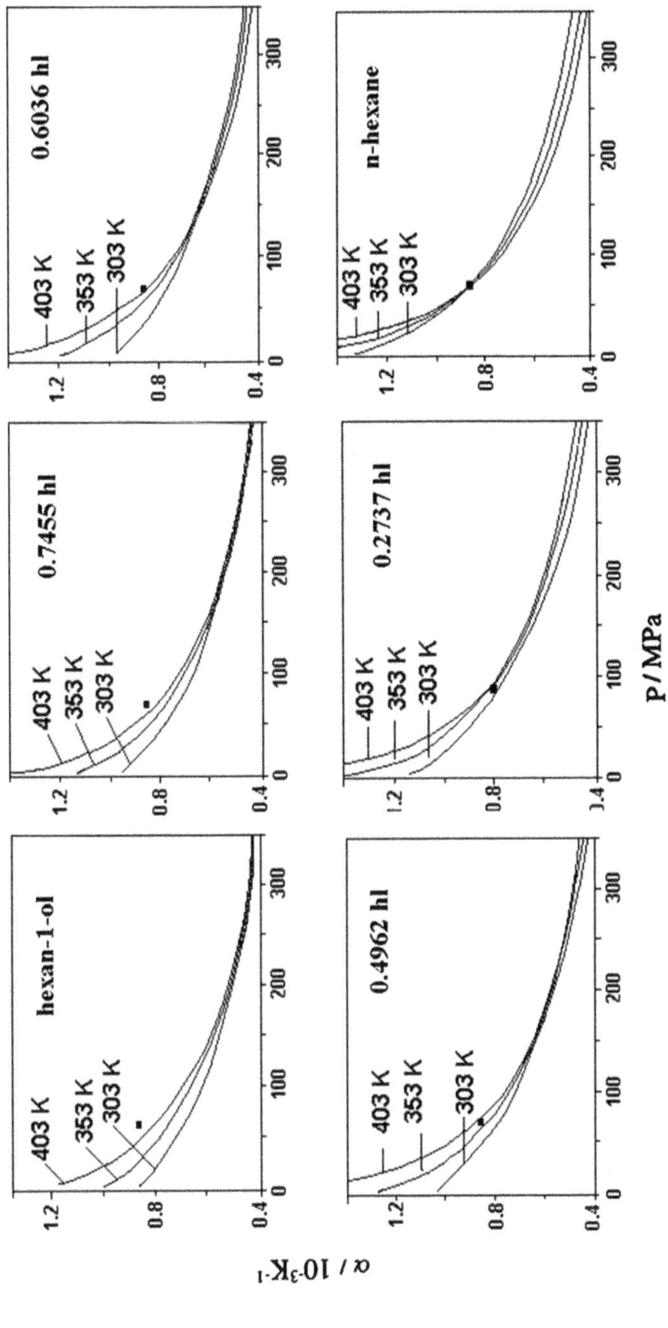

Figure 14.7 Comparison of α isotherms for hexan-1-ol[34] and *n*-hexane,[16] and for their binary mixtures.[35] ■ is the crossing point of α isotherms for *n*-hexane, hl refers to hexan-1-ol.

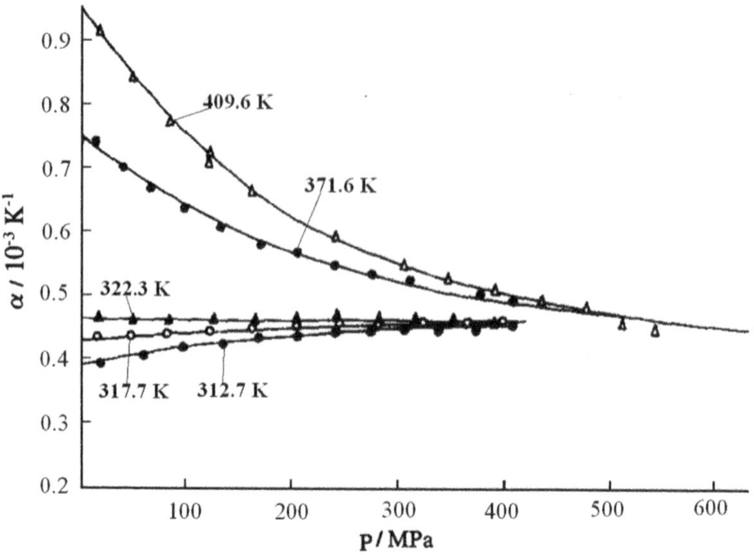

Figure 14.8 Selected isotherms of the coefficient of thermal expansion for liquid water.[12]

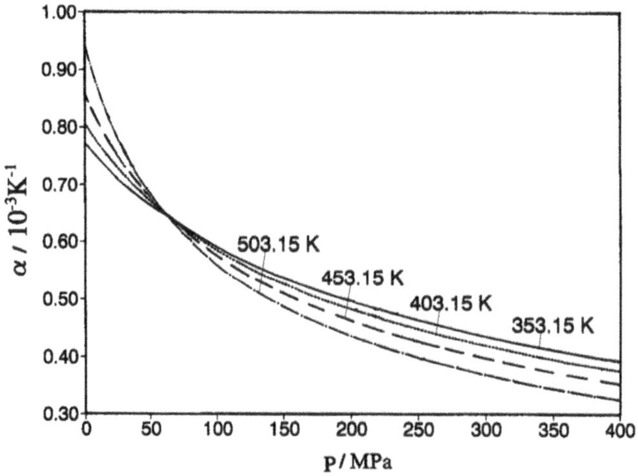

Figure 14.9 Selected isotherms of the coefficient of thermal expansion for liquid quinoline.[37]

of quinoline. With increasing temperature, the associating bonds are broken and quinoline behaves like a simple liquid. This is in accordance with a similar conclusion based on measurements of heat of dilution of quinoline in *n*-decane.[38]

Figure 14.10 presents selected thermal expansion isotherms for liquid *m*-cresol determined over wide pressure and temperature ranges.[39] It can be seen that there is no unique crossing point, the isotherms start to cross near

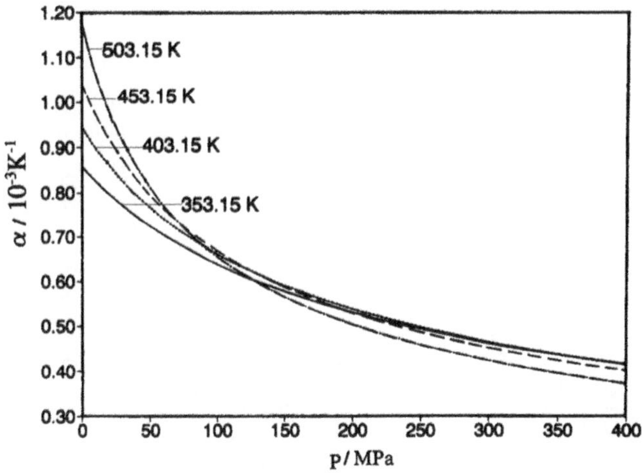

Figure 14.10 Selected isotherms of the coefficient of thermal expansion for liquid *m*-cresol.[39]

100 MPa, but the crossing points are temperature dependent. The isotherms cross at lower pressures as the temperature increases. Similarly to hexan-1-ol, this effect can be explained by the self-association equilibrium in the liquid *m*-cresol; the strong self-association of *m*-cresol was also observed in other studies.[38,40]

Figure 14.11 presents selected α isotherms of binary mixtures[41] of *m*-cresol with quinoline and compares them to those for the pure components (*m*-cresol[39] and quinoline[37]).

The most striking observation is that the mixture of *m*-cresol with quinoline near the 2 : 1 mole ratio behaves like a simple liquid without association: the isotherms exhibit a unique crossing point (■) at $p=(170\pm1.6)$ MPa and $\alpha=(5.27\pm0.01)\times10^{-4}$ K^{-1}. Such behaviour can be explained by the formation of very strong 2 : 1 complexes between *m*-cresol and quinoline. The position of this equilibrium is not much disturbed by changes in pressure and temperature over the investigated ranges and the mixture behaves like a simple liquid. Such a reasoning is confirmed by the known fact that in this binary system near the 2 : 1 mol ratio the excess enthalpy is at a maximum with a very large value of -7.7 kJ mol^{-1}.[42] For this reason, in this 2 : 1 mixture, the liquid phase is composed of strongly bound intermolecular complexes which behave like the molecules of a liquid without association. Thus, the macroscopic properties of such a mixture are similar to those of non-polar simple liquids. At other compositions, the equilibria among the 2 : 1 and 1 : 1 complexes between *m*-cresol and quinoline[38] and self-associated complexes of *m*-cresol[40] are shifted by pressure and temperature and preclude a unique crossing point of α isotherms.

Thermal expansion isotherms have also been determined calorimetrically for other polar liquids such as propionitrile, nitromethane, nitroethane, nitropropane, benzonitrile and nitrobenzene.[42] However, all the

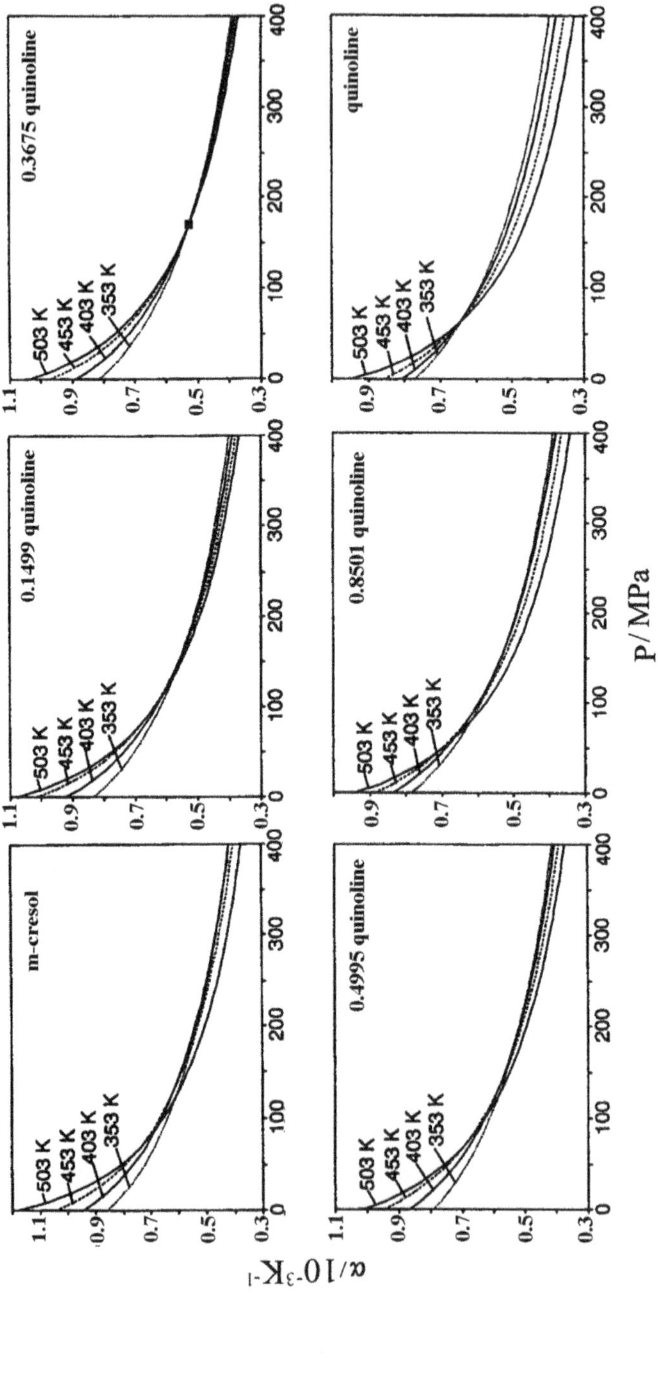

Figure 14.11 Comparison of α isotherms for *m*-cresol,[39] quinoline[37] and their selected binary mixtures.[41]

measurements have been performed over narrow pressure (5–55 MPa) and temperature (278.15–348.15 K) ranges and thus any conclusive comparison with other complex liquids discussed above cannot be done, although the α isotherms for benzonitrile and nitrobenzene exhibit near unique crossing points, near 25 MPa and 30 MPa, respectively.

14.4.3 Ionic Liquids

Ionic liquids have recently attracted an enormous research interest and have been intensively investigated with various techniques. Also, the calorimetric technique based on the Maxwell relation Equation (14.1) has already been used in investigation of a number of ionic liquids.[43,44]

Figure 14.12 presents thermal expansion isotherms for selected ionic liquids derived from fitting equations established on the basis of experimental data.[43,44] At first glance, two features are striking: α values decrease with temperature ($d\alpha/dT<0$) and the α isotherms are almost parallel. The first observation can be compared to the behaviour of α isotherms above the crossing point, observed for simple liquids, where the thermal expansion coefficient decreases with rising temperature. Such a comparison shows that the ionic liquids, even at low pressures, behave like a close-packed fluid in which thermal fluctuations are small and cohesion forces predominate.[45] This is confirmed by an observation that the isobaric heat capacity of ionic

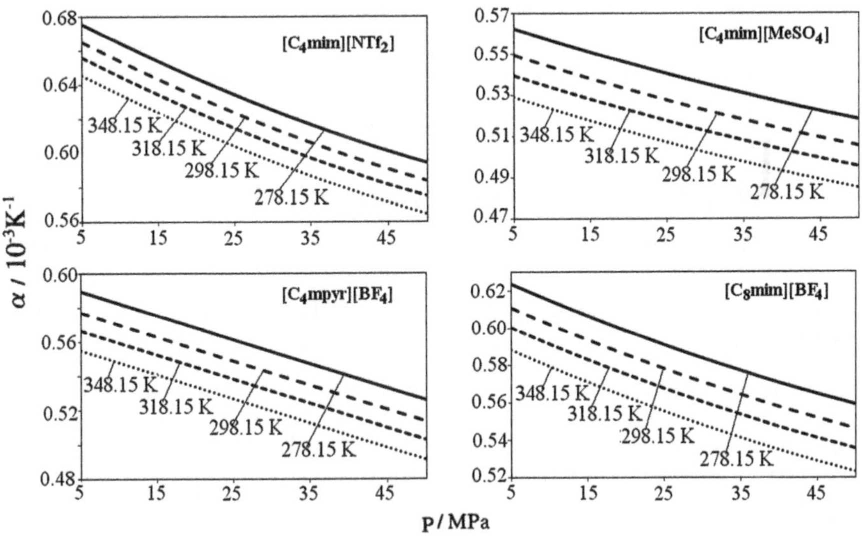

Figure 14.12 Isotherms of the coefficient of thermal expansion for selected ionic liquids.[43,44] Cations: [C_4mim] stands for 1-butyl-3-methylimidazolium, [C_4mpyr] stands for 1-butyl-3-methylpyridinium, [C_8mim] stands for 1-octyl-3-methylimidazolium. Anions: [NTf_2] stands for bis(trifluoromethylsulfonyl)imide, [$MeSO_4$] stands for methylsulfate and [BF_4] stands for tetrafluoroborate.

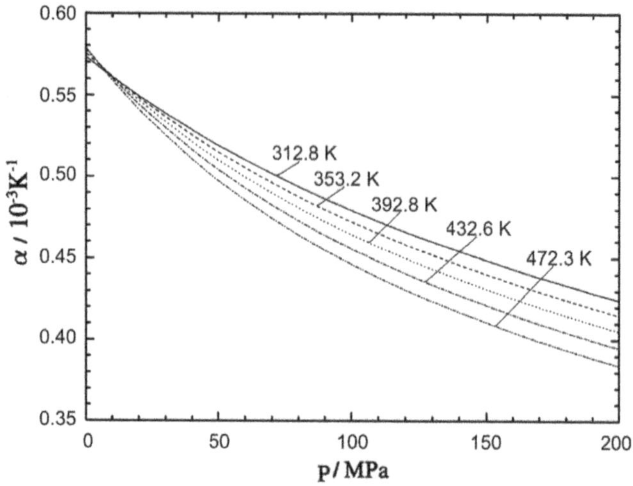

Figure 14.13 Isotherms of the coefficient of thermal expansion for 1-butyl-3-methylimidazolium hexafluorophosphate ([bmim+][PF_6−], ionic liquid).[47]

liquids rises with pressure, similarly to the simple liquids at high pressures.[46] The reported parallelism of α isotherms is most probably only apparent because the temperature and pressure ranges investigated were rather narrow, 278.15–348.15 K and 5–50 MPa, respectively. Figure 14.13 presents a set of α isotherms for an ionic liquid, obtained with a precise volumetric technique over much wider pressure and temperature ranges.[47] One can see that the general behaviour is very similar to the one observed for simple liquids, although the crossing point of α isotherms appears at a very low pressure.

14.4.4 Properties Near the Critical Point

Figure 14.14 presents thermomechanical coefficients determined with a volume controlled scanning transitiometer[7] at $T = 507.2$ K for liquid n-hexane, near its liquid–vapour critical point. Details of the derivation from experimental data are given in section 14.3. The most important observation is that the coefficient $\beta = \left(\frac{\partial P}{\partial T}\right)_V$ can be determined over the whole pressure range on the basis of Maxwell relation Equation (14.2), while at low pressures when approaching the saturation line Maxwell relation Equation (14.1) cannot be applied to determine α, because the compressibility is too high and the required pressure variations cannot be realised. It also worth noting that the value of $\beta = 0.0406$ MPa K^{-1} near the saturation line (see Table 14.1) is very close to the values obtained from a correlation equation for saturation pressure of n-hexane (0.040 MPa K^{-1})[16] and from experimental isochores (0.046 MPa K^{-1})[48] Another argument for the correctness of the presented results is that the values of κ_T derived directly from the measured volume

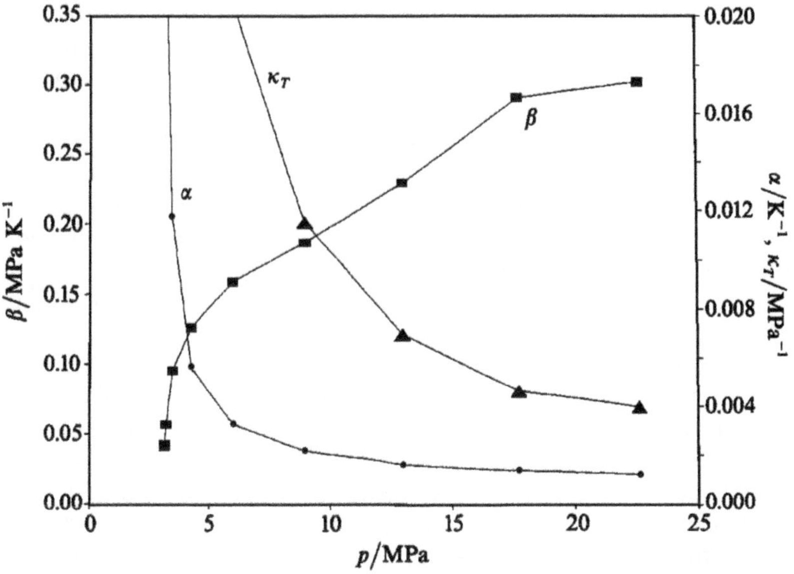

Figure 14.14 Thermomechanical coefficients α, κ_T and β for liquid n-hexane at $T = 507.2$ K (near its liquid–vapour critical point: $T_{cr} = 507.8$ K, $p_{cr} = 3.09$ MPa) determined with a volume controlled scanning transitiometer.[7]

and pressure variations (Equation (14.22) and Table 14.1) are very close to the values derived through the thermodynamic relation Equation (14.3) from the measured values of α and β. It is worth noting that the described transitiometric experiment confirms a known thermodynamic fact that both α and κ_T, second derivatives of G, go to infinite when approaching the critical point, while their ratio β keeps a finite value.

14.5 Conclusions

The present chapter is focused on the experimental aspects of the use of Maxwell relations in the investigation of liquids of various natures over wide pressure and temperature ranges, from near the solid–liquid line up to near the liquid–vapour critical point. For dense liquid phases, the preferred techniques are those based on isothermal variations of pressure, while for regions near the saturation line, and especially near the critical point, the preferred variable is volume. The direct measurements of the second order thermodynamic derivatives allow one to distinguish, on the (p, T) plane, the regions for a given liquid, where its properties are dominated by thermal fluctuations, repulsion forces, association and/or other specific interactions. The data collected for simple liquids reveal that they exhibit a characteristic pressure, at which the α isotherms cross. Below that point, at low pressures thermal fluctuations dominate the properties of the liquid phase. At high pressures, above the crossing point of α isotherms, the properties of the liquid are dominated by short-range repulsive forces, and fluctuations are negligible. For the alkanes series, the pressure of the crossing point of α isotherms decreases with an increasing alkane chain length, thus the region of dominating thermal fluctuations becomes smaller with increasing alkane chain length. For associated liquids the calorimetric techniques based on the Maxwell relations permit one to determine how the increasing temperature breaks the association bonds and the increasing pressure shifts the equilibrium towards associated species. Ionic liquids practically do not exhibit any region where the thermal fluctuations dominate. Even at low pressures, ionic liquids behave like a close-packed fluid with repulsive interactions, the behaviour of their thermal expansion isotherms is practically the same as for dense simple liquids above the crossing point of α isotherms.

The second order derivatives determined experimentally also form an important basis for verification of equations of state, simulations and modelling of the liquid phase.[28,29,31–33,45,49] The coefficient of thermal expansion and its temperature derivative are important properties in the determination of pressure effects on the heat capacity of liquids[50] and on other thermodynamic functions, for example for liquids of technological importance,[51,52] otherwise difficult to measure over wide pressure ranges.

References

1. S. L. Randzio, *Thermochim. Acta*, 1985, **89**, 215.
2. Ph. Pruzan, L. Ter Minassian, P. Figuiere and H. Szwarc, *Rev. Sci. Instrum.*, 1976, **47**, 66.
3. L. Ter Minassian and Ph. Pruzan, *J. Chem.Thermodyn.*, 1977, **9**, 375.
4. S. L. Randzio, *J. Phys. E: Sci. Instrum.*, 1983, **16**, 691.
5. S. L. Randzio, *J. Phys. E: Sci. Instrum.*, 1984, **17**, 1058.
6. S. L. Randzio, J.-P. E. Grolier and J. R. Quint, *Rev. Sci. Instrum.*, 1994, **65**, 960.
7. S. L. Randzio, J.-P. E. Grolier and J. R. Quint, *High Temp. – High Pressures*, 1998, **30**, 1025.
8. P. Kujawa and F. M. Winnik, *Macromolecules*, 2001, **34**, 4130.
9. S. L. Randzio, *Thermochim. Acta*, 2003, **398**, 75.
10. M. Chorazewski, J.-P. E. Grolier and S. L. Randzio, *J. Chem. Eng. Data*, 2010, **55**, 5489.
11. S. L. Randzio, *Chem. Soc. Rev.*, 1996, **25**, 383.
12. L. Ter Minassian, Ph. Pruzan and A. Soulard, *J. Chem. Phys.*, 1981, **75**, 3064.
13. S. L. Randzio, D. J. Eatough, E. A. Lewis and L. D. Hansen, *J. Chem. Thermodyn.*, 1988, **20**, 937.
14. D. Gonzalez-Salgado, J. L. Valencia, J. Troncoso, E. Carballo, J. Peleteiro, L. Romani and D. Bessieres, *Rev. Sci. Instrum.*, 2007, **78**, 055103/1.
15. S. L. Randzio and J. Suurkuusk, in *Biological Microcalorimetry*, ed. A. E. Beezer, Academic Press, London, 1980, pp. 311–341.

16. S. L. Randzio, J.-P. E. Grolier, J. R. Quint, D. J. Eatough, E. A. Lewis and L. D. Hansen, *Int. J. Thermophys.*, 1994, **15**, 415.
17. D. Bessieres, Th. Lafitte, J.-L. Daridon and S. L. Randzio, *Thermochim. Acta*, 2005, **428**, 25.
18. M. Zabransky, V. Ruzicka, V. Majer and E. S. Domalski, *Heat Capacity of Liquids. Critical Review and Recommended Values*. Monograph No 6, Vols. I and II, ACS and AIP for NIST, 1996.
19. P. Navia, J. Troncoso and L. Romaní, *J. Chem. Thermodyn.*, 2008, **40**, 1607.
20. L. Rodier-Renaud, S. L. Randzio, J.-P. E. Grolier, J. R. Quint and J. Jarrin, *J. Polym. Sci., Part B: Polym. Phys.*, 1996, **14**, 1229.
21. Ph. Pruzan, *J. Phys. (Paris), Lett.*, 1984, **45**, L273.
22. S. L. Randzio, *Thermochim. Acta*, 1987, **121**, 463.
23. Ph. Pruzan, *J. Chem. Thermodyn.*, 1991, **23**, 247.
24. C. Alba, L. Ter Minassian, A. Denis and A. Soulard, *J. Chem. Phys.*, 1985, **82**, 384.
25. P. Navia, J. Troncoso and L. Romani, *J. Chem. Eng. Data*, 2010, **55**, 2173.
26. A. H. Fuchs, Ph. Pruzan and L. Ter Minassian, *J. Phys. Chem. Solids*, 1979, **40**, 369.
27. L. Ter Minassian, K. Bouzar and C. Alba, *J. Phys. Chem.*, 1988, **92**, 487.
28. S. L. Randzio and U. K. Deiters, *Ber. Bunsenges. Phys. Chem.*, 1995, **99**, 1179.
29. S. L. Randzio, *Chem. Soc. Rev.*, 1995, **24**, 359.
30. N. F. Carnahan and K. E. Starling, *J. Chem. Phys.*, 1969, **51**, 635.
31. S. L. Randzio, *Phys. Lett.*, 1986, **117**, 473.
32. U. K. Deiters and S. L. Randzio, *Fluid Phase Equilib.*, 1995, **103**, 199.
33. J. Troncoso, P. Navia, L. Romani, D. Bessieres and T. Lafitte, *J. Chem. Phys.*, 2011, **134**, 094502-1.
34. S. L. Randzio, J.-P. E. Grolier and J. R. Quint, *Fluid Phase Equilib.*, 1995, **110**, 341.
35. S. L. Randzio, J.-P. E. Grolier and J. R. Quint, *Int. J. Thermophys.*, 1997, **18**, 753.
36. P. Navia, J. Troncoso and L. Romani, *J. Chem. Thermodyn.*, 2010, **42**, 23.
37. S. L. Randzio, D. J. Eatough, E. A. Lewis and L. D. Hansen, *Int. J. Thermophys.*, 1996, **17**, 405.
38. D. J. Eatough, S. L. Wolfley, L. J. Dungan, E. A. Lewis and L. D. Hansen, *Energy Fuels*, 1987, **1**, 94.
39. S. L. Randzio, E. A. Lewis, D. J. Eatough and L. D. Hansen, *Int. J. Thermophys.*, 1995, **16**, 883.
40. E. M. Woolley, J. G. Travers, B. O. Erno and L. G. Hepler, *J. Phys. Chem.*, 1971, **76**, 359.
41. S. L. Randzio, L. D. Hansen, E. A. Lewis and D. J. Eatough, *Int. J. Thermophys.*, 1997, **18**, 1183.
42. P. Navia, J. Troncoso and L. Romani, *J. Chem. Eng. Data*, 2010, **55**, 1537.
43. P. Navia, J. Troncoso and L. Romani, *J. Chem. Eng. Data*, 2010, **55**, 590.
44. P. Navia, J. Troncoso and L. Romani, *J. Chem. Eng. Data*, 2010, **55**, 595.

45. J. Troncoso, C. A. Cerdeirina, P. Navia, Y. A. Sanmamed, D. Gonzalez-Salgado and L. Romaní, *J. Phys. Chem. Lett.*, 2010, **1**, 211.
46. Y. A. Sanmamed, P. Navia, D. Gonzalez-Salgado, J. Troncoso and L. Romanı, *J. Chem. Eng. Data*, 2010, **55**, 600.
47. H. Machida, Y. Sato and R. L. Smith Jr., *Fluid Phase Equilib.*, 2008, **264**, 147.
48. D. S. Kurumov and B. A. Grigoriev, *Zh. Fiz. Khim.*, 1982, **56**, 551.
49. M. Taravillo, V. G Baonza, M. Caceres and J. Nunez, *J. Phys.: Condens. Matter*, 2003, **15**, 2979.
50. S. L. Randzio, Scanning Transitiometry and its use to Determine Heat Capacities of Liquids at High Pressures, in *Heat Capacities: Liquids, Solutions and Vapours*, ed. E. Wilhelm and T. Letcher, The Royal Society of Chemistry, 2010, ch. 8, pp. 153–184.
51. S. L. Randzio, *Thermochim. Acta*, 1987, **115**, 83.
52. M. Chorazewski, F. Dergal, T. Sawaya, I. Mokbel, J.-P. E. Grolier and J. Jose, *Fuel*, 2013, **105**, 440.

CHAPTER 15

Volumetric Properties and Thermodynamic Response Functions of Liquids and Liquid Mixtures

CARLOS LAFUENTE,*[a] IGNACIO GASCÓN,[a] CLAUDIO A. CERDEIRIÑA[b] AND DIEGO GONZÁLEZ-SALGADO[b]

[a] Departamento de Química Física, Universidad de Zaragoza, 50009, Zaragoza, Spain; [b] Departamento de Física Aplicada, Universidad de Vigo - Campus del Agua, 32004 Ourense, Spain
*Email: celadi@unizar.es

15.1 Introduction

The thermodynamic response functions or second-order derivatives of the thermodynamic potential are of wide relevance in characterising liquids and liquid mixtures. Some of these functions, like the isobaric expansibility α_p or compressibility, isothermal κ_T and isentropic κ_S, are volume derivatives, while others, like heat capacity (isobaric C_p and isochoric C_V) relate to the thermal response of the system. Specifically:

$$\kappa_T \equiv -\frac{1}{V}\left(\frac{\partial V}{\partial p}\right)_T, \ \kappa_S \equiv -\frac{1}{V}\left(\frac{\partial V}{\partial p}\right)_S, \ \alpha_p \equiv \frac{1}{V}\left(\frac{\partial V}{\partial T}\right)_p, \tag{15.1}$$

$$C_p \equiv T\left(\frac{\partial S}{\partial T}\right)_p, \ C_V \equiv T\left(\frac{\partial S}{\partial T}\right)_V, \tag{15.2}$$

Volume Properties: Liquids, Solutions and Vapours
Edited by Emmerich Wilhelm and Trevor M. Letcher

Published by the Royal Society of Chemistry, www.rsc.org

where V, S, T, p refer to the volume, entropy, temperature and pressure of the system.

Thermodynamic response functions are very sensitive to phenomena at the molecular level. Probably the best illustration of this fact is the diverging behaviour near critical points as a result of the long-range nature of fluctuations.[1,2] The anomalous thermodynamic properties of water at low temperatures is another example: the rise of C_p, κ_T and $-\alpha_p$ as temperature is lowered is attributed to an increase of local tetrahedral order in the hydrogen-bond network.[3] This has been claimed to be the mechanism leading to the hypothesised second, liquid–liquid critical point for deeply super-cooled water.[4] Molecular association *via* hydrogen bonding in other self-associated liquids (*e.g.*, alcohols or amines) is reflected in response functions, especially in the isobaric heat capacity.[5,6] These phenomena are also relevant in mixtures of aqueous solutions of alcohols[7] and also in alcohol–alkane mixtures.[8–11]

Some response functions are determined from others using the following two exact thermodynamic relations:

$$\kappa_T = \kappa_S + \frac{TV\alpha_p^2}{C_p}, \tag{15.3}$$

$$C_p = C_V + \frac{TV\alpha_p^2}{\kappa_T}. \tag{15.4}$$

Standard calorimetric techniques allow one to obtain the isobaric heat capacity as a function of temperature.[12] By combining density and speed of sound u data, the isentropic compressibility is readily determined *via* $\kappa_s = 1/\rho u^2$ if it is assumed that the effect of ultrasonic absorption is vanishingly small. Direct measurements of α_p and κ_T have been made in the past using dilatometric and piezothermal techniques, respectively. These two properties can also be determined with great accuracy from a so-called "expansion procedure".[13] With the aid of transitiometry, $C_p(T,p)$, $\alpha_p(T,p)$, and $\kappa_T(T,p)$ data can be obtained directly over wide temperature and pressure ranges.[14,15] Work on the temperature dependence of isochoric heat capacities using calorimetric techniques along isochores has been reported.[16]

It is quite usual and useful, nevertheless, to derive α_p and κ_T from primary ρpT data. The use of equations of state (EOS) with a physical background and a semi-phenomenological modified Tait equation are primary approaches. The reliability of calculated $\alpha_p(T,p)$ and $\kappa_T(T,p)$ data from those EOS rests on an appropriate parameterisation, which, in turn, requires as many experimental data points as possible. Furthermore, the higher the quality of data, the better is the performance of the EOS. Versions of the SAFT family of EOS as applied to alkanes or alcohols for which good data exist provide reliable $\alpha_p(T,p)$ and $\kappa_T(T,p)$ values.[17–20] Equally satisfactory is the modified Tait equation, which has been used by Cibulka and co-workers to correlate $\rho(T,p)$ data for a large number of pure liquids.[21–28] Along these lines, a more refined strategy is TRIDEN correlating system,[29] in which the Tait equation is combined with a modified Rackett equation,[30] for the liquid density at coexistence points and Wagner equation,[31] for vapour pressures.

Purely phenomenological multi-parameter EOSs[32] are designed with the objective of representing not only volumetric properties and the directly associated thermodynamic derivatives but also heat capacities, speeds of sound, *etc.* over wide ranges of properties. To do so, in addition to ρpT data, properties in the hypothetical ideal-gas state (as determined theoretically or from spectroscopic data) are required.

A different perspective emerges when a large database is lacking. This is the subject of the first part of this chapter, which deals with the determination of volumetric properties and thermodynamic response functions of "novel" liquids (or liquid mixtures) for which the necessary information is unknown. One central issue, discussed in Section 15.2.1, is the determination of α_p and κ_T from primary $\rho(T,p)$ data. The remainder of Section 15.2 summarises some approaches for pure fluids. In the second part of the chapter (Section 15.3), correlating schemes for ρpTx data of mixtures as well as the calculation of excess properties are discussed.

15.2 Pure Fluids

15.2.1 Derived Properties from ρpT Data

Two common situations are the derivation of $\kappa_T(p)$ and $\alpha_p(T)$ from only a few experimental datapoints of $\rho(p)$ along an isotherm and $\rho(T)$ along an isobar, respectively. In the former case, since it has a physical basis, the Tait equation proves fairly reliable in obtaining $\kappa_T(p)$ over the working pressure range. That is not the situation for $\rho(T)$ in the liquid state, for which a purely correlating, phenomenological equation must be used. This strategy proves reliable if one wishes to obtain α_p at a single temperature. It is, in general, inadequate when seeking the temperature dependence of α_p. The reason is that various functional forms that seemingly account well for $\rho(T)$ can give strikingly different results for $\alpha_p(T)$.

An alternative approach, which avoids the use of any fitting equation, is a so-called incremental procedure.[33] It consists of calculating the derivative $(\partial\rho/\partial T)_p$ by evaluating the density change $\Delta\rho$ corresponding to a ΔT temperature interval. The key point for ensuring the correctness of this calculation is the choice of an appropriate ΔT value. By "appropriate", we understand a ΔT interval over which ρ varies linearly within the resolution of the density data. On the other hand, ΔT must not be too small with a view to avoiding an undesired increase of the uncertainty in the $(\partial\rho/\partial T)_p \approx (\Delta\rho/\Delta T)_p$ value.

The incremental procedure has proved successful for obtaining $\alpha_p(T)$ of pure common organic liquids and mixtures (see Figure 15.1).[33,34] Quite remarkably, it has been used for elucidating the unusual temperature

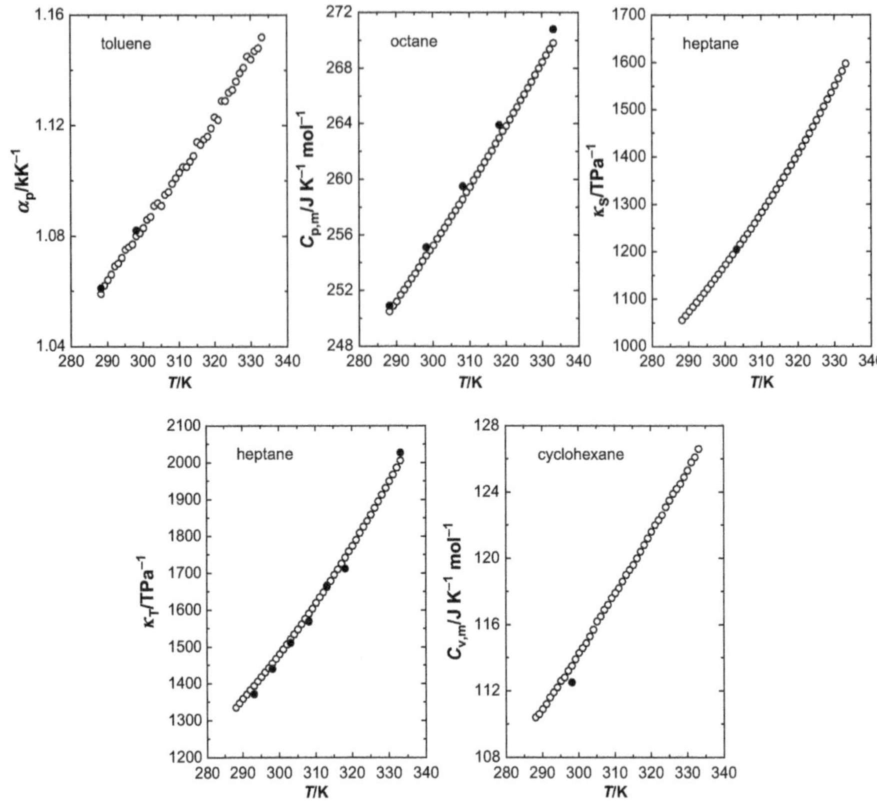

Figure 15.1 Plot of isobaric expansibility α_p, isobaric heat capacity $C_{p,m}$, isentropic compressibility κ_S, isothermal compressibility κ_T, and isochoric heat capacity $C_{v,m}$ against temperature T for several organic liquids: (●) literature data; (○) data obtained from measurements of density, speed of sound, and isobaric heat capacity at atmospheric pressure, see text for details.
The graphs have been drawn from data in reference 33.

dependence of $\alpha_p(T)$ for room temperature ionic liquids, which, in contrast to the commonly observed behaviour, is characterised by the term $(\partial\alpha_p/\partial T)_p$, being small and negative at atmospheric pressure.[35,36] This is a clear-cut example for which the use of a correlating $\rho(T)$ equation can be misleading but the incremental procedure proves quite correct.

15.2.2 Thermodynamic Response Functions

Very common and well-established experimental techniques are available for determination of the density $\rho(T)$, speed of sound $u(T)$ and isobaric heat capacity $C_p(T)$ at atmospheric pressure. As noted above, $\alpha_p(T)$ can be derived from $\rho(T)$, while the isentropic compressibility can be readily obtained from the Laplace equation: $\kappa_s = 1/\rho u^2$. Then, using Equations (15.3) and (15.4),

$\kappa_T(T)$ and $C_V(T)$ can be obtained. In this simple, straight forward way, a complete set of thermodynamic response functions as a function of temperature can be obtained. And, as has been shown (see Figure 15.1),[33] it has proved to be reasonably accurate for a number of common organic liquids.

Difficulties arise when such a thermodynamic characterisation is extended to high pressures. A natural approach involves use of $\rho(T,p)$, $u(T,p)$, and $C_p(T,p)$ data. Equations (15.3) and (15.4) indicate that knowledge of any set of three of the thermodynamic response functions under study allows one to obtain the two remaining ones. Then, since not only $\kappa_s(T,p)$ and $\alpha_p(T,p)$ but also $\kappa_T(T,p)$ are derived from $\rho(T,p)$ and $u(T,p)$ data, a test for thermodynamic consistency is feasible. To the best of our knowledge, no such task has been reported.

Because there are only a few laboratories in the world in which accurate high-pressure C_p data can be determined, alternative approaches have been adopted. One of them uses $\rho(T,p)$ and $u(T,p)$data as primary information.[37] From the derived values of $\kappa_s(T,p)$, $\alpha_p(T,p)$, and $\kappa_T(T,p)$, heat capacities can be calculated using Equations (15.3) and (15.4). This methodology provides results with reasonable accuracy for the volumetric second-order derivatives but it is, in general, not valid for heat capacities.[38] This last point is illustrated in Figure 15.2 for the isobaric heat capacity; a similar conclusion is reached for the isochoric heat capacity since its values depend strongly on the isobaric heat capacity through Equation (15.4). Secondly, from $\rho(T,p)$ and $C_p(T,p_{\text{ref}})$, where the reference pressure p_{ref} is usually

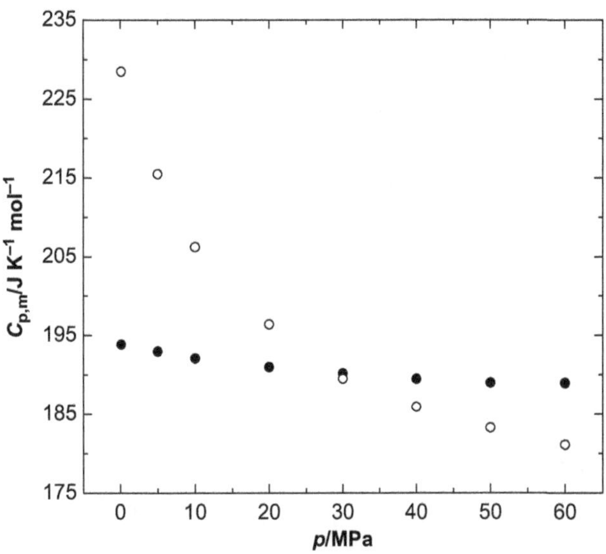

Figure 15.2 Isobaric heat capacity $C_{p,m}$ of hexane plotted against pressure p: (●) literature data; (○) data computed using $\rho(T,p)$ and $u(T,p)$ data. The graph has been drawn from data in reference 38.

atmospheric pressure, one can obtain $C_p(T,p)$ from the exact thermodynamic relation:

$$\left(\frac{\partial C_p}{\partial p}\right)_T = -\frac{T}{\rho}\left[\alpha_p^2 + \left(\frac{\partial \alpha_p}{\partial T}\right)_p\right]. \tag{15.5}$$

By combining the $C_p(T,p)$ obtained with the derived properties $\alpha_p(T,p)$ and $\kappa_T(T,p)$ and using Equations (15.3) and (15.4), the isentropic compressibility and the isochoric heat capacity can be evaluated.[38] This procedure can be used for providing data for $\alpha_p(T,p)$, $\kappa_T(T,p)$, and $\kappa_s(T,p)$ with acceptable accuracy. Results for the heat capacities usually show a great uncertainty[38] as is seen in Figure 15.3. Note that, although the uncertainty in the isobaric heat capacity is great, its effect on the isentropic compressibility is small (Figure 15.3). Minimal information is required by a method based on a "predictor–corrector" algorithm[39–41] that considers $\rho(T,p_{\mathrm{ref}})$, $C_p(T,p_{\mathrm{ref}})$, and

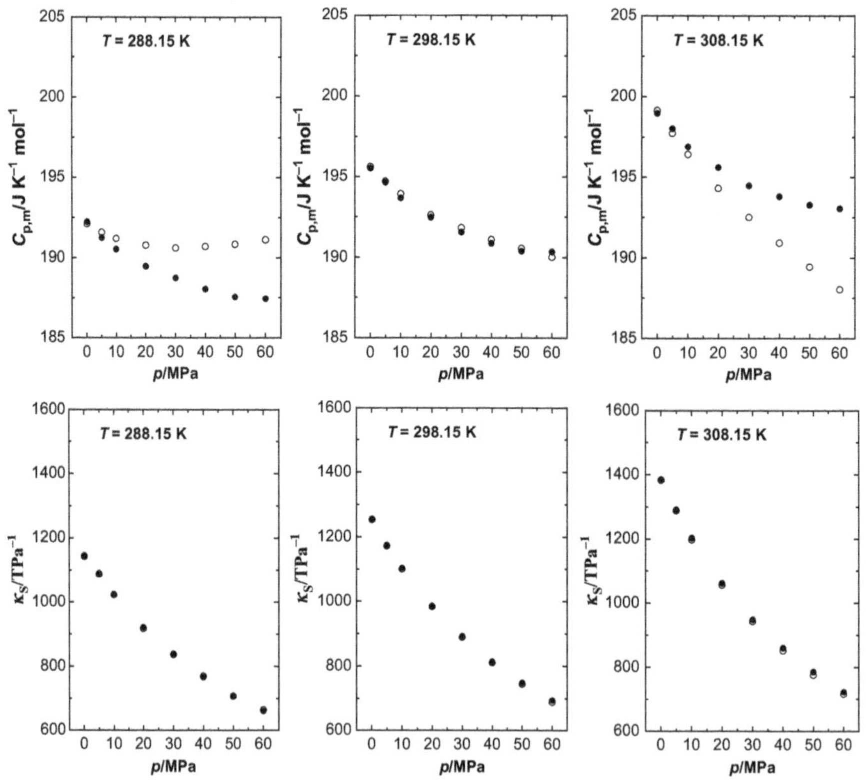

Figure 15.3 Isobaric heat capacity $C_{\mathrm{p,m}}$ (top) and isentropic compressibility κ_{S} (bottom) of hexane plotted against pressure p for several temperatures: (●) literature data; (○) data evaluated using Equation (15.5), see text for details.
The graphs have been drawn from data in reference 38.

$u(T,p)$ as input data, from which the remaining properties can be calculated. It is based on the following relation:

$$\rho(T,p) = \rho(T,p_{ref}) + \int_{p_{ref}}^{p} u^{-2}(T,p)dp + T\int_{p_{ref}}^{p} \left(\frac{\alpha_p^2(T,p)}{C_p(T,p)}\right) dp, \qquad (15.6)$$

where the first two terms of the right hand side of the equation can be easily computed from the input data. The computation of the third term is made using the "predictor–corrector" algorithm[39] that allows one to compute the values at high pressures starting from the value of the integrand at the lowest pressure (input data). Thus, with this evaluation, $\rho(T,p)$, $\alpha_p(T,p)$, and $C_p(T,p)$ are obtained. Furthermore calculating $\kappa_T(T,p)$ and deriving $\kappa_T(T,p)$, the isochoric heat capacity can be obtained. This method gives good results for volumetric properties but, in general, poor results for heat capacities,[38] as shown in Figure 15.4. A fourth approach[14] uses values of $\rho(T_{\text{ref}},p)$, $C_p(T,p_{\text{ref}})$ and $\alpha_p(T,p)$ as primary data. Here the density $\rho(T,p)$ is evaluated from the following thermodynamic relation:

$$\frac{M}{\rho(p,T)} = \frac{M}{\rho(p,T_{\text{ref}})} \exp\left[\int_{T_{ref}}^{T} \alpha_p(T,p)dT\right]. \qquad (15.7)$$

where M is the molar mass of the fluid. The isothermal compressibility is computed from these density values as was previously explained and Equation (15.5) is used to calculate the isobaric heat capacity from $\rho(T,p)$ and $\alpha_p(T,p)$. The isochoric heat capacity and the isentropic compressibility

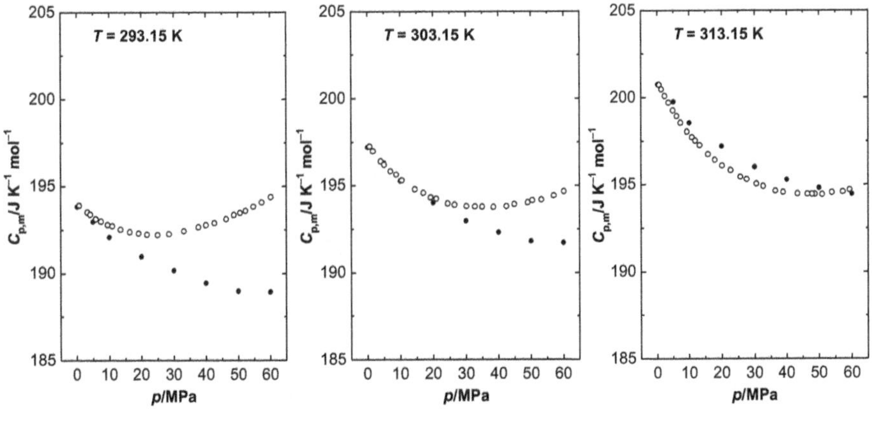

Figure 15.4 Isobaric heat capacity $C_{p,m}$ of hexane plotted against pressure p for several temperatures: (●) literature data; (○) data obtained from the "predictor–corrector" algorithm method, see text for details. The graphs have been drawn from data in reference 38.

can then be calculated using Equations (15.3) and (15.4). This procedure provides a reasonable description of the whole set of thermodynamic response functions.[14]

The same set of properties can be obtained using $\rho(T,p)$ and $C_p(T,p)$ as input data. In this case, $\alpha_p(T,p)$ and $\kappa_T(T,p)$ are derived from the density data and with $C_p(T,p)$ the isentropic compressibility and the isochoric heat capacity can be obtained. This method allows an accurate description of the behaviour of the thermodynamic response function with temperature and pressure,[42] as illustrated in Figure 15.5. This represents the best option if an accurate general characterisation is needed. In that sense, it can be concluded that, despite the central role that volumetric properties play in the evaluation of the response functions, the measurement of the isobaric heat capacity is, however, fundamental for an accurate overall description.

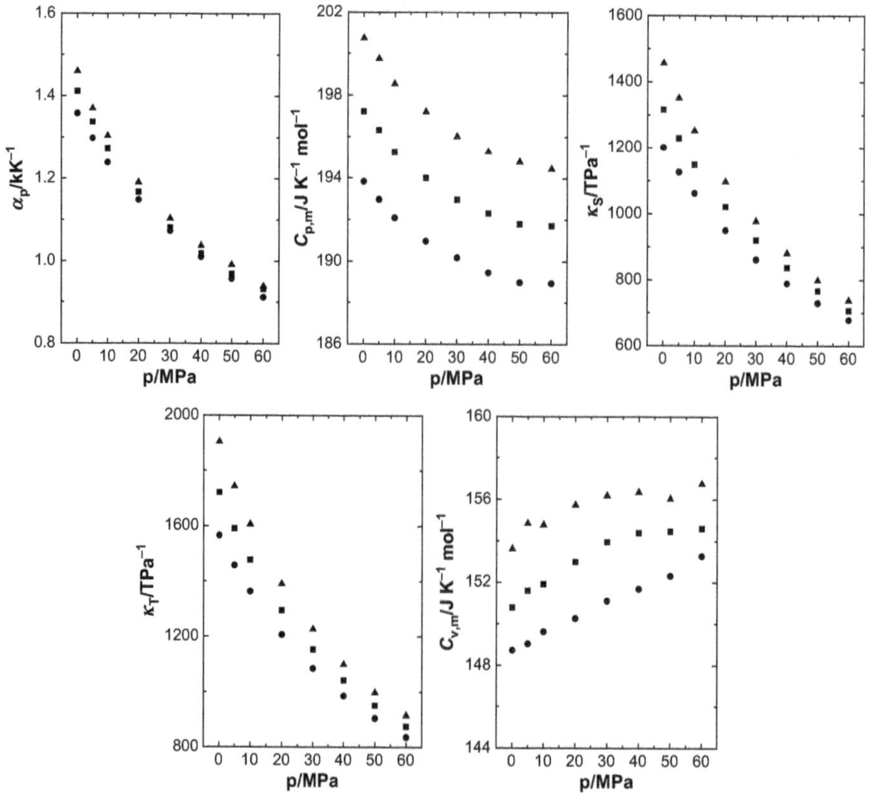

Figure 15.5 Isobaric expansibility α_p, isobaric heat capacity $C_{p,m}$, isentropic compressibility κ_S, isothermal compressibility κ_T, and isochoric heat capacity $C_{v,m}$ of hexane plotted against pressure: (●) $T=293.15$ K; (■) $T=303.15$ K; (▲) $T=313.15$ K.
The graphs have been drawn from data in reference 42.

15.3 Mixtures

Several methods can be used to correlate the densities of binary mixtures with temperature, pressure and composition. Here, we will focus on two different widely applied strategies: firstly, we will analyse the use of the Tait equation extended to mixtures and secondly we will describe the methods that correlate densities of mixtures through a correlation of their excess volumes.

15.3.1 Correlation Based on the Tait Equation

As Dymond and Malhotra wrote:[43] "the Tait equation, which is now widely used to fit liquid density data over wide pressure ranges, is a modification of the original equation of Tait, published 100 years ago, to fit his results on the compressibility of fresh water and seawater at different pressures". An extended or modified Tait equation, first proposed by Tammann,[44] includes temperature-dependent parameters.

The Tait equation has also been extended to include density correlations of binary mixtures:

$$\rho(T,p) = \frac{\rho_0(T,p_0(T))}{1 - C(T)\ln\left(\dfrac{B(T)+p}{B(T)+p_0}\right)}, \tag{15.8}$$

where p_0 is the reference pressure density (usually $p_0 = 0.1$ MPa), ρ_0 is the corresponding reference pressure density and B and C are adjustable parameters.

There are a number of different ways of calculating the parameters, B and C. Five different ways are discussed below.

(1) The values of parameters C and B are obtained for each specific composition of the mixture and temperature. Some authors[45–49] have taken the same parameter C for all the compositions.

In this case, obviously the reference pressure density is different for each composition and temperature, $\rho_0(x_1, T)$. This methodology implies that a large number of parameters are necessary for the characterisation of the whole binary system.

(2) The values of parameters C and B are obtained for each specific composition.[50–52] In this method, the parameter C is considered temperature independent while the dependence of parameter B with temperature is given by:

$$B(T) = b_0 + b_1 T + b_2 T^2 \tag{15.9}$$

Moreover, the reference pressure densities, ρ_0, are also fitted using a temperature-dependent polynomial equation:

$$\rho_0 = \sum_{i=0}^{n} \rho_{0i} T^i \tag{15.10}$$

(3) The parameters C and B are obtained using data from the entire temperature, pressure and composition ranges.[53,54] In this procedure, these parameters are functions of both temperature and composition:

$$C(x_1,T) = c_0 + c_x x_1 + c_T T + c_{xx} x_1^2 + c_x T x_1 T + c_{TT} T^2 \tag{15.11}$$

$$B(x_1,T) = b_0 + b_x x_1 + b_T T + b_{xx} x_1^2 + b_x T x_1 T + b_{TT} T^2 \tag{15.12}$$

Furthermore, the reference pressure densities, ρ_0, are correlated with temperature by means of a temperature-dependent Redlich–Kister polynomial through:

$$V_0^E = x_1 x_2 \sum_{i=0}^{n} (A_{0,i} + A_{T,i} T)(x_1 - x_2)^i \tag{15.13}$$

(4) The parameters C and B are also obtained using the whole temperature, pressure and composition ranges.[55] However, in this case, parameter C depends only on composition and parameter B only on temperature, as shown below:

$$C(x_2) = c_0 + c_1 x_1 + c_2 x_1^2 \tag{15.14}$$

$$B(T) = b_0 + b_1 T \tag{15.15}$$

The reference pressure densities, ρ_0, are correlated with temperature and composition using the following expression:

$$\rho_0 = x_1 \rho_{0,1} + x_2 \rho_{0,2} + x_1 x_2 \left[A_1 + A_2 x_1 + A_3 x_1^3 + A_5 x_1^5\right] \tag{15.16}$$

where is $\rho_{0,\mathrm{i}}$ the density of component i at the reference pressure. In order to represent these densities accurately with temperature, quadratic functions have been used.

(5) Finally, a modification of the Tait equation was adopted by Kubota *et al.*[56] to represent the pressure dependence of the molar volume of binary mixtures:

$$V = V_0\left(1 - C \ln \frac{B + p}{B + p_0}\right) \tag{15.17}$$

where V_0 is the molar volume at the reference pressure $p_0 = 0.1$ MPa,

The values of C and B are obtained at each specific temperature. In order to define the composition dependence of volumetric properties based on the Tait equation, two different correlations with composition, for each of the parameters C and B, were used depending on the concentration range.

The reference pressure molar volumes, V_0, are also correlated with composition:

$$V_0 = x_1 V_{0,1} + x_2 V_{0,2} + x_1 x_2 \sum_{i=0}^{n} a_\mathrm{i} (x_1 - x_2)^i \tag{15.18}$$

15.3.2 Density Correlation Using Excess Molar Volumes

The density of a binary mixture can be given as a function of temperature, pressure, and composition, using the following general expression, in terms of the excess molar volume:

$$\rho(T,p,x_i) = \frac{\sum_{i=1}^{2} M_i x_i}{\sum_{i=1}^{2} M_i x_i \,/\, \rho_i(T,p) + V^E(T,p,x_i)} \tag{15.19}$$

where M_i, and ρ_i and x_i are the molar mass, density and mole fraction of component i, respectively, and V^E is the excess molar volume of the mixture.

In this approach, the correlation equation used to describe the density of a binary mixture is based on an expression for the excess volume as a function of temperature, pressure and composition. Numerous equations exist to express excess functions as a function of composition at constant temperature and pressure. Usually, these equations are totally empirical and have the form of a polynomical expansion, with a number of terms determined by a statistical method. The most common procedure is to impose temperature and/or pressure dependencies on its parameters using one of the following models.

(1) V^E is given by a sum of the van Laar-type terms:[57,58]

$$V^E = x_1 x_2 \sum_{i=0}^{n} \frac{a_i(T,p)}{b_i(T,p)x_1 + x_2} \tag{15.20}$$

The model parameters a_i and b_i dependent linearly on temperature and pressure:

$$a_i = a_{i0} + a_{i1}(p - p_0) + a_{i2}(T - T_0) \tag{15.21}$$

$$b_i = b_{i0} + b_{i1}(p - p_0) + b_{i2}(T - T_0) \tag{15.22}$$

where p_0 and T_0 are pressure and temperature references, respectively.

(2) V^E has the form of a Redlich–Kister type equation with temperature- and pressure-dependent parameters:[59]

$$V^E = x_1 x_2 \sum_{i=0}^{n} a_i(T,p)(x_1 - x_2)^i \tag{15.23}$$

$$a_i(T,p) = a_{i0} + a_{i1}(p - p_0) + a_{i2}(T - T_0) + a_{i3}(p - p_0)(T - T_0) \tag{15.24}$$

where p_0 and T_0 are pressure and temperature references, respectively.

(3) V^E is given by a modified Redlich–Kister equation truncated at the second term, using a simple linear dependence with pressure and quadratic and linear dependences with temperature:[60]

$$V^E = x_1x_2[A(T, p) + B(T, p)(x_1 - x_2)] \quad (15.25)$$

$$A(p, T) = A_0(T) + A_1(T)p, \qquad B(p,T) = B_0(T) + B_1(T)p \quad (15.26)$$

$$A_0(T) = a_{00} + a_{01}T + a_{02}T^2, \qquad A_1(T) = a_{10} + a_{11}T \quad (15.27)$$

$$B_0(T) = b_{00} + b_{01}T + b_{02}T^2, \qquad B_1(T) = b_{10} + b_{11}T \quad (15.28)$$

(4) V^E is calculated through a modified Redlich–Kister equation, including a dependence on both pressure and temperature:[61]

$$V^E = x_1x_2(1 + Ap)\left[1 + \sum_{i=1}^{2} B_iT^i\right]\left[\sum_{j=0}^{3} C_j(x_1 - x_2)^j\right] \quad (15.29)$$

where A, B_i, and C_j are the adjustable parameters. This equation has not been used to correlate the densities of binary mixtures but only excess volumes.

15.3.3 Calculation of Excess Properties

The experimental determination of ρpT data of pure and mixed solvents can be used to calculate the following derived thermodynamic properties: isobaric expansibility, α_p, isothermal compressibility, κ_T, and internal pressure, p_i, from the following expressions:

$$\alpha_p = \frac{1}{V}\left(\frac{\partial V}{\partial T}\right)_p = -\frac{1}{\rho}\left(\frac{\partial \rho}{\partial T}\right)_p \quad (15.30)$$

$$\kappa_T = -\frac{1}{V}\left(\frac{\partial V}{\partial p}\right)_T = \frac{1}{\rho}\left(\frac{\partial \rho}{\partial p}\right)_T \quad (15.31)$$

$$p_i = \left(\frac{\partial U}{\partial V}\right)_T = T\frac{\alpha_p}{\kappa_T} - p = -T\frac{(\partial\rho/\partial T)_p}{(\partial\rho/\partial p)_T} - p \quad (15.32)$$

In the field of chemical thermodynamics, it is very usual to analyse the behaviour of liquid mixtures in terms of excess properties. Firstly, excess properties have been defined for extensive properties but now excess properties are also obtained for either molar or volume specific intensive properties. From volumetric properties of the mixture and of the pure components some interesting excess properties can be calculated, namely: excess molar volume, V^E, excess isobaric expansibility, α_p^E, excess isothermal compressibility, κ_T^E, and excess internal pressure, p_i^E.

A thermodynamic excess function, X^E, by definition represents the excess of the given thermodynamic property X of a real mixture over X^{id}, the value for an ideal mixture at the same temperature, pressure and composition:

$$X^E = X - X^{id} \tag{15.33}$$

In order to evaluate X^E a clear definition of X^{id} is necessary. According to Benson and Kiyohara,[62] values of the thermodynamic properties of an ideal mixture can be calculated from the values of the properties of the pure components as follows:

$$V^{id} = \sum_{i=1}^{n} x_i V_i \tag{15.34}$$

$$\alpha_p^{id} = \sum_{i=1}^{n} \phi_i \alpha_{p,i} \tag{15.35}$$

$$\kappa_T^{id} = \sum_{i=1}^{n} \phi_i \kappa_{T,i} \tag{15.36}$$

where $\phi_i = \frac{x_i V_i}{\sum_{i=1}^{n} x_i V_i}$ is the volume fraction of component i in the mixture referred to the unmixed state.

It can be outlined that the values of excess isobaric expansibility, α_p^E, and excess isothermal compressibility, κ_T^E, can be obtained directly from V^E values[63] using the following relations:

$$\alpha_p^E = \left[(\partial V^E/\partial T)_p - V^E \sum_{i=1}^{n} \phi_i \alpha_{p,i} \right] / V \tag{15.37}$$

$$\kappa_T^E = -\left[(\partial V^E/\partial p)_T + V^E \sum_{i=1}^{n} \phi_i \kappa_{T,i} \right] / V \tag{15.38}$$

where V is the molar volume of the mixture.

The calculation of excess internal pressure is different. Its non-Gibbs profile[64] made it necessary to define further the ideal value of this property, according to Marczak,[65]

$$p_i^{id} = \sum_{i=1}^{n} \psi_i p_{i,i} \tag{15.39}$$

where $\psi_i = \frac{x_i \kappa_{T,i}}{\sum_{i=1}^{n} x_i \kappa_{T,i}}$ is formally similar to the definition of the volume fraction; the only difference lies in the values of the molar compressibility,

K_T, substituted for volumes, thus the function may be called compression factor of component *i* in the mixture.

As an example, all these excess properties, calculated for the binary mixture (*n*-heptane + 1-chlorobutane)[66] are represented in Figures 15.6 and 15.7. In these figures, both the effect of temperature and pressure on the excess properties are shown.

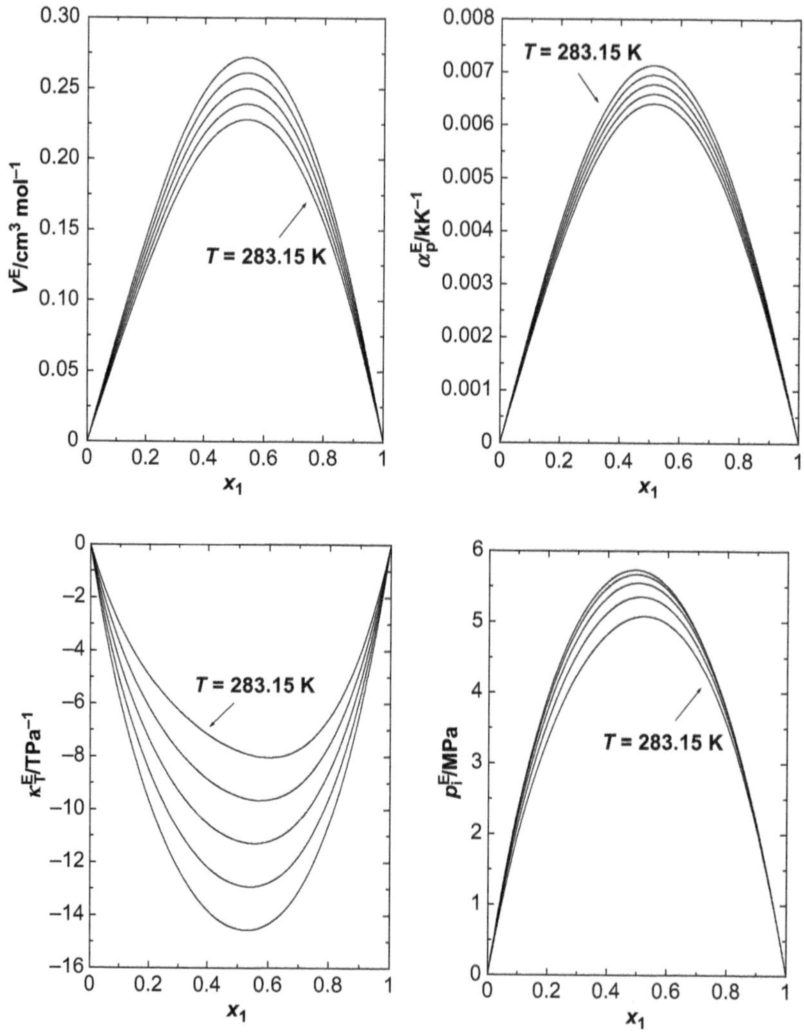

Figure 15.6 Excess properties at $p = 30.0$ MPa and at different temperatures, $T = 283.15$–323.15 K, for the binary mixture *n*-heptane + 1-chlorobutane.
The graphs have been drawn from data in reference 66.

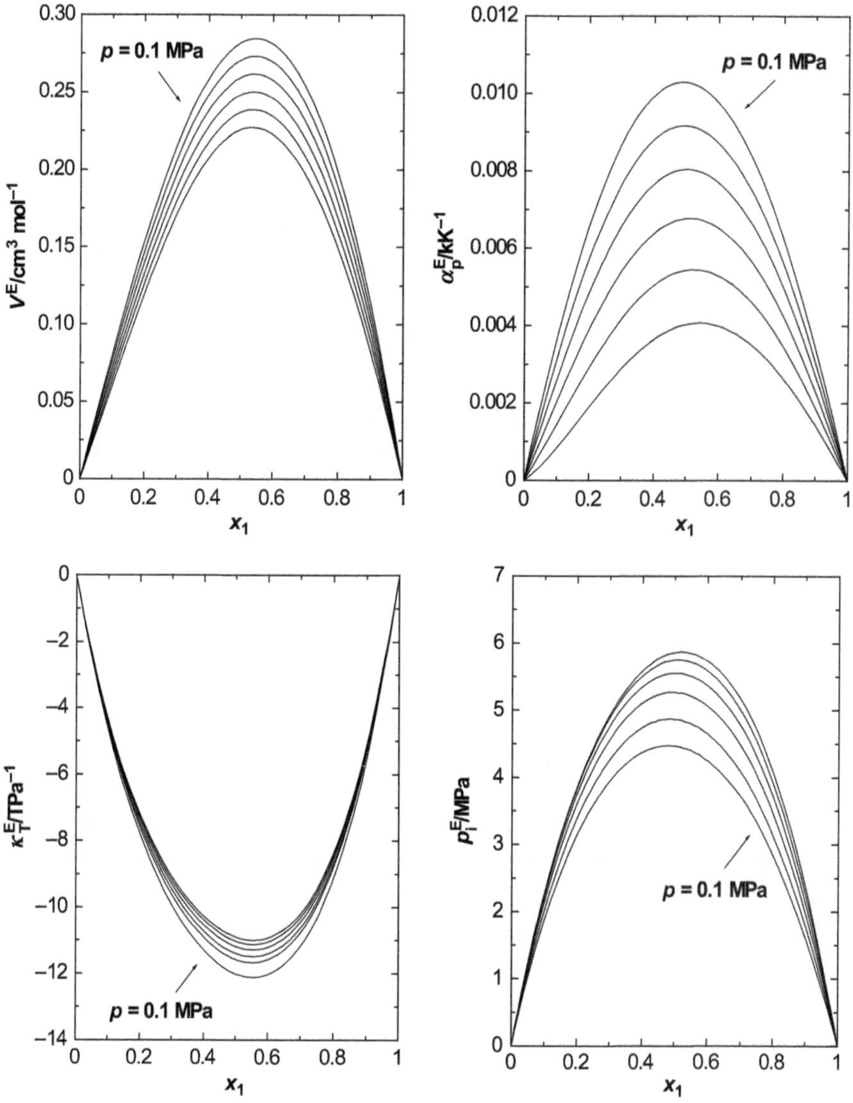

Figure 15.7 Excess properties at $T = 303.15$ K MPa and at different pressures, $p = 0.1$–50.0 MPa, for the binary mixture *n*-heptane + 1-chlorobutane. The graphs have been drawn from data in reference 66.

Finally, the isothermal effects of pressure on the excess molar Gibbs function, ΔG^E, entropic contribution to excess Gibbs function, $T\Delta S^E$, and excess molar enthalpy, ΔH^E, can be evaluated from the V^E data:

$$\Delta G^E = G^E(T,p,x_i) - G^E(T,p_0,x_i) = \int_{p_0}^{p} V^E(T,p,x_i)dp \qquad (15.40)$$

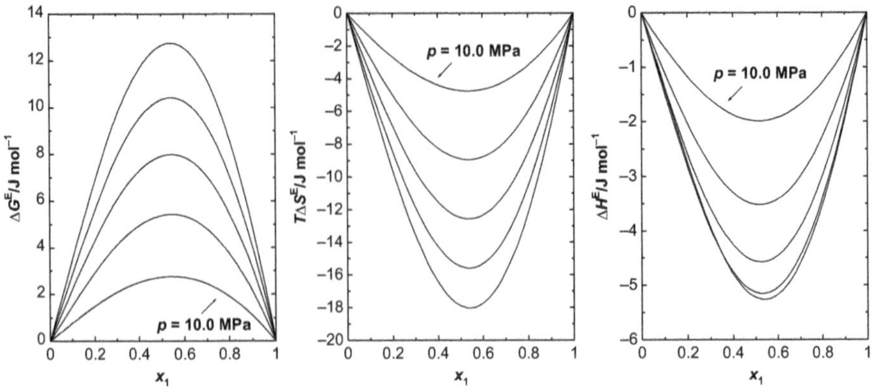

Figure 15.8 Effect of pressure, $p = 10.0$–50.0 MPa, on the isothermal excess properties at $T = 303.15$ K with respect to their values at $p_0 = 0.1$ MPa for the mixture *n*-heptane + 1-chlorobutane.
The graphs have been drawn from data in reference 66.

$$T\Delta S^{\mathrm{E}} = TS^{\mathrm{E}}(T,p,x_i) - TS^{\mathrm{E}}(T,p_0,x_i) = \int_{p_0}^{p} -T(\partial V^{\mathrm{E}}/\partial T)_p \mathrm{d}p \tag{15.41}$$

$$\Delta H^{\mathrm{E}} = H^{\mathrm{E}}(T,p,x_i) - H^{\mathrm{E}}(T,p_0,x_i) = \int_{p_0}^{p} \left[V^{\mathrm{E}} - T(\partial V^{\mathrm{E}}/\partial T)p\right] \mathrm{d}p \tag{15.42}$$

where p_0 is the reference pressure.

Usually values of the excess molar volume range between -1 and 1 cm^3 mol^{-1}, so the isothermal change in the excess molar Gibbs function with pressure is small. For a variation of pressure of 10 MPa the ΔG^{E} is between -10 and 10 J mol^{-1}.

The isothermal effects of pressure on the excess properties for the system (*n*-heptane + 1-chlorobutane)[34] are shown in Figure 15.8.

References

1. M. E. Fisher, *Rep. Prog. Phys.*, 1967, **30**, 615.
2. L. P. Kadanoff, W. Götze, D. Hamblen, R. Hetch, E. A. S. Lewis, V. V. Paciauskas, M. Rayl and J. Swift, *Rev. Mod. Phys.*, 1967, **39**, 395.
3. P. G. Debenedetti, *J. Phys.: Cond. Matter*, 2003, **15**, R1669.
4. P. G. Debenedetti and H. E. Stanley, *Phys. Today*, 2003, **56**, 40.
5. C. A. Cerdeiriña, D. González-Salgado, L. Romaní, M. D. Delgado, L. A. Torres and M. Costas, *J. Chem. Phys.*, 2004, **120**, 6648.
6. P. Navia, D. Bessieres and F. Plantier, *J. Chem. Thermodyn.*, 2013, **57**, 367.
7. G. C. Benson, P. J. Darcy and O. Kiyohara, *J. Solution Chem.*, 1980, **9**, 931.
8. M. Costas and D. Patterson, *Thermochim. Acta*, 1987, **120**, 161.
9. C. A. Cerdeiriña, C. A. Tovar, E. Carballo, L. Romaní, M. D. Delgado, L. A. Torres and M. Costas, *J. Phys. Chem. B*, 2002, **106**, 185.

10. C. A. Cerdeiriña, J. Troncoso, D. González-Salgado, G. García-Miaja, G. O. Hernández-Segura, D. Bessieres, M. Medeiros, L. Romaní and M. Costas, *J. Phys. Chem. B*, 2007, **111**, 1119.
11. C. Paz-Ramos, C. A. Cerdeiriña and M. Costas, *J. Phys. Chem. B*, 2011, **115**, 9626.
12. B. Wunderlich, *Thermal Analysis*, Academic Press, London, 1990.
13. M. Taravillo, F. J. Pérez, J. Núñez, M. Cáceres and V. G. Baonza, *J. Chem. Eng. Data*, 2007, **52**, 481.
14. S. L. Randzio, J.-P. E. Grolier, J. R. Quint, D. J. Eatough, E. A. Lewis and L. D. Hansen, *Int. J. Thermophys.*, 1994, **15**, 415.
15. S. L. Randzio, J.-P. E. Grolier and J. R. Quint, *Fluid Phase Equilib*, 1995, **110**, 341.
16. See *e.g.*, I. M. Abdulagatov, J. T. Safarov, F. S. Aliyev, M. A. Talibov, A. N. Shahverdiyev and E. P. Hassel, *Fluid Phase Equilib.*, 2008, **268**, 21.
17. F. Llovell and L. F. Vega, *J. Phys. Chem. B*, 2006, **110**, 11427.
18. T. Lafitte, D. Bessieres, M. M. Piñeiro and J.-L. Daridon, *J. Chem. Phys.*, 2006, **124**, 024509.
19. F. Llovell, C. J. Peters and L. F. Vega, *Fluid Phase Equilib*, 2006, **248**, 115.
20. T. Lafitte, M. M. Piñeiro, J.-L. Daridon and D. Bessieres, *J. Phys. Chem. B*, 2007, **111**, 3447.
21. I. Cibulka and M. Zikova, *J. Chem. Eng. Data*, 1994, **39**, 876.
22. I. Cibulka and L. Hnedkovsky, *J. Chem. Eng. Data*, 1996, **41**, 657.
23. I. Cibulka, L. Hnedkovsky and T. Takagi, *J. Chem. Eng. Data*, 1997, **42**, 2.
24. I. Cibulka, L. Hnedkovsky and T. Takagi, *J. Chem. Eng. Data*, 1997, **42**, 415.
25. I. Cibulka and T. Takagi, *J. Chem. Eng. Data*, 1999, **44**, 411.
26. I. Cibulka and T. Takagi, *J. Chem. Eng. Data*, 1999, **44**, 1105.
27. I. Cibulka, T. Takagi and K. Ruzicka, *J. Chem. Eng. Data*, 2001, **46**, 2.
28. I. Cibulka and T. Takagi, *J. Chem. Eng. Data*, 2001, **2002**, 1037.
29. E. C. Ihmels and J. Gmehling, *Ind. Eng. Chem. Res.*, 2001, **40**, 4470.
30. C. F. Spencer and R. P. Danner, *J. Chem. Eng. Data*, 1972, **17**, 236.
31. D. Ambrose, *J. Chem. Thermodyn.*, 1986, **18**, 45.
32. R. Span, *Multiparameter Equations of State*, Springer-Verlag, Berlin, 2000.
33. C. A. Cerdeiriña, C. A. Tovar, D. González-Salgado, E. Carballo and L. Romaní, *Phys. Chem. Chem. Phys.*, 2001, **3**, 5230.
34. J. Troncoso, C. A. Cerdeiriña, P. Losada-Pérez, E. Carballo and L. Romaní, *Fluid Phase Equilib.*, 2009, **280**, 144.
35. Y. A. Sanmamed, D. González-Salgado, J. Troncoso, C. A. Cerdeiriña and L. Romaní, *Fluid Phase Equilib.*, 2007, **252**, 96.
36. J. Troncoso, C. A. Cerdeiriña, P. Navia, Y. A. Sanmamed, D. González-Salgado and L. Romaní, *J. Phys. Chem. Lett.*, 2010, **1**, 211.
37. R. Gomes de Azevedo, J. Szydlowski, P. F. Pires, J. M. S. S. Esperanca, H. J. R. Guedes and L. P. N. Rebelo, *J. Chem. Thermodyn.*, 2004, **36**, 211.
38. J. L. Valencia, PhD Thesis, Universidad de Vigo, Spain, 2005.
39. L. A. Davis and R. B. Gordon, *J. Chem. Phys.*, 1967, **46**, 2650.

40. J. L. Daridon, B. Lagourette and J.-P. E. Grolier, *Int. J. Thermophys.*, 1998, **19**(1), 145.
41. D. González-Salgado, J. Troncoso, F. Plantier, J. L. Daridon and D. Bessieres, *J. Chem. Thermodyn.*, 2006, **38**, 893.
42. J. L. Valencia, D. González-Salgado, J. Troncoso, J. Peleteiro, E. Carballo and L. Romaní, *J. Chem. Eng. Data*, 2009, **54**, 904.
43. J. H. Dymond and R. Malhotra, *Int. J. Thermophys.*, 1988, **9**, 941.
44. G. Tammann, *Z. Phys. Chem.*, 1895, **17**, 620.
45. F. Audonnet and A. A. H. Padua, *Int. J. Thermophys.*, 2002, **23**, 1537.
46. L. Morávková, Z. Wagner and J. Linek, *Fluid Phase Equilib*, 2003, **209**, 81.
47. E. Widowati and M. Lee, *J. Chem. Thermodyn.*, 2013, **63**, 95.
48. J. Zhou, R. Zhu, H. Xu and Y. Tian, *J. Chem. Eng. Data*, 2010, **55**, 5569.
49. T. Ebina, M. Fukushima, D. Tomida and C. Yokoyama, *Int. J. Thermophys.*, 2009, **30**, 1466.
50. L. Lugo, M. J. P. Camuñas, E. R. López and J. Fernández, *Fluid Phase Equilib.*, 2001, **186**, 235.
51. J. Cendón, J. Vijande, J. L. Legido and M. M. Piñeiro, *J. Chem. Eng. Data*, 2006, **51**, 577.
52. G. Watson, C. K. Zéberg-Mikkelsen, A. Baylaucq and C. Boned, *J. Chem. Eng. Data*, 2006, **51**, 112.
53. L. Morávková, Z. Wagner and J. Linek, *J. Chem. Thermodyn.*, 2007, **39**, 1637.
54. O. Redlich and A. T. Kister, *Ind. Eng. Chem.*, 1948, **40**, 345.
55. I. M. Abdulagatov, A. Tekin, J. Safarov, A. Shahverdiyev and E. Hassel, *J. Solution Chem.*, 2008, **37**, 801.
56. H. Kubota, Y. Tanaka and T. Makita, *Int. J. Thermophys.*, 1987, **8**, 47.
57. J. J. Van Laar, *Z. Phys. Chem.*, 1910, **72**, 723.
58. T. Hofman, A. Gołdon, A. Nevines and T. M. Letcher, *J. Chem. Thermodyn.*, 2008, **40**, 580.
59. H. Guerrero, P. Cea, I. Gascón, F. M. Royo and C. Lafuente, *J. Phys. Chem. B*, 2013, **117**, 1084.
60. J. A. Amorin, O. Chiavone-Filho, M. L. L. Paredes and K. Rajagopal, *Fluid Phase Equilib.*, 2007, **259**, 89.
61. J. Cendón, J. Vijande, J. L. Legido and M. Piñeiro, *J. Chem. Eng. Data*, 2006, **51**, 577.
62. G. C. Benson and O. Kiyohara, *J. Chem. Thermodyn.*, 1979, **11**, 1061.
63. G. C. Benson and O. Kiyohara, *J. Solution Chem.*, 1980, **10**, 791.
64. J. C. R. Reis, M. J. Blandamer, M. I. Davis and G. Douhéret, *Phys. Chem. Chem. Phys.*, 2001, **3**, 1465.
65. W. Marczak, *Phys. Chem. Chem. Phys.*, 2002, **4**, 1889.
66. H. Guerrero, PhD Thesis, Universidad de Zaragoza, Spain, 2013.

CHAPTER 16

SAFT and Molecular Simulation Techniques: Application to Determination of Volumetric Excess Properties

FELIPE J. BLAS*[a,b] AND MANUEL M. PIÑEIRO[c]

[a] Departamento de Física Aplicada, Universidad de Huelva, E21071, Huelva, Spain; [b] Centro de Física Teórica y Matemática FISMAT, Universidad de Huelva, E31071, Huelva, Spain; [c] Departamento de Física Aplicada, Universidade de Vigo, E36310 Vigo, Spain
*Email: felipe@uhu.es

16.1 Introduction

The most important theoretical development in the field of complex liquids in the last 30 years has probably been the formulation of the so-called first-order thermodynamic perturbation theory of Wertheim.[1–4] The original Wertheim's theory is a statistical mechanical approach to account for the thermodynamic properties of hard spherical associating fluids. The statistical associating fluid theory (SAFT), which is a formalism based on Wertheim's theory, combines in general a chain reference contribution with an associating perturbation term (for systems with specific interactions, *i.e.*, hydrogen bonding, reactive systems, *etc.*) with the aim of analysing complex-chain and associating fluids. The implementation of the Wertheim theory as an equation of state (EoS) in the SAFT formalism has constituted a major advance towards a theoretical framework for

Volume Properties: Liquids, Solutions and Vapours
Edited by Emmerich Wilhelm and Trevor M. Letcher

Published by the Royal Society of Chemistry, www.rsc.org

modelling complex fluids, including systems characterised by anisotropic association interactions (such as those occurring in hydrogen bonding fluids), molecular shape, and electrostatic interactions (Coulombic and, in general, multipolar interactions), microscopic features exhibited by compounds as complex as surfactants, polymers, hydrogen bonding molecules such as water, and polyfunctional molecules, including amino acids and peptides, among many others. The SAFT approach has proven to be extremely successful for modelling associating and chain-like systems. Today, it is undoubtedly considered the state-of-the-art method for modelling thermodynamic properties and phase behaviour of complex fluids and has found application in small molecules such as carbon dioxide and water, through to complex copolymers, ionic liquids and amino acids, just to cite a few examples.

In the SAFT approach, molecules are modelled generally as associating chains formed of bonded spherical segments (normally referred to as monomers), with short ranged attractive sites. The Helmholtz energy in the context of SAFT is written as the sum of four separated contributions that account for different and specific microscopic features of the system under study:

$$A = A^{\text{ideal}} + A^{\text{mono}} + A^{\text{chain}} + A^{\text{assoc}}$$

where A^{ideal} is the ideal Helmholtz energy, A^{mono} the contribution to the Helmholtz energy due to the monomer–monomer dispersive (attractive and repulsive) interactions, A^{chain} the contribution due to the formation of bonds between monomeric segments, and A^{assoc} the contribution due to association.

Many different versions and extensions of the SAFT EoS can be found in the literature. Just to cite a few of the most popular: the simplest and perhaps most general version is that usually named SAFT-HS;[5,6] the most extensively used, denoted as SAFT-HR;[7,8] the version to treat Lennard-Jones (LJ) chain molecules, referred to as soft-SAFT;[9–12] the version proposed to deal with monomers with attractive potential of variable range, known as SAFT-VR;[13,14] and the PC-SAFT,[15,16] in which the reference system of the formalism is a hard-chain system, rather that a hard-sphere. There are a number of excellent reviews in which the SAFT formalism is described in detail, as for instance those of Müller and Gubbins,[17] Economou,[18] Paricaud *et al.*,[19] Tan *et al.*,[20] and more recently, the excellent account for the SAFT formalism by McCabe and Galindo.[21]

Another theoretical approach that can be used to determine the excess properties of a fluid mixture is molecular simulation (MS). From an experimental point of view, most experiments concerning the determination of thermophysical properties of solutions are carried out at constant temperature and pressure, so the natural thermodynamic ensemble that should be used to describe them by applying any statistical mechanical technique is the isothermal–isobaric (usually denoted as NPT) ensemble. Concerning MS, either Monte Carlo (MC) or molecular dynamics (MD) can be applied to the

NPT ensemble conditions, so both techniques are excellent candidates to be used as estimating tools for this objective. Concerning the thermodynamic theoretical foundations of these techniques and the technical details of their implementation, there is a large body of excellent monographs devoted to the description of MS techniques and their application to the determination of fluid phase equilibria, thermophysical, interfacial properties and many other related topics. Just to cite a few, and not intending to be exhaustive, the excellent books by Frenkel and Smit,[22] Allen and Tildesley,[23] Landau and Binder,[24] Rapaport,[25] Sadus[26] or Heyes[27] may serve as representative examples.

In general, writing and testing a proper (in technical terms) MS code intended to describe the properties of a molecular fluid are demanding tasks. This fact has significantly restricted the use of MS techniques. The need for extensive computer resources to run these costly central processing unit (CPU) time calculations has also traditionally acted as an activation barrier to performing MS calculations. Nevertheless, these two factors can no longer be considered as limitations. Concerning the hardware, the continuous improvement in computing performance nowadays allows the performance of many calculations on standard computers, and furthermore, computing centres have become readily available to many researchers. As for the software limitations, an increasing number of open source suite tools are available nowadays, allowing access to MS calculations to a wide community of scientists. Most of these tools have been developed to perform MD calculations, and this includes for instance Gromacs,[28] DL-Poly,[29] LAMMPS,[30] NAMD,[31] or AMBER[32] among others. Comparatively, the number of open source codes developed to perform MC calculations is more limited, and the Towhee[33] and ms2[34] codes may be cited as representative examples. All these software packages include not only the subroutines of the algorithms needed to perform the MS calculation, but also extensive compilations of molecular models, comprising inter- and intramolecular interaction potentials, molecular geometries and sets of characteristic parameters for many molecules. This allows almost any interested user to build a complete simulation setup, including an initial simulation box and input files describing in detail the calculation to be performed. Then, the MS can be run and the output information processed afterwards without the need to write a single code line. This has made MS calculations increasingly accessible, and nowadays it is a valuable and widespread resource in many scientific areas.

The importance of MS calculations does not need to be repeated, but let us recall two features that play an important role in the topic discussed here. First, the MS results can be considered the exact thermodynamic limit for the calculated properties of any interaction potential, provided that enough care has been put into the treatment of all the calculation variables, and this includes system size, treatment of long-range corrections, and so forth. This is very important to benchmark thermodynamic models derived from a certain interacting potential, as it occurs, for instance, with perturbation theory models, statistical models or many EoS. From this perspective, MS results can provide very useful hints about the performance limit of a given approach, and for the case studied here it will allow one to establish a connection between the molecular model considered and the trend of the excess magnitudes obtained. The second characteristic is that MS directly connects the molecular scale details of the molecular representation selected with the estimated macroscopic thermodynamic properties, allowing analysis of the influence of every individual feature of the molecular model on the excess properties. In addition, an MS calculation provides not only a value for all macroscopic thermodynamic properties compatible with the type of calculation and simulation ensemble selected, but also microscopic information about the fluid studied, including the structure (*e.g.*, radial distribution functions). Bearing this in mind, this tool has a much wider scope and range of applicability than other existing approaches. On the other hand, the long computing times needed must be considered because this implies a limitation to what type of systems can be considered feasible for MS calculations, and what range of thermodynamic conditions can be explored for a given system.

The thermodynamic potential corresponding to the NPT ensemble is the Gibbs energy (G), and thus the set of excess properties related include excess Gibbs energy, G^E, entropy, S^E, volume, V^E, heat capacity, $C_p{}^E$ and enthalpy, H^E. Focusing on the volumetric excess property, V^E, its calculation using MS is simple from the calculation of the density of the mixture and of those of the pure compounds. Nevertheless, the use of MS to estimate V^E presents important drawbacks. The first is the availability of a molecular model and force field (set of characteristic parameters of the interaction potentials) for the molecules considered. The determination of a molecular model for polyatomic molecules is a remarkably cumbersome task, and the collection of properly modelled molecules is still limited. If a parameterised model is not available the calculation cannot usually proceed, as the task of obtaining a molecular model is beyond the capability of most MS interested users. In addition, as the calculations are time consuming, when the molecule size grows, increasing the number of atoms or functional groups beyond a certain limit, the usual molecular representations become too complicated and need the use of approximations as united atom approaches or coarse graining to make calculations feasible. If molecular models are available, a second question arises related to the CPU time needed for the calculation. If the estimation of excess properties is the only objective, the cost of obtaining the results might be a reason to decline this option. Finally, the absolute value of excess properties is rather small, and it is obtained as the difference between magnitudes that are rather large. The statistical uncertainty associated with each value then adds up through error propagation and usually leads to an estimation of the final uncertainty result of a very large value, which sometimes even exceeds the excess property value itself. This implies that especial care has to be taken in the correct equilibration of the system during the simulation run, together with very careful uncertainty estimation calculations to minimise this undesirable effect. Thus, as with any other theoretical estimation tool, a careful evaluation of the advantages and limitations must be considered before proceeding. Despite this, very useful information and conclusions can be obtained by taking this route, as will be seen in the following section.

16.2 Classic Interaction Potentials

As stated before, MS calculations are often used to obtain the exact result of the properties of a thermodynamic system of particles interacting through a well-defined interaction potential. This has allowed the determination of the excess volume of several simple interaction potentials. Hard-sphere models have been widely used to describe fluid state behaviour, but few works have used them to estimate excess properties using MS. As an example, Morávek *et al.*[35] used MC to determine excess volumes of hard homonuclear dumbbells with different elongation, and compared the results with different EoS and also with the resolution of the Ornstein–Zernike integral equation.

The LJ[36] interacting potential has been extensively used in the description of dense fluids, as it is one of the most used effective pairwise potentials to represent dispersive interactions between atoms, functional groups, or molecules in condensed matter. A detailed description of its theoretical derivation can be found in the text by Kaplan.[37] The LJ interaction potential is characterised by two parameters, the segment diameter σ, and the potential well depth ε. Despite its simplicity, it can lead to an extremely rich number of scenarios describing fluid mixtures with very different behaviours. The simple mixture of isolated LJ spheres can lead to many different types of phase equilibria or thermodynamic profiles. If chains of LJ segments are considered instead of single spheres, or other types of interactions are combined with the LJ term, the number of possibilities diverges, and this is the reason why so many studies have explored different configurations based on the LJ potential.

For the determination of excess volumetric properties using MS, several authors have concentrated on the study of the LJ potential. Citing some examples in chronological order, Schaink and Hoheisel[38] used MD to determine phase equilibria and excess properties of LJ mixtures with non-additive cores, comparing the obtained results with perturbation theory estimations. Shukla,[39] also using MD, studied the mixture of LJ spheres and dimers with an added point quadrupole, analysing the dependence of excess properties on both dimer molecular elongation and quadrupole moment.

Blas[40–42] published a number of papers using MC and the soft-SAFT formalism to study the excess properties of the LJ potential. These works included the analysis of binary LJ segment mixtures for different size and dispersive energy ratios. The influence of the mixing rule was analysed by comparing the Lorentz–Berthelot classical combining rule with an association ($\varepsilon_{12}=\varepsilon_{11}$) and a solvation ($\varepsilon_{12}=\varepsilon_{22}$) scheme. The ability to describe the sign, symmetry and even the sigmoidal shape of the excess volume representations was discussed. The author also considered mixtures of LJ homonuclear chains, a simplified model of real chainlike molecules, such as *n*-alkanes, that includes the most important microscopic features of these systems, repulsive and attractive forces between chemical groups and the connectivity of segments to make up the chain. In particular, both MC simulations and SAFT are able to predict the general trends of the excess volume (V^E) and enthalpy (H^E) of real *n*-alkane mixtures, including their temperature and chain-length dependence. The last of these works[42] presents a generalisation of a study to deal with excess properties of self-associating mixtures of chains. Association is modelled by considering additional embedded off-centre square-well bonding sites. MC and SAFT allow the study of the effect of bonding energy and number of associating sites, as well as chain length, on V^E and H^E, whose behaviour is dominated by the interplay between the bond breaking of the structure formed by the self-associating molecules and the interstitial accommodation of the non-associating chains in the branched multimeric structure of the system. Comparison between MC and SAFT results demonstrates that the theory is able to provide a good description of the model.

Fujihara and Nakanishi[43] used MC to describe supercritical mixtures of LJ fluids, finding complex composition dependence and large absolute values for the computed V^E, together with a marked pressure dependence. The authors concluded that the trends obtained are representative of real fluids behaviour. Later, Fujihara *et al.*[44] again used MC, now focusing on mixtures of LJ monomers with small $\varepsilon_2/\varepsilon_1$ ratio for a wide range of temperature, including supercritical conditions.

Mukherjee and Bagchi[45] performed an interesting analysis showing how the fine tuning of LJ mixtures can lead to extremely opposite scenarios, and obtained NPT MD results for two binary mixtures of LJ monomers with opposite behaviour, first with strongly attractive behaviour encouraging structure formation, and in the second case with weaker mutual interactions than those of the pure monomers, so the mixture's structural behaviour was the opposite. In these conditions, the authors discussed the relationship between excess volume and viscosity, and diffusion coefficients. With a similar philosophy, Amore *et al.*[46] recently analysed symmetric LJ mixtures through MD, exploring miscible and immiscible systems and the nexus between chemical ordering and miscibility in dense liquid mixtures, finding that the sign of excess volume is not directly related to miscibility, but is governed instead by the interplay between repulsive and attractive interactions.

These cases are representative examples of how the analysis of an interaction potential using SAFT and MS can yield fruitful conclusions, in particular about the possible trends of the estimated excess properties, which, if later compared with experimental trends, represent useful hints in the development of thermodynamic models.

16.3 Molecular Models

Most of the SAFT-like works published have tried not only to understand the excess volumetric properties of systems from a molecular perspective, but also to obtain quantitative estimations of these properties. Vega and collaborators[47,48] have probably proposed the first model in the literature able to provide a nearly quantitative description of V^E (but considering also H^E) of mixtures of *n*-alkanes. In the first paper the authors developed a perturbation theory, based on Wertheim ideas, and in the second one they refined the model to make it possible to predict the V^E and H^E of *n*-alkanes. They reasoned that the original ideas of Wertheim were not able to provide a correct description of H^E of alkanes because the theory neglected conformational changes when *n*-alkanes are mixed. Combining a semi-phenomenological intramolecular contribution to the Helmholtz energy to account explicitly for conformations of *n*-alkanes, the authors were able to predict nearly quantitatively the behaviour of H^E for a number of mixtures of alkanes. Several years later, dos Ramos and Blas[49] adopted the same approach and generalised the soft-SAFT EoS to deal with conformations of linear chains. This allowed for quantitative predicting of the behaviour of both V^E and H^E for a number of mixtures of alkanes. In particular, the theory was able to account correctly for the most important features of H^E *viz.* positive values at low temperatures, negative values at high temperatures, and sigmoidal behaviour at intermediate temperatures. In addition, the theory shows that H^E of *n*-alkanes can be represented in a universal form, defining T_r in terms of the critical temperatures of the pure components. Using T_r, the authors found that H^E vanishes at the same value ($T_r \approx 0.51$–0.52) for all mixtures considered, in excellent agreement with experimental data. Sun *et al.*[50] generalised the SAFT-VRX approach of McCabe and Kiselev,[51,52] which was able to combine the accuracy of the SAFT-VR EoS away from the critical region with the asymptotic scaling behaviour seen near the critical point of real fluids, to predict the thermodynamic properties of binary mixtures of real substances. In this work, Sun and co-workers applied the new formalism to predict the phase equilibria and other thermodynamic properties of binary mixtures of alkanes and ($CO_2 + n$-alkane) systems. In particular, among the different properties considered, they studied the V^E of ($CO_2 +$ ethane) at several pressures. Agreement between experimental data from literature and theoretical predictions was excellent in all cases. As in most works dealing with excess volumetric properties, the molecular parameters obtained from the analysis of phase equilibria were transferred without further additional fitting. More recently, Blas and collaborators[53] have extended the study of mixtures of *n*-alkanes to binary systems formed from short alkanes. As mentioned below, the theoretical predictions were compared with MS results. The SAFT-VR approach is able to provide a good agreement with MS results, as well as with experimental data taken from the literature.

McCabe, Filipe, and collaborators have published two important series of papers devoted to the determination of excess volumetric properties of binary mixtures including an alkane (in some cases also considering cyclic compounds). In the first series of works,[54–57] the authors considered the phase behaviour, but also the most important excess volumetric properties, V^E, of (Xe + alkane) binary mixtures. It is important to recall that the authors established clearly, using different arguments, that the (Xe + alkane) mixtures can be considered as a particular case of mixtures of *n*-alkanes where Xe is the first member of the linear alkanes series. The authors have also considered the thermodynamic behaviour, including phase equilibria and excess volumetric properties, of mixtures of SF_6 with a short *n*-alkane, by considering SF_6 as the first member of the series.[58] In particular, they studied the excess volume of (ethane + SF_6) over a wide range of temperature and pressure. Agreement between experimental data and theoretical predictions is excellent in all cases. The use of the SAFT-VR in these works is doubly important: firstly, the theoretical formalism is used to check the ability of the EoS. Secondly, and perhaps more importantly from the experimentalist's point of view, it is used to interpret the experimental data obtained in the laboratory by the authors, which is quite often far from evident.

The second series of manuscripts published by these authors concerns systems in which at least one of the components has perfluoroalkyl functional groups. Perfluorinated compounds exhibit peculiar physical properties, such as chemical inertness and biocompatibility, among others. Today they are considered as key fluids in a wide range of fields, including medical applications (perfluoroalkanes can be used as oxygen carriers in blood substitute formulations[59] or as a fluid in eye surgery) and technological applications as solvents for biphasic synthesis, fire-extinguishers or lubricants, among many others. The work of Morgado *et al.*[60] represents the first paper of this series devoted to the determination, using the SAFT-VR EoS, of the excess volumetric properties, with special emphasis on V^E. Although the authors have also accounted for the phase equilibria of alkane + perfluoroalkane mixtures, one of the most interesting results is the study of the V^E for symmetric mixtures, including ($H_6 + F_5$) and ($H_5 + F_6$), ($H_6 + F_7$) and ($H_7 + F_6$), and ($H_6 + F_8$) and ($H_8 + F_6$), where H_n and F_n represent an alkane or perfluoroalkane with n carbon atoms, respectively. It is important to note that all molecular parameters for pure components had been obtained previously. This work served to extend the study to determine a number of thermodynamic properties, including liquid densities of perfluoroalkane and perfluoroalkylalkanes.[61,62] Perfluoralkylalkanes are non-ionic linear surfactant molecules in which a part is made up by alkyl groups (alkane-like) or CO_2-phobic part, and a second part is made up by perfluoroalkyl groups (perfluoroalkane-like) or CO_2-philic part. In addition to that, the authors also determined molar and molal volumes at infinite dilution of several perfluoroalkane and perfluoroalkylalkane molecules over a wide range of temperature and pressure.[63,64] In all cases, the authors obtained the experimental data and used the SAFT-VR approach to interpret the results. The importance of the results is due to the way in which the molecular formalism was used: all the molecular parameters were used in a transferable way, *i.e.*, the parameters for alkyl and perfluoroalkyl chemical groups in surfactants were the same as those used for describing pure alkanes and perfluoroalkanes in previous works.[60] This allowed for the prediction of the thermodynamic behaviour of perfluoroalkylalkanes without fitting any experimental data. This is probably one of the most powerful advantages of molecular modelling compared with other alternative macroscopic tools, since it allows one to obtain thermodynamic properties of a system under certain thermodynamic conditions from knowledge of the same system, or even a different system, in another state.

As already stated, the LJ potential is used to represent molecular models for different substances using MS; alkanes are a representative example of this. Very often they are modelled as flexible heteronuclear LJ chains, with a limited number of internal degrees of freedom governed by intramolecular bending and torsion potentials. Each LJ segment may represent either a single atom, in the models denoted as all-atom (AA), or a functional group, in the united-atom (UA) versions, but in all cases the LJ potential has shown to be extremely useful in being able to represent dispersive interactions. As an example of the calculation of their excess properties, recently dos Ramos *et al.*[53] used the Towhee MC code with the TraPPE[65] UA alkane model to determine the excess properties of mixtures of short alkanes in the gas, liquid and supercritical regions. These estimations were compared with SAFT VR results, and the simulation results captured the sign and trend of excess volume and enthalpy for all phases explored, but quantitatively the departure from experimental data was sometimes large. Also using LJ-based potentials, Palace Carvalho *et al.*[66] used the Towhee code and the TraPPE UA model to describe mixtures of Xe with short alkanes (up to butane). Here again intermolecular interactions are purely dispersive and described though LJ interactions, as Xe is modelled as a single sphere. The excess properties are well estimated in the case of the ethane mixture, but the quantitative differences increase as the alkane chain length grows longer, and the estimated excess volume is less negative than the experimental values, while the positive sign of the experimental excess enthalpy is not reproduced by the simulation results. In another paper,[67] the authors reproduced the same setup, now analysing the temperature trend of the studied excess properties for the same mixtures, estimating in detail as well the internal structure of the mixture through the radial distribution functions. The conclusions were much the same as that described in the previous paper concerning the quantitative ability of the employed technique to estimate excess properties.

The LJ interaction potential is a basic building block in the development of molecular models, describing dispersive interactions. If more complex interactions, such as those found in associating molecules, need to be modelled the electrostatic and polar effects have to be described, and the most direct solution is to consider point electric charges in the molecular structure. The combination of LJ interactions plus point electric charges leads to a wide variety of possibilities, useful to describe, for instance, alkanols, water, carbon dioxide, and many other widely used polar and associative substances. When dealing with associating molecules, it is well known that the MS calculations become more difficult, requiring large simulation boxes and very long runs, which are also especially CPU time-consuming due to the mathematical complexity of the treatment of coulombic interactions. The reason for this enhanced difficulty is that associating molecules establish strong hydrogen bond interactions, and this adds another higher order to the molecular scale due to the clusters formed within the bulk fluid. In a pure associating fluid this causes a somewhat regular microstructure that is, to some extent, well known but, in the cases of mixtures of associating molecules, the cross association leads to heterogeneities of nanometric scale within the fluid, whose evolution and eventual stability is an important challenge to thermodynamic models. For these nanostructure effects, associating molecules are the object of intensive studies using both SAFT and MS, despite the apparent simplicity of their molecular structures.

Alkanols represent perhaps the simplest associating homologous series of molecules, *i.e.*, chain-like systems with hydroxyl functional groups able to form hydrogen bonding. Unfortunately, little work has been done on predicting the excess volumetric properties of mixtures containing alcohols using SAFT. Khammar and Shaw[68] studied the speed of sound of several *n*-alkane + 1-alcohol binary mixtures using SAFT-VR Mie,[69,70] one of the variants of the original SAFT-VR EoS for fluids that interact through the Mie intermolecular potential. Watson *et al.*[71] determined volumetric and derivative properties experimentally, over a wide range of temperature and pressure, for the system (1-propanol + toluene) and compared their results with two of the most used SAFT versions, PC-SAFT and SAFT-VR. In both cases, the SAFT EoS provides an adequate description of the experimental data. Aparicio *et al.*[72] have performed *PVTx* measurements of the (*n*-methylpyrrolidone + methanol) binary mixture and obtained some excess properties, including excess volume, isothermal compressibility, enthalpy, and internal pressure. Experimental data were analysed using different EoS, including the original SAFT-HR[7,8] of Huang and Radosz. Finally, Kiselev *et al.*[73] extended the HRX-SAFT[74] formalism to deal with fluid mixtures and considered the application to mixtures of carbon dioxide, water, and methanol. Although the authors focused on the prediction of the phase equilibria of these mixtures, they did consider the H^E of the water + methanol binary mixture. Agreement between experimental data taken from the literature and estimations from theory is fairly good.

Alkanol solutions have been extensively studied using MS from many perspectives. Concerning excess properties, Guevara-Carrión *et al.*[75] studied the mixture methanol + ethanol, and focused on the calculation of transport properties through non-equilibrium MD, but also determined the V^E using a rigid non-polarisable model with point electric charges developed previously by themselves, and parameterised using only pure component VLE experimental data. The agreement with the experimental excess volumes was, in this case, only qualitative. As the mixture deviation from ideality is small, the V^E in this case represents less than 0.02% of the mixture volume. Under these conditions, V^E was obtained as the difference between large numerical quantities that yielded a vanishing small result. As already pointed out, the propagation of statistical uncertainties yields large errors. In any case, for this particular mixture V^E is overestimated. Duarte *et al.*[76] used MD and the OPLS-AA[77] force field to describe the system (ethanol + trifluoroethanol) in the high-pressure range. The trend of the excess properties in this solution is the result of the balance between weak dispersive forces between the hydrogenated and fluorinated functional groups and the hydrogen bonding between both molecules. An interesting result is that a 25% reduction of the F–H dispersive interaction in the MS produces a clear improvement in the quantitative agreement between the experimental and simulated H^E but produces no noticeable effect in V^E. This leads to the conclusion that the main reason causing the volume increase in these systems is not related to the weak dispersive interactions, as often stated, but should be related instead to the repulsive part of the intermolecular interacting potential.

Water, with its complex structure, deserves special treatment. Its importance and the difficulty of the modelling of all its properties need no comment. Undoubtedly this molecule is a huge scientific challenge, and the enormous research efforts that have been made concerning this singular molecule are apparent in the literature. As in the case of alcohols, there are not many SAFT studies of excess volumetric properties of systems involving water. Kiselev *et al.*[73] also considered the V^E of a (water + carbon dioxide) binary mixture. The authors found an excellent agreement between theoretical predictions from the HRX-SAFT[74] version and experimental data. Blas and co-workers[78,79] determined the phase equilibria and excess properties (V^E and H^E) of the (water + carbon dioxide) binary mixture using the original SAFT-VR approach. In particular, the authors tried to predict the properties of interest using the simplest model for water and carbon dioxide taken from previous literature SAFT-VR works, and only a single binary parameter was used to describe the global phase behaviour of the mixture, including the high-temperature and high-pressure regions of the phase diagram (vapour–liquid and liquid–liquid regions), as well as the excess behaviour over a wide range of temperature and pressure. Agreement between theoretical predictions and experimental data taken from the literature is remarkable, especially taking into account the simplicity of the models. More recently, another interesting work in which some excess volumetric properties have been determined is that of dos Ramos and McCabe[80] In this work, the phase behaviour, excess enthalpies and other thermodynamic properties of the (water + hydrogen sulphide) mixture were modelled using the SAFT-VR + D approach, a version of the SAFT-VR formalism modified to deal with dipolar fluids. The theory was able to predict the phase behaviour very accurately, as well as the H^E of the mixture without requiring any fitted binary interaction parameters; a remarkable achievement in spite of the complexity of the mixture.

Aqueous solutions of alcohols are a subject of particular interest and complexity. If we consider, for instance, the mixture water + methanol, the replacement of a proton in water by a methyl group, which is the structural difference between them, leads to a completely different behaviour. Methanol molecules arrange themselves in chains formed by hydrogen bonds, or form small rings, while water promotes the well-known tetrahedral pattern. In their solution, the competition for formation of hydrogen bonds leads to a rich microstructure or microscopic immiscibility in certain composition ranges, which is responsible for the anomalous trends of their transport and thermodynamic properties, a stringent test to any fluid model.

Many authors have focused on this (water + methanol) mixture using MS techniques. Koh *et al.*[81] studied the atmospheric excess properties of this system from experimental, theoretical and MS perspectives. Methanol was modelled according to the OPLS[82] force field, and different rigid non-polarisable models, including CC and TIP4P, were used for water (an excellent recent review by Vega and Abascal[83] summarises the latest advances using rigid non-polarisable models for water). González-Salgado and Nezbeda[84] also used MC to simulate OPLS methanol and TIP4P water, determining the atmospheric pressure excess properties, finding reasonable agreement for V^E, but the qualitative trend of the partial molar volume at low concentrations was not adequately reproduced. This study was later extended (Dopazo-Paz *et al.*[85]) considering other versions of the TIP4P model family, concluding that the increase in the electric charge value of the water molecular model in general reduces the values of the excess properties, but represents no qualitative differences in the fluid structure described by the radial distribution functions. With a similar approach, Vlcek and Nezbeda[86] also used MC to compare primitive *versus* realistic models (OPLS for methanol and SPC/E for water), finding qualitative agreement for V^E and H^E, but both approaches failed to describe the minimum of the partial molar volume of methanol at low concentrations. These results suggest that the non-ideality of the mixture results from the non-additivity of the repulsive interactions; this affirmation being supported by the nearly ideal composition dependence of the hydrogen bond balance in the mixture. In addition, primitive models reproduce the main features of their parent models to a great extent. Weerasinghe and Smith[87] followed a different strategy, and developed a force field to describe the Kirkwood–Buff (KB) integrals of the aqueous solution of methanol at different concentrations. The SPC/E water model and the methanol molecular geometry taken from OPLS were used as a starting point, fixing the LJ dispersive parameters, while the point electric charge values were refitted to describe the KB integrals. The resulting force field was evaluated by estimating a complete set of thermodynamic properties, which also included partial molar and excess volumes, estimated in both cases with remarkable accuracy. Zhong *et al.*[88] proposed to combine MD and the fluctuation charge method to account for polarisation effects in the water–methanol solution. Water was described using TIP4P-FQ and methanol using a model derived by Patel and Brooks.[89] With this setup, they analysed bulk solution and structural properties related to hydrogen bonding. Among the mixing properties, as in other cases, H^E was the property that presented more deviation in the estimations. Finishing this review of aqueous methanol solution excess property estimation, Perera *et al.*[90] proposed an MD study of (methanol + water or acetone), using SPC/E water, OPLS for methanol and acetone, comparing the trend of the vanishingly small V^E of (methanol + acetone) with the much larger negative excess values for the aqueous methanol.

Other authors have studied aqueous alcohol solutions, not restricted only to methanol. For instance, Wensink *et al.*[91] used MD to analyse the mixtures of TIP4P water with OPLS-AA methanol, ethanol or propanol. This work calculated excess properties, but also dynamic properties including viscosity using non-equilibrium MD. Concerning volumetric properties, a systematic underestimation of V^E was reported, which could be interpreted in the sense that water–alcohol interaction is not strong enough with the parameterisation used. However, simultaneously an overestimation of H^E was found, which would suggest the opposite reasoning. This leads to the conclusion that the potential used is too simple to describe simultaneously the whole set of mixing properties, and the authors proposed several alternatives such as the use of polarisable models, changes in the treatment of the repulsive and attractive terms separately, use of smeared electric charges instead of point charges, *etc.* This is an interesting case study where the analysis of estimated excess properties resulted in discussing the limit of applicability of a certain molecular model, suggesting the need for further refinements that eventually could be benchmarked to the same experimental data set. Guevara-Carrión *et al.*[92] focused on the objective of determining the self-diffusion coefficient and shear viscosity of water and the binary mixtures with methanol and ethanol using MD, together with estimating excess properties. Different rigid non-polarisable models were used to describe water (SPC/E and TIP4P/2005) and alkanols. MS were performed with the sm2[34] code, and regarding the estimation of V^E, the authors reported better agreement with experimental data using the TIP4P/2005 water model, if compared with SPC/E. Despite some qualitative deviations they conclude that rigid molecular models perform better than the OPLS-AA used in the case of ethanol by Wensink *et al.*[91] Finally, Gómez-Álvarez *et al.*[93] recently published an original paper where the experimental volumetric behaviour of (water + 2-propanol) was studied over a wide range of pressure and temperature. Then MC calculations were performed using the TIP4P/2005 and OPLS molecular models. The results reveal good agreement for both mixture density and also for its pressure (κ_T) and temperature (α_P) derivatives. These two response functions were estimated using the fluctuation method, and good correspondence with experimental data was found.

Some other works have been published concerning other aqueous solutions. An example is the work by Destrigneville *et al.*[94] on the (water + carbon dioxide) mixture. Using MC, TIP4P water and MSM3[95] carbon dioxide, they determined V^E over a wide temperature and pressure range, finding good quantitative agreement. At high pressure V^E *versus* P goes through a maximum around 30–40 MPa and then decreases asymptotically to zero. This trend allows one to make a good estimation in the wide region where the mixture shows non-ideal behaviour, spanning pressures of up to 100 MPa at temperatures between 700 and 1000 K. Nevertheless, the negative trend of the low pressure values was not adequately described with the molecular models used.

In addition to the systems previously mentioned, a few more works have been devoted to the prediction of excess volumetric properties using the SAFT formalism. Zúñiga-Moreno *et al.*[96] determined the liquid densities and excess volume of the (CO_2 + thiophene) binary mixture in a wide range of temperature and pressure. The authors used PC-SAFT EoS to model the volumetric behaviour of the mixture and the V^E. Agreement between experimental data and theoretical predictions at one temperature considered is excellent over the whole composition range. Finally, Giner *et al.*[97,98] published a series of papers experimentally determining the phase behaviour of cyclic (ether + haloalkane) mixtures. In addition, they used SAFT-VR EoS to model phase equilibria. In particular, they considered mixtures of tetrahydrofuran and tetrahydropyran[97] and 1,3-dioxolane and 1,4-dioxolane[98] with some isomers of chlorobutane (1-chlorobutane, 2-chlorobutane, 1-chloro-2-methylpropane, and 2-chloro-2-methylpropane). They also estimated the G^E behaviour for the mixtures studied.

Excess volumes have also been computed using MS for molecules of more complex structure and interactions, as it is the case for ionic liquids (ILs). This term refers to ionic salts that are liquid around room conditions. Since the moment when they were pointed out as potential green solvents, a real avalanche of scientific works analysing their synthesis, characterisation, chemical and physical properties, toxicity, chemical engineering applications, *etc.* has been published, and the subject is a dynamic research area. The molecular modelling of these substances is still a challenge due to their molecular size, complex structure that makes coarse graining essential, and very intense and dominant coulombic interactions. Undoubtedly, and although many relevant contributions have been already presented, there is still a long path to be explored. The number of studies concerning mixtures of ILs is, in comparison, still small and this includes the experimental determination of excess properties. Nevertheless, there are a few good works that used MS to estimate ILs excess properties. For instance, Hanke and Lynden-Bell[99] used MD to study the mixture of water and two 1,3-dialkyl imidazolium derivatives. The SPC/E model was used for water, and ILs were modelled as chains of beads interacting through Buckingham[37] potentials and point electric charges. V^E obtained were negative when the IL anion was Cl^-, which is fully miscible in water, but turned positive for a less hydrophilic anion $[BF_4]^-$. Nevertheless, these differences in the V^E sign do not correlate with relevant differences in the microscopic fluid structure, which is very similar in both cases, and apparently not very sensitive to the hydrophilic character of the anion in the IL. Shimizu *et al.*[100] used a similar approach using MD to model the mixture of two ILs with a common ion, investigating the reasons for the bizarre quasi-ideal behaviour of this type of mixture. The OPLS-AA model accurately estimated the very small V^E of this mixture at equimolar conditions, and the authors recall in this case again that the magnitude of the uncertainty exceeded its calculated value; it must be reminded that V^E represents approximately 0.5% of the mixture volume, so its precise determination is rather difficult. Bearing in mind the limitations of the estimation technique due to the statistics of the process, the fact that a property as elusive as V^E can be described with quantitative accuracy for a system composed of molecules as complex as ILs is a strong argument supporting the application of MS for this type of thermodynamic property.

16.4 Concluding Remarks

SAFT approach and molecular simulation techniques represent two approaches that have shown their potential as quantitative estimative tools for complex solution excess properties. Other traditional and widely used approaches to estimate these deviations from ideal solution behaviour properties lack generality, rely on empirical approaches and their efficiency is usually limited. SAFT models are grounded on a rigorous application of the perturbation theory to an underlying detailed molecular model. With different versions that consider different types of monomer interaction potentials, chain terms, electrostatic or polar terms, and so on, their use allows for the detailed discussion of the influence of every molecular model's particular features on excess properties. This underlines the thermodynamic soundness of the method. In particular, the characteristic parameters of SAFT models have solid physical meanings, and may be used in a transferable way with remarkable results if the calculations are performed with rigour. The ability to describe excess properties of fluid solutions of complex behaviour as associative molecules underlines the wide range of applicability and reliability of this theoretical approach.

The influence of every feature and variable of a given molecular model on the excess properties can be explored in detail using MS, and the technique has proved to be quantitative even for molecules interacting through complex strong and long-range interaction potentials, as in the case of ILs. The long calculation times needed, and the limitations imposed by the statistics of the simulation runs on the uncertainty of results must be kept in mind, but undoubtedly the conclusions drawn from MS estimation of excess properties provide very useful hints or the validation of a certain molecular model, or the determination of its confidence limits, which are essential in the further development of EoS and thermodynamic or statistical fluid models.

Acknowledgements

The authors acknowledge financial support from project number FIS2011-13119-E, Red de Simulación Molecular (RdSiMol) of Subprograma de Acciones Complementarias del Ministerio de Ciencia e Innovación. FJB acknowledges financial support from project number FIS2010-14866 and MMP from project number FIS2012-33621 (this one co-financed with EU FEDER funds), both from the Spanish Ministerio de Ciencia e Innovación. Additional support from Universidad de Huelva and Junta de Andalucía is also acknowledged. MMP also acknowledges CESGA (www.cesga.es, Santiago de Compostela, Spain) for providing access to computing facilities.

References

1. M. S. Wertheim, *J. Stat. Phys.*, 1984, **35**, 19.
2. M. S. Wertheim, *J. Stat. Phys.*, 1984, **35**, 35.
3. M. S. Wertheim, *J. Stat. Phys.*, 1986, **42**, 459.
4. M. S. Wertheim, *J. Stat. Phys.*, 1986, **42**, 477.
5. W. G. Chapman, G. Jackson and K. E. Gubbins, *Mol. Phys.*, 1988, **65**, 1057.
6. G. Jackson, W. G. Chapman and K. E. Gubbins, *Mol. Phys.*, 1988, **65**, 1.
7. S. H. Huang and M. Radosz, *Ind. Eng. Chem. Res.*, 1990, **29**, 2284.
8. S. H. Huang and M. Radosz, *Ind. Eng. Chem. Res.*, 1991, **30**, 1994.
9. J. K. Johnson, E. A. Müller and K. E. Gubbins, *J. Phys. Chem.*, 1994, **98**, 6413.
10. D. Ghonasgi and W. G. Chapman, *AIChE J.*, 1994, **40**, 878.
11. F. J. Blas and L. F. Vega, *Mol. Phys.*, 1997, **92**, 135.
12. F. J. Blas and L. F. Vega, *Ind. Eng. Chem. Res.*, 1998, **37**, 660.
13. A. Gil-Villegas, A. Galindo, P. J. Whitehead, S. J. Mills, G. Jackson and A. N. Burgess, *J. Chem. Phys.*, 1997, **106**, 4168.
14. A. Galindo, L. A. Davies, A. Gil-Villegas and G. Jackson, *Mol. Phys.*, 1998, **93**, 241.
15. J. Gross and G. Sadowski, *Ind. Eng. Chem. Res.*, 2001, **40**, 1244.
16. M. Kleiner, F. Tumakaka and G. Sadowski, in *Molecular Thermodynamics of Complex Systems*, ed. X. Lu and Y. Hu, Springer-Verlag Berlin, Berlin, 2009, vol. 131, p. 75.
17. E. A. Müller and K. E. Gubbins, *Ind. Eng. Chem. Res.*, 2001, **40**, 2211.
18. I. E. Economou, *Ind. Eng. Chem. Res.*, 2002, **41**, 953.
19. P. Paricaud, A. Galindo and G. Jackson, *Fluid Phase Equilib.*, 2002, **194–197**, 87.
20. S. P. Tan, H. Adidharma and M. Radosz, *Ind. Eng. Chem. Res.*, 2008, **47**, 8063.
21. C. McCabe and A. Galindo, in *Applied Thermodynamics of Fluids*, ed. A. R. H. Goodwin, J. V. Sengers and C. J. Peters, Royal Society of Chemistry, Cambridge, 2010, ch. 8, p. 215.

22. D. Frenkel and B. Smit, *Understanding Molecular Simulation*, Elsevier, San Diego, 2nd edn, 2002.
23. M. P. Allen and D. J. Tildesley, *Computer Simulation of Liquids*, Oxford University Press, Oxford, 1987.
24. D. P. Landau and K. Binder, *A Guide to Monte Carlo Simulation in Statistical Physics*, Cambridge University Press, Cambridge, 2nd edn, 2005.
25. D. C. Rapaport, *The Art of Molecular Dynamics Simulation*, Cambridge University Press, Cambridge, 2005.
26. R. J. Sadus, *Molecular simulation of fluids: theory, algorithms, and object-orientation*, Elsevier, Amsterdam, 1999.
27. D. M. Heyes, *The Liquid state: Applications of molecular simulations*, John Wiley & Sons, Chichester, 1998.
28. H. J. C. Berendsen, D. van der Spoel and R. vand Runen, *Comput. Phys. Commun.*, 1995, **91**, 43.
29. I. T. Todorov, W. Smith, K. Trachenko and M. T. Dove, *J. Mater. Chem.*, 2006, **16**, 1611.
30. S. Plimpton, *J. Comput. Phys.*, 1995, **117**, 1.
31. J. C. Phillips, R. Braun, W. Wang, J. Gumbart, E. Tajkhorshid, E. Villa, C. Chipot, R. D. Skeel, L. Kale and K. Schulten, *J. Comput. Chem.*, 2005, **26**, 1781.
32. D. A. Case, T. E. Cheatham, III, T. Darden, H. Gohlke, R. Luo, K. M. Merz, Jr., A. Onufriev, C. Simmerling, B. Wang and R. Woods, *J. Comput. Chem.*, 2005, **26**, 1668.
33. M. G. Martin, *Mol. Simul.*, 2013, **39**, 1184.
34. S. Deublein, B. Eckl, J. Stoll, S. V. Lishchuk, G. Guevara-Carrion, C. W. Glass, T. Merker, M. Bernreuther, J. Vrabec and H. Hasse, *Comput. Phys. Commun.*, 2011, **182**, 2350.
35. P. Morávek, J. Kolafa, T. Hujo and S. Labík, *J. Mol. Liq.*, 2006, **125**, 22.
36. J. E. Lennard-Jones, *Proc. R. Soc. London, Ser. A*, 1924, **106**, 463.
37. I. G. Kaplan, *Intermolecular interactions: Physical picture, computational methods and model potentials*, John Wiley and Sons, Chichester, 2006.
38. H. M. Schaink and C. Hoheisel, *J. Chem. Phys.*, 1992, **97**, 6561.
39. K. Shukla, *Fluid Phase Equilib.*, 1997, **128**, 29.
40. F. J. Blas, *J. Phys. Chem. B*, 2000, **104**, 9239.
41. F. J. Blas, *Mol. Phys.*, 2002, **100**, 2221.
42. F. J. Blas and I. Fujihara, *Mol. Phys.*, 2002, **100**, 2823.
43. I. Fujihara and K. Nakanishi, *J. Chem. Phys.*, 1997, **107**, 3121.
44. I. Fujihara, Y. Miyano, K. Sakamoto, K. Satoh and K. Nakanishi, *Fluid Phase Equilib.*, 2012, **336**, 1.
45. A. Mukherjee and B. Bagchi, *J. Phys. Chem. B*, 2001, **105**, 9581.
46. S. Amore, J. Horbach and I. Egry, *J. Chem. Phys.*, 2011, **134**, 044515.
47. L. G. MacDowell, C. Vega and A. Lopez Rodriguez, *J. Chem. Phys.*, 1999, **111**, 3183.
48. C. Vega, L. G. MacDowell and A. Lopez Rodriguez, *J. Chem. Phys.*, 1999, **111**, 3192.
49. M. C. Dos Ramos and F. J. Blas, *J. Phys. Chem. B*, 2005, **109**, 12145.

50. L. Sun, H. Zhao, S. B. Kiselev and C. McCabe, *J. Phys. Chem. B*, 2005, **109**, 9047.
51. C. McCabe and S. B. Kiselev, *Ind. Eng. Chem. Res.*, 2004, **43**, 2839.
52. C. McCabe and S. B. Kiselev, *Fluid Phase Equilib.*, 2004, **219**, 3.
53. M. C. dos Ramos, A. Villegas-Páez, M. M. Piñeiro and F. J. Blas, *Fluid Phase Equilib.*, 2014, **361**, 93.
54. L. G. Martins, E. J. M. Filipe and J. C. G. Calado, *J. Phys. Chem. B*, 2001, **105**, 10936.
55. E. J. M. Filipe, E. J. S. Gomes de Azevedo, L. G. Martins, V. A. M. Soares, J. C. G. Calado, C. McCabe and G. Jackson, *J. Phys. Chem. B*, 2000, **104**, 1315.
56. E. J. M. Filipe, L. G. Martins, J. C. G. Calado, C. McCabe and G. Jackson, *J. Phys. Chem. B*, 2000, **104**, 1322.
57. R. P. M. F. Bonifácio, E. J. M. Filipe, M. C. dos Ramos, F. J. Blas and L. G. Martins, *Fluid Phase Equilib.*, 2011, **303**, 193.
58. L. M. B. Dias, E. J. M. Filipe, C. McCabe, T. Cordeiro and J. C. G. Calado, *J. Phys. Chem. B*, 2007, **111**, 5284.
59. U. Gross, G. Papke and S. Rudiger, *J. Fluorine Chem.*, 1993, **61**, 11.
60. P. Morgado, C. McCabe and E. J. M. Filipe, *Fluid Phase Equilib.*, 2005, **228–229**, 389.
61. P. Morgado, H. Zhao, F. J. Blas, C. McCabe, L. P. N. Rebelo and E. J. M. Filipe, *J. Phys. Chem. B*, 2007, **111**, 2856.
62. P. Morgado, R. Tomás, H. Zhao, M. C. dos Ramos, F. J. Blas, C. McCabe and E. J. M. Filipe, *J. Phys. Chem. C*, 2007, **111**, 15962.
63. P. Morgado, H. Rodrigues, F. J. Blas, C. McCabe and E. J. M. Filipe, *Fluid Phase Equilib.*, 2011, **306**, 76.
64. P. Morgado, J. B. Lewis, C. M. C. Laginhas, L. F. Martins, C. McCabe, F. J. Blas and E. J. M. Filipe, *J. Phys. Chem. B*, 2011, **115**, 15013.
65. M. G. Martin and I. J. Siepmann, *J. Phys. Chem. B*, 1999, **103**, 4508.
66. A. J. Palace Carvalho, J. P. Prates Ramalho and L. F. G Martins, *J. Phys. Chem. B*, 2007, **111**, 6437.
67. L. F. G. Martins, A. J. Palace Carvalho, J. P. Prates Ramalho and E. J. M. Filipe, *J. Phys. Chem. B*, 2011, **115**, 9745.
68. M. Khammar and J. M. Shaw, *Fluid Phase Equilib.*, 2010, **288**, 145.
69. T. Lafitte, D. Bessières, M. M. Piñeiro and J. L. Daridon, *J. Chem. Phys.*, 2006, **124**, 024509.
70. T. Lafitte, M. M. Piñeiro, J. L. Daridon and D. Bessières, *J. Phys. Chem. B*, 2007, **111**, 3447.
71. G. Watson, T. Lafitte, C. K. Zéberg-Mikkerlsen, A. Baylacq, D. Bessieres and C. Boned, *Fluid Phase Equilib.*, 2006, **247**, 121.
72. S. Aparicio, B. García, R. Alcalde, M. J. Dávila and J. M. Leal, *J. Phys. Chem. B*, 2006, **110**, 6933.
73. S. B. Kiselev, J. F. Ely, S. P. Tan, H. Adidharma and M. Radosz, *Ind. Eng. Chem. Res.*, 2006, **45**, 3981.
74. S. B. Kiselev, J. F. Ely, H. Adidharma and M. Radosz, *Fluid Phase Equilib.*, 2001, **183**, 53.

75. G. Guevara-Carrión, C. Nieto-Draghi, J. Vrabec and H. Hasse, *J. Phys. Chem. B*, 2008, **112**, 16664.
76. P. Duarte, M. Silva, D. Rodrigues, P. Morgado, L. F. G. Martins and E. J. M. Filipe, *J. Phys. Chem B*, 2013, **117**, 9709.
77. W. L. Jorgensen, D. S. Maxwell and J. Tirado-Rives, *J. Am. Chem. Soc.*, 1996, **118**, 11225.
78. M. C. dos Ramos, F. J. Blas and A. Galindo, *J. Phys. Chem. C*, 2007, **111**, 15924.
79. M. C. dos Ramos, F. J. Blas and A. Galindo, *Fluid Phase Equilib.*, 2007, **261**, 359.
80. M. C. dos Ramos and C. McCabe, *Fluid Phase Equilib.*, 2010, **290**, 137.
81. C. A. Koh, H. Tanaka, J. M. Walsh, K. E. Gubbins and J. A. Zollweg, *Fluid Phase Equilib.*, 1993, **83**, 51.
82. W. L. Jorgensen, *J. Phys. Chem.*, 1986, **90**, 1276.
83. C. Vega and J. L. F. Abascal, *Phys. Chem. Chem. Phys.*, 2011, **13**, 19663.
84. D. González-Salgado and I. Nezbeda, *Fluid Phase Equilib.*, 2006, **240**, 161.
85. A. Dopazo-Paz, P. Gómez-Álvarez and D. González-Salgado, *Collect. Czech. Chem. Commun.*, 2010, **75**, 617.
86. L. Vlcek and I. Nezbeda, *J. Mol. Liq.*, 2007, **131–132**, 158.
87. S. Weerasinghe and P. E. Smith, *J. Phys. Chem. B*, 2005, **109**, 15080.
88. Y. Zhong, G. L. Warren and S. Patel, *J. Comput. Chem.*, 2008, **29**, 1142.
89. S. Patel and C. L. Brooks, *J. Chem. Phys.*, 2005, **122**, 024508.
90. A. Perera, L. Zoranic, F. Sokolic and R. Mazighi, *J. Mol. Liq.*, 2011, **159**, 52.
91. E. J. W. Wensink, A. C. Hoffmann, P. J. van Maaren and D. van der Spoel, *J. Chem. Phys.*, 2003, **119**, 7308.
92. G. Guevara-Carrión, J. Vrabec and H. Hasse, *J. Chem. Phys.*, 2011, **134**, 074508.
93. P. Gómez-Álvarez, D. González-Salgado, J.-P. Bazile, D. Bessières and F. Plantier, *Fluid Phase Equilib.*, 2013, **358**, 7.
94. C. M. Destrigneville, J. P. Brodholt and B. J. Wood, *Chem. Geol.*, 1996, **133**, 53.
95. J. P. Brodholt and B. J. Wood, *Am. Mineral.*, 1993, **78**, 558.
96. A. Zúñiga-Moreno, L. A. Galicia-Luna and F. F. Betancourt-Cárdenas, *Fluid Phase Equilib.*, 2005, **236**, 193.
97. B. Giner, I. Bandrés, M. C. López, C. Lafuente and A. Galindo, *J. Chem. Phys.*, 2007, **127**, 144513.
98. B. Giner, I. Gascón, H. Artigas, C. Lafuente and A. Galindo, *J. Phys. Chem. B*, 2007, **111**, 9588.
99. C. G. Hanke and R. M. Lynden-Bell, *J. Phys. Chem. B*, 2003, **107**, 10 873.
100. K. Shimizu, M. Tariq, L. P. N. Rebelo and J. N. Canongia Lopes, *J. Mol. Liq.*, 2010, **153**, 52.

CHAPTER 17

Calculation of Thermodynamic Functions from Volumetric Properties

JOSEF P. NOVÁK, KVĚTOSLAV RŮŽIČKA* AND MICHAL FULEM

Department of Physical Chemistry, Institute of Chemical Technology, Technická 5, 166 28 Prague, Czech Republic
*Email: ruzickak@vscht.cz

17.1 Introduction

Volumetric measurements give information about the variation of thermodynamic properties with pressure or density at constant temperature and are key inputs in the calculation of thermodynamic properties of real fluids (mass and enthalpic balances, phase and chemical equilibria, *etc.*).

Direct experimental volumetric data are, however, seldom applied to calculations of thermodynamic properties. Instead, analytical formulae in the form of equations of state (EOS) are used:

$$p = p(T,\rho) \tag{17.1}$$

or

$$A_m = A_m(T,\rho) \tag{17.2}$$

where A_m stands for molar Helmholtz energy.

Pure component critical data are sufficient input information to obtain EOS parameters in the case of simple (cubic) EOS; vapour pressures are additionally required for the calculation of vapour–liquid equilibria (VLE).

Volume Properties: Liquids, Solutions and Vapours
Edited by Emmerich Wilhelm and Trevor M. Letcher

Published by the Royal Society of Chemistry, www.rsc.org

For advanced mutiparameter EOS covering wide ranges of temperatures and pressures, a simultaneous fit of several experimental properties is necessary for establishing EOS parameters.[1]

The number of industrially important mixtures outnumbers industrially important pure compounds by several orders of magnitude. The prerequisite for solving chemical and phase equilibria for mixtures is knowledge of the chemical potential, which can be obtained from volumetric and thermal properties.

This chapter is devoted to the calculation of thermodynamic properties based on volumetric behaviour. The simple and rational way of performing such calculations is shown for thermodynamic as well as partial molar properties. Also, the role of empirical rules enabling the utilisation of state-of-the-art EOS for each component of a given mixture is highlighted. All the above mentioned calculations can be rationalised by utilising appropriate auxiliary quantities, as described in this chapter. The subject covered in this chapter represents extension and clarification of topics included in recent monographs devoted to EOS and their applications.[2,3]

17.2 *pVT* Description

A *pVT* description is indispensable for the calculation of thermodynamic properties of a real fluid. It can be represented by four basic approaches: (a) by pressure-explicit EOS; (b) by corresponding states principle; (c) by Helmhotz energy-explicit EOS; and (d) by empirical "laws" or "rules" for mixtures.

(a) The **pressure-explicit cubic EOS** in the general form:[4]

$$p = \frac{RT}{V_{\mathrm{m}} - b} - \frac{\theta}{V_{\mathrm{m}}^2 + \delta V_{\mathrm{m}} + \varepsilon} \tag{17.3}$$

is the most frequently used. A classical textbook example is the van der Waals EOS ($\theta = a$, $\delta = \varepsilon = 0$), while the Redlich–Kwong EOS ($\theta = a/\sqrt{T}$, $\delta = b$, $\varepsilon = 0$) and its modifications are among the most used and popular EOS with wide practical applications. For a recent review on the pressure-explicit EOS the readers are referred to references 5, 6 and 7. In the case of mixtures, the parameters are considered as functions of composition.[8–10] The limitation of this approach is that the same EOS must be used for all the components. Furthermore, mixing rules for parameters of more complex EOS are usually unavailable.

(b) The application of the **corresponding state principle** (CSP) explicit in the compressibility factor z, is less frequently employed.[11–13]

(c) The **Helmholtz energy-explicit EOS** (also called fundamental EOS) have become very popular over the past few decades[1,14–17] (see Section 17.3.4) and are primarily used in calculations of thermodynamic properties (see Section 17.3.7). In the case of mixtures, an extended modified Joffe's rule is used (see Section 17.3.6).

(d) As an alternative, **empirical "laws" and "rules"** based on the properties of pure components are used for mixtures. A significant improvement can be achieved by combining such 'laws' with the CSP, as discussed in section 17.3.5. This approach enables one to use a different EOS for each component and thus take advantage of state-of-the-art fundamental EOS. We note that this approach has not been covered in the books edited by Sengers *et al.*[2,3]

17.3 Thermodynamic Properties of a Real Fluid

Calculations of the thermodynamic properties of a real fluid can be rationalised by dividing the calculation into two parts (i) the calculation of the thermodynamic properties of an ideal gas and (ii) the calculation of the correction for non-ideal behaviour by means of departure or residual properties. Two sets of independent variables $T,p,x_1,x_2,\ldots$ and $T,V_m,x_1,x_2,\ldots$ or $T,\ \rho,\ x_1,x_2,\ldots$ ($\rho=1/V_m$) are employed. While the former is preferred in practice, the latter is predominantly used by the scientific community.

17.3.1 Thermodynamic Properties of an Ideal Gas

Let us consider a system in the state of an ideal gas at temperature T and pressure p, with standard pressure $p^{st}=100$ kPa (or 101.325 kPa) and standard volume and density defined as $V^{st}=RT/p^{st}$ and $\rho^{st}=p^{st}/RT$, respectively. As a basis of all subsequent calculations, we will assume that at the system temperature T and standard pressure p^{st} (such states will be denoted by superscript "0") the values of enthalpy $[H^0_{m,i}(T,p^{st})=H^0_{m,i}(T,\rho^{st})=H^0_{m,i}(T)]$ and entropy $[S^0_{m,i}(T,p^{st})=S^0_{m,i}(T,\rho^{st})]$ are known. Then it holds that:

$$\begin{aligned}
&U^0_{m,i}(T,p^{st})=U^0_{m,i}(T,\rho^{st})=U^0_{m,i}(T)=H^0_{m,i}(T)-RT\\
&A^0_{m,i}(T,p^{st})=U^0_{m,i}(T,p^{st})-TS^0_{m,i}(T,p^{st})\\
&G^0_{m,i}(T,p^{st})=H^0_{m,i}(T,p^{st})-TS^0_{m,i}(T,p^{st})
\end{aligned}\tag{17.4}$$

Thermodynamic properties of a pure compound at system T,p or T,ρ (denoted by superscript "*") will be equal to $Y^0_{m,i}(T,p^{st})$ or $Y^0_{m,i}(T,\rho^{st})$ for the properties derived from the first law of thermodynamics ($Y=U,\ H,\ C_p,\ C_V$) and the summation of values for pure components yields the value for the mixture $Y=\sum x_iY_i$. For properties based on the second law of thermodynamics ($Y=S,\ A,\ G$):

$$S^*_{m,i}(T,p)=S^0_{m,i}(T,p^{st})-R\ln(p/p^{st}),\quad S^*_{m,i}(T,\rho)=S^0_{m,i}(T,\rho^{st})-R\ln(\rho/\rho^{st})\tag{17.5}$$

and analogous relations hold for A and G. For mixtures, the mixing entropy must be added:

$$S^*_{\rm m}(T,p,\mathbf{x})=\sum x_i S^*_{{\rm m},i}(T,p)-R\sum x_i \ln x_i=\sum x_i S^0_{{\rm m},i}(T,p^{\rm st})-R\sum x_i \ln(px_i/p^{\rm st})$$
$$S^*_{\rm m}(T,\rho,\mathbf{x})=\sum x_i S^*_{{\rm m},i}(T,\rho)-R\sum x_i \ln x_i=\sum x_i S^0_{{\rm m},i}(T,\rho^{\rm st})-R\sum x_i \ln(\rho x_i/\rho^{\rm st})$$
(17.6)

Again, analogous equations hold for A and G.

17.3.2 Departure and Residual Properties

Molar departure and molar residual functions (denoted by subscripts "d" and "res", respectively) are defined as:

$$Y_{\rm d}(T,p,\mathbf{x})=Y_{\rm m}(T,p,\mathbf{x})-Y^*_{\rm m}(T,p,\mathbf{x}),\quad Y_{\rm res}(T,\rho,\mathbf{x})=Y_{\rm m}(T,\rho,\mathbf{x})-Y^*_{\rm m}(T,\rho,\mathbf{x}) \tag{17.7}$$

Alternatively, these functions were defined by Beattie and Stockmayer[18] as:

$$Y_{\rm d}=\int_0^p\left[\left(\frac{\partial Y_{\rm m}}{\partial p}\right)_{T,\mathbf{x}}-\left(\frac{\partial Y_{\rm m}}{\partial p}\right)^*_{T,\mathbf{x}}\right]{\rm d}p \tag{17.8}$$

$$Y_{\rm res}=\int_\infty^{V_{\rm m}}\left[\left(\frac{\partial Y_{\rm m}}{\partial V_{\rm m}}\right)_{T,\mathbf{x}}-\left(\frac{\partial Y_{\rm m}}{\partial V_{\rm m}}\right)^*_{T,\mathbf{x}}\right]{\rm d}V_{\rm m}=\int_0^\rho\left[\left(\frac{\partial Y_{\rm m}}{\partial \rho}\right)_{T,\mathbf{x}}-\left(\frac{\partial Y_{\rm m}}{\partial \rho}\right)^*_{T,\mathbf{x}}\right]{\rm d}\rho \tag{17.9}$$

The departure and residual properties diminish with decreasing pressure and density, as is obvious from Equations (17.8) and (17.9) with the exception of the departure volume [see Equation (17.13)].

It should be noted that the departure and residual functions are not distinguished in most papers and/or not used in a consistent way. Both departure and residual functions express differences between the system in the ideal gas state and real state, but their definition is different, as seen in Figure 17.1.

Relations valid for thermodynamic properties hold also for residual and departure functions:

$$G_{\rm d}=H_{\rm d}-TS_{\rm d},\quad G_{\rm res}=H_{\rm res}-TS_{\rm res},\quad H_{\rm d}=U_{\rm d}+pV_{\rm d},\quad H_{\rm res}=U_{\rm res}+p_{\rm res}/\rho \tag{17.10}$$

as well as for their derivatives:

$$(\partial G_{\rm d}/\partial p)_T=V_{\rm d},\quad (\partial A_{\rm res}/\partial V_{\rm m})_T=-p_{\rm res},\quad (\partial H_{\rm d}/\partial T)_p=C_{p,{\rm d}} \tag{17.11}$$

The departure volume $V_{\rm d}$ is according to the definition expressed as:

$$V_{\rm d}=V_{\rm m}-V^*_{\rm m}=zRT/p-RT/p=(z-1)RT/p \tag{17.12}$$

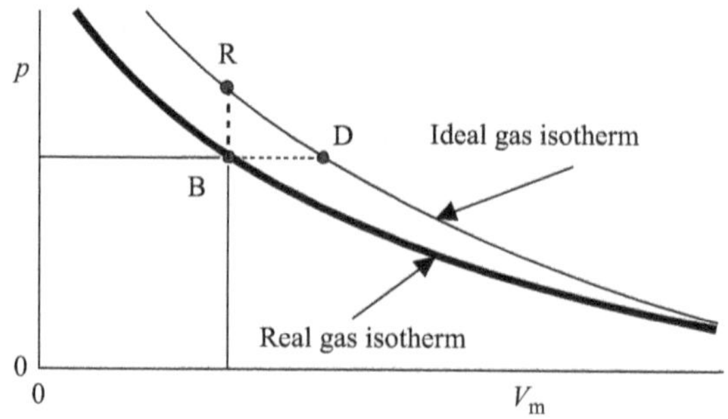

Figure 17.1 Graphical representation of departure and residual functions, Equations (17.8) and (17.6).

By using the pressure virial equation, we obtain:

$$V_d = B + C'p + \ldots \tag{17.13}$$

which means that $V_d \neq 0$ even at a limiting pressure of zero.

17.3.3 Application of *Q*-Quantities

Calculation of volumetric and other thermodynamic properties by means of EOS performed by substitution of relevant properties into Equations (17.8) and (17.9) represents a classical approach. A more efficient way is to introduce a few convenient "*Q*-quantities" defined as:[19,20]

$$Q_\rho = z + \rho\left(\frac{\partial z}{\partial \rho}\right)_{T,\mathbf{x}} = \frac{1}{RT}\left(\frac{\partial p}{\partial \rho}\right)_{T,\mathbf{x}} \tag{17.14}$$

$$Q_T = z + T\left(\frac{\partial z}{\partial T}\right)_{\rho,\mathbf{x}} = \frac{1}{R\rho}\left(\frac{\partial p}{\partial T}\right)_{\rho,\mathbf{x}} \tag{17.15}$$

which enable the calculation of volume derivatives:

$$\left(\frac{\partial V_m}{\partial T}\right)_{p,\mathbf{x}} = \frac{Q_T}{\rho T Q_\rho}, \qquad \left(\frac{\partial V_m}{\partial p}\right)_{T,\mathbf{x}} = -\frac{1}{\rho^2 RT Q_\rho} \tag{17.16}$$

and significantly facilitate the calculation of thermodynamic properties of real fluids. For the calculation of the *Q*-quantities, the compressibility factor z is considered as a function of temperature, density, and composition.

While the quantities z, Q_ρ, and Q_T are sufficient for the description of the volumetric behaviour of a system, the calculation of chemical potential and other thermodynamic properties requires additional auxiliary quantities Q_A, Q_U, and Q_C:

$$Q_A = \int_0^{\rho} (z-1)\,\mathrm{d}\ln\rho \tag{17.17}$$

$$Q_U = T\left(\frac{\partial Q_A}{\partial T}\right)_{\rho,\mathbf{x}} = T\int_0^{\rho}\left(\frac{\partial z}{\partial T}\right)_{\rho,\mathbf{x}} \mathrm{d}\ln\rho \tag{17.18}$$

$$Q_C = T^2\left(\frac{\partial^2 Q_A}{\partial T^2}\right)_{\rho,\mathbf{x}} = T^2\int_0^{\rho}\left(\frac{\partial^2 z}{\partial T^2}\right)_{\rho,\mathbf{x}} \mathrm{d}\ln\rho \tag{17.19}$$

This approach is general and can be used for any equation of state (virial, cubic, BWR-like, fundamental, *etc.*). A similar approach was independently developed by Schmidt and Wagner;[14] however their quantities $\alpha_{\tau\delta}$ are specific to the fundamental EOS (see Section 17.3.4).

It is apparent that for low densities:

$$\lim_{\rho\to 0} z = \lim_{\rho\to 0} Q_\rho = \lim_{\rho\to 0} Q_T = 1, \quad \lim_{\rho\to 0} Q_A = \lim_{\rho\to 0} Q_U = \lim_{\rho\to 0} Q_C = 0 \tag{17.20}$$

The residual and departure functions can be expressed as:[19,20]

$$\begin{aligned}
&U_\mathrm{d} = U_\mathrm{res} = -RTQ_U, \quad H_\mathrm{d} = H_\mathrm{res} = -RT(Q_U + 1 - z)\\
&C_{p,\mathrm{d}} = C_{p,\mathrm{res}} = -R\left[1 + 2Q_U + Q_C - Q_T^2/Q_\rho\right], \quad C_{V,\mathrm{d}} = C_{V,\mathrm{res}} = -R(2Q_U + Q_C)\\
&S_\mathrm{d} = -R(-\ln z + Q_A + Q_U), \quad S_\mathrm{res} = -R(Q_A + Q_U)\\
&A_\mathrm{d} = RT(Q_A - \ln z), \quad A_\mathrm{res} = RT\,Q_A\\
&G_\mathrm{d} = RT(Q_A - \ln z + z - 1), \quad G_\mathrm{res} = RT(Q_A + z - 1)\\
&\ln(f/p) = -\ln z + Q_A + z - 1,\\
&\mu_{JT} = \left[Q_T - Q_\rho\right]/\left(\rho C_{p\mathrm{m}} Q_\rho\right), \quad v_\mathrm{s}^2 = Q_\rho\left(C_{p\mathrm{m}}/C_{V\mathrm{m}}\right)(RT/M)
\end{aligned} \tag{17.21}$$

where μ_{JT} is the Joule–Thomson coefficient and v_s is the speed of sound.

For some types of calculations, derivatives of thermodynamic functions are also necessary. Their evaluation by means of the Q-quantities is summarised in Table 17.1.

As software packages capable of the calculation of analytical derivatives are commonly available, the only Q-quantity which must be derived for a given EOS is the Q_A defined by Equation (17.17). All other Q-quantities can be

Table 17.1 Derivatives of basic thermodynamic properties with respect to temperature, density, and pressure.

Quantity	$Y=Y(T,\rho,\mathbf{x})$ $(\partial Y/\partial T)_\rho$	$(\partial Y/\partial \rho)_T$	$Y=Y(T,p,\mathbf{x})$ $(\partial Y/\partial T)_p$	$(\partial Y/\partial p)_T$
z	$\frac{(Q_T-z)}{T}$	$\frac{(Q_\rho-z)}{\rho}$	$\frac{(z/T)(Q_T-Q_\rho)}{Q_\rho}$	$\frac{[(Q_\rho-z)/Q_\rho]}{\rho RT}$
p, V_m	$R\rho Q_T$	RTQ_ρ	$Q_T/(\rho TQ_\rho)$	$-(\rho^2RTQ_\rho)^{-1}$
U_m	C_{Vm}	$\frac{-RT(Q_T-z)}{\rho}$	$C_{Vm}+ RQ_T(Q_T-z)/Q_\rho$	$\frac{-R(Q_T-z)}{\rho Q_\rho}$
H_m	$C_{pm}+ RQ_T(Q_\rho-Q_T)/Q_\rho$	$\frac{-RT(Q_T-Q_\rho)}{\rho}$	C_{pm}	$\frac{-(Q_T-Q_\rho)}{\rho Q_\rho}$
S_m	C_{Vm}/T	$-RQ_T/\rho$	C_{pm}/T	$-Q_T/(\rho TQ_\rho)$
A_m	$-S_m$	zRT/ρ	$-S_m+RzQ_T/Q_\rho$	$z/(\rho Q_\rho)$
G_m	$-S_m+RQ_T$	$\frac{RTQ_\rho}{\rho}$	$-S_m$	$\frac{1}{\rho}$
$\ln f$	$-H_d/(RT^2)$ $+(QT-1)/T$	Q_ρ/ρ	$-H_d/(RT^2)$	$1/(\rho RT)$

obtained as its derivative. Q_U and Q_C are defined by Equations (17.18) and (17.19), respectively, and it can be shown that for z, and Q_ρ and Q_T:

$$z=1+\rho\left(\frac{\partial Q_A}{\partial \rho}\right)_{T,\mathbf{x}}$$

$$Q_\rho=1+2\rho\left(\frac{\partial Q_A}{\partial \rho}\right)_{T,\mathbf{x}}+\rho^2\left(\frac{\partial^2 Q_A}{\partial \rho^2}\right)_{T,\mathbf{x}} \tag{17.22}$$

$$Q_T=1+\rho\left(\frac{\partial Q_A}{\partial \rho}\right)_{T,\mathbf{x}}+\rho T\left(\frac{\partial^2 Q_A}{\partial \rho \partial T}\right)_{\mathbf{x}}$$

The relationships for Q_A derived from some common EOS are given below.

volume virial EOS $$Q_A=\frac{\rho}{RT}\left[B+\frac{1}{2}C\rho+\frac{1}{3}D\rho^2+\ldots\right] \tag{17.23}$$

van der Waals EOS $$Q_A=-\ln(1-b\rho)-\frac{a\rho}{RT} \tag{17.24}$$

Redlich-Kwong EOS $$Q_A=-\ln(1-b\rho)-\frac{a}{bRT^{3/2}}\ln(1+b\rho) \tag{17.25}$$

Redlich-Kwong-Soave EOS $$Q_A=-\ln(1-b\rho)-\frac{a}{RTb}\ln(1+b\rho) \tag{17.26}$$

Peng-Robinson EOS $$Q_A = -\ln(1-b\rho) - \frac{a}{\sqrt{8}RTb}\ln\frac{1+b\rho(1+\sqrt{2})}{1+b\rho(1-\sqrt{2})} \quad (17.27)$$

BWR EOS $$Q_A = \rho\left(B_0 - \frac{A_0}{RT} - \frac{C_0}{RT^3}\right) + \frac{\rho^2}{2}\left(b - \frac{a}{RT}\right) + \frac{\alpha a}{5RT}\rho^5 + \frac{c\rho^2}{RT^3}\left[\frac{1-\exp(-\gamma\rho^2)}{\gamma\rho^2} - \frac{1}{2}\exp(-\gamma\rho^2)\right] \quad (17.28)$$

17.3.4 Calculation of Thermodynamic Properties from Helmholtz Energy

The fundamental EOS describes the volumetric behaviour and thermodynamic properties by means of the dimensionless Helmholtz energy:[1,14–17]

$$A_m/(RT) = \alpha(\tau = T_c/T,\ \delta = \rho/\rho_c) = \alpha^0(\tau,\delta) + \alpha^r(\tau,\delta) \quad (17.29)$$

where τ is the reciprocal reduced temperature, δ is the reduced density; A_m is the molar Helmholtz energy; α^0 (τ,δ) and α^r (τ,δ) are the ideal and residual part of the Helmholtz energy, respectively; α^0 (τ,δ) is a function of the standard ideal gas enthalpy and entropy at selected reference state and of ideal gas heat capacity; and α^r $(\tau,\ \delta)$ reflects the non-ideal behaviour of the system.

Basic thermodynamic properties can be expressed as a combination of the dimensionless derivatives of α^0 and α^r:

$$\alpha^r_\delta = \delta\left(\frac{\partial\alpha^r}{\partial\delta}\right)_\tau, \quad \alpha^r_{\delta\delta} = \delta^2\left(\frac{\partial^2\alpha^r}{\partial\delta^2}\right)_\tau \quad (17.30)$$

$$\alpha^r_\tau = \tau\left(\frac{\partial\alpha^r}{\partial\tau}\right)_\delta, \quad \alpha^r_{\tau\tau} = \tau^2\left(\frac{\partial^2\alpha^r}{\partial\tau^2}\right)_\delta, \quad \alpha^r_{\tau\delta} = \tau\delta\left(\frac{\partial^2\alpha^r}{\partial\tau\partial\delta}\right). \quad (17.31)$$

Furthermore:

$$z = 1 + \alpha^r_\delta, \quad p = RT\rho z \quad (17.32)$$

$$U_m = RT[\alpha^0_\tau + \alpha^r_\tau], \quad H_m = RT[1 + \alpha^0_\tau + \alpha^r_\tau + \alpha^r_\delta] \quad (17.33)$$

$$S_m = -R[\alpha^0 - \alpha^0_\tau + \alpha^r - \alpha^r_\tau] \quad (17.34)$$

$$A_m = RT[\alpha^0 + \alpha^r], \quad G_m = RT[1 + \alpha^0 + \alpha^r + \alpha^r_\delta] \quad (17.35)$$

$$f = \rho RT\exp[\alpha^r + z - 1] = \rho RT\exp[\alpha^r + \alpha^r_\delta] \quad (17.36)$$

$$C_{Vm} = -R[\alpha^0_{\tau\tau} + \alpha^r_{\tau\tau}], \quad C_{pm} = C_{Vm} + R\frac{[1 + \alpha^r_\delta - \alpha^r_{\tau\delta}]^2}{1 + 2\alpha^r_\delta + \alpha^r_{\delta\delta}} \quad (17.37)$$

Comparison of the expression for A_{res} from equation (17.21) and (17.35) reveals that:

$$Q_A = \alpha^{r} \tag{17.38}$$

17.3.5 Amagat's Law and Other Empirical Rules

Amagat's law and other empirical rules represent an alternative to the "classical" approach, where the determination of thermodynamic properties is performed similarly, regardless of whether the system is a pure compound or a mixture (in the latter case, EOS parameters or critical properties are replaced by mixture parameters or pseudocritical properties). In many cases, the correction terms (*e.g.* binary interaction parameters) must be used to minimise differences between experimental and calculated values. For the evaluation of the correction terms, experimental data for a given mixture must be available. This obviously cannot be done for all practically important mixtures. It can easily be shown that for a complete description of a ten-component mixture (with composition of each component changing by $\Delta x_i = 0.1$), it would be necessary to perform experiments for 92 378 mixtures of different composition (9 points for each of 45 binaries, 36 points for each of 120 ternaries *etc.*), which would represent several decades of work by an experienced experimentalist. There is an alternative to this "single fluid" approach, which refers to the application of empirical "laws" or "rules". This allows the calculation of mixture properties by a combination of the mixture's pure component properties.

A well-known example of such treatment is Amagat's law:

$$Y(T,p,\mathbf{x}) = \sum x_i Y_i(T,p) + \Delta, \quad Y_d(T,p,\mathbf{x}) = \sum x_i Y_{d,i}(T,p) \tag{17.39}$$

where $Y_i(T,p)$ or $Y_{d,i}(T,p)$ is the system or departure thermodynamic property of a component i at temperature T, pressure p and in the same phase as the mixture (this "phase limitation" represents an important restriction in the application of Amagat's law). In Equation (17.39): $\Delta = 0$ for U_m, H_m, C_{Vm}, C_{pm}; $\Delta = -R\sum x_i \ln x_i$ for entropy S_m, and $\Delta = RT\sum x_i \ln x_i$ for the Helmoholtz energy A_m and Gibbs energy G_m [compare with Equations (17.5) and (17.6)]. Empirical "laws" or "rules" expressed for either the Gibbs energy G (for independent properties T,p) or the Helmholtz energy A (for independent properties T,ρ) are defined as follows:

Amagat's law:

$$G_m(T,p,\mathbf{x}) = G^{*}(T,p,\mathbf{x}) + \sum x_i G_{d,i}(T,p) \tag{17.40}$$

modified Amagat's law:[21]

$$G_m(T,p,\mathbf{x}) = G^{*}(T,p,\mathbf{x}) + \sum x_i G_{d,i}(T, p_i = p x_i) \tag{17.41}$$

Joffe's rule:[22]

$$G_m(T,p,\mathbf{x}) = G^{*}(T,p,\mathbf{x}) + \sum x_i G_{d,i}[T(T_{c,i}/T'_c), p(p_{c,i}/p'_c)] \tag{17.42}$$

Bartlett's rule:[23]

$$A_{\mathrm{m}}(T,\rho,\mathbf{x}) = A^{*}(T,\rho,\mathbf{x}) + \sum x_i A_{\mathrm{res},i}(T,\rho) \tag{17.43}$$

Dalton's law:

$$A_{\mathrm{m}}(T,\rho,\mathbf{x}) = A^{*}(T,\rho,\mathbf{x}) + \sum x_i A_{\mathrm{res},i}(T,\rho_i = \rho x_i) \tag{17.44}$$

and modified Joffe's rule:[24]

$$A_{\mathrm{m}}(T,\rho,\mathbf{x}) = A^{*}(T,\rho,\mathbf{x}) + \sum x_i A_{\mathrm{res},i}[T(T_{\mathrm{c},i}/T'_{\mathrm{c}}),\rho(\rho_{\mathrm{c},i}/\rho')_{\mathrm{c}}] \tag{17.45}$$

It should be stressed that the Amagat's, modified Amagat's, Bartlett's, and Dalton's laws require that the mixture as well as all its components are in the same phase, which usually causes failure if one of the components of the mixture is near its critical point. The Joffe's and modified Joffe's rule on the other hand use reduced independent variables, which practically ensures that all the components as well as the mixture are in the same phase. Another criterion of applicability is the expression for cross virial coefficients B_{ij} derived from the respective "laws". It can be shown that the Amagat's law and also the Bartlett's rule yield $B_{ij} = (B_{ii} + B_{jj})/2$ while the Dalton's and modified Amagat's law lead to $B_{ij} = 0$. The Joffe's and modified Joffe's rule result in more complicated, but also more realistic, expressions for B_{ij}.

The Amagat's law closely describes liquid mixtures and is therefore used extensively for such systems. Also, the Amagat's law was selected by Scatchard[25] as a reference system in the definition of non-ideal and excess functions used for thermodynamic descriptions of real liquid or solid mixtures.

17.3.6 Method of Lemmon, Jacobsen and Tillner-Roth for Mixtures

As mentioned in the previous section, the Joffe's rule Equation (17.42) and modified Joffe's rule Equation (17.45) uses reduced independent variables, thus ensuring that all the components as well as the mixture are in the same phase. An approach resembling (and extending) the modified Joffe's rule has been suggested by several authors.[16,26–28] The advantage of this approach lies in the exploitation of the state-of-art EOS for accurate description of mixture properties. The dimensionless Helmholtz energy is described as:

$$\alpha = \alpha^{0}(T,\rho,\mathbf{x}) + \alpha^{\mathrm{r}}(\tau,\delta,\mathbf{x}) \tag{17.46}$$

where

$$\alpha^{0}(T,\rho,\mathbf{x}) = \sum_{i=1}^{k} x_i\left[\alpha_i^{0}(T,\rho) + \ln x_i\right] \tag{17.47}$$

and $\alpha^0_i(T,\rho)$ is the dimensionless Helmholtz energy for the *i*-th pure component in the ideal gas state at T, ρ. The residual part α^{r} $(\tau, \delta, \mathbf{x})$ is (for a binary mixture) given by:

$$\alpha^{\mathrm{r}}(\tau,\delta,x_1)=x_1\alpha_1{}^{\mathrm{r}}(\tau,\delta)+x_2\alpha_2{}^{\mathrm{r}}(\tau,\delta)+x_1x_2F_{12}\alpha_{12}(\tau,\delta). \tag{17.48}$$

While for a pure compound $\tau=T_c/T$, $\delta=\rho/\rho_c=\rho\cdot V_c$; in the case of mixtures pseudocritical instead of critical parameters are used. Several different expressions have been proposed for pseudocritical parameters, *e.g.* Kunz and Wagner[28] used:

$$\begin{aligned} T_c' &= \sum_{i=1}^{k}\sum_{\ell=1}^{k} x_i x_\ell \beta_{T,i\ell}\gamma_{T,i\ell}\frac{x_i+x_\ell}{\beta_{T,i\ell}^2 x_i+x_\ell}T_{c,i\ell}, \\ V_c' &= \sum_{i=1}^{k}\sum_{\ell=1}^{k} x_i x_\ell \beta_{V,i\ell}\gamma_{V,i\ell}\frac{x_i+x_\ell}{\beta_{V,i\ell}^2 x_i+x_\ell}V_{c,i\ell} \end{aligned} \tag{17.49}$$

with

$$T_{c,i\ell}=\sqrt{T_{c,i}T_{c,\ell}},\quad V_{c,i\ell}=\frac{1}{8}\left(V_{c,i}^{1/3}+V_{c,\ell}^{1/3}\right)^3 \tag{17.50}$$

Parameters for a number of systems have been tabulated.[28] The last term $\alpha_{12}(\tau,\delta)$ in Equation (17.48) was introduced in order to achieve a better agreement between experimental and calculated properties; $\alpha_{12}(\tau,\delta)$ is either evaluated from experimental data (then $F_{12}=1$) or calculated from generalised formula[28] (then $F_{12}\neq 1$ and its value is optimised in the fitting procedure).

17.3.7 Application of Volumetric and Thermodynamic Properties in Engineering Calculations

While the Gibbs energy is essential for phase and chemical equilibria in chemical applications, in engineering other thermodynamic quantities are also required as can be seen in Table 17.2. This table summarises the

Table 17.2 Calculation of heat and work.

Condition	Q	W_{vol}	W_{t}
Isochoric	ΔU	0	$V(p_2-p_1)$
Isobaric	ΔH	$-p(V_2-V_1)$	0
Reversible adiabatic	0	ΔU	ΔH
Reversible isothermal	$T\Delta S$	ΔA	ΔG
Isoenthalpic	0	$-p_2 V_2+p_1 V_1$	0

relations for the calculation of heat and of volume work W_{vol} and technical work W_{t} (also sometimes called shaft- or pressure- or flow-work), defined as:

$$W_{\text{vol}} = -\int_{V_1}^{V_2} p dV, \quad W_{\text{t}} = \int_{p_1}^{p_2} V dp. \tag{17.51}$$

The volume work W_{vol} is connected to the unrepeated reversible volume change (at given conditions). W_{t} is connected to flowing systems and for such systems, the first law of thermodynamics has the form:[4]

$$\Delta H = H_2 - H_1 = Q + W_{\text{t}} \tag{17.52}$$

17.4 Partial Molar Quantities

The Gibbs energy G and chemical potential:

$$\mu_i = (\partial G/\partial n_i)_{T,p,n_{j\neq i}} = \overline{G_i} \tag{17.53}$$

are indispensible for many calculations involving phase or chemical equilibria. Equation (17.53) was applied by Lewis[29] for a general definition of a partial molar quantity:

$$\overline{Y_i} = (\partial Y/\partial n_i)_{T,p,n_{j\neq i}} \tag{17.54}$$

The most important partial molar quantities are $\overline{V_i}$ and $\overline{S_i}$ as they represent the influence of pressure and temperature on μ_i. The partial molar enthalpy $\overline{H_i}$ is used to express temperature dependence of quantities derived from the chemical potential (activity a_i, fugacity f_i , fugacity coefficient ν_i):

$$\overline{H_i} = -T^2(\partial(\mu_i/T)/\partial T)_{p,\mathbf{x}} \tag{17.55}$$

17.4.1 Application of *Q*-Quantities for Calculation of Partial Molar Properties

For the calculation of partial molar quantities, auxiliary dimensionless variable $Q_{A,i}$ is defined as:

$$Q_{A,i} = Q_{A,i}(T,\rho,x_1,\ldots x_{k-1}) = Q_A + \left(\frac{\partial Q_A}{\partial x_i}\right)_{T,\rho,x_{j\neq i}} - \sum_{j=1}^{k-1} x_j \left(\frac{\partial Q_A}{\partial x_j}\right)_{T,\rho,x_{\ell\neq j}} \tag{17.56}$$

or

$$Q_{A,i} = Q_{A,i}(T,\rho,x_1,\ldots x_k) = Q_A + \left(\frac{\partial Q_A}{\partial x_i}\right)_{T,\rho,x_{j\neq i}} - \sum_{j=1}^{k} x_j \left(\frac{\partial Q_A}{\partial x_j}\right)_{T,\rho,x_{\ell\neq j}} \tag{17.57}$$

can be used. Equation (17.57) does not respect relation $\sum_{j=1}^{k} x_j = 1$ but is simpler to use; both forms yield identical results.[30]

Three other quantities are necessary for the calculation of partial molar quantities, namely:

$$Q_{U,i} = T\left(\frac{\partial Q_{A,i}}{\partial T}\right)_{\rho,\mathbf{x}}, \quad Q_{C,i} = T^2\left(\frac{\partial^2 Q_{A,i}}{\partial T^2}\right)_{\rho,\mathbf{x}}, \quad \tilde{z}_i = 1 + \rho\left(\frac{\partial Q_{A,i}}{\partial \rho}\right)_{T,\mathbf{x}} \tag{17.58}$$

Using Equations (17.56) and (17.58), partial molar quantities can be expressed as:[19,20]

$$\overline{V_i} = \left[1 + (\tilde{z}_i - z)/Q_\rho\right]/\rho = V_\mathrm{m}\left[1 + (\tilde{z}_i - z)/Q_\rho\right] \tag{17.59}$$

$$\overline{U_i} = U^0_{\mathrm{m},i} + RT\left[-Q_{U,i} + \frac{(Q_T - z)(\tilde{z}_i - z)}{Q_\rho}\right]. \tag{17.60}$$

$$\overline{H_i} = H^0_{\mathrm{m},i} + RT\left[z - 1 - Q_{U,i} + \frac{(\tilde{z}_i - z)Q_T}{Q_\rho}\right] \tag{17.61}$$

$$\overline{S_i} = S^0_{\mathrm{m},i} - R\ln\frac{x_i RT\rho}{p^{\mathrm{st}}} + R\left[-Q_{U,i} - Q_{A,i} + \frac{(\tilde{z}_i - z)Q_T}{Q_\rho}\right] \tag{17.62}$$

$$\overline{A_i} = A^0_{\mathrm{m},i} + RT\ln\frac{x_i RT\rho}{p^{\mathrm{st}}} + RT\left[Q_{A,i} - \frac{z(\tilde{z}_i - z)}{Q_\rho}\right] \tag{17.63}$$

$$\overline{G_i} = \mu_i = G^0_{\mathrm{m},i} + RT\ln\frac{x_i RT\rho}{p^{\mathrm{st}}} + RT[Q_{A,i} + z - 1] \tag{17.64}$$

$$\ln f_i = \ln(x_i RT\rho) + Q_{A,i} + z - 1 \tag{17.65}$$

$$\ln \nu_i = -\ln z + Q_{A,i} + z - 1 \tag{17.66}$$

(Quantities with superscript "0" are defined in Section 17.3.1).

For a given EOS, the expression for $Q_{A,i}$ is influenced by the mixing rule used for the EOS parameters. For example, in the case of cubic EOS, the mixing rule $\sum\sum x_i x_j a_{ij}$ as well as $\left(\sum x_i\sqrt{a_i}\right)^2$ are commonly used. To keep a simple general expression for $Q_{A,i}$ (regardless of which mixing rule is applied), "partial molar parameters" can be defined as:

$$\overline{a_i} = a + \left(\frac{\partial a}{\partial x_i}\right)_{x_{j\neq i}} - \sum_{j=1}^{k-1} x_j\left(\frac{\partial a}{\partial x_j}\right)_{x_{\ell\neq j}} = a + \left(\frac{\partial a}{\partial x_i}\right)_{x_{j\neq i}} - \sum_{j=1}^{k} x_j\left(\frac{\partial a}{\partial x_j}\right)_{x_{\ell\neq j}}. \tag{17.67}$$

For most frequently used mixing rules we obtain:

$$a = \sum_{j=1}^{k} x_j a_j \rightarrow \overline{a_i} = a_i \tag{17.68}$$

$$a = \left(\sum_{j=1}^{k} x_j \sqrt{a_j} \right)^2 \rightarrow \overline{a_i} = -a + 2\sqrt{a_i a} \tag{17.69}$$

$$a = \sum_{j=1}^{k} \sum_{\ell=1}^{k} x_j x_\ell a_{j\ell} \rightarrow \overline{a_i} = -a + 2\sum_{j=1}^{k} x_j a_{ji} \tag{17.70}$$

$$a = \left(\sum_{j=1}^{k} x_j \sqrt[3]{a_j} \right)^3 \rightarrow \overline{a_i} = -2a + 3\sqrt[3]{(a^2 a_i)} \tag{17.71}$$

$$a = \left(\sum_{j=1}^{k} x_j \sqrt[4]{a_j} \right)^4 \rightarrow \overline{a_i} = -3a + 4\sqrt[4]{(a^3 a_i)} \tag{17.72}$$

$$a = \sum_{j=1}^{k} \sum_{\ell=1}^{k} \sum_{m=1}^{k} x_j x_\ell x_m a_{j\ell m} \rightarrow \overline{a_i} = -2a + 3\sum_{j=1}^{k} \sum_{m=1}^{k} x_j x_m a_{jim} \tag{17.73}$$

$$a = \sum_{j=1}^{k} \sum_{\ell=1}^{k} \sum_{m=1}^{k} \sum_{n=1}^{k} x_j x_\ell x_m x_n a_{j\ell mn} \rightarrow \overline{a_i} = -3a + 4\sum_{j=1}^{k} \sum_{m=1}^{k} \sum_{n=1}^{k} x_j x_m x_n a_{ijmn} \tag{17.74}$$

For example, in the case of the van der Waals equation of state it holds that:

$$Q_{A,i} = -\ln(1 - b\rho) + \frac{\rho(\bar{b}_i - b)}{1 - b\rho} - \frac{\bar{a}_i \rho}{RT} \tag{17.75}$$

and the fugacity can be expressed as:

$$\ln f_i = \ln(x_i RT\rho) - \ln(1 - b\rho) + \frac{\rho(\bar{b}_i - b)}{1 - b\rho} - \frac{\bar{a}_i \rho}{RT} + z - 1 \tag{17.76}$$

then, by using mixing rule Equation (17.69) for parameter a and Equation (17.68) for parameter b we obtain:

$$\ln f_i = \ln \frac{x_i RT\rho}{1 - b\rho} + \frac{\rho b_i}{1 - b\rho} - \frac{2\rho}{RT}\sqrt{a_i a} \tag{17.77}$$

17.4.2 Calculation of Partial Molar Properties with the Model of Lemmon *et al.*

In cases where a high precision is required, the methodology introduced in Section 17.3.6[16,26–28] can also be used for the calculation of partial molar properties. The key quantity for the calculation of partial molar properties is

$Q_{A,i}$, which in the context of the notation used by Wagner can be labelled $\tilde{\alpha}_i^{\mathrm{r}}$[see Equation (17.38)] and from Equation (17.48) we can obtain:

$$\tilde{\alpha}_i^{\mathrm{r}} = \alpha_i^{\mathrm{r}} + (1-x_i)^2 F_{12}\alpha_{12}^{\mathrm{r}} + \left(\frac{\bar{T}'_{\mathrm{c},i}}{T'_{\mathrm{c}}} - 1\right)\sum_{j=1}^{2} x_j\alpha_{j,\tau}^{\mathrm{r}} + \left(\frac{\bar{V}'_{\mathrm{c},i}}{V'_{\mathrm{c}}} - 1\right)\sum_{j=1}^{2} x_j\alpha_{j,\delta}^{\mathrm{r}}$$
$$+ x_1x_2F_{12}\left[\left(\frac{\bar{T}'_{\mathrm{c},i}}{T'_{\mathrm{c}}} - 1\right)\alpha_{12,\tau}^{\mathrm{r}} + \left(\frac{\bar{V}'_{\mathrm{c},i}}{V'_{\mathrm{c}}} - 1\right)\alpha_{12,\delta}^{\mathrm{r}}\right], \quad i=1,2 \tag{17.78}$$

where $\alpha_{i,\delta}^{\mathrm{r}}$ *etc.* are defined analogically to Equations (17.30) and (17.31).

$\bar{T}'_{\mathrm{c},i}$and $\bar{V}'_{\mathrm{c},i}$ are "partial molar pseudocritical parameters", defined analogically as "partial molar EOS parameters" [see Equations (17.68) to (17.74)]. In the case of a binary system and pseudocritical parameters Equation (17.49), we obtain:

$$\bar{T}'_{\mathrm{c},1} = x_1(2-x_1)T_{\mathrm{c},1} - x_2^2T_{\mathrm{c},2} + 2x_2^2\beta_{T,12}\gamma_{T,12}T_{\mathrm{c},12}\frac{1}{(\beta_{T,12}^2x_1 + x_2)^2}$$
$$\bar{T}'_{\mathrm{c},2} = -x_1^2T_{\mathrm{c},1} + x_2(2-x_2)T_{\mathrm{c},2} + 2x_1^2\beta_{T,12}\gamma_{T,12}T_{\mathrm{c},12}\frac{\beta_{T,12}^2}{(\beta_{T,12}^2x_1 + x_2)^2} \tag{17.79}$$

$$\bar{V}'_{\mathrm{c},1} = x_1(2-x_1)V_{\mathrm{c},1} - x_2^2V_{\mathrm{c},2} + 2x_2^2\beta_{V,12}\gamma_{V,12}V_{\mathrm{c},12}\frac{1}{(\beta_{V,12}^2x_1 + x_2)^2}$$
$$\bar{V}'_{\mathrm{c},2} = -x_1^2V_{\mathrm{c},1} + x_2(2-x_2)V_{\mathrm{c},2} + 2x_1^2\beta_{V,12}\gamma_{V,12}V_{\mathrm{c},12}\frac{\beta_{V,12}^2}{(\beta_{V,12}^2x_1 + x_2)^2}. \tag{17.80}$$

Then the fugacity can be expressed as:

$$\ln f_i = \ln(x_iRT\rho) + \tilde{\alpha}_i^{\mathrm{r}} + z - 1 \tag{17.81}$$

Equation (17.81) enables routine calculations of phase- and chemical-equilibria in real systems.

17.5 Conclusions

In this chapter, a rational way for the calculation of thermodynamic quantities, including partial molar quantities based on the system volumetric properties, was presented. While common thermodynamic properties (U, H, S, C_p, C_V) find use in the calculation of heat and work, partial molar properties (chemical potential, fugacity) are used in solving phase and chemical equilibria.

Special attention was devoted to the exploitation of dimensionless quantities for the description of *pVT* behaviour based either on the compressibility factor $z=z(T,\rho,\mathbf{x})$ or on the dimensionless Helmholtz energy $\alpha=\alpha(T,\rho,\mathbf{x})=A_{\mathrm{m}}/(RT)$. The empirical "laws" or "rules" suitable for the evaluation of the volumetric behaviour and thermodynamic properties of mixtures were also discussed. This approach, which makes use of the properties of pure components, has seldom been reported in the literature.

References

1. R. Span, *Multiparameter Equations of State - An Accurate Source of Thermodynamic Property Data*, Springer, Berlin, 2000.
2. *Experimental Thermodynamics, Volume 5: Equations of State for Fluids and Fluid Mixtures*, ed. J. V. Sengers, R. F. Kayser, C. J. Peters and H. J. White, Jr., Elsevier, Amsterdam, 2000.
3. *Applied Thermodynamics of Fluids*, ed. A. R. H. Goodwin, J. V. Sengers and C. J. Peters, RSC Publishing, Cambridge, 2010.
4. S. I. Sandler, *Chemical, Biochemical, and Engineering Thermodynamics*, Wiley, Hoboken, NJ, 2006.
5. J. O. Valderrama, *Ind. Eng. Chem. Res.*, 2003, **42**, 1603–1618.
6. A. Anderko, Cubic and Generalized van der Waals Equations, in *Experimental Thermodynamics, Volume 5: Equations of State for Fluids and Fluid Mixtures*, ed. J. V. Sengers, R. F. Kayser, C. J. Peters and H. J. White, Jr., Elsevier, Amsterdam, 2000, pp. 75–126.
7. I. G. Economou, Cubic and Generalized van der Waals Equations of State, in *Applied Thermodynamics of Fluids*, ed. A. R. H. Goodwin, J. V. Sengers and C. J. Peters, RSC Publishing, Cambridge, 2010, pp. 53–83.
8. M. Lencka and A. Anderko, *Chem. Eng. Commun.*, 1991, **107**, 173–188.
9. S. I. Sandler and H. Orbey, Mixing and Combining Rules, in *Experimental Thermodynamics, Volume 5: Equations of State for Fluids and Fluid Mixtures*, ed. J. V. Sengers, R. F. Kayser, C. J. Peters and H. J. White, Jr., Elsevier, Amsterdam, 2000, pp. 321–357.
10. A. R. H. Goodwin and S. I. Sandler, Mixing and Combining Rules, in *Applied Thermodynamics of Fluids*, ed. A. R. H. Goodwin, J. V. Sengers and C. J. Peters, RSC Publishing, Cambridge, 2010, pp. 84–134.
11. B. I. Lee and M. G. Kesler, *AIChE J.*, 1975, **21**, 510–527.
12. J. F. Ely and M. F. Marrucho, The Corresponding-States Principle, in *Experimental Thermodynamics, Volume 5: Equations of State for Fluids and Fluid Mixtures*, ed. J. V. Sengers, R. F. Kayser, C. J. Peters and H. J. White, Jr., Elsevier, Amsterdam, 2000, pp. 289–320.
13. A. Anderko and K. S. Pitzer, *AIChE J.*, 1991, **37**, 1379–1391.
14. R. Schmidt and W. Wagner, *Fluid Phase Equilib.*, 1985, **19**, 175–200.
15. R. T. Jacobsen, S. G. Penoncello, E. W. Lemmon and R. Span, Multiparameter Equations of State, in *Experimental Thermodynamics, Volume 5: Equations of State for Fluids and Fluid Mixtures*, ed. J. V. Sengers, R. F. Kayser, C. J. Peters and H. J. White, Jr., Elsevier, Amsterdam, 2000, pp. 849–881.
16. E. W. Lemmon and R. Span, Multiparameter Equations of State for Pure Fluids and Mixtures, in *Applied Thermodynamics of Fluids*, ed. A. R. H.

Goodwin, J. V. Sengers and C. J. Peters, RSC Publishing, Cambridge, 2010, pp. 394–432.
17. C. McCabe and A. Galindo, SAFT Associating Fluids and Fluid Mixtures, in *Applied Thermodynamics of Fluids*, ed. A. R. H. Goodwin, J. V. Sengers and C. J. Peters, RSC Publishing, Cambridge, 2010, pp. 215–279.
18. J. A. Beattie and W. H. Stockmayer, in *A Treatise on Physical Chemistry. Vol. II. States of Matter*, ed. H. S. Taylor and S. Glasstone, Van Nostrand, New York, 3rd edn, 1951, pp. 187–352.
19. J. P. Novák, A. Malijevský and J. Pick, *Chem. Listy*, 1979, **73**, 1178–1182(in Czech).
20. J. P. Novák, V. Růžička, Jr., A. Malijevský, J. Matouš and J. Linek, *Collect. Czech. Chem. Commun.*, 1985, **50**, 1–22.
21. J. P. Novák, P. Voňka, K. Růžička, J. Matouš and A. Malijevský, *Termodynamické vlastnosti plynů*, VŠCHT, Praha, 2007(in Czech).
22. J. Joffe, *Ind. Eng. Chem. Fundam.*, 1971, **10**, 532–533.
23. E. P. Bartlett, H. L. Cupples and T. H. Tremearne, *J. Am. Chem. Soc.*, 1928, **50**, 1275–1288.
24. V. Měřičková, J. P. Novák and J. Pick, *Collect. Czech. Chem. Commun.*, 1982, **47**, 371–383.
25. G. Scatchard, *Chem. Rev.*, 1931, **8**, 321–333.
26. E. W. Lemmon and R. T. Jacobsen, *Int. J. Thermophys.*, 1999, **20**, 825–835.
27. E. W. Lemmon and R. Tillner-Roth, *Fluid Phase Equilib.*, 1999, **165**, 1–21.
28. O. Kunz and W. Wagner, *J. Chem. Eng. Data*, 2012, **57**, 3032–3091.
29. G. N. Lewis and M. Randall, *Thermodynamics and the Free Energy of Chemical Substances*, McGraw-Hill, New York, 1923.
30. O. Redlich, A. T. Kister and C. E. Turnquist, *Chem. Eng. Prog., Symp. Ser.*, 1952, **48**, 49–61.

CHAPTER 18

Molar Volumes of Electrolyte Solutions

GLENN HEFTER

Chemistry Department, Murdoch University, Murdoch, WA 6150, Australia
Email: g.hefter@murdoch.edu.au

18.1 Introduction

The volumes of electrolyte solutions continue to attract the attention of both experimentalists and theoreticians several hundred years after the first serious measurements and speculations about the nature of such properties were made.[1] The reasons for this ongoing interest are not hard to find. First there is their practical utility for calculating solution densities[1,2] and the effects of pressure on chemical equilibria,[3,4] both of which are routinely required for various purposes in areas such as process engineering[5] and oceanography.[6] Second, the molar volumes of electrolyte solutions provide valuable insights into ion–solvent interactions (at infinite dilution), ion–ion interactions (at finite concentrations) and, under some circumstances, even solvent–solvent interactions.[1,7] Last, but certainly not least, there is a tangible quality to molar volumes that distinguishes them from other thermodynamic quantities: volumes can be visualised at the molecular level in a manner that is almost impossible for, say, enthalpies or Gibbs energies.[8] This is so in both the experimental and theoretical senses. The *contraction* in the total liquid volume (after dissipation of the heat generated) when a finite volume of solid sodium hydroxide is dissolved in a given volume of water is readily visible to the naked eye; indeed it makes a useful lecture demonstration. In contrast, the considerable enthalpy change accompanying the same reaction can only be inferred from the observed temperature rise.

Volume Properties: Liquids, Solutions and Vapours
Edited by Emmerich Wilhelm and Trevor M. Letcher

Published by the Royal Society of Chemistry, www.rsc.org

In the theoretical realm, it is much easier to form a 'mental picture' of the volume occupied by a dissolved electrolyte, and thus to develop straightforward mathematical models[9] to account for it, than it would be to describe, say, its heat capacity. Such conceptualisation is aided, for example, by the fact that a major proportion of the molar volumes of electrolytes in solution can be attributed to their geometric (crystallographic or intrinsic) size.[1,7] Regardless of the motivations, molar volumes of electrolyte solutions continue to be an area of active research.

Analogous to other partial molar thermodynamic quantities,[8,10] the partial molar volume of a solute dissolved in a solvent is defined as the partial derivative of the (total) solution volume, V, with respect to the amount (or concentration) of the solute at constant temperature and pressure:[1,2,7]

$$\bar{V}_2 = (\partial V/\partial n_2)_{T,p,n_1\ldots} \tag{18.1}$$

where the subscripts 1 and 2 denote solvent and solute properties, respectively.

The advantage of partial molar thermodynamic quantities is that the overall solution property at constant T and p is given by simple summation over all species present; thus for volumes:

$$V = n_1\bar{V}_1 + n_2\bar{V}_2 + \ldots.. \tag{18.2}$$

In practice it is more convenient to use the apparent molar volume, V_ϕ, which is defined as:

$$V_\phi = (V - n_1\bar{V}_1^\circ)/n_2 \tag{18.3}$$

at constant T and p, where $\bar{V}_1^\circ$ is the molar volume of the pure solvent. This utility arises because V_ϕ can be expressed in experimentally measurable or known quantities:

$$V_\phi = (M_2/\rho) - 1000(\rho - \rho_o)/\rho\rho_o m \tag{18.4}$$

where M_2 is the molar mass of the solute (in g mol^{-1}), ρ and ρ_o are, respectively, the density (in g cm^{-3}) of the solution and pure solvent and m is the concentration of the solute in 'molality' units [(mol solute)(kg solvent)$^{-1}$]. The factor of 1000 is required to render the units of density and concentration consistent. An analogous expression:

$$V_\phi = (M_2/\rho_o) - 1000(\rho - \rho_o)/\rho_o c \tag{18.5}$$

is obtained when the concentration c is expressed in 'molarity'[(mol solute)(L solution)$^{-1}$] units. Note that the values of V_ϕ obtained by Equations (18.4) and (18.5) are *identical*. It should also be noted that apparent (or partial) *molar* volumes are often referred to as apparent (or partial) *molal* volumes. In this context the adjectives '*molar*' and '*molal*' do ***not*** refer to the unit of concentration; they are merely alternative ways of indicating that the volume is being quoted on a 'per mole' basis. If desired (it usually isn't), the partial molar volume of the solute can be derived using the expression:

$$\bar{V}_2 = V_\phi + n_2(\partial V_\phi / \partial n_2)_{T,\ p,\ n_1 \ldots} \tag{18.6}$$

in either concentration unit.[1,2]

At infinite dilution, the apparent molar volume becomes exactly equivalent to the standard state partial molar volume of the solute, $\bar{V}_2{}^\circ$:

$$\bar{V}_2{}^\circ = \lim_{m \text{ or } c \to 0} V_\phi = V_\phi{}^\infty \tag{18.7}$$

Since at infinite dilution solute particles no longer interact with each other, it follows that the values of V° (the other descriptors are not required for the purposes of this review) of the solute reflect only solute–solvent interactions. That is, V° is a fundamental property of a given solute in a given solvent at any specific temperature and pressure. Therein lies its significance.

The concepts discussed above, although expressed in general terms, have been focussed implicitly on binary electrolyte solutions (consisting of one solute and one solvent) but can readily be extended to more complex systems. For example, for ternary electrolyte solutions (consisting of two solutes and one solvent) with solute concentrations in molality units it can be shown that:[2,10]

$$V_\phi\ (2,3) = (M_t/\rho) - 1000(\rho - \rho_o)/\rho\rho_o m_t \tag{18.8}$$

where $m_t = \Sigma\, m_i$ is the sum of the molalities of all the solutes ($i \neq 1$) present in the solution and $M_t = (\Sigma\, m_i M_i)/m_t$ is the mean molar mass of the electrolytes in solution. Further manipulations of V_ϕ for mixed electrolyte solutions can be found in Millero's comprehensive reviews.[1,2,10] Measurements of molar volumes of electrolytes in mixed solvents can be handled in an analogous manner by appropriate definition of the solvent.[11]

18.2 Experimental Methods

The experimental methods for determining apparent molar volumes of electrolyte solutions can, for convenience, be divided into three groups: those involving the measurement of density, direct measurements by dilatometry and some miscellaneous techniques. These methods have been reviewed on many occasions in the literature[12–14] and, to some extent, in Chapters 2, 3 and 4 of this book. Accordingly, only a brief description of the methods that are most relevant for electrolyte solutions will be given here.

In fact there are rather few methods that are truly appropriate for the accurate determination of V_ϕ, especially at the lower concentrations ($c <$ *ca.* 0.05 mol L^{-1}) that are needed for the optimal estimation of V° (Section 18.3).[7] For example, for density measurements it is necessary [*cf.* Equations (18.4) and (18.5)] to determine $(\rho - \rho_o)$ to a precision of $\sim 1 \times 10^{-6}$ (~ 1 μg cm^{-3}) to accurately define V_ϕ at low c. Few instruments or researchers can routinely achieve such precision.[12–14] This is undoubtedly a major cause for the 'splaying' of independently determined experimental V_ϕ values that is often observed for electrolyte solutions at low solute concentrations.[15]

18.2.1 Density Measurements

18.2.1.1 Vibrating Tube Densimeters

While there is a great variety of methods for measuring the densities of solutions, it is fair to say that most have been superseded by the development of the vibrating tube densimeter (vtd), which has revolutionised density measurements since first developed by Kratky *et al.*[16] A detailed description of these devices is given in Chapter 3 of this book and in various reviews[13,17] so only some brief comments are made here.

Although real vibrating tubes are difficult to describe in purely theoretical terms (Chapter 3), the working equation:

$$(\rho_A - \rho_B) = K({\tau_A}^2 - {\tau_B}^2) \tag{18.9}$$

holds with good accuracy over wide ranges of experimental conditions. The (empirical) tube constant K is determined by measurements using two reference fluids R1 and R2:

$$K = (\rho_{R1} - \rho_{R2})/({\tau_{R1}}^2 - {\tau_{R2}}^2) \tag{18.10}$$

As discussed in Chapter 4, the choice of reference fluids R1 and R2 is extremely limited for ppm precision. The most popular are water (w) and dry nitrogen gas (N) for which density data of suitable accuracy are readily available, the latter usually in the form of an equation of state. This choice is not ideal because of the large difference between ρ_w (≈ 1 g cm^{-3}) and ρ_N (≈ 1 mg cm^{-3}) and because the densities of most electrolyte solutions are greater than ρ_w; in other words they are outside the calibration range. There is no way around this difficulty until appropriate density standards with $\rho > \rho_w$ become available.[18] Fortunately, the behaviour of vtds appears to be sufficiently consistent so that the common calibration procedure (using w and N) does not introduce serious errors, at least under peri-ambient conditions. Under more extreme conditions of temperature and/or pressure, tube behaviour may not be fully ideal and the absence of reliable density standards with $\rho > \rho_w$ is a potentially significant source of uncertainty.[18]

Apart from calibration difficulties, the only common problems that affect vtd measurement precision are temperature control (which must be ± 0.01 K or better), the possible presence of microbubbles, and the effects of solution viscosity. With modern thermostats the first of these is no longer a problem, while the second can be minimised by good experimental technique. The last is undoubtedly important for highly viscous electrolytes such as ionic liquids (Chapters 3 and 19) but the effects of small changes in solution viscosity typical of most electrolyte solutions are not well established and are generally ignored.

As these drawbacks are relatively minor, vtds have become highly popular: they are easy to use, fast, relatively inexpensive, require relatively little sample and are readily applicable to almost any solution.[17] Vibrating tube densimeters are particularly suited to the measurement of V_ϕ because they can be set up to detect, by contiguous measurements, the difference in density between the solution and the solvent, $(\rho - \rho_o)$, which is critical for evaluating V_ϕ at low concentrations *via* Equations (18.4) and (18.5). Furthermore, vtds are particularly adaptable. At least one commercial manufacturer offers an apparatus capable of measuring densities up to 150 °C and 40 MPa with a precision of $\sim 10\times10^{-6}$.[19] A number of designs have been developed and utilised (see Section 18.4.1) for measurements up to temperatures above the critical point of water (647 K)[20] and up to at least 40 MPa. A vtd developed in the author's laboratory has a Pt–Rh alloy tube with Pt inlet and outlet tubing, which enables measurements to be made on chemically aggressive solutions up to 300 °C and 30 MPa.[18]

18.2.1.2 Buoyancy Methods

Of the many other techniques available for measuring densities, only buoyancy methods[12–14] continue to be attractive for determining V_ϕ values. This is mostly because such techniques are generally thought to be more sensitive, by up to an order of magnitude (*i.e.* to $\sim 0.1\times10^{-6}$ in ρ), than vtds. While this does not mean that such methods are necessarily that accurate, the extra sensitivity should permit more reliable determinations of V_ϕ in the important low concentration region (Section 18.3).

All buoyancy methods work on Archimedes' principle that a body (subscript b) of true (*in vacuo*) mass $m_b{}^*$ immersed in a fluid will experience an upward (buoyant) force that is exactly equal to the mass of fluid displaced by that body.[12–14] The density of the fluid is then obtained as:

$$\rho = (m_b{}^* - m_b)/V_b \tag{18.11}$$

where m_b is the apparent mass of the body immersed in the fluid and V_b is its volume (usually obtained by weighing in pure water). Buoyancy measurements come in many forms, including hydrostatic balances (direct and differential) and magnetic floats. The latter are particularly suited to measurements over small density ranges and thus for the determination of V_ϕ. Researchers should consult the literature for design details and the precautions that need to be taken to obtain results of the desired accuracy.[12–14] It may also be noted that buoyancy methods are readily adapted for measurements under extreme conditions, albeit with an inevitable decrease in accuracy (Chapter 2).

18.2.1.3 Pycnometry

This technique involves measurement of the mass of a known volume of liquid; it is simple, inexpensive and, with sufficient care, is capable of ppm precision.[12,13] However, it is unsuited to measuring small differences in ρ and its accuracy is considerably worse than its precision, especially at low solute concentrations.[7] Pycnometry is really only of use for measuring V_ϕ at relatively high electrolyte concentrations and can be considered as superseded by vtds: it should only be used when other (faster and better) methods are unavailable, or perhaps for systems of very high viscosity.

18.2.2 Dilatometry

Dilatometry[13,14,21,22] involves the direct measurement of the volume change Δv that occurs when a known volume v' of a concentrated electrolyte solution is added to a known volume v ($\gg v'$) of pure solvent:

$$\Delta v = v'' - (v' + v) \tag{18.12}$$

where v'' is the final volume of the diluted solution. Apparent molar volumes are calculated as:

$$V_{\phi,\mathrm{f}} = V_{\phi,\mathrm{i}} + \Delta v / n_2 \tag{18.13}$$

where the subscripts f and i denote the final (diluted) and initial (concentrated) states of the electrolyte solution and n_2 is the number of mole of the electrolyte added to the solvent. Because only the *change* in volume needs to be measured the technique does not require any unattainably accurate (absolute) volume measurements. The value of $V_{\phi,\mathrm{i}}$ can be measured to the desired precision by making the titrant solution sufficiently concentrated and using any reasonable densimetry method such as a vtd or even pycnometry. Providing Δv can be measured with sufficient accuracy, $V_{\phi,\mathrm{f}}$ can be determined with similar uncertainty to that of $V_{\phi,\mathrm{i}}$, apparently down to concentrations as low as $\sim$1 mmol L^{-1}. Of course, considerable effort is required to realise high quality data under such conditions.[21,22] Given that such measurements are slow, labour intensive and difficult to automate, a considerable investment of human resources is necessary.

18.2.3 Other Methods

There are numerous other methods for determining V_ϕ, density or other related properties.[12–14] Some of these approaches, such as volumometers, expansimeters and piezometers, directly measure the effects of T or p on V_ϕ and are as such of more interest to the measurement of isobaric expansivities and isothermal compressibilities,[23–25] which are outside the scope of this chapter. An interesting recent development is the technique known as pressure perturbation calorimetry, which provides a quick and convenient way of obtaining precise expansibilities of electrolyte solutions.[26] Similar techniques are discussed in Chapter 14.

18.3 Extrapolation to Infinite Dilution

While V_ϕ (or $\bar{V}_2$) values are undoubtedly of great practical utility, the standard state $V_\phi^{\infty}(= V^\circ)$ values are usually of most scientific interest because of the insights they provide at the molecular level with respect to ion–solvent interactions.[1,7] The values of V° are also essential for tabulation purposes[27] and for developing mathematical models of the concentration dependence of V_ϕ and ρ and equations of state for electrolyte solutions.[28] Furthermore, since V° values for electrolytes must be strictly additive, they also provide stern tests of the quality and thermodynamic consistency of the data (and the extrapolation procedures) that have been used to obtain them.[1,7] It is important to recognise that V° is a virtual quantity that can only be obtained from experimental data measured at finite concentrations *via* an appropriate (theoretically rigorous) extrapolation or calculation.

It is generally accepted that the properties of very dilute electrolyte solutions are well described by the Debye–Hückel (DH) theory.[28,29] For volumes the appropriate expression is obtained by taking the derivative with respect to pressure of the DH equation for activity coefficients. As shown by Redlich and Rosenfeld[30] more than eighty years ago, this produces an expression of the form:

$$V_\phi = V^\circ + S_V{}^{DH} c^{1/2} \tag{18.14}$$

An expression analogous to Equation (18.14) can be written in terms of molality, bearing in mind that the value of $S_V{}^{DH}$ will be slightly different. Equation (18.14) provides a theoretically sound basis for the extrapolation of $V_\phi(c)$ or $V_\phi(m)$ data to infinite dilution so as to obtain V°.

The use of the older, superficially similar but *purely empirical* Masson equation:[31]

$$V_\phi = V^\circ + S_V{}^{M} c^{1/2} \text{ (or } m^{1/2}) \tag{18.15}$$

where both the slope ($S_V{}^M$) and V° are derived from a presumed linear fit of $V_\phi(c)$ or $V_\phi(m)$ against $c^{1/2}$ (or $m^{1/2}$) is unacceptable unless there is no alternative, as occurs, for example, in some non-aqueous solvents where the required solvent data (see below) are not known.

The DH slope in Equation (18.14) is applicable to any fully dissociated electrolyte in any solvent at any T or p and has the form:

$$S_V{}^{DH} = kw^{3/2} \tag{18.16}$$

The second term on the right hand side of Equation (18.16) is a valency factor with:

$$w = 0.5\, \Sigma\, \nu_i z_i^2 \tag{18.17}$$

where ν_i is the number and z_i the charge number of the ions i produced when one 'molecule' (formula unit) of the electrolyte fully dissociates, summed over all cations and anions.

The expression for k in SI units can be written:[7]

$$k = N_A^{\,2}\, e^3 (8\pi^2\, \varepsilon_o^{\,3} RT)^{-1/2}\, \varepsilon^{-3/2}\, [(\partial \ln\varepsilon/\partial p)_T - \kappa_T/3] \qquad (18.18)$$

where N_A is Avogadro's constant, e is the unit (proton) charge, ε_o is the electrical permittivity of free space, ε is the relative permittivity (dielectric constant) of the solvent and $(\partial \ln\varepsilon/\partial p)_T$ its pressure dependence, κ_T is the isothermal compressibility of the solvent and other symbols have their usual meanings. Equation (18.18) shows that the value of k depends upon three physical properties (ε, $(\partial \ln\varepsilon/\partial p)_T$ and κ_T) of the pure solvent.[9] While these three quantities are well established for water over reasonably wide ranges of T and p,[28,32] this is generally not true for non-aqueous solvents.[7] Indeed for the latter some of these properties (particularly $(\partial \ln\varepsilon/\partial p)_T$) are unknown, although they can sometimes be estimated.[9] Values of k for water have been tabulated on many occasions, often over wide ranges of conditions.[28,32] As these values may differ significantly (due to slight differences in the values selected for the physical properties), it is important that researchers should use the best available values. Values of k for some non-aqueous solvents at 25 °C and 0.1 MPa have also been tabulated;[9] few data are available under other conditions.

As has been noted,[7] the electrolyte concentrations up to which the DH equation for volumes is applicable are not well defined and may even be below experimentally-accessible values, although there are many data that suggest the opposite.[7] To better utilise the more reliable data that can be obtained at higher concentrations, Redlich and Meyer[33] added an empirical term to Equation (18.14) giving

$$V_\phi = V^\circ + S_V^{\,DH} c^{1/2} + b_V c \qquad (18.19)$$

which is often referred to as the Redlich–Rosenfeld–Meyer (RRM) equation. This equation provides a good description of V_ϕ data up to concentrations of $\sim$1 mol L^{-1} for most electrolytes in aqueous solution under peri-ambient conditions. Plots of $(V_\phi - S_V^{\,DH} c^{1/2})$ *vs.* c are usually linear and readily yield V° and b_V as the intercept and slope respectively. The applicability of Equation (18.19) under other conditions, such as at high temperatures and pressures or in non-aqueous solvents, is less well established; nevertheless the RRM equation provides a reasonable starting point for the analysis of $V_\phi(c \text{ or } m)$ data and the estimation of V° values, providing the required solvent parameters are known for the conditions of interest.

The fitting of V_ϕ data at higher concentrations (>*ca.* 1 mol L^{-1}) can be accomplished by adding higher polynomial terms to Equation (18.19). For example, Trevani *et al.*[34] and Hnědkovský *et al.*[18] have used the equation:

$$V_\phi = V^\circ + A_V m^{0.5} + b_V m + c_V m^{1.5} + d_V m^2 \qquad (18.20)$$

where $A_V = S_V^{\,DH}$ (in molality units) and x_V are empirical parameters, to describe satisfactorily the volumetric properties of NaOH(aq) up to high concentrations, temperatures and pressures. Note that the addition of data at higher concentrations and the extra adjustable parameters required to fit them do ***not*** improve the accuracy of the estimation of V°; indeed they may lower it. The strategy generally accepted as best for fitting $V_{\phi}(m)$ over wide ranges of m is to evaluate V° at low m (with the upper and lower limits of m being defined by the quality of the data available) *via* Equations (18.14) or (18.19) and then fit all the data with an equation such as (or similar to) Equation (18.20) using the V° value as a fixed parameter. This of course applies regardless of the concentration unit being used.

While little progress has been made on theoretically rigorous descriptions of electrolyte solution behaviour over the past 50 or more years, many semi-empirical equations have been proposed that can precisely describe volumetric behaviour to high concentrations, temperatures and pressures. Undoubtedly the most popular of these equations are the suite of expressions due to Pitzer.[28,29] For volumes of 1 : 1 electrolytes, the original expression of Pitzer takes the form:

$$
\begin{aligned}
V_{\phi} &= V^{\circ} + (A_{\mathrm{V}}/b)\ln\left(1 + bm^{1/2}\right) + 2RT\left(B^{\mathrm{V}}m + C^{\mathrm{V}}m^{2}\right) \\
A_{\mathrm{V}} &= S_{\mathrm{V}}^{\mathrm{DH}} = 4RT(\partial A_{\phi}/\partial p)_{T} \\
A_{\phi} &= (2\pi N_{\mathrm{A}}\rho_{\mathrm{o}}/9000)^{1/2}(e^{2}/4\pi\varepsilon_{\mathrm{o}}\varepsilon k_{\mathrm{B}}T)^{3/2} \\
B^{\mathrm{V}} &= \beta^{\mathrm{V}(0)} + 2\beta^{\mathrm{V}(1)}f(\alpha m^{1/2}) \\
f(x) &= [1 - (1 + x)\exp(-x)]/x^{2}
\end{aligned}
\tag{18.21}
$$

where A_{ϕ} is the Debye–Hückel expression for osmotic coefficients, $b = 1.2$, $\alpha = 2.0$, and $\beta^{\mathrm{V}(0)}$, $\beta^{\mathrm{V}(1)}$, and C^{V} are the pressure derivatives of the corresponding empirical Pitzer parameters for activity or osmotic coefficients.[28,29] There are many variants on these equations with authors introducing various changes to produce better fits for data sub-sets, particularly at high electrolyte concentrations. However, the key point to note in the present context is that the Pitzer equations do ***not*** provide a more rigorous way of extrapolating V_{ϕ} data to infinite dilution to obtain V°.[35,36] In essence, this is because of correlations between V° and the three other empirical constants required to fit the V_{ϕ} values at higher m [Equation (18.21)]. Under some circumstances, the extra flexibility of the Pitzer equations [*cf.* say Equation (18.19)] may cause increased uncertainty in the evaluation of V° or, in some cases, produce wildly incorrect estimates of V°.[36] When V_{ϕ} data are plentiful and have been confirmed by independent measurements, and therefore have a low uncertainty, the V° values obtained from fitting Equations (18.19), (18.20) and (18.21) over appropriate concentration ranges are virtually identical. When these conditions are not met there is little reason for preferring one over another.[35,36] This situation is unlikely to improve until there is a breakthrough in the theory of moderately concentrated electrolyte solutions.

The extrapolation of V_ϕ to infinite dilution to obtain V° in non-aqueous solvents is more fraught than in water.[7] Equations (18.14) and (18.19) should still be applicable and indeed have been confirmed experimentally in a few cases.[37,38] On the other hand the sort of information required for Equation (18.21) is rarely available. It is also true that for many non-aqueous solvents the physical data required to evaluate the DH slope [*cf.* Equation (18.18)] are not known.[7,9] Under these circumstances there is no choice but to use the less accurate empirical Masson equation [Equation (18.15)]. A further problem in obtaining reliable V° values in non-aqueous solvents is the greater degree of ion pairing *cf.* water because of their (typically) lower dielectric constants. Volumetric measurements are especially sensitive to ion pairing because the charge neutralisation that accompanies ion pair formation significantly reduces electrostriction.[1,7]

18.4 Quantitative Studies of Volumes of Electrolyte Solutions

This chapter does not attempt to provide a comprehensive survey of all the quantitative data that have been reported on the molar volumes of electrolytes in aqueous and non-aqueous solvents; that would require a book in itself. Instead what follows is a limited selection of the data available in the open literature. Fortunately, volumetric studies of electrolyte solutions, unlike solubility or stability constant data, are relatively easy to access from the literature using modern computerised searching tools.

18.4.1 Aqueous Solutions

The literature data on the molar volumes of electrolyte solutions up to *ca.* 1970 were covered comprehensively by Millero.[1,2] These seminal works provided a detailed account of the history of electrolyte volume studies, an extended discussion of the effects of electrolyte concentration, as well as tabulations of standard molar volumes of a wide range of electrolytes and ions in aqueous solution from 0 to 200 °C. No major review of electrolyte volumes in aqueous solutions has appeared since Millero's work, but a number of additions and revisions have appeared[29,32,35] as new data and or fitting equations became available. Updated lists of V° values for ions, which are more convenient to tabulate than the salts, have been reported.[27] Two major compilations of density and V_ϕ data are available in book form.[39,40] However, it should be noted that the emphasis of both of these books is on data at relatively high concentrations; densities are generally not listed to the precision required for reliable calculation of V_ϕ (except in very concentrated solutions); data sources are sometimes obscure or confusing; and the listed values occasionally show disturbing anomalies. Density and or volumetric data are beginning to be captured in electronic data bases such as those being developed by Dechema in Germany and NIST in the USA. Perhaps the most comprehensive volumetric database currently available is at Murdoch University (JESS.murdoch.edu.au).[35,41] At the time of writing (April, 2014) this database has quantitative information for about 200 electrolytes in aqueous solution and has the additional advantage that the data are in a form that can be processed for automatic checking of thermodynamic consistency.

The absence of recent reviews belies a steady stream of publications on the measurement and interpretation of the molar volumes of electrolytes in aqueous solution. These include: measurements on new materials such as room temperature ionic liquids (RTILs);[42] on systems of industrial interest[43] or that are chemically 'difficult';[44] better quality or more extensive (higher concentration) studies on previously studied salts;[15] and mixed electrolyte systems.[45] One particularly important area of ongoing interest is the volumetric properties of electrolytes in aqueous solution at high temperatures and pressures. Apart from providing important insights into the nature of water as a solvent under extreme conditions, such data are required for geochemical modelling,[46] for engineering calculations in high temperature industrial and hydrometallurgical processes,[5,47] and for understanding proposed high-temperature waste disposal processes employing hydrothermal destruction. To give some idea of the scope of the measurements in this area Table 18.1 lists a small sample of the available literature. Of particular note is the work of Wood, Majer and co-workers[48–50] who have investigated the behaviour of a number of electrolytes at temperatures above the critical point of water and confirmed the (expected) existence of a thermodynamic discontinuity at the critical temperature.

Table 18.1 Representative volumetric measurements of electrolytes in aqueous solution at high temperatures (T) and pressures (p) over various ranges of concentration (m).

Electrolyte	T-Range/K	p-Range/MPa	m-Range/mol kg^{-1}	Reference
NaCl	323–600	0.1–40	0.003–5.0	48
LiCl	322–550	0.8–33	0.05–3.0	49
CsBr	604–725	18–38	0.002–0.5	50
$CaCl_2$	298–523	7–40	0.2–6.2	51
Na_2SO_4[a]	298–573	10–30	0.01–1.0	52
$MgCl_2$	298–623	10–30	0.005–3.0	53
KF[a]	298–627	10–30	0.1–3.0	54
LiCl	29–607	1–30	0.1–15	25
NaH_2PO_4[a]	473–598	15	0.1–2.1	55
NaOH	323–573	10	0.1–8.0	18
$NaAl(OH)_4$[b]	323–573	10	1.0–6.0	56
KCl[ab]	298–523	0.1–40	0.1–5.8	57

[a]Data for one or more related salts are also reported.
[b]Electrolyte mixtures are also reported.

18.4.2 Non-Aqueous and Mixed-Solvent Solutions

Millero summarised the rather limited amount of data for salts in non-aqueous solvents under near-ambient conditions up to *ca.* 1970.[1] The far more extensive studies subsequently available for non-aqueous electrolyte solutions up to the end of 2003 were compiled and critically reviewed by Marcus and Hefter.[7] Sufficient data were found to warrant collection into separate tables for eighteen solvents other than water, although the number of salts studied were $\leq$20 in six of those solvents. Data were insufficient to justify separate tables for a further 33 solvents.[7] More importantly, the overwhelming majority of the compiled data were for 1 : 1 electrolytes and very few of those were "recommended", *i.e.*, had been confirmed by at least two fully independent studies. Data were patchy for 2 : 1 salts, rare for 3 : 1 salts and non-existent for 1 : 2 and 1 : 3 electrolytes.[7] Clearly, a great deal remains to be done experimentally on the volumes of common ions, especially those with charges $>|1|$, in all non-aqueous solvents. Relatively few publications on the volumetric behaviour of electrolytes in non-aqueous solvents have appeared since Marcus and Hefter's review, an important exception being the extensive measurements on *N,N*-dimethylformamide (DMF) solutions by Płaczek *et al.*, which included a number of 2 : 1 and 3 : 1 salts.[38]

No comprehensive investigation of the molar volumes of electrolytes in any non-aqueous solvent at high or low temperatures appears to have been reported although some data reporting expansibilities exist at peri-ambient conditions.[1,7] Rather more measurements have been reported in mixed aqueous–organic solvents but the range of ions studied has usually been limited and very few measurements have been independently confirmed; the same is true of mixed organic solvents.[7]

18.5 Interpretation of Electrolyte Volumes

While molar volumes of electrolytes in solution have a tangible quality this does not mean that they are easy to interpret, even in a qualitative manner. This is essentially because molar volumes are composite quantities with many contributing effects. To simplify the discussion, it is customary to restrict theoretical description to the standard state (infinite dilution) volumes where, by definition, electrolytes are fully dissociated into their constituent ions and the observed volumes reflect only ion–solvent interactions. The behaviour of electrolyte volumes at finite (especially higher) concentrations where solute–solute interactions eventually become dominant is appropriately discussed in Section 18.3 and will not be further considered here.

18.5.1 Ionic Volumes

The thermodynamic properties, including volumes, of individual ions (so-called 'single ion' or 'absolute' ion properties) are undefined within the framework of thermodynamics.[7] This is unfortunate because such quantities are required for theoretical and computational purposes and for a variety of practical applications. To obtain such quantities from the determinable whole salt values it is necessary to adopt a suitable extra-thermodynamic ('extra' meaning 'outside of') assumption.[7] Such assumptions cannot be proven to be correct but their reasonableness (or otherwise) can be assessed by comparison with alternative assumptions (if they exist) and with known chemical behaviour. Fortunately, there are many assumptions available for determining ionic volumes, V°(ion). These assumptions have been reviewed on many occasions[7,58–61] and, while they are not in as good agreement as some researchers have suggested,[61] careful consideration of the available assumptions has indicated that a value of $V^{\circ}(H^{+}) = -5.5\ cm^3\ mol^{-1}$ for the aquated proton at 25 °C, proposed by Conway,[58,59] is reasonable and has been widely accepted.[7,27,60] Note that it is only necessary to define V° for one ion because all other V°(ion) values can then be derived by assuming additivity:

$$V^{\circ}(MX) = V^{\circ}(M^{+}) + V^{\circ}(X^{-}) \tag{18.22}$$

which is strictly obeyed at infinite dilution. Analogous expressions apply, with appropriate allowance for stoichiometry, for non-1 : 1 salts and for the differences between different salts, for example:

$$V^{\circ}(MX_n) = V^{\circ}(M^{n+}) + nV^{\circ}(X^{-}) \tag{18.23}$$

$$V^{\circ}(MX) = V^{\circ}(MB) + V^{\circ}(AX) - V^{\circ}(AB) \tag{18.24}$$

and so on.

The choice of which ion(s) to use in defining the scale depends upon the assumption but for comparison purposes among scales it is common to express the results obtained in terms of $V^{\circ}(H^{+})$, at least in aqueous solution.[7,58,59] It should also be noted that there is a significant difference between deriving V°(ion) values *via* an appropriate extra-thermodynamic assumption, which tries to provide estimates that are as realistic as possible, and those obtained *via* the *convention* that $V^{\circ}(H^{+}) \equiv 0\ cm^3\ mol^{-1}$. The latter is suitable only for tabulation purposes and makes no pretence to be realistic. Even experienced researchers appear to forget this distinction from time to time (see, for example, reference 29, p. 683).

Hefter and Marcus[60] have given a careful critique of all of the common methods for obtaining V°(ion) values in aqueous, mixed and non-aqueous solvents and have suggested those based on the 'TATB' (tetraphenylarsonium tetraphenylborate, $Ph_4As^{+}BPh_4^{-}$) or its phosphonium analogue (TPTB, $Ph_4P^{+}BPh_4^{-}$) were the "least objectionable". Their recommendation can be expressed as:

$$V^{\circ}(Ph_4As^{+}) = [V^{\circ}(TATB) + 8]/2;\ V^{\circ}(BPh_4^{-}) = [V^{\circ}(TATB) - 8]/2 \tag{18.25}$$

in all solvents or equivalently (for the phosphonium analogue):

$$V^\circ(Ph_4P^+) = [V^\circ(TPTB) + 2]/2;\ V^\circ(BPh_4^-) = [V^\circ(TPTB) - 2]/2 \quad (18.26)$$

18.5.2 Interpretation of Ionic Volumes

It is beyond the scope of this chapter to provide a detailed account of the various theoretical treatments of ionic volumes in solution. However, it is appropriate to note that this is still an active area of research, with the ongoing work of Marcus[32,62,63] being particularly noteworthy. One thing that is clear from the existing data is that V°(ion) values in any solvent, including water, are composite quantities that reflect a variety of contributing effects. Naturally, most of the data available and most of the interpretations to date have focussed on electrolytes in water. This is unfortunate because the properties of water and of aqueous electrolyte solutions are quite distinct from those in other solvents.[7]

A detailed description of the various factors contributing to V°(ion) values has been provided by Marcus and Hefter[7] who identified six distinct stages in transferring an ion from the ideal gaseous state to infinite dilution in a solvent. To simplify the discussion, these effects were grouped into four (hopefully orthogonal) categories:

$$V^\circ(\text{ion}) = V_{int} + V_{el} + V_{cov} + V_{str} \quad (18.27)$$

where V_{int} is the intrinsic (physical or geometric) size of the ion for which a cavity in the solvent must be provided when the ion dissolves; this quantity is always positive. In contrast, V_{el}, the electrostriction volume, which reflects the strong compressive effect on the surrounding solvent exerted by the very large electric field of the ion, is always negative. The quantity V_{cov} reflects the volume contribution of all the short range chemical interactions such as donor–acceptor and hydrogen-bonding interactions that occur between the ion and the solvent, while V_{str} represents the changes (excluding electrostriction) that occur to the solvent structure *via* short range interactions with the ions. Breakdowns of standard molar volumes in aqueous solution similar to that given in Equation (18.27), often based on the well-known Frank and Wen model of ion hydration, have been proposed[1] but usually contain effects that are unique to water and thus, although important, will not be considered in the present discussion.

The intrinsic volumes of ions are readily, but not necessarily meaningfully, calculated as $(4/3)\pi r_i^3$ from crystallographic radii, r_i (themselves the subject of longstanding debate)[64] but allowance has to be made for void spaces.[7] Recently Marcus has suggested obtaining V°(ion) in solution by using the partial molar volumes of salts in concentrated solution or even the molar volumes of ionic melts.[63]

The negative electrostriction volume, V_{el}, contributed by the compression of the solvent in the intense (of the order of 10^{11} V m^{-1}) electric field surrounding an ion has been identified for well over one hundred years, following the work of Drude and Nernst.[65] Using a simple electrostatic model these authors were able to show:

$$V_{el} = (N_A e^2/8\pi\varepsilon_0)\ \varepsilon^{-1}\ z_i^2\ r_i^{-1}(\partial \ln\varepsilon/\partial p)_T \tag{18.28}$$

where z_i is the charge (number) and r_i is the radius of the ion and all other symbols have been defined previously [see Equation (18.18)]. Considerable success in calculating V_{el} has been achieved[62] more recently using a shell model originally proposed by Marcus and Hefter.[9]

Unfortunately, not all of the contributions to $V°$(ion) in Equation (18.27) are amenable to straightforward theoretical description. To try to get around these difficulties, Marcus *et al.*[66] performed a statistical analysis of the then available values of $V°$(ion) in a cross-section of solvents with a view to identifying the significantly-contributing properties of ions and solvents. In addition to the expected major effect of crystallographic size for all ions there were significant contributions to $V°$(ion) for small ions from solvent compressibility (reflecting V_{el}) and H-bonding ability and some other parameters reflecting solvent structure and covalent effects. For large ions (which have small V_{el}), solvent and ion polarisability, which reflect covalent and structural effects, were also important.[66]

18.5.3 Ionic Transfer Volumes

Analogous to other thermodynamic transfer quantities, the volume of transfer of an ion is defined[7] as the difference in the standard partial molar volume of that ion in two distinct solvents:

$$\Delta_t V°(\text{ion, A}\rightarrow\text{B}) = V°(\text{ion, B}) - V°(\text{ion, A}) \tag{18.29}$$

and therefore represents the *differences* in ion–solvent interactions in the two solvents for that ion. The choice of reference solvent (A) is arbitrary but is normally water. Transfer volumes have a special advantage in that they enable comparisons to be made without the dominating effects of V_{int}, the latter being assumed to be more or less independent of the solvent. A comprehensive list of $\Delta_t V°$(ion, w→s) values for monovalent ions, based on the TATB or TPTB assumptions, has been given by Marcus and Hefter;[7] almost no reliable data were available for more highly charged ions. The values of $\Delta_t V°(Y^{\pm}$, w→s) for monovalent ions reveal some surprising characteristics. For example, the transfer volumes of small cations and small anions show no dependence on the dielectric constant (relative permittivity, ε) of the solvent whereas $\Delta_t V°(Ph_4X^{\pm}$, w→s) for the large 'tetraphenyl' ions do (at least at $\varepsilon < 50$). In a similar vein, $\Delta_t V°(Ph_4X^{\pm}$, w→s) have a much stronger dependence on solvent compressibility (κ_T) than do $\Delta_t V°(Y^{\pm}$, w→s). These two effects appear to be a consequence of the 'dielectric saturation' that occurs in the intense electric fields surrounding small ions, which largely equalises the electrostriction for all solvents.

Another surprising result is that while, as would be expected from Equation (18.28), ionic charge has a very large (negative) effect on $V°(M^{n+})$ these differences do not translate through to $\Delta_t V^\circ(M^{n+}, w \rightarrow s)$. For example, the ions Li^+, Mg^{2+} and Sc^{3+}, which have approximately equal (crystallographic) radii (69, 72 and 75 pm, respectively), have dramatically different V° values in water, being: −6.4, −32.2 and −58.4 $cm^3\ mol^{-1}$, respectively, at 25 °C.[27] The corresponding results in DMF,[38] the only solvent other than water for which reliable V° values exist for the higher charged cations, are −6.5, −27.8 and −45.0 $cm^3\ mol^{-1}$. Thus $\Delta_t V^\circ(M^{n+})_{w \rightarrow DMF} = -0.1$, +4.4 and +13.4 $cm^3\ mol^{-1}$ for Li^+, Mg^{2+} and Sc^{3+}, respectively.[38] Given that the isothermal compressibility, κ_T, of DMF (0.642 GPa^{-1}) is much greater than that of water (0.457 GPa^{-1}) at 25 °C,[20] the transfer volumes are increasingly *opposite* (with increasing charge) to what would be expected from electrostriction. Similar counter-intuitive effects have been discerned from the very limited database for other ions in other solvents.[7]

Further discussion of the effects contributing to $\Delta_t V^\circ(\text{ion}, w \rightarrow s)$ is given by Marcus and Hefter[7] but it is worth noting that, after expressing their concern about the quality of the available data, these authors concluded that the many unexpected features were probably a reflection of the "subtlety and complexity of the effects involved", a timely reminder that there are ongoing challenges for the measurement and interpretation of molar volumes in both aqueous and non-aqueous solvents.

References

1. F. J. Millero, *Chem. Rev.*, 1971, **71**, 147.
2. F. J. Millero, in *Water and Aqueous Solutions*, ed. R. A. Horne, Wiley, New York, 1971.
3. *Inorganic High Pressure Chemistry*, ed. R. van Eldik, Elsevier, Amsterdam, 1986.
4. *Organic High Pressure Chemistry*, ed. W. J. le Noble, Elsevier, Amsterdam, 1988.
5. J. F. Zamaitis, D. M. Clark, M. Rafal and N. C. Scrivner, *Handbook of Aqueous Electrolyte Thermodynamics, Theory and Application*, DIPPR Publication, New York, 1986.
6. F. J. Millero, *The Physical Chemistry of Natural Waters*, Wiley, New York, 2001.
7. Y. Marcus and G. Hefter, *Chem. Rev.*, 2004, **104**, 3405.
8. P. W. Atkins, *Physical Chemistry*, Oxford University Press, Oxford, UK, 3rd edn, 1986, p.161.
9. Y. Marcus and G. T. Hefter, *J. Solution Chem.*, 1999, **28**, 575.
10. F. J. Millero, in *Activity Coefficients in Electrolyte Solutions*, ed. R. M. Pytkowicz, CRC Press, Boca Raton, 1979, Vol. II, pp. 63–151.
11. See for example, G. T. Hefter, J.-P. E. Grolier and A. H. Roux, *J. Solution Chem.*, 1989, **18**, 229.
12. N. Bauer and S. Z. Lewin, in *Techniques of Chemistry, Volume I: Physical Methods of Chemistry, Part IV: Determination of Mass, Transport, and*

Electrical-Magnetic Properties, ed. A. Weissberger and B. W. Rossiter, Wiley, New York, 1972.
13. R. S. Davis and W. F. Koch, in *Physical Methods of Chemistry. Volume VI: Determination of Thermodynamic Properties*, ed. B. W. Rossiter and R. C. Baetzold, Wiley, New York, 2nd edn, 1992.
14. W. Wagner, R. Kleinrahm, H. W. Lösch and J. T. R. Watson, in *Experimental Thermodynamics Volume VI, Measurement of the Thermodynamic Properties of Single Phases*, ed. K. N. Marsh, W. A. Wakeham and A. R. H. Goodwin, Elsevier, Amsterdam, 2003, pp. 127–149.
15. See for example, S. Bochmann, P. M. May and G. T. Hefter, *J. Chem. Eng. Data*, 2011, **56**, 5081.
16. O. Kratky, H. Leopold and H. Stabinger, *Z. Angew. Phys.*, 1969, **27**, 273.
17. V. M. Majer and A. A. H. Pádua, in *Physical Methods of Chemistry. Volume VI: Determination of Thermodynamic Properties*, ed. B. W. Rossiter and R. C. Baetzold, Wiley, New York, 2nd edn, 1992, pp. 149–168.
18. L. Hnědkovský, E. Königsberger, I. Cibulka, L.-C. Königsberger, S. Schrödle, P. M. May and G. T. Hefter, *J. Chem. Eng. Data*, 2007, **52**, 2237.
19. http://www.anton-paar.com/
20. Y. Marcus, *The Properties of Solvents*, Wiley, Chichester, UK, 1998.
21. L. G. Hepler, J. M. Stokes and R. H. Stokes, *Trans. Faraday Soc.*, 1965, **61**, 20.
22. G. A. Bottomley and M. T. Bremers, *Aust. J. Chem.*, 1986, **39**, 1959.
23. L. A. Woolf, in *Physical Methods of Chemistry. Volume VI: Determination of Thermodynamic Properties*, ed. B. W. Rossiter and R. C. Baetzold, Wiley, New York, 2nd edn, 1992, pp. 168–191.
24. M. J. Blandamer, M. I. Davis, G. Douhéret and J. C. R. Reis, *Chem. Soc. Rev.*, 2001, **30**, 8.
25. I. M. Abdulagatov and N. D. Azizov, *Chem. Geol.*, 2006, **230**, 22 and references therein.
26. K. Boehm, J. Rsgen and H.-J. Hinz, *Anal. Chem.*, 2006, **78**, 984 and references therein.
27. Y. Marcus, *Ion Properties*, Dekker, New York, 1997.
28. *Activity Coefficients in Electrolyte Solutions*, ed. K. S. Pitzer, CRC Press, Boca Raton, USA, 2nd edn, 1991.
29. B. S. Krumgalz, R. Pogorelsky and K. S. Pitzer, *J. Phys. Chem. Ref. Data*, 1996, **25**, 663.
30. O. Redlich and P. Rosenfeld, *Z. Elektrochem.*, 1931, **37**, 705.
31. D. O. Masson, *Philos. Mag.*, 1929, **8**, 218.
32. Y. Marcus, *Chem. Rev.*, 2011, **111**, 2761.
33. O. Redlich and D. M. Meyer, *Chem. Rev.*, 1964, **64**, 221.
34. L. N. Trevani, E. C. Balodis and P. R. Tremaine, *J. Phys. Chem. B*, 2007, **111**, 2015.
35. P. M. May, D. Rowland, G. T. Hefter and E. Königsberger, *J. Chem. Eng. Data*, 2011, **56**, 5066.
36. D. Rowland, E. Königsberger, G. T. Hefter and P. M. May, *Appl. Geochem.*, in press.

37. T. Chen, G. Hefter, R. Buchner and G. Senanayake, *J. Solution Chem.*, 1998, **27**, 1067.
38. A. Płaczek, W. Grzybkowski and G. T. Hefter, *J. Phys. Chem. B*, 2008, **112**, 12366.
39. O. Söhnel and P. Novotny, *Densities of Aqueous Solutions of Inorganic Substances*, Elsevier, Amsterdam, 1985.
40. G. G. Asayev and I. D. Zaytsev, *Volumetric Properties of Electrolyte Solutions*, Begell House, New York, 1996.
41. P. M. May, D. Rowland, E. Königsberger and G. T. Hefter, *Talanta*, 2010, **81**, 142.
42. R. L. Gardas, D. H. Dagade, J. A. P. Coutinho and K. J. Patil, *J. Phys. Chem. B*, 2008, **112**, 3380.
43. A. Tromans, E. Königsberger, P. M. May and G. T. Hefter, *J. Chem. Eng. Data*, 2005, **50**, 2019.
44. H. Bianchi and P. R. Tremaine, *J. Solution Chem.*, 1995, **24**, 439.
45. See for example, A. A. Humffray, *Canad. J. Chem.*, 1987, **65**, 833 and references therein.
46. E. L. Shock and H. C. Helgeson, *Geochim. Cosmochim. Acta*, 1988, **52**, 2009 and references therein.
47. E. Königsberger, G. Eriksson, P. M. May and G. T. Hefter, *Ind. Eng. Chem. Res.*, 2005, **44**, 5805.
48. V. Majer, J. A. Gates, A. Inglese and R. H. Wood, *J. Chem. Thermodyn.*, 1988, **20**, 949.
49. V. Majer, A. Inglese and R. H. Wood, *J. Chem. Thermodyn.*, 1989, **21**, 321.
50. V. Majer and R. H. Wood, *J. Chem. Thermodyn.*, 1994, **26**, 1143.
51. C. S. Oakes, J. M. Simonson and R. J. Bodnar, *J. Solution Chem.*, 1995, **24**, 897.
52. M. Obšil, V. Majer, J.-P. E. Grolier and G. T. Hefter, *J. Chem. Soc., Faraday Trans.*, 1996, **92**, 4445 and references therein.
53. M. Obšil, V. Majer, G. T. Hefter and V. Hynek, *J. Chem. Thermodyn.*, 1997, **29**, 575.
54. V. Majer, M. Obšil, G. T. Hefter and J.-P. E. Grolier, *J. Solution Chem.*, 1997, **26**, 847.
55. G. E. Woolston, L. N. Trevani and P. R. Tremaine, *J. Chem. Eng. Data*, 2008, **53**, 1728.
56. L. Hnědkovský, E. Königsberger, L.-C. Königsberger, I. Cibulka, S. Schrödle, P. M. May and G. T. Hefter, *J. Chem. Eng. Data*, 2010, **55**, 1173.
57. D. Zezin, T. Driesner and C. Sanchez-Valle, *J. Chem. Eng. Data*, 2014, **59**, 736 and references therein.
58. B. E. Conway, *J. Solution Chem.*, 1978, **7**, 721.
59. B. E. Conway, *Ionic Hydration in Chemistry and Biophysics*, Elsevier, Amsterdam, 1981.
60. G. Hefter and Y. Marcus, *J. Solution Chem.*, 1997, **26**, 249.
61. F. J. Millero, *J. Phys. Chem.*, 1971, **75**, 280.
62. Y. Marcus, *J. Phys. Chem. B*, 2014, **118**, 2172 and references therein.

63. Y. Marcus, *J. Solution Chem.*, 2010, **39**, 1031.
64. See, for example, J. E. Huheey, *Inorganic Chemistry*, SI Units Edn., Harper & Row, London, 1975.
65. P. Drude and W. Nernst, *Z. Phys. Chem.*, 1894, **15**, 79.
66. Y. Marcus, G. T. Hefter and T. S. Pang, *J. Chem. Soc., Faraday Trans.*, 1994, **90**, 1899.

CHAPTER 19

Volumetric Behaviour of Room Temperature Ionic Liquids

YIZHAK MARCUS

Institute of Chemistry, The Hebrew University of Jerusalem, Jerusalem 91904, Israel
Email: ymarcus@vms.huji.ac.il

19.1 Introduction

Room temperature ionic liquids (RTILs) are stoichiometric combinations of cations and anions that have melting points, t_m, generally in the range 0 to 100 °C ($273 \leq T_m/K \leq 373$), hence they are liquid at ambient conditions or somewhat above. They have been proposed as "green" solvents because of their low vapour pressures and non-flammability, despite being rather viscous and expensive. In recent years, RTILs have become very popular subjects for research, the number of papers dealing with them doubling every few years; hence their volumetric properties cannot be dealt with in a chapter such as this in an exhaustive manner. There is a large list of cations and a corresponding list of anions that can be combined to form RTILs, and this chapter can deal with only a small fraction of the possible combinations for which there are data in the literature. The latter have been examined mainly from publications in the last decade, as the volumetric properties have not so far been summarised in a book devoted to RTILs.

The volumetric properties of RTILs, among other properties, depend on the water contents of the salts if they are not carefully dried; here only the dry salts are considered. Volumetric properties of water-saturated RTILs are available in the literature, *e.g.*, for imidazolium[1,2] and pyridinium[3] salts, compared with those of the dry salts.

Volume Properties: Liquids, Solutions and Vapours
Edited by Emmerich Wilhelm and Trevor M. Letcher

Published by the Royal Society of Chemistry, www.rsc.org

The cations of RTILs may be selected from a large list, as said above, but the most popular ones include 1-alkyl-3-methylimidazolium, *N*-alkylpyridinium, and tetraalkyl-ammonium or -phosphonium. Such RTILs are dealt with here in some detail, whereas those based on other cations are mentioned only cursorily. Similarly, the anions of RTILs dealt with here in detail include tetrafluoroborate, hexafluorophosphate, bis(trifluorosulfonyl)imide, alkylsulfate, small inorganic anions, and carboxylate anions.

The densities of RTILs at ambient pressures are generally linear with the temperature over a wide temperature range above the melting point:

$$\rho = a - 10^3 bT \tag{19.1}$$

The isobaric expansibility is therefore $\alpha_P = b/10^3\rho$, and the molar volume $V = M/\rho$, where M is the molar mass. The $\rho/(\text{g cm}^{-3})$ and $V/(\text{cm}^{-3}\ \text{mol}^{-1})$ data presented here generally pertain to 25 °C (298.15 K) and 0.1 MPa, but the values at other temperatures at ambient pressure are readily obtained from:

$$\rho(T) = \rho(298.15)[1 - \alpha_P(T - 298.15)] \tag{19.2}$$

The isothermal compressibility κ_T is obtained from the volume or density dependence on (fairly large values of the) pressure P at a given T:

$$\kappa_T(T) = -V^{-1}(\partial V/\partial P)_T = \rho^{-1}(\partial \rho/\partial P)_T = C(T)V_0(T)^{-1}[B(T) + P]^{-1} \tag{19.3}$$

where the last equality is a form of the Tait equation, for which $V_0(T) = V(P = 0.1\ \text{MPa}, T)$ and $B(T)$ and $C(T)$ are temperature-dependent substance-specific constants. The isothermal compressibility κ_T is related to the adiabatic compressibility κ_S as:

$$\kappa_T(T) = \kappa_S(T) + TV(T)\alpha_P(T)^2 C_P(T)^{-1} \tag{19.4}$$

where C_P is the molar heat capacity at constant pressure. The adiabatic compressibility κ_S is obtained at a given temperature T from the density and speed of ultrasound u at ambient pressures:

$$\kappa_S(T) = -V^{-1}(\partial V/\partial P)_S = \rho^{-1}(\partial \rho/\partial P)_S = \rho(T)^{-1}u(T)^{-2} \tag{19.5}$$

Various equations of state (EoSs) have been proposed for RTILs, yielding extensive *PVT* values, and these are discussed briefly in cases where they yield data not otherwise presented. Other methods for the prediction of the volumetric data, *e.g.*, from group contributions, are also briefly discussed.

19.2 The Volumetric Data

19.2.1 1-Alkyl-3-methylimidazolium Salts

RTILs consisting of 1-alkyl-3-methylimidazolium cations with a variety of anions have been studied extensively. The name of the cation is generally abbreviated as $C_n$mim, where n designates the number of carbon atoms in the 1-alkyl chain (sometimes also known as emim for C_2mim, bmim for C_4mim, *etc.*). The densities $\rho/(\mathrm{g\ cm^{-3}})$ at $T = 298.15$ K and 0.1 MPa of $C_n$mim salts with a given anion are presented in Table 19.1 for increasing values of n. Only those values are included that have been reported, or could be interpolated from reported data for these ambient conditions, with a precision that required at least 4 decimals related to the units used in Table 19.1. They have then been averaged and listed unless no more precise values are available. The densities of salts of $C_n$mim with the anions of alanine[4] and valine[5] are available from single sources and are reported as linear functions of n.

The molar masses of the $C_n$mim cations conform to the expression $M/(\mathrm{g\ mol^{-1}}) = 83.12 + 14.02n$, hence the molar volumes of the salts are readily calculated, using $M/(\mathrm{g\ mol^{-1}}) = 86.80$ for BF_4^-, 144.78 for PF_6^-, and 280.15 for $N(CF_3SO_2)_2^-$, the latter anion often being abbreviated as NTF_2. The molar volumes of the $C_n$mim salts with these three anions at $T = 298.15$ K conform to the following linear expressions:

$$V(\mathrm{C}_n\mathrm{mimBF_4})/(\mathrm{cm^3\ mol^{-1}}) = (121.7 \pm 1.7) + (16.8 \pm 0.2)n \quad (19.6)$$

$$V(\mathrm{C}_n\mathrm{mimPF_6})/(\mathrm{cm^3\ mol^{-1}}) = (138.1 \pm 2.8) + (17.3 \pm 0.2)n \quad (19.7)$$

$$V(\mathrm{C}_n\mathrm{mimNTF_2}/(\mathrm{cm^3\ mol^{-1}}) = (222.6 \pm 1.5) + (17.3 \pm 0.2)n \quad (19.8)$$

For the alkylsulfate salts the following expressions are valid:

$$V(\mathrm{C}_n\mathrm{mimMeSO_4})/(\mathrm{cm^3\ mol^{-1}}) = (139.5 \pm 1.0) + (17.0 \pm 0.4)n \quad (19.9)$$

$$V(\mathrm{C_2mimC}_m\mathrm{H}_{2m+1}\mathrm{SO_4})/(\mathrm{cm^3\ mol^{-1}}) = (156.3 \pm 0.4) + (17.1 \pm 0.1)m \quad (19.10)$$

Thus, the volume increment per CH_2 group in the alkyl chains of the salts is quite constant, $(17.0 \pm 0.3)\ \mathrm{cm^3\ mol^{-1}}$.

The expansibilities at ambient conditions vary little with the size of the alkyl chain and can be summarised as $\alpha_P/\mathrm{K^{-1}} = (0.62 \pm 0.03)\times10^{-3}$ for the BF_4^- and PF_6^- salts, as $(0.67 \pm 0.03)\times10^{-3}$ for the NTF_2^- salts, and as $(0.56 \pm 0.04)\times10^{-3}$ for the alkylsulfate salts. Similar values are reported for the salts of other anions. An uncertainty of 0.1×10^{-3} in $\alpha_P/\mathrm{K^{-1}}$ yields an uncertainty of $\leq$0.2% in the densities of the salts in the range of measurements 278 K to 318 K (5 to 45 °C), well within the average precision of the reported values. Therefore, these average expansibility values can serve well for the estimation of the densities and molar volumes of the $C_n$mim salts that have not been measured other than at room temperature.

The Tait expression in various forms has been employed to describe the pressure dependence of the densities [see Equation (19.1)] of $C_n$mim salts by a few authors.[2,6–9] The isothermal compressibilities κ_T presented in Table 19.2, however, were derived from the numerical tables of $\rho(T,P)$ provided by these authors and others, extrapolated to 0.1 MPa and interpolated to 298.15 K where necessary. They are seen to span a fairly narrow range of

Table 19.1 The densities ρ of RTIL 1-alkyl-3-methylimidazolium ($C_n$mim) salts at $T = 298.15$ K and 0.1 MPa.

Salt	ρ/(g cm^{-3})	Ref.
C_1mimBF_4	1.306	37
C_2mimBF_4	1.2803 ± 0.0004	41–43
C_3mimBF_4	1.237	44
C_4mimBF_4	1.2015 ± 0.0007	1,3,8,9,33,42,45–47
C_5mimBF_4	1.173	44
C_6mimBF_4	1.1453 ± 0.0002	42,48
C_8mimBF_4	1.1000 ± 0.0040	3,17,42
C_1mimPF_6	1.478	37
C_2mimPF_6	1.413	37
C_4mimPF_6	1.3647 ± 0.0010	1,7,8,17,40,46,49,50
C_6mimPF_6	1.2928 ± 0.0015	3,40,48,49
C_7mimPF_6	1.260	40
C_8mimPF_6	1.2291 ± 0.0090	3,17,40,49
$C_1mimN(CF_3SO_2)_2$	1.5620 ± 0.0040	37,39,51
$C_2mimN(CF_3SO_2)_2$	1.5183 ± 0.0008	1,7,21,33,45,46,49,52
$C_3mimN(CF_3SO_2)_2$	1.4757	6
$C_4mimN(CF_3SO_2)_2$	1.4371 ± 0.0010	1,7,33,45,46,49,52
$C_5mimN(CF_3SO_2)_2$	1.4045	6
$C_6mimN(CF_3SO_2)_2$	1.3708 ± 0.0002	46,49
$C_7mimN(CF_3SO_2)_2$	1.3477	21
$C_8mimN(CF_3SO_2)_2$	1.3203 ± 0.0012	21,46,49
$C_{10}mimN(CF_3SO_2)_2$	1.2783 ± 0.0002	46,49
$C_{12}mimN(CF_3SO_2)_2$	1.2447	49
$C_{14}mimN(CF_3SO_2)_2$	1.1882	49 (extrapolated)
$C_1mimCH_3SO_4$	1.3273	10
$C_2mimCH_3SO_4$	1.275 ± 0.011	53,54
$C_4mimCH_3SO_4$	1.2058	13
$C_2mimC_2H_5SO_4$	1.2390 ± 0.0012	1,7,11,42,43,46,52,55,56
$C_2mimC_4H_9SO_4$	1.1760 ± 0.0003	42,53
$C_2mimC_6H_{13}SO_4$	1.1300 ± 0.0004	42,53
$C_2mimC_8H_{17}SO_4$	1.0948 ± 0.0010	42,53
$C_4mimC_8H_{17}SO_4$	1.0668 ± 0.0030	8,13,47,52
C_4mimCl	1.080	37,40
C_6mimCl	1.0393	57
C_8mimCl	1.0097	57
C_6mimI	1.3814	33
C_4mimI_3	2.1486	33
$C_1mimSCN$	1.1574	46
$C_2mimSCN$	1.1162	43,58
$C_4mimSCN$	1.0695	48
C_3mimNO_3	1.184	44
C_4mimNO_3	1.153	44
C_5mimNO_3	1.127	44
C_6mimNO_3	1.103	44
$C_2mimHSO_4$	1.3674	43
$C_2mimAlCl_4$	1.2947	34
$C_6mimAlCl_4$	1.1953	34
$C_4mimFeCl_4$	1.380	37
$C_6mim\ FeCl_4$	1.333	37
$C_8mim\ FeCl_4$	1.280	37
$C_3mimCF_3BF_3$	1.310	40

Table 19.1 (*Continued*)

Salt	ρ/(g cm^{-3})	Ref.
$C_3mimC_2F_5BF_3$	1.380	40
$C_3mimC_3F_7BF_3$	1.440	40
$C_2mimN(CN)_2$	1.1016 ± 0.0004	41,43
$C_4mimN(CN)_2$	1.0588	45
$C_2mimCH_3SO_3$	1.2409 ± 0.0007	43,59
$C_3mimCH_3SO_3$	1.2031	59
$C_6mimCH_3SO_3$	1.1183	59
$C_8mimCH_3SO_3$	1.0818	59
$C_2mimCF_3SO_3$	1.3836	37,60
$C_4mimCF_3SO_3$	1.3006 ± 0.009	3,37,45
$C_4mimHCO_2$	1.0733	61
$C_2mimCH_3CO_2$	1.1437	49
$C_3mimCH_3CO_2$	1.1190	49
$C_4mimCH_3CO_2$	1.0665 ± 0.020	35,49,61
$C_5mimCH_3CO_2$	1.0773	35
$C_6mimCH_3CO_2$	1.0606	35
$C_4mimC_2H_5CO_2$	1.0350	61
$C_4mimC_3H_7CO_2$	1.0320	61
$C_2mimCH_3CHOHCO_2$	1.1461	62
$C_3mimH_2NCH_2CO_2$	1.1589	63
$C_{2\text{-}6}mimCH_3CHNH_2CO_2$	$1.1577 - 0.0193n$	4
$C_{2\text{-}6}mim(CH_3)_2CHCHNH_2CO_2$	$1.1438 - 0.0169n$	5

values, 0.33 to 0.54 GPa^{-1} (except for one outlying value for the ethylsulfate salt, which disagrees with the value derived from another author's data). The temperature dependence of κ_T appears to be variable among different authors, in the range from zero to 0.005 GPa^{-1} K^{-1}.

The adiabatic compressibilities of $C_1mimCH_3SO_4$[10] and $C_2mimC_2H_5SO_4$[11] are obtained from Equation (19.5) and the ρ and u data. At $T = 298.15$ K, $\kappa_S = 0.230$ GPa^{-1} for the former and 0.287 GPa^{-1} for the latter, with data also available at $T = 283.15$ to 345.15 K and 288.15 to 243.15 K, respectively. The C_P/J K^{-1} mol^{-1} = 423 value of $C_2mimC_2H_5SO_4$[12] permits the calculation of its $\kappa_T = 327$ GPa^{-1} according to Equation (19.4). For the C_1mim salt no C_P value was found so its κ_T could not be calculated. The adiabatic compressibilities of C_8mimCl, $C_4mimCH_3SO_4$, and $C_4mimC_8H_{17}SO_4$ at 298.15 K are, respectively, 0.337, 0.302, and 0.425 GPa^{-1}.[13] The adiabatic compressibility of $C_3mimNTF_2$ is similarly obtained from the ρ and u data in reference 6 yielding $\kappa_S = 0.443$ GPa^{-1} at $T = 298.15$ K.

Volumetric properties of further substituted imidazolium salts (*e.g.* methyl, also in the 2 position, abbreviated as $C_n$mmim[2,14]) are also available in the literature.

19.2.2 *N*-Alkyl Pyridinium Salts

Much less volumetric information is available for substituted pyridinium salts than for the imidazolium ones, but such RTILs are still quite popular.

Table 19.2 The isothermal compressibilities κ_T of RTIL 1-alkyl-3-methylimidazolium ($C_n$mim) salts at $T = 298.15$ K and 0.1 MPa.

Salt	κ_T/GPa^{-1}	dκ_T/dT/(GPa^{-1} K^{-1})	Ref.
C_2mimBF_4	0.337	0.0027	21
C_4mimBF_4	0.360	0.0013	7
	0.368		8 (at 313 K)
	0.390	0.0041	3
	0.382	0.0012	35
C_8mimBF_4	0.417	0.0034	3
C_4mimPF_6	0.412	0.0037	3
	0.348	0.0012	7
	0.381		8 (at 313K)
C_8mimPF_6	0.458	0.0039	3
$C_2mimN(CF_3SO_3)_2$	0.380	0.0015	7
$C_3mimN(CF_3SO_3)_2$	0.482	0.0000	6
	0.480	0.0037	21
$C_4mimN(CF_3SO_3)_2$	0.498	0.0010	7
$C_5mimN(CF_3SO_3)_2$	0.506	0.0004	6
$C_7mimN(CF_3SO_3)_2$	0.543	0.0046	21
$C_8mimN(CF_3SO_3)_2$	0.540	0.0052	21
$C_4mimCH_3SO_4$	0.379	0.0008	13
$C_2mimC_2H_5SO_4$	0.670	0.0010	7
	0.327		11 from κS
$C_4mimC_8H_{17}SO_4$	0.492	0.0022	13
	0.458		8 (at 313 K)
C_8mimCl	0.408	0.0035	13
$C_4mimN(CN)_2$	0.415	0.0032	21
$C_4mimCF_3SO_3$	0.432	0.0027	3

These pyridinium salts have an alkyl group attached to the nitrogen atom and may have a methyl group in the 3 or 4 position of the pyridine ring. The names of the cations are then abbreviated as C_nPy or C_n(3 or 4 M)Py. The densities ρ and isobaric expansibilities α_P at $T = 298.15$ K and 0.1 MPa of these pyridinium salts with some anions are presented in Table 19.3. Only those values are included that have been reported, or could be interpolated from reported data for these ambient conditions, with a precision that required at least 4 decimals (referring to units as given in Table 19.3) and have been averaged and listed unless no more precise values are available. For the C_nPyReO_4 salts with $n = 1, 2 \ldots 6$ the predictive expression

$$\rho/(\mathrm{g\ cm^{-3}}) = M/N_A v = (330.37 + 14.02n)/602.2(0.2272 + 0.0275n) \quad (19.11)$$

can be derived from the data given in reference 15 where M/(g mol^{-1}) is the molar mass and v/nm^3 is the molecular volume. However, some of the densities calculated according to $\rho = M/N_A v$ from reported v values[16] for other salts are appreciably larger than the experimental values for the same RTILs reported by other authors (see Table 19.3).

The compressibility of C_4PyBF_4, $\kappa_T = 0.378$ GPa^{-1}, has been calculated from data at increasing pressures.[17] The Tait expression was used to express the

Table 19.3 The densities ρ and expansibilities α_P of RTIL substituted pyridinium salts at $T = 298.15$ K and 0.1 MPa.

Salt	ρ/(g cm^{-3})	$10^3\alpha_P$/K^{-1}	Ref.
C_4PyBF_4	1.2138	0.50	17,64,65
C_8PyBF_4	1.1127	0.63	64
$C_2PyN(SO_2F)_2$	1.4587	0.60	43
$C_4PyN(SO_2F)_2$	1.3694	0.61	43
$C_6PyN(SO_2F)_2$	1.3019	0.61	43
C_2PyNTF_2	1.5368	0.59	66,67
C_3PyNTF_2	1.4933	0.59	68
C_4PyNTF_2	1.4521	0.63	67,69,70
C_5PyNTF_2	1.4214	0.63	67
C_6PyNTF_2	1.5635[b]		16
C_6PyNTF_2	1.3877	0.66	68
C_8PyNTF_2	1.3268	0.66	70
$C_{10}PyNTF_2$	1.2835	0.67	70
$C_{12}PyNTF_2$	1.2488	0.68	70
C_4PyReO_4	1.9025	0.80	15
C_5PyReO_4	1.8446	0.89	71
$C_2PyEtSO_4$	1.2223	0.53	18
$C_3OHPyNTF_2$	1.5464	0.60	72
$C_3OHPyPF_3(C_2F_5)_2$[a]	1.7368		73
$C_3(3M)PyNTF_2$	1.4486	0.71	2,18
$C_4(3M)PyBF_4$	1.3412		16,22
$C_4(3M)PyNTF_2$	1.5953[b]		16
$C_4(3M)PyNTF_2$	1.3936	0.63	72
$C_4(3M)PyNTF_2$	1.4128	0.65	2
$C_6(3M)PyNTF_2$	1.5379[b]		16
$C_6(3M)PyNTF_2$	1.3516		72
$C_4(3M)PyNTF_2$	1.4226	0.62	74
$C_4(4M)PyBF_4$	1.3434[b]		22
$C_2(4M)PyNTF_2$	1.4920	0.56	75
$C_4(4M)PyNTF_2$	1.4155	0.66	2,75

[a] *N*-(3-hydroxypropyl)pyridinium trifluorotris(pentafluoroethyl)phosphate.
[b] Estimated according to $\rho = M/N_A v$.

pressure and temperature dependence of the density in reference 18 yielding the isothermal compressibilities $\kappa_T = 0.543$ GPa^{-1} for $C_3(3M)PyNTF_2$ at ambient conditions. In reference 19 a figure was presented for the isothermal compressibility data of $C_3(3M)PyNTF_2$ from reference 18 as a function of the pressure and the temperature.

19.2.3 Quaternary Ammonium and Phosphonium Salts

RTILs based on quaternary ammonium and phosphonium salts are unsymmetrical (symmetrical ones generally have higher melting points), with *e.g.*, one long and three shorter alkyl chains, or based on pyrrolidinium or piperidinium rings. Most of the data available for the quaternary ammonium salts involve the bis(trifluoromethylsulfonyl)amide anion (NTF_2) with only few data for salts with other anions. These data, for $T = 298.15$ K and

Table 19.4 The densities ρ and expansibilities α_P of RTIL quaternary ammonium salts at $T = 298.15$ K and 0.1 MPa.

Salt	$\rho/(\mathrm{g\ cm^{-3}})$	$10^3\alpha_P/\mathrm{K}^{-1}$	Ref.
$BuMe_3NNTF_2$	1.3918	0.65	1,23,52,69
$HxMe_3NNTF_2$	1.3106	0.59	76
$DcMe_3NNTF_2$	1.2263	0.60	76
$BuEt_3NNTF_2$	1.2793	0.58	76
$HxEt_3NNTF_2$	1.2894	0.67	77
$HpEt_3NNTF_2$	1.2710	0.70	77
$OcEt_3NNTF_2$	1.2498	0.66	77
$DcEt_3NNTF_2$	1.2162	0.65	77
$DoEt_3NNTF_2$	1.1882	0.66	77
$TdEt_3NNTF_2$	1.1650	0.66	77
$MeOc_3NNTF_2$	1.1046	0.67	53
$BuPrMe_2NNTF_2$	1.3483	0.58	76
$HxPrMe_2NNTF_2$	1.2846	0.59	76
$DcPrMe_2NNTF_2$	1.2007	0.58	76
$MeEt_3NMeSO_4$	1.1729[a]	0.64	80
$PrMePyrrNTF_2$	1.4274	0.64	43
$BuMePyrrNTF_2$	1.3956	0.63	24,55,69,78
$HxMePyrrNTF_2$	1.32		78
$DcMePyrrNTF_2$	1.25		78
$(MeOEt)MePyrrNTF_2$	1.4547	0.64	55
$PrMePyrrN(SO_3F)_2$	1.3381	0.60	79
$EtMePyrrEtSO_4$	1.1960[a]	0.72	80
$BuMePyrrMeSO_4$	1.1669	0.66	80
$BuEtPyrrEtSO_4$	1.1665[a]	0.72	80
$BuMePyrrPF_3(C_2F_5)_3$	1.5832	0.58	55
$(MeOEt)MePyrr\ PF_3(C_2F_5)_3$	1.6309	0.68	55,73
$PrMePipNTF_2$	1.4121	0.69	18
$(MeOEt)MePipPF_3(C_2F_5)_3$	1.6077		73

[a]Extrapolated to 298.15 K.

0.1 MPa, are shown in Table 19.4, with the following common abbreviations: Me = methyl, Et = ethyl, Pr = 1-propyl, Bu = 1-butyl, Hx = 1-hexyl, Oc = 1-octyl, Dc = 1-decyl, Do = 1-dodecyl, Td = 1-tetradecyl, Pyrr = pyrrolidinium, and Pip = piperidinium (with the two alkyl groups attached to the nitrogen atom in the latter two rings). Where several references are given the values are the averages of closely agreeing data.

The densities of salts of the Bu_3MeN^+ cation with anions of amino acids are available in reference 20. The densities reported in reference 21 are consistently lower, where comparable, than the values reported by other authors and need to be questioned. Density values obtained from the $\rho = M/N_A\nu$ relationship with reported ν values[16,22] are consistently higher, where comparable, than the values reported by other authors and also need to be questioned.

For a few salts, the compressibilities have also been reported: $\kappa_T = 0.389\ \mathrm{GPa^{-1}}$ for $BuMe_3NNTF_2$ and $\kappa_T = 0.447\ \mathrm{GPa^{-1}}$ for $BuMePyrrNTF_2$,[23] and $\kappa_T = 0.510\ \mathrm{GPa^{-1}}$ for $PrMePipNTF_2$.[24] The speed of sound data[16] yield $\kappa_S = 0.445\ \mathrm{GPa^{-1}}$ for $BuMePyrrNTF_2$ which is not compatible with the κ_T value.[23]

Table 19.5 The densities ρ and expansibilities α_P of RTIL quaternary phosphonium salts at $T = 298.15$ K and 0.1 MPa.

Salt	$\rho/(g\ cm^{-3})$	$10^3\alpha_P/K^{-1}$	Ref.
$TdHx_3PCl$	0.8863	0.64	26,27,76,81
$TdHx_3PBr$	0.9552	0.64	81
$TdHx_3PCH_3CO_2$	0.8910	0.67	26,49
$TdHx_3PC_9H_{19}CO_2$	0.8806	0.70	81
$TdHx_3PN(CN)_2$	0.8991	0.68	24,27,81
$TdHx_3PMeSO_4$	0.9281	0.66	81
$TdHx_3PCF_3SO_3$	0.9824	0.67	49
$TdHx_3PNTF_2$	1.0660	0.67	26,49,81
$TdHx_3PiOc_2PO_4$	0.8853	0.71	81
$TdHx_3PPF_3(C_2F_5)_3$	1.1817	0.71	80
$TdBu_3PDoPhSO_3$[a]	0.9384	0.58	76
$EtBu_3Pet_2PO_4$	0.9978	0.64	76

[a]Dodecylbenzenesulfonate.

In the case of the quaternary phosphonium RTILs, most of the data are available for the cation $TdHx_3P^+$ with various anions, as shown in Table 19.5 for $T = 298.15$ K and 0.1 MPa. The densities reported in reference 21 are, as for the ammonium salts, consistently lower, where comparable, than the values reported by other authors and need to be questioned. The densities of salts of the Bu_4P^+ cation with anions of amino acids are available in reference 20. The series of salts C_nOc_3PCl, with $n = 3, 4 \ldots 10, 12$, and 14 have been studied,[25] resulting in the expressions $\rho/(g\ cm^{-3}) = 0.9015 - 0.001576n$ and $10^3\alpha_P/K^{-1} = 0.64 + 0.0033n$.

Isothermal compressibilities κ_T/GPa^{-1} at 298.15 K and 0.1 MPa could be derived from the reported data for $TdHx_3PCl$ (0.586), $TdHx_3PCH_3CO_2$ (0.585), and $TdHx_3PNTF_2$ (0.612),[26] and for $TdHx_3PN(CN)_2$ (0.478[24] or 0.492[27]), but a disagreeing (smaller) value was reported in reference 27 for $TdHx_3PCl$ (0.525).

19.3 Modelling and Correlations

Several attempts to predict the volumetric behaviour of RTILs have used various EoSs. The Sanchez–Lacombe EoS, based on a lattice fluid model, was employed in reference 8 and subsequently in reference 28 for the modelling of the *PVT* properties of RTILs. The EoS is:

$$\rho_r^2 + P_r + T_r[\ln(1 - \rho_r) + (1 - 1/r)\rho_r] = 0 \tag{19.12}$$

where the reduced temperature, $T_r = T/T^*$, pressure, $P_r = P/P^*$, and density, $\rho_r = \rho/\rho^*$, are expressed in terms of the substance-characteristic parameters T^*, P^*, and ρ^*, and include the additional segment number parameter $r = MP^*/RT^*\rho^*$. The parameters for C_4mim^+ salts with BF_4^-, PF_6^-, and $OcSO_4^-$ are listed in reference 8 and those for $TdHx_3P^+$ salts with Cl^-, Br^-, NTF_2^-, $N(CN)_2^-$, and $MeSO_3^-$ are listed in reference 28. The *PVT* values for these salts have then been modelled successfully over the useful temperature range 313 K to 473 K and pressure range 0.1 MPa to 50 MPa. An alternative

EoS, the modified cell model, was also applied successfully to the phosphonium salts.[28]

A cubic-plus-association EoS was used in reference 29 to model the *PVT* properties of a variety of RTILs: $C_n$mim$^+$ salts with BF_4^-, PF_6^-, NTF_2^-, and $MeSO_4^-$ as well as some quaternary ammonium salts with NTF_2^-. The compressibility factor $Z = PV/RT$ is taken as the sum of the cubic (Peng–Robinson) and association terms involving five parameters altogether.

A perturbed hard sphere scaled particle EoS was employed to model the *PVT* properties of $C_n$mim$^+$ salts with BF_4^- and PF_6^- in reference 30 and $C_n$mim$^+$ salts with NTF_2^-, RSO_4^-, and $CF_3SO_3^-$, as well as of pyridinium salts with NTF_2^-, and of $TdHx_3P^+$ salts with chloride and acetate.[31] The expression used is:

$$P/\rho kT = [1 + \eta + \eta^2 - \eta^3]/(1 - \eta)^3 - a(T)\rho/kT \tag{19.13}$$

where $\eta = b(T)\rho/4$ is the packing fraction. The parameters $a(T)$ and $b(T)$ depend on the critical temperature and density and 8 numerical parameters provided in the papers.

The statistical associating fluid theory (SAFT) EoS was applied to C_4mimBF_4[32] in its PC-modification and to C_nPyNTF$_2$ and C_n(3M)PyNTF$_2$ in reference 19 in its soft-modification for modelling their *PVT* properties, using five substance-specific parameters.

Apart from modelling *PVT* properties, the prediction of the densities (eventually their temperature dependences too) on the basis of group contributions has been attempted too. One method of doing this is by means of the parachors $Pa = M\gamma^{1/4}/\rho$, requiring surface tension data, γ.[33–35] Since, however, densities are more readily measured than surface tensions, there is no point in using this path. Again, modelling based on the critical temperature, the critical volume, and the normal boiling point of RTILs[36] to calculate their densities is of little use, because densities of liquids are measured so much more readily than the critical quantities.

More useful is the calculation of the molecular volumes v of RTILs by means of theoretical computations, *e.g.*, using the COSMO-RS method. This was applied successfully to yield $\rho/(\text{g cm}^{-3}) = M/(\text{g mol}^{-1})/602.2(v/\text{nm}^3)$ to a variety of imidazolium salts in reference 37. The molecular volumes of various cations of RTILs and common anions are reproduced in Table 19.6,

Table 19.6 Ionic volumes of the constituents of RTILs.[38,39]

Cation	v/nm^3	Anion	v/nm^3
$C_n$mim$^+$	$0.111 + 0.022n^a$	Cl^-	0.047
PrMePyrr$^+$	0.217	Br^-	0.056
C_nPy$^+$	$0.122 + 0.024n$	BF_4^-	0.073
C_4(4M)Py$^+$	0.221	PF_6^-	0.105
$C_nMe_3N^+$	$0.106 + 0.024n$	NTF_2^-	0.232
$C_nEt_3N^+$	$0.178 + 0.024n$	$N(CN)_2^-$	0.089
$TdHx_3P^+$	0.835		

^{a}For $2 \leq n \leq 8$.

with values taken or calculated from references 38 and 39. For the homologous series, per $-CH_3$ group the v is 0.035 nm^3 and per $-CH_2-$ group it is 0.028 nm^3.[39] These molecular volumes pertain to room temperature, and no provision for their temperature dependence was given. Some authors presented molecular volumes for the solid salts, calculated from the X-ray dimensions;[16] the volumes of the liquid salts are a few percent larger, hence the densities are accordingly lower than obtained from $M/N_A\nu_{solid}$.

The volumetric connectivity index was proposed in reference 40 for the estimation of the room temperature densities of RTILs. The expression used was:

$$\rho/(\mathrm{g\ cm^{-3}}) = a\Sigma(f_{Vi}f_{Vj})^{-1/2} + b + c \tag{19.14}$$

where a and b are listed anion parameters, c is non-zero for oxo-containing alkyl groups. The f_{Vi} and f_{Vj} are the connectivity indexes of connected groups, such as methyl and methylene groups to imidazolium, pyridinium, pyrrolidinium, and quaternary nitrogen and phosphorus atoms, as well as common RTIL anions, all listed.[40] A learning set of 62 RTIL density values was used to establish the parameter values that were then used to predict the densities of a further 80 RTILs with an average relative deviation of 0.63%.

References

1. J. Jaquemin, P. Husson, A. A. S. Padus and V. Majer, *Green Chem.*, 2006, **8**, 172.
2. R. L. Gardas, M. G. Freire, P. J. Carvalho, I. M. Marrucho, I. M. A. Fonseca, A. G. M. Ferreira and J. A. P. Coutinho, *J. Chem. Eng. Data*, 2007, **52**, 80.
3. F. S. Oliveira, M. G. Freire, P. J. Carvalho, J. A. P. Coutinho, J. N. Canongia Lopes, L. P. N. Rebelo and I. M. Marrucho, *J. Chem. Eng. Data*, 2010, **55**, 4514.
4. D.-W. Fang, W. Guan, J. Tong, Zh.-W. Wang and J.-Zh. Yang, *J. Phys. Chem. B*, 2008, **112**, 7499.
5. J. Tong, B. Song, C. X. Wang, L. Li, W. Guan, D.-W. Fang and J.-Zh. Yang, *Ind. Eng. Chem. Res.*, 2011, **50**, 2418.
6. J. M. S. S. Esperança, Z. P. Visak, N. V. Plechkova, K. R. Seddon, H. J. R. Guedes and L. P. N. Rebelo, *J. Chem. Eng. Data*, 2006, **51**, 2009.
7. J. Jaquemijn, P. Husson, V. Mayer and I. Cibulka, *J. Chem. Eng. Data*, 2007, **52**, 2204.
8. H. Machida, Y. Sato and R. L. Smith, Jr., *Fluid Phase Equilib.*, 2008, **264**, 147.
9. M. R. Curras, J. Vijande, M. M. Piñeiro, L. Lugo, J. Salgado and J. Garcia, *Ind. Eng. Chem. Res.*, 2011, **50**, 4065.
10. A. B. Pereiro, F. Santamaria, E. Tojo, A. Rodriguez and J. Tojo, *J. Chem. Eng. Data*, 2006, **51**, 952.
11. E. Gomez, B. Gonzazlez, N. Calvar, E. Tojo and A. Dominguez, *J. Chem. Eng. Data*, 2006, **51**, 2096.

12. A. Fernandez, J. S. Torreceilla, J. Garcia and F. Rodriguez, *J. Chem. Eng. Data*, 2007, **52**, 1979.
13. T. Singh and A. Kumar, *J. Solution Chem.*, 2009, **38**, 1043.
14. O. Ciocirlan and O. Iulian, *J. Chem. Eng. Data*, 2012, **57**, 3142.
15. D.-W. Fang, S.-L. Zang, W. Guan, J. Tong and J.-Zh. Yang, *J. Chem. Thermodyn.*, 2010, **42**, 860.
16. U. P. R. M. Preiss, J. M. Slattery and I. Krossing, *Ind. Eng. Chem. Res.*, 2009, **48**, 2290.
17. Zh. Gu and J. F. Brennecke, *J. Chem. Eng. Data*, 2002, **47**, 339.
18. R. L. Gardas, H. F. Costa, M. G. Freire, P. J. Carvalho, I. M. Marrucho, I. M. A. Fonseca, A. G. M. Ferreira and J. A. P. Coutinho, *J. Chem. Eng. Data*, 2008, **53**, 805.
19. M. B. Oliveira, F. Llovell, J. A. P. Coutinho and L. F. Vega, *J. Phys. Chem. B*, 2012, **116**, 9089.
20. R. L. Gardas, R. Ge, P. Goodrich, Ch. Hardacre, A. Hussain and D. W. Rooney, *J. Chem. Eng. Data*, 2010, **55**, 1505.
21. R. L. Gardas, M. G. Freire, P. J. Carvalho, I. M. Marrucho, I. M. A. Fonseca, A. G. M. Ferreira and J. A. P. Coutinho, *J. Chem. Eng. Data*, 2007, **52**, 1881.
22. T. Singh and A. Kumar, *J. Phys. Chem. B*, 2008, **112**, 12968.
23. J. Jaquemijn, P. Nancarrow, D. W. Rooney, M. F. Costa Gomes, P. Husson, V. Majer, A. A. H. Padua and Ch. Hardacre, *J. Chem. Eng. Data*, 2008, **53**, 2133.
24. A. Pereiro, H. I. M. Veiga, J. M. S. S. Esperança and A. Rodriguez, *J. Chem. Thermodyn.*, 2009, **41**, 1419.
25. G. Adamova, R. L. Gardas, L. P. N. Rebelo, A. J. Robertson and K. R. Seddon, *Dalton Trans.*, 2011, **40**, 12750.
26. J. M. S. S. Esperança, H. R. B. Guedes, M. Blesic and L. P. N. Rebelo, *J. Chem. Eng. Data*, 2006, **51**, 37.
27. F. A. M. M. Gonçalves, C. S. M. F. Costa, C. E. Ferreira, J. C. S. Bernardo, I. Johnson, I. M. A. Fonseca and A. G. M. Ferreira, *J. Chem. Thermodyn.*, 2011, **43**, 914.
28. L. I. N. Tome, R. L. Gardas, P. J. Carvalho, M. J. Pastoriza-Gallego, M. M. Piñeiro and J. A. P. Coutinho, *J. Chem. Eng. Data*, 2011, **56**, 2205.
29. J. Ma, J. Li, D. Fan, Ch. Peng, H. Liu and Y. Hu, *Chin. J. Chem. Eng.*, 2011, **19**, 1009.
30. S. M. Hosseini, J. Moghadassi, M. M. Papari and F. F. Nobandegani, *J. Mol. Liq.*, 2011, **160**, 67.
31. K. R. Harris and L. A. Woolf, *J. Chem. Eng. Data*, 2011, **56**, 4672.
32. O. Ciocirlan, O. Croitoru and O. Iulian, *J. Chem. Eng. Data*, 2011, **56**, 1526.
33. M. Deetlefs, K. R. Seddon and M. Shara, *Phys. Chem. Chem. Phys.*, 2006, **8**, 642.
34. J. Tong, Q.-Sh. Liu, W.-G. Xu, D.-W. Fang and J.-Zh. Yang, *J. Phys. Chem. B*, 2008, **112**, 4381.
35. W. Guan, X.-X. Ma, L. Li, J. Tong, D.-W. Fang and J.-Zh. Yang, *J. Phys. Chem. B*, 2011, **115**, 12915.

36. J. Valderrama and K. Zarricueta, *Fluid Phase Equilib.*, 2009, **275**, 145.
37. J. Palomar, V. R. Ferro, J. S. Torrecilla and F. Rodriguez, *Ind. Eng. Chem. Res.*, 2007, **46**, 6041.
38. J. M. Slattery, C. Daguenet, P. J. Dyson, T. J. S. Schubert and I. Krossing, *Angew. Chem., Int. Ed.*, 2007, **46**, 5384.
39. Ch. Ye and J. M. Shreeve, *J. Phys. Chem. B*, 2007, **111**, 1456.
40. Y. Xiong, J. Ding, D. Yu, Ch. Peng, H. Liu and Y. Hu, *Ind. Eng. Chem. Res.*, 2011, **50**, 14155.
41. A. Stoppa, R. Buchner and G. Hefter, *J. Mol. Liq.*, 2010, **153**, 46.
42. E. Rilo, M. Dominguez-Perez, L. M. Varela and O. Cabez, *J. Chem. Thermodyn.*, 2012, **54**, 68.
43. Sh. Seki, S. Tsuzuki, K. Hayamizu, Y. Umebayashi, N. Serizawa, K. Takei and H. Miyashiro, *J. Chem. Eng. Data*, 2012, **57**, 2211.
44. H. Shirota, T. Mandai, H. Fukazawa and T. Kazo, *J. Chem. Eng. Data*, 2011, **56**, 2453.
45. Ch. P. Fredlake, J. M. Crosthwaite, D. G. Hert, S. N. V. K. Aki and J. F. Brennecke, *J. Chem. Eng. Data*, 2004, **49**, 954.
46. J. Jaquemijn, R. Ge, P. Nancarrow, D. W. Rooney, M. F. Costa Gomes, A. A. H. Padua and Ch. Hardacre, *J. Chem. Eng. Data*, 2008, **53**, 716.
47. M. J. Davila, S. Aparicio, R. Alcalde, B. Garcia and J. M. Leal, *Green Chem.*, 2007, **9**, 221.
48. G. Vakili-Nezhaad, M. Vatani, M. Asghari and I. Ashour, *J. Chem. Thermodyn.*, 2012, **54**, 148.
49. M. Tariq, P. A. S. Forte, M. F. Costa Gomes, J. N. Canongia Lopes and L. P. N. Rebelo, *J. Chem. Thermodyn.*, 2009, **41**, 790.
50. A. Pal and B. Kumar, *J. Mol. Liq.*, 2011, **163**, 128.
51. A. Berthod, M. J. Ruiz-Angel and S. Carda-Broch, *J. Chromatogr. A*, 2008, **1184**, 6.
52. A. Wandschneider, J. K. Lehmann and A. Heintz, *J. Chem. Eng. Data*, 2008, **53**, 569.
53. A. J. L. Costa, J. M. S. S. Esperança, I. M. Marrucho and L. P. N. Rebelo, *J. Chem. Eng. Data*, 2011, **56**, 3433.
54. J.-Y. Wang, F.-Y. Zhao, Y.-M. Liu, X.-L. Wang and Y.-Q. Hu, *Fluid Phase Equilib.*, 2011, **305**, 114.
55. F. M. Gaciño, T. Regueira, L. Lugo, M. J. P. Comuñas and J. Fernandez, *J. Chem. Eng. Data*, 2011, **56**, 4984.
56. M. Larriba, S. Garcia, J. Garcia, J. S. Torrecilla and F. Rodriguez, *J. Chem. Eng. Data*, 2011, **56**, 3589.
57. H. Ning, M.-Q. Hou, Q.-Q. Mei, Y.-F. Liu, D.-Zh. Yang and B.-X. Han, *Sci. China, Ser. B: Chem.*, 2012, **55**, 1509.
58. W. Jin and J. Gui, *Int. J. Chem.*, 2011, **3**, 72.
59. M. Blesic, M. Swadzba-Kwasny, T. Belhocine, H. Q. Nimal Guanarantew, J. N. Canongia Lopes, M. F. Costa Gomes, A. A. H. Padua, K. R. Seddon and L. P. N. Rebelo, *Phys. Chem. Chem. Phys.*, 2009, **11**, 8939.
60. E. Vercher, A. V. Orchilles, F. J. Llopis, V. Gonzalez-Alfaro, and A. Martinez-Andreu, *J. Chem. Eng. Data*, 2011, **56**, 4633.

61. A. Xu, J. Wang, Y. Zhang and Q. Chen, *Ind. Eng. Chem. Res.*, 2012, **51**, 3458.
62. J.-Y. Wang, H.-C. Jiang, Y.-M. Liu and Y.-Q. Hu, *J. Chem. Thermodyn.*, 2011, **43**, 800.
63. J.-Zh. Yang, Q.-C. Zhang, B. Wang and J. Tong, *J. Phys. Chem. B*, 2006, **110**, 22521.
64. B. Mokhtarani, A. Sharifi, H. R. Mortaheb, M. Mirzaei, M. Mafi and F. Sadeghian, *J. Chem. Thermodyn.*, 2009, **41**, 1505.
65. M. Larriba, S. Garcia, P. Navarro, J. Garcia and F. Rodriguez, *J. Chem. Eng. Data*, 2012, **57**, 1318.
66. R. Kato and J. Gmehling, *Fluid Phase Equilib.*, 2004, **226**, 37.
67. Q.-Sh. Liu, M. Yang, P.-F. Yan, X.-M. Liu, Zh.-Ch. Tan and U. Weiz-Biermann, *J. Chem. Eng. Data*, 2010, **55**, 4928.
68. Q.-Sh. Liu, M. Yang, P.-P. Li, S.-S. Sun, U. Weiz-Biermann, Zh.-Ch. Tan and Q.-G. Zhang, *J. Chem. Eng. Data*, 2011, **56**, 4094.
69. H. Tokuda, K. Ishii, Md. Abu bin Hasan Susan, S. Tsuzuki, K. Hayamizu and M. Watanabe, *J. Phys. Chem. B*, 2006, **110**, 2833.
70. N. M. Yunus, M. I. Abdul Mutalib, Z. Man, M. A. Bustam and T. Murugesan, *J. Chem. Thermodyn.*, 2010, **42**, 491.
71. Q. Wang, D.-W. Fang, H. Wang, Y. Liu and Sh.-L. Zang, *J. Chem. Eng. Data*, 2011, **56**, 1714.
72. Y. Deng, P. Husson, A.-M. Delort, P. Bess-Hoggan, M. Sancelme and M. F. Costa Gomes, *J. Chem. Eng. Data*, 2011, **56**, 4194.
73. M. Součkova, J. Klomfar and J. Patek, *Fluid Phase Equilib.*, 2012, **333**, 38.
74. Q.-G. Zhang, Y. Wei, S.-S. Sun, C. Wang, M. Yang, Q.-Sh. Liu and Y.-A. Gao, *J. Chem. Eng. Data*, 2012, **57**, 2185.
75. Q.-Sh. Liu, P.-P. Li, U. Weiz-Biermann, X.-X. Liu and J. Chen, *J. Chem. Eng. Data*, 2012, **57**, 2999.
76. P. Kilaru, G. A. Baker and P. Scovazzo, *J. Chem. Eng. Data*, 2007, **52**, 2306.
77. K. Machanova, A. Boisset, Z. Sedlakova, M. Anouti, M. Bendova and J. Jacquemin, *J. Chem. Eng. Data*, 2012, **57**, 2227.
78. H. Jin, B. O'Hare, J. Dong, S. Arzhantzev, G. A. Baker, J. F. Wishart, A. J. Benesi and M. Maroncelli, *J. Phys. Chem. B*, 2008, **112**, 81.
79. T. Makino, M. Kanakubo, T. Umecky, A. Suzuki, T. Nishida and J. Takano, *J. Chem. Eng. Data*, 2012, **57**, 751.
80. B. Gonzalez, E. Gomez, A. Dominguez, M. Vilas and E. Tojo, *J. Chem. Eng. Data*, 2011, **56**, 14.
81. C. M. S. S. Neves, P. J. Carvalho, M. G. Freire and J. A. P. Coutinho, *J. Chem. Thermodyn.*, 2011, **43**, 948.

CHAPTER 20

Volumetric Behaviour of Molten Salts and Molten Salt Hydrates

YIZHAK MARCUS

Institute of Chemistry, The Hebrew University of Jerusalem,
Jerusalem 91904, Israel
Email: ymarcus@vms.huji.ac.il

20.1 Introduction

Molten salts are ionic liquids that have melting points above, say, 200 °C (but not including silicates and slags). Purely inorganic salts that melt below this temperature have a large measure of covalent bonding between the constituent atoms and are not treated here. This holds, *i.e.*, the covalent bonding, unless they include water in their crystals, *i.e.*, they are salt hydrates that melt congruently. Molten salts and salt hydrate melts received a great deal of attention in the 1960s and 1970s, partly related to thermal energy storage and to nuclear reactor technology. However, they have been replaced in the attention of investigators by room temperature ionic liquids (Chapter 19) and the interest in high melting salts has waned markedly recently. Nevertheless, they have interesting properties, from which much can be learned about their liquid structures. Insights can be obtained from their volumetric behaviour, among other properties including densities, molar volumes, expansibilities and compressibilities.

A great deal of experimental information on the densities of molten salts as a function of temperature is available in the Molten Salts Handbook by

Volume Properties: Liquids, Solutions and Vapours
Edited by Emmerich Wilhelm and Trevor M. Letcher

Published by the Royal Society of Chemistry, www.rsc.org

Janz[1] and subsequent compilations by Janz and coworkers.[2–5] Further information concerning high melting salts has hardly accrued since these publications, but density data on molten salt hydrates are dispersed in the literature.

In view of the large range of melting points of various salts (not including the salt hydrates that are unstable, above, say 120 °C, losing their water) a discussion of the resulting molar volumes of the salts that increase with temperature to various extents is difficult. Based on the concept of corresponding states, a corresponding temperature should be selected for such a discussion, as pointed out by Bloom,[6] among others. He suggested that any residual effects due to post-melting phenomena would probably have been overcome at, say, $1.1T_m$, where T_m/K is the melting point of the salt. This temperature was also suggested by Reiss *et al.*,[7] applying it successfully to the surface tension of uni-univalent salts, and by Yosim and Owens,[8] who compared the entropies of molten alkali halides. The present author, too, has taken up this concept of corresponding temperatures for molten salts.[9–12]

The densities of molten salts[1–5] are generally linear with the temperature over a wide temperature range above the melting point, up to and beyond $1.1T_m$ and are measured at ambient pressures. This is expressed by Equation (20.1):

$$\rho/(\mathrm{g\ cm^{-3}}) = a - 10^3 b(T/\mathrm{K}) \tag{20.1}$$

Therefore, the following quantities for a large number of molten salts at $1.1T_m$ are readily calculated: the densities, $\rho = a - 10^3 b(1.1T_m/\mathrm{K})$, the isobaric expansibilities, $10^3\alpha_P/\mathrm{K}^{-1} = b/\rho$, and the molar volumes, $V/(\mathrm{cm^{-3}\ mol^{-1}}) = M/\rho$, where the $M/(\mathrm{g\ mol^{-1}})$ are their molar masses. The isothermal compressibility κ_T is obtained from the volume or density dependence on (fairly large values of the) pressure P at a given T:

$$\kappa_T(T) = -V^{-1}(\partial V/\partial P)_T = \rho^{-1}(\partial\rho/\partial P)_T = \ln(10)C(T)V_0^{-1}[B(T)+P]^{-1} \tag{20.2}$$

where the last equality is a form of the Tait equation, with $V_0 = V(P = 0.1\ \mathrm{MPa})$ and $B(T)$ and $C(T)$ being temperature-dependent substance-specific constants. The adiabatic compressibility κ_S is obtained at a given temperature T from the density and speed of ultrasound u at ambient pressures:

$$\kappa_S(T) = -V^{-1}(\partial V/\partial P)_S = \rho^{-1}(\partial\rho/\partial P)_S = \rho(T)^{-1}u(T)^{-2} \tag{20.3}$$

The isothermal compressibility κ_T is related to the adiabatic compressibility κ_S as:

$$\kappa_T(T) = \kappa_S(T) + TV(T)\alpha_P(T)^2 C_P(T)^{-1} \tag{20.4}$$

Here C_P is the molar heat capacity at constant pressure. The pertinent quantities, namely $\rho(T)$, $u(T)$, $V(T)$, $\alpha_P(T)$, and $C_P(T)$, are available for a large number of molten salts.

20.2 Methodology

The general methods used for obtaining the densities of liquids are described in Chapter 2. The vibrating tube method (Chapter 3) is hardly applicable at the elevated temperatures needed for molten salts, and the most widely employed methods are the Archimedean sinker and pycnometry/dilatometry.

The Archimedean sinker method was illustrated by Janz and Lorenz[13] and is applicable up to at least 1000 °C. The design permitted the simultaneous measurement of the density and the surface tension, and the density values were corrected for the surface tension effect. The precision attained was ± 0.001 g cm^{-3} at the elevated temperatures regulated to ± 0.2 °C. The pycnometry/dilatometry method was illustrated by Zavodnaya and Vorob'ev,[14] who employed a pycnometer with a capillary tube of 2 mm diameter to record the volume changes as the temperature was increased, having been calibrated with a liquid of known densities at two temperatures within the range of measurements.

The thermal pressure coefficient, $(\partial P/\partial T)_V$, of molten salts yields the isothermal compressibility according to $\kappa_T = \alpha_P/(\partial P/\partial T)_V$. It was measured by Cleaver *et al.*[15] by completely filling the melt into the measuring vessel, equipped with a membrane attached to a strain gauge to measure the pressure, and raising the temperature gradually. Corrections for thermal expansion and dilation of the vessel were applied. The method could be employed in the range 300 to 400 °C and precision of $\pm 1.5\%$ was claimed.

Bockris and Richards[16] employed the double transducer method at temperatures appropriate for molten salts to measure the speed $u(T)$ of ultrasound waves. The acoustic path of *ca.* 7 cm in the melt between the two vertical quartz rods attached to transducers outside of the furnace could be measured accurately. The temperature was controlled within ± 0.15 °C and u was measured to $\pm 0.8\%$, yielding κ_S values with a maximal error of 1.5%. Denielou *et al.*[17] subsequently described a pulsed method for measuring u in molten salts. The signal from the sample was made to coincide with that from a delay line formed from a variable length of water. Knape and Torell[18] employed spontaneous Brillouin scattering to measure the temperature dependence of hypersonic wave speeds in salt melts, a method applicable in the range of 200 °C to 500 °C.

20.3 The Volumetric Data

The density data[1–5] according to Equation (20.1), or in few cases where $\rho(T)$ data up to $1.1T_m$ are available that do not conform to Equation (20.1), were used to obtain the values at $1.1T_m$ of ρ, $10^3\alpha_P$, and V of molten salts that are shown in Table 20.1. If desired, t_m/°C, and $b = \rho_{1.1Tm} \times \alpha_{P1.1Tm}$ and $a = \rho_{1.1Tm} + 1.1T_m b$ of Equation (20.1) are readily back calculated. Also included in Table 20.1 are the values of κ_S and κ_T of the molten salt at $1.1T_m$, obtained from the references indicated. When κ_T was not reported directly, it

Table 20.1 The corresponding temperatures $T = 1.1T_m$, the densities $\rho_{1.1Tm}$, the expansibilities $\alpha_{P1.1Tm}$, and the molar volumes $V_{1.1Tm}$ of molten salts (from references 1–5 unless otherwise noted) and the adiabatic compressibility $\kappa_{S1.1Tm}$ and isothermal compressibility $\kappa_{T1.1Tm}$ (from the indicated references).

Salt	T K	ρ g cm^{-3}	$10^3\alpha_P$ K^{-1}	V cm^3 mol^{-1}	κ_S GPa^{-1}	κ_T GPa^{-1}
$LiOH$[v]	820	1.3429	0.339	17.8		
LiF	1233	1.7536	0.278	14.8		0.093[a]
$LiCl$	971	1.4706	0.287	28.8	0.176[b]	0.216[b]
$LiBr$	905	2.4759	0.264	35.1	0.198[b]	0.235[b]
LiI	816	3.0415	0.303	44.6		
$LiNO_3$	578	1.7410	0.236	39.6	0.178[b]	0.197[b]
$LiBF_4$[o]	635	1.8392	0.249	50.9		
$LiClO_4$[p]	560	1.9944	0.307	53.3		0.230[c]
Li_2CO_3	980	1.8372	0.203	40.2	0.075[d]	0.091[e]
Li_2SO_4	1245	1.9573	0.208	56.2		0.099[f]
Li_2MoO_4[f]	1071	2.804	0.181	61.7		0.105[f]
Li_2WO_4[f]	1114	4.216	0.188	62.1		0.122[f]
$NaOH$	650	1.7570	0.273	22.8		
NaF	1395	1.8731	0.302	22.4		0.133[a]
$NaCl$	1181	1.4977	0.362	39.0	0.247[b]	0.343[b]
$NaBr$	1122	2.2582	0.361	45.6	0.280[b]	0.361[b]
NaI	1026	2.6544	0.356	56.5	0.313[b]	0.436[b]
$NaNO_2$	613	1.7684	0.422	39.0		0.180[c]
$NaNO_3$	637	1.9070	0.324	44.6	0.173[b]	0.198[b]
$NaClO_3$[q]	537	2.0718	0.466	51.4		0.29[c]
$NaPO_3$	991	2.2101	0.164	46.1		
$NaBF_4$[o]	749	1.9055	0.394	57.6		
$NaCH_3CO_2$[x]	662	1.2363	0.487	66.4	0.412[z]	
Na_2CO_3	1244	1.9213	0.234	55.2	0.102[d]	0.118[e]
Na_2SO_4	1273	2.0128	0.241	70.6		0.132[f]
Na_2CrO_4[f]	1177	2.193	0.221	73.8		0.138[f]
Na_2MoO_4	1056	2.7427	0.230	75.1		0.121[f]
Na_2WO_4	1068	3.8025	0.237	77.3		0.133[f]
$Na_2S_2O_7$[w]	781	1.7329	0.328	128.2		
Na_3AlF_6[g]	1407	1.9834	0.495	105.8		0.718[g]
KOH	696	1.7069	0.258	32.9		
KF	1238	1.8399	0.353	31.6		0.186[a]
KCl	1147	1.4670	0.395	50.8	0.308[b]	0.442[b]
KBr	1108	2.0440	0.403	58.3	0.330[b]	0.465[b]
KI	1049	2.3571	0.407	70.4	0.380[b]	0.572[b]
$KSCN$	495	1.5623	0.512	54.5		0.265[c]
KNO_3	672	1.8235	0.400	55.4	0.193[b]	0.234[b]
KHF_2[u]	563	1.9213	0.296	40.7		
KBF_4[o]	927	1.6950	0.481	74.3		
KCH_3CO_2[x]	635	1.3357	0.480	72.7	0.466[z]	
K_2CO_3	1289	1.8456	0.238	74.9	0.146[d]	0.178[e]
K_2SO_4	1484	1.8121	0.267	96.2		0.207[f]
K_2MoO_4	1319	2.5147	0.113	94.7		
K_2WO_4	1323	2.2877	0.442	117.0		0.228[f]
$K_2S_2O_7$[w]	795	1.5633	0.410	160.4		
$K_2Cr_2O_7$	738	2.2395	0.313	131.4		0.187[c]

Table 20.1 (*Continued*)

Salt	T K	ρ g cm^{-3}	$10^3\alpha_P$ K^{-1}	V cm^3 mol^{-1}	κ_S GPa^{-1}	κ_T GPa^{-1}
RbOH[t]	722	2.5476	0.306	40.2		
RbF	1217	2.7526	0.371	38.0		0.176[h]
RbCl	1095	2.1537	0.410	56.1		0.429[i]
RbBr	1062	2.6008	0.412	63.6		0.499[i]
RbI	1012	2.7927	0.408	76.1		
$RbNO_3$	641	2.4261	0.400	60.8	0.179[j]	0.208[k]
$RbBF_4$[o]	941	2.1003	0.495	82.0		
$RbCH_3CO_2$[y]	571	1.9460	0.494	73.4	0.386[z]	
Rb_2CO_3[r]	1221	2.7689	0.245	83.4	0.158[d]	0.209[e]
Rb_2SO_4	1481	2.4571	0.271	108.7		0.251[f]
$Rb_2S_2O_7$[w]	814	2.1132	0.400	164.2		
CsF	1074	3.5227	0.363	43.1		0.228[h]
CsCl	1010	2.6932	0.396	62.5	0.306[b]	0.461[b]
CsBr	1000	3.0216	0.406	70.4	0.362[b]	0.584[b]
CsI	989	3.0709	0.386	84.8	0.51[c]	0.69[c]
$CsNO_3$	756	2.3415	0.500	83.2	0.268[j]	0.308[k]
$CsBF_4$[o]	911	2.3928	0.497	91.8		
$CsCH_3CO_2$[y]	514	2.3723	0.414	80.5	0.355[z]	
Cs_2CO_3[r]	1172	3.3734	0.254	96.6	0.170[d]	0.229[e]
Cs_2SO_4	1421	2.2833	0.257	158.5		0.275[f]
$Cs_2S_2O_7$[w]	850	2.1513	0.422	205.4		
CuCl	765	3.6215	0.218	27.3		
AgCl	801	4.2525	0.221	33.7	0.071[l]	0.101[c]
AgBr	778	4.9330	0.211	38.1	0.083[l]	0.108[c]
AgI	912	4.9424	0.204	47.5	0.139[l]	0.141[e]
$AgNO_3$	534	3.3185	0.325	51.2		0.113[j]
Ag_2SO_4[f]	1031	4.734	0.156	65.9		0.072[f]
InCl	548	2.8755	0.505	52.2		
TlCl	773	4.5192	0.398	53.1		
TlBr[s]	806	5.8840	0.326	48.3		
TlI	715	6.1421	0.304	53.9		
$TlNO_3$	531	4.8158	0.363	55.3		
BeF_2	908	1.9899	0.26	23.6		
$BeCl_2$	784	1.4139	0.778	56.5		
MgF_2	1690	2.3494	0.223	26.5		0.123[h]
$MgCl_2$	1079	1.6601	0.193	57.4	0.550[m]	0.769[e]
$MgBr_2$	1086	2.5689	0.186	71.7	0.448[m]	0.717[e]
MgI_2	1015	2.9808	0.218	93.3	0.391[m]	0.839[e]
CaF_2	1860	2.4517	0.160	31.8		0.064[h]
$CaCl_2$	1161	2.0357	0.209	54.5	0.124[m]	0.384[e]
$CaBr_2$	1103	3.0660	0.163	65.2	0.161[m]	0.528[e]
CaI_2	1163	3.3598	0.223	87.5	0.263[m]	0.718[e]
SrF_2	1840	3.4022	0.221	36.9		0.068[h]
$SrCl_2$	1263	2.6598	0.218	59.6	0.116[m]	0.434[e]
$SrBr_2$	1008	3.6393	0.205	68.0	0.116[m]	0.403[e]
SrI_2	867	4.0351	0.219	84.6	0.151[m]	0.453[e]
BaF_2	1752	4.0405	0.247	43.4		0.089[h]
$BaCl_2$	1359	3.0895	0.220	67.4	0.130[m]	0.547[e]
$BaBr_2$	1235	3.8567	0.223	77.1	0.163[m]	0.679[e]
BaI_2	1114	4.1338	0.236	94.6	0.157[m]	0.635[e]

Table 20.1 (*Continued*)

Salt	T K	ρ g cm^{-3}	$10^3\alpha_P$ K^{-1}	V cm^3 mol^{-1}	κ_S GPa^{-1}	κ_T GPa^{-1}
$NiCl_2$	1309	2.6349	0.251	49.2		
$ZnCl_2$	650	2.4917	0.186	54.7	0.434[m]	0.547[e]
$ZnBr_2$	738	3.4091	0.280	66.1	0.513[m]	0.679[e]
ZnI_2	791	3.7805	0.360	84.4	0.576[m]	0.635[e]
$CdCl_2$	925	3.3221	0.250	55.2	0.285[m]	0.448[e]
$CdBr_2$	924	3.9849	0.271	68.3	0.405[m]	0.563[e]
CdI_2	726	4.3221	0.258	84.7	0.424[m]	0.668[e]
$SnCl_2$	571	3.3301	0.360	56.9		
$PbCl_2$	851	4.8350	0.310	57.5		0.537[n]
$PbBr_2$	711	5.6159	0.296	65.4		0.623[n]
PbI_2	751	5.5827	0.285	82.6		
GaI_3	512	3.5610	0.668	126.5	0.663[k]	0.793[k]
$InCl_3$	945	1.9600	1.059	112.5		
$InBr_3$	780	3.0128	0.498	117.7		
InI_3	531	3.7455	0.398	132.3	0.547[k]	0.596[k]
$BiCl_3$	553	3.8453	0.598	82.0		
$BiBr_3$	540	4.6205	0.523	97.1		
BiI_3	750	4.5500	0.484	129.6		
YCl_3	1048	2.4800	0.209	78.7		
LaF_3	1943	4.4679	0.153	43.8		
$LaCl_3$	1260	3.1101	0.251	78.9		
$LaBr_3$	1167	4.1180	0.243	91.9		
LaI_3	1156	4.1739	0.266	124.5		
CeF_3	1906	4.4690	0.209	44.7		
$CeCl_3$	1205	3.1396	0.293	78.5		
$PrCl_3$	1059	3.266	0.230	76.6		
$NdCl_3$	1134	3.2094	0.290	78.1		
NdI_3	1156	4.1699	0.257	125.9		
$GdCl_3$	970	3.4978	0.192	75.4		
$DyCl_3$	1048	3.5520	0.192	75.7		
UCl_3	1221	3.9536	2.009	87.1		
ThF_4[h]	1521	5.3595	0.127	51.7		0.088[h]
$ThCl_4$	1147	3.2172	0.435	116.2		
UF_4[h]	1440	6.3555	0.156	49.4		0.13[h]
$NbCl_5$	526	1.9345	1.402	139.7		
$TaCl_5$	53.8	2.3498	1.299	152.7		

[a]Reference 19. [b]Reference 16. [c]Reference 20. [d]Reference 21. [e]Reference 11. [f]Reference 22. [g]Reference 23. [h]Reference 24. [i]Reference 25. [j]Reference 18. [k]Reference 26. [l]Reference 27. [m]Reference 28. [n]Reference 20. [o]Reference 30. [p]Reference 31. [q]Reference 32. [r]Reference 33. [s]Reference 55. [t]Reference 61. [u]Reference 62. [v]Reference 63. [w]Reference 64. [x]Reference 65. [y]Reference 66. [z]Reference 67.

was calculated[11] from κ_S by means of Equation (20.4), with the $10^3\alpha_P$ and V in Table 20.1 and the relevant heat capacities, C_P.[34]

A few of the compressibilities in Table 20.1 (those from references 19 and 22) pertain to T_m, rather than to $1.1T_m$, because no temperature dependence of κ_T was reported. Where the densities were calculated from data in reference 22, as noted in Table 20.1, it was assumed that the expansibilities have a negligible dependence on the temperature.

The volumetric properties of molten salt hydrates, $C_pA_q \cdot nH_2O$, were gleaned from a large number of references, as noted in Table 20.2. The table records the corresponding temperatures $1.1T_m$, the densities $\rho_{1.1Tm}$, the expansibilities $\alpha_{P1.1Tm}$, the molar volumes $V_{1.1Tm}$ and the molar electrostriction $\Delta_{el}V_{1.1Tm}$, the latter being a negative quantity.

The molar electrostriction is the difference between the molar volume of the salt hydrate melts $V_{1.1Tm}$ on the one hand and the intrinsic volume of the ions, $V_{i\ intr}$, plus n times the molar volume of the water V_W^* on the other.

Table 20.2 The corresponding temperatures $T = 1.1T_m$, the densities $\rho_{1.1Tm}$, the expansibilities $\alpha_{P1.1Tm}$, the molar volumes $V_{1.1Tm}$ and the electrostriction $-\Delta V_{1.1Tm}$ of molten salt hydrates.

Salt	T K	ρ g cm^{-3}	$10^3\alpha_P$ K^{-1}	V cm^3 mol^{-1}	$-\Delta_{el}V$
$LiClO_3 \cdot 3H_2O$	309	(1.689)[a]	(0.654)	(71.1)	(21.9)
$LiClO_4 \cdot 3H_2O$	405	1.4927[b]	0.542	107.5	11.7
$LiI \cdot 3H_2O$	383	1.9903[b]	0.619	94.4	10.5
$LiNO_3 \cdot 3H_2O$	333	1.4012[b]	0.554	87.8	5.2
$NaOH \cdot H_2O$	371	1.6228[c]	0.518	35.7	−0.7
$NaCH_3CO_2 \cdot 3H_2O$	364	1.2421[c]	0.715	90.3	
$Na_2S_2O_5 \cdot 5H_2O$	354	1.6546[c]	0.453	150.0	
	354	1.6334[d]	0.472	151.9	
$KF \cdot 4H_2O$	321	1.4346[e]	0.088	103.4	−7.0
$Mg(NO_3)_2 \cdot 6H_2O$	399	1.5027[f]	0.638	170.6	14.1
	399	1.4906[g]	0.644	172.0	12.7
$CaCl_2 \cdot 6H_2O$	334	1.4990[h]	0.437	146.1	22.8
$CaBr_2 \cdot 6H_2O$	337	1.9082[h]	0.445	161.2	21.7
	337	1.9218[i]	0.459	160.1	22.8
$CaI_2 \cdot 6H_2O$	347	2.2124[h]	0.426	181.7	29.6
$Ca(NO_3)_2 \cdot 4H_2O$	347	1.7073[j]	0.471	138.7	11.8
		1.7200[k]	0.417	137.3	13.3
$Al(NO_3)_3 \cdot 9H_2O$	379	1.5120[f]	0.446	248.1	27.1
	379	1.5613[g]	0.607	240.3	34.9
$Al_2(SO_4)_3 \cdot 16H_2O$	406	1.7153[l]	0.374	367.5	94.3
$NH_4Al(SO_4)_2 \cdot 12H_2O$	404	1.3510[g]	0.631	335.6	29.3
$Cr(NO_3)_3 \cdot 9H_2O$	373	1.6329[f]	0.606	245.1	30.5
	373	1.6341[m]	0.560	244.9	30.7
$Mn(NO_3)_2 \cdot 6H_2O$	337	1.7782[f]	0.569	161.4	24.2
$FeCl_3 \cdot 6H_2O$	341	1.5823[n]	1.117	170.8	21.5
$Fe(NO_3)_3 \cdot 9H_2O$	348	1.6653[k]	0.616	242.6	33.1
	348	1.6384[o]	0.702	246.6	29.1
$Co(NO_3)_2 \cdot 6H_2O$	361	1.7469[f]	0.570	166.6	18.3
$Ni(NO_3)_2 \cdot 6H_2O$	363	1.7416[f]	0.562	167.0	17.5
$Zn(NO_3)_2 \cdot 6H_2O$	341	1.7922[n]	0.644	166.0	18.9
	341	1.8207[k]	0.587	163.4	21.5
$Cd(NO_3)_2 \cdot 4H_2O$	366	2.2475[j]	0.481	137.3	12.3
		2.2449[k]	0.485	137.4	12.2

[a]Reference 36, but see the discussion of the data. [b]Reference 37. [c]Reference 38. [d]m Reference 39. [e]Reference 40, values extrapolated from concentrated aqueous solutions. [f]Reference 41. [g]Reference 42. [h]Reference 43. [i]Reference 44, values extrapolated from concentrated aqueous solutions. [j]Reference 45. [k]Reference 46. [l]Reference 47. [m]Reference 48. [n]Reference 49. [o]Reference 50.

The intrinsic volume of the ions is the sum of the stoichiometric coefficients p and q times the intrinsic volumes of the ions C^{k+} and A^{l-}, respectively. Therefore:

$$V_{1.1T\mathrm{m}} = pV_{\mathrm{i\ intr}}(\mathrm{C}^{k+}) + qV_{\mathrm{i\ intr}}(\mathrm{A}^{l-}) + nV_{\mathrm{W}}^{*} + \Delta_{\mathrm{el}}V_{1.1T\mathrm{m}} \tag{20.5}$$

It is assumed that the intrinsic ionic volumes $V_{\mathrm{i\ intr}}$ are independent of the temperature and for monatomic ions they are given by the Mukerjee expression:[51]

$$V_{\mathrm{i\ intr}}/\mathrm{cm^3\ mol^{-1}} = (4\pi N_{\mathrm{A}}/3)[1.213(r_{\mathrm{i}}/\mathrm{nm})]^3 \tag{20.6}$$

where r_{i} are the crystal ionic radii.[52] The $V_{\mathrm{i\ intr}}$ values for some of the polyatomic ions are given by the present author in previous publications.[53,54] The molar volume of water at the corresponding temperature $T = 1.1T_{\mathrm{m}}$ is calculated from:

$$V_{\mathrm{W}}^{*}/\mathrm{cm^3\ mol^{-1}} = 22.05 - 0.0318(T/\mathrm{K}) + 6.2\times10^{-5}(T/\mathrm{K})^2 \tag{20.7}$$

20.4 Internal Pressures

A quantity relevant to fluids that is closely related to their volumetric behaviour is their internal pressure, P_{int}. This is defined thermodynamically as:

$$P_{\mathrm{int}} = (\partial U/\partial V)_T = T(\partial P/\partial T)_V - P \tag{20.8}$$

where U is the molar internal energy of the substance (a negative quantity) and $(\partial P/\partial T)_V$ is the isochoric thermal pressure coefficient. Noting that $(\partial P/\partial T)_V = -(\partial V/\partial T)_P/(\partial V/\partial P)_T$, $\alpha_{\mathrm{P}} = (\partial V/\partial T)_P/V$, and $\kappa_{\mathrm{T}} = -(\partial V/\partial P)_V/V$, the internal pressure can be expressed as:

$$P_{\mathrm{int}} = T\alpha_{\mathrm{P}}/\kappa_{\mathrm{T}} - P \tag{20.9}$$

The internal pressure of most fluids is $P_{\mathrm{int}} \geq 100$ MPa at ambient conditions and saturation vapour pressures, hence the last term, $-P$, in Equations (20.8) and (20.9) can generally be neglected.

Cleaver *et al.*[15] obtained P_{int} values for molten sodium nitrate between 340 and 400 °C by direct $(\partial P/\partial T)_V$ measurements according to Equation (20.8). Cerisier and El-Hazime[56] reported values of P_{int} calculated according to Equation (20.9) at T_{m} and Marcus[10] reported such values for the corresponding temperature $T = 1.1T_{\mathrm{m}}$ shown in Table 20.3. The P_{int} values at 800 °C were calculated from the κ_{T} and α_{P} values reported by Bockris and Richards[16] for some alkaline earth halides, and the P_{int} values for some 2:1 molten salts reported by Denielou *et al.*[22] at various temperatures are shown in Table 20.4.

Values of P_{int}, such as those reported by Murgulescu and Terzi[57] and by Sanguri and Singh,[58] obtained in an indirect evaluation of the κ_{T} and α_{P} values from surface tension and reduced volume data, rather than directly from Equation (20.9) and measured values, cannot be correct.

Table 20.3 The internal pressure of 1 : 1 molten salts.

Salt	T_m/K	P_{int}/GPa at T_m [56]	at 1073 K [16]	at $1.1(T_m/K)$[10]
LiF	1121	2.94		3.24
LiCl	887	1.32	1.30	1.46
LiBr	823	1.00		1.10
NaF	1265	2.43		2.67
NaCl	1073	1.32	1.35	1.24
NaBr	1023	1.10	0.99	1.13
NaI	935	0.87	0.86	0.96
KF	1129	1.73		1.90
KCl	1043	1.08	1.09	1.23[a]
KBr	1008	0.95	0.93	0.96
KI	958	0.77	0.73	0.75
RbCl	988	1.18		1.30
RbBr	953	1.05		1.16
RbI	913	0.86		0.95
CsF	955			1.29
CsCl	918		0.850	0.80
CsBr	909		0.740	0.67
CsI	894			0.58
$LiNO_3$	525	0.89	0.920	0.8
$NaNO_3$	603	1.19	1.180	1.31
$NaBF_4$	567			0.96
$NaCH_3CO_2$	679			0.74
KNO_3	611	1.11	1.150	1.22
KCH_3CO_2	582			0.59
$RbNO_3$	583	0.82		0.90
$CsNO_3$	687	0.95		1.05
$AgNO_3$	476			1.23

[a]The value 1630 in reference 10 is a misprint.

The P_{int} values for the molten salts are generally 4 to 20 times larger than those for liquids at ambient temperatures,[59] due to the strong coulombic forces and depending only partly on the higher temperatures involved. On the other hand, whereas for liquids at ambient temperatures the P_{int} values are commensurate with those of the cohesive energy density, this is not the case for molten salts. For these the cohesive energy density is generally >10 times the internal pressure.[10,59]

20.5 Correlations of Volumetric Data

The volumetric properties recorded in Tables 20.1 and 20.2 yield some interesting correlations with other properties of the salts and their constituent ions that shed light on the structures of the molten salts and salt hydrates.

The best fit of the *molar volumes* of 1 : 1 salts, is:[12]

$$V_{1.1Tm}(\text{calc})/(\text{cm}^{-3}\ \text{mol}^{-1}) = (4\pi N_A/3)[3.3r_C{}^3 + 1.78r_A{}^3] = (-1.1 \pm 2.0) + (1.001 \pm 0.039)V_{1.1Tm}(\text{expt})/(\text{cm}^{-3}\ \text{mol}^{-1}) \quad (20.10)$$

Table 20.4 The internal pressure of 1 : 2[28] and 2 : 1[22] molten salts.

Salt	t/°C	P_{int}/GPa
$MgCl_2$	800	2.5
$MgBr_2$	800	2.1
MgI_2	800	1.8
$CaCl_2$	800	15.3
$CaBr_2$	800	9.5
CaI_2	800	8.4
$SrCl_2$	900	17.1
$SrBr_2$	800	14.1
SrI_2	800	8.8
$BaCl_2$	1000	18.7
$BaBr_2$	900	12.9
BaI_2	800	11.8
$ZnCl_2$	600	3.7
$ZnBr_2$	600	3.6
ZnI_2	600	4.0
Li_2SO_4	857	2.2
Na_2SO_4	882	1.9
K_2SO_4	1068	1.5
Rb_2SO_4	1066	1.4
Cs_2SO_4	1001	1.2
Ag_2SO_4	664	2.0
Li_2CrO_4	482	1.5
Na_2CrO_4	797	1.7

The factor $1.78 = 1.213^3$ multiplying the anion radius cubed $r_A{}^3$ is the Mukerjee factor of Equation (20.6). The factor 3.3 of the cation radius cubed $r_C{}^3$ is the fitting coefficient, reliable to ± 0.1, yielding a correlation coefficient $r_{corr} = 0.9884$ and a standard error of the fit of $\sigma_{fit} = 2.9\ cm^{-3}\ mol^{-1}$ as the second equality of Equation (20.10). The ionic molar volumes pertaining to Equation (20.10) are shown in Tables 20.3 and 20.4. The expression is valid for 31 salts, including the alkali metal halides, alkali metal nitrates and tetrafluoroborates ($CsNO_3$, $LiBF_4$, and $NaBF_4$ are outliers), NaOH, KOH, $AgNO_3$, and InCl. However 1 : 1 salts with considerable covalent bonding between the ions, such as the silver halides, CuCl, and the Tl(I) salts do not conform and have appreciably smaller molar volumes than expected from Equation (20.10). It is suspected that in the cases of $CsNO_3$, $LiBF_4$, and $NaBF_4$ the reported densities, leading to the deviating molar volumes, may be incorrect, since there is no obvious reason for them not to conform to the correlation.

The non-symmetrical 1 : 2, 1 : 3, and 2 : 1 molten salts pose a problem in attempts to correlate the molar volumes of the salts listed in Table 20.1 with ionic contributions. This is because of the presence of vacancies in the cation quasi-lattice for the 1 : 2 and 1 : 3 salts and in the anion quasi-lattice of the 2 : 1 salts. For such salts the correlation expression is:

$$V_{1.1Tm} = k_C V_{1.1Tm}(C) + k_A V_{1.1Tm}(A) \tag{20.11}$$

For the 1:2 and 1:3 salts, the anions A are the halides, with $k_A = 0.72z_C$, *i.e.*, 1.44 for the 1:2 salts and 2.16 for the 3:1 salts, and volumes $V_{1.1Tm}(A)$ from Tables 20.3 and 20.4. An exception occurs for the very small magnesium salts, for which $k_A = 1.81$. The cation volumes are given by:

$$k_C V_{1.1Tm}(C)/(\mathrm{cm}^{-3}\ \mathrm{mol}^{-1}) = -7.8 + 291(r_C/\mathrm{nm}) \tag{20.12}$$

an exception being the small zinc cation, for which the term −7.8 is omitted. Equations (20.11) and (20.12) are valid for 36 1:2 and 1:3 salts yielding the relationship between the calculated and experimental molar volumes with a correlation coefficient $r_{corr} = 0.9921$ and a standard error of the fit of $\sigma_{fit} = 2.7\ \mathrm{cm}^{-3}\ \mathrm{mol}^{-1}$:

$$V_{1.1Tm}(\mathrm{calc})/(\mathrm{cm}^{-3}\ \mathrm{mol}^{-1}) = (0.7 \pm 2.3) + (0.984 \pm 0.031)V_{1.1Tm}(\mathrm{expt})/(\mathrm{cm}^{-3}\ \mathrm{mol}^{-1}) \tag{20.13}$$

This pertains[12] to the halides of the alkaline earth metals, transition- and post-transition metals, lanthanides, YCl_3 and UCl_3, but not to GaI_3 and the indium trihalides, where some covalent bonding should occur. A notable outlier is CaI_2, and again it is suspected that the density data leading to its molar volume are incorrect.

For the alkali metal salts of divalent anions, Equation (20.11) applies with $k_C = 1.68$ and the $V_{1.1Tm}(C)$ values from Table 20.5. The following values of $k_A V_{1.1Tm}(A)$ apply: 38.1 for carbonate, 57.4 for sulfate, 58.5 for molybdate, and 66.9 for tungstate, and 123.1 for pyrosulfate, which are roughly proportional to the radii r_A of the anions. The fit of the experimental data of 18 salts is:[12]

$$V_{1.1Tm}(\mathrm{calc})/(\mathrm{cm}^{-3}\ \mathrm{mol}^{-1}) = (2.4 \pm 5.1) + (0.976 \pm 0.066)V_{1.1Tm}(\mathrm{expt})/(\mathrm{cm}^{-3}\ \mathrm{mol}^{-1}) \tag{20.14}$$

with a correlation coefficient of $r_{corr} = 0.9737$ and a standard deviation of the fit of $\sigma_{fit} = 4.5\ \mathrm{cm}^{-3}\ \mathrm{mol}^{-1}$, considerably worse than Equations (20.10) and (20.13). The experimental $V_{1.1Tm}$ of Cs_2SO_4 and Li_2WO_4 are outliers, the former being much too large and the latter too small, but $V_{1.1Tm}$ of Ag_2SO_4 is included in the fit.

Table 20.5 Ionic contributions to the molar volumes of 1 : 1 molten salts at $1.1T_m$: $V(C) = 3.3(4\pi N_A/3)r_C^3$ and $V(A) = 1.78(4\pi N_A/3)r_A^3$.

Cation	$V(C)/\mathrm{cm}^3\ \mathrm{mol}^{-1}$	Anion	$V(A)/\mathrm{cm}^3\ \mathrm{mol}^{-1}$
Li^+	2.7	F^-	10.6
Na^+	8.8	Cl^-	26.6
K^+	21.9	Br^-	33.8
Rb^+	27.0	I^-	47.8
Cs^+	40.9	OH^-	10.6
Ag^+	12.7	NO_3^-	35.9
		BF_4^-	54.6

The molar volumes of the molten salt hydrates $C_pA_q \cdot nH_2O$ at their corresponding temperatures, $1.1T_m$, can be estimated from the intrinsic volumes and the molar volume of the water according to Equation (20.5), noting that the electrostriction volumes are proportional to the number of water molecules per formula unit: $\Delta_{el}V_{1.1Tm} = (3.3 \pm 0.5)n$. This holds for 21 molten salt hydrates:[54]

$$V_{1.1Tm}(\mathrm{calc})/(\mathrm{cm}^{-3}\ \mathrm{mol}^{-1}) = (-0.8 \pm 2.6) + (1.008 \pm 0.014)V_{1.1Tm}(\mathrm{expt})/(\mathrm{cm}^{-3}\ \mathrm{mol}^{-1}) \quad (20.15)$$

with a correlation coefficient of $r_{corr} = 0.9975$ and a standard deviation of the fit of $\sigma_{fit} = 4.5\ \mathrm{cm}^{-3}\ \mathrm{mol}^{-1}$. There are some outliers from this fit: $LiClO_3 \cdot 3H_2O$, $KF \cdot 4H_2O$, and $Al_2(SO_4)_3 \cdot 16H_2O$.

The *expansibilities* of molten salts at $1.1T_m$ correlate well with the inverse of their cohesive energies, *ce*:

$$10^3\alpha_P \times ce = (245 \pm 35)z_Cz_A\ \mathrm{K}^{-1}\ \mathrm{kJ\ mol}^{-1} \quad (20.16)$$

for 64 1 : 1, 1 : 2, 1 : 3 and 2 : 1 salts, the z_i being the ionic charge numbers. The cohesive energies *ce* at $1.1T_m$ of 1 : 1 salts[10] and of 1 : 2 and 2 : 1 salts[11] have been published. Those of the 1 : 3 lanthanide halides are not available, but their lattice energies are,[60] ranging from 3900 to 4700 kJ mol^{-1}. Their correction from zero K to $1.1T_m$ is minor (estimated at only -20 kJ mol^{-1}), so that they could be included in the correlation [Equation (20.16)]. The silver halides and post-transition metal 1 : 2 salts are outliers from this relationship, having smaller and larger expansibilities, respectively, than expected from Equation (20.16). The expansibility of UCl_3 is very much larger than expected, indicating that the reported density data are probably incorrect.

The expansibilities of molten salt hydrates can be correlated by means of the expression proposed by Sharma *et al.*:[38]

$$\rho/\rho_m = 1 - 0.1806[(T/T_m) - 1] \quad (20.17)$$

where ρ_m is the density at the melting point. This expression may be recast in the form of the expansibility at any temperature, including $T = 1.1T_m$, *as*:

$$\alpha_P = \rho_m{}^2T_m{}^{-2}(6.537T_m - T) \quad (20.18)$$

requiring knowledge of the density at the melting point, ρ_m.

An inverse correlation was found for the *isothermal compressibilities*, κ_T/GPa^{-1}, of a series of related molten salts and their cohesive energy densities, *ced*/GPa $= ce/V_{1.1Tm}$. For 27 alkali, alkaline earth, and divalent transition- and post-transition metal halides:[11]

$$\kappa_T \times ced = (4.37 \pm 0.80)z_C{}^2 \quad (20.19)$$

For 10 alkali metal nitrates and carbonates the correlation is:

$$\kappa_T \times ced = (2.51 \pm 0.19)z_A \quad (20.20)$$

Here $LiNO_3$ is an outlier but $AgNO_3$ is included. The anions of these salts have similar planar triangular shapes, so their compressibility follows the same expression. The tetrahedrally shaped sulfate anion allows a somewhat smaller compressibility of the molten salts than that given by Equation (20.20), namely $(1.78 \pm 0.21)z_A$ for the 5 alkali metal sulfates.[11]

20.6 Modelling the Volumetric Properties

The generally accepted model of highly ionic molten salts represents them in terms of two intertwining quasi-lattices, one for the cations and one for the anions. This results from the strong coulombic attractive forces between ions with charges of unlike signs that tend to be near neighbours, and the corresponding repulsion between ions with charges of like signs, that occupy next-nearest neighbour sites. This repulsion balances the attraction and maintains the volume of the molten salts, as expressed by Equation (20.10) for the symmetrical 1:1 salts and Equation (20.11) for the unsymmetrical 1:2, 1:3, and 2:1 salts. In melts that are nominally salts but have a considerable degree of covalent bonding between the ions of opposite charges the repulsion between next-nearest-neighbouring ions of the same charge is less effective in maintaining the volume. Only partial net charges are then operative between the constituent ions and such melts have, indeed, smaller molar volumes than are observed for highly ionic melts and calculated by means of Equations (20.10) and (20.11).

A feature of molten 1:1 salts that pertains to their expansibilities and compressibilities is the existence in them of holes beyond the free volume available for the oscillations of the ions in their quasi-lattice sites. The existence of such holes was deduced by Bockris and Richards[16] from the expansion of the salts on melting. The inter-ionic distances in the melts are not larger than they are in the crystalline salts (on the contrary, diffraction measurements show them to be somewhat smaller), hence the expansion on melting yields coordination numbers in the molten salts that are smaller than in the corresponding crystals, due to the existence of the holes. These authors[16] then showed that the expansibility α_P and compressibility κ_T are predominantly those due to the holes rather than those due to the quasi-lattices. Bockris *et al.* subsequently[28] extended the hole-model to unsymmetrical salts, including the site vacancies and association of small divalent cations with large anions, to estimate their expansibilities and compressibilities.

If no experimental compressibility data are available, these could possibly be estimated from the scaled particle theory (SPT) expression presented by Stillinger:[35]

$$\kappa_T = (V/RT)(1-y)^4/(1+2y)^2 \tag{20.21}$$

where y is the packing fraction[29] of the ions considered as hard spheres. The ions have a diameter σ (the mean of the cation and anion values) and it is noted that in symmetrical salts there are 2 particles per formula unit. Hence the packing fraction is:

$$y = (\pi N_A/3)\sigma^3/V \tag{20.22}$$

However, when this approach was applied[24] to the estimation of the compressibilities of molten alkali metal fluorides, the compressibilities were underestimated, resulting in 70% of the experimental values.[11] A modified compressibility expression derived from the SPT was shown by Yosim and Owens[8] to yield values for the alkali metal halides that are roughly only one half of the experimental values, hence the assumptions leading to this modified expression were deemed not to be valid.

The extension of the volumetric properties of molten salts to high pressures (the data in Table 20.1 pertain to ambient pressure, *i.e.* 0.1 MPa) can be made by means of the Tait expression [*cf.* Equation (20.2)] according to Goldmann and Tödheide.[68] It utilises the isothermal compressibility κ_T at ambient pressure P_0 and at temperature T (available from the references in Table 20.1) and assumes a constant value, $C(T) = 0.1$ and the expression $B(T)/\text{MPa} = 400 + 8.9(\kappa_T/\text{MPa})^{-1}$. The resulting expression is:

$$\rho(P, T) = \rho(P_0, T)/[1 - 0.1\ln\{B(T) + P\}/B(T)] \tag{20.23}$$

The model that describes the behaviour of molten salt hydrates considers the hydrated multivalent cations to be single particles having a radius $r = r_C + 2r_{H2O}$, of low electrical field strength.[69] Thus, in molten $Ca(NO_3)_2 \cdot 4H_2O$ the cation $[Ca(H_2O)_4]^{2+}$ has a z/r value of 5.3 nm^{-1}, comparable with that of Cs^+, 5.9 nm^{-1}. This view is confirmed by the consideration of $Ca(NO_3)_2 \cdot 4H_2O$ as a component in the thermodynamic sense in its mixtures with KNO_3 at 100 °C, as it yields a straight line in a plot of the volume per mole of mixture against the mole fraction of KNO_3.[70] These early examples of molten salt hydrates puts them in the same class of fluids as anhydrous molten salts, rather than regarding them as concentrated aqueous solutions. This should be valid for congruently melting salt hydrates, particularly with hydrated multivalent cations. How this concept can be applied to salts such as $KF \cdot 4H_2O$ and $Na_2S_2O_3 \cdot 5H_2O$ (see Table 20.2) is not clear, however.

References

1. G. J. Janz, *Molten Salts Handbook*, Academic Press, New York, 1967.
2. G. J. Janz, G. L. Gardner, U. Krebs and R. P. T. Tomkins, *J. Phys. Chem. Ref. Data*, 1974, **3**, 1.
3. G. J. Janz, R. P. T. Tomkins, C. B. Allen, J. R. Downey, G. L. Garner, U. Krebs and S. K. Singer, *J. Phys. Chem. Ref. Data*, 1975, **4**, 871.
4. G. J. Janz, R. P. T. Tomkins, C. B. Allen, J. R. Downey, G. L. Garner and S. K. Singer, *J. Phys. Chem. Ref. Data*, 1977, **6**, 409.
5. G. J. Janz and R. P. T. Tomkins, *J. Phys. Chem. Ref. Data*, 1980, **9**, 831.
6. H. Bloom, *The Chemistry of Molten Salts*, Benjamin, New York, 1967.

7. H. Reiss, S. W. Mayer and J. L. Katz, *J. Chem. Phys.*, 1961, **35**, 820.
8. S. J. Yosim and B. B. Owens, *J. Chem. Phys.*, 1964, **41**, 2032.
9. Y. Marcus, *Thermochim. Acta*, 2009, **495**, 81.
10. Y. Marcus, *J. Chem. Thermodyn.*, 2010, **42**, 60.
11. Y. Marcus, *J. Chem. Thermodyn.*, 2013, **61**, 7.
12. Y. Marcus, *Thermochim. Acta*, 2013, **559**, 111.
13. G. J. Janz and M. R. Lorenz, *J. Electrochem. Soc.*, 1961, **108**, 1052.
14. G. E. Zavodnaya and V. A. Vorob'ev, *Zh. Prikl. Khim.*, 1966, **39**, 1864.
15. B. Cleaver, B. C. J. Neil and P. N. Spencer, *Rev. Sci. Instrum.*, 1971, **42**, 578.
16. J. O'M. Bockris and N. E. Richards, *Proc. R. Soc. London*, 1957, **241**, 44.
17. L. Denielou, J. P. Petitet and C. Tequi, *Can. J. Chem.*, 1974, **52**, 400.
18. H. E. G. Knape and L. M. Torell, *J. Chem. Phys.*, 1975, **62**, 4111.
19. K. Yajima, H. Moriyama and J. Oishi, *J. Phys. Chem.*, 1984, **88**, 4390.
20. B. Cleaver and P. N. Spencer, *High Temp. - High Pressures*, 1955, **7**, 539.
21. Z. Hong-Min, T. Saito, Y. Sato, T. Yamamura, K. Shimakage and T. Ejima, *J. Jpn. Inst. Met.*, 1991, **55**, 936.
22. L. Denielou, J.-P. Petitet and C. Tequi, *J. Phys.*, 1976, **37**, 1017.
23. R. Fernandez and T. Østvold, *Acta Chem. Scand.*, 1989, **43**, 151.
24. S. Hara and K. Ogino, *ISIJ Int.*, 1989, **29**, 477.
25. E. L. Herie, *J. Phys. Chem.*, 1965, **69**, 2785.
26. B. Cleaver and P. Zani, *High Temp. - High Pressures*, 1978, **10**, 437.
27. K.-I. Takesawam, S. Takeda, S. Harada and S. Tamaki, *J. Phys. Soc. Jpn.*, 1989, **58**, 538.
28. J. O'M. Bockris, A. Pilla and J. L. Barton, *Rev. Chim., Acad. Repub. Pop. Roum.*, 1962, **7**, 59.
29. T. Veraiah, *Curr. Sci.*, 1966, **23**, 589.
30. S. Cantor, D. P. M. Dermott and L. O. Gilpatrick, *J. Chem. Phys.*, 1970, **52**, 3600.
31. G. Petersen, W. M. Ewing and G. P. Smith, *J. Chem. Eng. Data*, 1961, **6**, 540.
32. A. N. Campbell and E. T. van der Kouwe, *Can. J. Chem.*, 1968, **46**, 1279.
33. H.-M. Zhu, T. Saito and Y. Sato, *J. Jpn. Inst. Met.*, 1991, **55**, 937.
34. Y. Marcus, in *Heat Capacities*, ed. E. Wilhelm and T. Letcher, Royal Society Chemistry, Cambridge, 2010, ch. 22, pp. 472–489.
35. F. H. Stillinger, *J. Chem. Phys.*, 1961, **35**, 1581.
36. K. Gawron and J. Schröder, *Energy Res.*, 1977, **1**, 351.
37. V. P. Mashovets, N. M. Baron and G. E. Zavodnaya, *Russ. J. Phys. Chem.*, 1969, **43**, 971.
38. S. K. Sharma, C. K. Jotshi and A. Singh, *Can. J. Chem. Eng.*, 1987, **65**, 171.
39. C. Bhattacharjee, S. Ismail and K. Ismail, *J. Chem. Eng. Data*, 1986, **31**, 117.
40. P. Novotny and O. Söhnel, *J. Chem. Eng. Data*, 1988, **33**, 49.
41. S. K. Jain, *J. Chem. Eng. Data*, 1977, **22**, 383.
42. A. Minevich, Y. Marcus and L. Ben-Dor, *J. Chem. Eng. Data*, 2004, **49**, 1451.

43. S. K. Jain, S. Prashar and S. K. Jain, *Indian J. Chem., Sect. A: Inorg., Bio-inorg., Phys., Theor. Anal. Chem.*, 1999, **38A**, 778.
44. G. Scatchard, *Intl. Crit. Tables*, 1928, **3**, 73.
45. S. K. Jain, *J. Chem. Eng. Data*, 1973, **18**, 397.
46. K. V. Ramana, R. C. Sharma and H. C. Gaur, *J. Chem. Eng. Data*, 1986, **31**, 288.
47. D. Ornek, T. Gurkan and C. Oztin, *Ind. Eng. Chem. Res.*, 1998, **37**, 2687.
48. S. K. Jain, *J. Chem. Eng. Data*, 1978, **23**, 216.
49. S. K. Jain, *J. Chem. Eng. Data*, 1978, **23**, 170.
50. S. Gupta, R. C. Sharma and H. C. Gaur, *J. Chem. Eng. Data*, 1981, **26**, 187.
51. P. Mukerjee, *J. Phys. Chem.*, 1961, **65**, 740.
52. Y. Marcus, *Ion Properties*, Dekker, New York, 1997.
53. Y. Marcus, *J. Phys. Chem. B*, 2012, **116**, 7232.
54. Y. Marcus, *J. Chem. Eng. Data*, 2013, **58**, 488.
55. E. R. Buckle, P. E. Tsaoussoglou and A. R. Ubbelohde, *Trans. Faraday Soc.*, 1964, **60**, 684.
56. P. Cerisier and D. El-Hazime, *J. Chim. Phys.*, 1983, **80**, 255.
57. I. G. Murgulescu and M. Terzi, *Rev. Roum. Chim.*, 1979, **24**, 113.
58. V. Sanguri and N. Singh, *J. Indian Chem. Soc.*, 2011, **88**, 163.
59. Y. Marcus, *Chem. Rev.*, 2013, **113**, 6536.
60. *Handbook of Chemistry and Physics*, ed. D. Lide, CRC Press, Baton Rouge, 82nd edn, 2001–2002.
61. D. Bogart, *J. Phys. Chem.*, 1954, **58**, 1168.
62. O. Matal, J. Zaloudek, F. Ciganek, P. Slavicek, B. Zmij and M. Nejedly, *Z. Naturforsch.*, 2001, **56a**, 707.
63. A. N. Kruglov, V. P. Kochergin and V. Ya. Poluyanova, *Izv. Vizsh. Ucheb. Zaved. Khim. Khim. Tekhnol.*, 1978, **21**, 1085.
64. G. Hatem, F. Abdoun, M. Gaune-Escard, K. M. Eriksen and R. Fehrmann, *Thermochim. Acta*, 1998, **319**, 33.
65. E. J. Hazlewood, E. Rhodes and A. R. Ubbelohde, *Trans. Faraday Soc.*, 1966, **82**, 3101.
66. D. Leonesi, A. Cingolani and G. Berchiesi, *Z. Naturforsch.*, 1976, **31A**, 1609.
67. V. P. Denisovets and VF. D. Prisyazhii, *Ukr. Khim. Zh.*, 1988, **54**, 1247.
68. G. Goldmann and K. Tödheide, *Z. Naturforsch., A*, 1976, **31A**, 769.
69. C. A. Angell, *J. Electrochem. Soc.*, 1965, **112**, 1224.
70. J. Braunstein, L. Orr and W. Macdonald, *J. Chem. Eng. Data*, 1967, **12**, 415.

CHAPTER 21

Partial Molar Volumes of Proteins in Solution

TIGRAN V. CHALIKIAN

Department of Pharmaceutical Sciences, Leslie Dan Faculty of Pharmacy, University of Toronto, 144 College Street, Toronto, Ontario M5S 3M2, Canada
Email: chalikan@phm.utoronto.ca

21.1 Introduction

An understanding of the molecular origins and driving forces of protein recognition events, including reactions of folding and binding, relies increasingly on the combination of structural, computational, and thermodynamic studies.[1–8] Without exemption, these reactions can be viewed as an exchange of solute–solvent for solute–solute interactions with concomitant changes in thermodynamic properties.[9–14] Solute–solvent interactions exert a mutually modifying effect on the protein and its waters of hydration. The latter are structurally, kinetically, and thermodynamically distinct from waters in the bulk.[15–17] Thermodynamic distinctions involve, among others, the differential packing density of water molecules solvating protein groups and water in the bulk. In this connection, volumetric (densimetric) measurements have proven useful for thermodynamic characterization of protein hydration as well as changes in hydration accompanying recognition events involving proteins.

Among thermodynamic variables of state, volume is the most easily comprehensible and intuitively appealing. The partial molar volume, V°, of a protein reflects the entire range of its intra- and intermolecular (solute–solvent) interactions and the packing of the polypeptide chain inside the

Volume Properties: Liquids, Solutions and Vapours
Edited by Emmerich Wilhelm and Trevor M. Letcher

Published by the Royal Society of Chemistry, www.rsc.org

protein. In recognition of this notion, the partial molar volumes of proteins, as well as their low molar mass model compounds, have been extensively studied as a function of environmental conditions, while attempts have been made to develop theoretical frameworks for molecular interpretation of volumetric results.

In this chapter, we present an overview of experimental and, partly, theoretical developments in the field. Throughout the review, we emphasize the unique insights into the molecular forces driving protein recognition events that can be gained from volumetric studies. The review is not designed to be comprehensive or systematic with respect to the body of published works. Instead, it is biased towards our own results while also touching upon related studies from select other research groups.

21.2 Theoretical Considerations

The partial molar volume, V°, of a solute is defined as the partial derivative of the volume of the solution, V, with respect to the number of moles, N_i, of the i-th solute:

$$V_i^\circ = (\partial V/\partial N_i)_{T,P,N_{j\neq i}} \tag{21.1}$$

where T is the temperature; P is the pressure; and N_j is the number of moles of the j-th solute.

If the derivative is taken at a vanishingly small concentration of a solute, the property determined is defined as the partial molar volume at infinite dilution:

$$V_i^\circ{}_\infty = \lim_{C_i\to 0} (\partial V/\partial N_i)_{T,P,N_{j\neq i}} \tag{21.2}$$

where C_i is the concentration of the i-th solute.

In more conventional terms, partial molar volume at infinite dilution is the apparent volume occupied by 1 mole of solute in solution at a negligibly small concentration. For over a century, researchers have invested considerable effort into understanding the constituents of the partial molar volume of solutes.[18–21] Scaled particle theory (SPT) offers a practical scheme for deconvoluting the partial molar volume of a solute into contributions:[22–24]

$$V^\circ = V_M + V_T + V_I + \beta_{T0}RT \tag{21.3}$$

where V_M is the intrinsic volume of a solute; V_T is the thermal volume; V_I is the interaction volume; β_{T0} is the coefficient of isothermal expansibility of the solvent; R is the universal gas constant; and T is the temperature.

Intrinsic volume, V_M, is the geometric volume of a solute which is not accessible to any part of solvent molecules. For a small, solvent-accessible solute, V_M is approximated well by its van der Waals volume, V_W. The latter can be calculated using an additive scheme and group contributions given in the literature.[20,25] For larger molecules such as proteins, the intrinsic volume, V_M, coincides with the volume enclosed by the molecular surface as defined by Richards.[26,27] The molecular surface of a protein has two components: (i) part of the protein surface which contacts a rolling probe solvent molecule; and (ii) a re-entrant surface, which corresponds to a series of patches formed by the interior-facing domain of the probe when it simultaneously contacts more than one atom on the protein surface.[26]

Thermal volume, V_T, in Equation (21.3) represents the "void" volume around the solute due to steric, vibrational, and structural effects. The steric component of V_T reflects the imperfect packing of solute and solvent molecules in the solution. The vibrational component results from thermally-induced mutual vibrational motions of solute and solvent molecules that, theoretically, should subside to zero at $T = 0$ K. The structural component reflects the open (tetrahedral) structure of water and related packing effects around solute molecules.[21]

The interaction volume, V_I, in Equation (21.3) reflects the effect of solute–solvent interactions. More generally, the interaction volume, V_I, reflects the volume effect of the difference between real solute–solvent interaction potential and the hard-sphere potential. For practical estimates, the interaction volume, V_I, can be viewed as resulting from the differential partial molar volume of water of solute hydration and bulk water:

$$V_i = \sum_i n_{hi}(V_{hi} - V_0) \tag{21.4}$$

where n_{hi} is the number of water molecules solvating the i-th solvent-exposed domain of the solute; and V_{hi} and V_o are the partial molar volumes of water solvating the i-th solute domain and bulk water, respectively.

The sum of the intrinsic, V_M, and thermal, V_T, contributions constitutes the so-called cavity volume, V_C, that is the volume of the cavity inside the solvent large enough to accommodate the solute molecule. Based on Equation (21.3), the cavity volume, V_C, can be viewed as the partial molar volume of the hard-sphere analogue of the solute molecule minus the ideal term $\beta_{T0}RT$. The hard-sphere analogue of a solute is a molecule having the same geometry but characterized by the hard-sphere solute–solvent interaction potentials. SPT-based hard-sphere calculations reveal a good agreement between the computed and experimental partial molar volumes of nearly spherical non-polar molecules.[22] Thus, from the volumetric perspective, non-polar solutes behave as hard-sphere molecules.

Molecular dynamics simulations have been used in conjunction with the Kirkwood–Buff theory to compute the partial molar volumes of a number of small solutes varying in shape and chemical nature.[28] Resulting volumetric data have been compared with experimental data, results of Monte Carlo simulation with hard-sphere potentials, and scaled particle theory-based computations.[28] Based on the comparison, it has been concluded that the partial molar volume of a small solute computed with the Lennard-Jones potential in the absence of the Coulombic term nearly coincides with the cavity volume, V_C.[28] This notion is consistent with the picture in which the partial molar volume of a non-polar solute of an arbitrary shape coincides with its cavity volume, V_C, plus the ideal term $\beta_{T0}RT$.

It has been suggested that the thermal volume, V_T, can be computed as the volume of void around the solute with the thickness, δ, being constant and independent of the chemical type of the solute or its geometry.[20,21] With this assumption, the cavity volume for a spherical solute can be presented as follows:[20]

$$V_C = V_M + V_T \approx (4N_A\pi/3)\,(r_W + \delta)^3 \tag{21.5}$$

where r_W is the van der Waals radius of the solute; and N_A is Avogadro's number.

There have been various empirical and theoretical estimates of the thickness of the empty layer, δ.[20,21,28–31] For small molecules, the different estimates of δ are in good agreement with each other, being within a range of 0.04 nm to 0.06 nm.[20,21,28] However, the values of δ estimated for proteins diverge significantly, ranging from 0.0226 to 0.1 nm.[29–31] In one study, an empirical analysis based on cross-correlating the partial molar volumes of 12 globular proteins with their structural data revealed that, for a native globular protein, the thickness of the thermal volume, δ, is ~0.1 nm, twice as high as that for small molecules.[29] This result raised a question about a possible dependence of δ on the size of a solute.

In a recent work, molecular dynamics simulations have been employed in conjunction with the Kirkwood–Buff theory to compute the partial molar volumes for organic solutes of varying size in water.[32] The thermal volume, V_T, for each compound was determined by subtracting the van der Waals volume, V_W, from V_C. The spherical approximation of solute geometry was applied to evaluating the thickness of the thermal volume, δ. The resulting data reveal a sigmoid increase in the thickness of the thermal volume, δ, from ~0.05 nm to ~0.1 nm as the solute radius increases from ~0.2 nm to ~0.7 nm (see Figure 21.1). This finding, which offers computational support to the empirical estimate of ~0.1 nm for δ of a globular protein, may be related to the de-wetting of large non-polar surfaces.[33]

21.3 Insights From Small Analogues of Proteins

In an attempt to model the hydration properties of proteins in their native and unfolded conformations, the volumetric properties of low molar mass model compounds mimicking protein groups, including amino acids, peptides, amines, carboxylic acids, and α,ω-aminocarboxylic acids, have been extensively studied.[18,34–45] Small molecules offer a number of advantages for hydration studies. Firstly, the molecular structure of a model compound can be varied easily to enable one to analyze the volumetric contribution of a specific functional group. Secondly, by judicious manipulation of the number and position of polar and/or charged groups in a parent molecule, one can modulate the solubility of its derivative. Thirdly, the simplicity of low molar mass compounds affords the task of empirical

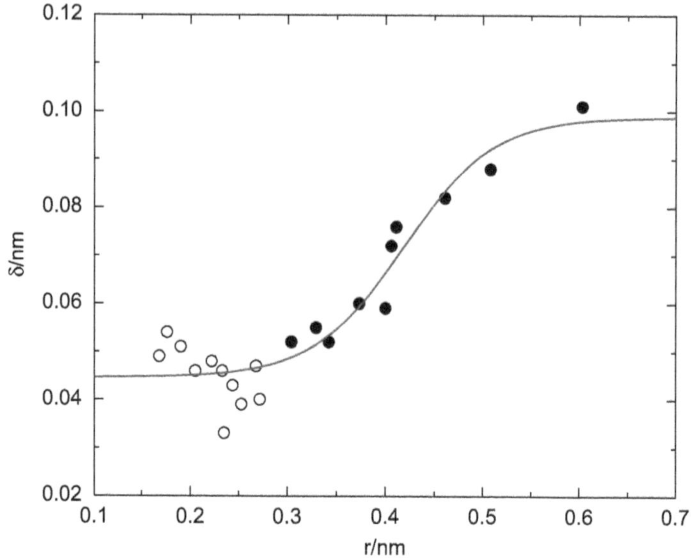

Figure 21.1 The dependence of the thickness of thermal volume, δ, on the van der Waals radii, r, of solutes from reference 32 (closed symbols) and reference 28 (open symbols), both computed in an identical manner. The sigmoidal fit of the data (solid line) only serves to guide the eye of the reader rather than conveying any analytical dependence.

Table 21.1 Partial molar volume contributions of the methylene group (–CH_2–), glycyl unit (–CH_2CONH–), and peptide bond (–CONH–).

	$V(-R)$ / (cm^3 mol^{-1})			
Functional Group	$t = 18$ °C	25 °C	40 °C	55 °C
-CH_2-[a]	15.5 ± 0.1	15.7 ± 0.1	15.9 ± 0.1	16.4 ± 0.1
-CH_2CONH-[b]	36.8 ± 0.1	37.4 ± 0.1	39.0 ± 0.2	39.3 ± 0.1
-CONH-	21.3 ± 0.1	21.7 ± 0.1	23.1 ± 0.1	22.9 ± 0.1

[a]From reference 43.
[b]From reference 44.

interpretation of volumetric data in terms of intra-solute and solute–solvent interactions.

The volumetric contribution of the glycyl residue, –CH_2CONH–, and methylene group, –CH_2–, have been determined from systematic studies of oligoglycines, NH_3^+–$(CH_2CONH)_n$–CH_2–COO^-, and α,ω-aminocarboxylic acids, NH_3^+–$(CH_2)_n$–COO^-.[43,44,46,47] Table 21.1 presents the group contributions of the independently hydrated –CH_2– and –CH_2CONH– groups at 18, 25, 40, and 55 °C. The volume contribution of the peptide group, –CONH–, computed as the difference between the group numbers of the –CH_2CONH– and –CH_2– group numbers is also presented in Table 21.1.

The group contribution of a specific amino acid residue in an extended polypeptide chain can be obtained by adding the group contribution of the glycyl unit (–CH_2CONH–) to that of the respective side chain. The latter can be obtained as the difference in partial molar volume, $V°$, between the corresponding amino acid and glycine in zwitterionic amino acids, amino acids with blocked termini, or oligopeptides. The most complete set of partial molar volumes of model compounds that can be used for determining the pH- and temperature-dependent volume contributions of amino acid side chains has been obtained on a set of *N*-acetyl amino acid amides, *N*-acetyl amino acid methylamides, and *N*-acetyl amino acids with neutralized carboxyl termini.[45] Table 21.2 contains the volume contributions for the 20 amino acid side chains as a function of temperature. For amino acids containing titratable groups (aspartic and glutamic acids, histidine, lysine, and arginine), the data presented in Table 21.2 refer to the neutral state of the side chain. Table 21.2 also presents the side chain contributions calculated from the partial molar volumes of zwitterionic amino acids, glycyl dipeptides, and diglycyl tripeptides.[36,37,42,48] Comparison between the several sets of data shown in Table 21.2 reveals that the volume contribution of an amino acid side chain is essentially independent of the model compound (in most cases, within $\pm$1–2 cm^3 mol^{-1}). This observation attests to the relative insensitivity of the partial molar volume observable to the microenvironment of individual atomic groups, a feature that renders additive calculations of the partial molar volumes of solutes reliable.

The ionic state of a protein influences all of its thermodynamic properties including partial molar volume.[49–51] Proper consideration of the volume effects of ionization is important in volumetric studies of protein folding and binding events, since these reactions are generally associated with a change in the state of ionization of abnormally titrating amino acid side chains.[52–54] In recognition of this need, changes in volume accompanying titration of amino and carboxyl termini as well as ionizable amino acid side chains in low molar mass model compounds have been extensively studied.[45,50,55,56] A recent work has produced the most complete set of ionization volumes and pK_a values for titratable protein groups determined at as a function of temperature.[45]

Several additive schemes have been devised for calculating the partial molar volume of extended polypeptide chains, including those of fully unfolded proteins.[45,57,58] These schemes take into account the contributions of amino acid side chains, the peptide backbone, and the amino and carboxyl terminal groups. The additive schemes proposed by Kharakoz[59] and Lee *et al.*[45] enable one to model the pH-dependent behaviour of partial molar volume by taking into account changes in the ionization/neutralization equilibria of the titratable amino acid side chains and the amino and carboxyl termini.

The data in Table 21.2 can be used in conjunction with the additive approach to calculate the partial specific volumes of proteins in their fully

Table 21.2 Partial molar volume contributions of amino acid side chains, *V(–R)*/(cm^3 mol^{-1}), as a function of temperature.

	V(–R)/(cm^3 mol^{-1})			
SC	*t* = 18 °C	25 °C	40 °C	55 °C
Ala	17.0 ± 0.1	16.8 ± 0.3	17.2 ± 0.3	17.9 ± 0.3
	18.2[c]	17.23[a]	18.4[c]	18.4[c]
		18.3[c]		
		16.14[d]		
		17.9[e]		
		17.5[f]		
Ala[g]	17.8 ± 0.1	17.2 ± 0.1	17.3 ± 0.2	17.9 ± 0.3
Val	47.5 ± 0.1	47.7 ± 0.1	47.7 ± 0.6	49.0 ± 0.2
	48.2[c]	47.46[a]	48.6[c]	49.0[c]
		48.3[c]		
		46.11[d]		
		49.2[e]		
		49.1[f]		
Leu	64.8 ± 0.2	65.4 ± 0.2	66.0 ± 0.3	67.4 ± 0.6
	64.3[c]	64.64[a]	65.1[c]	66.0[c]
		64.5[c]		
		63.39[d]		
		66.1[e]		
		65.7[f]		
Ile	63.5 ± 0.6	63.8 ± 0.1	63.8 ± 0.5	65.1 ± 0.4
	64.8[c]	62.61[a]	66.1[c]	67.3[c]
		65.2[c]		
		64.4[f]		
Pro	35.1 ± 0.2	35.6 ± 0.1	36.3 ± 0.4	36.7 ± 0.3
	32.9[c]	39.44[a]	33.5[c]	34.0[c]
		33.1[c]		
Phe	78.7 ± 0.1	79.9 ± 0.3	80.9 ± 0.2	82.4 ± 0.2
	80.3[c]	79.0[a]	82.2[c]	83.4[c]
		80.9[c]		
		79.22[d]		
		79.9[e]		
		79.1[f]		
Phe[g]	78.9 ± 0.1	79.9 ± 0.1	81.4 ± 0.1	82.7 ± 0.2
Trp	101.1 ± 0.1	102.1 ± 0.1	103.6 ± 0.2	104.7 ± 0.2
	98.5[c]	100.6[a]	99.9[c]	101.0[c]
		98.9[c]		
Trp[g]	99.6 ± 0.1	101.6 ± 0.1	102.6 ± 0.1	103.8 ± 0.5
Met	62.1 ± 0.2	62.8 ± 0.1	63.0 ± 0.4	64.3 ± 0.3
	64.7[c]	62.38[a]	66.4[c]	67.8[c]
		65.2[c]		
Cys[g]	29.2 ± 0.1	29.7 ± 0.1	30.3 ± 0.2	31.6 ± 0.4
	31.3[c]	30.0[b]	31.9[c]	32.3[c]
		31.5[c]		
Tyr	81.6 ± 0.3	82.2 ± 0.2	83.5 ± 1.0	85.1 ± 0.5
	82.4[c]	81.2[a]	84.0[c]	85.1[c]
		82.9[c]		
Ser[g]	17.3 ± 0.1	17.1 ± 0.1	17.4 ± 0.1	17.8 ± 0.3
	17.9[c]	17.43[a]	18.1[c]	18.2[c]
		18.0[c]		
		16.34[d]		

Table 21.2 (*Continued*)

SC	t = 18 °C	25 °C	40 °C	55 °C
Thr[g]	33.3 ± 0.3 33.9[c]	33.1 ± 0.1 33.64[a] 34.0[c] 32.27[d]	33.5 ± 0.2 34.4[c]	34.7 ± 0.3 34.7[c]
Asn[g]	34.0 ± 0.3 37.7[c]	34.0 ± 0.2 33.9[b] 38.1[c] 33.88[d]	34.5 ± 0.6 39.1[c]	35.4 ± 0.3 40.2[c]
Gln	49.9 ± 0.1 50.1[c]	50.8 ± 0.1 50.42[a] 50.3[c]	51.2 ± 0.2 50.9[c]	51.5 ± 0.2 51.5[c]
Asp	30.7 ± 0.4 31.4[c]	31.7 ± 0.1 31.6[a] 31.7[c]	32.4 ± 0.4 32.4[c]	32.3 ± 0.5 32.9[c]
Glu	46.7 ± 0.5 47.6[c]	47.7 ± 0.7 46.66[a] 48.0[c]	48.0 ± 0.2 48.7[c]	48.6 ± 0.4 49.3[c]
His[h]	56.9 ± 0.4 59.1[c]	57.0 ± 0.6 55.11[a] 59.5[c]	57.9 ± 0.2 60.3[c]	58.5 ± 0.4 61.0[c]
Lys	69.6 ± 0.2 64.3[c]	70.1 ± 0.4 64.5[c]	70.7 ± 0.3 65.1[c]	70.6 ± 0.3 65.8[c]
Arg	66.0 ± 0.5 71.6[c]	67.4 ± 0.3 72.1[c]	67.6 ± 0.3 73.0[c]	66.3 ± 0.7 73.7[c]

[a]Zwitterionic amino acids from reference 36.
[b]Zwitterionic amino acids from reference 42.
[c]GlyXGly tripeptides from reference 37.
[d]GlyX dipeptides from reference 48.
[e]XGlyGly tripeptides from reference 165.
[f]GlyGlyX tripeptides from reference 165.
[g]Calculated from *N*-acetyl amino acid data.
[h]Calculated from *N*-acetyl amino acid methylamide data.

unfolded conformations as a function of pH and temperature. The calculations are performed based on the primary amino acid structure of the polypeptide using the following additive scheme:[45]

$$\nu^o = \Bigg[V^o(\text{trigly}) - \Delta V_{\text{NH}_2}\,/\,(1 + 10^{\text{pH}-\text{p}K_{\text{a}}\text{NH}_2}) + \Delta V_{\text{COOH}}/(1 + 10^{\text{p}K_{\text{a}}\text{COOH}-\text{pH}}) + (n-3)V(-\text{CH}_2\text{CONH}-) + \sum_{i=1}^{n} V_{\text{i}}(-\text{R}) + \sum_{j=1}^{l} \Delta V_{\text{j}}/(1 + 10^{\text{pH}-\text{p}K_{\text{aj}}}) - \sum_{k=1}^{m} \Delta V_{\text{k}}/(1 + 10^{\text{p}K_{\text{ak}}-\text{pH}})\Bigg]/M \tag{21.6}$$

where ν° is the partial specific volume of the protein in question (expressed per gram rather that per mole of protein); V°(trigly) is the partial molar volume of zwitterionic triglycine; ΔV_{NH_2} and ΔV_{COOH} are the protonation volumes of the amino and carboxyl termini, respectively; pK_{aNH_2} and

pK_{aCOOH} are the dissociation constants of the amino and carboxyl termini, respectively; $V_i(-R)$ is the volume contribution of the i-th amino acid side chain [for an ionizable side chain, $V_i(-R)$ corresponds to its neutral state]; n is the number of amino acid residues in the polypeptide chain; l is the number of basic side chains; pK_{aj} and ΔV_j are the dissociation constant and the protonation volume of the j-th basic side chains; m is the number of acidic side chains; pK_{ak} and ΔV_k are the dissociation constant and the protonation volume of the k-th acidic side chain; and M is the molar mass of the polypeptide.

The list of temperature-dependent dissociation constants, pK_a, and protonation volumes, ΔV, for the ionizable amino acid side chains and terminal groups can be found in reference 45. Figure 21.2 presents the calculated temperature dependences of the partial specific volumes, $\nu°$, of the extended conformations of ubiquitin, apocytochrome c, ribonuclease A, lysozyme, apomyoglobin, and α-chymotrypsinogen A. Figure 21.2 also shows the experimentally determined temperature-dependent data on $\nu°$ for apocytochrome c, a protein which is unfolded under the physiological conditions. The observed disparity between the calculated and experimental partial specific volumes of apocytochrome c is consistent with the picture in which the protein, although unfolded, is not fully extended but retains a considerable number of water-inaccessible atomic groups.[45]

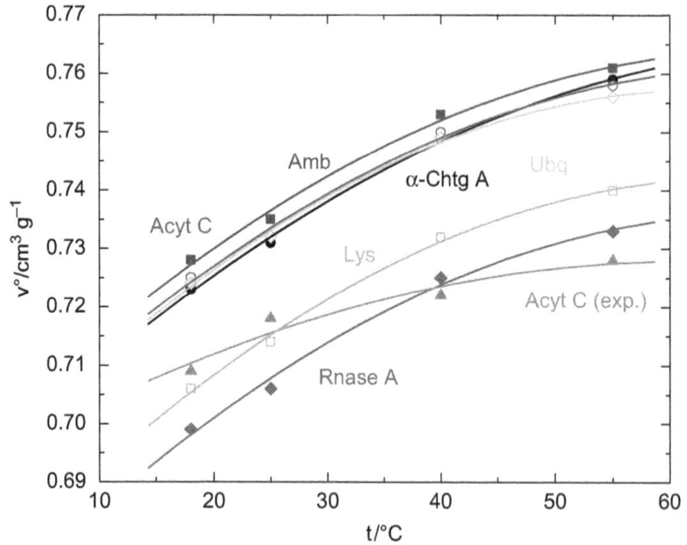

Figure 21.2 Calculated temperature dependences of the partial specific volumes, $\nu°$, of the fully extended conformations of ubiquitine (open diamond), apocytochrome (open circle), ribonuclease A (closed diamond), lysozyme (open square), apomyoglobin (closed square), and α-chymotrypsinogen A (closed circle) at pH 7. Experimental temperature dependence of $\nu°$ for apocytochrome c (closed triangles).

21.4 Partial Molar Volumes of Proteins

Table 21.3 lists some available data on the partial specific volumes, ν°, of globular proteins as a function of temperature. Globular proteins exhibit little diversity with respect to their partial specific volumes which, at 25 °C, group compactly around (0.725 ± 0.013) cm^3 g^{-1}. According to Equation (21.3), the similarity of the partial molar volumes of globular proteins reflects similarities in their interior packing and hydration properties. In fact, globular proteins are surprisingly homogeneous with respect to the chemical nature of buried and solvent-exposed atomic groups. The fractions of buried non-polar, polar, and charged groups are (62 ± 1)%, (31 ± 2)%, and (7 ± 2)%, respectively,[60] while the fractions of the same groups exposed to the solvent are (53 ± 5)%, (33 ± 7)%, and (14 ± 4)%, respectively.[29]

Inspection of data in Table 21.3 reveals an increase in the partial specific volumes, ν°, of the proteins with an increase in temperature consistent with positive partial specific expansibilities, $e^\circ = (\partial \nu^\circ/\partial T)_P$. Bull and Breese were first to determine the partial specific expansibility, e°, of globular proteins and changes in e° accompanying their heat-induced conformational transitions.[61] The average partial specific expansibility, e°, of lysozyme, BSA, ovalbumin, β-lactoglobulin, and methemoglobin at room temperature has been found to be $(3.4 \pm 0.4)10^{-4}$ cm^3 g^{-1} K^{-1}.[61] This value is in excellent agreement with $(3.5 \pm 0.8)10^{-4}$ cm^3 g^{-1} K^{-1}, the value of e° that can be computed from the data presented in Table 21.3 as the average temperature slope of the partial specific volumes, ν°, of the globular proteins at 25 °C.[29]

According to Equation (21.3), the partial specific volume of a protein consists of the intrinsic, $\nu_M = V_M/M$, thermal, $\nu_T = V_T/M$, and interaction,

Table 21.3 Protein partial specific volumes, ν°/(cm^3 g^{-1}), as a function of temperature.[a] [29]

		ν° / (cm^3 g^{-1})				
Proteins	M/kDa	t = 18 °C	25 °C	35 °C	45 °C	55 °C
Conalbumin	75.5	0.726	0.729	0.733	0.735	0.739[b]
BSA	68.0	0.735	0.739	0.744	0.751	0.756[b]
Hemoglobin	68.0	0.743	0.745	0.747	0.750	
Ovalbumin	46.0	0.735	0.737	0.740	0.743	0.746[b]
Pepsin	35.5	0.730	0.733	0.736	0.738	0.741
α-Chymo-trypsinogen A	25.7	0.727	0.730	0.733		
α-Chymotrypsin	25.3	0.717	0.721	0.724	0.727	
Trypsin	23.0	0.718	0.720	0.724	0.727	0.731[b]
Trypsinogen	23.0	0.721	0.725	0.730	0.734	0.739
β-Lactoglobulin	18.4	0.731	0.734	0.737	0.741	
Myoglobin	17.8	0.742	0.745	0.748	0.750	0.753[b]
α-Lactalbumin	14.3	0.711	0.713	0.717	0.720	
Lysozyme	14.3	0.699	0.702	0.704	0.707	0.710
Ribonuclease A	13.6	0.702	0.704	0.707	0.710	
Cytochrome C	12.4	0.735	0.738	0.742	0.746	0.750

[a]Estimated average error of measurements is ± 0.003 cm^3 g^{-1}.
[b]Extrapolated values.

$\nu_I = V_I/M$, contributions. The intrinsic volume of a protein, V_M, is the sum of the van der Waals volumes of all of its atoms, V_W, and the volume of the void space, V_V, inside the protein originating from the imperfect packing of the components of the polypeptide chain. The volume contribution of a specific atom inside the protein core can be determined using the Voronoi polyhedra procedure or one of its more recent modifications.[62] The average van der Waals volumes of specific protein atoms reported by different research groups generally agree with each other within a few percent.[60,63–66]

Empirical analysis of available structural data reveals that the intrinsic volume, V_M, of an average globular protein is related to its molar mass, M, *via* $V_M/(\text{nm}^3) = (1.2 \pm 0.5) + (1.04 \pm 0.02) \times 10^{-3}\ M$,[29,51] while the average density of a protein interior is (1.47 ± 0.05) g cm^{-3}.[67] However, the interior packing is not uniform throughout the protein with respect to the filling of the available space.[68–70] Based on the analysis of the three-dimensional structure of subtilisin BPN', carboxypeptidase A, α-chymotrypsin, lysozyme, and ribonuclease A, Beardsley and Kauzmann have outlined a general pattern in which the protein interior consists of a central low-density core surrounded by a higher density subsurface shell.[69] An important general trend is a loose packing of the active site groove of a globular enzyme, such as lysozyme and ribonuclease A, which suggests that the configurational flexibility of local residues plays a role in facilitating substrate binding.[68,69] Consistent with this notion, it has been proposed that the local packing density of protein micro-domains essentially determines the protein flexibility profile as reflected in the crystallographic B-factors (atomic mean-square displacements) of amino acid residues.[71]

MD simulations of barnase and T4 phage lysozyme have revealed that the likelihood of finding atomic size packing defects (cavities) is higher in non-polar regions containing buried hydrophobic side chains.[72] In general, the packing density in non-polar regions of proteins is higher than in polar regions since the regular backbone-to-backbone hydrogen bonding networks are open-structured and do not favor maximizing tight van der Waals interactions.[69,72] On the other hand, polar groups that form hydrogen bonds occupy a smaller volume than the same groups when they do not form hydrogen bonds.[73]

Provided that the intrinsic volume of a protein is defined as its molecular volume, the thermal volume can be considered as consisting of an empty space around the protein molecule of a thickness, δ, of $\sim$0.1 nm.[29,32] Consequently, the value of ν_T can be approximated by the product of δ and the solvent-accessible surface area, S_A, of a protein. The solvent-accessible surface area, S_A, of a globular protein correlates with its molar mass: $S_A/(\text{nm}^2) = (0.047 \pm 0.002)\ M^{0.76 \pm 0.03}$.[74,75] Thus, for an average protein, the thermal volume, ν_T, is given by $\nu_T/(\text{cm}^3\ \text{g}^{-1}) = (N_A/M)\ \delta\ (0.0047\ M^{0.76}) = 2.83\ M^{-0.24}$. The interaction volume, ν_I, is the difference between the partial specific volume, $\nu°$, of a protein and the sum of the intrinsic, ν_M, and thermal, ν_T, contributions. Since the average value of the partial specific volume, $\nu°$, of a globular protein is 0.72 cm^3 g^{-1}, for an average globular protein one obtains

$\nu_I/(\text{cm}^3\ \text{g}^{-1}) = \nu° - \nu_M - \nu_T = 0.094 - 0.723/M - 2.83\ M^{-0.24}$. These simple considerations can be used for rough estimates of the intrinsic, thermal, and interaction contributions to the partial specific volume of an "average" globular protein. Such estimates yield highly negative values of ν_I which are suggestive of a smaller partial molar volume (greater density) of water solvating a protein compared to that of bulk water. This expectation is consistent with the experimental observations which indicate a ~10–15% increase in the density of water of protein hydration relative to bulk water.[76–78]

In one study, the hydration contributions to volume and compressibility of 1 nm^2 of polar, non-polar, and charged type of solvent accessible surfaces have been determined as a function of temperature by a linear regression analysis of the values of the partial molar volume and compressibility of 12 globular proteins and X-ray derived data on their intrinsic volumes and solvent accessible surface areas.[29] The comparison of these protein-derived volumetric unit contributions with similar results derived based on low molar mass compounds revealed the following features: the hydration contributions to volume and compressibility of charged protein surface groups are similar to those of charged groups in small organic molecules. By contrast, the hydration contributions to volume and compressibility of polar protein surface groups are qualitatively different from those of polar groups in low molar mass compounds. The disparity may reflect the presence of networks of water molecules adjacent to polar protein surface areas, with these networks involving waters from second and third coordination spheres.[29] For non-polar protein surface groups, the ability of low molar mass compounds to successfully model protein properties depends on the temperature domain being examined. The volumetric data have been rationalized to suggest that non-polar groups on protein surfaces are hydrated independently at low temperatures, while, at higher temperatures, some of the solvating waters become influenced by neighboring polar groups.[29]

21.5 Conformational Transitions

Volumetric measurements have been applied to characterization of protein transitions induced by pH, salts, denaturants, temperature, and pressure. Changes in volume accompanying protein folding/unfolding transitions under the conditions of constant temperature and pressure have been measured using densimetric and dilatometric techniques.[79–82] Densimetric and pressure-perturbation calorimetric techniques have been used to study temperature-induced transitions,[83,84] while pressure-induced transitions have been studied by various spectroscopic techniques.[85–88] Each experimental technique samples a specific type of protein transition. A transition volume, ΔV, measured by one technique may not be directly related to a volume change obtained on the same protein with another technique. The underlying reason is due not only to the difference in the type of the perturbant of the native structure (pH, denaturant, temperature, pressure), which may influence the nature of the unfolded state being sampled, but also non-zero values of changes in expansibility, $\Delta E = (\partial \Delta V/\partial T)_P$, and isothermal compressibility, $\Delta K_T = -(\partial \Delta V/\partial P)_T$ associated with the transition. These observables are needed for extrapolating data obtained at elevated temperatures and pressures to room temperature and ambient pressure. Therefore, the volumetric differences between the folded and unfolded states may change in magnitude and even in sign upon a change in temperature and pressure.

As a general statement, a change in volume, ΔV, associated with a conformational transition of a globular protein is small, in absolute value being in the order of ~1% (or less) of the protein's partial molar volume.[89–91] Significantly, the sign of ΔV for protein transitions measured at atmospheric pressure can be either positive or negative.[81–83,92–95] By contrast, changes in volume associated with pressure-induced denaturation of globular proteins at elevated pressures are always negative.[85–88,96–104] This observation reflects the fact that, according to Le Chatelier's principle, a protein may be denatured by pressure only if the partial volume of its pressure-induced denatured state is smaller than that of the native state at and above the denaturation pressure. Therefore, the volume change, ΔV, associated with pressure-induced protein denaturation is invariably negative. If the atmospheric pressure value of ΔV is positive, the protein can still be denatured by pressure if the change in isothermal compressibility, $\Delta K_T = -(\partial \Delta V/\partial \mathrm{P})_T$, associated with the pressure-induced transition is positive.[58] Even if the initial value of ΔK_T at atmospheric pressure is negative, ΔK_T may depend on pressure and change its sign from negative to positive at elevated pressures. In both scenarios, a protein will denature above the pressure at which the sign of ΔV becomes negative.

According to Equation (21.3), a change in volume accompanying a protein transition, ΔV, is the sum of changes in the intrinsic, ΔV_M, thermal, ΔV_T, and interaction, ΔV_I, contributions:[51,90,105,106]

$$\Delta V = \Delta V_M + \Delta V_T + \Delta V_I \tag{21.7}$$

Protein unfolding leads to elimination of intraglobular voids with a concomitant decrease in intrinsic volume, V_M. Thermal volume, V_T, increases upon protein unfolding due to an increase in the solvent accessible surface area, S_A, of a protein, although an increase in ΔV_T is not necessarily proportional to ΔS_A.[106] Finally, protein unfolding brings previously buried polar and charged groups into contact with water which causes a decrease in interaction volume, V_I, although the hydration patterns of folded and unfolded protein states may be different.[29,51,106] Thus, near-zero values of ΔV suggest that an increase in the thermal contribution, V_T, nearly compensates decreases in the intrinsic, V_M, and interaction, V_I, contributions.[90,106]

Protein folding/unfolding transitions are generally associated with a change in the state of ionization of abnormally titrating groups and, consequently, are coupled with changes in buffer ionization/neutralization equilibria.[52,53] Buffer ionization, therefore, influences measured changes in volume accompanying protein denaturation. In fact, the contribution due to ionization of biologically relevant buffers is comparable in magnitude with the volume of protein transition.[107] Thus, the impact of buffer ionization on the volume of protein denaturation can be very significant with the potential to affect not only its magnitude but also the sign. In agreement with this notion, pressure perturbation calorimetric (PPC) studies of lysozyme and ribonuclease A at pH 3.0 in four buffers differing in their ionization volumes have identified buffer ionization as an important determinant of protein transition volume that needs to be carefully taken into account.[107] Figure 21.3 presents the PPC-determined changes in volume, ΔV, accompanying the heat-induced unfolding of lysozyme and ribonuclease A in pyrophosphate, citrate, glycine, and HEPES plotted against the volume of buffer ionization, ΔV_{buff}.[107] The span of the buffer-dependent changes in ΔV_M for lysozyme is 23 cm^3 mol^{-1}, while that for ribonuclease A is 36 cm^3 mol^{-1}. Given the magnitude of the relative contributions of buffer ionization to the net value of the transition volume, ΔV, the choice of the buffer should have an impact on the stability of a protein with the potential to alter the nature of its pressure dependence from pressure-induced denaturation to pressure-induced stabilization and *vice versa*. The importance of these results is general and applies to all volumetric investigations, in particular, extending to derivation of pressure–temperature phase diagrams of protein stability.

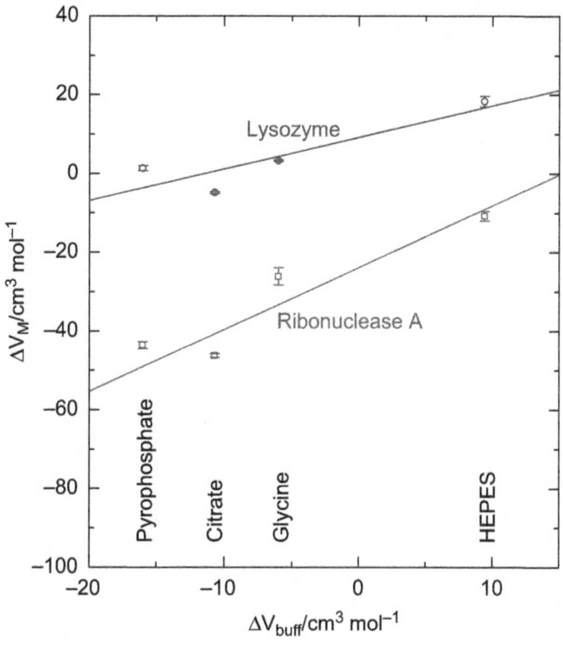

Figure 21.3 Dependence of the heat-induced unfolding transition volumes, ΔV_M, on the volume of buffer ionization, ΔV_{buff}, for lysozyme and ribonuclease A.

21.6 High Pressure Studies

Since the pioneering study by Bridgman of the effect of pressure on ovalbumin published in 1914,[108] it has been known that hydrostatic pressure has the ability to modulate the native/unfolded equilibria of globular proteins and may, ultimately, lead to protein denaturation. The volume difference between the native and pressure-induced denatured forms of the protein can be determined from:

$$\Delta V = -RT(\partial \ln K/\partial P)_T \quad (21.8)$$

where K is the equilibrium constant for the reaction of interconversion between the folded and unfolded protein states.

To assess the pressure dependence of the equilibrium constant, K, a sigmoid pressure denaturation profile sampled by some spectroscopic observable (typically, NMR, fluorescence, light absorbance) is analyzed within the framework of the two-state approximation of the folding/unfolding transition:

$$K = \alpha/(1-\alpha) = \exp[-(\Delta G_o - P\Delta V)/RT] \quad (21.9)$$

where α is the fraction denatured; and ΔG_o is the protein stability at ambient pressure.

While high pressure volumetric investigations have provided important new insights into the thermodynamics and kinetics of protein folding,[87,88,97,98,100,109] inspection of Equation (21.9) reveals limitations of such studies. Firstly, the analysis and resulting volumetric data become physically meaningless for a non-two-state transition involving intermediate states. Secondly, only those transitions that exhibit negative values of ΔV will be detected and characterized. And, finally, it is difficult to detect any pressure dependence of the estimated values of ΔV even if they are strongly pressure-dependent. In fact, in one study, an attempt has been made to take into account the pressure dependence of ΔV as reflected in a change in compressibility, $\Delta K_T = -(\partial \Delta V/\partial P)_T$, associated with the transition.[103] However, the resolution of spectroscopically-detected protein denaturation profiles is, generally, not sufficiently high for reliable evaluation of ΔK_T. This limitation is unfortunate since conformational transitions of proteins are characterized by significant changes in compressibility.[51,75] Therefore, ignoring the effect of compressibility may cause significant and unpredictable error in the values of ΔV, especially if extrapolated to a specific pressure within the experimental pressure range.

Notwithstanding, volumetric investigations at elevated pressures have provided important thermodynamic and kinetic insights into the molecular mechanisms of protein transitions and conformational dynamics of proteins. A compilation of literature data on volume changes accompanying pressure-induced protein transitions is given in reference 99.

Of particular importance is the stream of high pressure studies coming from the laboratory of Kazuyuki Akasaka, in which NMR measurements have been employed to characterize the low populated sub-conformations of globular proteins that may be of biological importance.[109] The native state of many proteins may contain, in addition to the main, low-energy conformation, a high-energy sub-conformation which, despite its being sparingly populated, may, nevertheless, play an important role in the biological function of the protein. At physiological conditions, the high energy sub-state, being extremely low-populated, cannot be characterized either structurally or thermodynamically. However, if the biologically-relevant high-energy sub-state of the protein is characterized by a smaller partial molar volume relative to the ground state, an increase in pressure, typically to 100 to 200 MPa, will shift the equilibrium towards the high-energy state according to Le Chatelier's principle. The pressure-induced increase in the population of the high-energy state enables one to characterize it both structurally (NMR) and thermodynamically. For example, an increase in pressure causes ubiquitin to initially populate a low-populated native sub-species, N_2, which subsequently undergoes a transition to an intermediate state, I, which, in turn, denatures to the unfolded state, U.[109,110]

Pressure-induced denaturation of a globular protein has been analyzed by comparing values of the volume and compressibility exhibited by the twenty amino acid side chains and the peptide bond when buried inside the protein and when solvent-exposed in the fully unfolded conformation.[111] The analysis revealed that the transfer of a peptide group from the protein interior to water becomes increasingly favorable as the pressure increases. This observation identifies solvation of the peptide backbone of the protein as the main driving force in pressure-induced denaturation. Polar side chains do not appear to exhibit any significant pressure-dependent alteration in their affinity for the protein interior or water. The transfer of non-polar groups from the protein interior to water becomes more unfavorable as the pressure increases. Consequently, the pressure-induced denatured state of a protein is not fully unfolded but contains a sizeable population of clustered solvent-inaccessible non-polar side chains. Intra-globular voids in the native state, which disappear in the denatured state, negatively contribute to the transition volume, ΔV, thereby representing another factor facilitating the pressure-induced protein denaturation.[102,111]

21.7 Ligand Binding

A detailed understanding of the thermodynamic forces participating in the control of ligand–protein recognition events is required for characterizing and, even to a larger degree, predicting the affinity and specificity of such events.[7,112–116] Amongst these forces, hydration/dehydration of interacting surfaces and, possibly, more distant loci of the protein occupy a special place due to their sizeable contribution to the binding thermodynamics and the difficulties involved in quantifying changes in hydration and the related thermodynamic impact.[13,113–115,117–124] Changes in hydration and related changes in thermodynamic observables have been implicated as a major factor contributing to the ubiquitous phenomenon of enthalpy–entropy compensation in protein recognition.[115,125,126]

Characterization of the role of hydration in protein–ligand association requires the use of physical parameters that can discriminate between the water solvating protein and ligand sites and water in the bulk. Volumetric observables such as volume, compressibility, and expansibility offer one way for tackling the problem of hydration in protein recognition.[51,105,127,128] To this end, volume changes accompanying protein–ligand binding reactions have been measured in an attempt to characterize the role of hydration as a driving force.[106,129–134]

Table 21.4 lists changes in volume, ΔV, determined for some ligand–protein and protein–protein binding events. Measured changes in volume, ΔV, can be used in conjunction with structural data to quantify changes in hydration accompanying the binding event.[106,130] According to Equation (21.3), a change in volume accompanying ligand binding, $P+L \leftrightarrow PL$, is the sum of changes in the intrinsic, ΔV_M, thermal, ΔV_T, and interaction, ΔV_I, contributions. The intrinsic term, ΔV_M, is the difference between the intrinsic volume of the protein–ligand complex (PL) and the sum of the intrinsic volumes of a free protein (P) and a free ligand (L). The thermal term, ΔV_T, is the difference between the thermal volume, V_T, of the protein–ligand complex and the sum of the thermal volumes of a free protein and a free ligand. The interaction term, ΔV_I, reflects the difference between the solute–solvent interactions of the protein–ligand complex and the sum of the solute–solvent interactions of a free protein and a free ligand.

A change in intrinsic volume, ΔV_M, can be evaluated using conventional computation techniques based on the three-dimensional structures of the protein–ligand complex and the free protein and the free ligand.[26,27,135] A change in thermal volume can be estimated as $\Delta V_T = \delta_C S_C - \delta_P S_P - \delta_L S_L$, where S_C, S_P, and S_L are the solvent-accessible surface areas of the complex, the free protein, and the ligand, respectively; and δ_C, δ_P, and δ_L are the

Table 21.4 Changes in volume accompanying protein association events.

Reaction	$\Delta V/(\mathrm{cm^3\ mol^{-1}})$	Δn_h
Glucose–hexokinase[a]	76 ± 30	332 ± 20
$(GlcNAC)_3$–lysozyme[b]	-44 ± 2	79 ± 44
2′-CMP–ribonuclease A[c]	22 ± 15	210 ± 40
BPTI–trypsinogen[d]	-278 ± 45	110 ± 40
OMTKY3–chymotrypsin[e]	170 ± 100	454 ± 22
Operator DNA-*cro*[f]	324 ± 7	811 ± 10
cAMP–EPAC1[g]	-59 ± 4	103 ± 2

[a]From reference 132.
[b]From reference 134.
[c]From reference 130.
[d]From reference 133.
[e]From reference 131.
[f]From F. Han, I. V. Beletskaya and T. V. Chalikian, (unpublished).
[g]From reference 166.

average thicknesses of the thermal volume for the complex, the free protein, and the ligand, respectively. In such calculations, the values of δ_C, δ_P, and δ_L depend on the size of the solute; for proteins and protein complexes, the thickness of the thermal volume, δ, is $\sim$0.1 nm, while for low molar mass ligands it is within the range of 0.05 nm to 0.06 nm.[32] A change in interaction volume, ΔV_I, thus can be calculated as the difference between the net binding change in volume, ΔV, and the sum of the intrinsic, ΔV_M, and thermal, ΔV_T, contributions; $\Delta V_I = \Delta V - \Delta V_M - \Delta V_T$.

A change in interaction volume, ΔV_I, is the only component of ΔV that reflects redistribution of water molecules between the bulk and hydration phases. Based on Equation (21.4), the value of ΔV_I is given by the sum:

$$\Delta V_I = \sum_i \Delta n_{hi}(V_{hi} - V_0) \tag{21.10}$$

where Δn_{hi} is the number of water molecules taken up by the i-th domain of the ligand or the protein upon their association; and V_{hi} and V_0 are the partial molar volumes of water of hydration of the i-th solute domain and bulk water, respectively. Under the simplifying assumption of the uniformity of the hydration shells of the ligand and the protein, Equation (21.10) reduces to $\Delta V_I = \Delta n_h(V_h - V_0)$, from which $\Delta n_h = \Delta V_I/(V_h - V_0)$. The partial molar volume of water solvating proteins and many small organic molecules, V_h, is roughly 10% smaller than that of bulk water.[76,77,136]

Application of this approach to protein–ligand and protein–protein binding events has produced the respective numbers of water molecules, Δn_h, released to the bulk. The estimated numbers of water molecules released to the bulk upon each protein association reaction are listed in the third column of Table 21.4.

21.8 Proteins in Binary Solvents

Interactions of a protein with the components of the surrounding solvent represent a major force modulating and guiding the processes of folding and binding.[9,137,138] The solvent is invariably water to which more or less significant amounts of water-miscible co-solvents are added. Organic co-solvents are often referred to as osmolytes, since they are used by cells of a variety of organisms to counteract the osmotic loss of cellular water. Binary solvents, representing a concentrated mixture of a co-solvent and water as the principal solvent, are of special interest from both the biological and the physico-chemical perspectives.[139–141] The added co-solvent may exert a stabilizing or destabilizing influence on the native conformation of a protein or may be neutral.

The task of understanding the differential thermodynamics of solute–principal solvent *versus* solute–co-solvent interactions has attracted considerable attention from both experimental and theoretical research groups.[142–150] The volumetric component of these efforts has been relatively modest. In fact, until recently, volumetric measurements have not been applied in any systematic manner to investigations of solvation in binary solvents. The deficiency was related to demanding experimental protocols related to extremely high co-solvent concentrations and the lack of conceptual formalisms that can be used for microscopic interpretations of macroscopic experimental data.

The volumetric properties of a solute, including partial molar volume, are influenced by the volume of a cavity that should be created in the solvent to accommodate the solute and the manifold of solute–solvent interactions. In the simplest model, the partial molar volume, V°, of a solute in a binary solvent can be presented as follows:

$$V^{\circ} = V_{\mathrm{C}} + n_1(V_{1\mathrm{h}} - V_{10}) + n_3(V_{3\mathrm{h}} - V_{30}) + \beta_{\mathrm{T0}}RT \tag{21.11}$$

where V_{C} is the volume of the cavity in the solvent enclosing a solute; $V_{1\mathrm{h}}$ and V_{10} are the partial molar volumes of water in the solvation shell of a solute and in the bulk, respectively; and $V_{3\mathrm{h}}$ and V_{30} are the partial molar volumes of the co-solvent in the solvation shell of a solute and in the bulk, respectively.

More rigorously, the partial molar volume of a solute in a binary solvent can be defined within the framework of the Kirkwood–Buff theory.[10,117,151] The Kirkwood–Buff theory is an exact statistical mechanical theory that links the radial distribution functions, $g_{ij}(r)$, between the different solution components with their thermodynamic properties, including partial molar volumes and compressibilities.[10,117,152,153] These properties are expressed *via* the Kirkwood–Buff integrals $G_{ij} = \int_0^{\infty}[g_{ij}(r) - 1]4\pi r^2 dr$. Significantly, the Kirkwood–Buff integrals can also be evaluated in an alternative (inversion) procedure from experimental data (density, compressibility, and activity coefficient derivatives) in the absence of the knowledge of specific radial distribution functions $g_{ij}(r)$.[154]

For the limit of a very dilute solution of the solute, its partial molar volume in a binary mixture of the principal solvent and co-solvent is given by the expression:[10,117]

$$V^{\circ} = -\mathrm{C}_1V_1G_{12} - \mathrm{C}_3V_3G_{23} + \beta_{\mathrm{T0}}RT = -N_{21}V_1 - N_{23}V_3 + \beta_{\mathrm{T0}}RT \tag{21.12}$$

where $N_{21} = N_{\mathrm{A}}C_1G_{21}$ and $N_{23} = N_{\mathrm{A}}C_3G_{23}$ are, respectively, the excess numbers of water and co-solvent molecules within the region, including the solute proper and its immediate surroundings.[144,155] For a one-component solvent, Equation (21.12) simplifies to $V^{\circ} = -G_{12} + \beta_{\mathrm{T0}}RT$.

Recently, an expression for partial molar volume has been derived based on the statistical thermodynamic approach and the solvent exchange model.[156–158] In this model, the interaction of a co-solvent with a hydrated solute is viewed as an association reaction in which r water molecules become released to the bulk. Following a chain of simple steps, one derives an equation for a change in volume accompanying the transfer of a solute from water to a concentrated co-solvent solution:[156,157]

$$\Delta V^{\circ} = \Delta V_{\mathrm{C}} - \gamma_1 n\Delta V^{\circ}{}_1 + \Delta V(n/r)(\mathrm{a}_3/\mathrm{a}_1{}^r)\mathrm{k}/[1 + (\mathrm{a}_3/\mathrm{a}_1{}^r)\mathrm{k}] \tag{21.13}$$

where ΔV_C is the differential cavity volume in a concentrated co-solvent solution and water; n is the number of water molecules solvating the solute in pure water; k is the equilibrium constant for the reaction in which a co-solvent molecule replaces r water molecules at the binding site; $\Delta V = \Delta V_0 + \gamma_1 r \Delta V^\circ_1 - \gamma_3 \Delta V^\circ_3$ is the change in volume associated with replacement of water with co-solvent at the binding site in a concentrated co-solvent solution; ΔV_0 is the exchange volume in an ideal solution; ΔV°_1 and ΔV°_3 are the excess partial molar volumes of water and co-solvent in a concentrated solution; and γ_1 and γ_3 are the correction factors reflecting the influence of the bulk solvent on the properties of solvating water and co-solvent, respectively. The values of γ_1 and γ_3 may change from 0 (the properties of the solvation shell change in parallel with those of the bulk) to 1 (the properties of the solvation shell are independent of those of the bulk).

To analyze experimental volumetric data within the framework of Equation (21.13), one needs to estimate a change in cavity volume, V_C, which accompanies the transfer of a solute from water to a co-solvent solution. The cavity volume, V_C, in a mixed solvent can be computed based on the concepts of scaled particle theory (SPT) in which the solute and solvent molecules are approximated by a mixture of hard spheres.[24] Although SPT has been subsequently extended to include solutes of arbitrary shapes,[159–161] the "spherical" version of SPT is simpler and, consequently, has been used more extensively in practical calculations. Within the framework of SPT, the cavity volume can be computed from the equation:[24]

$$V_C = (\beta_{T0}/\alpha_0 T)H_C + N_A \pi d_S{}^3/6 \tag{21.14}$$

where β_{T0} and α_0 are, respectively, the coefficients of isothermal compressibility and thermal expansibility of the principal solvent; d_S is the hard-sphere diameter of a solute molecule. The enthalpy of cavity formation, H_C, is given by the relationship:[162]

$$H_C = [\alpha_0 RT^2/(1-\xi_3)][\xi_3 + 3\xi_2 d_S/(1-\xi_3) + 3\xi_1 d_S{}^2/(1-\xi_3) + 9\xi_2{}^2 d_S{}^2/(1-\xi_3)^2] \tag{21.15}$$

where $\xi_k = (\pi N_A/6)\sum_{i=1}^{m} C_i d_i^k$; k has the values of 1, 2, and 3; m is the number of solvent components; C_i and d_i are, respectively, the molar concentration and the molecular diameter of the i-th solvent component.

Systematic investigations of the volumetric properties of N-acetyl amino acid amides and oligoglycines in urea[157] and glycine betaine (GB)[163] have yielded concentration-dependent changes in the volume contributions of the amino acid side chains and the peptide group accompanying their transfer from water to concentrated solutions of the two co-solvents. Table 21.5 lists the volume contributions of the amino acid side chains and the glycyl residue at various urea concentrations at 25 °C.[157]

Urea is a quintessential denaturant which destabilizes the native protein structure. In contrast, GB is a stabilizing co-solvent (osmolyte). Both co-solvents have been extensively used in protein folding studies. The differential action of urea and GB on protein structure originates from the

Table 21.5 Partial molar volume contributions of amino acid side chains, *V(–R)*/(cm^3 mol^{-1}), as a function of urea concentration.[157]

	V(–R)/(cm^3 mol^{-1})				
SC M/(mol L^{-1}) = 0		**2**	**4**	**6**	**8**
Ala	16.8 ± 0.3	17.0 ± 0.1	17.0 ± 0.1	17.0 ± 0.3	16.9 ± 0.4
Ala[a]	17.1 ± 0.2	17.1 ± 0.1	17.4 ± 0.1	17.4 ± 0.3	17.4 ± 0.4
Val	47.7 ± 0.1	47.9 ± 0.1	48.0 ± 0.2	48.0 ± 0.1	48.2 ± 0.4
Leu	65.4 ± 0.2	65.4 ± 0.1	65.5 ± 0.1	65.6 ± 0.1	65.8 ± 0.4
Ile	63.2 ± 0.3	63.0 ± 0.2	63.0 ± 0.3	63.2 ± 0.3	63.3 ± 0.4
Pro	35.6 ± 0.1	35.4 ± 0.3	35.4 ± 0.1	35.3 ± 0.1	35.3 ± 0.3
Phe	79.9 ± 0.3	80.0 ± 0.1	80.3 ± 0.3	80.6 ± 0.3	80.7 ± 0.3
Phe[a]	79.9 ± 0.1	80.4 ± 0.1	80.6 ± 0.1	80.9 ± 0.2	80.9 ± 0.4
Trp	102.1 ± 0.1	102.1 ± 0.2	102.4 ± 0.1	102.4 ± 0.2	102.5 ± 0.6
Trp[a]	100.5 ± 0.1	102.4 ± 0.8	102.6 ± 0.3	102.9 ± 0.4	103.1 ± 0.4
Met	62.8 ± 0.1	63.2 ± 0.1	63.4 ± 0.4	63.6 ± 0.1	63.5 ± 0.4
Cys[a]	29.7 ± 0.1	30.5 ± 0.1	30.9 ± 0.2	31.0 ± 0.3	31.2 ± 0.2
Tyr	82.2 ± 0.2	82.7 ± 0.1	83.0 ± 0.1	83.1 ± 0.1	83.1 ± 0.4
Ser[a]	17.1 ± 0.1	17.4 ± 0.1	17.8 ± 0.1	17.9 ± 0.3	17.9 ± 0.4
Thr[a]	33.1 ± 0.1	33.5 ± 0.5	34.0 ± 0.5	34.3 ± 0.3	34.5 ± 0.4
Asn[a]	34.0 ± 0.2	34.8 ± 0.3	35.3 ± 0.2	35.5 ± 0.2	35.7 ± 0.3
Gln	50.8 ± 0.1	51.1 ± 0.2	51.3 ± 0.1	51.5 ± 0.7	51.5 ± 0.4
Asp	31.7 ± 0.1	32.1 ± 0.5	32.3 ± 0.3	32.8 ± 0.2	33.1 ± 0.4
Glu	47.7 ± 0.7	47.9 ± 0.3	48.3 ± 0.2	48.7 ± 0.2	48.7 ± 0.5
His[b]	57.0 ± 0.6	57.5 ± 0.6	58.0 ± 0.3	58.2 ± 0.4	58.5 ± 0.6
Lys	70.1 ± 0.4	70.6 ± 0.5	70.8 ± 1.0	70.8 ± 0.6	71.1 ± 0.6
Arg	67.4 ± 0.3	68.7 ± 0.6	69.8 ± 0.9	70.7 ± 0.6	71.9 ± 0.7
-CH_2CONH-[c]	37.5 ± 0.2	37.0 ± 0.6	37.0 ± 0.7	36.9 ± 0.6	36.8 ± 0.5
-CH_2CONH-[d]	34.9 ± 0.4	35.1 ± 0.3	35.6 ± 0.1	35.7 ± 0.4	35.8 ± 0.2

[a]Calculated from *N*-acetyl amino acid data.
[b]Calculated from *N*-acetyl amino acid methylamide data.
[c]Calculated from the data on oligoglycines.
[d]Calculated as the difference between the data on *N*-acetyl glycine methylamide and *N*-methyl acetamide.

difference in their interactions with protein groups.[140,146] The analyses of the urea- and GB-dependences of the volume contributions of the amino acid side chains and the peptide bond within the framework of Equation (21.13) have yielded the equilibrium constants, *k*, for the association reactions of each functional group with urea and GB.[157,163] Comparison between the values of *k* for a particular group enables one to characterize its differential affinity for urea and GB, thereby providing molecular insights into the stabilizing/destabilizing action of the two co-solvents.[157,163]

In a recent study, the partial molar volumes and compressibilities of cytochrome *c*, ribonuclease A, lysozyme, and ovalbumin have been measured in aqueous solutions of GB at concentrations between 0 M and 4 M.[164] The fact that globular proteins do not undergo any conformational transitions in the presence of GB provides an opportunity to study the interactions of GB with proteins in their native states within the entire range of experimentally accessible GB concentrations. Figure 21.4 presents the partial molar volumes

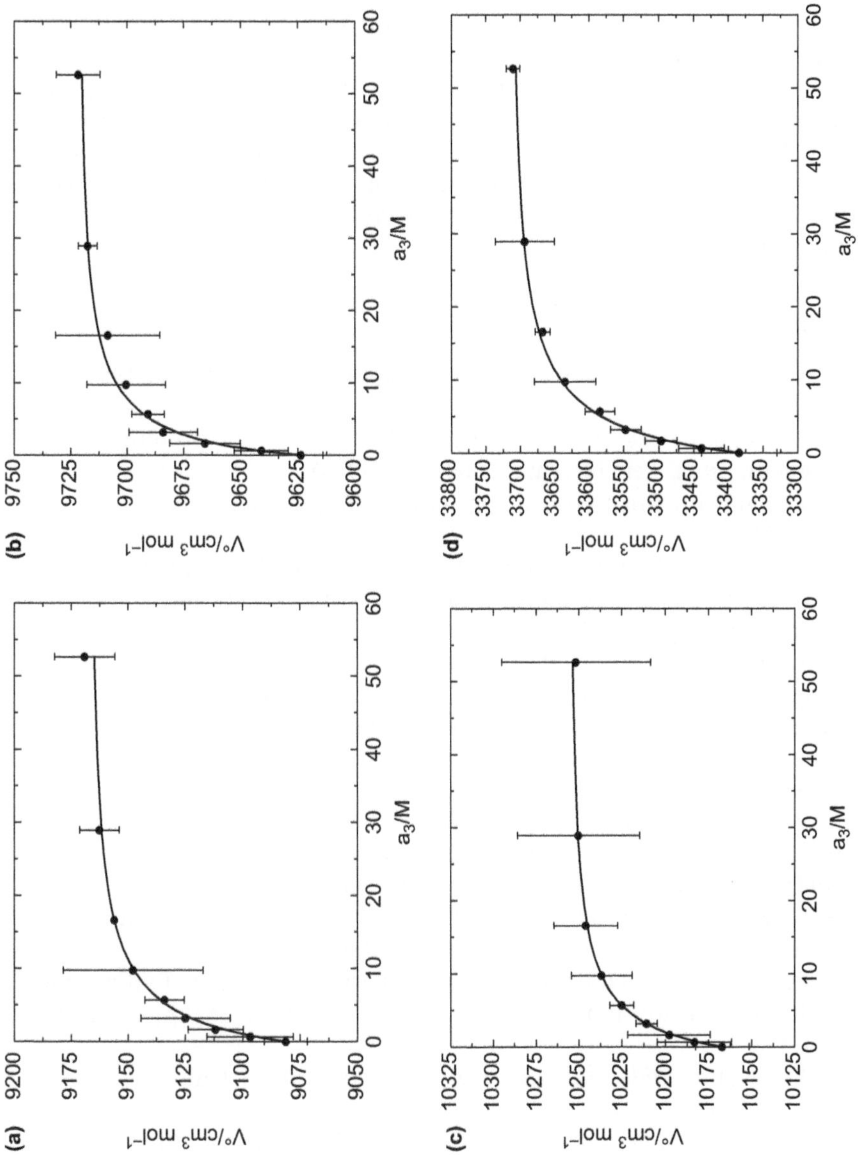

Figure 21.4 The partial molar volumes of cytochrome *c* (a), ribonuclease A (b), lysozyme (c), and ovalbumin (d) as a function of glycine betaine activity.

of cytochrome *c* (panel a), ribonuclease A (panel b), lysozyme (panel c), and ovalbumin (panel d) plotted *versus* the GB activity.[164] Inspection of Figure 21.4 reveals that, for all the proteins, the partial molar volumes increase hyperbolically with an increase in GB activity and level off at ~20 M. These volumetric data have been analyzed within the framework of Equation (21.13), in which each instance of GB interaction with a protein is viewed as a binding reaction that is accompanied by release of four water molecules.[164] The analysis has yielded the association constants, k, as well as changes in volume, ΔV_0, accompanying each GB–protein association event in an ideal solution. By comparing these parameters with the similar characteristics determined for low molar mass analogues of proteins, an inference has been drawn that no significant cooperative effects are involved in GB interactions with the proteins.[164]

References

1. K. A. Dill, Dominant forces in protein folding, *Biochemistry*, 1990, **29**, 7133–7155.
2. K. A. Dill, S. B. Ozkan, M. S. Shell and T. R. Weikl, The protein folding problem, *Annu. Rev. Biophys.*, 2008, **37**, 289–316.
3. R. L. Baldwin, Energetics of protein folding, *J. Mol. Biol.*, 2007, **37**, 283–301.
4. Y. W. Chen, F. Ding, H. F. Nie, A. W. Serohijos, S. Sharma, K. C. Wilcox, S. Y. Yin and N. V. Dokholyan, Protein folding: Then and now, *Arch. Biochem. Biophys.*, 2008, **469**, 4–19.
5. E. A. Freire, Thermodynamic approach to the affinity optimization of drug candidates, *Chem. Biol. Drug Design*, 2009, **74**, 468–472.
6. E. Shakhnovich, Protein folding thermodynamics and dynamics: Where physics, chemistry, and biology meet, *Chem. Rev.*, 2006, **106**, 1559–1588.
7. J. Wereszczynski and J. A. Mccammon, Statistical mechanics and molecular dynamics in evaluating thermodynamic properties of biomolecular recognition, *Q. Rev. Biophys.*, 2012, **45**, 1–25.
8. H. X. Zhou and M. K. Gilson, Theory of free energy and entropy in noncovalent binding, *Chem. Rev.*, 2009, **109**, 4092–4107.
9. Y. Levy and J. N. Onuchic, Water mediation in protein folding and molecular recognition, *Annu. Rev. Biophys. Biomol. Struct.*, 2006, **35**, 389–415.
10. A. Ben-Naim, *Molecular theory of solutions*, Oxford University Press, Oxford, 2006.
11. A. Ben-Naim, *Molecular theory of water and aqueous solutions. Part 1: Understanding water*, World Scientific Publishing Co. Pte. Ltd., Singapore, 2011.
12. A. Ben-Naim, *Molecular theory of water and aqueous solutions. Part 2: The role of water in protein folding, self-assembly and molecular recognition*, World Scientific Publishing Co. Pte. Ltd., Singapore, 2012.

13. J. E. Ladbury, Just add water! The effect of water on the specificity of protein-ligand binding sites and its potential application to drug design, *Chem. Biol.*, 1996, **3**, 973–980.
14. R. U. Lemieux, How water provides the impetus for molecular recognition in aqueous solution, *Acc. Chem. Res.*, 1996, **29**, 373–380.
15. P. Ball, Water as an active constituent in cell biology, *Chem. Rev.*, 2008, **108**, 74–108.
16. D. P. Zhong, S. K. Pal and A. H. Zewail, Biological water: A critique, *Chem. Phys. Lett.*, 2011, **503**, 1–11.
17. B. Bagchi, Water dynamics in the hydration layer around proteins and micelles, *Chem. Rev.*, 2005, **105**, 3197–3219.
18. E. J. Cohn, T. L. McMeekin, J. T. Edsall and M. H. Blanchard, Studies in the physical chemistry of amino acids, peptides and related substances. 1. The apparent molal volume and the electrostriction of the solvent, *J. Am. Chem. Soc.*, 1934, **56**, 784–794.
19. S. Terasawa, H. Itsuki and S. Arakawa, Contribution of hydrogen bonds to partial molar volumes of nonionic solutes in water, *J. Phys. Chem.*, 1975, **79**, 2345–2351.
20. J. T. Edward and P. G. Farrell, Relation between van der Waals and partial molal volumes of organic molecules in water, *Can. J. Chem.*, 1975, **53**, 2965–2970.
21. D. P. Kharakoz, Partial molar volumes of molecules of arbitrary shape and the effect of hydrogen bonding with water, *J. Solution Chem.*, 1992, **21**, 569–595.
22. R. A. Pierotti, Aqueous solutions of non-polar gases, *J. Phys. Chem.*, 1965, **69**, 281–288.
23. F. H. Stillinger, Structure in aqueous solutions of non-polar solutes from the standpoint of scaled particle theory, *J. Solution Chem.*, 1973, **2**, 141–158.
24. R. A. Pierotti, Scaled particle theory of aqueous and non-aqueous solutions, *Chem. Rev.*, 1976, **76**, 717–726.
25. A. Bondi, Van der Waals Volumes + Radii, *J. Phys. Chem.*, 1964, **68**, 441–451.
26. F. M. Richards, Areas, volumes, packing, and protein structure, *Annu. Rev. Biophys. Bioeng.*, 1977, **6**, 151–176.
27. F. M. Richards, Calculation of molecular volumes and areas for structures of known geometry, *Methods Enzymol.*, 1985, **115**, 440–464.
28. N. Patel, D. N. Dubins, R. Pomes and T. V. Chalikian, Parsing partial molar volumes of small molecules: a molecular dynamics study, *J. Phys. Chem. B*, 2011, **115**, 4856–4862.
29. T. V. Chalikian, M. Totrov, R. Abagyan and K. J. Breslauer, The hydration of globular proteins as derived from volume and compressibility measurements: cross correlating thermodynamic and structural data, *J. Mol. Biol.*, 1996, **260**, 588–603.
30. T. Imai, A. Kovalenko and F. Hirata, Partial molar volume of proteins studied by the three-dimensional reference interaction site model theory, *J. Phys. Chem. B*, 2005, **109**, 6658–6665.

31. M. Bano and J. Marek, How thick is the layer of thermal volume surrounding the protein? *Biophys. Chem.*, 2006, **120**, 44–54.
32. N. Patel, D. N. Dubins, R. Pomes and T. V. Chalikian, Size dependence of cavity volume: a molecular dynamics study, *Biophys. Chem.*, 2012, **161**, 46–49.
33. D. Chandler, Interfaces and the driving force of hydrophobic assembly, *Nature*, 2005, **437**, 640–647.
34. H. Hoiland, Partial molar volumes of biochemical model compounds in aqueous solution, in *Thermodynamic data for biochemistry and biotechnology*, ed. H.-J. Hinz, Springer-Verlag, Berlin, Heidelberg, New York, Tokyo, 1986, pp. 17–44.
35. F. J. Millero, A. L. Surdo and C. Shin, Apparent molal volumes and adiabatic compressibilities of aqueous amino acids at 25 °C, *J. Phys. Chem.*, 1978, **82**, 784–792.
36. A. K. Mishra and J. C. Ahluwalia, Apparent molal volumes of amino acids, N-acetylamino acids, and peptides in aqueous solutions, *J. Phys. Chem.*, 1984, **88**, 86–92.
37. M. Hackel, H. J. Hinz and G. R. Hedwig, Partial molar volumes of proteins: amino acid side-chain contributions derived from the partial molar volumes of some tripeptides over the temperature range 10–90 °C, *Biophys. Chem.*, 1999, **82**, 35–50.
38. M. Hackel, H. J. Hinz and G. R. Hedwig, The partial molar volumes of some tetra- and pentapeptides in aqueous solution: a test of amino acid side-chain group additivity for unfolded proteins, *Phys. Chem. Chem. Phys.*, 2000, **2**, 4843–4849.
39. M. A. Schwitzer and G. R. Hedwig, Thermodynamic properties of peptide solutions. Part 17. Partial molar volumes and heat capacities of the tripeptides GlyAspGly and GlyGluGly, and their salts K[GlyAspGly] and Na[GlyGluGly] in aqueous solution at 25 °C, *J. Solution Chem.*, 2005, **34**, 801–821.
40. T. Vogl, H. J. Hinz and G. R. Hedwig, Partial molar heat capacities and volumes of Gly–X–Gly tripeptides in aqueous solution: model studies for the rationalization of thermodynamic parameters of proteins, *Biophys. Chem.*, 1995, **54**, 261–269.
41. G. I. Makhatadze, V. N. Medvedkin and P. L. Privalov, Partial molar volumes of polypeptides and their constituent groups in aqueous solution over a broad temperature range, *Biopolymers*, 1990, **30**, 1001–1010.
42. D. P. Kharakoz, Volumetric properties of proteins and their analogues in diluted water solutions. 1. Partial volumes of amino acids at 15–55 °C, *Biophys. Chem.*, 1989, **34**, 115–125.
43. T. V. Chalikian, A. P. Sarvazyan and K. J. Breslauer, Partial molar volumes, expansibilities, and compressibilities of α,ω-aminocarboxylic acids in aqueous solutions between 18 and 55 °C, *J. Phys. Chem.*, 1993, **97**, 13017–13026.
44. T. V. Chalikian, A. P. Sarvazyan, T. Funck and K. J. Breslauer, Partial molar volumes, expansibilities, and compressibilities of

oligoglycines in aqueous solutions at 18–55 °C, *Biopolymers*, 1994, **34**, 541–553.

45. S. Lee, A. Tikhomirova, N. Shalvardjian and T. V. Chalikian, Partial molar volumes and adiabatic compressibilities of unfolded protein states, *Biophys. Chem.*, 2008, **134**, 185–199.
46. F. Shahidi and P. G. Farrell, Partial molar volumes of organic compounds in water. 4. Aminocarboxylic acids, *J. Chem. Soc., Faraday Trans. 1*, 1978, **74**, 858–868.
47. S. Cabani, G. Conti, E. Matteoli and M. R. Tine, Volumetric properties of amphionic molecules in water. 1. Volume changes in the formation of zwitterionic structures, *J. Chem. Soc., Faraday Trans. 1*, 1981, **77**, 2377–2384.
48. G. R. Hedwig, J. D. Hastie and H. Høiland, Thermodynamic properties of peptide solutions. 14. Partial molar expansibilities and isothermal compressibilities of some glycyl dipeptides in aqueous solution, *J. Solution Chem.*, 1996, **25**, 615–633.
49. J. Rasper and W. Kauzmann, Volume changes in protein reactions. 1. Ionization reactions of proteins, *J. Am. Chem. Soc.*, 1962, **84**, 1771–1777.
50. W. Kauzmann and A. Bodanszky, and J. Rasper, Volume changes in protein reactions. 2. Comparison of ionization reactions in proteins and small molecules, *J. Am. Chem. Soc.*, 1962, **84**, 1777–1788.
51. T. V. Chalikian, Volumetric properties of proteins, *Annu. Rev. Biophys. Biomol. Struct.*, 2003, **32**, 207–235.
52. C. Tanford, Protein denaturation, *Adv. Protein Chem.*, 1968, **23**, 121–282.
53. C. Tanford, Protein denaturation. C. Theoretical models for the mechanism of denaturation, *Adv. Protein Chem.*, 1970, **24**, 1–95.
54. B. M. Baker and K. P. Murphy, Dissecting the energetics of a protein–protein interaction: The binding of ovomucoid third domain to elastase, *J. Mol. Biol.*, 1997, **268**, 557–569.
55. T. V. Chalikian, D. P. Kharakoz, A. P. Sarvazyan, C. A. Cain, R. J. Mcgough, I. V. Pogosova and T. N. Gareginian, Ultrasonic study of proton transfer reactions in aqueous solutions of amino acids, *J. Phys. Chem.*, 1992, **96**, 876–883.
56. N. Taulier and T. V. Chalikian, Volumetric effects of ionization of amino and carboxyl termini of α,ω-aminocarboxylic acids, *Biophys. Chem.*, 2003, **104**, 21–36.
57. G. R. Hedwig and H. J. Hinz, Group additivity schemes for the calculation of the partial molar heat capacities and volumes of unfolded proteins in aqueous solution, *Biophys. Chem.*, 2003, **100**, 239–260.
58. D. P. Kharakoz, Partial volumes and compressibilities of extended polypeptide chains in aqueous solution: additivity scheme and implication of protein unfolding at normal and high pressure, *Biochemistry*, 1997, **36**, 10276–10285.
59. D. P. Kharakoz, Partial volumes and compressibilities of extended polypeptide chains in aqueous solution: Additivity scheme and

implication of protein unfolding at normal and high pressure, *Biochemistry*, 1997, **36**, 10276–10285.
60. Y. Harpaz, M. Gerstein and C. Chothia, Volume changes on protein folding, *Structure*, 1994, **2**, 641–649.
61. H. B. Bull and K. Breese, Temperature dependence of partial volumes of proteins, *Biopolymers*, 1973, **12**, 2351–2358.
62. A. Poupon, Voronoi and Voronoi-related tessellations in studies of protein structure and interaction, *Curr. Opin. Struct. Biol.*, 2004, **14**, 233–241.
63. C. Chothia, Structural invariants in protein folding, *Nature*, 1975, **254**, 304–308.
64. J. Pontius, J. Richelle and S. J. Wodak, Deviations from standard atomic volumes as a quality measure for protein crystal structures, *J. Mol. Biol.*, 1996, **264**, 121–136.
65. J. Tsai, R. Taylor, C. Chothia and M. Gerstein, The packing density in proteins: standard radii and volumes, *J. Mol. Biol.*, 1999, **290**, 253–266.
66. J. Tsai and M. Gerstein, Calculations of protein volumes: sensitivity analysis and parameter database, *Bioinformatics*, 2002, **18**, 985–995.
67. M. L. Quillin and B. W. Matthews, Accurate calculation of the density of proteins, *Acta Crystallogr., Sect. D: Biol. Crystallogr.*, 2000, **56**, 791–794.
68. F. M. Richards, The interpretation of protein structures: total volume, group volume distributions and packing density, *J. Mol. Biol.*, 1974, **82**, 1–14.
69. D. S. Beardsley and W. J. Kauzmann, Local densities orthogonal to beta-sheet amide planes: patterns of packing in globular proteins, *Proc. Natl. Acad. Sci. U. S. A.*, 1996, **93**, 4448–4453.
70. P. J. Fleming and F. M. Richards, Protein packing: dependence on protein size, secondary structure and amino acid composition, *J. Mol. Biol.*, 2000, **299**, 487–498.
71. B. Halle, Flexibility and packing in proteins, *Proc. Natl. Acad. Sci. U. S. A.*, 2002, **99**, 1274–1279.
72. J. P. Kocher, M. Prevost, S. J. Wodak and B. Lee, Properties of the protein matrix revealed by the free energy of cavity formation, *Structure*, 1996, **4**, 1517–1529.
73. D. Schell, J. Tsai, J. M. Scholtz and C. N. Pace, Hydrogen bonding increases packing density in the protein interior, *Proteins: Struct., Funct., Bioinf.*, 2006, **63**, 278–282.
74. S. Miller, J. Janin, A. M. Lesk and C. Chothia, Interior and surface of monomeric proteins, *J. Mol. Biol.*, 1987, **196**, 641–656.
75. N. Taulier and T. V. Chalikian, Compressibility of protein transitions, *Biochim. Biophys. Acta*, 2002, **1595**, 48–70.
76. T. V. Chalikian, Structural thermodynamics of hydration, *J. Phys. Chem. B*, 2001, **105**, 12566–12578.
77. D. I. Svergun, S. Richard, M. H. J. Koch, Z. Sayers, S. Kuprin and G. Zaccai, Protein hydration in solution: Experimental observation by

x-ray and neutron scattering, *Proc. Natl. Acad. Sci. U. S. A.*, 1998, **95**, 2267–2272.

78. F. Merzel and J. C. Smith, Is the first hydration shell of lysozyme of higher density than bulk water? *Proc. Natl. Acad. Sci. U. S. A.*, 2002, **99**, 5378–5383.
79. J. A. Ybe and P. C. Kahn, Slow folding kinetics of ribonuclease A by volume change and circular dichroism. Evidence for two independent reactions, *Protein Sci.*, 1994, **3**, 638–649.
80. K. Foygel, S. Spector, S. Chatterjee and P. C. Kahn, Volume changes of the molten globule transitions of horse heart ferricytochrome C. A thermodynamic cycle, *Protein Sci.*, 1995, **4**, 1426–1429.
81. T. V. Chalikian, V. S. Gindikin and K. J. Breslauer, Volumetric characterizations of the native, molten globule and unfolded states of cytochrome c at acidic pH, *J. Mol. Biol.*, 1995, **250**, 291–306.
82. R. Filfil and T. V. Chalikian, Volumetric and spectroscopic characterizations of the native and acid-induced denatured states of staphylococcal nuclease, *J. Mol. Biol.*, 2000, **299**, 827–842.
83. T. V. Chalikian, J. Volker, D. Anafi and K. J. Breslauer, The native and the heat-induced denatured states of alpha-chymotrypsinogen A: thermodynamic and spectroscopic studies, *J. Mol. Biol.*, 1997, **274**, 237–252.
84. L. N. Lin, J. F. Brandts, J. M. Brandts and V. Plotnikov, Determination of the volumetric properties of proteins and other solutes using pressure perturbation calorimetry, *Anal. Biochem.*, 2002, **302**, 144–160.
85. A. Zipp and W. Kauzmann, Pressure denaturation of metmyoglobin, *Biochemistry*, 1973, **12**, 4217–4228.
86. S. A. Hawley, Reversible pressure-temperature denaturation of chymotrypsinogen, *Biochemistry*, 1971, **10**, 2436–2442.
87. K. Heremans, High pressure effects on proteins and other biomolecules, *Annu. Rev. Biophys. Bioeng.*, 1982, **11**, 1–21.
88. I. Daniel, P. Oger and R. Winter, Origins of life and biochemistry under high-pressure conditions, *Chem. Soc. Rev.*, 2006, **35**, 858–875.
89. A. A. Zamyatnin, Amino acid, peptide, and protein volume in solution, *Annu. Rev. Biophys. Bioeng.*, 1984, **13**, 145–165.
90. T. V. Chalikian and K. J. Breslauer, On volume changes accompanying conformational transitions of biopolymers, *Biopolymers*, 1996, **39**, 619–626.
91. H. Durchshlag, *Specific volumes of biological macromolecules and some other molecules of biological interest*, ed. H.-J. Hinz, Springer-Verlag, Berlin, Heidelberg, New York, Tokyo, 1986, pp. 45–128.
92. T. V. Chalikian, V. S. Gindikin and K. J. Breslauer, Spectroscopic and volumetric investigation of cytochrome c unfolding at alkaline pH: characterization of the base-induced unfolded state at 25 degrees C, *FASEB J.*, 1996, **10**, 164–170.
93. N. Taulier and T. V. Chalikian, Characterization of pH-induced transitions of beta-lactoglobulin: ultrasonic, densimetric, and spectroscopic studies, *J. Mol. Biol.*, 2001, **314**, 873–889.

94. N. Taulier, I. V. Beletskaya and T. V. Chalikian, Compressibility changes accompanying conformational transitions of apomyoglobin, *Biopolymers*, 2005, **79**, 218–229.
95. Y. Tamura and K. Gekko, Compactness of thermally and chemically denatured ribonuclease A as revealed by volume and compressibility, *Biochemistry*, 1995, **34**, 1878–1884.
96. J. F. Brandts, R. J. Oliveira and C. Westort, Thermodynamics of protein denaturation. Effect of pressure on denaturation of ribonuclease A, *Biochemistry*, 1970, **9**, 1038–1047.
97. K. Heremans and L. Smeller, Protein structure and dynamics at high pressure, *Biochim. Biophys. Acta*, 1998, **1386**, 353–370.
98. F. Meersman, L. Smeller and K. Heremans, Protein stability and dynamics in the pressure-temperature plane, *Biochim. Biophys. Acta*, 2006, **1764**, 346–354.
99. C. A. Royer, Revisiting volume changes in pressure-induced protein unfolding, *Biochim. Biophys. Acta*, 2002, **1595**, 201–209.
100. J. L. Silva, D. Foguel and C. A. Royer, Pressure provides new insights into protein folding, dynamics and structure, *Trends Biochem. Sci.*, 2001, **26**, 612–618.
101. L. Mitra, J. B. Rouget, B. Garcia-Moreno, C. A. Royer and R. Winter, Towards a quantitative understanding of protein hydration and volumetric properties, *ChemPhysChem*, 2008, **9**, 2715–2721.
102. J. Roche, J. A. Caro, D. R. Norberto, P. Barthe, C. Roumestand, J. L. Schlessman, A. E. Garcia, B. Garcia-Moreno and C. A. Royer, Cavities determine the pressure unfolding of proteins, *Proc. Natl. Acad. Sci. U. S. A.*, 2012, **109**, 6945–6950.
103. K. E. Prehoda, E. S. Mooberry and J. L. Markley, Pressure denaturation of proteins: Evaluation of compressibility effects, *Biochemistry*, 1998, **37**, 5785–5790.
104. R. Winter, D. Lopes, S. Grudzielanek and K. Vogtt, Towards an understanding of the temperature/pressure configurational and free-energy landscape of biomolecules, *J. Non-Equilib. Thermodyn.*, 2007, **32**, 41–97.
105. T. V. Chalikian and K. J. Breslauer, Thermodynamic analysis of biomolecules: a volumetric approach, *Curr. Opin. Struct. Biol.*, 1998, **8**, 657–664.
106. T. V. Chalikian and R. Filfil, How large are the volume changes accompanying protein transitions and binding? *Biophys. Chem.*, 2003, **104**, 489–499.
107. S. Lee, H. Heerklotz and T. V. Chalikian, Effects of buffer ionization in protein transition volumes, *Biophys. Chem.*, 2010, **148**, 144–147.
108. P. W. Bridgman, The coagulation of albumen by pressure, *J. Biol. Chem.*, 1914, **19**, 511–512.
109. K. Akasaka, Probing conformational fluctuation of proteins by pressure perturbation, *Chem. Rev.*, 2006, **106**, 1814–1835.

110. R. Kitahara and K. Akasaka, Close identity of a pressure-stabilized intermediate with a kinetic intermediate in protein folding, *Proc. Natl. Acad. Sci. U. S. A.*, 2003, **100**, 3167–3172.
111. T. V. Chalikian and R. B. Macgregor, Jr. Origins of pressure-induced protein transitions, *J. Mol. Biol.*, 2009, **394**, 834–842.
112. D. H. Williams, E. Stephens, D. P. O'Brien and M. Zhou, Understanding noncovalent interactions: Ligand binding energy and catalytic efficiency from ligand-induced reductions in motion within receptors and enzymes, *Angew. Chem., Int. Ed.*, 2004, **43**, 6596–6616.
113. J. E. Ladbury and M. A. Williams, The extended interface: measuring non-local effects in biomolecular interactions, *Curr. Opin. Struct. Biol.*, 2004, **14**, 562–569.
114. T. S. G. Olsson, M. A. Williams, W. R. Pitt and J. E. Ladbury, The thermodynamics of protein–ligand interaction and solvation: insights for ligand design, *J. Mol. Biol.*, 2008, **384**, 1002–1017.
115. T. S. G. Olsson, J. E. Ladbury, W. R. Pitt and M. A. Williams, Extent of enthalpy–entropy compensation in protein–ligand interactions, *Protein Sci.*, 2011, **20**, 1607–1618.
116. C. E. A. Chang, W. Chen and M. K. Gilson, Ligand configurational entropy and protein binding, *Proc. Natl. Acad. Sci. U. S. A.*, 2007, **104**, 1534–1539.
117. A. Ben-Naim, *Statistical thermodynamics for chemists and biochemists*, Plenum Press, New York, London, 2002.
118. C. S. Poornima and P. M. Dean, Hydration in drug design. 1. Multiple hydrogen-bonding features of water molecules in mediating protein–ligand interactions, *J. Comput.-Aided Mol. Des.*, 1995, **9**, 500–512.
119. C. S. Poornima and P. M. Dean, Hydration in drug design. 2. Influence of local site surface shape on water binding, *J. Comput.-Aided Mol. Des*, 1995, **9**, 513–520.
120. C. S. Poornima and P. M. Dean, Hydration in drug design. 3. Conserved water molecules at the ligand-binding sites of homologous proteins, *J. Comput.-Aided Mol. Des.*, 1995, **9**, 521–531.
121. D. J. Huggins, M. Marsh and M. C. Payne, Thermodynamic properties of water molecules at a protein–protein interaction surface, *J. Chem. Theory Comput.*, 2011, **7**, 3514–3522.
122. N. Shimokhina, A. Bronowska and S. W. Homans, Contribution of ligand desolvation to binding thermodynamics in a ligand–protein interaction, *Angew. Chem., Int. Ed.*, 2006, **45**, 6374–6376.
123. T. Beuming, Y. Che, R. Abel, B. Kim, V. Shanmugasundaram and W. Sherman, Thermodynamic analysis of water molecules at the surface of proteins and applications to binding site prediction and characterization, *Proteins*, 2012, **80**, 871–883.
124. C. Clarke, R. J. Woods, J. Gluska, A. Cooper, M. A. Nutley and G. J. Boons, Involvement of water in carbohydrate–protein binding, *J. Am. Chem. Soc.*, 2001, **123**, 12238–12247.

125. R. Lumry and S. Rajender, Enthalpy–entropy compensation phenomena in water solutions of proteins and small molecules: a ubiquitous property of water, *Biopolymers*, 1970, **9**, 1125–1227.
126. D. H. Leung, R. G. Bergman and K. N. Raymond, Enthalpy–entropy compensation reveals solvent reorganization as a driving force for supramolecular encapsulation in water, *J. Am. Chem. Soc.*, 2008, **130**, 2798–2805.
127. A. P. Sarvazyan, Ultrasonic velocimetry of biological compounds, *Annu. Rev. Biophys. Biophys. Chem.*, 1991, **20**, 321–342.
128. T. V. Chalikian, A. P. Sarvazyan and K. J. Breslauer, Hydration and partial compressibility of biological compounds, *Biophys. Chem.*, 1994, **51**, 89–107.
129. G. B. Ogunmola, W. Kauzmann and A. Zipp, Volume changes in binding of ligands to methemoglobin and metmyoglobin, *Proc. Natl. Acad. Sci. U. S. A.*, 1976, **73**, 4271–4273.
130. D. N. Dubins, R. Filfil, R. B. Macgregor and T. V. Chalikian, Role of water in protein–ligand interactions: Volumetric characterization of the binding of 2′-CMP and 3′-CMP to ribonuclease A, *J. Phys. Chem. B*, 2000, **104**, 390–401.
131. R. Filfil and T. V. Chalikian, The thermodynamics of protein–protein recognition as characterized by a combination of volumetric and calorimetric techniques: the binding of turkey ovomucoid third domain to alpha-chymotrypsin, *J. Mol. Biol.*, 2003, **326**, 1271–1288.
132. R. Filfil and T. V. Chalikian, Volumetric and spectroscopic characterizations of glucose–hexokinase association, *FEBS Lett.*, 2003, **554**, 351–356.
133. R. Filfil, A. Ratavosi and T. V. Chalikian, Binding of bovine pancreatic trypsin inhibitor to trypsinogen: spectroscopic and volumetric studies, *Biochemistry*, 2004, **43**, 1315–1322.
134. I. Son, Y. L. Shek, D. N. Dubins and T. V. Chalikian, Volumetric characterization of tri-*N*-acetylglucosamine binding to lysozyme, *Biochemistry*, 2012, **51**, 5784–5790.
135. M. L. Connolly, Solvent-accessible surfaces of proteins and nucleic acids, *Science*, 1983, **221**, 709–713.
136. M. Gerstein and C. Chothia, Packing at the protein–water interface, *Proc. Natl. Acad. Sci. U. S. A.*, 1996, **93**, 10167–10172.
137. S. N. Timasheff, Water as ligand: preferential binding and exclusion of denaturants in protein unfolding, *Biochemistry*, 1992, **31**, 9857–9864.
138. S. N. Timasheff, Protein hydration, thermodynamic binding, and preferential hydration, *Biochemistry*, 2002, **41**, 13473–13482.
139. P. H. Yancey, M. E. Clark, S. C. Hand, R. D. Bowlus and G. N. Somero, Living with water stress: evolution of osmolyte systems, *Science*, 1982, **217**, 1214–1222.
140. J. A. Schellman, Protein stability in mixed solvents: a balance of contact interaction and excluded volume, *Biophys. J.*, 2003, **85**, 108–125.

141. P. H. Yancey, Organic osmolytes as compatible, metabolic and counteracting cytoprotectants in high osmolarity and other stresses, *J. Exp. Biol.*, 2005, **208**, 2819–2830.
142. S. N. Timasheff, The control of protein stability and association by weak interactions with water: how do solvents affect these processes? *Annu. Rev. Biophys. Biomol. Struct.*, 1993, **22**, 67–97.
143. J. Rosgen, B. M. Pettitt and D. W. Bolen, An analysis of the molecular origin of osmolyte-dependent protein stability, *Protein Sci.*, 2007, **16**, 733–743.
144. S. Shimizu, Estimating hydration changes upon biomolecular reactions from osmotic stress, high pressure, and preferential hydration experiments, *Proc. Natl. Acad. Sci. U. S. A.*, 2004, **101**, 1195–1199.
145. S. Shimizu and N. Matubayasi, Preferential hydration of proteins: A Kirkwood-Buff approach, *Chem. Phys. Lett.*, 2006, **420**, 518–522.
146. E. J. Guinn, L. M. Pegram, M. W. Capp, M. N. Pollock and M. T. Record, Jr. Quantifying why urea is a protein denaturant, whereas glycine betaine is a protein stabilizer, *Proc. Natl. Acad. Sci. U. S. A.*, 2011, **108**, 16932–16937.
147. M. W. Capp, L. M. Pegram, R. M. Saecker, M. Kratz, D. Riccardi, T. Wendorff, J. G. Cannon and M. T. Record, Interactions of the osmolyte glycine betaine with molecular surfaces in water: thermodynamics, structural interpretation, and prediction of m-values, *Biochemistry*, 2009, **48**, 10372–10379.
148. D. J. Felitsky and M. T. Record, Application of the local-bulk partitioning and competitive binding models to interpret preferential interactions of glycine betaine and urea with protein surface, *Biochemistry*, 2004, **43**, 9276–9288.
149. D. R. Canchi, P. Jayasimha, D. C. Rau, G. I. Makhatadze and A. E. Garcia, Molecular mechanism for the preferential exclusion of TMAO from protein surfaces, *J. Phys. Chem. B*, 2012, **116**, 12 095–12 104.
150. D. R. Canchi and A. E. Garcia, Co-solvent effects on protein stability, *Annu. Rev. Phys. Chem.*, 2013, **64**, 273–293.
151. J. G. Kirkwood and F. P. Buff, The statistical mechanical theory of solutions. 1. *J. Chem. Phys.*, 1951, **19**, 774–777.
152. D. G. Hall, Kirkwood-Buff theory of solutions. Alternative derivation of part of it and some applications, *Trans. Faraday Soc.*, 1971, **67**, 2516–2524.
153. A. Ben-Naim, Solute and solvent effects on chemical equilibria, *J. Chem. Phys.*, 1975, **63**, 2064–2073.
154. A. Ben-Naim, Inversion of Kirkwood-Buff theory of solutions. Application to water–ethanol system, *J. Chem. Phys.*, 1977, **67**, 4884–4890.
155. R. Chitra and P. E. Smith, Preferential interactions of co-solvents with hydrophobic solutes, *J. Phys. Chem. B*, 2001, **105**, 11513–11522.
156. S. Y. Lee and T. V. Chalikian, Volumetric properties of solvation in binary solvents, *J. Phys. Chem. B*, 2009, **113**, 2443–2450.

157. S. Lee, Y. L. Shek and T. V. Chalikian, Urea interactions with protein groups: A volumetric study, *Biopolymers*, 2010, **93**, 866–879.
158. T. V. Chalikian, Volumetric measurements in binary solvents: Theory to experiment, *Biophys. Chem.*, 2011, **156**, 3–12.
159. M. Irisa, K. Nagayama and F. Hirata, Extended scaled particle theory for dilute solutions of arbitrary shaped solutes. An application to solvation free energies of hydrocarbons, *Chem. Phys. Lett.*, 1993, **207**, 430–435.
160. M. Irisa, T. Takahashi, K. Nagayama and F. Hirata, Solvation free energies of non-polar and polar solutes reproduced by a combination of extended scaled particle theory and the Poisson–Boltzmann equation, *Mol. Phys.*, 1995, **85**, 1227–1238.
161. G. Graziano, Dimerization thermodynamics of large hydrophobic plates: A scaled particle theory study, *J. Phys. Chem. B*, 2009, **113**, 11232–11239.
162. N. Desrosiers and J. E. Desnoyers, Enthalpies, heat capacities, and volumes of transfer of tetrabutylammonium ion from water to aqueous mixed solvents from point of view of scaled particle theory, *Can. J. Chem.*, 1976, **54**, 3800–3808.
163. Y. L. Shek and T. V. Chalikian, Volumetric characterization of interactions of glycine betaine with protein groups, *J. Phys. Chem. B*, 2011, **115**, 11481–11489.
164. Y. L. Shek and T. V. Chalikian, Interactions of glycine betaine with proteins: insights from volume and compressibility measurements, *Biochemistry*, 2013, **52**, 672–680.
165. T. V. Chalikian, V. S. Gindikin and K. J. Breslauer, Hydration of diglycyl tripeptides with non-polar side chains: a volumetric study, *Biophys. Chem.*, 1998, **75**, 57–71.
166. I. Son, R. Selvaratnam, D. N. Dubins, G. Melacini and T. V. Chalikian, Ultrasonic and densimetric characterization of the association of cyclic AMP with the cAMP-binding domain of the exchange protein EPAC1, *J. Phys. Chem. B*, 2013, **117**, 10779–10784.

CHAPTER 22

Partial Molar Volumes of Proteins in Solution: Prediction by Statistical–Mechanical, 3D-RISM–KB Molecular Theory of Solvation

ANDRIY KOVALENKO

National Institute for Nanotechnology, and Department of Mechanical Engineering, University of Alberta, 11421 Saskatchewan Dr, Edmonton, Alberta, T6G 2M9, Canada
Email: andriy.kovalenko@nrc-cnrc.gc.ca

22.1 Introduction

The partial molar volume (PMV) of proteins is one of the most fundamental thermodynamic quantities to characterize the conformational stability of a protein,[1] especially in the analysis of pressure-induced denaturation of protein from the viewpoint of Le Chatelier's principle.[2] The thermodynamic stability of a native protein is modulated under pressure, and as a result the protein changes its structure to one with a smaller PMV value.[3] Pressure-induced denaturation (unfolding or structural transition) of proteins has been continuously attracting the attention of biochemists and biophysicists[2–6] since Bridgman[7] observed that egg white coagulated under a hydrostatic pressure of several hundred MPa. In addition to being of

Volume Properties: Liquids, Solutions and Vapours
Edited by Emmerich Wilhelm and Trevor M. Letcher

Published by the Royal Society of Chemistry, www.rsc.org

scientific interest, the pressure effect has found industrial applications, particularly in the food processing industry.[8]

The question, "Why and how does pressure change the structure of a protein?" can be converted into "Why and how is the partial molar volume (PMV) of the high-pressure structure (HPS) less than that of the low-pressure structure (LPS) of the protein?" through the following thermodynamic relation:[2]

$$\left(\frac{\partial \ln K}{\partial P}\right)_T = -\frac{\Delta \overline{V}}{N_A k_B T}, \tag{22.1}$$

where $\Delta \overline{V} = \overline{V}_{HPS} - \overline{V}_{LPS}$ is the difference in PMV between the LPS and HPS of the protein, K is the equilibrium constant between the two structures at pressure P and temperature T, and N_A and k_B are the Avogadro number and the Boltzmann constant, respectively. The latter question is more amenable to study than the former. Based on a high-pressure NMR measurement, Akasaka and co-workers[9–11] proposed a "volume theorem" of protein folding that the PMV of a protein generally decreases in parallel with the loss of its conformational order. In other words, PMV could be an order parameter in protein folding and conformational changes, which would provide an insight into the mechanisms of protein functioning.

Although the molecular mechanism of the pressure denaturation of proteins has not been completely understood, it is considered that pressure-induced effects in a protein may be attributed to the penetration of water molecules into the protein interior, which stimulates the structural change.[3–5,12–17] Thus, thermodynamic stability of a protein can be modulated not only by temperature and pressure thermodynamic variables but also by solvent conditions. Native protein is known to be marginally stable along temperature and pressure axes, as well as with the chemical conditions of the solvent.[18]

Several studies have investigated the change in the protein volume associated with pressure denaturation in the presence of co-solvents.[19–25] Staphylococcal nuclease (SNase) has been widely employed as a model for understanding protein folding because it is a small (149 amino acids), single-domain monomeric protein that contains no sulfhydryl residues or disulfide linkages. Winter and co-workers[26] studied the effect of such chaotropic and kosmotropic co-solvents as glycerol, sucrose, sorbitol, K_2SO_4, $CaCl_2$, and urea on the secondary structure and thermodynamic properties upon unfolding and denaturation of SNase by using FT-IR spectroscopy. The data showed that different co-solvents have a profound effect on the denaturation pressure and the changes in the Gibbs energy and PMV with unfolding.

The most common way to analyze the relationship between the PMV and the structure of a molecule is to empirically decompose the whole PMV into several additive contributions.[27–29] In the simplest way, the PMV of amino acids is broken up into contributions from constituent atomic groups.[30–32] Such analysis, however, seems to be too crude if one considers that the PMV of a solute molecule in solution is very sensitive to its geometry. According to Stillinger,[28] the PMV comprises the ideal gas term, the contribution from the excluded volume of the solute molecule, and from the packing of the solvent molecules around the solute. Chalikian and Breslauer[29] further decomposed the PMV into five contributions including the geometric and hydration terms for the molecular description:

$$\overline{V} = V_{\text{id}} + V_{\text{W}} + V_{\text{V}} + V_{\text{T}} + V_{\text{I}}, \tag{22.2}$$

where the components are identified as follows: V_{id} is the ideal contribution to PMV from the translational degrees of freedom of the solute biomolecule at a given number density and temperature. The next two components are the geometric volume contributions, the van der Waals volume V_{W} and the void volume V_{V} owing to structural voids within the solvent-inaccessible core. The last two components are the solvation effects on the PMV. The V_{T} term is the so-called thermal volume resulting from thermally induced molecular fluctuations between the solute and solvent molecules. It has been introduced to explain the distinction between the PMV and the molecular volume ($V_{\text{W}} + V_{\text{V}}$) for small non-polar molecules.[29] Finally, V_{I} is the interaction volume representing the change in the solvent volume by the intermolecular electrostatic interaction between the solute and solvent molecules, which is directly related to the electrostriction effect. Although this decomposition provides a molecular picture of the PMV, it is problematic for experiments to resolve the last two contributions because they reflect the solvation structure and are closely related to each other. A direct solution to this challenge of predicting these solvation effects without empirical assumption or ambiguity is enabled with statistical mechanics-based integral equation theory of liquids,[33] and in particular with the reference interaction site model (RISM) molecular theory of solvation,[34] as discussed below.

22.2 Evaluation of PMV of Proteins by MD Simulations

Since Levy and co-workers[35] investigated bovine pancreatic trypsin inhibitor under a high pressure by molecular dynamics (MD) simulation, there have been many theoretical studies of proteins under high pressure using molecular simulations to provide information about the response of native proteins to external pressure.[36] Because the PMV change upon unfolding, $\Delta\overline{V} = \overline{V}_{\text{HPS}} - \overline{V}_{\text{LPS}}$, is only a few percent of the protein PMV, obtaining its value from MD simulation is a challenging task.[37] Given that the observation of pressure unfolding of a protein is not feasible by direct MD simulations, a possible way to quantify the thermodynamics of the process is based on the thermodynamic cycle that relates the free energy of unfolding at a high pressure P_{H} to that at ambient (reference) low pressure P_{L}:[37]

$$\begin{array}{ccc} G_{\text{LPS}}(P_{\text{L}}), \overline{V}_{\text{LPS}}(P_{\text{L}}) & \xrightarrow{\text{LPS}(p_{\text{L}})\rightarrow\text{HPS}(p_{\text{L}})} & G_{\text{HPS}}(P_{\text{L}}), \overline{V}_{\text{HPS}}(P_{\text{L}}) \\ \downarrow & & \downarrow \\ G_{\text{LPS}}(P_{\text{H}}), \overline{V}_{\text{LPS}}(P_{\text{H}}) & \xrightarrow{\text{LPS}(p_{\text{H}})\rightarrow\text{HPS}(p_{\text{H}})} & G_{\text{HPS}}(P_{\text{H}}), \overline{V}_{\text{HPS}}(P_{\text{H}}) \end{array}, \tag{22.3}$$

which translates to the relation for the pressure stability of a protein at a high pressure of interest:

$$\Delta G_{\text{LPS}\to\text{HPS}}(P_{\text{H}}) = \Delta G_{\text{LPS}\to\text{HPS}}(P_{\text{L}}) + \int_{p_{\text{L}}}^{p_{\text{H}}} \text{d}P(\overline{V}_{\text{HPS}}(P) - \overline{V}_{\text{LPS}}(P)) \tag{22.4}$$

which requires knowledge of the PMVs of folded (LPS) and unfolded (HPS) states of the protein as a function of pressure.

The protein PMV is given by the pressure derivative of its Gibbs energy in solution and can be calculated as a change in the system volume upon addition or removal of one molecule (or mole) of a protein at constant temperature, pressure, and constant number of moles of water solvent (and other solution components):

$$\overline{V}_{\text{p}} = \left(\frac{\partial(nV)}{\partial n_{\text{p}}}\right)_{T,P,n_{\text{w}}}, \tag{22.5}$$

where nV is the total volume of the aqueous solution, and n_{p} and n_{w} are the number of moles of protein and water in the solution ($n = n_{\text{p}} + n_{\text{w}}$). In typical MD simulations that include one copy of a protein in a bath of a sufficient number of water molecules, this translates to:

$$\overline{V}_{\text{p}} = N_{\text{A}}(\langle V_{\text{p+w}}\rangle - \langle V_{\text{w}}\rangle), \tag{22.6}$$

where $\langle\ldots\rangle$ indicates an average in the isothermal–isobaric (N,P,T) ensemble, and $V_{\text{p+w}}$ and V_{w} are the volumes of systems containing one protein molecule and water, and a pure water system containing the same number of water molecules, respectively.[38] Provided the structures of both folded HPS and unfolded LPS of the protein are accessible to MD simulations, Equation (22.6) is applied to obtain PMVs of those states required in Equation (22.4). In this way, MD simulations of protein–water systems provide a net account of contributions both from the excluded volume of a complex-shaped protein, $V_{\text{W}} + V_{\text{V}}$, and from the packing of water molecules, which is determined by the local shape of the protein and its interactions with water, $V_{\text{T}} + V_{\text{I}}$ in Equation (22.4). Dividing the PMV change $\Delta\overline{V}$ into the five partial contributions according to Equation (22.2), the last two terms in particular, pose a significant complication and a further challenge to the evaluation of $\Delta\overline{V} = \overline{V}_{\text{HPS}} - \overline{V}_{\text{LPS}}$ from the MD simulation using Equation 22.6.

22.3 Calculation of the PMV of Proteins by the Molecular Theory of Solvation

An alternative route to the theoretical evaluation of the PMV is based on the statistical mechanics of liquids.[39] The calculation of the PMV of a spherical ion in aqueous solution was first performed by using the scaled particle theory of the Gibbs energy of cavity formation in a liquid.[40] Although successfully providing a microscopic picture for an "intrinsic volume", defined as the volume of a hypothetical ion without charge,[41] that approach had an inherent weakness of not treating the molecular interactions explicitly. Another method is based on the Kirkwood–Buff (KB) theory,[42] which provides a general framework for evaluating the thermodynamics properties of a liquid mixture involving the PMV and the isothermal compressibility in terms of the pair density correlation functions of the constituent molecular species. Expressions in terms of the site–site pair correlation functions rather than the molecular pair correlation functions are obtained[43] by coupling the KB theory[42] with the RISM molecular theory of solvation.[34]

Consider a solute biomolecule at infinite dilution in a solvent system comprising a mixture of solvent with other components such as co-solvent, electrolyte ions, and ligand molecules. With a straightforward extension of the RISM–KB theory[43] to a multicomponent solvent system, the PMV of the solute at infinite dilution $\overline{V}_{\mathrm{u}}$ and the isothermal compressibility of the (solvent) system χ_T^{v} can be expressed as:

$$\overline{V}^{\mathrm{u}} = k_{\mathrm{B}} T \chi_T^{\mathrm{v}} \left(1 - \sum_{\alpha \in \mathrm{u}} \sum_{\gamma \in \mathrm{v}} \rho_\gamma^{\mathrm{v}} c_{\alpha\gamma}^{\mathrm{uv}}(k=0) \right), \tag{22.7}$$

$$k_{\mathrm{B}} T \chi_T^{\mathrm{v}} = \left[\rho^{\mathrm{v}} \left(1 - \sum_{\alpha \in \mathrm{v}} \sum_{\gamma \in \mathrm{v}} \rho_\gamma^{\mathrm{v}} c_{\alpha\gamma}^{\mathrm{vv}}(k=0) \right) \right]^{-1}, \tag{22.8}$$

where indices α,γ enumerate interaction sites of species, that is, all sites on the solute biomolecule denoted as "u" and all sites on all sorts of molecules of the solvent system denoted as "v"; $c_{\alpha\gamma}^{\mathrm{uv}}(r)$ is the radial direct correlation function between site α on biomolecule "u" and site γ on species of solvent system "v"; $c_{\alpha\gamma}^{\mathrm{uv}}(r)$ is the radial direct correlation function between sites α and γ, both on species of solvent system "v"; ρ_γ^{v} is the number density of site γ (of the corresponding solvent species) and ρ^{v} is the total number density of molecules of solvent system "v"; and $c(k=0) = 4\pi \int_0^\infty \mathrm{d}r\ c(r)$ means the Fourier transform of the correlation function $c(r)$ at wave vector $k=0$.

The solute–solvent and solvent–solvent direct correlation functions $c_{\alpha\gamma}^{\mathrm{uv}}(r)$ and $c_{\alpha\gamma}^{\mathrm{uv}}(r)$ inserted in Equations (22.7) and (22.8) are obtained by solving the integral equations of the RISM theory (summarized in the next section). From Equation (22.7), the ideal contribution to the PMV is identified as $k_{\mathrm{B}} T \chi_T^{\mathrm{v}}$, which accounts for the centre of mass of the solute biomolecule comprising multiple interaction sites (atoms), the result originating from the intramolecular geometric constraints on atom pair distances in RISM theory, Equation (22.15). Application of the RISM–KB theory to amino acids in aqueous solution shows that it can account for the effect of the chemical specificities of 20 amino acids on their PMV; however, compared to the experimental data, the calculated PMV showed a systematic underestimation—increasing nearly proportionally to the number of atoms in the amino acid.[43] The deviation was attributed in some part to an "ideal fluctuation volume" originating from the flexibility of the solute molecules,[43] but mainly to the imperfectness of RISM in describing excluded volume effects of macromolecules with a large number of sites.[44] This drawback was mitigated by adding a Verlet type bridge correction (calculated for a reference solute stripped of charges) to RISM theory, which resulted in a significant improvement of the PMV calculated for the 20 amino acids and oligopeptides of glutamic acids in extended and α-helix conformations.[44]

The drawback of underestimation of the solute excluded volume by RISM operating with radial correlation functions is overcome to a large extent in the three-dimensional (3D) version of RISM theory which performs the orientational averaging of the correlation functions for solvent species but not for the solute macromolecule and introduces 3D maps of solvent site correlation functions around the solute of arbitrary shape.[34] Coupled with the KB method, the 3D-RISM–KB theory generalizes the PMV expression [Equation (22.7)] as follows:[45]

$$\overline{V}^{\mathrm{u}} = k_{\mathrm{B}} T \chi_T^{\mathrm{v}} \left(1 - \sum_{\gamma \in \mathrm{v}} \rho_\gamma^{\mathrm{v}} c_\gamma^{\mathrm{uv}}(\mathbf{k}=0) \right), \tag{22.9}$$

where $c_\gamma^{\mathrm{uv}}(\mathbf{r})$ is the 3D direct correlation function of site γ on species of solvent system "v" around the solute macromolecule, and $c(\mathbf{k}=0) = \int d\mathbf{r}\, c(\mathbf{r})$ is the 3D Fourier transform at the 3D wave vector $\mathbf{k}=0$.

The volume components in Equation (22.2) are obtained from the 3D-RISM–KB theory and geometric volume calculation.[46–48] The ideal volume V_{id} is the ideal gas contribution to the PMV and is naturally included in the 3D-RISM–KB equations. The van der Waals term V_{W} is the volume occupied by the van der Waals spheres of interaction sites, representing chemical groups in a united-atom or atoms in all-atom classical force field. The void volume V_{V} is defined as void space inside the solute molecule or at its surface which the solvent probe cannot access. The geometric terms V_{W} and V_{V} can be calculated, *e.g.*, using Alpha Shapes methods and software.[49] The thermal volume V_{T} is defined by $V_{\mathrm{T}} = \overline{V}^0 - V_{\mathrm{id}} - V_{\mathrm{W}} - V_{\mathrm{V}}$ in terms of the PMV of the same protein but with the solute–solvent electrostatic interactions switched off by completely removing the protein partial atomic charges, $\overline{V}^0$, which is essentially the solvent-packing contribution to the PMV. Finally, the interaction volume V_{I} is defined as the effect of the electrostatic interaction between the solute and solvent, $V_{\mathrm{I}} = \overline{V} - \overline{V}^0$. Both the PMVs $\overline{V}$ and $\overline{V}^0$ at full and zero partial atomic charges of the protein are obtained from 3D-RISM theory.

22.4 Statistical–Mechanical, 3D-RISM Molecular Theory of Solvation

Molecular theory of solvation, also known as the reference interaction site model (RISM),[33,34] is based on the first principles of statistical mechanics and the Ornstein–Zernike (OZ) type integral equation theory of molecular liquids.[33] It provides a firm platform to handle complex chemical and biomolecular systems in solution. As distinct from molecular simulations which explore the phase space of a molecular system by direct sampling, RISM theory operates with spatial distributions rather than trajectories of molecules and is based on analytical summation of the free energy diagrams which yields the solvation structure and thermodynamics in the statistical–mechanical ensemble. It yields the solvation structure by solving the RISM integral equations for the correlation functions and then the solvation thermodynamics analytically as a single integral of a closed form in terms of the correlation functions obtained. Its three-dimensional (3D) version, 3D-RISM theory gives the 3D distribution maps of solvent around a solute macromolecule of arbitrary shape.[50–59] An important component enabling utility of 3D-RISM theory for complex systems in solution has been the closure relation proposed by Kovalenko and Hirata (KH approximation).[54,57,59] For simple and complex solvents and solutions of a given composition, including buffers, salts, polymers, ligands and other cofactors at a finite concentration, the 3D-RISM–KH molecular theory of solvation properly accounts for chemical functionalities by representing in a single formalism both electrostatic and non-polar features of solvation, such as hydrogen bonding, structural solvent molecules, salt bridges, solvophobicity, and other electrochemical, associative and steric effects. For real systems, solving the 3D-RISM–KH integral equations is far less computationally expensive than running molecular simulations which must be long enough to sample relevant exchange and binding events. This enables handling of complex systems and processes occurring over large space and long time scales; these are problematic and frequently not feasible for molecular simulations. The 3D-RISM–KH theory successfully describes both simple and complex associating liquids with different chemical functionalities,[59–67] including ionic liquids[63] and polyelectrolyte gels,[64] in a range of fluid thermodynamic conditions,[34,65–67] and in various local environments such as interfaces with metal,[54,57] metal oxide,[68] zeolite,[58,69] clay,[70] and in confinement of carbon nanotubes,[61] synthetic organic rosette nanotubes,[59,71–73] and biomolecular systems.[59,73,74–88] The latter range from structural water, xenon, and ions bound to the lysozyme protein,[74,75] selective binding and permeation of water, ions, and protons in channels,[59,64,75–78] salt-induced conformational transitions of DNA,[78–80] aggregation of Aβ oligomers and fibrils and prion proteins,[82–84] binding modes of inhibitors of pathologic conversion and aggregation of prion proteins,[64,84] binding modes of thiamine against extracytoplasmic thiamine binding lipoprotein MG2[85,86,89] binding modes and ligand efflux pathways in multidrug transporter AcrB,[87] to biomolecular systems as large as the *Gloebacter violaceus* pentameric ligand-gated ion channel (GLIC) homologue in a lipid bilayer[59,64,85] and the GroEL/ES chaperonin complex.[88] The 3D-RISM–KH theory provided an insight into soft matter phenomena such as the structure and stability of gels formed by oligomeric polyelectrolyte gelators in different solvents.[64]

22.4.1 Integral Equations for the 3D Solvation Structure

The solvation structure is represented by the probability density $\rho_\gamma^{\mathrm{v}} g_\gamma^{\mathrm{uv}}(\mathbf{r})$ of finding interaction site γ of solvent molecules at 3D space position $\mathbf{r}$ around the solute molecule (which can be both a macromolecule and supramolecule), as given by the average number density ρ_γ^{v} in the solution bulk times the normalized density distribution, or 3D distribution function, $g_\gamma^{\mathrm{uv}}(\mathbf{r})$. The values of $g_\gamma^{\mathrm{uv}}(\mathbf{r})>1$ and $g_\gamma^{\mathrm{uv}}(\mathbf{r})<1$ indicate areas of density enhancement or depletion, respectively, relative to the average density at a distance from the solute in the solution bulk where $g_\gamma^{\mathrm{uv}}(\mathbf{r})\rightarrow 1$. The 3D distribution functions of solvent interaction sites around the solute molecule are determined from the 3D-RISM integral equation:[50–59]

$$h_\gamma^{\mathrm{uv}}(\mathbf{r}) = \sum_\alpha \int \mathrm{d}\mathbf{r}' c_\alpha^{\mathrm{uv}}(\mathbf{r}-\mathbf{r}')\chi_{\alpha\gamma}^{\mathrm{vv}}(r'), \tag{22.10}$$

where $h_\gamma^{\mathrm{uv}}(\mathbf{r})$ and $c_\gamma^{\mathrm{uv}}(\mathbf{r})$ are the 3D total and direct correlation functions of solvent site γ, respectively; $\chi_{\alpha\gamma}^{\mathrm{vv}}(r)$ is the site-site susceptibility of pure solvent which is an input to the 3D-RISM theory; and indices α and γ enumerate all sites on all sorts of solvent species. The diagrammatic analysis relates the distribution function to the total correlation function as:[33]

$$g_\gamma^{\mathrm{uv}}(\mathbf{r}) = h_\gamma^{\mathrm{uv}}(\mathbf{r}) + 1, \tag{22.11}$$

and so the latter has the meaning of the normalized probability density of spatial correlations between the solute and solvent molecules. As follows from the diagrammatic expansion of the direct correlation function,[33] the leading term of its asymptotes beyond the short-range region $D^{(\mathrm{sr})}$ of the solute–solvent repulsive core and attractive well (first solvation shell maximum):

$$c_\gamma^{\mathrm{uv}}(\mathbf{r}) \approx -u_\gamma^{\mathrm{uv}}(\mathbf{r})\Big/(k_\mathrm{B}T) \quad \text{for} \quad \mathbf{r}\notin D^{(\mathrm{sr})}, \tag{22.12}$$

is given by the interaction potential $u_\gamma^{\mathrm{uv}}(\mathbf{r})$ between the solute molecule and solvent interaction site γ, scaled by the Boltzmann factor $k_\mathrm{B}T$. Inside the repulsive core, the direct correlation function strongly deviates from the asymptotes [Equation (22.12)] and assumes the values related to the solvation Gibbs energy of the solute molecule immersed in the solvent.

22.4.2 Closures to the Integral Equations

To obtain the correlation functions, the 3D-RISM integral Equation (22.10) has to be complemented with another relation between the two functions $h_\gamma^{\mathrm{uv}}(\mathbf{r})$ and $c_\gamma^{\mathrm{uv}}(\mathbf{r})$ which is called a closure and also involves the 3D solute–solvent site interaction potential $u_\gamma^{\mathrm{uv}}(\mathbf{r})$ specified with a molecular force field. The exact closure has a non-local functional form which can be expressed as an infinite diagrammatic series in terms of multiple integrals of the total correlation function.[33] However, these diagrams are cumbersome and the series suffers from the usual drawback of poor convergence, which makes it impossible to truncate the series at reasonably low-order diagrams and thus renders the exact closure computationally intractable. Therefore, it is replaced in practice with amenable approximations with analytical features that properly represent physical characteristics of the system, such as the long-range asymptotes of the correlation functions related to electrostatic forces and their short-range features related to the solvation structure and to thermodynamics. Examples of closure relations to Ornstein–Zernike type integral equations (including RISM) suitable for liquids of polar and charged species are the so-called hypernetted chain (HNC) closure and the mean spherical approximation (MSA),[33] since they both enforce the long-range asymptotes like Equation (22.12), which is critical for systems with charges. Such other well-known approximations as the Percus–Yevick, Modified Verlet, Martynov–Sarkisov, and Ballone–Pastore–Galli–Gazzillo; closures[33] reproduce solvation features for species with short-range repulsion but do not properly account for electrostatics in the interaction potential. By generalization, the 3D-HNC and 3D-MSA closures to the 3D-RISM integral Equation (22.10) can be constructed.[50–53,57] In particular, the 3D-HNC closure has the form:

$$g_\gamma^{\mathrm{uv}}(\mathbf{r}) = \exp\left(-u_\gamma^{\mathrm{uv}}(\mathbf{r})\Big/(k_\mathrm{B}T) + h_\gamma^{\mathrm{uv}}(\mathbf{r}) - c_\gamma^{\mathrm{uv}}(\mathbf{r})\right). \tag{22.13}$$

While enforcing the asymptotes [Equation (22.12)], the 3D-HNC closure [Equation (22.13)] strongly overestimates association effects and therefore the 3D-RISM–HNC equations diverge for macromolecules with considerable site charges solvated in polar solvents or electrolyte solutions, which is almost always the case for biomolecules in aqueous solution. The 3D-MSA closure is free from that shortcoming but suffers from another serious deficiency of producing nonphysical negative values of the distribution function in the areas adjacent to the associative peaks. Those drawbacks are overcome with the approximation proposed by Kovalenko and Hirata (KH closure), the 3D version of which reads:[54,57,59]

$$g_\gamma^{\mathrm{uv}}(\mathbf{r}) = \begin{cases} \exp\left(-u_\gamma^{\mathrm{uv}}(\mathbf{r})\Big/(k_\mathrm{B}T) + h_\gamma^{\mathrm{uv}}(\mathbf{r}) - c_\gamma^{\mathrm{uv}}(\mathbf{r})\right) & \text{for} \quad g_\gamma^{\mathrm{uv}}(\mathbf{r}) \le 1 \\ 1 - u_\gamma^{\mathrm{uv}}(\mathbf{r})\Big/(k_\mathrm{B}T) + h_\gamma^{\mathrm{uv}}(\mathbf{r}) - c_\gamma^{\mathrm{uv}}(\mathbf{r}) & \text{for} \quad g_\gamma^{\mathrm{uv}}(\mathbf{r}) > 1 \end{cases}. \tag{22.14}$$

The 3D-KH closure relation [Equation (22.14)] couples, in a nontrivial way, the HNC approximation and the MSA linearization; the latter automatically applied to spatial regions of solvent density enrichment $g_\gamma^{\mathrm{uv}}(\mathbf{r})>1$ such as strong peaks of association and long-range tails of near-critical fluid phases, and the former to spatial regions of density depletion $g_\gamma^{\mathrm{uv}}(\mathbf{r})<1$ (including the repulsive core), while keeping the right asymptotes [Equation (22.12)] peculiar to both HNC and MSA. The distribution function and its first derivative are continuous at the joint boundary $g_\gamma^{\mathrm{uv}}(\mathbf{r})=1$ by construct. The KH approximation consistently accounts for both electrostatic and non-polar (associative and steric) effects of solvation in simple and complex liquids, non-electrolyte and electrolyte solutions, and complex macromolecular and supramolecular solutes in various chemical,[54–69] soft matter,[64] synthetic organic,[59,71–73] and biomolecular[59,74–88] systems.

The 3D-KH closure underestimates the height of strong associative peaks of the 3D site distribution functions because of the MSA linearization applied to them.[54,89] However, it somewhat widens the peaks and so 3D-RISM–KH quite accurately reproduces the coordination numbers of the solvation structure in different systems, including micromicelles in water–alcohol solutions,[59,67] solvation shells of metal–water,[54,57] metal oxide–water,[68] and mixed organic solvent–clay[70] interfaces, and structural water solvent localized in biomolecular confinement.[75,88] For example, the coordination numbers of water strongly bound to the MgO surface are calculated from the 3D-RISM–KH theory with a 90% accuracy and the peak positions within a 0.05 nm deviation, compared to MD simulations.[68] The 3D solvation map $g_\gamma^{uv}(\mathbf{r})$ of function-related structural water in the GroEL chaperonin complex (shown to be strongly correlated to the rate of protein folding inside the chaperonin cavity) obtained in an expensive MD simulation with explicit solvent involving ~1 million atoms is reproduced from the 3D-RISM–KH theory in a relatively short calculation on a workstation with an accuracy of over 90% correlation for the 3D density map and about 98% correlation for the 3D density maxima.[88]

Note that different approximate 3D bridge functions, or bridge corrections, to the 3D-HNC approximation can be constructed, such as the 3D-HNCB closure which reproduced MD simulation results for the height of the 3D distribution peaks of water solvent around the bovine pancreatic trypsin inhibitor (BPTI) protein.[89] A related promising approach is the partial series expansions of order *n* (PSE-*n*) of the HNC closure.[90] The PSE-*n* closures interpolate between the KH and HNC approximations and therefore combine numerical stability with enhanced accuracy, as demonstrated in aqueous solutions of alkali halide ions.[90,91] However, a particular appeal of the 3D-KH closure is that it provides adequate accuracy and existence of solutions if expected from physical considerations for a wide class of solution systems "from the wild", including various chemical and biomolecular solutes in solution with different solvents, co-solvents, and electrolytes,[54–88] and moreover, with ligand molecules or fragments considered as part of solvent for efficient 3D mapping of binding affinities in problems such as molecular recognition and fragment based drug design.[75,86,87] Construction of bridge functions for such systems with different solvent components other than water constitutes a significant challenge not addressed so far, and the 3D-KH closure constitutes a reliable choice for a given system.

22.4.3 Integral Equation for Site–Site Radial Correlation Functions of the Solvent System

The site–site susceptibility of the solvent system breaks up into the intra- and intermolecular terms:

$$\chi_{\alpha\gamma}^{vv}(r) = \omega_{\alpha\gamma}^{vv}(r) + \rho_\alpha^v h_{\alpha\gamma}^{vv}(r), \tag{22.15}$$

where the intramolecular correlation function $\omega_{\alpha\gamma}^{vv}(r)$ normalized as $\int d\mathbf{r}\, \omega_{\alpha\gamma}^{vv}(r) = 1$ represents the geometry of solvent molecules. [$\omega_{\alpha\gamma}^{vv}(r) = 0$ for sites α and γ on different species.] For rigid species with site separations $l_{\alpha\gamma}^{vv}$ it has the form $\omega_{\alpha\gamma}^{vv}(r) = \delta(r - l_{\alpha\gamma}^{vv})/(4\pi(l_{\alpha\gamma}^{vv})^2)$ specified in the reciprocal k-space as:

$$\omega_{\alpha\gamma}^{vv}(r) = j_0(kl_{\alpha\gamma}^{vv}), \tag{22.16}$$

and $h_{\alpha\gamma}^{vv}(r)$ is the intermolecular, radial total correlation function between sites α and γ enumerating all sites on all sorts of molecules in bulk solvent. The site–site total correlation functions $h_{\alpha\gamma}^{vv}(r)$ used Equation (22.15) and then in Equation (22.10) are obtained in advance of the 3D-RISM calculation from the dielectrically consistent RISM theory[92] using the HNC closure[33] or otherwise coupled with the KH closure (DRISM–KH approach).[57,59] The latter can be applied to the bulk solution of a given composition, including polar solvent, co-solvent, electrolyte, and ligands at a given concentration. The DRISM integral equation reads:[90]

$$\tilde{h}_{\alpha\gamma}^{vv}(r) = \tilde{\omega}_{\alpha\mu}^{vv}(r) * c_{\mu\nu}^{vv}(r) * \tilde{\omega}_{\nu\gamma}^{vv}(r) + \tilde{\omega}_{\alpha\mu}^{vv}(r) * c_{\mu\nu}^{vv}(r) * \rho_{\nu}^{v} \tilde{h}_{\nu\gamma}^{vv}(r), \tag{22.17a}$$

where $c_{\alpha\gamma}^{vv}(r)$ is the site–site direct correlation function of the bulk solvent system, and both the intramolecular correlation function $\tilde{\omega}_{\alpha\gamma}^{vv}(r)$ and the total correlation function $\tilde{h}_{\alpha\gamma}^{vv}(r)$ are renormalized due to a dielectric bridge correction in a particular analytical form[90] that ensures the given phenomenological value as well as consistency of the dielectric constant determined through the three different routes related to the solvent–solvent, solvent–ion, and ion–ion effective interactions in electrolyte solution:

$$\tilde{\omega}_{\alpha\gamma}^{vv}(r) = \omega_{\alpha\gamma}^{vv}(r) + \rho_{\alpha}^{v} \chi_{\alpha\gamma}^{vv}(r), \tag{22.17b}$$

$$\tilde{h}_{\alpha\gamma}^{vv}(r) = h_{\alpha\gamma}^{vv}(r) - \chi_{\alpha\gamma}^{vv}(r). \tag{22.17c}$$

The renormalized dielectric correction, enforcing the given phenomenological value of the dielectric constant and the proper orientational behavior and consistency of the dielectric response in the electrolyte solution, is obtained in the analytical form specified in the reciprocal *k*-space as follows:[90]

$$\chi_{\alpha\gamma}^{vv}(k) = j_0(kx_\alpha) j_0(ky_\alpha) j_1(kz_\alpha) h_c(k) j_0(kx_\gamma) j_0(ky_\gamma) j_1(kz_\gamma), \tag{22.18}$$

where $j_0(x)$ and $j_1(x)$ are the zeroth-order and first-order spherical Bessel functions, $\mathbf{r}_\alpha = (x_\alpha, y_\alpha, z_\alpha)$ are the Cartesian coordinates of partial charge q_α of site α on species s with respect to its molecular origin, both sites α and γ are on the same species s, and its dipole moment $\mathbf{d}_s = \sum_{\alpha \in s} q_\alpha \mathbf{r}_\alpha$ is oriented along the *z*-axis, $\mathbf{r}_\alpha = (0,0,d_s)$. Note that the renormalized dielectric correction given by Equation (22.18) is nonzero only for polar solvent species of electrolyte solution which possess a dipole moment and thus are responsible for the dielectric response in the DRISM approach. The envelope function $h_c(k)$ has the value at $k = 0$ determining the dielectric constant of the solution and is assumed to be in a smooth non-oscillatory form that quickly falls off at wave vectors k larger than the inverse characteristic size l of liquid molecules:[90]

$$h_c(k) = A\exp\left(-l^2k^2/4\right),$$

where $A = \rho_{\mathrm{polar}}^{-1}\,(\varepsilon/y - 3)$ is the amplitude related to the dielectric constant ε of the electrolyte solution. Equation (22.18) is extended to mixed solvents[93] with the total number density of polar species in solution:

$$\rho_{\mathrm{polar}} = \sum_{s\in\mathrm{polar}} \rho_s$$

and the solution dielectric susceptibility:

$$y = \frac{4\pi}{9k_\mathrm{B}T}\sum_{s\in\mathrm{polar}} \rho_s (d_s)^2.$$

The parameter l specifies the characteristic separation from a liquid molecule below which the dielectric correction [Equation (22.18)] is switched off so as not to distort the short-range solvation structure. It can be chosen to be about $l = 1$ Å for water as a solvent; however, in solvents of larger molecules, such as octanol, or in the presence of such co-solvents it should be increased to about $l = 1.0$ nm so as to avoid "ghost" associative peaks appearing in the radial distributions if the dielectric correction [Equation (22.18)] interferes with the intramolecular structure of the large solvent species.

The HNC closure to the DRISM integral Equation (22.17) reads:[33]

$$g_{\alpha\gamma}^{\mathrm{vv}}(r) = \exp\left(-u_{\alpha\gamma}^{\mathrm{vv}}(r)\Big/(k_\mathrm{B}T) + h_{\alpha\gamma}^{\mathrm{vv}}(r) - c_{\alpha\gamma}^{\mathrm{vv}}(r)\right), \tag{22.19}$$

The DRISM integral Equation (22.17) with the KH closure:[57,59]

$$g_{\alpha\gamma}^{\mathrm{vv}}(r) = \begin{cases} \exp\left(-u_{\alpha\gamma}^{\mathrm{vv}}(r)\Big/(k_\mathrm{B}T) + h_{\alpha\gamma}^{\mathrm{vv}}(r) - c_{\alpha\gamma}^{\mathrm{vv}}(r)\right) & \text{for} \quad g_{\alpha\gamma}^{\mathrm{vv}}(r) \le 1 \\ 1 - u_{\alpha\gamma}^{\mathrm{vv}}(r)\Big/(k_\mathrm{B}T) + h_{\alpha\gamma}^{\mathrm{vv}}(r) - c_{\alpha\gamma}^{\mathrm{vv}}(r) & \text{for} \quad g_{\alpha\gamma}^{\mathrm{vv}}(r) > 1 \end{cases} \tag{22.20}$$

retains the same dielectrically consistent asymptotes [Equation (22.18)] as the originally derived DRISM–HNC theory[90] but extends the description to solutions with strong associative species in a wide range of composition and thermodynamic conditions not amenable to the HNC closure due to its overestimation of associative forces and phase transition phenomena.

22.4.4 Analytical Expressions for the Solvation Thermodynamics

Further to the analytical expressions for the PMV and compressibility, Equations (22.7)–(22.9), RISM theory with both the HNC and KH closures yields analytical expressions for the solvation Gibbs energy. Much as the HNC approximation,[33] the 3D-KH closure (Equation 22.14) to the 3D-RISM integral Equation (22.10) has an exact differential of the solvation Gibbs energy,[34] and allows one to analytically perform Kirkwood's thermodynamic integration gradually switching on the solute–solvent interaction. This gives the solvation Gibbs energy of the solute macro- or supramolecule in multi-component solvent in a closed analytical form in terms of the 3D site total and direct correlation functions $h_\gamma(\mathbf{r})$ and $c_\gamma(\mathbf{r})$:[54,57,59]

$$\mu^{\mathrm{u}}_{\mathrm{solv}} = \sum_{\gamma} \int_V \mathrm{d}\mathbf{r}\, \Phi^{\mathrm{uv}}_{\gamma}(\mathbf{r}), \tag{22.21a}$$

$$\Phi^{\mathrm{uv}}_{\gamma}(\mathbf{r}) = \rho^{\mathrm{v}}_{\gamma} k_{\mathrm{B}} T \left[\frac{1}{2} (h^{\mathrm{uv}}_{\gamma}(\mathbf{r}))^2 \Theta(-h^{\mathrm{uv}}_{\gamma}(\mathbf{r})) - c^{\mathrm{uv}}_{\gamma}(\mathbf{r}) - \frac{1}{2} h^{\mathrm{uv}}_{\gamma}(\mathbf{r}) c^{\mathrm{uv}}_{\gamma}(\mathbf{r}) \right], \tag{22.21b}$$

where $\Theta(x)$ is the Heaviside step function. The integrand $\Phi^{\mathrm{uv}}_{\gamma}(\mathbf{r})$ in Equation (22.21a) is interpreted as the solvation Gibbs energy density (3D-SFED) coming from interaction site γ of solvent molecules around the solute. The solvation Gibbs energy of the solute macromolecule $\mu^{\mathrm{u}}_{\mathrm{solv}}$ is obtained by summation of the 3D-SFED partial contributions from all solvent species and spatial integration over the whole space. (Note that the integration volume V in the form of Equation (22.21) comprises both the solvation shells and the solute–solvent molecular repulsive cores, since the direct correlation functions $c^{\mathrm{uv}}_{\gamma}(\mathbf{r})$ inside the latter are related to the free energy of creation of a cavity to accommodate the solute excluded volume.) In this way, $\Phi^{\mathrm{uv}}_{\gamma}(\mathbf{r})$ characterizes the intensity of effective solvation forces in different 3D spatial regions of the solvation shells and indicates where they contribute the most/least to the entire solvation Gibbs energy. The solvation Gibbs energy [Equation (22.21a)] and 3D-SFED [Equation (22.21b)] can be further decomposed into partial contributions of interaction sites and/or chemical groups of the solute macro- or supramolecule.[94,95]

The solvation Gibbs energy [Equation (22.21)] can be split up into the energetic and entropic contributions:

$$\mu^{\mathrm{u}}_{\mathrm{solv}} = \varepsilon^{\mathrm{uv}} + \varepsilon^{\mathrm{vv}} - Ts_{\mathrm{V}}, \tag{22.22}$$

by calculating the solvation entropy at constant volume as:

$$s_{\mathrm{V}} = -\frac{1}{T} \left(\frac{\partial \mu^{\mathrm{u}}_{\mathrm{solv}}}{\partial T} \right)_{\mathrm{V}}, \tag{22.23}$$

the internal energy of the solute ("u")–solvent ("v") interaction as:

$$\varepsilon^{\mathrm{uv}} = k_{\mathrm{B}} T \sum_{\gamma} \rho^{\mathrm{v}}_{\gamma} \int \mathrm{d}\mathbf{r}\, g^{\mathrm{uv}}_{\gamma}(\mathbf{r})\, u^{\mathrm{uv}}_{\gamma}(\mathbf{r}), \tag{22.24}$$

and the remaining term $\varepsilon^{\mathrm{vv}}$ giving the energy of solvent reorganization around the solute molecule.

22.4.5 Analytical Treatment of the Electrostatic Asymptotes

To treat electrostatic forces properly in electrolyte solution with a polar molecular solvent and ionic species in 3D-RISM/DRISM theory, we analytically handle the electrostatic asymptotes of the radial site–site total and direct correlation functions of the bulk solvent in the DRISM–HNC or DRISM–KH integral equations, Equations (22.17), (22.19) or (22.20), and the 3D site total and direct correlation functions in the 3D-RISM–HNC or 3D-RISM–KH integral equations, Equations (22.10), (22.13) or (22.14), as well as in the expressions for the solvation Gibbs energy [Equation (22.21)] and its derivatives [Equations (22.22)–(22.24)].[55–59,62] The spatial convolution in the 3D-RISM integral Equation (22.10) is calculated by using the 3D fast Fourier transform (3D-FFT) technique. The DRISM integral equations are made discrete on a uniform radial grid with resolution of 0.001–0.01 nm, and the 3D-RISM integral equations are made discrete on a uniform 3D rectangular grid with a resolution of 0.02–0.05 nm in a 3D box of a size which includes 2–3 solvation shells around the solute macromolecule (supramolecule). The analytical forms for the non-periodic electrostatic asymptotes in the direct and reciprocal space are separated out from all the correlation functions before, and then added back after, doing the 3D-FFT transform. Note that, although the solvent susceptibility $\chi^{vv}_{\alpha\gamma}(r)$ has a long-range electrostatic part, no aliasing occurs in the backward 3D-FFT of the short-range part of $h^{uv}_{\gamma}(\mathbf{k})$ on the 3D box super cell since the short-range part of the convolution product $h^{uv}_{\gamma}(\mathbf{k})$ contains merely 2–3 oscillations (solvation shells) and vanishes outside the 3D box for physical reasons.[58,59] Accordingly, the electrostatic asymptote terms in the thermodynamic integral (Equation 22.21) are handled analytically and reduced to one-dimensional integrals, which are easy to compute.[55–59,62] Note that in the RISM and 3D-RISM expressions for the PMV $\bar{V}^{u}$, Equations (22.7) and (22.9), and for the solvent system compressibility χ^{v}_{T}, Equation (22.8), the long-range electrostatic asymptotes in both $c^{uv}_{\gamma}(\mathbf{k}=0)$ and $c^{uv}_{\alpha\gamma}(\mathrm{k}=0)$ cancel out upon summation over all interaction sites of all species of the solvent system (without the protein and counterion(s) at infinite dilution) due to its global electroneutrality.

22.4.6 Accelerated Numerical Solution

The 3D-RISM integral equations typically converge to a root mean square accuracy of 10^{-5}–10^{-8} and the DRISM integral equations to an accuracy of 10^{-8}–10^{-12} by using the modified direct inversion in the iterative subspace (MDIIS) accelerated numerical solver.[55–58,96] MDIIS is an iterative procedure accelerating convergence of integral equations of liquid state theory by optimizing each iterative guess in a Krylov subspace of typically 10–20 successive iterations and then making the next iterative guess by mixing the optimized solution with the approximated optimized residual.[55,57,59,96] It is closely related to Pulay's DIIS approach for quantum chemistry equations[97] and other similar algorithms like the generalized minimal residual (GMRes) solver.[98] The MDIIS solver combines the simplicity and relatively small memory usage of an iterative approach with the efficiency of a direct method. Compared to damped (Picard) iterations, MDIIS provides substantial acceleration with quasi-quadratic convergence throughout practically the entire range of root mean square residual values, and is robust and stable. Of particular importance is that the MDIIS method ensures convergence (provided a solution exists) for complex charged systems with strong associative and steric effects, which is usually not achievable by Picard iterations and constitutes a challenging task in the case of 3D integral equations on large 3D grids.

22.5 Predictions of the Molecular Theory of Solvation for PMV and Pressure-Induced Structural Transitions of Proteins

Examples are given below, illustrating the accuracy of the 3D-RISM–KB method in predicting PMV and revealing molecular mechanisms of pressure denaturation of proteins.

22.5.1 PMV of Amino Acids from 3D-RISM/RISM *Versus* Experimental Data

Table 22.1 makes a comparison of the PMV predicted from the 3D-RISM–HNC[45] and RISM–HNC[43] theory to the experimental values[99–102] for the 20 amino acids in the zwitterionic form, classified into five categories in the conventional manner: aliphatic, non-polar, aromatic, polar, and charged. As is evident, compared to RISM–KB, the predictions from the 3D-RISM–KB method give a much better agreement with the experimental data. The comparison is also visualized in the theory-*versus*-experiment plot of Figure 22.1. The 3D-RISM–KB results show a remarkable correlation with the experiment, while the RISM-KB trend systematically underestimates the PMV. The most conspicuous deviation from the experiment is seen in the amino acids with larger residues. For the largest residue size, the 3D-RISM–KB approach somewhat underestimates the PMV; and, as noted in Section 22.3, this can be attributed to an "ideal fluctuation volume" originating from the flexibility of the larger amino acids.[43]

22.5.2 PMV Changes Associated with the Helix–Coil Transition of an Alanine-rich Peptide AK16 in Aqueous Solution

Takekiyo *et al.*[103,104] estimated the volume change associated with the helix-to-coil transition of an alanine-rich oligopeptide Ac–YGAAKAAAAKAAAAKA–NH_2, so-called AK16 peptide,[105] from the pressure dependence of infrared

Table 22.1 Partial molar volume ($cm^3\ mol^{-1}$) of the 20 amino acids in zwitterionic form in aqueous solution.

	Molar mass (g mol^{-1})	RISM–HNC[43]	3D-RISM–HNC[45]	Experiment
Aliphatic				
Ala	89.10	50.3	63.9	60.5[99]
Val	117.15	63.8	90.6	90.8[99]
Leu	131.17	73.7	105.5	107.8[99]
Ile	131.17	72.0	104.6	105.7[99]
Non-polar				
Gly	75.07	40.4	48.8	43.2[99]
Pro	115.13	61.5	84.7	82.5[99]
Cys	121.16	59.2	74.3	73.4[100]
Met	149.21	82.7	106.9	104.8[99]
Aromatic				
His	155.16	71.1	94.9	98.9[101]
Phe	165.25	84.0	111.8	121.5[99]
Tyr	181.24	84.7	113.1	124.3[99]
Trp	204.27	90.5	123.9	143.4[99]
Polar				
Asn	132.12	58.2	76.8	77.3[101]
Gln	146.15	69.2	93.5	93.6[101]
Ser	105.10	52.6	66.5	60.6[102]
Thr	119.12	59.7	81.9	76.9[102]
Asn	132.12	58.2	76.8	77.3[101]
Charged				
Lys	147.20	83.4	115.5	108.5[100]
Arg	175.23	84.8	123.6	127.3[100]
Asp	132.10	58.9	69.0	73.8[100]
Glu	146.12	68.1	82.9	85.9[100]

(IR) band intensities. It is about +1 $cm^3\ mol^{-1}$ per residue, which is in fact different from the values for polypeptides. What is the dominant factor in the volume change associated with the helix–coil transition?

The PMV values for the α-helix and coil structure of AK16, along with the volume components according to Equation (22.2) calculated by the 3D-RISM–KB theory, are shown in Table 22.2.[48] The PMV values and the standard deviation for the coil structure are the average for 64 random coil structures generated in MD simulation. Figure 22.2 shows the Ramachandran plot with the distributions of φ,ψ angles of the 16 residues of the 64 random coil structures employed. The volume changes associated with the helix-to-coil transition, determined as the difference of the average values, are also listed. The theoretical result demonstrates that the PMV increases by 13.0 $cm^3\ mol^{-1}$ accompanying the conformational transition; this agrees well with the experimental data (refined by using the integrated intensity instead of the absorbance of IR spectra), 14.9 ± 0.5 $cm^3\ mol^{-1}$.[103,104]

In the structure transition, V_{id} does not change by definition, and V_W is also unchanged because the van der Waals spheres cannot overlap with each

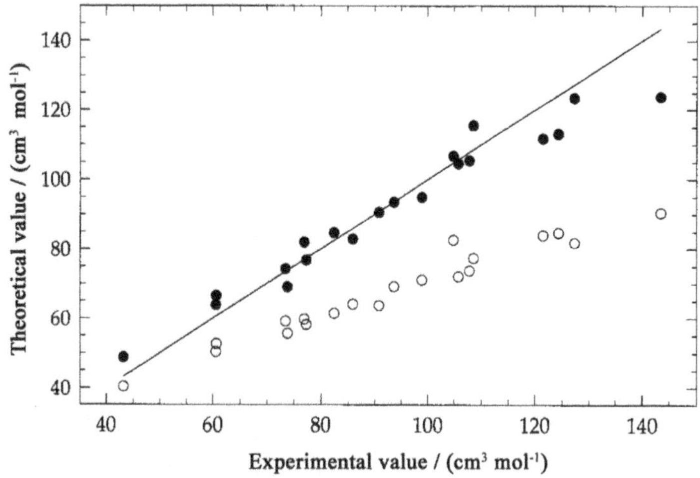

Figure 22.1 Partial molar volume of the 20 amino acids obtained from the 3D-RISM–KH and RISM–KH theory (filled and open circles, respectively) plotted against experimental data. The straight line indicates ideal match.

Table 22.2 Partial molar volume $\Delta\overline{V}$ (cm^3 mol^{-1}) and its components according to Equation (22.2) for AK16 in aqueous solution at $T = 298.15$ K.

Volume/cm^3 mol^{-1}	$\overline{V}$	V_{id}	V_W	V_V	V_T	V_I
α-Helix	1017.2	1.3	645.1	159.9	203.8	7.1
Coil						
Average	1030.2	1.3	645.5	187.5	198.4	− 2.4
Change in the helix-to-coil transition	13.0	0.0	0.4	27.5	− 5.4	− 9.6
Standard deviation	7.8	0.0	0.2	18.8	14.3	4.4

other under the present condition. The component V_V is the only contribution that increases, while V_T and V_I decrease in the helix-to-extended structure transition. On average, the increasing term of $\overline{V}$ is primarily due to the growth of void space in the peptide, which is canceled in part by the PMV reduction due to the increase in the peptide–water interaction. The change in V_T is less than in V_V and V_I.

It might seem strange that ΔV_V is positive because the peptide loses the void space with helix unfolding. In fact, ΔV_V takes a negative value when the peptide is completely extended.[46] (See also Table 22.3.) However, a coil is not always extended. In Figure 22.2, the dihedral angles take various values corresponding not only to extended conformation but also to helix, sheet, turn, and the other metastable conformations. Besides, the average radius of gyration of coils, 0.75 nm, is close to that of the helix, 8.4 Å; for the fully extended structure the distance is 1.80 nm. When forming a coil structure, the peptide can gain additional voids created among the main chain and side chains while losing the voids within the helical main chain.

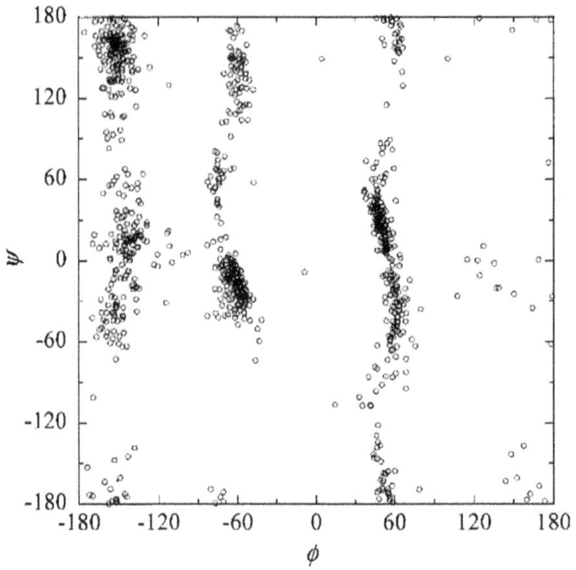

Figure 22.2 The φ,ψ map for the 64 random coil structures of AK16.

Table 22.3 Partial molar volume $\Delta\overline{V}$ ($cm^3\ mol^{-1}$) and its components for the extended structure of AK16 in aqueous solution at $T = 298.15$ K. Change in the partial molar volume $\Delta\overline{V}$ and in its components associated with the helix-to-extended structure transition of AK16.

Volume/$cm^3\ mol^{-1}$	$\overline{V}$	V_{id}	V_W	V_V	V_T	V_I
$\overline{V}$ of extended conformation	1013.4	1.3	645.1	105.0	270.8	− 8.9
Change $\Delta\overline{V}$ in helix-to-extended structure transition	− 3.9	0.0	0.0	− 54.9	67.1	− 16.0

It should be noted that the standard deviations are very large compared to the average volume changes. Figure 22.3 shows the histograms for the components of the volume change. In the case of ΔV_T the standard deviation exceeds the average, which means that ΔV_T is negative for some of the coils, and positive for the others. Applying pressure destabilizes most of the coil structures, but stabilizes the other few coil structures. A similar conclusion is true for the volume components. Although ΔV_I is relatively limited to within a narrow volume range, the volumes ΔV_V and ΔV_T vary widely and can take either sign. Therefore, the above explanation for the positive $\Delta\overline{V}$ being due to the growth of void space, and being cancelled in part by the volume reduction due to the intermolecular interaction, has some exceptions when it is applied to a particular coil structure.

Figure 22.4 shows the histogram for the sum of the volume changes $\Delta V_V + \Delta V_T$ as well as the correlation between these terms, which were reported to tend to cancel each other. The distribution becomes apparently narrower, and $\Delta V_V + \Delta V_T$ can no longer take negative values. As seen, there is

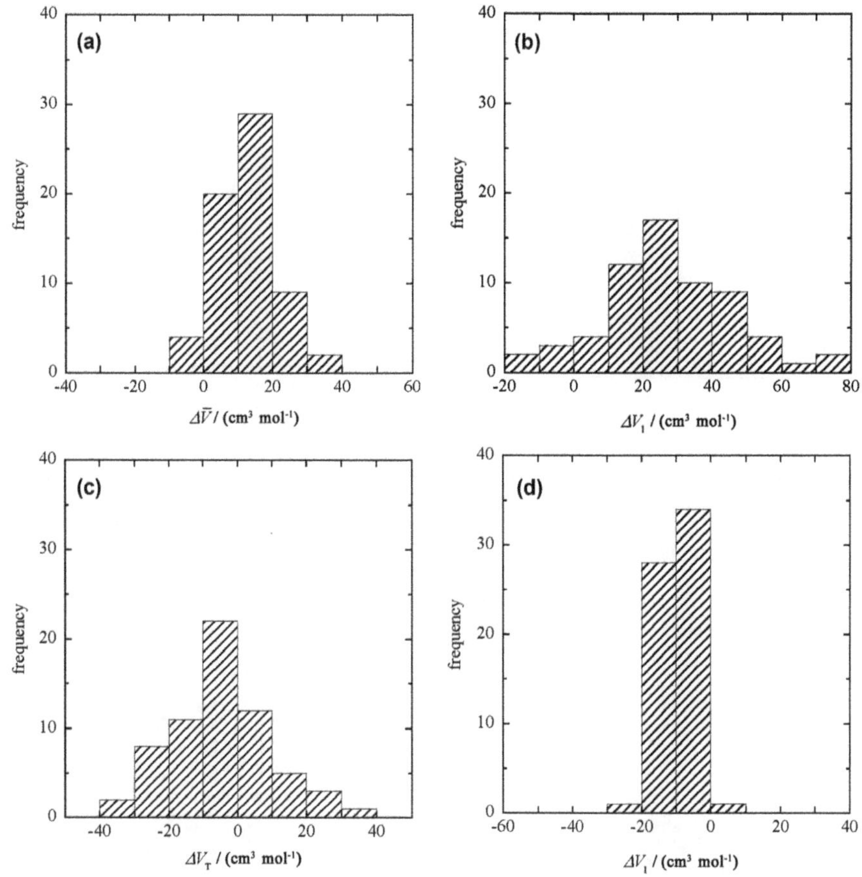

Figure 22.3 Histograms of the volume changes associated with the helix-to-coil transition of AK16. (a) Partial molar volume $\Delta\overline{V}$, (b) void volume ΔV_V, (c) thermal volume ΔV_T, and (d) interaction volume ΔV_I.

an anti-correlation between $\Delta V_V + \Delta V_T$. The compensation can be explained as follows. As ΔV_V is defined by the void space within the solute molecule, ΔV_T can be related by primitive consideration to a kind of void space between the solute and solvent molecules created by imperfect packing of the solvent molecules.[106,107] When a conformational change creates an additional void space within the solute molecule, some solvent surface layer is involved in the internal void space. Thus, the void space between the solute and solvent molecules is converted into the void space within the solute molecule. The reverse also occurs. However, the competition is not even, and $\Delta V_V + \Delta V_T$ become larger with an increase in ΔV_V which dominates.

It was demonstrated that the volume components related to hydration, ΔV_T and ΔV_I, have apparent correlation with solvent accessible surface area (ASA).[47] Figure 22.5 contains plots of these correlations for AK16. The V_T term is directly proportional to the ASA, with coefficient 0.0210 nm.

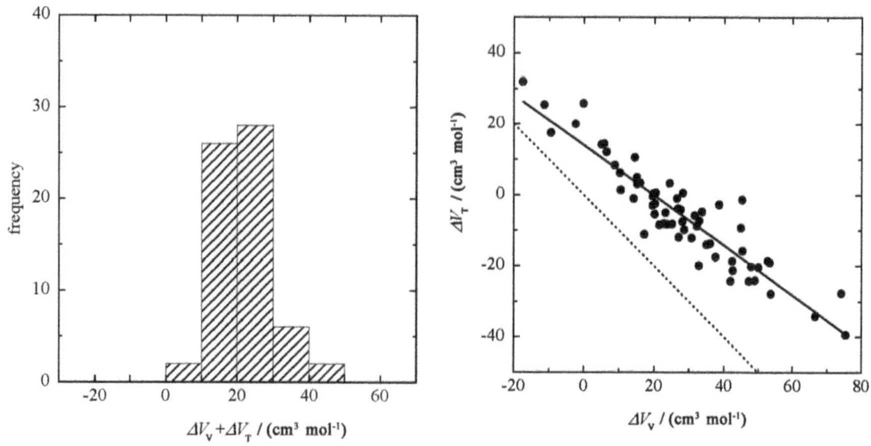

Figure 22.4 Histogram of the sum of the void and thermal volume changes $\Delta V_V + \Delta V_T$ (left panel), associated with the helix-to-coil transition of AK16. Correlation between ΔV_V and ΔV_T (right panel). Dotted line indicates $\Delta V_V + \Delta V_T = 0$.

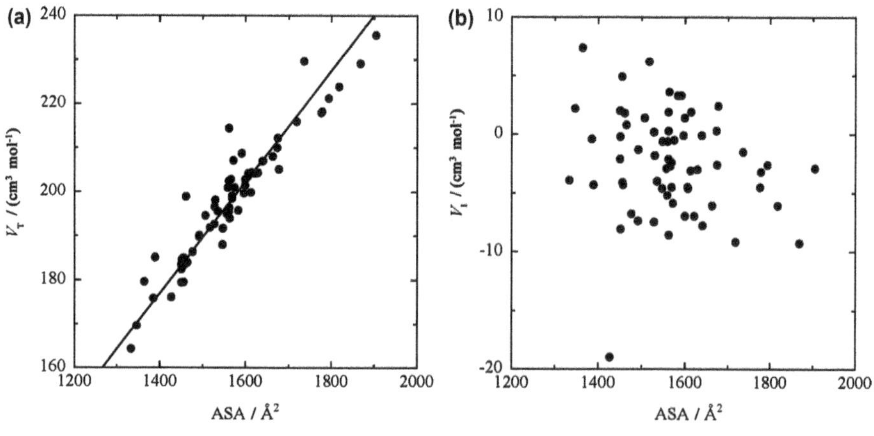

Figure 22.5 Correlations of the thermal volume ΔV_T and interaction volume ΔV_I of random coil structures to the accessible surface area (ASA). Solid line is least-squares fit.

The latter is very close to the value 0.0226 nm obtained for several proteins with different sizes and a globular structure.[47] The coincidence of the results for various structures of single peptide AK16 and for different peptides of similar structure suggests a more general conclusion that ΔV_T is directly proportional to the ASA, regardless of size and structure of the solute biomolecule.

On the other hand, V_I has no apparent correlation with ASA, though negative correlation between V_I and ASA was found for globular species.[47] One might attribute the negative ΔV_I to the increase of the interaction (hydrogen bond) between main-chain peptide groups and water molecules due to their exposure to the aqueous phase. However, V_I is also found to be independent of the ASA of peptide group. Although charged groups are expected to have a great influence on the quantity of V_I, no apparent correlation between V_I and ASA of charged groups (*e.g.* NH_3^+ of lysine) was found. This is because the electrostatic interaction between the peptide and water molecules is modified through more complicated processes other than the change of ASA. For example, a salt bridge (in this case, between peptide group and NH_3^+ group) reduces the electrostriction of water molecules, even when it is exposed to the solvent. When the groups with the same sign are put close to each other, the electrostriction is enhanced. Such effects, naturally included in 3D-RISM calculations, are difficult to explain using primitive ASA analysis.

22.5.3 PMV Changes Associated with the Pressure-Induced Structural Transition of Ubiquitin

The 3D structure of ubiquitin at high pressure (300 MPa) as well as at low pressure (3 MPa), resolved by Akasaka and co-workers,[108] had been one of the first systems studied experimentally at such high pressures. The HPS of ubiquitin at 300 MPa is significantly different in some local structures from the LPS at 3 MPa, although the secondary structures remain the same, as seen in Figure 22.6. Ubiquitin thus presents a good case for study of pressure-induced structural transitions of proteins.

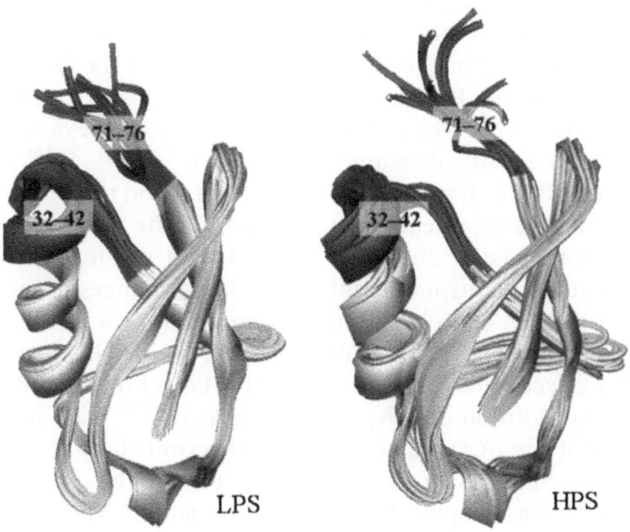

Figure 22.6 Solid ribbon representation of low-pressure (3 MPa) and high-pressure (300 MPa) structures of ubiquitin (LPS and HPS, respectively). Each display consists of 10 superimposed structures included in PDB files 1V80 (LPS) and 1V81 (HPS). The dark gray residues are the segments undergoing significant displacement in the LPS-to-HPS transition.

Table 22.4 Partial molar volume and its components ($cm^3\ mol^{-1}$) for low-pressure (3 MPa) and high-pressure (300 MPa) structures (LPS and HPS, respectively) of ubiquitin in aqueous solution at $T=298.15$ K.

Volume/($cm^3\ mol^{-1}$)	$\overline{V}$	V_W	V_V	V_T	V_I
LPS	5788.4	3768.3	1546.9	645.5	−173.6
HPS	5741.2	3764.7	1445.6	680.5	−150.8
Difference	−47.2	−3.6	−101.3	34.9	22.8

Table 22.4 shows the 3D-RISM–KB results for the PMV and its components of the HPS and LPS of ubiquitin.[109] The total PMV decrease $\Delta\overline{V}$ is primarily caused by the decrease in the void volume ΔV_V which is partially canceled by the increases in the two hydration terms, thermal ΔV_T and interaction ΔV_I. The van der Waals volume change ΔV_W is negligible, as the pressure does not cause any unusual van der Waals overlaps. The decrease in the void volume indicates a partial loss of the structural voids in the protein. The increase in the thermal volume implies the generation of additional empty space around the protein, primarily due to the extension of the protein surface. The increment of the interaction volume typically represents the reduction of the protein–water attractive interactions causing the so-called electrostriction.[41] However, other complicated mechanisms can be involved in the interaction volume change.[43,47]

Which parts in the protein have the most substantial effect on the volume changes? A presumable candidate is the protein segment where the structure is most drastically perturbed by the pressure. Figure 22.7 shows the RMSD per residue between LPS and HPS. It is apparent that the segment of residues 71–76, which is the *N*-terminal "tail" of ubiquitin (see Figure 22.6), undergoes the most significant displacement through the structural transition. Table 22.5 lists the PMV change and its components for the fragments A{71–76} and B{1–70}, as well as the contribution from the overlapping part ΔAB, with the total PMV decomposed as $\overline{V}=\overline{V}^{A}-\overline{V}^{B}-\overline{V}^{\Delta AB}$. The result shows that the volume changes of both fragment A and the overlap part DAB are minor, while the significant part is located in the residual segment B.

As is seen from Figure 22.7, the segment with residues 32–42 has the secondary outstanding displacement. The volume changes of the fragments C{32–42} and D{1–31, 43–76} and the overlap (ΔCD) contributions are given in Table 22.5. It is seen from this analysis that the overlap contribution to the PMV change accounts for the major part (71%) of the total PMV change, although the contribution of the excised segment $\overline{V}^{C}$ itself is minor. The PMV decrease of the ΔCD part is primarily determined by the reduction of the void volume, much as for the total PMV decrease. The thermal volume of the ΔCD part increases upon the structural transition. The interaction volume change contributes less to the PMV change.

The finding that, in the overlap part DCD, the void volume decreases and the thermal volume increases upon the LPS-to-HPS transition implies that solvent water molecules penetrate into the boundary region between the segments C and D through the structural transition. That is because the

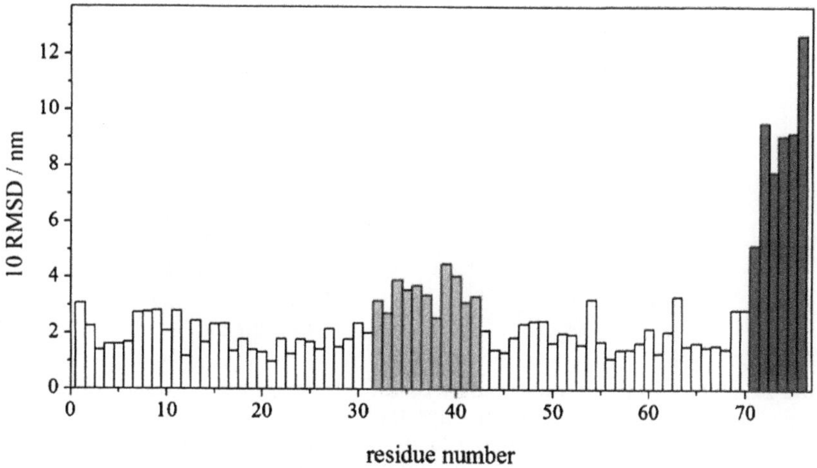

Figure 22.7 Root mean square displacement (RMSD) per residue between low- and high-pressure structures of ubiquitin. Dark gray and light gray indicate the first and second highest RMSD segments, respectively.

Table 22.5 Contributions of ubiquitin fragments and overlapping parts to the partial molar volume change $\Delta\overline{V}$ and its components (cm^3 mol^{-1}), associated with the pressure-induced structural transition of ubiquitin.

Volume/(cm^3 mol^{-1})	$\overline{V}$	V_W	V_V	V_T	V_I
Fragment A{71–76}	− 1.8	0.2	− 4.3	3.5	− 1.1
Fragment B{1–70}	− 37.4	− 3.2	− 95.0	41.5	19.4
Overlaping part ΔAB	− 8.0	− 0.6	− 1.9	− 10.0	4.5
Fragment C{32–42}	6.8	− 0.3	7.5	− 4.3	3.9
Fragment D{1–31, 43–76}	− 20.5	− 2.3	− 48.6	20.0	10.3
Overlapping part ΔCD	− 33.5	− 1.0	− 60.2	19.2	8.5

water penetration can eliminate the void space in the region and simultaneously can expand the thermally induced empty space between the protein and water by creating the additional surface. It is most important that the former rather than latter effect governs the total PMV change. The water penetration into the region thus reduces the PMV of ubiquitin. The water penetration is confirmed in terms of the 3D distribution of water around the protein (3D water map). Figure 22.8 shows an isosurface of the 3D distribution functions of water oxygen for LPS and HPS obtained in this calculation by the 3D-RISM theory. It is apparent from the figure that a structural channel is created in the boundary region between the segments C and D of HPS so that the water distribution is enhanced there. This change in the water distribution obviously supports the water penetration predicted above by the volume analysis. As the PMV is calculated from the correlation functions obtained by the 3D-RISM theory, the change in the water distribution naturally affects the PMV change through the statistical-mechanical relation in this calculation. Therefore, the consistency in the two theoretical

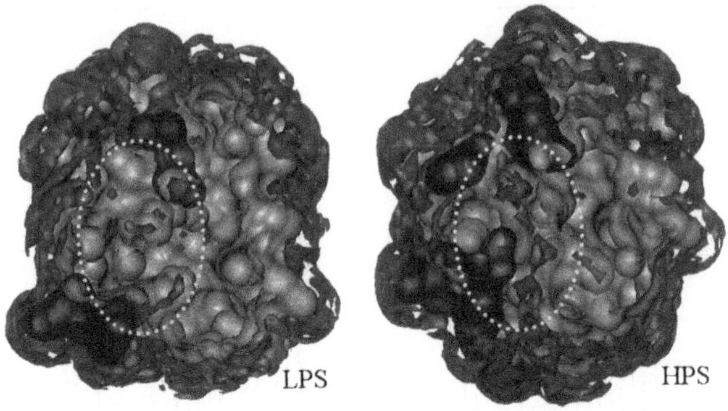

Figure 22.8 Isosurface representation of the 3D water map around low-pressure (3 MPa) and high-pressure (300 MPa) structures (LPS and HPS) of ubiquitin. Lighter surfaces show the area with water density enhancement >2. Dark gray and light gray surfaces indicate the molecular surfaces of ubiquitin segments C (residues 32–42) and D (residues 1–31 and 43–76). Top-view representation with the upper parts (the frontal parts in the figure) are clipped in order to bring the boundary region between C and D (marked with the dotted circle) into view. LPS is represented by the sixth structure of 1V80 with the largest void volume in the boundary region $V_V(\Delta CD)$, and HPS is represented by the second structure of 1V81 with the smallest $V_V(\Delta CD)$.

results supports the concept that the water penetration decreases the PMV by eliminating the structural voids against the volume expansion due to additional hydration. This constitutes the molecular mechanism of the PMV reduction associated with the pressure-induced structural transition of ubiquitin.

The consistency between the volume analysis and the water distribution analysis in the 3D-RISM–KB method is crucial in determining the cause of the pressure-induced structural change. Based on observing just the water distribution change implying water penetration, one could not conclude whether the water penetration was the cause or the result of the structural transition. The two findings that (i) the water penetration reduces the PMV of the protein and (ii) the PMV reduction induces the structural transition lead together to the conclusion that the water penetration is the driving force of the pressure-induced structural transition of ubiquitin.

22.5.4 Co-solvent Effect on the PMV Change of Staphylococcal Nuclease Associated with Pressure Denaturation

Pressure denaturation of staphylococcal nuclease (SNase) protein in the presence of co-solvents has been observed in experiments.[19–26] The molecular mechanisms of the effect of urea and glycerol co-solvents at 0.5 mol L^{-1} concentration on the PMV change associated with pressure denaturation of SNase was studied by using MD simulation to generate structures, followed by the 3D-RISM–KB analysis of the PMV.[110] The MD simulation was performed along the following thermodynamic cycle to obtain pressure-induced denatured SNase:

$$\begin{array}{ccc} 298\,\mathrm{K}, 400\,\mathrm{MPa} & \leftarrow & 500\,\mathrm{K}, 400\,\mathrm{MPa} \\ \Uparrow & & \uparrow \\ 298\,\mathrm{K}, 0.1\,\mathrm{MPa} & \rightarrow & 500\,\mathrm{K}, 0.1\,\mathrm{MPa} \end{array}.$$

The MD simulations were run along the pathway shown with the single line arrows, as the denaturation pathway along the pressure axis on the T–P phase diagram of the protein (double-line arrow in this diagram) is a prohibitively time-consuming task for MD.[14,36,37] The PMV was calculated from the 3D-RISM–KB theory for 100 conformations picked out from the 1 ns trajectory (a conformation every 10 ps) for the native SNase after 200 ps of equilibration and for another set of 100 conformations from the last 1 ns trajectory for the denatured SNase. The snapshots of the 100th native and denatured conformations are shown in Figure 22.9. In the native state, SNase has a hydrophobic core shown by the gray circle in part (a). As the simulation progresses, water molecules gradually penetrate the inside of the core and, as a result of penetration, this protein forms two sub-domains, separated as shown in part (b). This denatured conformation is in qualitative agreement with the experiment using small-angle neutron scattering.[14] The experimental results suggested that the denatured SNase under high pressure forms two separate sub-domains.

Table 22.6 exhibits the 3D-RISM–KB predictions for the PMV in the native and denatured states, $\overline{V}_{\mathrm{N}}$ and $\overline{V}_{\mathrm{D}}$, averaged over the conformations, and the PMV change $\overline{V}_{\mathrm{D}} - \overline{V}_{\mathrm{N}}$ associated with SNase denaturation at $T = 298$ K and 400 MPa. The negative PMV change associated with denaturation agrees with the experimental observations for pressure-induced denaturation in the range of -27 to -112 cm^3 mol^{-1},[3,111] as illustrated in Figure 22.10.

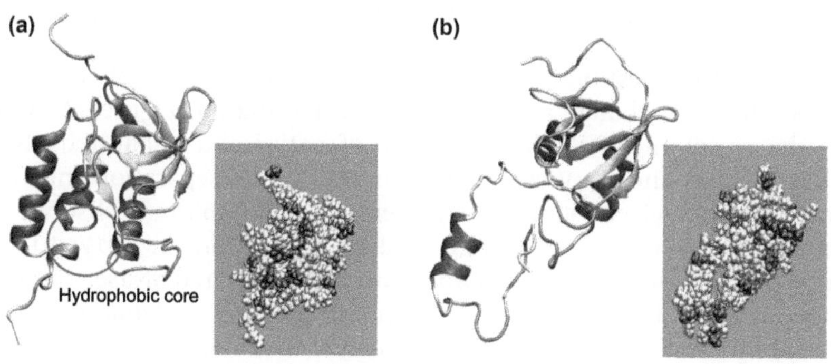

Figure 22.9 Conformations of native SNase (a) and denatured SNase under a high pressure of 400 MPa (b). The grey circle in part (a) shows the hydrophobic core. Cartoon representation and the space-filling model (with the hydrophobic and polar residues shown in gray and white, respectively).

Table 22.6 Partial molar volume ($cm^3\ mol^{-1}$) of the native (at 298 K, 0.1 MPa) and denatured (at 298 K, 400 MPa) states of SNase, and the PMV change associated with SNase denaturation in water and in water–urea and water–glycerol mixtures, predicted by the 3D-RISM-KB theory.

PMV/$cm^3\ mol^{-1}$	In water	In water–urea	In water–glycerol
Native, $\overline{V}_N$	12 800	12 810	12 816
Denatured, $\overline{V}_D$	12 619	12 632	12 640
Change, $\overline{V}_D - \overline{V}_N$	−181	−178	−176

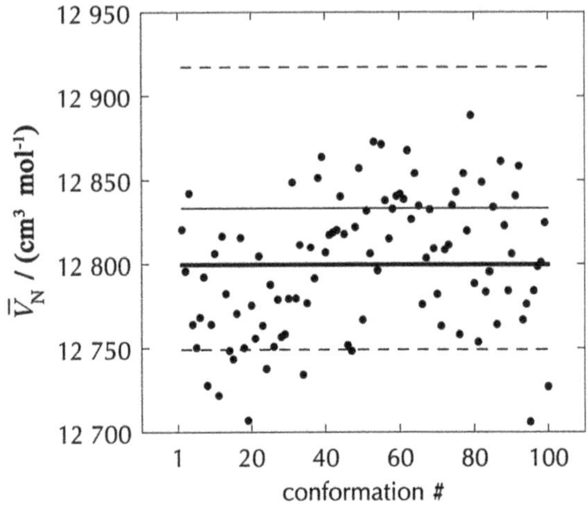

Figure 22.10 Partial molar volume of native SNase in water. Calculated PMV for each native conformation (symbols) and their average (bold solid line). Experimental data (thin solid line) and confidence interval (dashed lines).[111]

Although the difference is very small, both the urea and glycerol co-solvents decrease the PMV change associated with pressure-induced denaturation, compared to that in pure water. This is in agreement with the experimental results.[26]

To answer the question how and why both urea and glycerol co-solvents decrease the PMV change associated with denaturation, Figure 22.11 presents the co-solvent effects on the PMV of native and denatured SNase, $\overline{V}_N$(Water→Mix) and $\overline{V}_D$(Water→Mix), as well as its separation into electrostatic and non-electrostatic parts. There are two important observations related to the effect of urea and glycerol co-solvents on the PMVs: (i) The co-solvents increase the PMV of both native and denatured SNase compared to those in water; (ii) The co-solvents increase the PMV of denatured SNase more than that of the native SNase. To analyze the origin of these changes, the co-solvent effect on the PMV is further decomposed (see the right panel of Figure 22.11) into nonelectrostatic and electrostatic contributions, $\overline{V}$(Water→Mix) = $\Delta\overline{V}^{el}$(Water→Mix) − $\Delta\overline{V}^{nonel}$(Water→Mix). The PMV non-electrostatic part $\Delta\overline{V}^{nonel}$ was obtained by applying 3D-RISM to the solvated

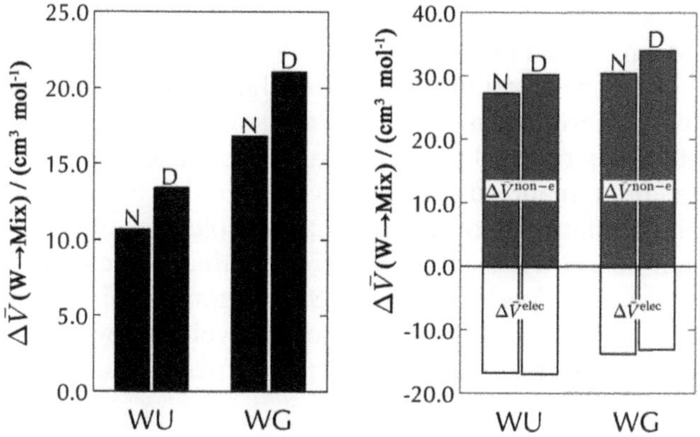

Figure 22.11 Left panel: Co-solvent effect in water–urea (WU) and water–glycerol (WG) mixtures on the partial molar volume of native (N) and denatured (D) SNase protein. Right panel: decomposition of the co-solvent effect on the partial molar volume into electrostatic (in white) and non-electrostatic (in grey) parts.

native and denatured conformations with all its partial site charges switched off, and the rest of the PMV change for the protein with full charges was attributed to the electrostatic part $\Delta\overline{V}^{\text{el}}$. In this decomposition, $\Delta\overline{V}^{\text{el}}$ is negative in each case, which suggests that the electrostriction effect is larger in the mixture than in pure water. However, $\Delta\overline{V}^{\text{nonel}}$ is positive and larger in magnitude than $\Delta\overline{V}^{\text{el}}$. Moreover, $\Delta\overline{V}^{\text{nonel}}$ in the denatured state is larger than that in the native state. The division thus clearly shows that the non-electrostatic part of PMV dominates the co-solvent effect on the PMV change associated with denaturation, resulting in the above observations.

In the decomposition given in Figure 22.11, the non-electrostatic part $\Delta\overline{V}^{\text{nonel}}$ is further subdivided into four contributions: the ideal term V_{id}, van der Waals volume V_{W}, void volume V_{V}, and thermal volume V_{WT}. The ideal term change from water to the solvent mixtures is negligible $V_{\text{id}} = 1.21$ in water, 1.21 in water–urea, and 1.22 cm^3 mol^{-1} in the water–glycerol mixture. The van der Waals volume and the void volume do not change for the same protein conformations in every solvent. The non-electrostatic part difference $\Delta\overline{V}^{\text{nonel}}$ thus comes only from the difference between the thermal volume (packing effect) in the solvent mixture and in pure water. The molecular volumes of both the urea and the glycerol molecules are larger than the water molecule, and so the empty space between the protein surface and a layer of co-solvent molecules is larger than that between the protein and water molecules. This increases the thermal volume V_{T} in the solvent mixture more than in the water and makes $\Delta\overline{V}^{\text{nonel}}$ positive for both native and denatured proteins. Furthermore, the protein denatured under high pressure allows solvent access not only to its surface but also inside the protein core and exposes the hydrophobic residues that are hidden in the native state. These processes contribute to making the solvent-accessible surface area (SASA) of the protein larger. Calculation with a probe radius of 0.14 nm gives the average values of SASA for the 100 native and 100 denatured proteins as 99.51 nm^2 and 114.62 nm^2, respectively. Analysis of the 3D maps of water and co-solvent distributions shows that both water and co-solvent (urea or glycerol) have access to the protein surface and are actually spread over it in both the native and the denatured states. The empty space between the denatured protein and solvents is thus expected to be larger than that for the protein in the native state. Therefore, the molecules of co-solvent which are larger than water increase the PMV of denatured SNase more than that of the native SNase, in proportion to the SASA difference. This argument on the packing effect and the decomposition of the volume by the 3D-RISM-KB method supports the PMV changes observed experimentally.[26]

22.5.5 Xenon–Lysozyme Binding: Molecular Mechanism of Pressure Reversal of Anesthesia

The pressure reversal of general anesthesia has been considered to provide a key to elucidating the molecular mechanism of general anesthesia.[112–117] Early workers believed that anesthetics acted non-specifically on hydrophobic lipid components of cells. However, recent studies have demonstrated that anesthetics interact with specific sites of transmembrane proteins such as ion channels which regulate signal transduction.[118–120] In those studies, the anesthetic–protein binding was considered as a physico-chemical or thermodynamic process at the molecular level. If this is the case, the pressure reversal of general anesthesia should be explained in terms of thermodynamics.

It was proposed that the volume expansion of proteins is due mainly to the release of "freezing" water molecules on the protein surface.[112,115] Water molecules around an ionic species become denser compared to the bulk, due to the strong electrostatic interactions, and consequently the PMV is reduced, the phenomenon known as electrostriction.[41] It was suggested that the binding of anesthetics releases such "electrostricted" water molecules from the vicinity of ionic residues of protein. However, no experimental evidence for the electrostriction hypothesis has been presented and the molecular mechanism of the volume increase, as well as the dehydration accompanying the anesthetic binding, remains unanswered.

The substrate binding site of hen egg white lysozyme can be a model site of the pressure reversal of general anesthesia.[121] It was demonstrated using docking simulation that the binding of xenon, one of the simplest anesthetics, to the site was energetically destabilized under high pressure. The xenon–lysozyme system is thus a pressure reversal model for 3D-RISM–KB study, since the 3D structure of lysozyme in solution was resolved[122] and xenon binding sites were estimated by docking simulation.[121] In the model of the xenon–lysozyme complex illustrated in Figure 22.12, one binding site of xenon corresponds to the binding pocket of native ligands (substrate binding site). The other site (internal site) is located in a cavity of lysozyme.

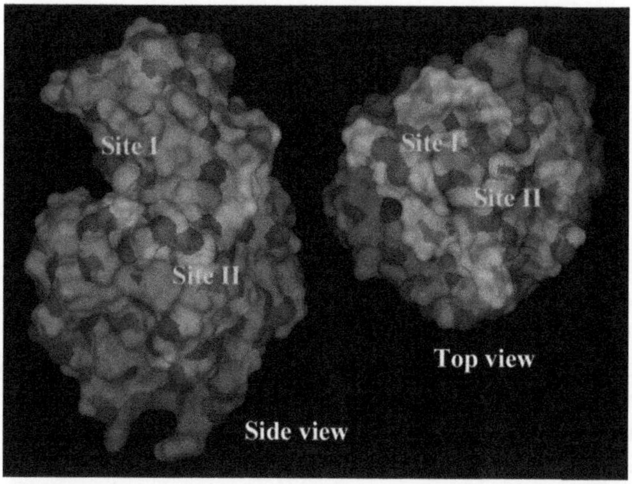

Figure 22.12 Structure of the xenon–lysozyme complex. Site I is the "substrate binding site". Site II is the "internal site" located in a cavity of lysozyme.

Table 22.7 Partial molar volume and its components (cm^3 mol^{-1}) for xenon, lysozyme, and two complexes in aqueous solution at 298.15 K.

$V/cm^3\ mol^{-1}$	$\overline{V}$	V_{id}	V_W	V_V	V_T	V_I
Xenon	28.6	1.3	13.7	0.0	13.6	0.0
Lysozyme	9413.8	1.3	6082.4	2580.4	804.4	− 54.8
Xenon + lysozyme[a]	9442.4	2.7	6096.1	2580.4	818.1	− 54.8
Complex (Substrate binding site)	9451.9	1.3	6096.1	2603.8	802.3	− 51.6
$\Delta V_{binding}$[b]	9.5	− 1.3	0.0	23.4	− 15.7	3.2
Complex (internal site)	9440.6	1.3	6096.0	2600.3	789.9	− 56.0
$\Delta V_{binding}$[b]	− 1.8	− 1.3	0.0	19.9	− 19.1	− 1.2

[a] $\overline{V}_{xenon} + \overline{V}_{lysozyme}$.
[b] $\overline{V}_{complex} - (\overline{V}_{xenon} + \overline{V}_{lysozyme})$.

The first finding is that the PMV increased by 10 cm^3 mol^{-1} at the substrate binding site. Based on Equation (22.1), this is in concordance with the pressure reduction of the binding constant of xenon. The PMV values for xenon, lysozyme, and two complexes (with one xenon molecule binding to the substrate binding site or internal site) in aqueous solution calculated using the 3D-RISM–KB theory are given in Table 22.7. The PMV changes upon the bindings, $\Delta V_{binding} = \overline{V}_{complex} - (\overline{V}_{xenon} + \overline{V}_{lysozyme})$, are shown as well. Unlike the effect at the substrate binding site, the binding to an internal site results in a slight decrease or negligible change to the PMV. Furthermore, the pressure effect on binding to an internal site [from Equation (22.1)] is small.

The calculated PMV increase of 10 cm^3 mol^{-1} accompanying the xenon–lysozyme binding is far lower than an experimental result for the diethylether–albumin binding which shows a PMV expansion of 295 cm^3 mol^{-1} (reference 115). It is most likely that the discrepancy comes from the difference in the number of anesthetic molecules interacting with protein. It is assumed that only a few xenon molecules bind to lysozyme because docking simulation, as well as X-ray analysis, could detect only a few binding sites which are energetically distinguished from others.

The second finding is that the PMV increase for the substrate binding site primarily consists of a large increase in the void volume, which is partially compensated by a decrease in the thermal volume. The change in the ideal volume is purely the entropic effect caused by the binding. In other words, the decrease in the ideal volume is due to the decrease of the number of molecules from 2 to 1. The effect is almost negligible compared to the PMV increase. The interaction volume contributes positively. For the internal site, an increase in the void volume is almost completely compensated by a decrease in the thermal volume. The contribution of the interaction volume for the internal site is also negligible.

The PMV change associated with the binding of xenon is related to the exchange between xenon and water molecules at the site. The water maps analysis of the hydration change from the xenon binding to the substrate binding site before and after binding of xenon shows that the release of two water molecules makes a 40 $cm^3\ mol^{-1}$ cavity belonging to the lysozyme molecule. Thus, one xenon molecule of volume 28.6 $cm^3\ mol^{-1}$ (See Table 22.7) bound to the site expels two water molecules of volume $\sim$40 $cm^3\ mol^{-1}$ from the site. Consequently, the xenon binding increases the PMV by about 10 $cm^3\ mol^{-1}$. Similarly, the xenon binding to the internal site reduces the number of molecules within the binding site from 2 to 1. In this case, one water molecule is exchanged for one xenon molecule. The volume contribution of the decrease in the hydration number is about +20 $cm^3\ mol^{-1}$, which is approximately the volume of one water molecule. These results imply that the xenon binding to the internal site only interchanges one xenon with one water molecule, so that it does not significantly change the PMV.

What is the difference between the two sites? When a xenon molecule binds to the lysozyme site, some thermal voids between xenon and water vanish because the xenon molecule gets partially dehydrated. Instead, some structural voids between the xenon and the lysozyme are created. In the same way, some thermal voids between lysozyme and water transform to the structural voids between xenon and lysozyme. If the packing between xenon and lysozyme are similar to the packing of water, the increase in the void volume balances the decrease in the thermal volume. The binding of xenon to the internal site expels only one water molecule so that the voids between xenon and water and between lysozyme and water entirely transform to the structural voids between xenon and lysozyme, as illustrated in Figure 22.13. On the other hand, at the substrate binding site two water molecules are exchanged for one xenon molecule, as shown above. Such an exchange makes larger voids between xenon and lysozyme (Figure 22.13). Therefore, the difference of the packing between xenon and lysozyme from the packing of hydration water, or interfacial water, creates the increase in the void

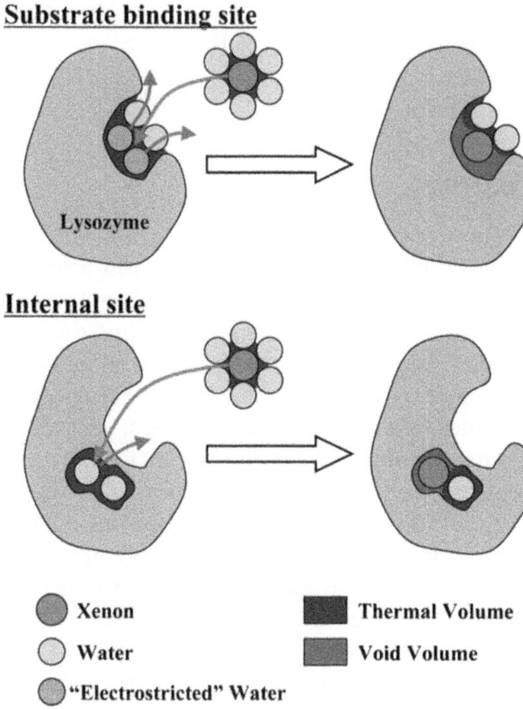

Figure 22.13 Illustration of the molecular mechanism of xenon–lysozyme binding and the volume change.

volume which exceeds the decrease in the thermal volume; this is the origin of the PMV increase of anesthetic binding at the molecular level.

This molecular mechanism explains that the loose binding of ligands such as anesthetics to protein sites (typically protein surface) expands the PMV, while the tight binding of specific ligands to their own sites does not increase the PMV or can reduce the PMV depending on the degree of packing. This is consistent with experimental results[117] that the binding of halothane (anesthetic) to luciferase increases the PMV and the binding of myristate (specific inhibitor) decreases the PMV.

The interaction volume change ΔV_{I} is directly related to the electrostriction hypothesis.[112,115] The substrate binding site is proximate to the active site of lysozyme and has two charged residues, Glu35 and Asp52. Before a xenon molecule binds to the site, the charged groups are strongly hydrated. When a xenon molecule comes to the site, two water molecules are released from the site. The dehydration reduces the electrostriction. The contribution of electrostriction to the PMV is represented by the change $\Delta V_{\mathrm{I}} = +3.2\ \mathrm{cm^3\ mol^{-1}}$. This confirms and identifies the release of electrostricted water; however, electrostricted water release is found to be a secondary contribution to the PMV increase which will cause pressure reversal of anesthesia.

For the internal site, the interaction volume change ΔV_I is almost negligible. The water molecules confined in the cavity are isolated from bulk water, and the detailed hydration structure in such a cavity affects neither the interaction volume nor the PMV.

22.6 Conclusions

One of the most fundamental thermodynamic quantities characterizing the conformational stability of a protein in solution under pressure is the partial molar volume (PMV). According to experimental suggestions by Akasaka and co-workers,[9–11] the PMV might be considered to be an order parameter in protein folding and conformational changes, which has substantial implications to the understanding of protein functions. Based on the first principles of statistical mechanics, the method of three-dimensional reference interaction site model (3D-RISM) molecular theory of solvation[34,59] coupled with the Kirkwood–Buff (KB) theory[42] provides a firm approach to calculation and analysis of the partial molar volume of biomolecules in solution.[45] The method is readily applicable to complex systems with a co-solvent. An advantage and a predictive capability of the 3D-RISM–KB method is that it simultaneously evaluates both the partial molar volume and three-dimensional water maps (in general, solvent mixture maps), consistently obtained in the same statistical–mechanical approach. The 3D-RISM–KB method readily yields a decomposition of the partial molar volume into physically meaningful components. Coupled with 3D water maps, this provides an insight into molecular mechanisms of pressure-induced transitions of proteins in solution, such as pressure-induced denaturation, helix–coil transition, and local transitions in essential parts of the protein structure, as has been illustrated in a number of biomolecular systems in solution.

Acknowledgements

This work was supported by the National Institute for Nanotechnology, National Research Council of Canada, University of Alberta, and Alberta Prion Research Institute.

References

1. T. V. Chalikian, *Annu. Rev. Biophys. Biomol. Struct.*, 2003, **32**, 207.
2. C. Balny, P. Masson and K. Heremans, *Biochim. Biophys. Acta*, 2002, **1595**, 3.
3. C. A. Royer, *Biochim. Biophys. Acta*, 2002, **1595**, 201.
4. J. L. Silva and G. Weber, *Annu. Rev. Phys. Chem.*, 1993, **44**, 89.
5. C. Balny, *Biochim. Biophys. Acta*, 2006, **1764**, 632.
6. F. Meersman, C. M. Dobson and K. Heremans, *Chem. Soc. Rev.*, 2006, **35**, 908.

7. P. W. Bridgman, *J. Biol. Chem.*, 1914, **19**, 511.
8. M. F. San Martín, G. V. Barbosa-Cánovas and B. G. Swanson, *Crit. Rev. Food Sci. Nutr.*, 2002, **42**, 627.
9. R. Kitahara, H. Yamada, K. Akasaka and P. E. Wright, *J. Mol. Biol.*, 2002, **320**, 311.
10. R. Kitahara and K. Akasaka, *Proc. Natl. Acad. Sci. U. S. A.*, 2003, **100**, 3167.
11. K. Akasaka, *Biochemistry*, 2003, **42**, 10 875.
12. G. Hummer, S. Garde, A. E. García, M. E. Paulaitis and L. R. Pratt, *Proc. Natl. Acad. Sci. U. S. A.*, 1998, **95**, 1552.
13. H. Lesch, J. Schlichter, J. Friedrich and J. M. Vanderkooi, *Biophys. J.*, 2004, **86**, 467.
14. A. Paliwal, D. Asthagiri, D. P. Bossev and M. E. Paulaitis, *Biophys. J.*, 2004, **87**, 3479.
15. M. D. Collins, G. Hummer, M. L. Quillin, B. W. Matthews and S. M. Gruner, *Proc. Natl. Acad. Sci. U. S. A.*, 2005, **102**, 16668.
16. Y. Harano and M. Kinoshita, *J. Chem. Phys.*, 2006, **125**, 024910.
17. F. Meersman, C. M. Dobson and K. Heremans, *Chem. Soc. Rev.*, 2006, **35**, 908.
18. C. Scharnagl, M. Reif and J. Friedrich, *Biochim. Biophys. Acta*, 2005, **1749**, 187.
19. G. Pappenberger, C. Saudan, M. Becker, A. E. Merbach and T. Kiefhaber, *Proc. Natl. Acad. Sci. U. S. A.*, 2000, **97**, 17.
20. K. Ruan, C. Xu, Y. Yu, J. Li, R. Lange, N. Bec and C. Balny, *Eur. J. Biochem.*, 2001, **268**, 2742.
21. J. N. Webb, S. D. Webb, J. L. Cleland, J. F. Carpenter and T. W. Randolph, *Proc. Natl. Acad. Sci. U. S. A.*, 2001, **98**, 7259.
22. S. Perrett and J. M. Zhou, *Biochim. Biophys. Acta*, 2002, **1595**, 210.
23. J. Skerjanc, V. Dolecek and S. Lapanje, *Eur. J. Biochem.*, 1970, **17**, 160.
24. J. Skerjanc and S. Lapanje, *Eur. J. Biochem.*, 1972, **25**, 49.
25. J. C. Lee and S. N. Timasheff, *Biochemistry*, 1974, **13**, 257.
26. H. Herberhold, C. A. Royer and R. Winter, *Biochemistry*, 2004, **43**, 3336.
27. W. Kauzmann, *Adv. Protein Chem.*, 1959, **14**, 1.
28. F. H. Stillinger, *J. Solution. Chem.*, 1973, **2**, 141.
29. T. V. Chalikian and K. Breslauer, *J. Biopolym.*, 1996, **39**, 619.
30. A. A. Zamyatnin, *Annu. Rev. Biophys. Bioeng.*, 1984, **13**, 145.
31. J. T. Edsall, in *Proteins, Amino Acids, and Peptides*, ed. E. J. Cohn and J. T. Edsall, Reinhold, New York, 1943, p. 155.
32. F. J. Millero, A. L. Surdo and C. Shin, *J. Phys. Chem.*, 1978, **82**, 784.
33. J.-P. Hansen and I. McDonald, *Theory of Simple Liquids*, Elsevier, Amsterdam, the Netherlands, 3rd edn, 2006.
34. Molecular Theory of Solvation, ed. F. Hirata, *Understanding Chemical Reactivity Series*, Vol. 24, Kluwer Academic Publishers, Dordrecht, the Netherlands, 2003.
35. D. B. Kitchen, L. H. Reed and R. M. Levy, *Biochemistry*, 1992, **31**, 10083.
36. E. Paci, *Biochim. Biophys. Acta*, 2002, **1595**, 185.

37. S. Sarupria, T. Ghosh, A. E. García and S. Garde, *Proteins: Struct., Funct., Bioinf.*, 2010, **78**, 1641.
38. R. De Vane, C. Ridley, R. W. Larsen, B. Space, P. B. Moore and S. I. Chan, *Biophys. J.*, 2003, **85**, 2801.
39. F. Hirata, F. Imai and M. Irisa, *Rev. High Pressure Sci. Technol.*, 1998, **8**, 96.
40. F. Hirata and K. Arakawa, *Bull. Chem. Soc. Jpn.*, 1973, **46**, 3367.
41. F. J. Millero, *Chem. Rev.*, 1971, **71**, 147.
42. J. G. Kirkwood and F. P. Buff, *J. Chem. Phys.*, 1951, **19**, 774.
43. T. Imai, M. Kinoshita and F. Hirata, *J. Chem. Phys.*, 2000, **112**, 9469.
44. M. Kinoshita, T. Imai, A. Kovalenko and F. Hirata, *Chem. Phys. Lett.*, 2001, **348**, 337.
45. Y. Harano, T. Imai, A. Kovalenko, M. Kinoshita and F. Hirata, *J. Chem. Phys.*, 2001, **114**, 9506.
46. T. Imai, Y. Harano, A. Kovalenko and F. Hirata, *Biopolymers*, 2001, **59**, 512.
47. T. Imai, A. Kovalenko and F. Hirata, *J. Phys. Chem. B*, 2005, **109**, 6658.
48. T. Imai, T. Takekiyo, A. Kovalenko, F. Hirata, M. Kato and Y. Taniguchi, *Biopolymers*, 2005, **79**, 97.
49. H. Edelsbrunner, M. Facello, P. Fu and J. Liang, Measuring proteins and voids in proteins, in *Proc. 28th Annu. Hawaii Int. Conf. System Sci.*, IEEE Computer Society Press, Los Alamitos, CA, 1995, vol. 5, pp. 256–264.
50. D. Chandler, J. McCoy and S. Singer, *J. Chem. Phys.*, 1986, **85**, 5971.
51. D. Chandler, J. McCoy and S. Singer, *J. Chem. Phys.*, 1986, **85**, 59771.
52. D. Beglov and B. Roux, *J. Phys. Chem. B*, 1997, **101**, 7821.
53. A. Kovalenko and F. Hirata, *Chem. Phys. Lett.*, 1998, **290**, 237.
54. A. Kovalenko and F. Hirata, *J. Chem. Phys.*, 1999, **110**, 10095.
55. A. Kovalenko and F. Hirata, *J. Chem. Phys.*, 2000, **112**, 10391.
56. A. Kovalenko and F. Hirata, *J. Chem. Phys.*, 2000, **112**, 10403.
57. A. Kovalenko, in *Molecular theory of solvation*, ed. F. Hirata, Understanding Chemical Reactivity Series, Vol. 24, Kluwer Academic Publishers, Dordrecht, the Netherlands, 2003, pp. 169–275.
58. S. Gusarov, B. S. Pujari and A. Kovalenko, *J. Comput. Chem.*, 2012, **33**, 1478.
59. A. Kovalenko, *Pure Appl. Chem.*, 2013, **85**, 159.
60. S. Gusarov, T. Ziegler and A. Kovalenko, *J. Phys. Chem. A*, 2006, **110**, 6083.
61. D. Casanova, S. Gusarov, A. Kovalenko and T. Ziegler, *J. Chem. Theory Comput.*, 2007, **3**, 458.
62. J. W. Kaminski, S. Gusarov, T. A. Wesolowski and A. Kovalenko, *J. Phys. Chem. A*, 2010, **114**, 6082.
63. M. Malvaldi, S. Bruzzone, C. Chiappe, S. Gusarov and A. Kovalenko, *J. Phys. Chem. B*, 2009, **113**, 3536.
64. A. Kovalenko, A. E. Kobryn, S. Gusarov, O. Lyubimova, X. Liu, N. Blinov and M. Yoshida, *Soft Matter*, 2012, **8**, 1508.
65. A. Kovalenko and F. Hirata, *Chem. Phys. Lett.*, 2001, **349**, 496.
66. A. Kovalenko and F. Hirata, *J. Theor. Comput. Chem.*, 2002, **1**, 381.

67. K. Yoshida, T. Yamaguchi, A. Kovalenko and F. Hirata, *J. Phys. Chem. B*, 2002, **106**, 5042.
68. V. Shapovalov, T. N. Truong, A. Kovalenko and F. Hirata, *Chem. Phys. Lett.*, 2000, **320**, 186.
69. S. R. Stoyanov, S. Gusarov and A. Kovalenko, in *Industrial Applications of Molecular Simulations*, ed. M. Meunier, CRC Press, Boca Raton, 2011, pp. 203–230.
70. J. Fafard, O. Lyubimova, S. Stoyanov, G. Kenne Dedzo, S. Gusarov, A. Kovalenko and C. Detellier, *J. Phys. Chem. C*, 2013, **117**, 18556.
71. J. G. Moralez, J. Raez, T. Yamazaki, R. K. Motkuri, A. Kovalenko and H. Fenniri, *J. Am. Chem. Soc.*, 2005, **127**, 8307.
72. R. S. Johnson, T. Yamazaki, A. Kovalenko and H. Fenniri, *J. Am. Chem. Soc.*, 2007, **129**, 5735.
73. T. Yamazaki, H. Fenniri and A. Kovalenko, *Chem. Phys. Chem.*, 2010, **11**, 361.
74. N. Yoshida, S. Phongphanphanee, Y. Maruyama, T. Imai and F. Hirata, *J. Am. Chem. Soc.*, 2006, **128**, 12042.
75. N. Yoshida, T. Imai, S. Phongphanphanee, A. Kovalenko and F. Hirata, *J. Phys. Chem. B*, 2009, **113**, 873.
76. S. Phongphanphanee, N. Yoshida and F. Hirata, *J. Am. Chem. Soc.*, 2008, **130**, 1540.
77. S. Phongphanphanee, T. Rungrotmongkol, N. Yoshida, S. Hannongbua and F. Hirata, *J. Am. Chem. Soc.*, 2010, **132**, 9782.
78. Y. Maruyama, N. Yoshida and F. Hirata, *Interdiscip. Sci.: Comput. Life Sci.*, 2011, **3**, 290.
79. Y. Yonetani, Y. Maruyama, F. Hirata and H. Kono, *J. Chem. Phys.*, 2008, **128**, 185102.
80. Y. Maruyama, N. Yoshida and F. Hirata, *J. Phys. Chem. B*, 2010, **114**, 6464.
81. M. C. Stumpe, N. Blinov, D. Wishart, A. Kovalenko and V. S. Pande, *J. Phys. Chem. B*, 2011, **115**, 319.
82. T. Yamazaki, N. Blinov, D. Wishart and A. Kovalenko, *Biophys. J.*, 2008, **95**, 4540.
83. N. Blinov, L. Dorosh, D. Wishart and A. Kovalenko, *Biophys. J.*, 2010, **98**, 282.
84. N. Blinov, L. Dorosh, D. Wishart and A. Kovalenko, *Mol. Simul.*, 2011, **37**, 718.
85. A. Kovalenko and N. Blinov, *J. Mol. Liq.*, 2011, **164**, 101.
86. D. Nikolic, N. Blinov, D. Wishart and A. Kovalenko, *J. Chem. Theory Comput.*, 2012, **8**, 3356.
87. T. Imai, N. Miyashita, Y. Sugita, A. Kovalenko, F. Hirata and A. Kidera, *J. Phys. Chem. B*, 2011, **115**, 8288.
88. M. C. Stumpe, N. Blinov, D. Wishart, A. Kovalenko and V. S. Pande, *J. Phys. Chem. B*, 2011, **115**, 319.
89. J. S. Perkyns, G. C. Lynch, J. J. Howard and B. M. Pettitt, *J. Chem. Phys.*, 2010, **132**, 064106.

90. S. M. Kast and T. Kloss, *J. Chem. Phys.*, 2008, **129**, 236101.
91. I. S. Joung, T. Luchko and D. A. Case, *J. Chem. Phys.*, 2013, **138**, 044103.
92. J. S. Perkyns and B. M. Pettitt, *J. Chem. Phys.*, 1992, **97**, 7656.
93. B. Kvamme, *Int. J. Thermophys.*, 1995, **16**, 743.
94. T. Yamazaki and A. Kovalenko, *J. Chem. Theory Comput.*, 2009, **5**, 1723.
95. T. Yamazaki and A. Kovalenko, *J. Phys. Chem. B*, 2011, **115**, 310.
96. A. Kovalenko, S. Ten-no and F. Hirata, *J. Comput. Chem.*, 1999, **20**, 928.
97. P. Pulay, *Chem. Phys. Lett.*, 1980, **73**, 393.
98. Y. Saad and M. H. Schultz, *SIAM J. Sci. Stat. Comput.*, 1986, **7**, 856.
99. M. Kikuchi, M. Sakurai and K. Nitta, *J. Chem. Eng. Data*, 1995, **40**, 935.
100. A. A. Zamyatnin, *Annu. Rev. Biophys. Bioeng.*, 1984, **13**, 145.
101. Y. Yasuda, N. Tochio, M. Sakurai and K. Nitta, *J. Chem. Eng. Data*, 1998, **43**, 205.
102. M. Mizuguchi, M. Sakurai and K. Nitta, *J. Solution Chem.*, 1997, **26**, 579.
103. T. Takekiyo, A. Okuno, T. Imai, A. Shimizu, M. Kato and Y. Taniguchi, in *Portable Synchrotron Light Sources and Advanced Applications*, ed. H. Yamada, N. Mochizuki-Oda and M. Sasaki, American Institute of Physics, Melville, New York, 2004, pp. 184–187.
104. T. Takekiyo, A. Shimizu, M. Kato and Y. Taniguchi, *Biochim. Biophys. Acta*, 2005, **1750**, 1.
105. D. T. Clerke, A. J. Doig, B. J. Stapley and G. R. Jones, *Proc. Natl. Acad. Sci. U. S. A.*, 1999, **96**, 7232.
106. S. Terasawa, H. Itsuki and S. Arakawa, *J. Phys. Chem.*, 1975, **79**, 2345.
107. J. T. Edward and P. G. Farrell, *Can. J. Chem.*, 1975, **53**, 2965.
108. R. Kitahara, S. Yokoyama and K. Akasaka, *J. Mol. Biol.*, 2005, **347**, 277.
109. T. Imai, S. Ohyama, A. Kovalenko and F. Hirata, *Protein Sci.*, 2007, **16**, 1927.
110. T. Yamazaki, T. Imai, F. Hirata and A. Kovalenko, *J. Phys. Chem. B*, 2007, **111**, 1206.
111. R. Filfil and T. V. Chalikian, *J. Mol. Biol.*, 2000, **299**, 827.
112. H. Eyring, J. W. Woodbury and J. S. D'Arrigo, *Anesthesiology*, 1973, **38**, 415.
113. K. W. Miller, W. D. M. Paton, R. A. Smith and E. B. Smith, *Mol. Pharmacol.*, 1973, **9**, 131.
114. N. P. Franks and W. R. Lieb, *Nature*, 1982, **300**, 487.
115. I. Ueda and T. Mashimo, *Physiol. Chem. Phys.*, 1982, **14**, 157.
116. G. W. J. Moss, W. R. Lieb and N. P. Franks, *Biophys. J.*, 1991, **60**, 1309.
117. I. Ueda, H. Matsuki, H. Kamaya and P. R. Krishna, *Biophys. J.*, 1999, **76**, 483.
118. N. P. Franks and W. R. Lieb, *Nature*, 1994, **367**, 607.
119. K. W. Miller, *Br. J. Anaesth.*, 2002, **89**, 17.
120. J. A. Campagna, K. W. Miller and S. A. Forman, *N. Engl. J. Med.*, 2003, **348**, 2110.
121. H. Isogai, T. Seto and S. Nosaka, *Int. Congr. Ser.*, 2005, **1283**, 318.
122. M. Refaee, T. Tezuka, K. Akasaka and M. P. Williamson, *J. Mol. Biol.*, 2003, **327**, 857.

Subject Index